6th International Conference on Mathematical Modeling in Physical Sciences (IC-MSquare 2017)

Journal of Physics: Conference Series Volume 936

Pafos, Cyprus
28 - 31 August 2017

Editor:

Dimitrios S. Vlachos

ISBN: 978-1-5108-5845-9
ISSN: 1742-6588

Printed from e-media with permission by:

Curran Associates, Inc.
57 Morehouse Lane
Red Hook, NY 12571

Printed by Curran Associates, Inc. (2018)

For permission requests, please contact the Institute of Physics at the address below.

Institute of Physics
Dirac House, Temple Back
Bristol BS1 6BE UK

Phone: 44 1 17 929 7481
Fax: 44 1 17 920 0979

techtracking@iop.org

Additional copies of this publication are available from:

Curran Associates, Inc.
57 Morehouse Lane
Red Hook, NY 12571 USA
Phone: 845-758-0400
Fax: 845-758-2633
Email: curran@proceedings.com
Web: www.proceedings.com

Table of contents

Volume 936

6th International Conference on Mathematical Modelling in Physical Sciences (IC-MSQUARE 2017)

28–31 August 2017, Pafos, Cyprus

Accepted papers received: 11 December 2017
Published online: 19 February 2018

Preface

Papers

IOP Conf. Series: Journal of Physics: Conf. Series **936** (2017) 011001 doi:10.1088/1742-6596/936/1/011001

Preface

The 6th International Conference on Mathematical Modeling in Physical Sciences (IC-MSQUARE) took place at Pafos, Cyprus, from Monday, 28th of August, to Thursday, 31th of August 2017.

The Conference was attended by more than 120 participants and hosted about 140 oral, poster, and virtual presentations while counted more than 500 pre-registered authors. The 6th IC-MSQUARE consisted of different and diverging workshops and thus covered various research fields where Mathematical Modeling is used, such as Theoretical/Mathematical Physics, Neutrino Physics, Non-Integrable Systems, Dynamical Systems, Computational Nanoscience, Biological Physics, Computational Biomechanics, Complex Networks, Stochastic Modeling, Fractional Statistics, DNA Dynamics, Macroeconomics etc.

The scientific program was rather heavy, however, according to all attendees, the program was excellent with high level of talks and the scientific environment was fruitful, thus all attendees had a creative time.

We would like to thank the hosting Neapolis University of Pafos for providing their excellent infrastructure for the needs of the Conference. We also would like to thank the Members of the International Advisory and Scientific Committees as well as the Members of the Organizing Committee.

The Conference Chairman

Dimitrios Vlachos
University of Peloponnese

IOP Conf. Series: Journal of Physics: Conf. Series **936** (2017) 011001 doi:10.1088/1742-6596/936/1/011001

Dimitrios S. Vlachos, University of Peloponnese, Greece

IOP Conf. Series: Journal of Physics: Conf. Series **936** (2017) 011002 doi:10.1088/1742-6596/936/1/011002

Peer review statement

All papers published in this volume of *Journal of Physics: Conference Series* have been peer reviewed through processes administered by the proceedings Editors. Reviews were conducted by expert referees to the professional and scientific standards expected of a proceedings journal published by IOP Publishing.

IOP Conf. Series: Journal of Physics: Conf. Series **936** (2017) 012001 doi:10.1088/1742-6596/936/1/012001

Mathematical modeling of polymer flooding using the unstructured Voronoi grid

T F Kireev[1, 2], G T Bulgakova[1] and I F Khatmullin[2]

[1]Ufa State Aviation Technical University, Ufa, Russian Federation

[2]UNTC LLC, Ufa, Russian Federation

E-mail: kireevtf@gmail.com

Abstract. Effective recovery of unconventional oil reserves necessitates development of enhanced oil recovery techniques such as polymer flooding. The study investigated the model of polymer flooding with effects of adsorption and water salinity. The model takes into account six components that include elements of the classic black oil model. These components are polymer, salt, water, dead oil, dry gas and dissolved gas. Solution of the problem is obtained by finite volume method on unstructured Voronoi grid using fully implicit scheme and the Newton's method. To compare several different grid configurations numerical simulation of polymer flooding is performed. The oil rates obtained by a hexagonal locally refined Voronoi grid are shown to be more accurate than the oil rates obtained by a rectangular grid with the same number of cells. The latter effect is caused by high solution accuracy near the wells due to the local grid refinement. Minimization of the grid orientation effect caused by the hexagonal pattern is also demonstrated. However, in the inter-well regions with large Voronoi cells flood front tends to flatten and the water breakthrough moment is smoothed.

1. Introduction

While the polymer flooding, the aqueous solution of high molecular weight polymer (about 1 million g/mol for partially hydrolyzed polyacrylamide) is injected into a reservoir and its mobility is several times lower than water mobility, providing the increase in oil recovery. This method, as well as most reservoir development technologies, requires a numerical modeling.

The software products known in the oil and gas industry include only the preselected sets of mathematical models and calculation tools. For a user it is not possible to make changes in these models. In addition, such software is expensive and for small companies it is more practical to develop the in-house software and make calculations by themselves.

The studies of polymer flooding modeling began in the late 1960s and are actively continued at present time [1, 2].

In the late 1980s, for the first time the unstructured grid was applied for the oil reservoir modeling [3-5]. These grids provide a number of advantages over classical rectangular grids: they allow better description of reservoir heterogeneity and more accurate calculation of the problem solution in the near-well area, reduction of the grid orientation effect and/ in some cases, the calculating speed increase from reducing the number of grid cells.

In the presented work we review a model of polymer flooding with the reservoir water salinity consideration, which was generated from the classical black-oil model. The novelty of the work is in the application of the unstructured Voronoi grid for space representation. The flow under consideration in a porous medium consists of six-components: water, oil, free gas, dissolved gas, polymer and salt brine. The viscosity of water phase increases with the polymer concentration increase and the polymer adsorption on the rock surface as defined by the adsorption isotherms with subsequent decrease of rock permeability to water.

2. Basic equations

The equations of polymer flooding model are:

$$\frac{\partial}{\partial t}\left[\phi\left(\frac{S_w}{B_w}\right)\right]+\nabla\cdot\left(\frac{\vec{u}_w}{B_w}\right)+q_w\delta_{well}=0, \tag{1}$$

$$\frac{\partial}{\partial t}\left[\phi\left(\frac{S_o}{B_o}\right)\right]+\nabla\cdot\left(\frac{\vec{u}_o}{B_o}\right)+q_o\delta_{well}=0, \tag{2}$$

$$\frac{\partial}{\partial t}\left[\phi\left(\frac{R_s S_o}{B_o}+\frac{S_g}{B_g}\right)\right]+\nabla\cdot\left(\frac{R_s\vec{u}_o}{B_o}+\frac{\vec{u}_g}{B_g}\right)+q_g\delta_{well}=0, \tag{3}$$

$$\frac{\partial}{\partial t}\left[\phi\left(\frac{S_w C_p}{B_w}\right)\right]+\frac{\partial}{\partial t}\left[(1-\phi)\rho_r C_a\right]+\nabla\cdot\left(\frac{\vec{u}_w C_p}{B_w}\right)+q_w C_p\delta_{well}=0, \tag{4}$$

$$\frac{\partial}{\partial t}\left[\phi\left(\frac{S_w C_s}{B_w}\right)\right]+\nabla\cdot\left(\frac{\vec{u}_w C_s}{B_w}\right)+q_w C_s\delta_{well}=0, \tag{5}$$

$$S_w+S_o+S_g=1. \tag{6}$$

where: ϕ = rock porosity; S_w = water saturation; S_o = oil saturation; S_g = gas saturation; B_w, B_o, B_g = water, oil and free gas volume factors, respectively; R_s = solution gas-oil ratio; p_b = bubble point pressure, which is constant at $S_g=0$ and equal to reservoir pressure at $S_g>0$; $\vec{u}_w$, $\vec{u}_o$, $\vec{u}_g$ = water, oil and free gas flow rates, respectively; C_p = mass of polymer dissolved in a unit volume of water; C_s = mass of salt dissolved in a unit volume of water; C_a = mass of polymer adsorbed on the surface of unit rock mass (adsorption isotherm); ρ_r = rock density; q_w, q_o, q_g = the flow rate of water, oil and free gas, respectively, produced from a reservoir unit volume per unit time; δ_{well} = Dirac delta function localized for the well location.

The first three equations (1) - (3) describe the classical black-oil model. They are the equations of mass conservation for the water, oil and gas phases, respectively. Equations (4) and (5) reflect the mass conservation laws for polymer and salt components, respectively. The velocities of fluids flowing in a porous media are described by the Darcy's law (7) - (9):

$$\vec{u}_w=\frac{kk_{rw}}{\mu_{w\_eff}R}\nabla p, \tag{7}$$

$$\vec{u}_o=\frac{kk_{ro}}{\mu_o}\nabla p, \tag{8}$$

$$\vec{u}_g=\frac{kk_{rg}}{\mu_g}\nabla p. \tag{9}$$

The sources simulating the wells are described by equations (10) – (12):

IOP Conf. Series: Journal of Physics: Conf. Series **936** (2017) 012001 doi:10.1088/1742-6596/936/1/012001

$$q_w = \frac{2\pi kh}{\ln\left(r_{eq} / r_w\right)} \frac{k_{rw}}{\mu_{w\_eff} B_w R} (p - p_{bh}), \tag{10}$$

$$q_o = \frac{2\pi kh}{\ln\left(r_{eq} / r_w\right)} \frac{k_{ro}}{\mu_o B_o} (p - p_{bh}), \tag{11}$$

$$q_g = \frac{2\pi kh}{\ln\left(r_{eq} / r_w\right)} \left(\frac{k_{rg}}{\mu_g B_g} + \frac{R_s k_{ro}}{\mu_o B_o} \right) (p - p_{bh}). \tag{12}$$

where: p = reservoir pressure; k = rock permeability; k_{rw}, k_{ro}, k_{rg} = relative permeability to water, oil and free gas, respectively; μ_{w_eff} = effective water viscosity; μ_o, μ_g = viscosity of oil and free gas, respectively; R = resistance factor, reflecting the rock permeability reduction by the polymer adsorption; h = reservoir thickness; r_{eq} = equivalent wellbore radius [6]; r_w = wellbore radius; p_{bh} = bottomhole pressure. The resistance factor is calculated as:

$$R = 1 + (R_{res} - 1) \frac{C_a}{C_{a,\max}},$$

where R_{res} = residual resistance factor, $C_{a,\max}$ = maximum adsorption isotherm.

The basic assumptions of the model are:

- the fluids temperature remains constant;
- water does not mix and does not exchange masses with oil and gas phases;
- water, polymer and salt are completely mixed into a homogeneous solution;
- the gas phase contains free gas and oil-dissolved gas;
- volume factors of water, oil and gas do not depend on polymer and salt concentrations;
- porosity and volume factors are the functions of reservoir pressure;
- solution gas-oil ratio depends on reservoir pressure and bubble point;
- relative permeability is a function of water saturation and gas saturation;
- effective water viscosity is a two-dimensional function of polymer and salt concentrations (equal to pure water viscosity in the absence of polymer and salt admixtures);
- the adsorption isotherm is a function of polymer concentration.

3. The problem statement

Let us assume that the fluid flow occurs in a thin reservoir, which properties are weakly varied along a vertical axis. Neglecting gravitational effects and variations of all quantities along the vertical axis, a planar two-dimensional reservoir representation can be used $\Omega \subset \mathbb{R}^2$. The reservoir boundary is described by a closed contour $\Gamma = \partial\Omega$. The reservoir is penetrated by several vertical wells, which operate at the given constant bottomhole pressure. The rock permeability is given as non-uniform and isotropic. The reservoir thickness, rock density, residual resistance factor, and viscosity of pure water, oil and gas are known and constant. The functions $\phi(p)$, $B_w(p)$, $B_o(p)$, $B_g(p)$, $k_{rw}(S_w)$, $k_{ro}(S_w, S_g)$, $k_{rg}(S_g)$, $R_s(p, p_b)$, $\mu_{w\_eff}(C_p, C_s)$, $C_a(C_p)$ are given and the values of required functions are known at the reference time.

The reservoir boundary is considered as a non-flow (barrier) boundary:

$$\left.\frac{\partial p}{\partial \vec{n}}\right|_{\Gamma} = 0.$$

Thus, the values of p, S_w, S_g, p_b, C_p, C_s, q_w, q_o, q_g should be calculated for the subsequent time period $t > 0$.

4. Grid generation

A two-dimensional Voronoi grid is generated from a given set of points on the plane called 'grid nodes'. This is such partition of the plane where each grid cell forms a set of points being closer to one particular node than to any other node (Figure 1). Let's denote the grid nodes as P_i, $i = 1...N$. Then, the corresponding cells of the Voronoi grid G_i can be defined as:

$$G_i = \left\{ x \in \Omega \middle| \, |x, P_i| \le |x, P_j|, j = 1...\mathrm{N},\ \mathrm{i} \ne \mathrm{j} \right\},\ \mathrm{i} = 1...\mathrm{N}.$$

For representing the reservoir as a grid, the grid nodes should be arranged on contour Γ and as a certain pattern within the contour. First of all, the entire $\mathbb{R}^2$ plane is covered by the Voronoi segmentation for using the sweeping line algorithm [3]. After that, the algorithm [5] cuts off the parts of the cells beyond the contour Γ (Figure 1). Around the well locations we can increase the density of grid nodes and arrange them radially from the center of the well. This approach allows reflecting the radial nature of reservoir pressure distribution in the wellbore vicinity. If in any area to increase a node density, it defines a refined grid ensuring more accurate solution of the problem. Fine grids should be used for the reservoir regions with high pressure gradient and strong permeability variation. The classical rectangular grid, used for local grid refinement, will require the modification of mass conservation equations.

We have to point to one important property of the Voronoi grid: a segment connecting the nodes of any two neighboring cells is perpendicular to a face between these cells (Figure 2). And thus, the reservoir pressure gradient in the normal direction to the face between two cells can be approximated by the finite difference of pressure values at the nodes of these cells – in the same way as in the rectangular grid case. This approach is called 'two-point flux approximation'.

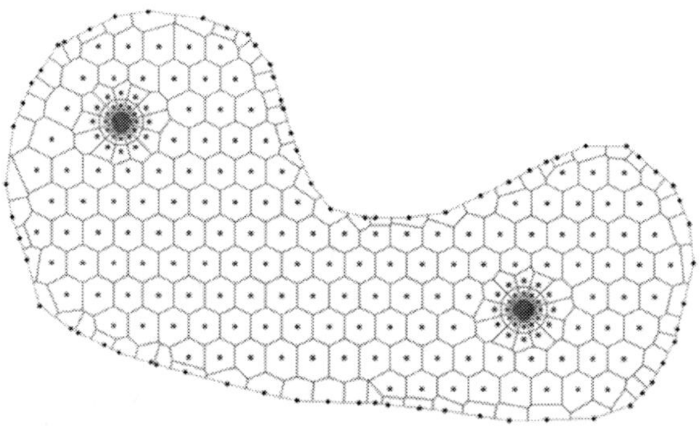

Figure 1. Reservoir representation as a Voronoi grid with a hexagonal pattern and local grid refinement in the well vicinity.

5. Solution of the problem

Though the computational grid is two-dimensional, the cells were considered as three-dimensional of height h equal to the reservoir thickness. Let's make discretization of equation (4). Integrating equation (4) by the cell volume i, we obtain:

$$\int_{V_i} \left\{ \frac{\partial}{\partial t}\left[\phi\left(\frac{S_w C_p}{B_w} \right) \right] + \frac{\partial}{\partial t}\left[(1-\phi)\rho_r C_a \right] \right\} dV + \int_{V_i} \nabla \cdot \left(\frac{\vec{u}_w C_p}{B_w} \right) dV + \int_{V_i} q_w C_p \delta_{well} dV = 0.$$

Let subscript i to determine the average value inside cell i. Replacing the integrands by the means and applying the Gauss-Ostrogradsky formula, the resulted equation is:

$$V_i \frac{\partial}{\partial t}\left[\phi\left(\frac{S_w C_p}{B_w}\right) + (1-\phi)\rho_r C_a\right]_i + \int_{S_i} \frac{C_p}{B_w}\vec{u}_w \cdot \vec{n} dS + q_{w,i} C_{p,i} = 0.$$

The integral over the surface S_i of cell i in the last equation can be substituted by a sum of the surface integrals over each face of this cell:

$$V_i \frac{\partial}{\partial t}\left[\phi\left(\frac{S_w C_p}{B_w}\right) + (1-\phi)\rho_r C_a\right]_i + \sum_{j\in\Psi(i)} \frac{C_{p,ij}}{B_{w,ij}} \int_{S_{ij}} \vec{u}_w \cdot \vec{n} dS + q_{w,i} C_{p,i} = 0,$$

$$V_i \frac{\partial}{\partial t}\left[\phi\left(\frac{S_w C_p}{B_w}\right) + (1-\phi)\rho_r C_a\right]_i + \sum_{j\in\Psi(i)} \frac{C_{p,ij} s_{ij}}{B_{w,ij}} \vec{u}_w \cdot \vec{n}_{ij} + q_{w,i} C_{p,i} = 0.$$

Where $\Psi(i)$ = set of cells adjacent to i-th; S_{ij} = face between i-th and j-th cells, s_{ij} = area of this face; $\vec{n}_{ij}$ = normal to face S_{ij}, directed toward the cell j (Figure 2); d_{ij} = distance between nodes P_i and P_j. The subscript ij for all other parameters reflects the mean value of the parameter on face S_{ij}.

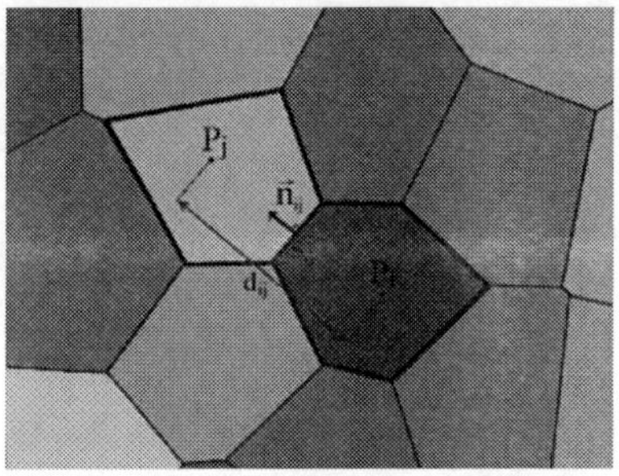

Figure 2. Computation of flow between grid cells.

p_i, p_j are the reservoir pressure values at nodes P_i and P_j, respectively. The fluid flow velocity (7) normal to a face is described by two-point approximation:

$$V_i \frac{\partial}{\partial t}\left[\phi\left(\frac{S_w C_p}{B_w}\right) + (1-\phi)\rho_r C_a\right]_i + \sum_{j\in\Psi(i)} C_{p,ij} s_{ij} \frac{k_{ij} k_{rw,ij}}{\mu_{w\_eff,ij} B_{w,ij} R_{ij}} \left(\frac{p_i - p_j}{d_{ij}}\right) + q_{w,i} C_{p,i} = 0.$$

The time derivative is described by a finite-difference approximation:

$$\frac{V_i}{\Delta t}\left[\left(\phi\frac{S_w C_p}{B_w} + (1-\phi)\rho_r C_a\right)^{n+1} - \left(\phi\frac{S_w C_p}{B_w} + (1-\phi)\rho_r C_a\right)^{n}\right]_i +$$

$$\sum_{j\in\Psi(i)} \left[\frac{s}{d}\frac{k k_{rw} C_p}{\mu_{w\_eff} B_w R}\right]_{ij}^{n+1} \left(p_i^{n+1} - p_j^{n+1}\right) + q_{w,i}^{n+1} C_{p,i}^{n+1} = 0.$$

The value of rock permeability on the cell face k_{ij} is calculated as the cell-average harmonic permeability.

The values of relative permeability, volume factor, viscosity, resistance factor and polymer concentration are calculated for the face in the upstream direction. The equivalent radius of the well grid cell is calculated as:

$$r_{eq} = \left(e^{-2\pi kh} \prod_{j \in \Psi(i)} d_{ij}^{T_{ij}} \right)^{1/\sum_{j \in \Psi(i)} T_{ij}},$$

$$T_{ij} = \frac{k_{ij} s_{ij}}{d_{ij}}.$$

After equations (1) - (3) discretization and a similar procedure for (5) gives a system of five nonlinear equations for the i-th grid cell.

For the grid with N cells the number of equations equal to $5N$. The unknown variables are p_i, $S_{w,i}$, 'gas variable in the i-th cell', $C_{p,i}$ and $C_{s,i}$ for $i = 1...N$. If the oil in the cell is saturated with gas ($S_{g,i} > 0$), then the 'gas variable' equals to gas saturation $S_{g,i}$, in the contrary case – to gas-in-oil solubility factor $R_{s,i}$. Oil saturation $S_{o,i}$ is a known quantity as per equation (6).

If all $5N$ variables are known for a time layer n, then the system of equations, becomes a closed system. Its solution will be the values of unknown variables for the time layer $n+1$. This system of equations in vector form is expressed as:

$$\vec{F}(\vec{x}^{n+1}) = 0. \tag{13}$$

The no-flow boundary condition is valid providing that in the mass conservation equations the flows across the cell boundary are summed only by faces of two neighboring cells. The system is solved by Newton's iteration method:

$$\vec{x}^{\nu+1} = \vec{x}^{\nu} - \mathbf{J}^{-1}(\vec{x}^{\nu}) \cdot \vec{F}(\vec{x}^{\nu}), \;\; \mathbf{J}(\vec{x}^{\nu}) = \frac{d\vec{F}(\vec{x}^{\nu})}{d\vec{x}^{\nu}}.$$

The largest part of the computational time is related to finding the inverse Jacobian matrix $\mathbf{J}^{-1}$, which order linearly depends on the number of cells in the computational grid.

6. Numerical experiment

The relative permeability curves to water and to gas were built using method [7], and relative permeability to oil was calculated from model [8].

The correlation between the viscosity of polymer solution and local concentrations of polymer and salt were assumed as linear. Volumetric coefficients of water and degassed oil and a linear isotherm of polymer adsorption are also linear. The reservoir porosity and permeability are constant.

The injection of polymer solution was simulated for 1750 days (57 months). The polymer and salt concentrations in the injected solution are 2 kg/m3 and 0.01 kg/m3, respectively. After the polymer, pure water was injected during 2700 days (90 months). For numerical simulation 6 different grid configurations were used: a hexagonal locally refined Voronoi grid with 121, 444 and 2678 grid cells and a rectangular grid with 121, 441 and 2601 grid cells (Figure 3).

Throughout the entire period of polymer rim injection, an oil production rate shows the decline (Figure 5), caused by the increased high viscosity areas in the reservoir. After the beginning of pure water injection, in 1750 days the size of these areas decreases due to low viscosity of pure water and partial polymer adsorption by rock surfaces that leads to a sharp increase in the oil production rate. After 2400 days the injected polymer is entirely adsorbed, the oil production profile achieves a plateau. After

3000 days, the injected water reaches a production well causing the oil production rate decrease and water production rate increase.

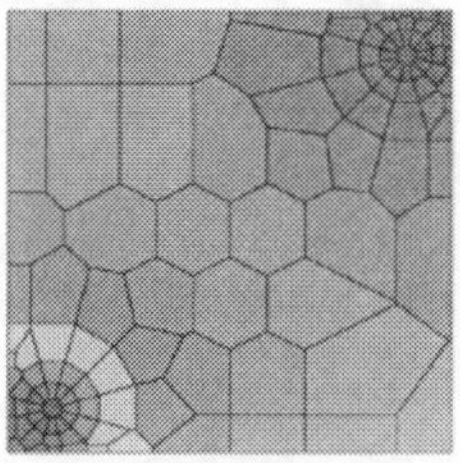

(a) 121 grid cells

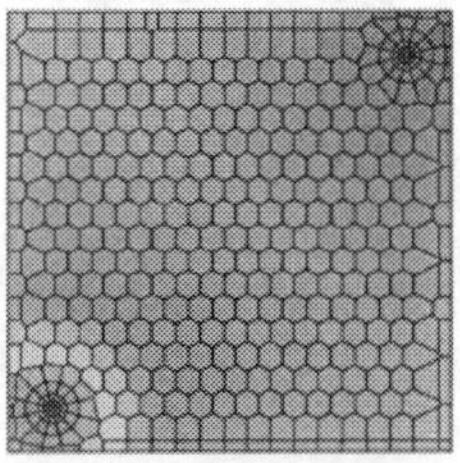

(b) 444 grid cells

(c) 2678 grid cells

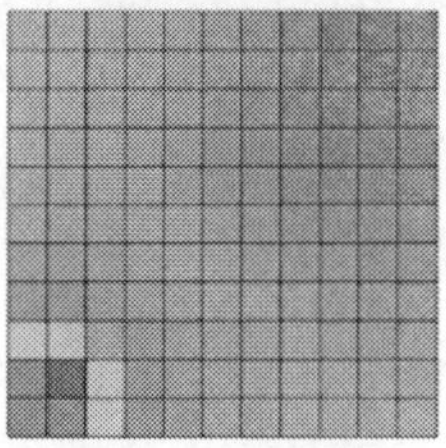

(d) 121 grid cells

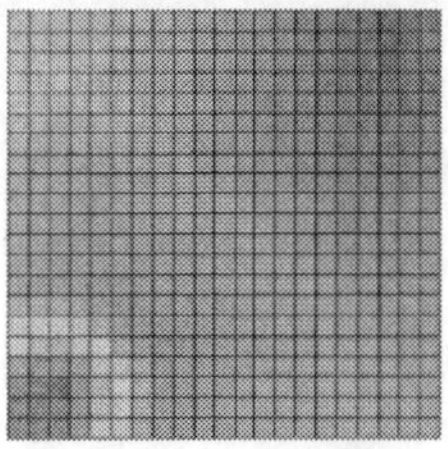

(e) 441 grid cells

(f) 2601 grid cells

Figure 3. Grid configuration and reservoir pressure after 60 months of waterflooding. The hexagonal locally refined Voronoi grid is shown in (a), (b), (c). A rectangular grid is shown in (d), (e), (f).

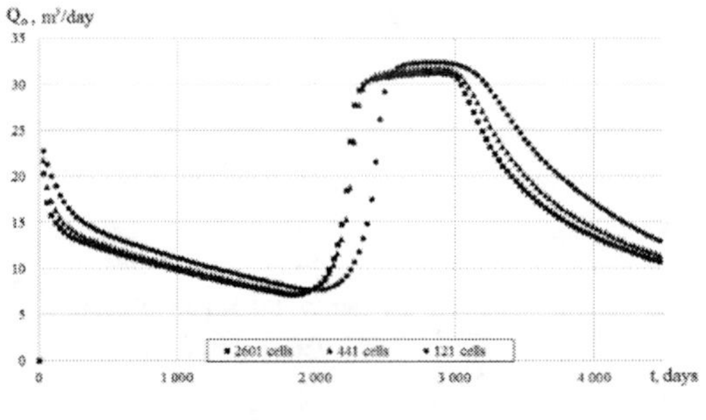

Figure 4. Oil flow rate calculated on the rectangular grid.

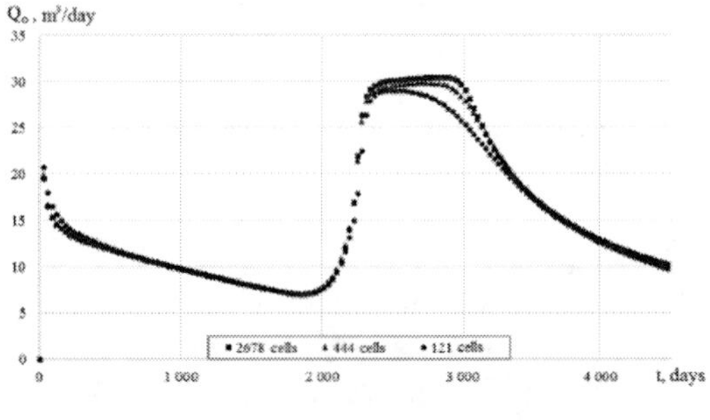

Figure 5. Oil flow rate calculated on the hexagonal locally refined Voronoi grid.

(a) Rectangular grid with 2601 grid cells.

(b) Hexagonal locally refined Voronoi grid with 2678 grid cells.

Figure 6. Water saturation after 80 month of waterflooding.

Figure 3 demonstrates that the Voronoi grid gives more realistic picture of the reservoir pressure distribution in the well vicinity than a rectangular grid with the same number of grid cells. From analysis of Figures 4 and 5 the following conclusion is made: unlike the rectangular grid, changes in the grid cell number the Voronoi grid yields smaller dispersion in oil production. Using this grid, the production rates calculated on a small number of grid cells are close to the rates for a large number of grid cells. Thus, the Voronoi grid keeps the calculation accuracy for well indicators giving a significant reduction in the number of grid cells. This can be explained by the fact that in the regions with high pressure gradient the solution quality does not change much, since the grid cells in these regions always remain fine and the cells with wells do not change their location. However, with a lower number of grid cells the fluid front is worse described at a distance from the wells, which is expressed in the oil production profile smoothing at water breakthrough into the production well (Figure 5).

Figure 6 illustrates the grid orientation effect, causing the distortion of water front shape. Compared with the rectangular grid, the Voronoi grid gives more realistic picture of the water front movement when using the hexagonal pattern; however, it required longer computational time due to the Jacobian matrix structure complication.

7. Conclusion

A tool was developed for evaluating the polymer flooding performance, which can be used for defining optimal polymer properties and injection volumes.

The Voronoi grid with a given number of cells allows achieving solution accuracy by redistributing the grid cell density through the reservoir regions. With local grind refinement near the wells, the computation of well technological parameters is characterized by higher accuracy, but with coarse estimation of the fluid front movement. The grid orientation effect can be reduced from using different grid pattern (rectangular, hexagonal or octagonal). The Voronoi grid describes much more accurately the structure of complex reservoirs, horizontal well geometry and wells with hydraulic fracture than the classical rectangular one.

Summarizing, we can state that the Voronoi grid provides more detail solution to the problem as well as flexibly in balancing between accuracy and computational time.

8. Acknowledgments

This study was supported by the Russian Foundation for Basic Research (Project 17-41-020226 r_a) and UNTC LLC.

References

[1] Zubkov P T and Fedorov K M *Izvestiia RAN Mekhanika zhidkosti i gaza* 1994(**1**) 98-103
[2] Nikiforov A I and Sadovnikov R V *Math Models Comput Simul* 2017 **9(2)** 221-231
[3] Fortune S *Algorithmica* 1987 **2(1-4)** 153–174
[4] Kim S K, Bae G O and Lee K K Geosci J *Geosciences Journal* 2015 **19(3)** 527–535
[5] Palagi C L and Aziz K *SPE Advanced Technology Series* 1994 **2(2)** 69–77
[6] Abou-Kassem J H and Aziz K *Society of Petroleum Engineers* 1985 **25(4)** 573–579
[7] Corey A T *Producers monthly* 1954 **19(1)** 38–41
[8] Stone H L *Journal of Canadian Petroleum Technology* 1973 **12(4)** 53–61

IOP Conf. Series: Journal of Physics: Conf. Series **936** (2017) 012002 doi:10.1088/1742-6596/936/1/012002

Surface preparation effect on oxidation kinetics of Ni-base superalloy

Wojciech J Nowak[1], **Bartek Wierzba, Jan Sieniawski**
Department of Materials Science, Faculty of Mechanical Engineering and Aeronautics, Rzeszow University of Technology, al. Powstancow Warszawy 12, 35-959 Rzeszow, Poland

E-mail: w.nowak@prz.edu.pl

Abstract. In the present study an oxidation behavior of commercially available Ni-base superalloy IN 713C with different surface finishing, namely polished up to 1 μm diamond paste (resulting in R_a=0.023) and ground by SiC paper with 80 grit (R_a=1.23), were investigated at 950°C in air. It was found that ground sample developed very thin, protective α-Al_2O_3 scale. Moreover, locally formation of θ-Al_2O_3 in the forms of spikes was observed. For the sample with polished surface formation of outer chromia scale accompanied with internal oxidation of aluminum was observed. Moreover, for specimen which does not form continuous alumina scale or sub-scale, the GD-OES depth profiles showed an enrichment of boron at the oxide scale/alloy interface. It was proposed that the formation of protective alumina scale on ground samples is caused by introducing the defects into the near-surface region of the material, which are an easy diffusion paths for the elements forming a protective oxide scales.

1. Introduction

Ni-based alloys are optimized to provide a very good strength, also at elevated temperatures. To increase high temperature oxidation resistance an elements like Al and Cr are added to form protective oxide scales (chromia and alumina respectively) during exposures at high temperatures in atmospheres containing oxygen [1, 2]. The alternative method to provide a resistance against environmental reactions (oxidation, nitridation, carburization or sulfidation) is applying a protective coatings such as MCrAlY [3] (where M is mainly Ni or Co) or β-NiAl [4, 5], which are an alumina forming materials. However, coatings production is time consuming, results in additional component costs, and can negatively affect alloy mechanical properties, such as fatigue strength.
Therefore, mechanical surface treatment seems to be an alternative method for promoting protective oxide scales formation without changing the material composition and/or manufacturing methods.
The influence of metallic samples preparation, formation was already investigated in literature. Giggins and Pettit [6] investigated the influence of grain size and surface deformation on selective oxidation of chromium in Ni-Cr alloys. The authors observed that oxidation behavior of the samples with deformed surface, due to grit blasting, was similar to that with fine grained structure. This phenomenon was explained by producing fine grains via recrystallization of cold worked layer during sample exposure at elevated temperature. The influence of grain size and deformation on selective oxidation of chromium was observed up to 30 wt.% of Cr. This was found to be correlated with enhanced diffusion of chromium at the grain boundaries of the alloy.

[1] To whom any correspondence should be addressed.

The effect of grain boundaries density, introduced into the material by cold working with subsequent recrystalization, on the oxidation and carburization of Alloy 800 (Fe-base, 35%-Ni, 20%Cr and minor amount of C, Mn, S, Si, Cu, Al and Ti) was already investigated by Leistikow et al. [7]. The authors found that samples containing higher grain boundaries densities showed the smallest mass gain. With decreasing density of grain boundaries the tendency of internal oxidation increased. This observation confirm the role of grain boundaries as an easy diffusion path for oxide forming elements. Therefore, from the crystallographic point of view, the grain boundaries present in the polycrystalline materials are also classified as a defects.

In the recent work of Sudbrack et al. [8] an effect of surface preparation of two single crystal, Ni-based superalloys LDS-1101+HF and CMSX-4+Y on their oxidation behavior was investigated. The samples surface finish includes low-stress ground, polishing up to 9 μm and up to 3 μm as well as electrochemical thinning. The authors observed that the type of the oxide scale formed on samples strongly depends on its surface preparation.

The aim of the present work is to investigate the influence of different surface finishing of commercially based Ni-base superalloy IN 713C on its oxidation behavior at 950°C during air exposure. Moreover, the model for internal oxidation process will be proposed. The model will be based on the generalized Darken method and multiphase approximation. The two-phase zones (internal oxidation) will be mathematically described.

2. Experimental

From the rod of convectionally cast Ni-based superalloy IN 713 C (Ni-bal., Cr-14, Ta-2, Mo-4, Nb-5.1, Al-6, Ti-1, C-0.2, B-0.01 given in wt. %) a coupons with 20 mm diameter and 2 mm height were cut. The surfaces of the samples were prepared with three different surface finishing: polishing up to 1 μm and grinding using 80 grit SiC paper. Surface of such prepared samples were investigated by the profilometer HOMMEL to determine the roughness parameters and using scanning electron microscopy Hitachi S3400N (SEM) to investigate the surface topography. Such prepared samples were oxidized for 48 hours in laboratory air at 950°C using tube furnace Carbolite STF 16/450. The samples were balanced before and after oxidation tests using microbalance RADWAG WAA 100/C/1 for oxidation kinetics determination. After exposure samples were investigated using glow discharge optical emission spectrometry (GD-OES). The GD-OES depth profiles were quantified using the procedure described in references [9, 10, 11]. After GD-OES depth-profiling, all samples were sputtered with very thin gold layer using cathodic evaporation, electrolithycally coated with nickel and mounted in epoxy resin. Metallographic cross section of the alloy after oxidation as well as in the as-cast condition were prepared by series of grinding and polishing steps, finishing with a fine polishing in SiO_2 suspension with 0.25 μm granulation. The cross sections were analyzed using optical microscope Nikon Epiphot 300 and scanning electron microscope Hitachi S3400N (SEM).

3. Oxidation modeling

In this chapter the internal oxidation model will be proposed. The model will be based on the generalized Darken method with drift determination and multiphase approximation of components during the diffusion process. The main law is the mass conservation for average component of the ternary diffusion couple, e.g. Ni-Cr-O.

$$\frac{\partial \overline{c}_i}{\partial t} = -\frac{\partial \overline{J}_i}{\partial x}, \quad i = 1,2,3 \tag{1}$$

The average concentration can be defined from the Gibbs concentration triangle, when the values of the phase ratio f^α and f^β are known ($f^\alpha + f^\beta = 1$):

$$\overline{c}_i = f^\alpha c_i^\alpha + f^\beta c_i^\beta \tag{2}$$

where $\overline{J}_i$ is the average flux of the component, c_i^α and c_i^β are the concentration in α and β phase, respectivelly.

IOP Conf. Series: Journal of Physics: Conf. Series **936** (2017) 012002 doi:10.1088/1742-6596/936/1/012002

$$\bar{J}_i = f^\alpha J_i^\alpha + f^\beta J_i^\beta, \quad i = 1,2,3 \tag{3}$$

The fluxes in α and β phase are defined after Darken [12]:

$$J_i^\alpha = -D_i^\alpha \frac{\partial c_i^\alpha}{\partial x} + c_i^\alpha v^\alpha, \quad i = 1,2,3 \tag{4}$$

and

$$J_i^\beta = -D_i^\beta \frac{\partial c_i^\beta}{\partial x} + c_i^\beta v^\beta, \quad i = 1,2,3 \tag{5}$$

D_i^α and D_i^β are the intrinsic diffusion coefficients of the components. The drift velocity can be calculated from the volume continuity equation in each phase:

$$v^\alpha = \frac{1}{c^\alpha} \sum_{i=1}^{3} D_i^\alpha \frac{\partial c_i^\alpha}{\partial x} \quad \text{and} \quad v^\beta = \frac{1}{c^\beta} \sum_{i=1}^{3} D_i^\beta \frac{\partial c_i^\beta}{\partial x} \tag{6}$$

Finally, the mass conservation for average concentration in ternary system can be rewritten in the following form:

$$\frac{\partial \bar{c}_i}{\partial t} = -\frac{\partial}{\partial x}\left(-f^\alpha D_i^\alpha \frac{\partial c_i^\alpha}{\partial x} - f^\beta D_i^\beta \frac{\partial c_i^\beta}{\partial x} + f^\alpha c_i^\alpha v^\alpha + f^\beta c_i^\beta v^\beta \right), \quad i = 1,2,3 \tag{7}$$

when the phase diagram is known, the tie-lines and thermodynamic properties, the reactive diffusion can be calculated.

4. Results and discussion

Figure 1 shows the SEM/BSE image of IN 713C in the as-received conditions. The alloys shows the microstructure based on the γ-Ni matrix strengthened by the γ'-Ni_3Al precipitates [13]. Due to the high Nb content an relatively high carbon content the Nb-carbides formation is observed (see Fig. 1). The Nb-carbides formation is also one of the strengthening mechanisms [13]. A similar microstructure of IN 713C has been observed by Binczyk and Śleziona [14] or Matysiak et al. [15]. Moreover the surface topography captured using SEM/BSE 3D mode revealed that the surface of the sample prepared with 1 μm polishing differs substantially from the sample with ground surface. This observation is strongly supported by the roughness profiles presented in Fig. 2. The average R_a value based on five measurement is presented in Table 1. The measurements results reveal that the R_a values differ of about two orders of magnitude between the 1 μm and 80 grit SiC-paper.

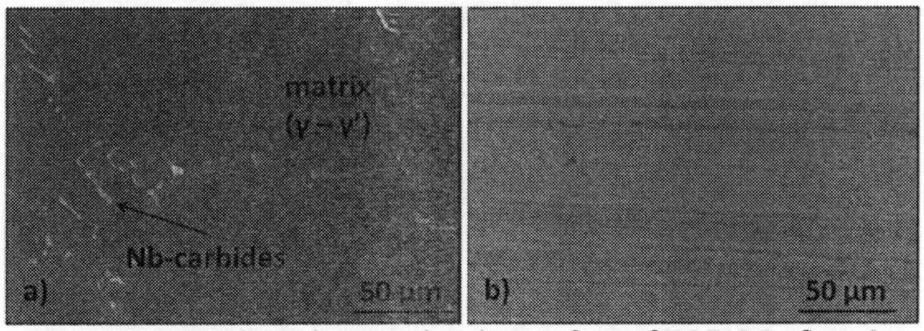

Figure 1. SEM/BSE images showing surface of IN 713C after: a) polishing up to 1 μm polishing and b) grinding using 80 grit SiC paper

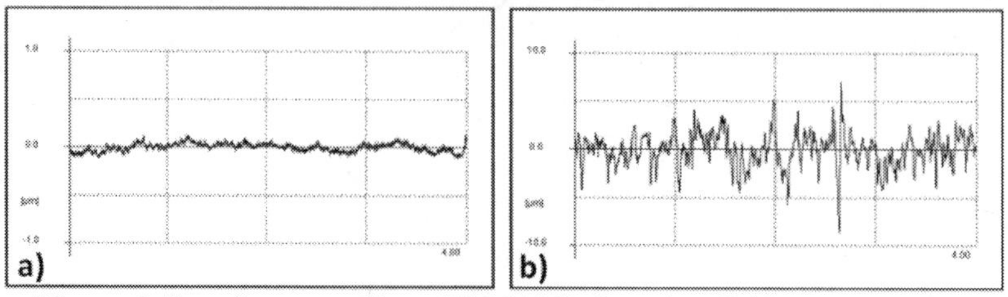

Figure 2. Roughness profiles of IN 713C after: a) polishing up to 1 μm polishing and b) grinding using 80 grit SiC paper

Table 1 The average values of the R_a values measured on polished and ground IN 713C

Average measured value	1 um	80 grit
R_a	0.023	1.240
Standard deviation	0.002	0.212

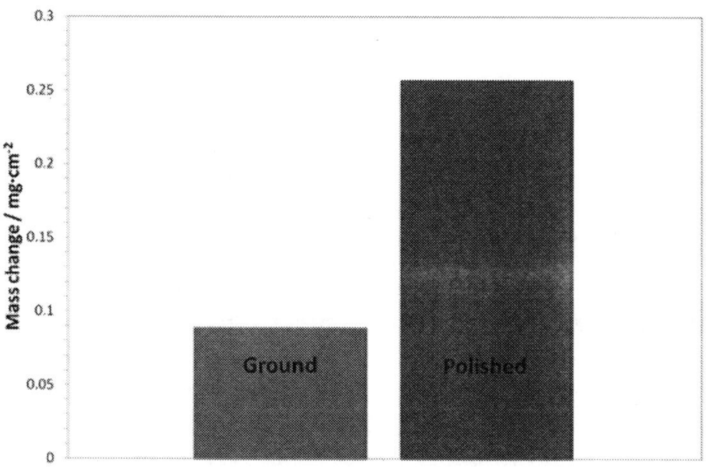

Figure 3. Mass change plot of ground and polished IN 713C after air exposure at 950°C for 48 hours.

The GD-OES depth profile of polished IN 713C (Fig. 4 a) showed that the outer part of the oxide scale is enriched in Ni. Below outer Ni-peak a co-enrichment in Cr and Nb is present. Moreover whole outer oxide scale is slightly enriched in Ti. Also an enrichment in boron is observed at the oxide scale/alloy interface. Below the boron peak a zone of boron depletion is present. An enrichment of aluminum accompanied with decreasing oxygen profile suggests presence of zone of internal Al-oxidation. The GD-OES depth profile of ground specimen revealed an enrichment of aluminum in the outer part of the oxide scale accompanied with slight enrichment in Ti and Nb (Fig. 4 b). For ground specimen no boron peak is found. A similar effect of hampering boron diffusion through alumina scale was observed by Nowak et al. [16].

Considering the SEM images of the cross-sections (Fig. 5 a and b) one can observe that on polished specimen non uniform oxide scale is formed, i.e. regions with an outer Cr-rich oxide scale and internally oxidized aluminum and regions with relatively thin oxide scale without internal oxidation zone are present. Contrary, polished sample developed thin, aluminum rich oxide scale. However, as shown if Fig. 5 b, two types of oxides morphology are present: a dense and flat oxide scale and oxides in the form of spikes. The first one are typical for the α-Al_2O_3 while the spikes are characteristic for θ-Al_2O_3. It is known from the literature that at 950°C a metastable θ-Al_2O_3 can form [17]. An nodules observed on the surface of oxidized ground IN 713C might be correlated with the presence of θ-Al_2O_3. Formation of θ-Al_2O_3 might be supported by the larger zone of γ'-depletion below the spiky oxides, which in turn is correlated with higher growth rate o of θ-Al_2O_3 as compared with α-Al_2O_3 and the same consumption of Al from its in the γ' is larger.

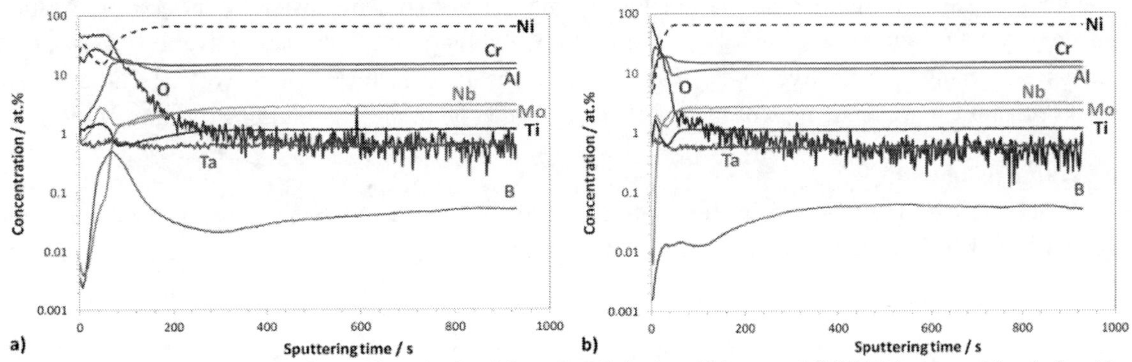

Figure 4. The GD-OES depth profiles obtained for: a) polished and b) ground IN 713C exposed in air for 48 hours at 950°C.

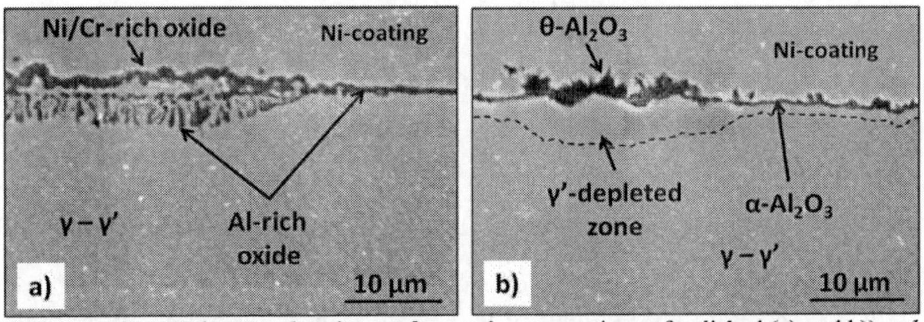

Figure 5. SEM/SE images showing surface and cross sections of polished (a) and b)) and ground (c) and d)) IN 792 after isothermal oxidation test at 950°C for 48 hours in air. The dashed line indicates the zone of γ' depletion

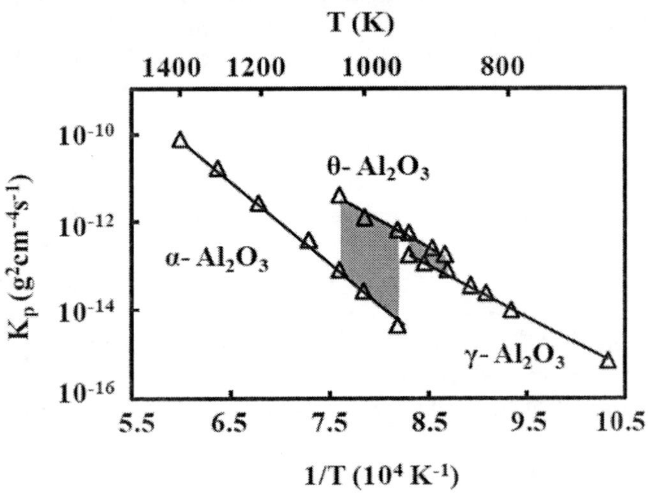

Figure 6. Arrhenius plot of parabolic rate constant (k_p) for the oxidation of pure NiAl [17]

5. Summary and conclusions

As shown in the present work, the type of formed oxide scale can be affected by different surface treatment. When the surface was ground (R_a=1.23), the alloy formed a protective, continuous alumina layer at all tested temperatures, since polished specimens (R_a=0.02) developed outer Ni/Cr-mixed oxide and internally oxidized aluminium. Therefore, one can conclude that increasing of R_a value by two orders of magnitude forced material to form a protective alumina scale. It is known that at lower temperatures the grain boundary diffusion plays bigger role than lattice diffusion. The reason for that is lower activation energy of the boundary process caused by more disordered structures in the boundaries [18]. Considering the latter, the hypothesis is that with different surface preparation a different amount and type of defects are introduced which also lowers an activation energy of the elements forming a protective oxide scales. It was observed in the present work, that the different surface treatment alters the type of the oxide scale formed on nominally alumina forming alloy.

6. Acknowledgment

This work has been supported by the National Science Centre (NCN) in Poland, decision number 2015/19/P/ST8/03995 in the frame of grant "POLONEZ 1".

7. Literature

[1] R. Prescott and M. J. Graham, "The Formation of Aluminum Oxide Scales on High-Temperature Alloys", Oxidation of Metals, Vol. 38, Nos. 3/4, 1992
[2] C. S. Giggins, F. S. Pettit, "Oxidation of Ni-Cr-Al Alloys Between 1100° and 1200°C", Solid State Science Vol. 118, No. 11, pp 1782-1790, (1971)
[3] K. Yuan, R. Eriksson, R. L. Peng, X.-H. Li, S. Johansson, Y.-D. Wang, "Modelling of microstructural evolution and lifetime prediction of MCrAlY coatings on nickel based superalloys during high temperature oxidation", Surface and Coating Technology, Volume 232, pp 204-215, (2013)
[4] M.J. Pomeroy, "Coatings for gas turbine materials and long term stability issues", Materials and Design 26, 223-231, (2005)
[5] K. H. Stern, "Metallurgical and Ceramic Protective Coatings", Chapman and Hall, 2-6 Boundary Row, London SE1 8HN, UK, (1996)
[6] C. S. Giggins and F. S. Pettit, "The effect of Alloy Grain-Size and Surface Deformation on the selective Oxidation of Chromium in Ni-Cr Alloys at temperatures of 900°C and 1100°C", Transactions of the Metallurgical Society of AIME, Vol. 245, 2509-2514, 1969
[7] S. Leistikow, I. Wolf and H. J. Grabke, "Effect of cold work on the oxidation behavior and carburization resistance of Alloy 800", Werkstoffe und Korrosion 38, 556-562 (1987)
[8] C. K. Sudbrack, D. L. Beckett, R. A. Mackay, "Effect of Surface Preparation on the 815°C Oxidation of Single Crystal Ni-Based Superalloys", JOM, Vol. 67, No. 11, 2015
[9] J.P. Pfeifer, H. Holzbrecher, W.J. Quadakkers, J. Speier: Quantitative analysis of oxide films on ODS*-alloys using MCs+-SIMS and e-beam SNMS**, *J. Anal.Chem.,* **346**, 1993, 186 – 191.
[10] W.J Quadakkers, A. Elschner, W. Speier, H. Nickel: Composition and growth mechanisms of alumina scales on FeCrAl-based alloys determined by SNMS, *App. Surf. Sci.,* **52**, 1991, 271 – 287
[11] Wojciech J. Nowak: Characterization of oxidized Ni-base superalloys by GD-OES, *JAAS*, 2017, DOI: 10.1039/C7JA00069C
[12] Danielewski M, Wierzba B, "Thermodynamically consistent bi-velocity mass transport phenomenology", Acta Materialia 58, 2010, 6717
[13] J.R. Davis: ASM Specialty Handbook, Heat Resistant Materials. ASM International - Materials Park, Ohio, 1997
[14] F. Binczyk , J. Śleziona: Phase transformations and microstructure of IN-713C nickel superalloy, *ARCHIVES OF FOUNDRY ENGINEERING* **9**, 2009, 109-112

[15] H. Matysiak, M. Zagorska, A. Balkowiec, B. Adamczyk-Cieslak, R. Cygan, J. Cwajna, J. Nawrocki, and K. J. Kurzydłowski: The Microstructure Degradation of the IN 713C Nickel-Based Superalloy After the Stress Rupture Tests, *Journal of Materials Engineering and Performance* **23**, 2014, 3305-3313

[16] W.J. Nowak, D. Naumenko, A. Jałowicka, D.J. Young, V. Nischwitz and W.J. Quadakkers: Effect of alloy composition on the oxidation-induced boron depletion in cast Ni-base superalloy components, *Mat. Corr.* **68** No. 2, 2017, 171- 185

[17] H.J.Grabke, Oxidation of NiAl and FeAl, Intermetallics, **7** (10), 1999, 1153-1158

[18] D. Young, High Temperature Oxidation and Corrosion of Metals, Second ed., Elsevier, 2016

Dissipative gravitational bouncer on a vibrating surface

J. S. Espinoza Ortiz[1], **R. E. Lagos**[2]

[1]Departamento de Física, UAEF, Universidade Federal de Goiás, Catalão, Goiás 75704/020, Brasil.
[2]Departamento de Física, IGCE, UNESP, Rio Claro, São Paulo 13500/970, Brasil.

E-mail: `jsespino@physics.org`

Abstract. We study the dynamical behavior of a particle flying under the influence of a gravitational field, with dissipation constant λ (Stokes-like), colliding successive times against a rigid surface vibrating harmonically with restitution coefficient α. We define re-scaled dimensionless dynamical variables, such as the relative particle velocity Ω with respect to the surface's velocity; and the real parameter τ accounting for the temporal evolution of the system. At the particle-surface contact point and for the $k'th$ collision, we construct the mapping described by $(\tau_k\,, \Omega_k)$ in order to analyze the system's nonlinear dynamical behavior. From the dynamical mapping, the fixed point trajectory is computed and its stability is analyzed. We find the dynamical behavior of the fixed point trajectory to be stable or unstable, depending on the values of the re-scaled vibrating surface amplitude Γ, the restitution coefficient α and the damping constant λ. Other important dynamical aspects such as the phase space volume and the one cycle vibrating surface (decomposed into absorbing and transmitting regions) are also discussed. Furthermore, the model rescues well known results in the limit $\lambda = 0$.

1. Introduction

The dynamical behavior of a particle bouncing on a vertically vibrating surface has been extensively studied, both theoretically and experimentally [1, 2, 3]. The model was established as an alternative to the Fermi-Ulam model for the cosmic ray acceleration mechanism introduced by Fermi [4, 5]. A fundamental characteristic of the model lies on its dynamical map exactly iterated at any time, under the assumption that the ball is considered small so the surface is not affected by repeated impacts of the ball. There are different types of ball-particle collision mechanisms to be considered, displaying diverses periodic, non-periodic and chaotic motion regimes. However, most of the map studies assume that the distance traveled by the free falling ball between impacts is large compared to the overall displacement of the vibrating surface, and an approximate equation for the time interval between impacts, is obtained [6]. In this work, we study a model that considers the explicit equation of motion of the ball, as in [7]. In addition, we assume the particle to be under the influence of a Stokes-like damping force. In section 2 we present the dynamical equations governing the particle-surface system, and the mapping is presented in section 3. We compute and discuss the stability of the fixed points in section 4. General aspects related to the dynamics of the system are present in section 5 and the conclusions follow in section 6.

2. The bouncer

Let us consider an unitary mass particle flying under the influence of a gravitational field and a Stokes-like damping force. The particle collides repetitive times with a vibrating rigid platform governed by the amplitude $s(t) = a\sin(\omega t)$. The nature of the collision is defined by the restitution coefficient α, a real positive parameter. The particle dynamics is described by Newton's second law: $\ddot{x}(t) = -\lambda\dot{x}(t) - g$, recasted, in convenient dimensionless variables, as

$$\ddot{X}(\tau) = -\tilde{\lambda}V(\tau) - 2\,, \tag{1}$$

where $t = T\tau$, $x = \frac{g}{2}T^2X$, $\dot{x} = \frac{g}{2}T\dot{X}$, $\ddot{x} = \frac{g}{2}\ddot{X}$ and $\tilde{\lambda} = \lambda T$ with $T = \frac{2\pi}{\omega}$. Also we define $\triangle\tau = \tau - \tau_o \geq 0$ and integrate the equation with initial conditions $X(\tau_o)$ and $V(\tau_o)$, yielding

$$\begin{aligned} V(\tau) &= -\tfrac{2}{\tilde{\lambda}} + \left[V(\tau_o) + \tfrac{2}{\tilde{\lambda}}\right]\exp(-\tilde{\lambda}\triangle\tau)\,, \\ X(\tau) &= X(\tau_o) - \tfrac{2}{\tilde{\lambda}}\triangle\tau - \tfrac{1}{\tilde{\lambda}}\left[V(\tau_o) + \tfrac{2}{\tilde{\lambda}}\right]\left[\exp(-\tilde{\lambda}\triangle\tau) - 1\right]. \end{aligned} \tag{2}$$

Notice, the vibrating surface is now given by $S(\tau) = \frac{\Gamma}{2\pi}\sin(2\pi\tau)$ where $\Gamma = \frac{4\pi}{g}\frac{a}{T^2}$.

3. The Mapping

The temporal variables τ_1 and τ_0 respectively are, the time at which the first collision occurs and the time at which the system initiates its dynamics. We define the particle-surface distance as $D(\tau) = X(\tau) - S(\tau)$ and its derivative (relative velocity) denoted by $\overline{\Omega}(\tau)$. $\Omega_o = V(\tau_o) - \Gamma\cos(2\pi\tau_o)$ is the relative velocity at τ_o (the beginning of the motion), where $X(\tau_o) = S(\tau_o)$. Thus, for $\tau_o \leq \tau \leq \tau_1$, we have

$$\begin{aligned} D(\tau) &= \tfrac{\Gamma}{2\pi}\left[\sin(2\pi\tau_o) - \sin(2\pi\tau)\right] - \tfrac{2}{\tilde{\lambda}}\triangle\tau + \tfrac{1}{\tilde{\lambda}}\left[\Omega_o + \Gamma\cos(2\pi\tau_o) + \tfrac{2}{\tilde{\lambda}}\right]\left[1 - \exp(-\tilde{\lambda}\triangle\tau)\right], \\ \overline{\Omega}(\tau) &= -\Gamma\cos(2\pi\tau) - \tfrac{2}{\tilde{\lambda}} + \left[\Omega_o + \Gamma\cos(2\pi\tau_o) + \tfrac{2}{\tilde{\lambda}}\right]\exp(-\tilde{\lambda}\triangle\tau)\,. \end{aligned} \tag{3}$$

We denote Ω_1 as the relative velocity just after the collision. We construct the dynamical mapping $\mathcal{M} : (\tau_o, \Omega_o) \to (\tau_1, \Omega_1)$ by imposing at τ_1 the relative distance and the relative velocity to satisfy

$$\begin{aligned} D(\tau_1) &= 0\,, \\ \Omega_1 &= -\alpha\,\overline{\Omega}(\tau_1)\,. \end{aligned} \tag{4}$$

Figure 1 schematically depicts the dynamics in the (τ, Ω)-space.

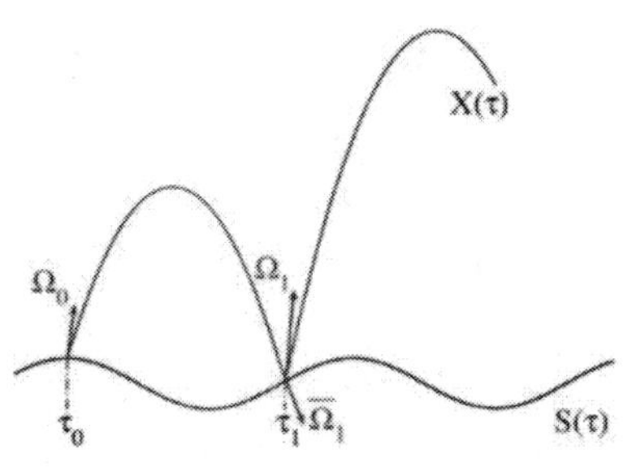

Figure 1. The dynamic of the system represented in the $(\tau\,,\Omega)$ space.

4. Fixed points and their stability

The mapping fixed points are defined by $(\tau_\star, \Omega_\star)$, satisfying the condition $\Omega_1 = \Omega_o = \Omega_\star$ and $\tau_o = \tau_\star$, $\tau_1 = \tau_\star + m$ (the integer m represents a platform cycle), then conditions given in Eqn.(4) become:

$$\begin{aligned} D(\tau_\star + m) &= 0\,, \\ \Omega_\star &= -\alpha\,\overline{\Omega}(\tau_\star + m)\,, \end{aligned} \tag{5}$$

yielding

$$\begin{aligned} \Gamma\cos(2\pi\tau_\star) = \beta(\alpha, m, \tilde{\lambda}) &= \frac{2\,m}{(1+\alpha)}\frac{[1+\alpha\exp(-\tilde{\lambda}m)]}{[1-\exp(-\tilde{\lambda}m)]} - \frac{2}{\tilde{\lambda}}\,, \\ \Omega_\star &= \frac{2\,\alpha}{(1+\alpha)}m\,. \end{aligned} \tag{6}$$

We discuss the stability of the fixed points by allowing small displacements around their neighborhood, $\tau_o = \tau_\star + \delta\tau_o$, $\Omega_o = \Omega_\star + \delta\Omega_o$, $\tau_1 = \tau_\star + m + \delta\tau_1$, $\Omega_1 = \Omega_\star + \delta\Omega_1$. We define the convenient parameter $\sigma = 2\pi^2 X(\tau_\star) = \pi\Gamma\sin(2\pi\tau_\star)$. Linearizing Eqn. (4) in the standard fashion, we find

$$\begin{pmatrix} \delta\tau_1 \\ \delta\Omega_1 \end{pmatrix} = \begin{pmatrix} [1-(1+\alpha)\sigma x] & \frac{1}{2}(1+\alpha)x \\ 2\alpha\sigma\,[(\alpha - z) - (1+\alpha)(1-\sigma x)] & \alpha\,[z - \sigma(1+\alpha)x] \end{pmatrix} \begin{pmatrix} \delta\tau_o \\ \delta\Omega_o \end{pmatrix}, \tag{7}$$

where $\tilde{\lambda}m\,x = \left(1 - e^{-\tilde{\lambda}m}\right)$ and $z = \alpha\,e^{-\tilde{\lambda}m}$.

The eigenvalues $\chi_\pm$ of the monodromy matrix are computed by $\chi_\pm = \left(\frac{\Sigma}{2}\right) \pm \sqrt{\left(\frac{\Sigma}{2}\right)^2 - \Pi}$, where Σ (trace) and Π (determinant) are given respectively by

$$\begin{aligned} \Sigma &= (1+\alpha z) - (1+\alpha)^2\sigma x\,, \\ \Pi &= \alpha z\,. \end{aligned} \tag{8}$$

Now, we analyze the nature of these solutions and their consequences on the stability of the dynamical mapping:

4.0.1. Complex eigenvalues: $\left(\frac{\Sigma}{2}\right)^2 - \Pi < 0$, yields the condition

$$\frac{(1-\sqrt{\alpha z})^2}{(1+\alpha)^2 x} < \sigma < \frac{(1+\sqrt{\alpha z})^2}{(1+\alpha)^2 x}\,. \tag{9}$$

A convenient parametric solution of the above condition is the linear interpolation $\sigma(q) = \frac{[1+\alpha z - 2\sqrt{\alpha z}(1-2q)]}{(1+\alpha)^2 x}$ with $0 \le q \le 1$. It follows: (i) if $0 \le q \le 1/2$, then $\Sigma = 2\Re\,\chi_\pm > 0$; (ii) if $1/2 \le q \le 1$, then $\Sigma = 2\Re\,\chi_\pm < 0$.

4.0.2. Real eigenvalues: $\left(\frac{\Sigma}{2}\right)^2 - \Pi > 0$, yields: (i) $\sigma < \frac{(1-\alpha z)^2}{(1+\alpha)^2 x}$ and the solution are $\chi_\pm > 0$; (ii)$\sigma > \frac{(1+\alpha z)^2}{(1+\alpha)^2 x}$ and the solutions are $\chi_\pm < 0$.

We compact our results defining: $\sigma_o = \frac{(1+\alpha z)}{(1+\alpha)^2 x}$, $\Delta = \frac{2\sqrt{\alpha z}}{(1+\alpha)^2 x}$ and $\sigma_o^\pm = \sigma_o \pm \Delta$. The table below displays the several eigenvalues regimes, for a given value of σ :

$\sigma < \sigma_o^-$	$\sigma_o^- < \sigma < \sigma_o$	$\sigma_o < \sigma < \sigma_o^+$	$\sigma > \sigma_o^+$
Real: $\chi_\pm > 0$	Complex: $\Re\,\chi_\pm > 0$	Complex: $\Re\,\chi_\pm < 0$	Real: $\chi_\pm < 0$

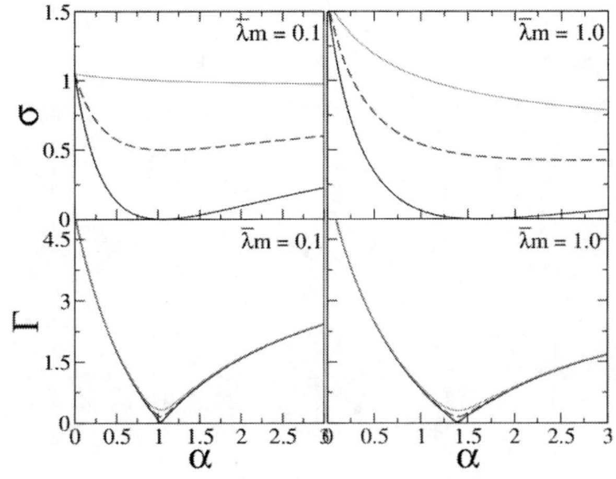

Figure 2. We plott σ and Γ versus α, for $m = 5$ and $\lambda = 2\times 10^{-2}, 2\times 10^{-1}$. Values of σ above the continuous green line and below the continuous black line define real eigenvalues. The region between the continuous-black-line and the continuous-green-line define the complex eigenvalues.

Notice that, for given $\tilde{\lambda}$ and m, Γ is computed, from Eqn. (6) as

$$\Gamma = \sqrt{\left(\tfrac{\sigma}{\pi}\right)^2 + \beta^2}\,. \tag{10}$$

Figure 2 displays σ and Γ as functions of the restitution coefficient α, for $m = 5$ and for $\tilde{\lambda}$ equal to 2×10^{-2} and 2×10^{-1}. For the σ's values above the continuous green line eigenvalues are real and negative, below the continuous black line the eigenvalues are real and positive. For values in between green and black line the eigenvalues are complex.

Finally, we analyze the stability of the system according to the eigenvalues' modules:

Complex χ	$\lvert\chi_\pm\rvert = \alpha\, e^{-\tilde{\lambda}m/2} < 1$, stable/focus $\lvert\chi_\pm\rvert = \alpha\, e^{-\tilde{\lambda}m/2} > 1$, unstable/source	Depending on $\alpha, \tilde{\lambda}$ and m
Real χ	$\lvert\chi_\pm\rvert < 1$, stable/node $\lvert\chi_-\rvert < 1 < \lvert\chi_+\rvert$, hyperbolic/saddle $\lvert\chi_\pm\rvert > 1$, unstable/node	Depending on $\Gamma, \alpha, \tilde{\lambda}$ and m

5. Global aspects of the dynamics

5.1. The phase space volume

Both variables Ω and $\Delta\tau$ will remain bounded, as long as the parameters α and $\tilde{\lambda}$ are appropriately chosen. In order to estimate the upper bounds Ω_{max} and $\Delta\tau_{max}$, we consider the reduced phase space, where the time variable compact (see just above eqn.5). A rough upper bound can be derived from Eqns.(3-4),

$$\begin{aligned} \Omega_1 &\leq \alpha\left\{\Gamma + \tfrac{2}{\tilde{\lambda}} - \left[\Omega_o + \Gamma + \tfrac{2}{\tilde{\lambda}}\right]\exp\left(-\tilde{\lambda}\Delta\tau\right)\right\}, \\ D(\tau) &\leq \tfrac{\Gamma}{\pi} - \tfrac{2}{\tilde{\lambda}}\Delta\tau + \tfrac{1}{\tilde{\lambda}}\left[\Omega_o + \Gamma + \tfrac{2}{\tilde{\lambda}}\right]\left[1 - \exp(-\tilde{\lambda}\Delta\tau)\right]. \end{aligned} \tag{11}$$

We maximize $\Omega_0 = \Omega_1 = \Omega_{max}$, from Eqn. (6) yielding $\Gamma = \frac{2}{\tilde{\lambda}}\left[\frac{\tilde{\lambda}m}{(1+\alpha)}\,\frac{(1-\alpha e^{-\tilde{\lambda}m})}{(1-e^{-\tilde{\lambda}m})} - 1\right]$ and

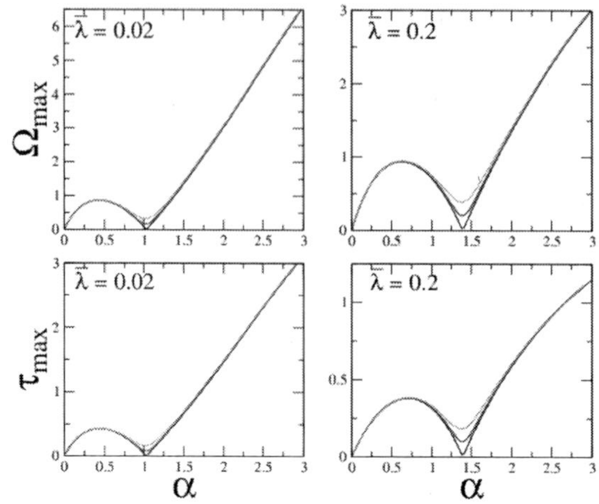

Figure 3. The maximum values for both the relative velocity and time versus the restitution coefficient, for $m = 5$ and two values of the damping force.

by linearizing the latter, we obtain $m_{max} = \frac{(1+\alpha)\,\Gamma/2}{1+\frac{\tilde{\lambda}}{2}\,(1+\alpha)\,\Gamma}$, thus

$$\Omega_{max} = \frac{\alpha\,\Gamma}{1+\frac{\tilde{\lambda}}{2}\,(1+\alpha)\,\Gamma} \quad , \quad \Delta\tau_{max} = \frac{1}{\tilde{\lambda}} \ln\left(\frac{1+\frac{\tilde{\lambda}}{2}(\Omega_{max}+\Gamma)}{1+\frac{\tilde{\lambda}}{2}\Gamma}\right). \tag{12}$$

In Figure 3 we plot the maximum values of both the relative velocity and the time as a function of the restitution coefficient. Both plots feature a critical value α_c ($\approx$ 1 or 1.5 , $\tilde{\lambda}$ dependence). For $\alpha < \alpha_c$ the dynamical variables become bounded .

5.2. Absorbing and transmitting regions

Following [7], one cycle of the vibrating surface can be decomposed into a transmitting and an absorbing region, according to whether or not long flights are allowed. For any take-off time $0 < \tau_0 < 1$, we define (θ, Ω_A) at the point where the collision occurs with relative velocity zero; and from eqn.(3) we obtain

$$\begin{aligned} \frac{\Gamma}{2\pi}\left[\sin\left(2\pi\tau_o\right) - \sin\left(2\pi\theta\right)\right] - \frac{2}{\tilde{\lambda}}\left(\theta - \tau_0\right) + \frac{1}{\tilde{\lambda}}\left[\Omega_A + \Gamma\cos\left(2\pi\tau_o\right) + \frac{2}{\tilde{\lambda}}\right]\left[1 - e^{-\tilde{\lambda}(\theta-\tau_0)}\right] &= 0\,, \\ -\Gamma\cos\left(2\pi\theta\right) - \frac{2}{\tilde{\lambda}} + \left[\Omega_A + \Gamma\cos\left(2\pi\tau_o\right) + \frac{2}{\tilde{\lambda}}\right] e^{-\tilde{\lambda}(\theta-\tau_0)} &= 0\,. \end{aligned} \tag{13}$$

By linearizing and considering $\Gamma \gg \tilde{\lambda}$, we obtain approximate expressions for Ω_A and θ

$$\Omega_A/\Gamma = e^{\tilde{\lambda}(\theta-\tau_o)}\cos(2\pi\theta) \; - \; \cos(2\pi\tau_0)\,. \tag{14}$$

$$\{\sin(2\pi\tau_0) - \sin(2\pi\theta)\} + 2\pi\left\{\cos(2\pi\theta) - \tilde{\lambda}\left[\sin(2\pi\tau_0) - \sin(2\pi\theta)\right]\right\}(\theta - \tau_0) + \tilde{\lambda}\pi\cos(2\pi\theta)(\theta - \tau_0)^2 = 0\,. \tag{15}$$

When $\Omega_0 < \Omega_A$ (absorbing region) the particle rides on the membrane allowing some small bouncing (chattering) or simply stick to the membrane (locking) until the velocity is reverse and $\Omega_A > \Omega_o$. In the latter case, namely $\Omega_o > \Omega_A$ (transmitting region) the ball is launched with a strong impulse.

6. Conclusions

We presented a model strictly derived from the equation of motion of a ball colliding against a vibrating surface, influenced by both gravitational and a Stokes-like damping forces. We constructed the associated mapping governing the system's dynamical motion. We computed the mapping fixed points and discussed its stability. The bounded motion regime exist even for for $\alpha > 1$ and the complexity of the absorbing-transmitting region depends on the dissipative constant $\tilde{\lambda}$. We emphasize that our model rescues well known results in the limit $\lambda = 0$, as in [7] .

In order to obtain a quantitative picture of the many dynamical regimes in parameter space we have ongoing numerical calculations. Our results can be applied to a variety of areas. In particular, granular flows gives rise to a variety of phenomena such as convection, surface excitation and fluidization. In all cases the character of collision between a grain and a wall of the container strongly affects the dynamical behavior of such dissipative media [3]. Another interesting issue refers to the study of the dissipative effects on regular or chaotic trajectories [8], including relativistic effects [2] .

Acknowledgments

The authors would like to thank the support of the Goiás Research Foundation - FAPEG.

References

[1] Pustil'nikov L D 1983 Theor Math Phys **57** 1035-1038
[2] Pinto Rafael S and Letelier Patricio S 2011 Phys Lett A **375** 3273-3278
[3] Falcon E, Laroche C, Fauve S and Coste C 1998 Eur Phys J B **3** 45-47
[4] Lieberman M A and Lichtenberg A J 1971 Phys Rev A **5** 1852-1866
[5] Fermi E 1949 Phys Rev **75** 1169-1174
[6] Oliveira Diego F M and Leonel Edson 2011 New Journal of Physics **13** 123012
[7] Luck J M and Mehta Anita 1993 Phys. Rev. E **48**(5) 3988-3997
[8] Jousseph C A , Kruger T S , Manchein C, Lopes S R , Beims M.W. 2016 Physica A **456** 68-74

IOP Conf. Series: Journal of Physics: Conf. Series **936** (2017) 012004 doi:10.1088/1742-6596/936/1/012004

Quantized Vector Potential and the Photon Wave-function

C Meis[1,3*] and P R Dahoo[2,3]

1 CEA Saclay. National Institute for Nuclear Science and Technology, Université Paris Saclay – 91191 Gif-sur-Yvette, France.
2 LATMOS /IPSL, UVSQ Université Paris-Saclay, UPMC Univ. Paris 6, CNRS, F-78280, Guyancourt, France.
3 Chair Materials Simulation and Engineering, UVSQ, UPSUD, INSTN-CEA, Université Paris-Saclay. France.

E-mail: constantin.meis@cea.fr

Abstract. The vector potential function $\vec{\alpha}_{k\lambda}(\vec{r},t)$ for a k-mode and λ-polarization photon, with the quantized amplitude $\alpha_{0k}(\omega_k)=\xi\omega_k$, satisfies the classical wave propagation equation as well as the Schrodinger's equation with the relativistic massless Hamiltonian $\widetilde{H}=-i\hbar c\vec{\nabla}$ and finally an equivalent quantum equation for the vector potential amplitude operator $\widetilde{\alpha}_0=-i\xi c\vec{\nabla}$. Thus, $\vec{\alpha}_{k\lambda}(\vec{r},t)$ behaves as a wave function for the photon within a non-local representation that can be suitably normalized. It is deduced that the probability for detecting a k-mode photon around a point on the propagation axis depends on the square of the angular frequency. Taking into account the left and right circularly polarized states and weighting $\vec{\alpha}_{k\lambda}(\vec{r},t)$ by $\omega_k\sqrt{2\varepsilon_0}$ we define a six components function as a general function for a k-mode photon. The square of the modulus of the defined general function gives the energy density at a given coordinate which depends on the fourth power of the angular frequency. The amplitudes of the electric and magnetic fields of a single k-mode photon free of cavity are also calculated and it is shown that they are proportional to the square of the angular frequency. In this way, the influence of the photon electric and/or magnetic fields on the energy levels of atoms and molecules might be used for a non-destructive photon detection.

Introduction

The challenging difficulties encountered for establishing a wave function for the photon are well developed in the literature [1-5]. Indeed, the photon as a relativistic massless particle constitutes a particular case and a wave function within the quantum mechanical concept can hardly be defined. However, in recent years, the experimental evidence of the entangled states with single photons sources has demonstrated the permanent violation of Bell's inequality entailing that the existence of hidden variables within a local quantum mechanical representation is excluded [6, and references therein]. At the same time it revealed the necessity of developing a non-local representation for the photon, through a real wave function, in which new variables are not explicitly excluded. Hence, we explore here the possibility of establishing a photon wave function through the vector potential with quantized amplitude. We also analyze the resulting expressions of the intrinsic electric and magnetic fields of a free photon opening perspectives for new experiments that might be used for a non-destructive photon detection method.

Vector potential amplitude quantization

A detailed unit analysis of the general solution of Maxwell's equations for the vector potential has shown that it is proportional to the frequency. Consequently, for a k-mode photon with angular frequency ω_k the vector potential amplitude α_{0k} can be written [7-10]

$$\alpha_{0k} = \xi \omega_k \tag{1}$$

where ξ is a constant.

In a first approximation the value of the constant ξ has been evaluated [9,11] by normalizing the mean energy of a plane electromagnetic wave over a period to Planck's expression for the energy of a single photon $\hbar\omega_k$, where $\hbar = h/2\pi$ is Planck's reduced constant, getting

$$|\xi| \propto \left| \frac{1}{(2\pi)^{3/2}} \sqrt{\frac{\hbar}{8\alpha_{FS}\varepsilon_0 c^3}} \right| = \left| \frac{\hbar}{4\pi e c} \right| = 1.747\ 10^{-25}\ V\, m^{-1} s^2 \tag{2}$$

with $\alpha_{FS} = 1/137$ the Fine Structure constant, ε_0 the vacuum electric permittivity, c the speed of light in vacuum and e the electron charge.

Hence, in the plane wave representation the vector potential for a k-mode and λ-polarization photon can be written

$$\vec{\alpha}_{k,\lambda}(\vec{r},t) = \omega_k \left(\xi \hat{\varepsilon}_\lambda e^{i(\vec{k}\cdot\vec{r} - \omega_k t + \theta)} + cc \right) = \omega_k \vec{\Xi}_{k\lambda}(\omega_k, \vec{r}, t) \tag{3}$$

where cc is the complex conjugate and λ takes two values corresponding to the left (L) and right (R) circular polarization expressed by the unit vector ε_λ.

The general equation for the vector potential of an electromagnetic wave considered as a superposition of plane waves now writes

$$\vec{A}(\vec{r},t) = \sum_{k,\lambda} \omega_k \left[\xi \hat{\varepsilon}_\lambda e^{i(\omega_k t - \vec{k}\cdot\vec{r} + \varphi)} + \xi^* \hat{\varepsilon}_\lambda^* e^{-i(\omega_k t - \vec{k}\cdot\vec{r} + \varphi)} \right] = \sum_{k,\lambda} \omega_k \vec{\Xi}_{k\lambda}(\omega_k, \vec{r}, t) \tag{4}$$

According to the relation (1) the quantized vector potential amplitude is an intrinsic property of a single photon state defining the wave function (3).

The vector potential as a wave function for the photon

It can be easily shown [7,9,12] that the vector potential function (3) satisfies the classical wave propagation equation in vacuum

$$\vec{\nabla}^2 \vec{\alpha}_{k,\lambda}(\vec{r},t) - \frac{1}{c^2}\frac{\partial^2}{\partial t^2} \vec{\alpha}_{\kappa,\lambda}(\vec{r},t) = 0 \tag{5}$$

as well as the *vector potential – energy* (*wave – particle*) equation

$$i\begin{pmatrix}\xi\\ \hbar\end{pmatrix}\frac{\partial}{\partial t}\vec{\alpha}_{k;\lambda}(\vec{r},t)=\begin{pmatrix}\tilde{\alpha}_0\\ \tilde{H}\end{pmatrix}\vec{\alpha}_{k,\lambda}(\vec{r},t) \tag{6}$$

where the vector potential amplitude operator $\tilde{\alpha}_0 = -i\xi c\vec{\nabla}$ and the relativistic Hamiltonian for a massless particle $\tilde{H} = -i\hbar c\vec{\nabla}$ have the eigenvalues $\xi\omega_k$ and $\hbar\omega_k$ respectively [7-9]. Consequently, the vector potential function $\vec{\alpha}_{k\lambda}(\vec{r},t)$ with the quantized amplitude $\xi\omega_k$ behaves as a real wave function for the photon within a non-local representation. Obviously, when considering the propagation of a *k*-mode photon along e.g. the *z* axis, unlike the momentum operator $\tilde{P}_z$ which is well defined in quantum mechanics $\vec{P}_z = \hbar\vec{k}_z = \frac{h}{\lambda_k}\hat{e}_z \rightarrow \tilde{P}_z = -i\hbar\frac{\partial}{\partial z}$, it is not evident how to define a position operator and this difficulty has been widely commented in the literature [1,4,6,9,13]. In fact, Heisenberg's uncertainty for the position and momentum for a photon with a wavelength λ_k propagating along the *z*-axis can be written

$$\delta z\,\delta P_z \geq h \rightarrow \delta z\,\delta\left(\frac{1}{\lambda_k}\right) \geq 1 \tag{7}$$

The physical meaning of Heisenberg's relation is that the photon as an integral entity cannot be localized precisely on the propagation axis *z* except with a significant uncertainty which is of the order of the wavelength λ_k. This fact is a natural consequence that lays in the intrinsic wave-particle nature of the photon and has been widely commented in the literature [6, 13, 15-17]. As quoted by many authors, despite the fact that the *point-particle* model used in the QED formalism is a practical mathematical tool, in reality the photon cannot be defined within a length shorter than its wavelength. Thus, what we can essentially say is that when a *k*-mode photon is emitted at a coordinate z_e at an instant t_e and propagates in vacuum along the *z*-axis the probability $P_k(z)$ to be localized at time *t* and at the coordinate $z = z_e + c\,(t-t_e)$ has rather to be considered within a volume having roughly the value λ_k^3 [6, 16, 17] around the point *z* and according to (3) it is proportional to the square of the angular frequency

$$P_k(\vec{r}) \propto |\vec{\alpha}_{k\lambda}(\vec{r},t)|^2 \propto \xi^2\omega_k^2 \tag{8}$$

Considering the normalization factor $1/\xi\omega_{max}$ for $\vec{\alpha}_{k\lambda}(\vec{r},t)$ and taking $\omega_{max} \sim 10^{25}$ rad Hz, which corresponds to highly energetic gamma photons with wavelengths much shorter than the electron radius, then the detection probability becomes $P_k(\vec{r}) \propto \omega_k^2/\omega_{max}^2$ and tends to 1 for $\omega_k = \omega_{max}$. It is also worthy noticing that, disregarding the physical units, equation (8) also tends naturally to one for the same value $\omega_k = \omega_{max} = 10^{25}$ rad Hz and can be practically used as a direct indication. Another normalization possibility consists of considering $\omega_{max}\sim 10^{43}$ rad Hz corresponding roughly to Planck's energy 10^{19} GeV. However, we do not actually have any experimental evidence of photons with

IOP Conf. Series: Journal of Physics: Conf. Series **936** (2017) 012004 doi:10.1088/1742-6596/936/1/012004

frequencies as high as 10^{43} Hz.

Now, taking into account the left (L) and right (R) circularly polarized states ($\lambda = L,R$) and weighting the vector potential by the factor $\omega_k\sqrt{2\varepsilon_0}$ we define the six components general function

$$\Phi_{k,(L,R)}(\vec{r},t) = \omega_k\sqrt{2\varepsilon_0}\begin{pmatrix}\vec{\alpha}_{k,L}(\vec{r},t)\\ \vec{\alpha}_{k,R}(\vec{r},t)\end{pmatrix} \tag{9}$$

Obviously, for both L and R polarizations $\Phi_{k,(L,R)}(\vec{r},t)$ satisfies also the wave propagation equation (5) and the *vector potential – energy* equation (6). The square of the modulus of the defined general function $\left|\Phi_{k,(L,R)}(\vec{r},t)\right|^2$ gives the energy density of the electromagnetic field composed of a single mode

$$W_k(\vec{r},t) = \left|\Phi_{k,(L,R)}(\vec{r},t)\right|^2 = 2\varepsilon_0\xi^2\omega_k^4 \tag{10}$$

It is worth noticing that the last expression depends on the fourth power of the angular frequency as in the case of the energy density of a radiating dipole [13, 14].

We can also deduce the amplitude of the electric field $|\vec{\varepsilon}_k|$ of a free single k-mode photon [18] which can be expressed through the square of the angular frequency

$$|\vec{\varepsilon}_k| = \left|-\frac{\partial}{\partial t}\vec{\alpha}_{k\lambda}(\vec{r},t)\right| \propto \xi\omega_k^2 \tag{11}$$

Considering a plane wave representation the corresponding magnetic field $|\vec{\beta}_k|$ of the photon is equally proportional to the square of the angular frequency

$$|\vec{\beta}_k| \propto \sqrt{\varepsilon_0\mu_0}\,\xi\omega_k^2 \tag{12}$$

where μ_0 is the vacuum magnetic permeability.

It is of fundamental interest to investigate experimentally the last results, *e.g.* the coupling of left and right circularly polarized photons gives rise to a linearly polarized state whose spatial region around $\lambda_k/4$, where the electric field is maximum, corresponds (in the visible-UV range) to a considerable distance of many nanometers and can be crossed by a large number of atoms or molecules. Under these conditions, high resolution laser spectroscopy might be employed for the detection of the Stark shifts induced by the photon electric field on the atomic or molecular energy levels. Such an experimental investigation could confirm the dependence of the single photon electric/magnetic fields on the square of the angular frequency. This might also open perspectives on the development of non-destructive photon detection through the measurement of the influence of the photon electric or magnetic fields upon the atomic or molecular energy levels.

IOP Conf. Series: Journal of Physics: Conf. Series **936** (2017) 012004 doi:10.1088/1742-6596/936/1/012004

Conclusion

We have seen that the vector potential function $\vec{\alpha}_{k\lambda}(\vec{r},t)$ with the quantized amplitude $\alpha_{0k}(\omega_k)=\xi\omega_k$ satisfies the propagation equation and the coupled *vector potential – energy* equation. It can therefor play the role of a real wave function for the photon in a non-local representation that can be suitably normalized. A *k*-mode photon as an integral physical entity subsists over a period (thus over a wavelength λ_k) through the oscillation of the quantized vector potential and propagates along the wave function. Weighting $\vec{\alpha}_{k\lambda}(\vec{r},t)$ by $\omega_k\sqrt{2\varepsilon_0}$ we may define a six components general function $\Phi_{k,(L,R)}(\vec{r},t)$ for the photon including both circular polarizations. The square of the modulus of $\Phi_{k,(L,R)}(\vec{r},t)$ gives the energy density on a point of the propagation axis. The oscillation of the vector potential gives birth to intrinsic electric and magnetic fields within a single free photon state over a wavelength.

References

[1] Bialynicki-Birula I, *Photon Wave Function*, Progress in Optics XXXVI, Wolf E, Elsevier (1996).
[2] Chandrasekar N, Quantum Mechanics of Photons, *Adv. Studies Th. Phys.*, 6, (2012).
[3] Sipe J E, Photon wave functions, *Phys. Rev. A* 52, 1875, (1995).
[4] Smith B J and Raymer M G, Photon wave functions, wave packet quantization of light and coherence theory, *New Journal of Physics*, 9, 414, (2007).
[5] Khokhlov D L, Spatial and Temporal Wave Functions of Photons, *Appl.Phys. Res.* 2(2), (2010).
[6] Grynberg G, Aspect A & Fabre C, *Quantum optics*, Cambridge University Press (2010).
[7] Meis C, *Phys. Res. Int.* Vol.2014, ID 187432 (2014).
[8] Meis C, *Found. Phys.*,**27**, (1997) 865.
[9] Meis C, *Light and Vacuum*, 2nd Edition, World Scientific (2017).
[10] Meis C, The vacuum energy density and the cosmological constant. QED beyond the standard Description, *International Review of Physics,* 10 (4), (2016).
[11] Meis C, *Physics Essays*, vol. 22, 1, (2009).
[12] Meis C and Dahoo P R, *Journal of Physics*: Conference Series 738 (2016) 012099 doi:10.1088/1742-6596/738/1/012099
[13] Garrison J C and Chiao R, *Quantum Optics,* Oxford University Press (2008).
[14] Read F H, *Electromagnetic radiation*, John Wiley & Sons, New York (1980).
[15] Loudon R, The Quantum Theory of Light, Oxford University Press (2000).
[16] Mandel L and Wolf E, *Optical Coherence and Quantum Optics.* Cambridge University Press (1995).
[17] Mandel L, *Phys. Rev.* 144, 1071 (1966).
[18] Meis C and Dahoo P R, *Int. Jour. of Quantum Inf.* (accepted, august 2017).

IOP Conf. Series: Journal of Physics: Conf. Series **936** (2017) 012005 doi:10.1088/1742-6596/936/1/012005

Understanding dynamics of Martian winter polar vortex with "improved" moist-convective shallow water model

M. Rostami and V. Zeitlin

LMD-ENS, 24 Rue Lhomond, 75005 Paris, France

E-mail: zeitlin@lmd.ens.fr

Abstract. We show how the properties of the Mars polar vortex can be understood in the framework of a simple shallow-water type model obtained by vertical averaging of the adiabatic "primitive" equations, and "improved" by inclusion of thermal relaxation and convective fluxes due to the phase transitions of CO_2, the major constituent of the Martian atmosphere. We perform stability analysis of the vortex, show that corresponding mean zonal flow is unstable, and simulate numerically non-linear saturation of the instability. We show in this way that, while non-linear adiabatic saturation of the instability tends to reorganize the vortex, the diabatic effects prevent this, and thus provide an explanation of the vortex form and longevity.

1. Introduction

A characteristic feature of the general circulation of planetary atmospheres is the presence of polar vortices and/or circumpolar jets. They clearly appear in the data after the flow fluctuations are removed by time and/or zonal (longitudinal) averaging. As such, the polar vortices obey the laws of hydrodynamics, and can be subject to hydrodynamic instabilities according to well-known instability criteria. A surprising property of the polar vortex in the winter hemisphere on Mars is that it keeps its annular form for the long time, while according to the above-mentioned criteria, this structure is *unstable*. In order to understand this apparent paradox, we employ the simplest atmospheric model which is obtained by vertical averaging of the adiabatic "primitive" equations of the atmosphere. We, however, "improve" the model by adding radiative cooling and convective fluxes due to the phase transitions of CO_2, which is the major constituent of the Mars atmosphere. The approach we are using was successfully tested in modelling atmospheric vortices [1], [2], and recently allowed to understood the peculiar form of the Saturn's North pole jet plus vortex system [3]. It consists, first, in analysing stability properties of the vortex configuration with the help of the adiabatic version of the model. Once unstable modes are identified, they are used to initialize numerical simulations with a high-resolution finite-volume code, and compare adiabatic saturation of the instability to the diabatic one, with inclusion of convective fluxes, as explained in [4].

2. The model in adiabatic approximation, and the vortex configuration

The rotating shallow water (RSW) equations are obtained from the adiabatic atmospheric primitive equations by vertical averaging between material surfaces in so-called pseudo-height

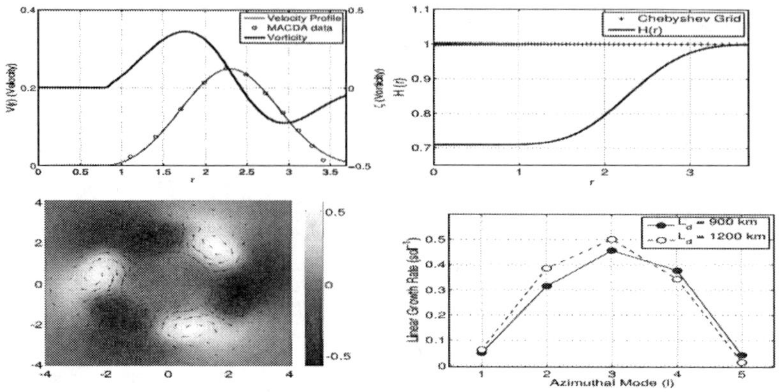

Figure 1. Upper left: Fit of azimuthal velocity and vorticity of Martian North polar vortex superimposed on vertically integrated MACDA data. Upper right: Pressure $H(r)$ of the background vortex, and Chebyshev grid points. Lower left: Pressure (colours) and velocity (arrows) in the $x-y$ plane of the most unstable mode with azimuthal wavenumber $l=3$. Lower right: dependence of the growth rate on l for two different values of L_d.

pressure coordinates in the absence of forcing and dissipation, and under hypothesis of horizontally constant potential temperature. Although spherical coordinates are natural to use for large-scale atmospheric motions, we will work with *cylindrical coordinates* on the polar *tangent plane*. The RSW equations thus read:

$$\frac{D\boldsymbol{v}}{Dt} + (f + \frac{v}{r})(\hat{\boldsymbol{z}} \times \boldsymbol{v}) = -g\boldsymbol{\nabla}_H h, \quad \frac{Dh}{Dt} + h\boldsymbol{\nabla}_H.\boldsymbol{v} = 0. \tag{1}$$

Here (r,θ) are the coordinates in the plane, $\hat{\boldsymbol{z}}$ is the unit vector normal to the plane, $\boldsymbol{v} = u\hat{r} + v\hat{\theta}$, g is gravitational acceleration, f is the Coriolis parameter, and h is thickness of the layer. $D/Dt = \partial_t + \vec{v}.\boldsymbol{\nabla}_H$ is the material derivative, and $\boldsymbol{\nabla}_H = \hat{r}\partial_r + \hat{\theta}(1/r)\partial_\theta$.

Equations (1) have exact solutions verifying the conditions of *cyclo-geostrophic equilibrium*, i.e. equilibrium between the pressure, and centrifugal and Coriolis forces, which link azimuthal velocity $V(r)$ to the thickness $H(r)$ of the vortex.

We start with the observed mean velocity profile of the Mars' polar vortex, and make a simple analytic fit of it. We then reconstitute corresponding pressure distribution via cyclo-geostrophic equilibrium, and analyse stability properties of thus obtained solution of (1). The radial distribution of the mean azimuthal velocity obtained by vertical averaging of the MACDA data [5], together with our fit, are presented in the upper left panel of Figure 1. We use f^{-1} as a unit of time, and scaling all lengths with the deformation radius $L_d = \sqrt{gH_0}/f$, where H_0 is the equivalent depth of the atmospheric layer.

3. Linear stability analysis of the Martian polar vortex

Linearising (1) about the above-described vortex, and looking for solutions of the linearised equation in the form of azimuthal harmonics $\propto e^{i(l\theta - \omega t)}$, leads to a linear eigenproblem for the frequencies ω. Eigenvalues with non-zero imaginary part correspond to unstable modes. The resulting eigenproblem is discretised in radial direction on a stretched Chebyshev grid, and solved by the pseudo-spectral collocation method. The linear stability analysis shows that the Martian polar vortex is *unstable*. The results are presented in Figure 1, where the typical structure of the unstable mode is displayed, together with dependence of the growth rates on l. The most unstable mode, thus, has azimuthal wavenumber $l = 3$. This mode will be used for initialisation of nonlinear simulations below.

4. Improving the model with diabatic effects

As already said we are interested in *nonlinear saturation* of just displayed instability. The previous studies [1],[2], [3], however, show that diabatic effects, which are absent in (1) can be important. We, thus, improve the mode by including them. The first, and generic for the atmospheres, diabatic effect is thermal relaxation to an equilibrium state through radiative cooling/heating. In terms of pressure it may be represented as a relaxation of the pressure to the equilibrium value $H(r)$. Hence

$$\frac{Dh}{Dt} + h\boldsymbol{\nabla}_H.\boldsymbol{v} = -\frac{h - H(r)}{\tau_r}. \tag{2}$$

The relaxation time τ_r is of the order of one Martian (solar) day.

The second diabatic effect is specific to Mars, with its almost pure CO_2 atmosphere. At winter temperatures at the pole, CO_2 undergoes a gas - solid deposition phase transition, if pressure (i.e. h in the present context) is high enough. The deposition process leads to diminishing the mass of gaseous CO_2, and can be thought of as a relaxational sink in the equation for h of the form $\frac{h - h_{th}}{\tau_c} H(h - h_{th})$, where h_{th} is a threshold value over the equilibrium pressure profile, and $\tau_c \ll \tau_r$. To be efficient the deposition needs nuclei, which are provided by dust particles. We can consider that deposition starts when the bulk number of nuclei in the atmospheric column N reaches some threshold N_{th}, and increases with the number of nuclei. Hence, for the sink in the h equation we get

$$C = \frac{h - h_{th}}{\tau_c} H(h - h_{th})(N - N_{th})H(N - N_{th}). \tag{3}$$

The dust particles are simply advected by the flow. Upon vertical averaging, this gives the conservation equation (we neglect the changes in N due to deposition): $\frac{DN}{Dt} + N\boldsymbol{\nabla}_H.\boldsymbol{v} = 0$.

At the same time, the deposition process leads to latent heat release, and related vertical convective flux. This flux also acts as a sink of h, cf [4], and is proportional to C. Hence, the resulting equation for h combining thermal relaxation and deposition effects, is:

$$\frac{Dh}{Dt} + h\boldsymbol{\nabla}_H.\boldsymbol{v} = -\frac{h - H}{\tau_r} - \beta C, \tag{4}$$

where the coefficient β can be incorporatet in τ_c.

5. Nonlinear saturation of the instability. Adiabatic vs diabatic scenarios

Nonlinear saturation of the instability is studied by direct numerical simulations initialised with the main vortex configuration with superimposed most unstable mode of weak amplitude (several per cent). Simulations are made with finite-volume numerical scheme used in the above-cited references. A useful diagnostics of the outputs of such simulations is provided by potential vorticity (PV) $q = \frac{\zeta + f}{h}$, which is a Lagrangian invariant of the adiabatic system. Here $\zeta_i = \partial_x v - \partial_y u$ is relative vorticity. We present in Figure 2 a comparison of adiabatic and diabatic saturations of the instability in terms of PV. A pronounced difference between adiabatic and diabatic saturations is that, while in the former the vortex loses its annular structure, it keeps it in the latter. The $l = 2$ component is manifest in the diabatic evolution with thermal relaxation (2nd row in the Figure). The vortex structure at the late stages in this simulation is in striking resemblance with the data (cf. [5]). As follows from te Figure, the diabatic evolution with

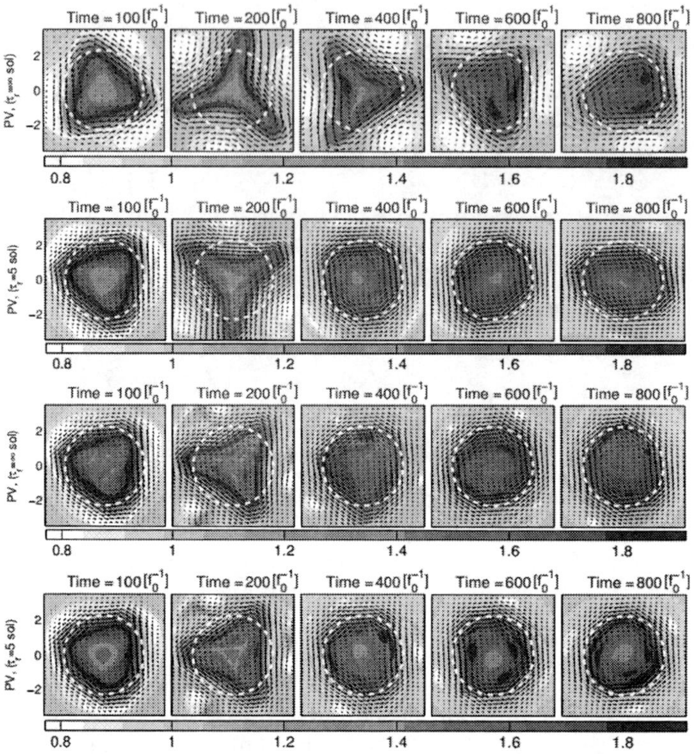

Figure 2. Evolution of PV of the vortex perturbed by the most unstable modein *adiabatic* - 1st row, *diabatic with thermal relaxation only* - 2nd row, *diabatic with deposition without thermal relaxation* - 3rd row, *diabatic with thermal relaxation and deposition* - 4th row, simulations. Dashed line indicates initial radius of maximum wind.

deposition and without thermal relaxation also mainatins the annular structure of the vortex, but rather favours appearance of the $l = 3$ mode, and not of the $l = 2$ one. Yet, the zones of enhanced PV in the annulus, which are typical for the Martian polar vortex, cf. [5], but absent without deposition, appear.

6. Summary and discussion

Thus, the simplest possible vertically integrated model of Martian atmosphere allows, after inclusion of roughly parametrised effects of radiative cooling and depostion phase transition, to understand some salient features of Martian winter polar vortex, such as maintenance of annular structure, loss of axial symmetry, and appearance of zones of enhanced potential vorticity. Such crude representation of dynamics and thermodynamics of Martian polar atmosphere can, obviously, give only a first idea of the dynamical role of diabatic effects, which is nevertheless to be born in mind in more detailed modelling.

References

[1] Lahaye, N and Zeitlin, V., 2016, *J. Atmos. Sci.*, **73**, 505-23.
[2] Rostami, M. and Zeitlin, V., 2017, *Geoph. Astroph. Fluid Dyn.*, **111**, 1–31.
[3] Rostami, M., Zeitlin, V., and Spiga, A. 2017, *Icarus*, to be published.
[4] Bouchut, F., J. , Lambaerts, G. Lapeyre, and V. Zeitlin, 2009, *Phys. Fluids*, **21**, 116-604
[5] Montabone, L. *et al*, 2014, *Geosci. Data Journal*, **1**, 129–139.

IOP Conf. Series: Journal of Physics: Conf. Series **936** (2017) 012006 doi:10.1088/1742-6596/936/1/012006

Correction to the Wills-Harrison approach: Influence on the Fe-based liquid alloys thermodynamics

N Dubinin[1,2,*]
[1]Institute of Metallurgy of the Ural Branch of Russian Academy of Science, 101 Amundsen st., Ekaterinburg, 620016, Russia
[2] Ural Federal University, 19 Mira st., Ekaterinburg, 620002, Russia

E-mail: ned67@mail.ru

Abstract. Concentration dependencies of the Helmholtz free energy of mixing for liquid Fe-Co and Fe-Ni alloys are calculated near their melting temperatures in the framework of modified by us earlier the Wills-Harrison model at different values of the probability that not only diagonal couplings between *d* states in different atoms are possible in transition metal. It is found that an account of non-diagonal *d-d* couplings leads to an improvement of theoretical results for thermodynamic properties of liquid transition-metal binary alloys.

1. Introduction

The knowledge of physical-chemical properties of melts containing transition metals plays an important role for investigations of metallurgical processes and understanding the characteristics of produced steels and alloys. Among different theoretical approaches for studying the melts of transition metals (for example, at last years, [1-8]; see also the review [9]), the Wills-Harrison (WH) approach [10] is one of the most successful. The WH model is based on the Harrison-Froyen [11] approximation which introduces some elements of the muffin-tin orbital theory [12] into transition-metal pseudopotential theory [13, 14], the rectangular model for density of *d* states [15] and the simple-metal pseudopotential theory which works in the framework of the nearly-free-electron (NFE) approximation [16].

In works [10, 11] the diagonal matrix elements with respect to the magnet quantum number, m, between *d* states of neighboring atoms are taken into account only. It is truly in the case of the rotational symmetry with axis along the interatomic distance. Since the *d*-electron contribution to the potential energy, E_d, is non-pairwise in principle, the named above symmetry should be disturbed and non-diagonal *d-d* couplings may arise.

Recently, we suggested the correction of the WH model taking into account non-diagonal couplings between *d* states in transition metals [17] and applied this correction to study the WH effective pair interaction in pure liquid Fe, Co, Ni [17] and equiatomic Co-Ni liquid alloy [18]. In the present work, using the variational method [19] with the additive hard-sphere (HS) reference system [20] we investigate how this correction influences on the thermodynamic properties of binary liquid alloys of transition metals.

[*]To whom any correspondence should be addressed.

2. Theory

The Helmholtz free energy per atom, F, of the binary liquid alloy described by the WH approximation in the framework of the HS variational method is calculated by minimization of

$$F_{\text{var-HS-WH}} = F_{\text{HS}} + U_e - TS_e + \langle E_d \rangle_{\text{HS}} + \frac{2\pi}{\Omega}\sum_{i,j=1}^{2} c_i c_j \int_{\sigma_{ij}}^{\infty} \varphi_{ij\text{NFE}}(r) g_{ij\text{HS}}(r) r^2 \mathrm{d}r \qquad (1)$$

with respect to the partial HS diameters, σ_{ij}. Here, U_e is the s-electron contribution to the internal energy; S_e is the electron entropy; T is the temperature; $\langle E_d \rangle_{\text{HS}}$ is the average of E_d over the HS reference system; Ω is the mean atomic volume; c_i is the concentration of the i-th component; $g_{ij}(r)$ is the partial pair correlation function; $\varphi_{ij}(r)$ is the partial pair potential (hereafter, all formulas are written in atomic units (a.u.)):

$$\varphi_{ij\text{NFE}}(r) = \frac{z_{si} z_{sj}}{r} + \frac{\Omega}{8\pi^2}\int_0^{\infty} F_{ij}(q)\frac{\sin(qr)}{qr} q^2 \mathrm{d}q \ , \qquad (2)$$

where z_{si} is the effective number of valence s electrons per ion in the pure metal of the i-th kind; $F_{ij}(q)$ is the partial energy wave-number characteristic:

$$F_{ij}(q) = -\frac{\Omega q^2 \omega_i(q)\omega_j(q)}{8\pi[(\varepsilon(q)-1)^{-1} + 1 - f(q)]} \ , \qquad (3)$$

where $\omega_i(q)$ is the form-factor of the unscreened i-th-kind ion pseudopotential; $\varepsilon(q)$ is the Hartree dielectric function; $f(q)$ is the exchange-correlation function used here in the Vashishta-Singwi form [21].

We use suggested specially for transition metals the Bretonnet-Silbert (BS) [22] local model pseudopotential generalized to binary alloys [23]:

$$\omega_{i\text{BS}}(r) = \begin{cases} \sum_{n=1}^{2} B_{ni} \exp\left(\dfrac{r}{na_i}\right), & r \le R_{\text{C}i} \\ -z_{si}/r, & r \ge R_{\text{C}i} \end{cases} \ , \qquad (4)$$

$$\omega_{i\text{BS}}(q) = \frac{4\pi a_i^3}{\Omega}\left[\frac{B_{1i}J_{1i}(q)}{(1+a_i^2q^2)^2} + \frac{8B_{2i}J_{2i}(q)}{(1+4a_i^2q^2)^2}\right] - -\frac{4\pi z_{si}\cos(qR_{\text{C}i})}{\Omega q^2} \ , \qquad (5)$$

where $R_{\text{C}i}$ and a_i are pseudopotential parameters;

$$B_{1i} = (z_{si}/R_{\text{C}i})[1 - 2a_i/R_{\text{C}i}]\exp(R_{\text{C}i}/a_i) \ ; \qquad (6)$$

$$B_{2i} = (2z_{si}/R_{\text{C}i})[a_i/R_{\text{C}i} - 1]\exp(0.5R_{\text{C}i}/a_i) \ ; \qquad (7)$$

$$J_{1i}(q)=2-\frac{1}{e^{(R_{Ci}/a_i)}}\left\{\left[R_{Ci}\frac{1+a_i^2q^2}{a_i}+1-a_i^2q^2\right]\frac{\sin(qR_{Ci})}{a_iq}+\left[2+\frac{R_{Ci}(1+a_i^2q^2)}{a_i}\right]\cos(qR_{Ci})\right\}\ ; \quad (8)$$

$$J_{2i}(q)=2-\frac{1}{e^{(R_{Ci}/2a_i)}}\left\{\left[R_{Ci}\frac{1+4a_i^2q^2}{2a_i}+1-4a_i^2q^2\right]\frac{\sin(qR_{Ci})}{2a_iq}+\left[2+\frac{R_{Ci}(1+4a_i^2q^2)}{2a_i}\right]\cos(qR_{Ci})\right\}\ . \quad (9)$$

The expression for the structure-dependent d-electron energy of the binary mixture was obtained in [24]:

$$\langle E_d\rangle_{\text{HS}}=-\frac{\sqrt{12}K_b\bar{z}_d}{2}\left(\frac{10-\bar{z}_d}{10}\right)\left(\frac{4\pi}{\Omega}\sum_{i,j=1}^{2}\frac{c_ic_j(r_{di}r_{dj})^3}{7\sigma_{ij}^7}+\frac{2}{\pi}\sum_{i,j=1}^{2}\sqrt{c_ic_j}\,r_{di}^3r_{dj}^3\int_0^\infty\left[S_{ij\text{HS}}(q)-\delta_{ij}\right]q\mathrm{d}q\int_{\sigma_{ij}}^\infty\frac{\sin(qr)}{r^9}\mathrm{d}r\right)^{\frac{1}{2}}$$

$$+\bar{z}_dK_c\left(\frac{2\pi}{\Omega}\sum_{i,j=1}^{2}\frac{c_ic_j(r_{di}r_{dj})^3}{5\sigma_{ij}^5}+\frac{1}{\pi}\sum_{i,j=1}^{2}\sqrt{c_ic_j}\,r_{di}^3r_{dj}^3\int_0^\infty\left[S_{ij\text{HS}}(q)-\delta_{ij}\right]q\mathrm{d}q\int_{\sigma_{ij}}^\infty\frac{\sin(qr)}{r^7}\mathrm{d}r\right), \quad (10)$$

where $\bar{z}_d=c_1z_{d1}+c_2z_{d2}$; z_{di} is the effective number of valence d electrons per ion in the pure metal of the i-th kind, $z_{di}=z_i-z_{si}$ (z_i is the total number of valence electrons per i-th kind ion); r_{di} is the d-state radius of the free atom of the i-th kind; $S_{ij\text{HS}}(q)$ is the partial structure factor in the Ashcroft-Langreth analytical form [25]; δ_{ij} is the Kronecker symbol; K_b and K_c are coefficients arising in the result of summarizing the d-d matrix elements. When the orbital quantum number is equal to two (that corresponds to the case of the d-band metal), these coefficients within the traditional WH model [10] are

$$K_b=28.06/\pi\ , \quad (11)$$

$$K_c=225/\pi^2\ . \quad (12)$$

In [17] were obtained the expressions of K_b and K_c for a general case, when the non-diagonal d-d couplings can take place in principle:

$$K_b=\left[\frac{1}{5}\left(\left(1-\frac{4p}{5}\right)y_0^2+\left(2-\frac{6p}{5}\right)(y_2^2+y_1^2)+\frac{4p}{5}y_0(y_1+y_2)+\frac{8p}{5}y_1y_2\right)\right]^{\frac{1}{2}}, \quad (13)$$

$$K_c=-\frac{2}{5}\left[\left(1-\frac{4p}{5}\right)y_0x_0+\left(2-\frac{6p}{5}\right)(y_1x_1+y_2x_2)+\frac{2p}{5}(y_0(x_1+x_2)+x_0(y_1+y_2))+\frac{4p}{5}(y_1x_2+y_2x_1)\right]$$

(14)

where p is the probability that any from all 25 equiprobable *d-d* couplings can be realized between two neighboring atoms in metal; $(1-p)$ is the probability that any from 5 diagonal couplings only are possible; y_m and x_m ($y_m = y_{|m|}$ and $x_m = x_{|m|}$) are the combinatoric coefficients:

$$y_m = -\frac{(-1)^{|m|}180}{\pi(2+|m|)!(2-|m|)!}, \tag{15}$$

$$x_m = -\frac{1}{8}\left(1+\frac{4m^2-1}{9}\right)y_m . \tag{16}$$

At $p=0$ Eqs. (13) and (14) are transformed to Eqs. (11) and (12), respectively.

3. Results and Discussion

Earlier, in the framework of the WH model we calculated the concentration dependencies of the Helmholtz free energy of mixing, ΔF, for liquid Fe-Co [24] and Fe-Ni [26] alloys at $T = 1863\text{K}$. Here, these dependencies at different values of p are considered.

Values of r_{di}, z_{si}, R_{Ci}, and a_i are taken the same as values of r_d, z_s, R_C, and a for corresponding pure metals from the works of Wills and Harrison [10] for r_{di} and Jakse and Bretonnet [27] for other parameters (Table 1). The mean atomic volume for each alloy composition is calculated by means the minimization of $F_{\text{var-HS-WH}}$ with respect to Ω.

The calculated concentration dependencies of ΔF in comparison with experiment [28] are presented in Figs. 1, 2. It is observed that the better agreement with experimental data among used values of p is achieved at $p = 0.75$.

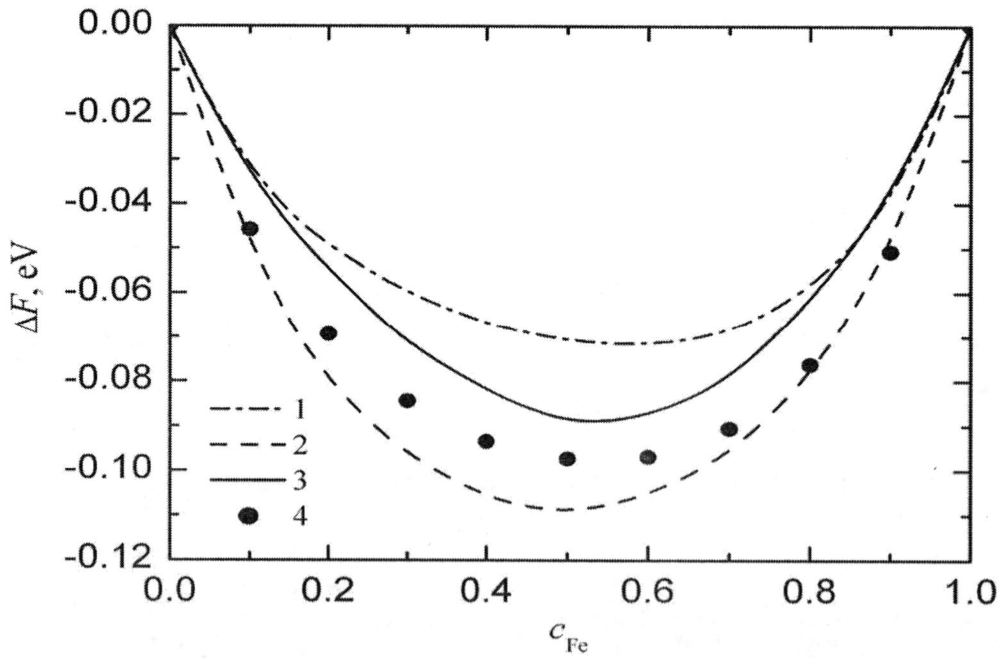

Figure 1. The Helmholtz free energy of mixing of liquid Fe-Co alloy at $T = 1863\text{K}$
(1): $p=0$; (2): $p=1$; (3): $p=0.75$; (4): experiment

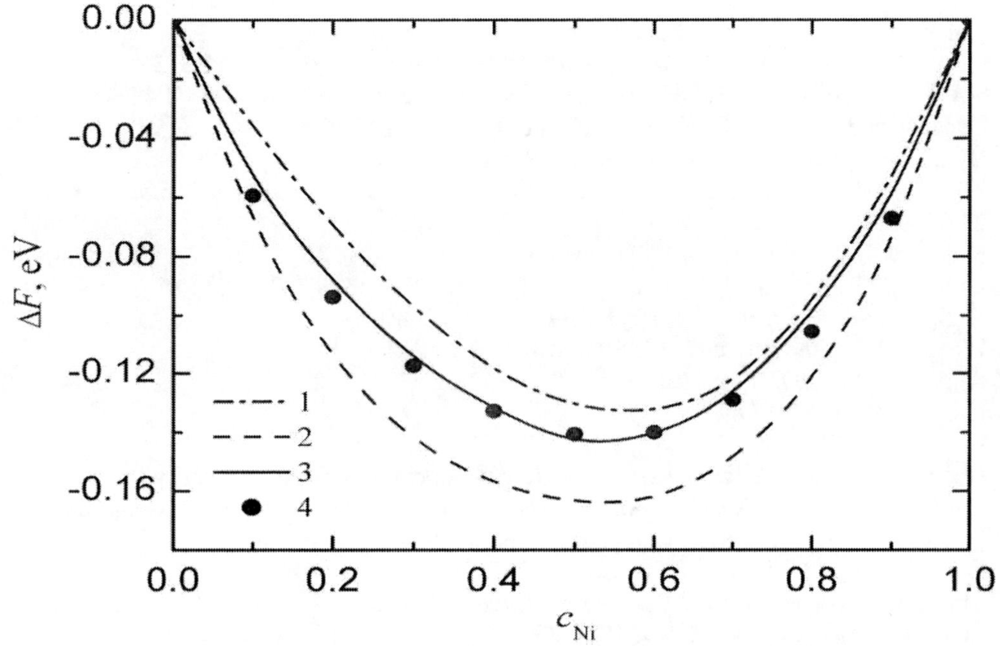

Figure 2. The Helmholtz free energy of mixing of liquid Fe-Ni alloy at $T = 1863$K (1): $p = 0$; (2): $p = 1$; (3): $p = 0.75$; (4): experiment

Table 1. Input parameters for the calculation

	Fe	Co	Ni
r_{di} (a.u.)	1.512	1.437	1.342
z_{si}	1.4	1.4	1.4
z_{di}	6.6	7.6	8.6
R_{Ci} (a.u.)	1.540	1.641	1.030
a_i (a.u.)	0.363	0.393	0.207

4. Conclusion

The results obtained show that the account of non-diagonal *d-d* couplings within the Wills-Harrison model leads to an improvement of theoretical results for thermodynamic properties of liquid transition-metal binary alloys. In the further study it is necessary to find a method determining values of the probability p for different transition metals and alloys *a priory*.

Acknowledgments
This work is supported by the federal target project "R&D for Priority Areas of the Russian Science-and-Technology Complex Development for 2014-2020", government contract № 14.578.21.0200 on the subject "Development of ceramic components and parts production by selective laser melting technology, using innovative diagnostic processes of products and methods" (Application Code «2016–14–579–0009–3076»).

IOP Conf. Series: Journal of Physics: Conf. Series **936** (2017) 012006 doi:10.1088/1742-6596/936/1/012006

References

[1] Son L D, Ryltsev R E and Sidorov V E 2007 *J. Non-Cryst. Solids* **353** 3722
[2] Bolshov L A, Bogdanov V I and Gorbunov V A 2008 *J. Phys.: Conf. Series* **98** 032019
[3] Gorbunov V A and Kurkina L I 2008 *Bull. Rus. Acad. Sci.: Physics* **72** 515
[4] Son L D, Ryltcev R E, Sidorov V E and Sordelet D 2007 *Mat. Sci. Eng. A* **449-451** 582
[5] Ornat M and Paja A 2008 *J. Phys.: Condens. Matter* **20** 375102
[6] Ryltsev R E and Son L D 2011 *Physica B* **406** 3625
[7] Zhang L, Martinez E, Caro A, Liu X Y and Demkowicz M J 2013 *Mod. Simul. Mater. Sci. Eng.* **21** 025005
[8] Kulikova T V, Mayorova A V, Shunyaev K Yu and Ryltsev R E 2015 *Physica B* **466-467** 90
[9] Dubinin N E, Vatolin N A and Filippov V V 2014 *Rus. Chem. Rev.* **83** 987
[10] Wills J M and Harrison W A 1983 *Phys. Rev. B* **28** 4363
[11] Harrison W A and Froyen S 1980 *Phys. Rev. B* **21** 3214
[12] Andersen O K 1975 *Phys. Rev. B* **12** 3060
[13] Harrison W A 1969 *Phys. Rev.* **181** 1036
[14] Moriarty A 1972 *Phys. Rev. B* **5** 2066
[15] Friedel J 1969 in *The Physics of Metals* (New York: Cambridge University Press)
[16] Harrison W A 1966 *Pseudopotential in the Theory of Metals* (New York: Benjamin)
[17] Dubinin N E 2012 *J. Phys.: Conf. Series* **338** 012004
[18] Dubinin N E 2013 *Russian Metallurgy* **2** 157
[19] Mansoori G A and Canfield F B 1969 *J. Chem. Phys.* **51** 4958
[20] Lebowitz J L 1964 *Phys. Rev. A* **133** 895
[21] Vashishta P and Singwi K 1972 *Phys. Rev. B* **6** 875
[22] Bretonnet J L and Silbert M 1992 *Phys. Chem. Liq.* **24** 169
[23] Dubinin N E, Son L D and Vatolin N A 2007 *Defect Diffus. Forum* **263** 105
[24] Dubinin N E, Son L D and Vatolin N A 2008 *J. Phys.: Condens. Matter* **20** 114111
[25] Ashcroft N W and Langreth D C 1967 *Phys. Rev.* **156** 685
[26] Dubinin N E 2009 *J. Phys.: Conf. Ser.* **144** 012115
[27] Jakse N and Bretonnet J L 1995 *J. Phys.: Condens. Matter* **7** 3803
[28] Hultgren R R, Desai P D, Hawkins D T, Gleiser M, Kelley K K and Wagman D D 1973 *Selected Values of the Thermodynamic Properties of the Elements and Binary Alloys* (Ohio: Amer. Soc. Met.)

IOP Conf. Series: Journal of Physics: Conf. Series **936** (2017) 012007 doi:10.1088/1742-6596/936/1/012007

Diffusion of neutrons in the toroidal nuclear electrogenerator

V Yu Tertychny-Dauri

Saint-Petersburg National Research University of Information Technologies, Mechanics & Optics
Department of Physics and Engineering
49 Kronverkskiy Prospect, Saint-Petersburg 197101, Russia

E-mail: tertychny-dauri@mail.ru

Abstract. Diffusion of neutrons in the toroidal nuclear electrogenerator (nuclegen) is investigated. Important conclusion about the practically complete absence of the diffusion of neutrons and charged fission particles through the external boundary under toroidal motion is substantiated by the solution of corresponding equation diffusion with the aid of Fourier's standard division method.

1. Introduction

It is well known, the process of formation of free neutrons inside a fission medium (thereby of formation of charged fission particles) for realization and maintenance of nuclear chain reaction is a decisive significance. Therefore the question about diffusion of neutrons and charged particles flex over the whole volume of external magnetic reflecting field which created by toroidal nuclear electrogenerator (nuclegen) is presented sufficiently important.

At critical stationary generator the coefficient of neutrons reproduction k from generation to generation isn't change ($k = 1$). Hence it follows that for any point of active zone we may write the equation of neutrons balance

$$F - L - A - R = 0, \tag{1}$$

where F is the "formation", L is the "leakage", A is the "absorption" and R is the "reflection". Here through R in relation (1) is indicated the external magnetic pressure is making in respect to the neutron flex the reflected and restored function and reflected effect of itself toroidal surface also.

2. Diffusion problem and its solution by Furier's method

The main equation for diffusion effects (in this case neutrons diffusion) is the Fick's law [1], [2], which establishes that resulting current (flex) of diffusive substance is proportional to the gradient of this substance and directed to the side of scope with the least density

$$J = -\,D\,\nabla\Phi, \tag{2}$$

where $\nabla = (\partial/\partial x,\, \partial/\partial y,\, \partial/\partial z)$ is the operator of dimensional gradient, J is the resulting flex of neutrons, $\Phi = nv$ is the scalar neutron flex, D is the coefficient of diffusion determined by relation (2). Here n is the number of neutrons in the unit of volume (the dimentional density), v is the average neutrons density.

Denote the leakage L in equation (1) through L too. The value L is express itself in the equality

$$L = -D\,\Delta\Phi, \tag{3}$$

where $\Delta = (\partial^2/\partial x^2,\ \partial^2/\partial y^2,\ \partial^2/\partial z^2)$ is the Laplace operator. If to count up that in equation (1) "formation", "absorption", "reflection" each separately are proportional to the flex Φ then taking into account relations (2), (3) equation (1) can be written for some coefficient of proportion R in the form [3], [4]:

$$D\,\Delta\Phi + R\Phi = 0,$$

or

$$\Delta\Phi + a^2\Phi = 0, \qquad \Phi\,|_\Gamma \in \Omega, \tag{4}$$

where $a^2 = R/D$, $\Phi\,|_\Gamma \in \Omega$ is the boundary condition on the Γ of some scope Ω concerning the function Φ. Solve equation (4) with given boundary condition of scope Ω into supposition that the flex Φ is twice continuously differentiable on dimensional variables. With that end in view we introduce cylindrical coordinates (r, φ, z). Have for the arbitrary function $F(r, \varphi, z)$:

$$\Delta\Phi = \frac{1}{r}\frac{\partial}{\partial r}\left(r\,\frac{\partial F}{\partial r}\right) + \frac{1}{r^2}\frac{\partial^2 F}{\partial \varphi^2} + \frac{\partial^2 F}{\partial z^2}.$$

Owing to the symmetry of toroid the flex doesn't depend on angular coordinate φ and the diffusive problem is becoming twodimensional: $\Phi = \Phi(r, z)$, i. e.

$$\frac{\partial^2\,\Phi(r,z)}{\partial r^2} + \frac{1}{r}\frac{\partial\,\Phi(r,z)}{\partial r} + \frac{\partial^2\,\Phi(r,z)}{\partial z^2} + a^2\,\Phi(z,z) = 0, \tag{5}$$

where $r_1 \le r \le r_2$, $-r_0 \le z \le r_0$, $r_0 = (r_2 - r_1)/2$. Make use of the Furier's method variables division for the solving of equation (5) supposed that the flex $\Phi(r, z)$ is presented in the form of two functions product as

$$\Phi(r, z) = R(r)\,Z(z). \tag{6}$$

Bring expression (6) in equation of diffusion (5). After termwise division on $R(r)\,Z(z)$ we receive the equation

$$\frac{1}{R(r)}\left[\frac{d^2\,R(r)}{dr^2} + \frac{1}{r}\frac{d\,R(r)}{dr}\right] + a^2 = -\,\frac{1}{Z(z)}\frac{d^2\,Z(z)}{dz^2}. \tag{7}$$

The left part of equation (7) depends only from r and the right — only from z. Function (6) can be the solution of equation (7) as soon as in the case if both parts of this equation are constant by the arbitrary changing of variables r and z, i. e. must be carried out the following relations

$$\frac{1}{R(r)}\left[\frac{d^2\,R(r)}{dr^2} + \frac{1}{r}\frac{d\,R(r)}{dr}\right] = -\,a_r^2, \tag{8}$$

$$\frac{1}{Z(z)}\frac{d^2\,Z(z)}{dz^2} = -\,a_z^2, \tag{9}$$

where a_r^2, a_z^2 are positive numbers connected by equality $a^2 = a_r^2 + a_z^2$ owing to equation (7).

3. Solution of equation on $R(r)$

Consider equation (8). At first we termwise its over $r^2\,R(r)$:

$$r^2\,\frac{d^2\,R(r)}{dr^2}+r\,\frac{d\,R(r)}{dr}+r^2\,R(r)\,a_r^2=0,$$

and then introduce the new variable $\rho = ra_r$. As a result we obtain the Bessel's equation of zero order

$$\rho^2\,\frac{d^2\,R(\rho)}{d\rho^2}+\rho\,\frac{d\,R(\rho)}{d\rho}+\rho^2\,R(\rho)=0. \tag{10}$$

The general solution of equation (10) is presented in the form

$$R(\rho)=C_1\,J_0(\rho)+C_2\,Y_0(\rho),$$

where C_1, C_2 are the arbitrary constants, $J_0(\rho)$, $Y_0(\rho)$ are the Bessel's functions of zero order first and second types determined with the help of power serieses

$$J_0(\rho)=1+\sum_{k=1}^{\infty}\frac{(-1)^k\,\rho^{2k}}{(k!)^2\,2^{2k}},$$

$$Y_0(\rho)=\lim_{n\to 0}\frac{J_n(\rho)\cos n\pi-J_{-n}(\rho)}{\sin n\pi}.$$

Here $J_{-n}(\rho)=(-1)^n\,J_n(\rho)$, $J_n(\rho)$ is the Bessel's function of n–th order with the constant $a^{(n)}$:

$$J_n(\rho)=a^{(n)}\,\rho^n\sum_{k=0}^{\infty}\frac{(-1)^k\,\rho^{2k}}{k!\,(n+1)(n+2)\cdots(n+k)\,2^{2k}},$$

$$a^{(n)}=\frac{1}{2^n\,\Gamma(n+1)},\qquad \Gamma(n)=\int_0^{\infty}e^{-x}\,x^{n-1}\,dx,$$

where $\Gamma(n)$ is the gamma-function of Euler. Note that by small ρ

$$Y_0(\rho)\approx-\frac{2}{\pi}\ln\left(\frac{2}{\gamma\rho}\right),$$

where $\ln\gamma = C = 0,5772...$ is the Euler's constant. Sometimes the function $Y(\rho)$ is called the Neumann's (or the Weber's) function of zero order. If come back to the variable r then we can write as a result

$$R(r)=C_1\,J_0(a_r r)+C_2\,Y_0(a_r r). \tag{11}$$

4. Solution of equation on $Z(z)$ and the general solution

Consider now equation (9). Its solution is determined easily

$$Z(z)=C_3\sin a_z z+C_4\cos a_z z, \tag{12}$$

where C_3, C_4 are the arbitrary constants. Take from conditions of symmetry

$$Z=Z_0,\qquad \left.\frac{d\,Z(z)}{dz}\right|_{z=0}=0. \tag{13}$$

Obviosly then in solution (12) by conditions (13) we shall behaving $C_3 = 0$, $C_4 = Z_0$ and consequently $Z(z) = Z_0\cos a_z z$.

IOP Conf. Series: Journal of Physics: Conf. Series **936** (2017) 012007 doi:10.1088/1742-6596/936/1/012007

Multiply these two solutions: (11) and $Z(z)$. As a result we obtain the expression for flex (6):

$$\Phi(r,z) = \left[\, C_1\, J_0(a_r r) + C_2\, Y_0(a_r r) \,\right] Z_0 \cos a_z z. \tag{14}$$

Solution (14) must sutisfy the main boundary condition:

$$\Phi(r,z) = 0, \qquad \forall\, r, z \in T_\delta, \tag{$*$}$$

where

$$T_\delta = \{\, r, z:\; [r - (r_1 + r_2)/2]^2 + z^2 = (r_0 + \delta)^2,\; \varphi: 0 \le \varphi \le 2\pi \,\},$$

T_δ — δ–toroid, and axial conditions of symmetry

$$\left.\frac{\partial\, \Phi(r,z)}{\partial r}\right|_{r = r_1 + r_0} = 0, \qquad \left.\frac{\partial\, \Phi(r,z)}{\partial z}\right|_{z=0} = 0. \tag{$**$}$$

These conditions answer the neutron flex $\Phi = (\Phi_r,\, \Phi_z)$ for which $\Phi_r = 0$, $\Phi_z = 0$. The starting equation of diffusion (4) with boundary ($*$) and axial ($**$) conditions answer the flex $\Phi = (\Phi_r, \Phi_z, \Phi_\varphi)$, where

$$\Phi_z = 0, \qquad \Phi_z = 0, \qquad \Phi_\varphi = \Phi_0 = \text{const}. \tag{15}$$

Parameter δ attends in condition ($*$). It's named the length of extrapolation. Neutron flex doesn't turn into zero on the boundary as far as always there is (although insignificant) the leakage across external boundary. It means that in immediate nearness from the boundary inside of active zone must be neutrons which leave its limits. On the boundary part of neutrons leaves outside but another part diffuses inside and raises the significance of density for neutron flex. In consequence of this neutron flex on boundary be not equal zero but is supposed be equal zero from the outside on some little distance δ.

5. Conclusions

Flex (15) corresponds to the absence of the leakage in stationary ($k = 1$) nuclear electrogenerator, the availability of zero parameter ($a = 0$) in equation (4) and the carrying-out of neutron balance equation: $F - A - R = 0$. The fulfilment of this condition means the important peculiarity and advantage of toroidal vacuum-nuclear systems, expressed into practically complete (without losses) circular motion of nuclear flexes inside of the toroidal active zone of nuclegen.

References

[1] Harrer J M 1963 *Nuclear Reactor Control Engineering* (Princeton, N. J.: D. van Nostrand Co., Inc.) 587 p
[2] Keepin G R 1965 *Physics of Nuclear Kinetics* (Massachusetts, London: Addison-Wesley Publishing Company) 428 p
[3] Tertychny-Dauri V Yu 2006 Optimal stabilization in the problems of adaptive nuclear kinetics *Different. Equat.* **42** 374-384
[4] Tertychny-Dauri V Yu 2008 *Galamech* Vol 3 *Hyperreactive Mechanics* (Moscow: Fizmatlit) 576 p

IOP Conf. Series: Journal of Physics: Conf. Series **936** (2017) 012008 doi:10.1088/1742-6596/936/1/012008

Explanation of Rotation Curves in Galaxies and Clusters of them, by Generalization of Schwarzschild Metric and Combination with MOND, eliminating Dark Matter

Spyridon Vossos[1,3] and Elias Vossos[2] [1]Chemist, [2]Physicist, School of Science, National and Kapodistrian University of Athens, Panepistimiopolis, GR15784 Zografos, Athens, Hellas

E-mail: spyrolia@hotmail.com

Keywords: dark matter, Relativistic Doppler shift, Einstein equations, Euclidean closed linear transformations, complex spacetime, Euclidean complex relativistic mechanics, Euclidean metric, Galaxies: kinimatics and dynamics of, Galaxy: NGC 3198, Galilean metric, general relativity, gravitation, gravitational red shift, human senses, isometry, linear spacetime transformation, Lorentz matrix, Lorentz metric, Lorentz transformation, Minkowski space, Modified Newtonian Dynamics (MOND), modified gravitation, Schwarzschild metric, simple interpolating function, Solar system: kinimatics and dynamics of, special relativity, standard interpolating function, real spacetime, time dilation, time contraction, universal rotation curve, universal speed, Vossos matrix, Vossos transformation.

PACS: 02.10.Ud, 02.40.Dr, 03.30.+p, 04.20.-q, 04.50.Kd, 04.80.Cc, 95.35.+d, 96.12.De, 96.15.De, 98.20.+d

Abstract. *Schwarzschild Metric* is the first and the most important solution of *Einstein vacuum field equations*. This is associated with *Lorentz metric* of flat spacetime and produces the relativistic potential (Φ) and the field strength (g) outside a spherically symmetric mass or a non-rotating black hole. It has many applications such as gravitational red shift, the precession of Mercury's orbit, Shapiro time delay etc. However, it is inefficient to explain the rotation curves in large galaxies and clusters of them, causing the necessity for dark matter. On the other hand, *Modified Newtonian Dynamics* (MOND) has already explained these rotation curves in many cases, using suitable interpolating function (μ) in *Milgrom's Law*. In this presentation, we initially produce a *Generalized Schwarzschild potential* and the corresponding Metric of spacetime, in order to be in accordance with any isotropic metric of flat spacetime (including *Galilean Metric* of spacetime which is associated with *Galilean Transformation* of spacetime). From this Generalized *Schwarzschild potential* (Φ), we calculate the corresponding field strength (g), which is associated with the interpolating function (μ). In this way, a new relativistic potential is obtained (let us call *2nd Generalized Schwarzschild potential*) which describes the gravitational interaction at any distance and for any metric of flat spacetime. Thus, not only the necessity for Dark Matter is eliminated, but also MOND becomes a pure Relativistic Theory of Gravitational Interaction. Then, we pass to the case of flat spacetime with *Lorentz metric* (*Minkowski space*), because the experimental data have been extracted

[3]Spyridon Vossos, 39 Epidavrou Str., GR-15233 Chalandri, Athens, Hellas, e-mail: spyrolia@hotmail.com.

using the *Relativistic Doppler Shift* and the gravitational red shift of Classic Relativity (CR). Thus, we Explain the Rotation Curves in Galaxies (e.g. NGC 3198) and Clusters of them as well as the Solar system, eliminating Dark Matter .This relativistic potential and the corresponding metric of spacetime have been obtained by the light of *Euclidean Closed Linear Transformations of Complex Spacetime endowed with the Corresponding Metric*. Of course, may also be applied by scientists who prefer the hyperbolic geometry of Classic Relativity (CR).

1. Introduction

1.1. Schwarzschild Metric, Relativistic Potential and Field Strength

Schwarzschild Metric

$$g_{00} = -\left(1-\frac{r_S}{r}\right) \; ; \; g_{rr} = \frac{1}{\left(1-\frac{r_S}{r}\right)} \tag{1}$$

is the first and the most important solution of *Einstein vacuum field equations*. It is associated with *Lorentz metric* of flat spacetime, e.g.

$$\eta = \begin{bmatrix} -1 & 0 & 0 & 0 \\ 0 & 1 & 0 & 0 \\ 0 & 0 & 1 & 0 \\ 0 & 0 & 0 & 1 \end{bmatrix} \tag{2}$$

if a contravariant four-vector is expressed in *Cartesian Coordinates* (CCs)

$$X = \begin{bmatrix} ct \\ x \\ y \\ z \end{bmatrix} \tag{3}$$

and produces the corresponding Relativistic Potential

$$\Phi = \frac{c^2}{2}\ln\left(1-\frac{r_S}{r}\right) \tag{4}$$

where

$$r_S = \frac{2GM}{c^2} \tag{5}$$

is *Schwarzschild radius*. It is referred to the space outside a spherically symmetric mass or a non-rotating black hole. Thus, it has many applications such as gravitational red shift, the precession of Mercury's orbit, Shapiro time delay etc [1]. According *Schwarzschild Metric* the field strength is

$$g = \frac{GM}{r^2}\left(1-\frac{r_s}{r}\right)^{-\frac{1}{2}} \tag{6}$$

which far away from *Schwarzschild radius* becomes

$$g = \frac{GM}{r^2}\left(1+\frac{r_s}{2r}\right) \tag{7}$$

and at even larger distances is equal to the *Newtonian field strength*

$$g = \frac{GM}{r^2} \tag{8}$$

In case of Uniform Circular Motion (UCM), it is

$$g = \frac{\upsilon^2}{r} \tag{9}$$

so

$$\upsilon^2 = \frac{GM}{r} \tag{10}$$

However, the above field strength is inefficient to explain the rotation curves in large galaxies and clusters of them, causing the necessity for dark matter [2-6].

1.2. Modified Newtonian Dynamics (MOND)

On the other hand, *Modified Newtonian Dynamics* (MOND) has already explained these rotation curves in many cases, using suitable interpolating function (μ) in *Milgrom's Law* [7-11]. The *Modified Newtonian field strength* is

$$g = \frac{GM}{r^2}\frac{1}{\mu} \tag{11}$$

The interpolating function (μ) is an unspecified function having the properties

$$\lim_{a \to 0} \mu = \frac{a}{a_0} \tag{12}$$

where a is the acceleration of a body and a_o is a new fundamental constant with value (see [8] p. 1)

$$a_0 \approx 1.2 \times 10^{-10} \frac{m}{s^2} \tag{13}$$

and $\mu = 1$, if $a >> a_o$. Two common choices are the *Simple interpolating function*

$$\mu = \frac{1}{1 + \frac{a_0}{a}} \tag{14}$$

and the *Standard interpolating function*

$$\mu = \frac{1}{\sqrt{1 + \left(\frac{a_0}{a}\right)^2}} \tag{15}$$

Using *Simple* μ of MOND (14) and (11), for a=g, it emerges

$$g = \frac{GM}{r^2}\frac{1}{2}\left(1 + \sqrt{1 + \frac{4a_0 r^2}{GM}}\right) \tag{16}$$

Let us define *Milgrom radius*

$$r_0 = \sqrt{\frac{GM}{4a_0}} \tag{17}$$

We may return to the *Newtonian* results, if we take the limit $a_o \to 0$ or $r_o \to +\infty$. With that (16) becomes

$$g = \frac{GM}{r^2}\frac{1}{2}\left(1 + \sqrt{1 + \left(\frac{r}{r_0}\right)^2}\right) \tag{18}$$

and we obtain *Simple interpolating function*

$$\mu = \frac{2}{1 + \sqrt{1 + \left(\frac{r}{r_0}\right)^2}} \tag{19}$$

In case of Uniform Circular Motion (UCM), it is

$$\upsilon^2 = \frac{GM}{2r}\left(1+\sqrt{1+\left(\frac{r}{r_0}\right)^2}\right) \tag{20}$$

Using *Standard* μ of MOND (15) and (11), for $\alpha = g$, it emerges

$$g = \frac{GM}{r^2}\frac{1}{\sqrt{2}}\sqrt{1+\sqrt{1+\frac{1}{4}\left(\frac{r}{r_0}\right)^4}} \tag{21}$$

and we obtain *Standard interpolating function*

$$\mu = \frac{\sqrt{2}}{\sqrt{1+\sqrt{1+\frac{1}{4}\left(\frac{r}{r_0}\right)^4}}} \tag{22}$$

In case of Uniform Circular Motion (UCM), it is

$$\upsilon^2 = \frac{GM}{r}\frac{1}{\sqrt{2}}\sqrt{1+\sqrt{1+\frac{1}{4}\left(\frac{r}{r_0}\right)^4}} \tag{23}$$

In Figure 1, we show the plots of $1/\mu_{Simpl}$ and $1/\mu_{Stand}$ wrt the distance (r/r_o) from the center of gravity, where μ_{Simpl} and μ_{Stand} are *simple and standard interpolating functions*, respectively. We observe that they are similar to the contribution of dark matter (see e.g. Figure 3). Besides *Simple* μ produces stronger gravity than *Standard* μ.

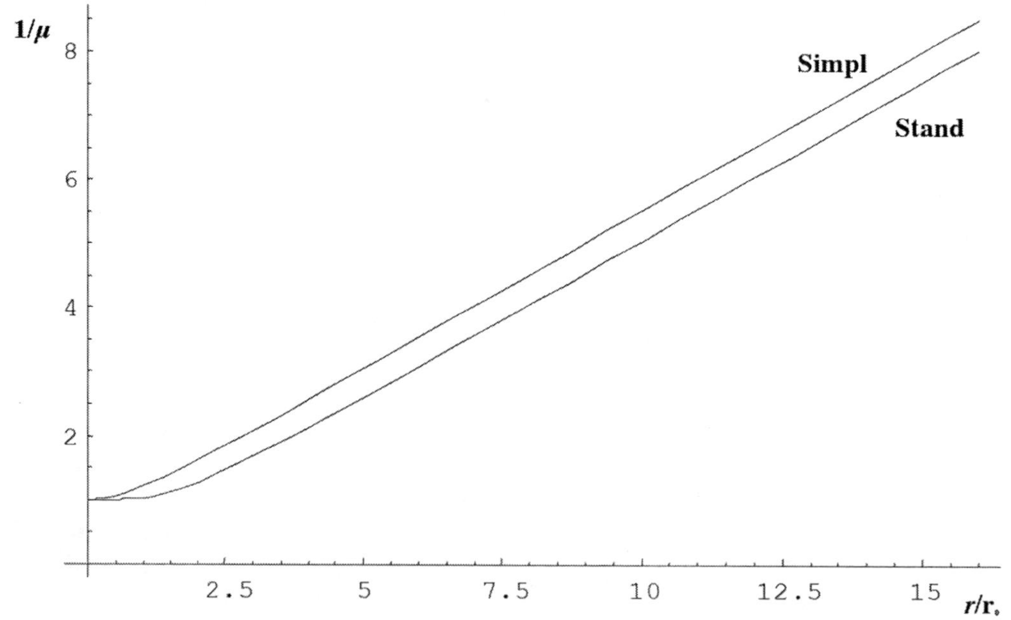

Figure 1. Plot of $1/\mu_{Simpl}$ and $1/\mu_{Stand}$ wrt the distance (r/r_o) from the center of gravity, where μ_{Simpl} and μ_{Stand} are *simple* and *standard interpolating functions*, respectively.

Both *Interpolating functions* give the same velocity at infinite distance from the center of gravity

$$\upsilon_\infty^{\ 2} = \sqrt{GMa_0} \tag{24}$$

IOP Conf. Series: Journal of Physics: Conf. Series **936** (2017) 012008 doi:10.1088/1742-6596/936/1/012008

so

$$\beta_\infty = \frac{\upsilon_\infty}{c} = \frac{\sqrt[4]{GMa_0}}{c} \tag{25}$$

Milgrom radius is connected with *Schwarzschild radius* via the formula

$$r_0 = \sqrt{\frac{GM}{4a_0}\frac{r_S c^2}{2GM}} = \frac{c^2}{4\sqrt{GMa_0}} r_S = \frac{c^2}{4\upsilon_\infty^{\ 2}} r_S = \frac{1}{4\beta_\infty^{\ 2}} r_S \tag{26}$$

Thus, we have another characteristic parameter of a body or cluster

$$N_0 = \frac{c^2}{4\sqrt{GMa_0}} = \frac{c^2}{4\sqrt{Ga_0}}\frac{1}{\sqrt{M}} \tag{27}$$

Three fundamental physical constants are combined giving an extremely large value

$$N = \frac{c^2}{4\sqrt{Ga_0}} = 2.510664\times10^{26} Kg^{-\frac{1}{2}} \tag{28}$$

Thus, only Clusters of Galaxies and entire Universe have mass that makes r_s having similar value to r_0. In any other case

$$r_s << r_0 \tag{29}$$

Some values of the aforementioned characteristic parameters for the Earth, the Sun (see [12] p 8 and [13] pp 1-1, 14-2) and galaxy NGC3198 (see [4] p 56 and [2] p 3) are contained in Table 1.

Table 1. Characteristic parameters (mass M, radius R, *Schwarzschild radius* r_s, *Milgrom radius* r_0, r_0/r_s, velocity at infinite distance υ_∞ and β_∞) for the Earth, the Sun and galaxy NGC 3198.

	Earth	**Sun**	**NGC 3198**
M / Kg	5.9742×10^{24}	1.9891×10^{30}	6.76294×10^{40}
R / m	6378140	6.9599×10^{8}	2.47×10^{20}
/ Kpc	2.066999×10^{-13}	2.2555×10^{-11}	8
r_s / m	8.873578523×10^{-3}	2,954.44328	1.0044505×10^{14}
/ Kpc	$2.875710057\times10^{-22}$	9.57462903×10^{-17}	0.0000680703
r_0 / m	9.114261212×10^{11}	5.25908150×10^{14}	9.6972671×10^{19}
/ Kpc	2.953709437×10^{-8}	0.0000170434	3.1426474
r_0/r_s	1.027123520×10^{14}	1.78005837×10^{11}	965,430
υ_∞ / m s^{-1}	14.7899	355.272	152,556
β_∞	4.933392321×10^{-8}	1.18506×10^{-6}	0.000508873

The work that is needed for moving one unit of mass from a point A with distance r from the center of gravity to another point B with larger distance r_1 (the Potential Difference between point A and point B), is

$$w = V_{(r_1)} - V_{(r)} = \int_r^{r_1} g\,dr = GM\int_r^{r_1} \frac{dr}{r^2 \mu_{(r)}} \tag{30}$$

In case of *Simple* μ, we have

$$w = \Delta V = 2a_0 r_0 \int_r^{r_1} \frac{1}{\left(\frac{r}{r_0}\right)^2}\left(1+\sqrt{1+\left(\frac{r}{r_0}\right)^2}\right) d\left(\frac{r}{r_0}\right) = \frac{GM}{2}\left[-\frac{1}{r}-\frac{1}{r}\sqrt{1+\left(\frac{r}{r_0}\right)^2} + \sqrt{\frac{GM}{4a_0}} ArcSinh\left(\frac{r}{r_0}\right)\right]_r^{r_1} \tag{31}$$

because the integral

IOP Conf. Series: Journal of Physics: Conf. Series **936** (2017) 012008 doi:10.1088/1742-6596/936/1/012008

$$I_{Simpl} = \int \frac{1}{x^2}\left(1+\sqrt{1+x^2}\right)\mathrm{d}x = -\frac{1}{x} - \frac{\sqrt{1+x^2}}{x} + \mathrm{ArcSinh}\,x = -\frac{1}{x} - \frac{\sqrt{1+x^2}}{x} + \ln\left(x+\sqrt{1+x^2}\right) \quad (32)$$

Taking the limit $r \to +\infty$, we find that if we wish to move the mass at infinite distance from the center of gravity, we must pay infinite work. So, we cannot define the infinite as the point with zero potential. In addition, it appears that nothing cannot be moved out of the Universe unless we pay infinite work.

In case of *Standard* μ, the work that is needed for moving one unit of mass from a point A with distance r from the center of gravity to another point B with larger distance r_1, is

$$w = \Delta V = 2\sqrt{2}a_0 r_0 \int_r^{r_1} \frac{1}{\left(\frac{r}{r_0}\right)^2}\sqrt{1+\sqrt{1+\frac{1}{4}\left(\frac{r}{r_0}\right)^4}}\,\mathrm{d}\left(\frac{r}{r_0}\right) \quad (33)$$

Unfortunately, the integral

$$I_{Stand} = \int \frac{1}{x^2}\sqrt{1+\sqrt{1+\frac{x^4}{4}}}\,\mathrm{d}x \quad (34)$$

has no analytical solution.

The values of Circular Velocities are contained in Table 1 and the Mass of the Galaxy NGC 3198 that is enclosed within the circular orbit wrt the distance from the center of Galaxy are contained in Table 2 (see [5] p 2). In addition, the corresponding Rotation Curves and Mass Distribution in Galaxy NGC 3198 are shown in Figure 2 and Figure 3 respectively.

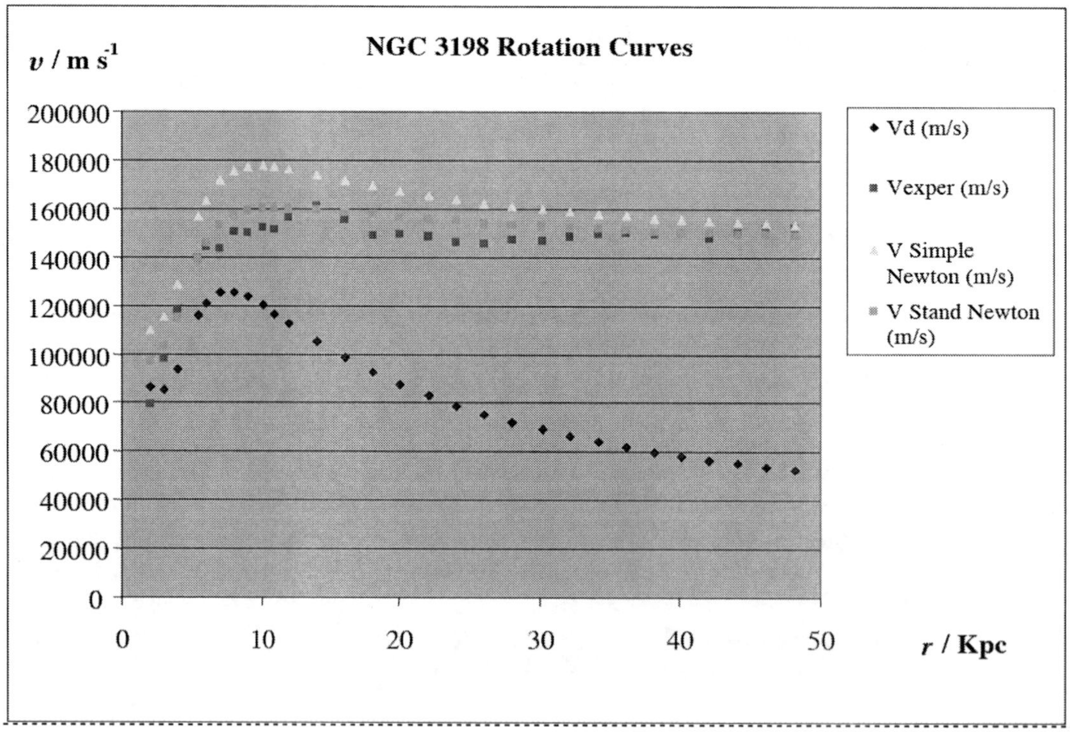

Figure 2. Universal Rotation Curves in Galaxy NGC 3198. Circular Velocities (experimental V_{exp}, calculated by *Schwarzschild* or *Newtonian field strength* V_d , calculated by *Simple* μ in *Milgrom's Law* V_{simp}, calculated by *Standard* μ in *Milgrom's Law* V_{stand}) wrt the distance from the center of Galaxy NGC 3198.

Table 2. Circular Velocities (experimental V_{exp}, calculated using *Schwarzschild* or *Newtonian field strength* V_d , *Simple* μ in *Milgrom's Law* V_{simp}, *Standard* μ in *Milgrom's Law* V_{stand}) and Luminous Mass Distribution (Luminous Mass of the galaxy that is enclosed within the circular orbit M_d) wrt the distance from the center of Galaxy NGC 3198. $(\Delta V_{exp})_r \approx 8\%$, $(\Delta V_d)_r$, $(\Delta V_{simp})_r$, and $(\Delta V_{stand})_r$ are the corresponding relative errors.

r / Kpc	M_d /10^{40} kg	V_{exp} / Km s^{-1}	V_d / Km s^{-1}	$(\Delta V_d)_r$ %	V_{simp} / Km s^{-1}	$(\Delta V_{simp})_r$ %	V_{stand} / Km s^{-1}	$(\Delta V_{stand})_r$ %
2.0	0.687	79.0	86.2	9	109.6	39	97.2	23
3.0	1.012	97.8	85.4	-13	115.6	18	102.8	5
4.0	1.620	118.0	93.6	-21	128.8	9	114.8	-3
5.5	3.404	139.4	115.7	-17	156.6	12	139.3	0
6.0	3.698	144.2	120.8	-16	163.4	13	145.5	1
7.0	5.089	143.3	125.4	-12	171.8	20	153.1	7
8.0	5.825	150.3	125.5	-17	175.76	17	157.0	4
9.0	6.346	149.9	123.5	-18	177.3	18	158.9	6
10.1	6.735	152.1	120.1	-21	177.6	17	159.9	5
11.0	6.914	151.1	116.6	-23	176.9	17	159.9	6
12.1	7.093	156.2	112.6	-28	176.14	13	159.9	2
14.1	7.214	161.0	105.2	-35	173.81	8	159.2	-1
16.1	7.237	155.3	98.6	-37	171.5	10	158.4	2
18.1	7.191	148.7	92.7	-38	169.3	14	157.4	6
20.1	7.115	149.1	87.5	-41	167.3	12	156.5	5
22.1	7.005	148.4	82.8	-44	165.4	11	155.5	5
24.1	6.901	146.2	78.7	-46	163.7	12	154.7	6
26.1	6.806	145.5	75.1	-48	162.3	12	153.9	6
28.1	6.716	147.3	71.9	-51	161.0	9	153.2	4
30,2	6.628	146.5	68.9	-53	159.8	9	152.6	4
32,2	6.544	148.4	66.3	-55	158.7	7	152.0	2
34.2	6.456	149.3	63.9	-57	157.7	6	151.4	1
36.2	6.392	149.9	61.8	-59	156.9	5	151.0	1
38.2	6.316	149.3	59.8	-60	156.0	5	150.4	1
40.2	6.252	150.0	58.0	-61	155.3	4	150.0	0
42.1	6.191	147.6	56.4	-62	154.6	5	149.6	1
44.2	6.159	149.8	54.9	-63	154.2	3	149.5	0
46.2	6.114	151.5	53.5	-65	153.6	1	149.1	-2
48.2	6.072	151.9	52.2	-66	153.2	1	148.8	-2

The Circular Velocities V_d have been calculated using (10) of *Schwarzschild* or *Newtonian field strength* , V_{simp} using (20) of *Simple* μ in *Milgrom's Law* and V_{stand} using (23) of *Standard* μ in *Milgrom's Law*. We observe that in case of Galaxy NGC 3198, *Schwarzschild* or *Newtonian field strength* produces maximum relative error about 66% at extra large distances, while *Simple* μ in *Milgrom's Law* gives better results, producing maximum relative error about 39% near to the galactic center and *Standard* μ in *Milgrom's Law* gives even better results, producing maximum relative error about 23%

at the center of the galaxy. It is noted that the relative error of experimental Circular Velocities is $(\Delta V_{exp})_r \approx 8\%$. Thus *Standard* μ in *Milgrom's Law* gives the best results of MOND.

Table 3. Mass Distribution (Luminous Mass M_d, Total Mass M_{tot} and Dark Mass M_{dark} of the galaxy that is enclosed within the circular orbit) wrt the distance from the center of Galaxy NGC 3198.

r / Kpc	M_d /10^{40} kg	M_{tot}/10^{40} kg	M_{dark} /10^{40} kg	R / Kpc	M_d /10^{40} kg	M_{tot} /10^{40} kg	M_{dark} /10^{40} kg
2.0	0.687	0.577	-0.110	22.1	7.005	22.501	15.496
3.0	1.012	1.327	0.315	24.1	6.901	23.816	16.915
4.0	1.620	2.575	0.955	26.1	6.806	25.546	18.740
5.5	3.404	4.941	1.537	28.1	6.716	28.188	21.472
6.0	3.698	5.287	1.589	30.2	6.628	29.966	23.338
7.0	5.089	6.646	1.557	32.2	6.544	32.785	26.241
8.0	5.825	8.355	2.530	34.2	6.456	35.245	28.789
9.0	6.346	9.350	3.003	36.2	6.392	37.606	31.214
10.1	6.735	10.803	4.067	38.2	6.316	39.367	33.051
11.0	6.914	11.611	4.697	40.2	6.252	41.817	35.565
12.1	7.093	13.649	6.556	42.1	6.191	42.404	36.212
14.1	7.214	16.897	9.683	44.2	6.159	45.856	39.697
16.1	7.237	17.952	10.716	46.2	6.114	49.025	42.911
18.1	7.191	18.503	11.312	48.2	6.072	51.418	45.346
20.1	7.115	20.659	13.544				

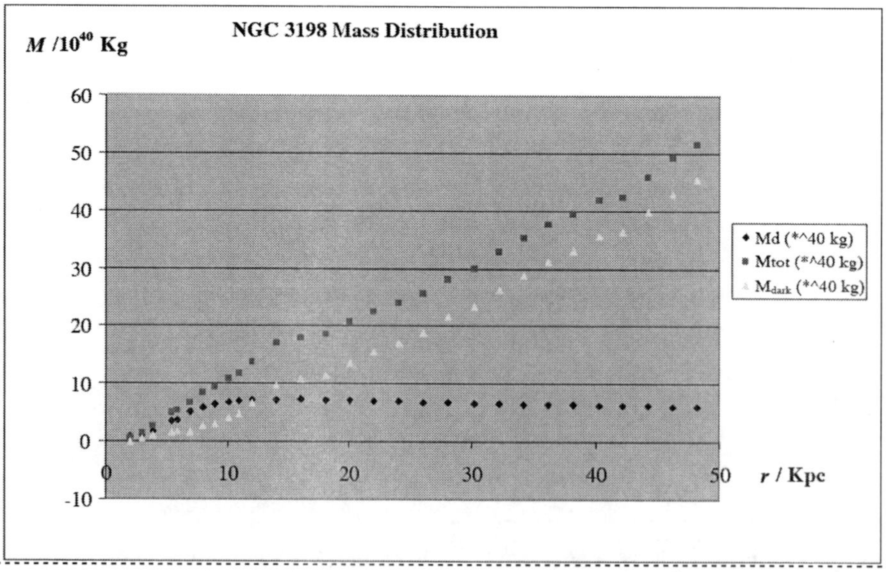

Figure 3. Mass Distribution in Galaxy NGC 3198. Luminous Mass M_d, Total Mass M_{tot} and Dark Mass M_{dark} of the galaxy that is enclosed within the circular orbit wrt the distance from the center of Galaxy NGC 3198.

IOP Conf. Series: Journal of Physics: Conf. Series **936** (2017) 012008 doi:10.1088/1742-6596/936/1/012008

2. Euclidean Closed Linear Transformations of Complex Spacetime endowed with the Corresponding Metric

2.1. Theoretical Background

In case of *Closed Isometric* Linear Spacetime Transformation (LSTT), 3D-space is isotropic [12]. So for Relativistic Inertial observers (RIOs), the representation of the non-degenerate inner product in basis $[c\hat{t},\hat{x},\hat{y},\hat{z}]$ is the real matrix

$$g_{\mathrm{I}}=\begin{bmatrix} g_{\mathrm{I}00} & 0 & 0 & 0 \\ 0 & g_{\mathrm{I}11} & 0 & 0 \\ 0 & 0 & g_{\mathrm{I}22} & 0 \\ 0 & 0 & 0 & g_{\mathrm{I}33} \end{bmatrix}=g_{\mathrm{I}11}\begin{bmatrix} \dfrac{1}{\omega_{\mathrm{I}}^{2}} & 0 & 0 & 0 \\ 0 & 1 & 0 & 0 \\ 0 & 0 & 1 & 0 \\ 0 & 0 & 0 & 1 \end{bmatrix} \tag{35}$$

where

$$\omega_{\mathrm{I}}^{2}=\frac{g_{\mathrm{I}11}}{g_{\mathrm{I}00}} \tag{36}$$

The Universal Speed (U_{I}) of the corresponding SR is given by the equation

$$U_{\mathrm{I}}^{2}=-\frac{\mathrm{c}^{2}}{\omega_{\mathrm{I}}^{2}} \tag{37}$$

and the transformation of a contravariant four-vector is

$$\mathrm{d}X'=\Lambda_{(\omega_{\mathrm{I}},\vec{\beta})}\,\mathrm{d}X \tag{38}$$

where

$$\Lambda_{(\omega_{\mathrm{I}},\vec{\beta})}=\gamma_{(\mathrm{i}\,\omega_{\mathrm{I}}\vec{\beta})}\begin{bmatrix} 1 & \omega_{\mathrm{I}}^{2}\beta_{x} & \omega_{\mathrm{I}}^{2}\beta_{y} & \omega_{\mathrm{I}}^{2}\beta_{z} \\ -\beta_{x} & 1 & \omega_{\mathrm{I}}\beta_{z} & -\omega_{\mathrm{I}}\beta_{y} \\ -\beta_{y} & -\omega_{\mathrm{I}}\beta_{z} & 1 & \omega_{\mathrm{I}}\beta_{x} \\ -\beta_{z} & \omega_{\mathrm{I}}\beta_{y} & -\omega_{\mathrm{I}}\beta_{x} & 1 \end{bmatrix}=\gamma_{(\mathrm{i}\,\omega_{\mathrm{I}}\vec{\beta})}\begin{bmatrix} 1 & \omega_{\mathrm{I}}^{2}\beta^{T} \\ -\beta & \mathrm{I}_{3}+\omega_{\mathrm{I}}\,\mathrm{A}_{(\beta)} \end{bmatrix} \tag{39}$$

$$\beta^{i}=\frac{\mathrm{d}x^{i}}{\mathrm{d}x^{0}}\;;\quad \beta=\begin{bmatrix}\beta_{x}\\ \beta_{y}\\ \beta_{z}\end{bmatrix}\;;\quad \mathrm{A}_{(\beta)}=\begin{bmatrix} 0 & \beta_{z} & -\beta_{y} \\ -\beta_{z} & 0 & \beta_{x} \\ \beta_{y} & -\beta_{x} & 0 \end{bmatrix} \tag{40}$$

and γ-function is

$$\gamma_{(\vec{\delta})}=\frac{1}{\sqrt{1-\left|\vec{\delta}\right|^{2}}} \tag{41}$$

The specific value $\omega_{\mathrm{I}}=0$ ($g_{\mathrm{I}11}\to 0$, $g_{\mathrm{I}00}\neq 0$) gives *Galilean Transformation* (GT) with Infinite Universal Speed (U_{I}) and the corresponding metric of the spacetime (let us call *Galilean metric*)

$$g_{\Gamma}=\begin{bmatrix} g_{\mathrm{I}00} & 0 & 0 & 0 \\ 0 & 0 & 0 & 0 \\ 0 & 0 & 0 & 0 \\ 0 & 0 & 0 & 0 \end{bmatrix} \tag{42}$$

If the *coefficients of metric of time and space have different signs*, then ω_{I} *is an imaginary number*

$$\omega_{\mathrm{I}}=\pm\xi_{\mathrm{I}}\mathrm{i} \tag{43}$$

where

$$\xi_I = \sqrt{\frac{g_{I11}}{-g_{I00}}} \tag{44}$$

is positive number. The corresponding SR has real Universal Speed (U_I)

$$U_I = \frac{1}{\xi_I}c \tag{45}$$

The specific value ω_I=±i or equivalently ξ=1 (g_{I11}= - g_{I00}) gives *Vossos Transformation* (VT) with Universal Speed

$$U_I\text{=c} \tag{46}$$

and the corresponding metric of the spacetime (let us call *Vossos metric*)

$$g_B = g_{iiI}\begin{bmatrix} -1 & 0 & 0 & 0 \\ 0 & 1 & 0 & 0 \\ 0 & 0 & 1 & 0 \\ 0 & 0 & 0 & 1 \end{bmatrix} = g_{iiI}\eta \tag{47}$$

which for g_{iiI}=1 becomes the *Lorentz metric*. Thus, we have the SR theory of *Euclidean Complex Relativistic Mechanics* (ECRMs) [14].

In case that *the metric depends on the position* of the event in spacetime (GR), the Closed LSTT is applied locally, not globally. The limit of vanishing acceleration leads to the corresponding SR. Thus

$$\lim_{\vec{a}\to 0} g_{11} = g_{I11} \tag{48}$$

$$\lim_{\vec{a}\to 0} g_{00} = \lim_{\vec{a}\to 0}\frac{g_{11}}{\omega^2} = g_{I00} = \frac{g_{11I}}{\omega_I^{\ 2}} \tag{49}$$

where

$$\omega^2 = \frac{g_{11}}{g_{00}} \tag{50}$$

The Local Invariant Speed (U) of the corresponding GR is given by the equation

$$U^2 = -\frac{c^2}{\omega^2} \tag{51}$$

and the transformation of a contravariant four-vector is

$$dX' = \Lambda_{(\omega,\vec{\beta})} dX \tag{52}$$

where

$$\Lambda_{(\omega,\vec{\beta})} = \gamma_{(i\omega\vec{\beta})}\begin{bmatrix} 1 & \omega^2\beta_x & \omega^2\beta_y & \omega^2\beta_z \\ -\beta_x & 1 & \omega\beta_z & -\omega\beta_y \\ -\beta_y & -\omega\beta_z & 1 & \omega\beta_x \\ -\beta_z & \omega\beta_y & -\omega\beta_x & 1 \end{bmatrix} = \gamma_{(i\omega\vec{\beta})}\begin{bmatrix} 1 & \omega^2\beta^T \\ -\beta & I_3 + \omega A_{(\beta)} \end{bmatrix} \tag{53}$$

In case of GR, the coefficients of metric of time and space combined producing ω which is *a local parameter of GR*. Thus, parameter ω is a real or an imaginary number.
If *the metric coefficients of time and space have different signs*, then ω *is an imaginary number*

$$\omega = \pm\xi i \tag{54}$$

where

$$\xi = \sqrt{\frac{g_{11}}{-g_{00}}} \tag{55}$$

is positive number. The corresponding GR has real Local Invariant Speed (U)

$$U = \frac{1}{\xi}c \tag{56}$$

In addition, any complex *Cartesian Coordinates* (CCs) of the theory may be turned to the corresponding real CCs, in order to be perceived by human senses.

2.2. Relativistic potential around a center of gravity (1st Generalized Schwarzschild potential)

The purpose is the creation of a spacetime metric around a center of gravity that is in accordance with any acceptable SR having Real, Imaginary or Infinite universal speed (U_I) (let us call 1st *Generalized Schwarzschild metric*). Thus, we define a relativistic potential Φ around a center of gravity (let us call *1st Generalized Schwarzschild potential*)

$$\Phi = \frac{U_I^2}{2}\ln\left(1 - \frac{r_{SI}}{r}\right) \tag{57}$$

where r is the distance between the center of the gravity and the spatial position of the event and r_{SI} is the radius (let us call 1st *Generalized Schwarzschild radius*)

$$r_{SI} = \frac{2GM}{U_I^2} \tag{58}$$

Thus the above equation combined with (5) and (37) gives

$$r_{SI} = -\omega_I^2 r_S \tag{59}$$

and (34) becomes

$$\Phi = \frac{-c^2}{2\omega_I^2}\ln\left(1 + \frac{\omega_I^2 r_S}{r}\right) = -\frac{c^2}{2}\frac{r_S}{r} + \ldots = -\frac{GM}{r} + \ldots \tag{60}$$

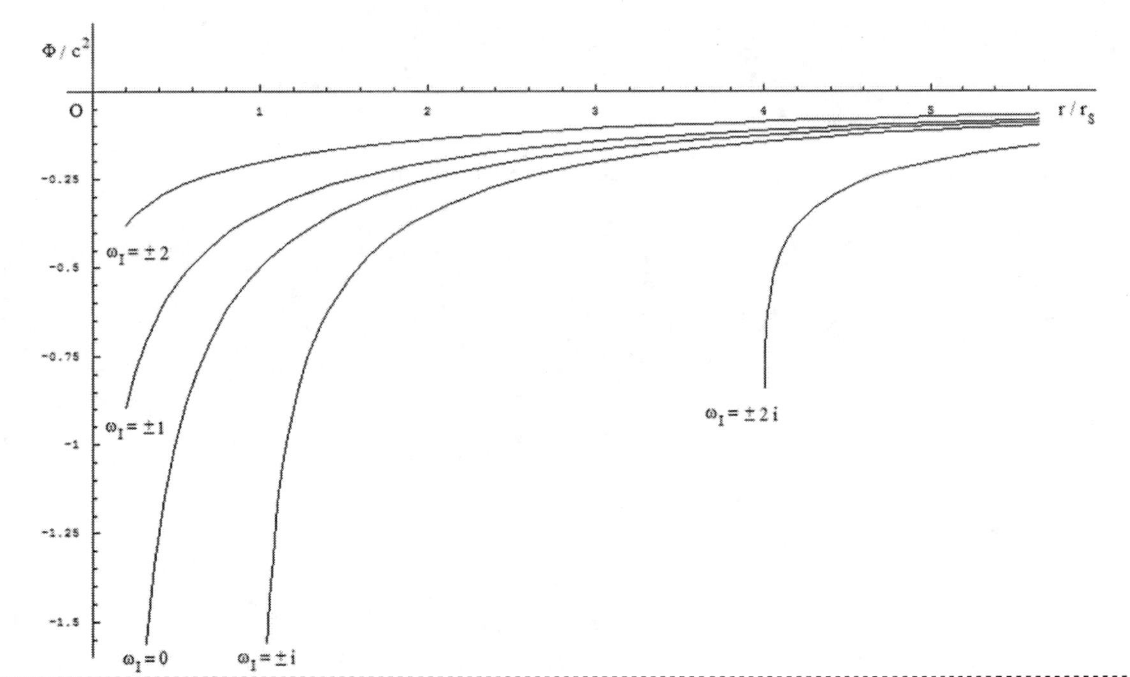

Figure 4. Parametric plot of 1st *Generalized Schwarzschild potential* (Φ) wrt the distance (r/r_S), for different values of characteristic parameter (ω_I) of the specific SR.

We observe that if $\omega_I = \pm i$ (VT), it emerges the original *Schwarzschild potential*. Moreover, if $\omega_I \to 0$ (GT), we compute

$$\lim_{\omega_\mathrm{I}\to 0}\Phi=\frac{-\mathrm{c}^2}{2}\lim_{\omega_\mathrm{I}\to 0}\left[\frac{1}{\omega_\mathrm{I}^{\,2}}\ln\left(1+\frac{\omega_\mathrm{I}^{\,2}r_\mathrm{S}}{r}\right)\right]=\frac{-\mathrm{c}^2}{4}\lim_{\omega_\mathrm{I}\to 0}\left[\frac{1}{\omega_\mathrm{I}}\frac{\frac{2\omega_\mathrm{I}r_\mathrm{S}}{r}}{1+\frac{\omega_\mathrm{I}^{\,2}r_\mathrm{S}}{r}}\right]=-\frac{\mathrm{c}^2}{2}\frac{r_\mathrm{S}}{r}=-\frac{\mathrm{G}M}{r} \tag{61}$$

Thus, it has produced the *Relativistic Potential of Newtonian Physics* (NP).

In Figure 4, we show the plot of *1st Generalized Schwarzschild potential* Φ wrt the distance r/r_s for different values of the characteristic parameter (ω_I) of the specific SR. We observe that for $\omega_\mathrm{I}\to\pm\infty$, the relativistic potential becomes zero–function. The curve with $\omega_\mathrm{I}=0$ ($g_{\mathrm{I}11}=0$) describes the *Relativistic Potential of Newtonian Physics* (NP) which corresponds to *Galilean metric* and *Galilean Transformation* (GT), while the curve with $\omega_\mathrm{I}=\pm i$ ($g_{\mathrm{I}11}=-g_{\mathrm{I}00}$) describes *Schwarzschild potential,* which corresponds to *Vossos metric* of flat spacetime and *Vossos Transformation* (VT).

2.3. Metric of spacetime around a center of gravity (1st Generalized Schwarzschild metric)

The 1st *Type of generalized Schwarzschild metric* is

$$\mathrm{d}S^2=\mathrm{d}X^\mathrm{T}g\,\mathrm{d}X=\begin{bmatrix}\mathrm{c}\mathrm{d}t & \mathrm{d}r & \mathrm{d}\theta & \mathrm{d}\phi\end{bmatrix}\cdot\begin{bmatrix}-(-g_{\mathrm{I}00})e^A & 0 & 0 & 0\\ 0 & g_{\mathrm{I}11}e^B & 0 & 0\\ 0 & 0 & g_{\mathrm{I}11}r^2 & 0\\ 0 & 0 & 0 & g_{\mathrm{I}11}r^2\sin^2\theta\end{bmatrix}\cdot\begin{bmatrix}\mathrm{c}\mathrm{d}t\\ \mathrm{d}r\\ \mathrm{d}\theta\\ \mathrm{d}\phi\end{bmatrix} \tag{62}$$

where

$$A=\frac{2}{U_\mathrm{I}^{\,2}}\Phi \tag{63}$$

This combined with *Einstein equations* in vacuum gives

$$\mathrm{d}S^2=g_{\mathrm{I}00}\left(1-\frac{r_\mathrm{SI}}{r}\right)\mathrm{c}^2\mathrm{d}t^2+\frac{g_{\mathrm{I}11}}{1-\frac{r_\mathrm{SI}}{r}}\mathrm{d}r^2+g_{\mathrm{I}11}r^2\,\mathrm{d}\theta^2+g_{\mathrm{I}11}r^2\sin^2\theta\,\mathrm{d}\phi^2 \tag{64}$$

Thus, the *isotropic form* of *1st generalized Schwarzschild metric* has

$$g_{00}=g_{\mathrm{I}00}\left(\frac{1-\frac{\xi_\mathrm{I}^{\,2}r_\mathrm{S}}{4\tilde r}}{1+\frac{\xi_\mathrm{I}^{\,2}r_\mathrm{S}}{4\tilde r}}\right)^2 \quad ; \quad \tilde g_{11}=g_{\mathrm{I}11}\left(1+\frac{\xi_\mathrm{I}^{\,2}r_\mathrm{S}}{4\tilde r}\right)^4 \tag{65}$$

where

$$\tilde r=\frac{\xi_\mathrm{I}^{\,2}r_\mathrm{S}}{4}\left(\frac{2r}{\xi_\mathrm{I}^{\,2}r_\mathrm{S}}\left(1+\sqrt{1-\xi_\mathrm{I}^{\,2}\frac{r_\mathrm{S}}{r}}\right)-1\right) \quad ; \quad r=\tilde r\left(1+\frac{\xi_\mathrm{I}^{\,2}r_\mathrm{S}}{4\tilde r}\right)^2 \tag{66}$$

In case that $\xi_\mathrm{I}\to 0$, it emerges $\tilde r\to r$, $g_{00}\to g_{\mathrm{I}00}$, $\tilde g_{11}\to 0$. We also have

$$\tilde\xi=\frac{\xi_\mathrm{I}\left(1+\frac{\xi_\mathrm{I}^{\,2}r_\mathrm{S}}{4\tilde r}\right)^3}{\left(1-\frac{\xi_\mathrm{I}^{\,2}r_\mathrm{S}}{4\tilde r}\right)} \tag{67}$$

and the Local Invariant speed is

$$\widetilde{U}=\frac{\left(1-\dfrac{\xi_{\rm I}^{2}r_{\rm S}}{4\widetilde{r}}\right)}{\xi_{\rm I}\left(1+\dfrac{\xi_{\rm I}^{2}r_{\rm S}}{4\widetilde{r}}\right)^{3}}\mathrm{c} \tag{68}$$

Besides

$$g_{00}=g_{\rm I00}\left(1-\xi_{\rm I}^{2}\frac{r_{\rm S}}{r}\right)\ ;\quad g_{rr}=\frac{g_{\rm I11}}{1-\xi_{\rm I}^{2}\dfrac{r_{\rm S}}{r}} \tag{69}$$

2.4. Field strength (acceleration) around a center of gravity with 1st Generalized Schwarzschild potential for SR having real, imaginary or infinite universal speed

From (60), we can calculate the field strength g (see e.g. [1] p. 230). The field is radial and we define

$$g=\sqrt{g_{\rm I11}}\left|grad\Phi\right|=\sqrt{g_{\rm I11}}\frac{d\Phi}{dl}=\sqrt{g_{\rm I11}}\frac{d\Phi}{dr}\frac{dr}{dl} \tag{70}$$

where l is the radial ruler distance (see eq. 64, 59)

$$\mathrm{d}l^{2}=\frac{g_{\rm I11}}{\left(1+\omega_{\rm I}^{2}\dfrac{r_{\rm S}}{r}\right)}\mathrm{d}r^{2}+g_{\rm I11}r^{2}\,\mathrm{d}\theta^{2}+g_{\rm I11}r^{2}\sin^{2}\theta\,\mathrm{d}\phi^{2} \tag{71}$$

So

$$\frac{dr}{dl}=\frac{1}{\sqrt{g_{\rm I11}}}\left(1+\omega_{I}^{2}\frac{r_{s}}{r}\right)^{\frac{1}{2}} \tag{72}$$

and

$$g=\frac{GM}{r^{2}}\left(1+\omega_{I}^{2}\frac{r_{s}}{r}\right)^{-\frac{1}{2}} \tag{73}$$

In case that $\boldsymbol{\omega_I=\pm i}$ ($g_{\rm I11}=-g_{\rm I00}$), we have the SR which corresponds to VT and it emerges the original *Schwarzschild metric, potential* and *field strength*.

3. 2nd *Generalized Schwarzschild metric* and the corresponding Relativistic potential and Field strength and Combination with MOND

3.1. 2nd Generalized Schwarzschild metric and the corresponding Relativistic potential and Field strength

Now, we can modify (60). Thus, we define a new relativistic potential Φ around a center of gravity (let us call *Modified Generalized Schwarzschild potential)* as

$$\Phi=h_{(r)}\frac{-\mathrm{c}^{2}}{2\omega_{\rm I}^{2}}\ln\left(1+f_{(r)}\frac{\omega_{\rm I}^{2}r_{\rm S}}{r}\right) \tag{74}$$

In case that

$$h_{(r)}=1\ ;\ f_{(r)}=1$$

it emerges *1st Generalized Schwarzschild potential*. Now, we examine the case that only $f_{(r)}=1$. Let us call this potential as *2nd Generalized Schwarzschild potential*

$$\Phi=h_{(r)}\frac{-\mathrm{c}^{2}}{2\omega_{\rm I}^{2}}\ln\left(1+\frac{\omega_{\rm I}^{2}r_{\rm S}}{r}\right) \tag{75}$$

We demand the above relativistic potential having value equal to the *1st Generalized Schwarzschild potential* at the distance of *1st Generalized Schwarzschild radius*. Thus, we have

$$h_{(r_{\mathrm{SI}})} = h_{(-\omega_{\mathrm{I}}{}^2 r_{\mathrm{S}})} = 1 \tag{76}$$

The metric of spacetime around a center of gravity (*Modified Generalized Schwarzschild metric*) is

$$\mathrm{d}S^2 = \mathrm{d}X^{\mathrm{T}} g\,\mathrm{d}X = [c\mathrm{d}t \quad \mathrm{d}r \quad \mathrm{d}\theta \quad \mathrm{d}\phi] \cdot \begin{bmatrix} -(-g_{\mathrm{I}00})e^{A} & 0 & 0 & 0 \\ 0 & g_{\mathrm{I}11}e^{B} & 0 & 0 \\ 0 & 0 & g_{\mathrm{I}11}r^2 & 0 \\ 0 & 0 & 0 & g_{\mathrm{I}11}r^2\sin^2\theta \end{bmatrix} \cdot \begin{bmatrix} c\mathrm{d}t \\ \mathrm{d}r \\ \mathrm{d}\theta \\ \mathrm{d}\phi \end{bmatrix} \tag{77}$$

where

$$A = \frac{2}{U_{\mathrm{I}}{}^2}\Phi \tag{78}$$

This combined with (74) gives

$$A = h_{(r)} \ln\left(1 + \frac{\omega_{\mathrm{I}}{}^2 r_{\mathrm{S}}}{r}\right) = \ln\left(1 + \frac{\omega_{\mathrm{I}}{}^2 r_{\mathrm{S}}}{r}\right)^{h_{(r)}} \tag{79}$$

Thus

$$g_{00} = g_{\mathrm{I}00}\left(1 + \frac{\omega_{\mathrm{I}}{}^2 r_{\mathrm{S}}}{r}\right)^{h_{(r)}} \tag{80}$$

The coefficients $(-g_{00\mathrm{I}})$ and $g_{\mathrm{I}rr}$ may be written as

$$-g_{\mathrm{I}00} = e^{\ln(-g_{\mathrm{I}00})}, \; g_{\mathrm{I}11} = e^{\ln g_{\mathrm{I}11}} \tag{81}$$

Thus, we have

$$\mathrm{d}S^2 = [c\mathrm{d}t \quad \mathrm{d}r \quad \mathrm{d}\theta \quad \mathrm{d}\phi] \cdot \begin{bmatrix} -e^{A+\ln(-g_{\mathrm{I}00})} & 0 & 0 & 0 \\ 0 & e^{B+\ln g_{\mathrm{I}11}} & 0 & 0 \\ 0 & 0 & g_{\mathrm{I}11}r^2 & 0 \\ 0 & 0 & 0 & g_{\mathrm{I}11}r^2\sin^2\theta \end{bmatrix} \cdot \begin{bmatrix} c\mathrm{d}t \\ \mathrm{d}r \\ \mathrm{d}\theta \\ \mathrm{d}\phi \end{bmatrix} \tag{82}$$

We may define

$$A_1 = A + \ln(-g_{\mathrm{I}00}), \; B_1 = B + \ln g_{\mathrm{I}11} \tag{83}$$

So, we obtain the formula of the original *Schwarzschild metric* except for containing A_1 and B_1

$$\mathrm{d}S^2 = [c\mathrm{d}t \quad \mathrm{d}r \quad \mathrm{d}\theta \quad \mathrm{d}\phi] \cdot \begin{bmatrix} -e^{A_1} & 0 & 0 & 0 \\ 0 & e^{B_1} & 0 & 0 \\ 0 & 0 & g_{\mathrm{I}11}r^2 & 0 \\ 0 & 0 & 0 & g_{\mathrm{I}11}r^2\sin^2\theta \end{bmatrix} \cdot \begin{bmatrix} c\mathrm{d}t \\ \mathrm{d}r \\ \mathrm{d}\theta \\ \mathrm{d}\phi \end{bmatrix} \tag{84}$$

This metric combined with *Einstein equations* in vacuum (see e.g. [1] p 229, eq 11.8) gives

$$\frac{\mathrm{d}B_1}{\mathrm{d}r} = -\frac{\mathrm{d}A_1}{\mathrm{d}r} \tag{85}$$

or

$$B_1 = -A_1 + C_1 \tag{86}$$

So

$$B = -A + \ln\left(\frac{1}{-g_{\mathrm{I}00}g_{\mathrm{I}11}}\right) + \ln e^{C_1} \tag{87}$$

Taking into account (79), we have

$$B=\ln\left(\frac{e^{C_1}}{-g_{\mathrm{I}00}g_{\mathrm{I}11}\left(1+\frac{\omega_{\mathrm{I}}^2 r_{\mathrm{S}}}{r}\right)^{h_{(r)}}}\right) \tag{88}$$

and

$$g_{rr}=\frac{e^{C_1}}{-g_{\mathrm{I}00}\left(1+\frac{\omega_{\mathrm{I}}^2 r_{\mathrm{S}}}{r}\right)^{h_{(r)}}} \tag{89}$$

At infinite distance, we have metric of the corresponding SR. So

$$g_{\mathrm{I}11}=\frac{e^{C_1}}{-g_{\mathrm{I}00}}$$

and

$$g_{rr}=\frac{g_{\mathrm{I}11}}{\left(1+\frac{\omega_{\mathrm{I}}^2 r_{\mathrm{S}}}{r}\right)^{h_{(r)}}} \tag{90}$$

Thus, the 2nd *generalized Schwarzschild metric* in spherical coordinates is

$$\mathrm{d}S^2=g_{\mathrm{I}00}\left(1+\frac{\omega_{\mathrm{I}}^2 r_{\mathrm{S}}}{r}\right)^{h_{(r)}}\mathrm{c}^2\mathrm{d}t^2\,\frac{g_{\mathrm{I}11}}{\left(1+\frac{\omega_{\mathrm{I}}^2 r_{\mathrm{S}}}{r}\right)^{h_{(r)}}}\mathrm{d}r^2+g_{\mathrm{I}11}r^2\,\mathrm{d}\theta^2+g_{\mathrm{I}11}r^2\sin^2\theta\,\mathrm{d}\phi^2 \tag{91}$$

with spatial part

$$\mathrm{d}l^2=\frac{g_{\mathrm{I}11}}{\left(1+\frac{\omega_{\mathrm{I}}^2 r_{\mathrm{S}}}{r}\right)^{h_{(r)}}}\mathrm{d}r^2+g_{\mathrm{I}11}r^2\,\mathrm{d}\theta^2+g_{\mathrm{I}11}r^2\sin^2\theta\,\mathrm{d}\phi^2 \tag{92}$$

Now, we can calculate this radial field strength defining

$$\vec{g}=-\sqrt{g_{\mathrm{I}11}}\nabla\Phi=-\sqrt{g_{\mathrm{I}11}}\frac{d\Phi}{dl}\hat{r}=-\sqrt{g_{\mathrm{I}11}}\frac{d\Phi}{dr}\frac{dr}{dl}\hat{r} \tag{93}$$

and

$$g=\sqrt{g_{\mathrm{I}11}}\frac{d\Phi}{dr}\frac{dr}{dl} \tag{94}$$

The positive value of field strength means gravity, while negative value means antigravity. Moreover, it is

$$\frac{dr}{dl}=\frac{1}{\sqrt{g_{\mathrm{I}11}}}\left(1+\frac{\omega_{\mathrm{I}}^2 r_{\mathrm{S}}}{r}\right)^{\frac{h_{(r)}}{2}} \tag{95}$$

and

$$\frac{d\Phi}{dr}=\frac{-\mathrm{c}^2}{2\omega_{\mathrm{I}}^2}\left(-\frac{\omega_{\mathrm{I}}^2 r_{\mathrm{S}}}{r^2}\frac{h_{(r)}}{1+\frac{\omega_{\mathrm{I}}^2 r_{\mathrm{S}}}{r}}+\frac{\mathrm{d}h}{\mathrm{d}r}\ln\left(1+\frac{\omega_{\mathrm{I}}^2 r_{\mathrm{S}}}{r}\right)\right) \tag{96}$$

So

$$g=\frac{-c^2}{2\omega_I^2}\left(-\frac{\omega_I^2 r_S}{r^2}h_{(r)}\left(1+\frac{\omega_I^2 r_S}{r}\right)^{\frac{h_{(r)}}{2}-1}+\frac{dh}{dr}\left(1+\frac{\omega_I^2 r_S}{r}\right)^{\frac{h_{(r)}}{2}}\ln\left(1+\frac{\omega_I^2 r_S}{r}\right)\right) \quad (97)$$

or

$$g=\frac{GM}{r^2}h_{(r)}\left(1+\frac{\omega_I^2 r_S}{r}\right)^{\frac{h_{(r)}}{2}-1}-\frac{c^2}{2\omega_I^2}\frac{dh}{dr}\left(1+\frac{\omega_I^2 r_S}{r}\right)^{\frac{h_{(r)}}{2}}\ln\left(1+\frac{\omega_I^2 r_S}{r}\right) \quad (98)$$

In case of UCM, it is

$$\upsilon^2=\frac{GM}{r}h_{(r)}\left(1+\frac{\omega_I^2 r_S}{r}\right)^{\frac{h_{(r)}}{2}-1}-\frac{c^2}{2\omega_I^2}r\frac{dh}{dr}\left(1+\frac{\omega_I^2 r_S}{r}\right)^{\frac{h_{(r)}}{2}}\ln\left(1+\frac{\omega_I^2 r_S}{r}\right) \quad (99)$$

3.2. Combination of Galilean-Newtonian 2nd Generalized Schwarzschild metric with MOND

We firstly examine the case which flat spacetime follows *Galilean Metric* ($\omega_I \to 0$). Thus

$$\Phi=-\frac{GM}{r}h_{(r)} \quad (100)$$

or

$$\Phi=-\frac{GM}{r}\frac{\sqrt{4a_0 r_0}}{\sqrt{GM}}h_{(r)}=-\frac{2}{\frac{r}{r_0}}\sqrt{GMa_0}\,h_{(r)}=-\frac{2\upsilon_\infty^2}{\frac{r}{r_0}}h_{(r)} \quad (101)$$

Besides

$$g=\frac{GM}{r^2}h_{(r)}-\frac{GM}{r}\frac{dh}{dr}=\frac{GM}{r}\left(\frac{h_{(r)}}{r}-\frac{dh}{dr}\right) \quad (102)$$

and

$$\upsilon^2=\frac{GM}{r}h_{(r)}-GM\frac{dh}{dr} \quad (103)$$

Using *Interpolating function* of MOND and (11), it emerges

$$\frac{1}{\mu_{(r)}}=h_{(r)}-r\frac{dh}{dr} \quad (104)$$

Thus, if we know function $h_{(r)}$, we may calculate *Interpolating function* $\mu_{(r)}$, using the above equation. But, if we know *Interpolating function* $\mu_{(r)}$, we may calculate $h_{(r)}$, solving the above differential equation and it emerges

$$\frac{h_{(r)}}{r}=-\int\frac{dr}{r^2\mu_{(r)}} \quad (105)$$

So, calculating the above integral, we easily obtain function $h_{(r)}$. The value of integration constant is obtained using the condition (76). Besides (105) may be written as

$$GM\frac{h_{(r)}}{r}=-\int\frac{GM\,dr}{r^2\mu_{(r)}}=-\int g dr \quad (106)$$

It emerges that the work we need to pay for moving one mass unit from a point A with distance r from the center of gravity to another point B with larger distance r_1 (the Potential Difference between point A and point B), is

$$w = \Delta V = V_{(r_1)} - V_{(r)} = \int_r^{r_1} g dr = -GM\frac{h_{(r_1)}}{r_1} + GM\frac{h_{(r)}}{r} \quad (107)$$

We observe that

$$V_{(r)} = \Phi = -\frac{GM}{r} h_{(r)} \quad (108)$$

Finally, from (91) we have the *(Modified) Galilean metric*

$$\mathrm{d}S^2 = -c^2 \mathrm{d}t^2 + 0\mathrm{d}r^2 + 0r^2\mathrm{d}\theta^2 + 0r^2\sin^2\theta \mathrm{d}\phi^2 \quad (109)$$

This is independent by distance from the center of gravity.

3.2.1 *Combination of Galilean-Newtonian 2nd Generalized Schwarzschild metric with* MOND *Simple μ*

Replacing (19) to (105), we obtain

$$\frac{h_{(r)}}{r} = -\int \frac{1}{2r^2}\left(1 + \sqrt{1 + \left(\frac{r}{r_0}\right)^2}\right)\mathrm{d}r \quad (110)$$

or

$$\frac{h_{(r)}}{r} = -\frac{1}{2r_0}\int \frac{1}{\left(\frac{r}{r_0}\right)^2}\left(1 + \sqrt{1 + \left(\frac{r}{r_0}\right)^2}\right)\mathrm{d}\left(\frac{r}{r_0}\right) \quad (111)$$

and using (32), we have

$$h_{(r)} = \frac{1}{2}\left(1 + \sqrt{1 + \left(\frac{r}{r_0}\right)^2} - \frac{r}{r_0}\mathrm{ArcSinh}\left(\frac{r}{r_0}\right)\right) + rC_1 \quad (112)$$

Thus

$$\frac{\mathrm{d}h}{\mathrm{d}r} = C_1 - \frac{1}{2r_0}\mathrm{ArcSinh}\left(\frac{r}{r_0}\right) \quad (113)$$

Because 1st *Generalized Schwarzschild radius* is equal to the zero in *Galilean-Newtonian Physics*, the value of integration constant C_1 is calculated as following

$$h_{(-\omega_I^2 r_S)} = \frac{1}{2}\left(1 + \sqrt{1 + \left(\frac{-\omega_I^2 r_S}{r_0}\right)^2} + \frac{\omega_I^2 r_S}{r_0}\mathrm{ArcSinh}\left(\frac{-\omega_I^2 r_S}{r_0}\right)\right) - \omega_I^2 r_S C_1 = 1$$

Thus

$$C_1 = \frac{1}{2r_S\omega_I^2}\left(-1 + \sqrt{1 + \left(\frac{-\omega_I^2 r_S}{r_0}\right)^2} + \frac{\omega_I^2 r_S}{r_0}\mathrm{ArcSinh}\left(\frac{-\omega_I^2 r_S}{r_0}\right)\right) \quad (114)$$

Taking the limit $\omega_I \to 0$, it emerges

$$C_1 = 0 \quad (115)$$

So

$$h_{(r)} = \frac{1}{2}\left(1 + \sqrt{1 + \left(\frac{r}{r_0}\right)^2} - \frac{r}{r_0}\mathrm{ArcSinh}\left(\frac{r}{r_0}\right)\right) \quad (116)$$

and

IOP Conf. Series: Journal of Physics: Conf. Series **936** (2017) 012008 doi:10.1088/1742-6596/936/1/012008

$$\frac{\mathrm{d}h}{\mathrm{d}r} = -\frac{1}{2r_0}\mathrm{ArcSinh}\left(\frac{r}{r_0}\right) \leq 0 \tag{117}$$

In Figure 5, we show the plot of function h_ω wrt the distance (r/r_0) for the Combination of *Galilean-Newtonian* 2nd *Generalized Schwarzschild metric* with MOND *Simple* μ.

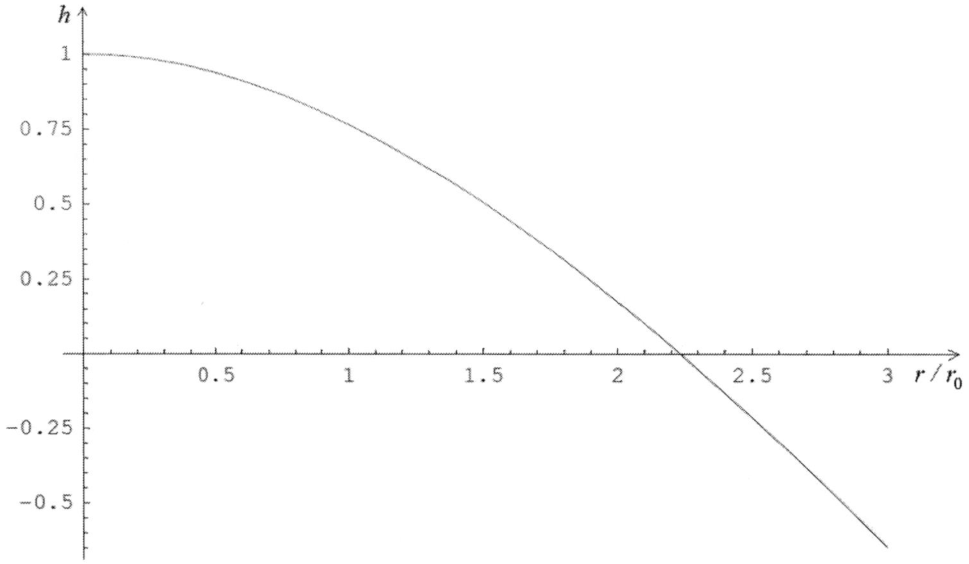

Figure 5. Plot of function h_ω wrt the distance (r/r_0) for the Combination of *Galilean-Newtonian* 2nd *Generalized Schwarzschild metric* with MOND *Simple* μ .

We observe that function h_ω is decrescent, its graph cuts the distance axis at $r \approx 2.2 r_0$ and goes to the minus infinite ($-\infty$) at infinite distance from the center of gravity.

Now, using (101) we calculate the corresponding Relativistic Potential

$$\Phi = -\frac{\upsilon_\infty^2}{\frac{r}{r_0}}\left(1+\sqrt{1+\left(\frac{r}{r_0}\right)^2} - \frac{r}{r_0}\mathrm{ArcSinh}\left(\frac{r}{r_0}\right)\right) \tag{118}$$

In Figure 6, we show the plot of Relativistic Potential (Φ/υ_∞^2) wrt the distance (r/r_0) for the Combination of *Galilean-Newtonian* 2nd *Generalized Schwarzschild metric* with MOND *Simple* μ.

We observe that this Modified Relativistic Potential Φ is increscent, its graph cuts the distance axis at $r \approx 2.3 r_0$ and goes slightly to the plus infinite ($+\infty$) at infinite distance from the center of gravity, in contrast to the pure *Newtonian Potential* that goes to zero.

IOP Conf. Series: Journal of Physics: Conf. Series **936** (2017) 012008 doi:10.1088/1742-6596/936/1/012008

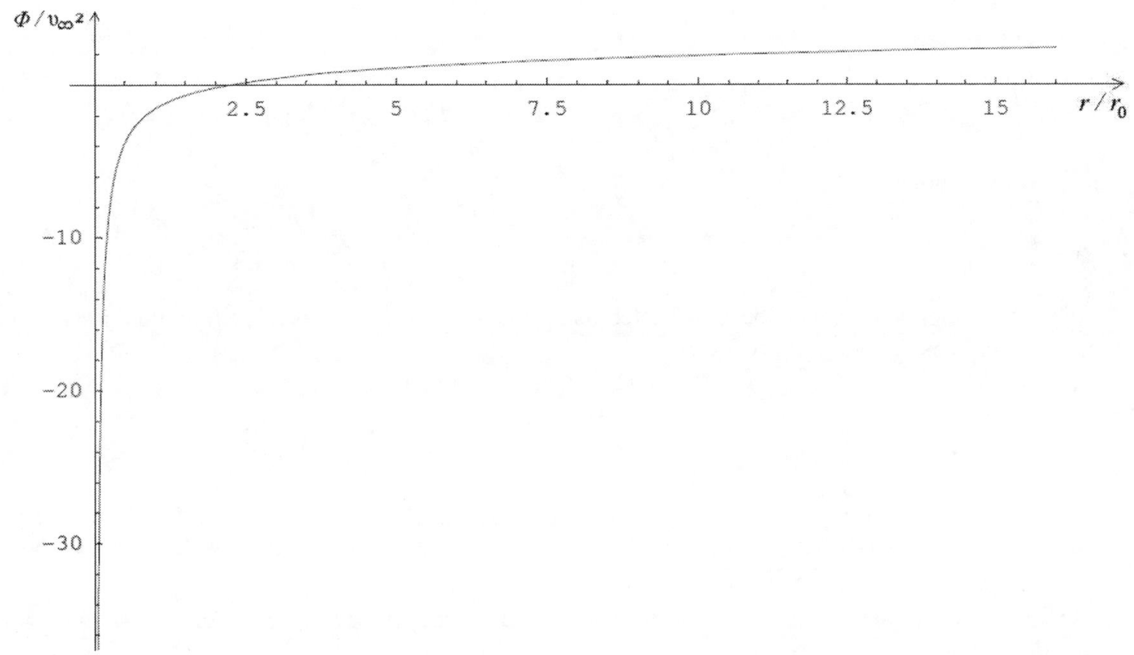

Figure 6. Plot of Relativistic Potential (Φ/υ_∞^2) wrt the distance (r/r_0) for the Combination of *Galilean-Newtonian 2nd Generalized Schwarzschild metric* with MOND *Simple* μ .

3.2.2 *Combination of Galilean-Newtonian 2nd Generalized Schwarzschild metric with* MOND *Standard Interpolating Function*

Replacing (22) to (105) we have results

$$\frac{h_{(r)}}{r} = -\int \frac{1}{\sqrt{2}r^2}\sqrt{1+\sqrt{1+\frac{1}{4}\left(\frac{r}{r_0}\right)^4}}\,\mathrm{d}r \tag{119}$$

or

$$\frac{h_{(r)}}{r} = -\frac{1}{\sqrt{2}r_0}\int \frac{1}{\left(\frac{r}{r_0}\right)^2}\sqrt{1+\sqrt{1+\frac{1}{4}\left(\frac{r}{r_0}\right)^4}}\,\mathrm{d}\left(\frac{r}{r_0}\right) \tag{120}$$

Unfortunately, the integral (34) has not yet analytical solution. So, we cannot go further our analysis.

3.3. Combination of Lorentzian-Einsteinian 2nd Generalized Schwarzschild metric with MOND

In case of *Lorentzian-Einsteinian Physics* the 2nd *generalized Schwarzschild metric* in spherical coordinates (91) becomes

$$\mathrm{d}S^2 = -\left(1-\frac{r_S}{r}\right)^{h_{(r)}} \mathrm{c}^2\mathrm{d}t^2 + \frac{1}{\left(1-\frac{r_S}{r}\right)^{h_{(r)}}}\mathrm{d}r^2 + r^2\,\mathrm{d}\theta^2 + r^2\sin^2\theta\,\mathrm{d}\phi^2 \tag{121}$$

2nd *Generalized Schwarzschild potential* (75) is written as

$$\Phi = h_{(r)} \frac{c^2}{2} \ln\left(1 - \frac{r_S}{r}\right) \tag{122}$$

and field strength

$$g = \frac{GM}{r^2} h_{(r)} \left(1 - \frac{r_S}{r}\right)^{\frac{h_{(r)}}{2}-1} + \frac{c^2}{2} \frac{dh}{dr} \left(1 - \frac{r_S}{r}\right)^{\frac{h_{(r)}}{2}} \ln\left(1 - \frac{r_S}{r}\right) \tag{123}$$

as well as υ^2 in UCM

$$\upsilon^2 = \frac{GM}{r} h_{(r)} \left(1 - \frac{r_S}{r}\right)^{\frac{h_{(r)}}{2}-1} + \frac{c^2}{2} r \frac{dh}{dr} \left(1 - \frac{r_S}{r}\right)^{\frac{h_{(r)}}{2}} \ln\left(1 - \frac{r_S}{r}\right) \tag{124}$$

The experimental data have been extracted using the *Relativistic Doppler Shift* and the gravitational red shift of Classic Relativity (CR)

$$f_E = f_S \gamma (1 - \beta \cos\varphi) \tag{125}$$

$$\frac{f_S}{f_E} = \sqrt{\frac{g_{00(S)}}{g_{00(E)}}} \tag{126}$$

where f_s and f_E is the frequency of the radiation which has emitted by the Star, measured on the Star and on the Earth respectively and β is the velocity factor of the Star wrt the Earth and ϕ is the angle between the position vector $(\vec{r})$ and the velocity of the Star wrt Earth $(\vec{\upsilon})$.

In case of Combination of *Lorentzian-Einsteinian 2nd Generalized Schwarzschild metric with* MOND *Simple* μ, it is

$$g_{00} = -\left(1 - \frac{r_S}{r}\right)^{h_{(r)}} \tag{127}$$

but there is not significant change to the corresponding values (see Table 5). So, the experimental data may be used without problem. Moreover, for $h>0$, there is time dilation of the accelerated observer wrt unmoved RIO at infinite distance form the center of gravity, while time contraction happens for $h<0$.

3.3.1 *Combination of Lorentzian-Einsteinian 2nd Type of Generalized Schwarzschild metric with MOND Simple Interpolating Function*

We firstly use function $h_{(r)}$ that has produced by the combination of *Galilean-Newtonian 2nd Type of Generalized Schwarzschild metric* with MOND *Simple* μ. We simply change integration constant C_1, in order to correspond to *Lorentzian-Einsteinian Physics*. Thus, if we replace ω_r=-1 to (114), we obtain

$$C_1 = -\frac{1}{2r_S}\left(-1 + \sqrt{1 + \left(\frac{r_S}{r_0}\right)^2} - \frac{r_S}{r_0} \text{ArcSinh}\left(\frac{r_S}{r_0}\right)\right) \tag{128}$$

Moreover (112) becomes

$$h_{(r)} = \frac{1}{2}\left(\frac{r}{r_S} + 1 + \sqrt{1 + \left(\frac{r}{r_0}\right)^2} - \frac{r}{r_S}\sqrt{1 + \left(\frac{r_S}{r_0}\right)^2} - \frac{r}{r_0}\text{ArcSinh}\left(\frac{r}{r_0}\right) + \frac{r}{r_0}\text{ArcSinh}\left(\frac{r_S}{r_0}\right)\right) \tag{129}$$

Due to (26), the above equation may be written as

$$h_{(r)} = \frac{1}{2}\left(\frac{1 + 4\beta_\infty^2 \text{ArcSinh}\left(4\beta_\infty^2\right) - \sqrt{1 + 16\beta_\infty^4}}{4\beta_\infty^2}\frac{r}{r_0} + 1 + \sqrt{1 + \left(\frac{r}{r_0}\right)^2} - \frac{r}{r_0}\text{ArcSinh}\left(\frac{r}{r_0}\right)\right) \tag{130}$$

We observe that this function $h_{(r)}$ has a different value than the corresponding *Galilean-Newtonian* because of this additional term

IOP Conf. Series: Journal of Physics: Conf. Series **936** (2017) 012008 doi:10.1088/1742-6596/936/1/012008

$$\frac{1+\mathrm{ArcSinh}\left(4\beta_{\infty}^{2}\right)-\sqrt{1+16\beta_{\infty}^{4}}}{4\beta_{\infty}^{2}}\ \frac{r}{r_0} \tag{131}$$

Besides (113) becomes

$$\frac{\mathrm{d}h}{\mathrm{d}r}=-\frac{1}{2}\left(-\frac{1}{r_S}+\frac{1}{r_S}\sqrt{1+\left(\frac{r_S}{r_0}\right)^2}-\frac{1}{r_0}\mathrm{ArcSinh}\left(\frac{r_S}{r_0}\right)+\frac{1}{r_0}\mathrm{ArcSinh}\left(\frac{r}{r_0}\right)\right) \tag{132}$$

Due to (26) the above equation may be written as

$$\frac{\mathrm{d}h}{\mathrm{d}r}=-\frac{1}{2}\left(-\frac{1}{r_S}+\frac{1}{r_S}\sqrt{1+16\beta_{\infty}^{4}}-\frac{1}{r_0}\mathrm{ArcSinh}\left(4\beta_{\infty}^{2}\right)+\frac{1}{r_0}\mathrm{ArcSinh}\left(\frac{r}{r_0}\right)\right) \tag{133}$$

3.3.2 *Combination of Lorentzian-Einsteinian 2nd Generalized Schwarzschild metric with MOND Standard Interpolating Function*

The *Standard* μ in *Milgrom's Law* gives the best results, explaining the rotation curves in galaxies. Unfortunately, the integral (34) has not yet analytical solution and we cannot go further our analysis.

4. Experimental Validation

4.1. The Combination of *Lorentzian-Einsteinian 2nd Generalized Schwarzschild metric* with MOND *Simple Interpolating Function* in Galaxy NGC 3198

The values of Circular Velocities (experimental V_{exp} and calculated by the Combination of *Lorentzian-Einsteinian 2nd Generalized Schwarzschild metric* with MOND *Simple* μ $V_{simp,Lor}$), the Luminous Mass of the galaxy that is enclosed within the circular orbit (M_L), the corresponding *Schwarzschild* radius (r_s), *Milgrom* radius (r_o), the value of function h and the value of time coefficient of metric (g_{00}) wrt the distance from the center of Galaxy NGC 3198, are contained in Table 5 (see [5] p 2). The Circular Velocities $V_{simp,Lor}$ have been calculated using (124) and the values of time coefficient of metric (g_{00}) using (127).

In Figure 7, we show the plot of function h_{00} wrt the distance from the center of NGC 3198 for the Combination of *Lorentzian-Einsteinian 2nd Generalized Schwarzschild metric* with MOND *Simple* μ.

In Figure 8, we show the plot of Relativistic Potential ($\Phi/\upsilon_{\infty}^{2}$) wrt the distance from the center of NGC 3198 (r/r_o) for the Combination of *Lorentzian-Einsteinian 2nd Generalized Schwarzschild metric* with MOND *Simple* μ, as well as the original *Schwarzschild potential*.

In Figure 9, we show the plot of coefficients of metric g_{00}, g_{rr} wrt the distance from the center of NGC 3198 for the Combination of *Lorentzian-Einsteinian 2nd Generalized Schwarzschild metric* with MOND *Simple* μ.

We observe that in case of Galaxy NGC 3198, the Combination of *Lorentzian-Einsteinian 2nd Generalized Schwarzschild metric* with MOND *Simple* μ gives almost the same Circular Velocities as these those calculated by *Simple* μ in *Milgrom's Law* (16). It is noted that the relative error of experimental Circular Velocities is $(\Delta V_{exp})_r \approx 8\%$. In addition, the values of time coefficient metric (g_{00}) show that it emerges time contraction at distances larger than 3.0 Kpc. this does not produce gravitational bleu shift to the radiation emitted by the star surface, because there exist the self-gravitation of the star which is very-very stronger. Finally, the values at distance 13.8 Mpc which is the distance of Galaxy NGC 3198 from Earth give us the image of what happens at extremely large distances (compare with the corresponding values in Table 1).

IOP Conf. Series: Journal of Physics: Conf. Series **936** (2017) 012008 doi:10.1088/1742-6596/936/1/012008

Table 4. Circular Velocities (experimental V_{exp} and calculated by the Combination of *Lorentzian-Einsteinian* 2nd *Generalized Schwarzschild metric* with MOND *Simple* μ ($V_{simp,Lor}$), the Luminous Mass of the galaxy that is enclosed within the circular orbit (M_l), the corresponding *Schwarzschild* radius (r_s), *Milgrom* radius (r_o), the value of function h and the value of time coefficient of metric (g_{00}) wrt the distance from the center of Galaxy NGC 3198. The relative errors of the experimental Velocities are $(\Delta V_{exp})_r \approx 8\%$.

r / Kpc	M_l /10^{40} kg	V_{exp} / Km s^{-1}	r_s / 10^{14} m	r_o /10^{20} m	h	g_{00}	$V_{simp,Lor}$ / Km s^{-1}
2.0	0.687	79.0	0.1020	0.3091	0.1768	-0.9999999708	109.6
3.0	1.012	97.8	0.1502	0.3750	-0.1867	-1.0000000303	115.6
4.0	1.620	118.0	0.2406	0.4746	-0.2964	-1.0000000578	128.8
5.5	3.404	139.4	0.5056	0.6880	-0.1856	-1.0000000553	156.6
6.0	3.698	144.2	0.5493	0.7171	-0.1863	-1.0000000605	163.4
7.0	5.089	143.3	0.7558	0.8412	-0.2690	-1.0000000941	171.8
8.0	5.825	150.3	0.8652	0.9000	-0.4181	-1.0000001465	175.7
9.0	6.346	149.9	0.9426	0.9394	-0.6070	-1.0000002060	177.3
10.1	6.735	152.1	1.0003	0.9677	-0.8507	-1.0000002731	177.6
11.0	6.914	151.1	1.0269	0.9805	-1.0825	-1.0000003275	176.9
12.1	7.093	156.2	1.0534	0.9931	-1.3798	-1.0000003893	176.1
14.1	7.214	161.0	1.0715	1.0016	-1.9952	-1.0000004914	173.8
16.1	7.237	155.3	1.0748	1.0031	-2.6769	-1.0000005791	171.5
18.1	7.191	148.7	1.0680	0.9999	-3.4247	-1.0000006549	169.3
20.1	7.115	149.1	1.0567	0.9946	-4.2299	-1.0000007207	167.3
22.1	7.005	148.4	1.0404	0.9869	-5.0999	-1.0000007781	165.4
24.1	6.901	146.2	1.0250	0.9796	-6.0130	-1.0000008288	163.7
26.1	6.806	145.5	1.0108	0.9728	-6.9645	-1.0000008741	162.3
28.1	6.716	147.3	0.9975	0,9664	-7.9533	-1.0000009149	161.0
30,2	6.628	146.5	0.9844	0,9600	-9.0290	-1.0000009538	159.8
32,2	6.544	148.4	0.9719	0,9539	-10.0944	-1.0000009874	158.7
34.2	6.456	149.3	0.9589	0,9475	-11.2033	-1.0000010180	157.7
36.2	6.392	149.9	0.9494	0.9428	-12.3175	-1.0000010469	156.9
38.2	6.316	149.3	0.9380	0.9371	-13.4845	-1.0000010731	156.0
40.2	6.252	150.0	0.9286	0.9324	-14.6652	-1.0000010978	155.3
42.1	6.191	147.6	0.9196	0.9278	-15.8182	-1.0000011197	154.6
44.2	6.159	149.8	0.9148	0.9254	-17.0560	-1.0000011440	154.2
46.2	6.114	151.5	0.9080	0.9220	-18.2900	-1.0000011650	153.6
48.2	6.072	151.9	0.9018	0.9189	-19.5417	-1,0000011849	153.2
13800	6.763	-	1.0045	0.9697	-17,741.1	-1.0000041848	152.6

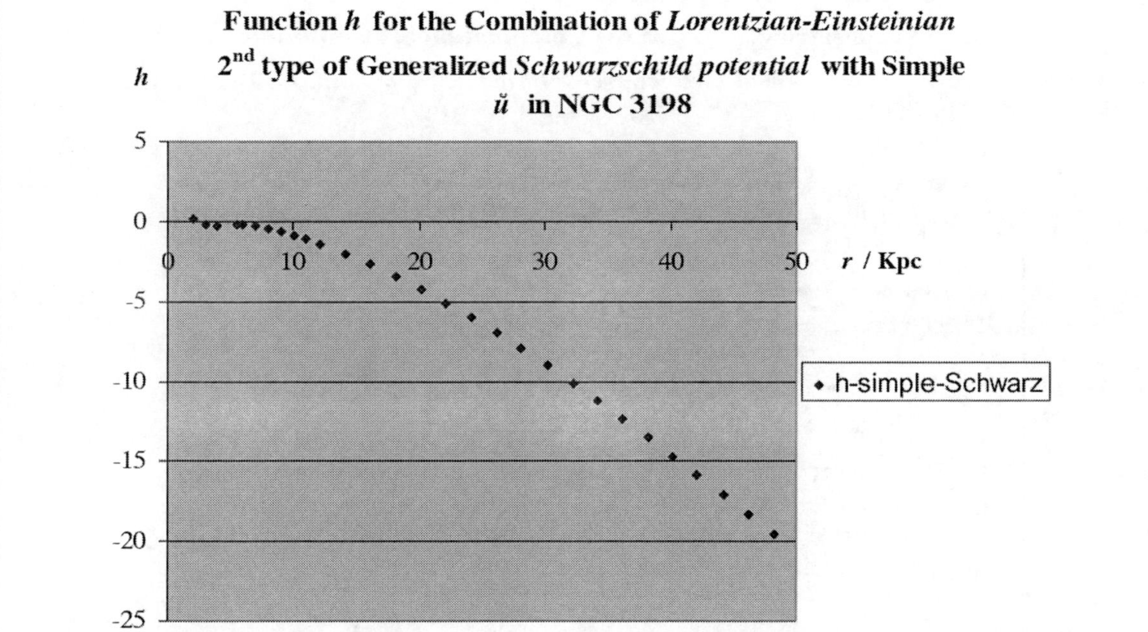

Figure 7. Plot of function h_ω wrt the distance (r) from the center of NGC 3198 for the Combination of *Lorentzian-Einsteinian 2nd Generalized Schwarzschild metric* with MOND *Simple* μ.

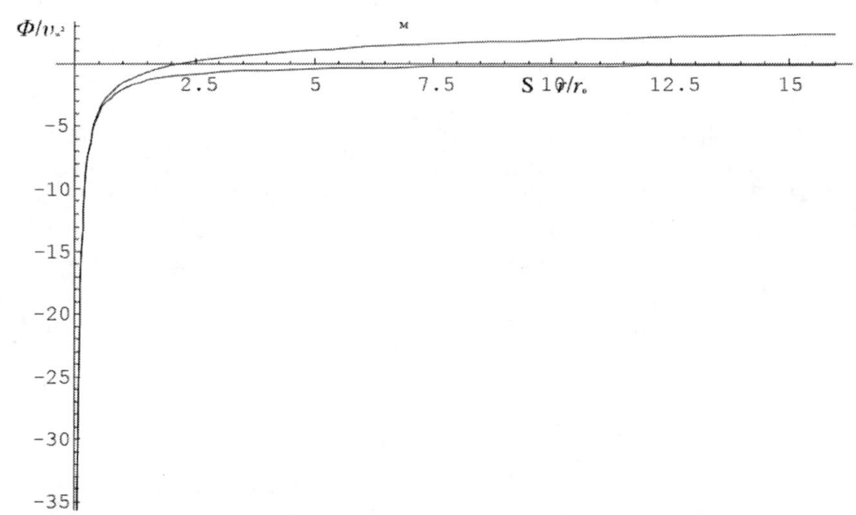

Figure 8. Plot of Relativistic Potential (Φ/υ_∞^2) wrt the distance $(r/r_\circ)$ from the center of NGC 3198 for the Combination of *Lorentzian-Einsteinian 2nd Generalized Schwarzschild metric* with MOND *Simple Interpolating Function* (M), as well as the original *Schwarzschild potential* (S).

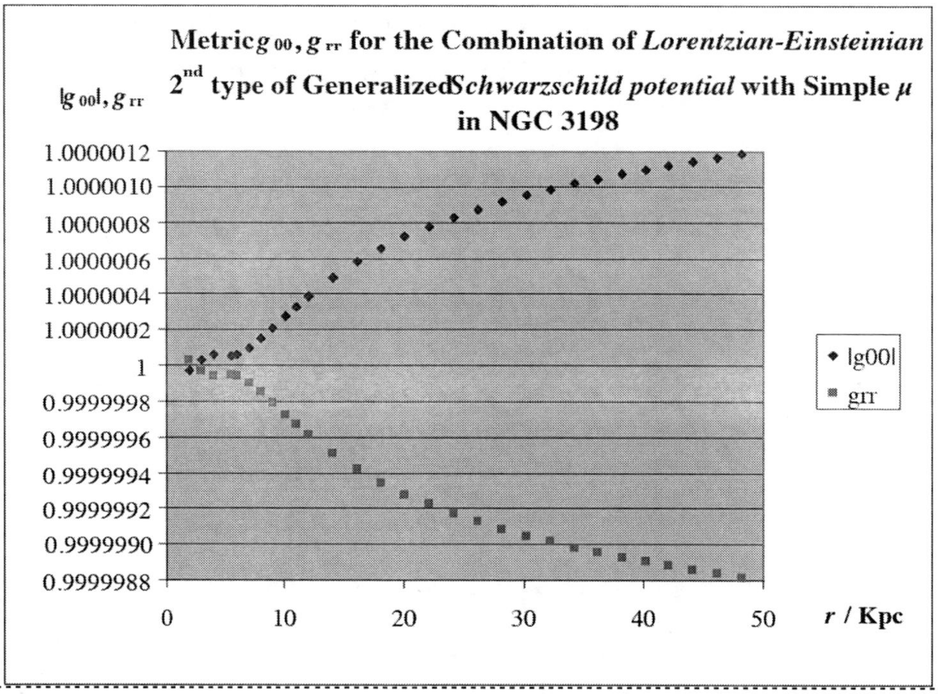

Figure 9. Plot of coefficients of metric g_{00}, g_{rr} wrt the distance from the center of NGC 3198 for the Combination of *Lorentzian-Einsteinian* 2nd *Generalized Schwarzschild metric* with MOND *Simple* μ.

4.2. The Combination of Lorentzian-Einsteinian 2nd Generalized Schwarzschild metric with MOND Simple Interpolating Function in the Solar System

In order to find out what is the effect of the modification at medium mass and size systems, we now examine our Solar System.

The mean values of Rotational Velocities are contained in Table 6 and the Mass of the Solar System that is enclosed within the orbit wrt the mean distance the planet from the Sun, are contained in Table 7 (see [13] p 14-3). The Circular Velocities $V_{simpLor}$ have been calculated using (124) and the values of time coefficient of metric (g_{00}) using (127). In addition, the corresponding Rotation Curves and Mass Distribution in the Solar System are shown in Figure 10 and Figure 11 respectively.

In Figure 12, we show the plot of function $h_{(r)}$ wrt the distance of the planet from the Sun for the Combination of *Lorentzian-Einsteinian* 2nd *Type of Generalized Schwarzschild metric* with MOND *Simple* μ.

In Figure 13, we show the plot of coefficients of metric g_{00}, g_{rr} wrt the distance of the planet from the Sun for the Combination of *Lorentzian-Einsteinian* 2nd *Generalized Schwarzschild metric* with MOND *Simple* μ.

We observe that in case of Solar System, the Combination of *Lorentzian-Einsteinian* 2nd *Generalized Schwarzschild metric* with MOND *Simple* μ, gives almost the same Rotational Velocities as those calculated by *Schwarzschild field strength* V_{Schwar} as well as by *Simple* μ in *Milgrom's Law* (16). Besides, the values of the coefficients of metric (g_{00}) for the Combination of *Lorentzian-Einsteinian* 2nd *Generalized Schwarzschild metric* with MOND *Simple* μ are almost the same as those computed by the original *Schwarzschild metric*. Thus, there are not significant changes to the *Relativistic Doppler Shift*, the gravitational red shift as well as the precession of Mercury's orbit (g_{00}=0.9999999490). Finally, the values at distance 13.8 Mpc which is the distance of Galaxy NGC 3198 from Earth give us the image of what happens at extra large distances (compare with the corresponding values in Table 1).

Table 5. Rotational Velocities (experimental V_{exp} and calculated by the Combination of *Lorentzian-Einsteinian 2nd Type of Generalized Schwarzschild metric* with MOND *Simple Interpolating Function* $V_{simp,Lor}$), the Luminous Mass of the Solar System that is enclosed within the circular orbit (M_d), the corresponding *Schwarzschild* radius (r_s), *Milgrom* radius (r_0), the value of function *h* and the value of time coefficient of metric (g_{00}) wrt the mean distance from the Sun.

Name	***r*** / ua	M_d /10^{24} kg	V_{Schwar} / Km s^{-1}	***h***	g_{00}	$V_{simp,Lor}$ / Km s^{-1}
Sun Surface	0.00465	1,989,100	436.747	0.9999999999991	-0,9999957553	436.747
Mercury	0.38710	1,989,100.0000	47.880	0.9999999976629	-0,9999999490	47.880
Venus	0.72333	1,989,100.3302	35.027	0.9999999900082	-0,9999999727	35.027
Earth	1.00000	1,989,105.1992	29.790	0.9999999781689	-0,9999999803	29.790
Mars	1.52369	1,989,111.1715	24.134	0.9999999507762	-0,9999999870	24.134
Jupiter	5.20283	1,989,111.8134	13.060	0.9999994561830	-0,9999999962	13.060
Saturn	9.53876	1,991,010.6134	9.650	0.9999981701195	-0,9999999979	9.650
Uranus	19.19139	1,991,579.1134	6.804	0.9999925682555	-0,9999999990	6.804
Neptune	30.06107	1,991,665.7384	5.437	0.9999817262519	-0,9999999993	5.437
Pluto	39.52940	1,991,768.5184	4.741	0.9999684583836	0,9999999995	4.741
NGC 3198	2.846×10^{12}	1,991,768.5334	0.0177	-8,174,346,104.103	-1,0000000000	0.00000

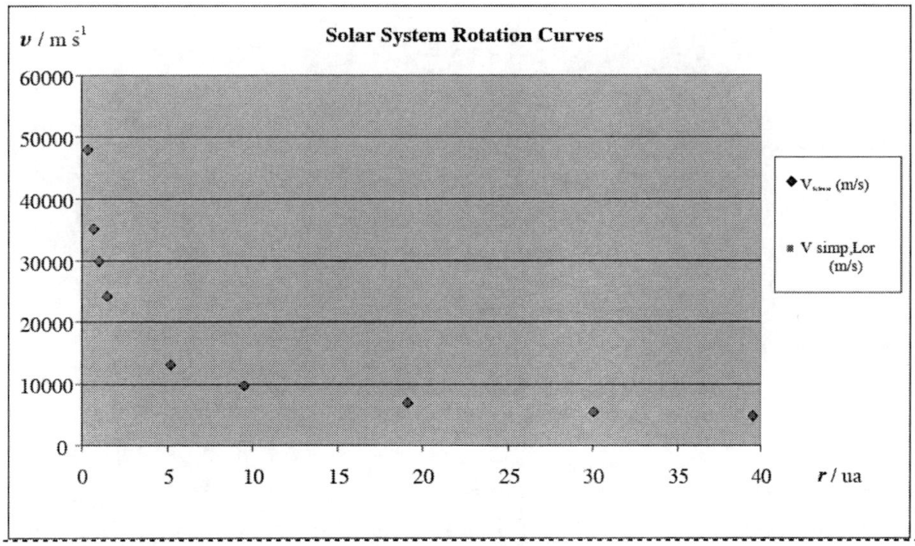

Figure 10. Rotation Curves in the Solar System. Rotational Velocities (calculated by *Schwarzschild field strength* V_{Schwar} and by the Combination of *Lorentzian-Einsteinian* 2nd *Generalized Schwarzschild metric* with MOND *Simple Interpolating Function* $V_{simp,Lor}$) wrt the distance from the Sun (ua=astronomical unit).

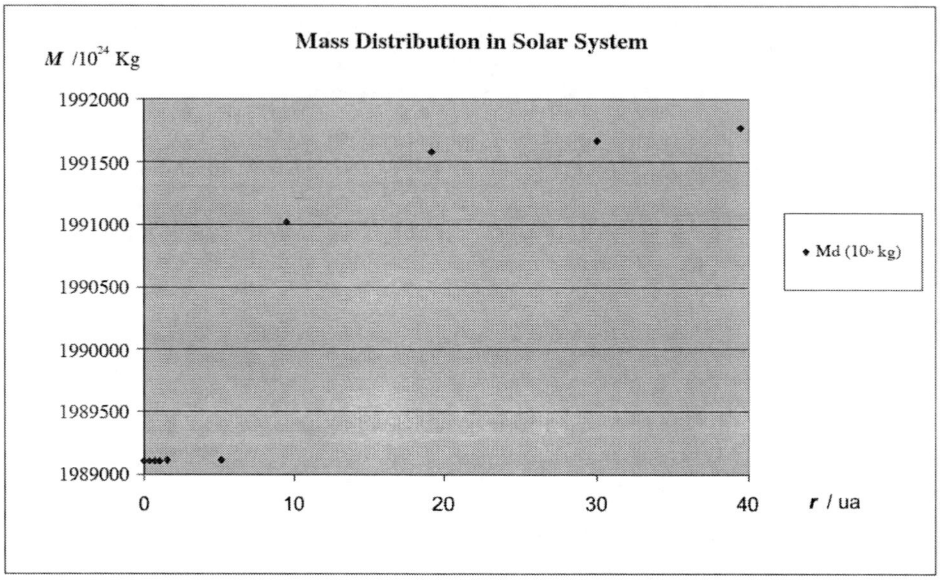

Figure 11. Mass Distribution in the Solar System. Mass M_c that is enclosed within the orbit of the planet wrt the distance from the Sun (ua=astronomical unit).

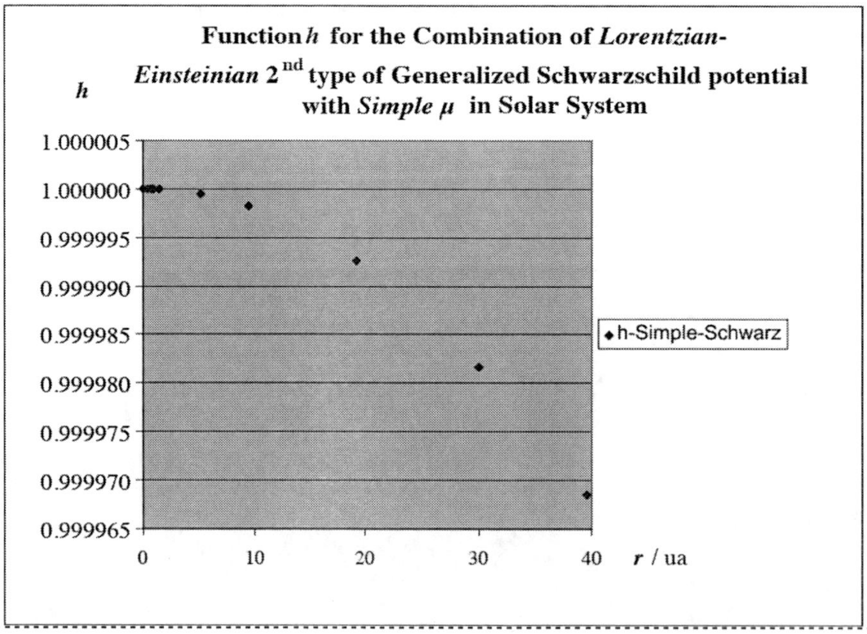

Figure 12. Plot of function h_{ω} wrt the distance of the planet (r) from the Sun, for the Combination of *Lorentzian-Einsteinian* 2nd *Generalized Schwarzschild metric* with MOND *Simple* μ (ua=astronomical unit).

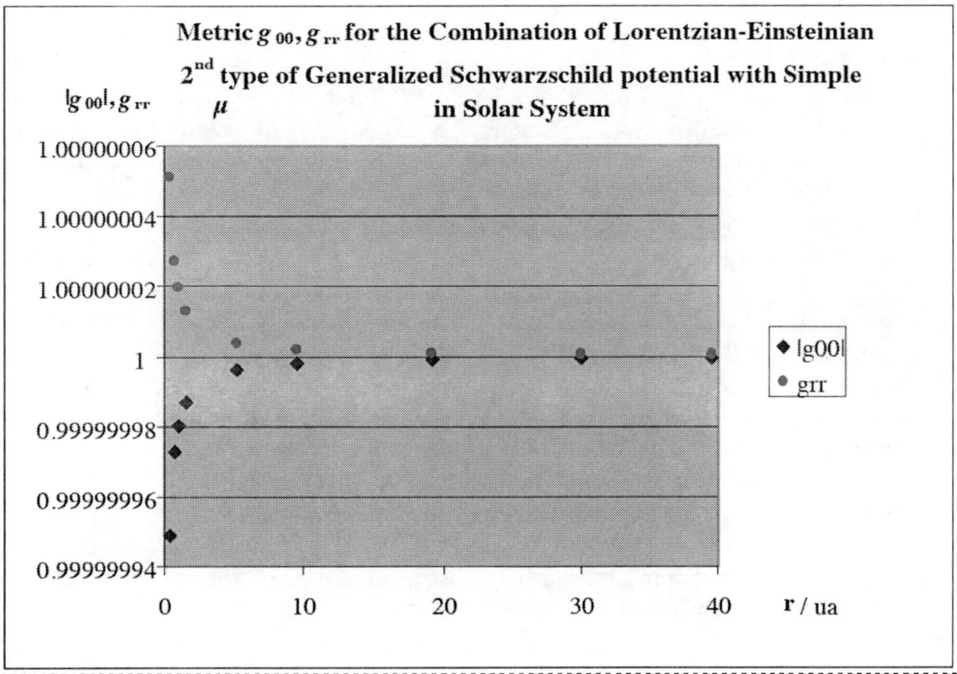

Figure 13. Plot of coefficients of metric g_{00}, g_{rr} wrt the distance of the planet from the Sun for the Combination of *Lorentzian-Einsteinian 2nd Generalized Schwarzschild metric* with MOND *Simple* μ (ua=astronomical unit).

5. Conclusion

A new modified relativistic potential (2nd *Generalized Schwarzschild potential*) is obtained. This describes the gravitational interaction at any distance and for any metric of flat spacetime (any theory of physics). In case of *Galilean-Newtonian physics*, the *Modified Newtonian Dynamics* (MOND) has already been developed, explaining the Rotation Curves in many Galaxies, using suitable Interpolating function (μ) in *Milgrom's Law*. Solving suitable integral, we may obtain the corresponding new Interpolating function h_ω that is contained in 2nd *Generalized Schwarzschild potential* and metric. In this paper, we present the results of *simple interpolating function*, although the results of *standard interpolating function* are better. Unfortunately, we have not yet analytical solution of the corresponding integral, but also a completely new h_ω may be proposed.

We then studied the case of flat spacetime with *Lorentz metric* (*Minkowski space*), because the experimental data have been extracted using the *Relativistic Doppler Shift* and the gravitational red shift of *Lorentzian-Einsteinian Relativity*. We use h_ω which has developed by *Galilean-Newtonian physics* by adding suitable term, in order to achieve *Schwarzschild potential* near to *Schwarzschild radius*. The experimental validation was performed on the Rotation Curves of both the Galaxy NGC 3198 and the Solar system.

This new modified relativistic potential and the corresponding metric of spacetime have been obtained by the light of *Euclidean Closed Linear Transformations of Complex Spacetime endowed with the Corresponding Metric*. Of course, may also be applied by scientists who prefer the hyperbolic geometry of Classic Relativity (CR).

Acknowledgments
The authors would like to thank Prof. Ch. G. Massouros for the valuable remarks and advice.

References

[1] Rindler W 2006 *Relativity: Special, General and Cosmological* (New York: Oxford University Press) pp 229-245 [ISBN: 978-0-19-856732-5]
[2] Heymans C 2017 *The Dark Universe* (IOP Publishing, Bristol, UK) pp 3-4 [ISBN: 978-0-7503-1373-5] [doi:10.1088/978-0-7503-1373-5] http://iopscience.iop.org/book/978-0-7503-1373-5.pdf
[3] Bershady MA, Verheijen MAW, Swaters RA, Andersen DR, Westfall KB and Martinsson T 2010 DiskMass Survey. I. Overview *Astrophysical Journal* 716(1) 198-233. [doi:10.1088/0004-637X/716/1/198] http://iopscience.iop.org/article/10.1088/0004-637X/716/1/198/pdf
[4] Begeman KG 1989 H, rotation curves of spiral galaxies. I - NGC 3198 *Astron Astrophys* 223(1-2) 47-60 http://articles.adsabs.harvard.edu/cgi-bin/nph-iarticle_query?1989A%26A...223...47B&defaultprint=YES&filetype=.pdf
[5] Karukes EV, Salucci P and Gentile G 2015 The Dark Matter Distribution in the Spiral NGC3198 out of 0.22 Rvir *Astron Astrophys* 578 A13 pp 1-8 [doi: 10.1051/0004-6361/201425339] https://www.aanda.org/articles/aa/pdf/2015/06/aa25339-14.pdf
[6] Sánchez-Gil CM, Alfaro EJ and Pérez E 2015 Corrugated Velocity Patterns in the Spiral Galaxies NGC 278, NGC 1058, NGC 2500 & UGC 3574 *Mon Not R Astron Soc* 000 1-19 https://academic.oup.com/mnras/article-abstract/454/4/3376/996897/Corrugated-velocity-patterns-in-the-spiral?redirectedFrom=fulltext
[7] Milgrom M 1992 Dynamics with a non-standard inertia-acceleration relation an alternative to dark matter in Galactic Systems *Annals of Physics* 229(2) pp 384-415 [doi:10.1006/aphy.1994.1012] Open access version: https://arxiv.org/pdf/astro-ph/9303012.pdf
[8] McGaugh S and Milgrom M 2013 Andromeda Dwarfs in Light of MOND *Astrophysical Journal* 766(1), 22 pp 1-7 [doi: 10.1088/0004-637X/766/1/22] http://iopscience.iop.org/article/10.1088/0004-637X/766/1/22/pdf
[9] Milgrom M and Sanders RH 2016 Perspective on MOND emergence from Verlinde's 'emergent gravity' and its recent test by weak lensing https://arxiv.org/pdf/1612.09582.pdf
[10] Sanders RH 1986 Finite length-scale anti-gravity and observations of mass discrepancies in galaxies *Astron Astrophys* 154(1-2) pp 135-144 http://articles.adsabs.harvard.edu/cgi-bin/nph-iarticle_query?1986A%26A...154..135S&data_type=PDF_HIGH&whole_paper=YES&type=PRINTER&filetype=.pdf
[11] Bottema R, Pestaña JLG, Rothberg B and Sanders RH 2002 MOND rotation curves for spiral galaxies with Cepheid-based distances *Astron Astrophys* 393(2) pp 453-460 [doi: 10.1051/0004-6361:20021021] https://www.aanda.org/articles/aa/pdf/2002/38/aah3496.pdf
[12] Vossos S and Vossos E 2016 Euclidean Closed Linear Transformations of Complex Spacetime and generally of Complex Spaces of dimension four endowed with the Same or Different Metric *J Phys.: Conf. Ser.* 738 012048 [doi:10.1088/1742-6596/738/1/012048] http://iopscience.iop.org/article/10.1088/1742-6596/738/1/012048/meta
[13] Lide D R ed 2009 *CRC Handbook of Chemistry and Physics* 89th Edition (Internet Version) (FL: CRC Press/Taylor and Francis, Boca Raton)
[14] Vossos S and Vossos E 2015 Euclidean Complex Relativistic Mechanics: A New Special Relativity Theory *J. Phys.: Conf. Ser.* 633 012027 [doi:10.1088/1742-6596/633/1/012027] http://dx.doi.org/10.1088/1742-6596/633/1/012027

Accelerated procedure to solve kinetic equation for neutral atoms in a hot plasma

Mikhail Z. Tokar

Forschungszentrum Jülich GmbH, Institut für Energie- und Klimaforschung - Plasmaphysik, 52425 Jü, Germany

E-mail: m.tokar@fz-juelich.de

Abstract. The recombination of plasma charged components, electrons and ions of hydrogen isotopes, on the wall of a fusion reactor is a source of neutral molecules and atoms, recycling back into the plasma volume. Here neutral species participate, in particular, in charge-exchange (c-x) collisions with the plasma ions and, as a result, atoms of high energies with chaotically directed velocities are generated. Some fraction of these hot atoms hit the wall. Statistical Monte Carlo methods normally used to model c-x atoms are too time consuming for reasonably small level of accident errors and extensive parameter studies are problematic. By applying pass method to evaluate integrals from functions, including the ion velocity distribution, an iteration approach to solve one-dimensional kinetic equation [1], being alternative to Monte Carlo procedure, has been tremendously accelerated, at least by a factor of 30-50 [2]. Here this approach is developed further to solve the 2-D kinetic equation, applied to model the transport of c-x atoms in the vicinity of an opening in the wall, e.g., the entrance of the duct guiding to a diagnostic installation. This is necessary to determine firmly the energy spectrum of c-x atoms penetrating into the duct and to assess the erosion of the installation there. The results of kinetic modeling are compared with those obtained with the diffusion description for c-x atoms, being strictly relevant under plasma conditions of low temperature and high density, where the mean free path length between c-x collisions is much smaller than that till the atom ionization by electrons. It is demonstrated that the previous calculations [3], done with the diffusion approximation for c-x atoms, overestimate the erosion rate of Mo mirrors in a reactor by a factor of 3 compared to the result of the present kinetic study.

1. Introduction

In devices for thermonuclear fusion of hydrogen isotopes deuterium and tritium these are in the form of a hot strongly ionized plasma. To avoid too strong interactions of this plasma with the machine wall there is normally a peripheral edge layer, the so called scrape-off layer (SOL), where plasma particles and energy is transported along magnetic field lines into a special volume, the so called divertor [4]. In a future reactor like DEMO [5, 6] processes in the divertor will significantly restrain this flow in the main part of the SOL [2] and a significant fraction of plasma particles, diffusing from the plasma core, will reach the vessel wall before they are exhausted into divertor. Arising plasma fluxes to the wall, with a density of $10^{20}m^{-2}s[2]$, saturate the wall with fuel particles in a time much shorter than the discharge duration. Thus, a comparable influx of neutral species will recycle back into the plasma. These neutral particles are not confined by the magnetic field and can penetrate deep, at several centimeters into the SOL plasma. Here charge-exchange interactions with plasma ions generate atoms of energies much higher than that of primary recycling particles. A noticeable fraction, up to 60%, of such secondary hot neutrals escape to the wall, can get into diagnostic ports and destroy installations there, see figure 1.

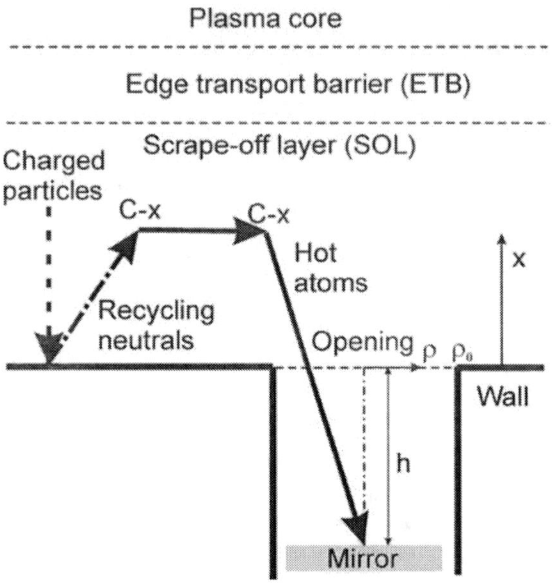

Figure 1. Processes near an opening in the reactor wall.

Statistical Monte Carlo methods [7] are normally used to model c-x atoms at the edge of fusion devices. A crucial obstacle to apply these approaches routinely for extensive parameter studies, e.g., with the aim to optimize the duct geometry, is too large calculation time for a reasonably small level of accident errors. In a one-dimensional geometry an alternative approach based on iteration procedure to solve the kinetic equation represented in an integral form has been elaborated in Ref.[1]. Being, unlikely to Monte Carlo, free from statistical noise and permitting the convergence of iterations to the error level defined by the machine accuracy, this method, nonetheless, is also time consuming. However, by applying the pass method to evaluate integrals from functions, including the velocity distribution of plasma ions, a tremendous acceleration of kinetic calculations, at least by a factor of 30-50, has been achieved in Ref.[2]. This has allowed to perform a thorough comparison of calculations for 1-D transfer of c-x atoms in a hot plasma in the DEMO SOL done both with the kinetic description and by applying the so called diffusion approximation. Strictly speaking, the latter is relevant under plasma conditions of low temperature and high density, e.g., in the divertor, where the mean free path length between c-x collisions is much smaller than that till the atom ionization by electrons. Nonetheless, as it has been demonstrated in Ref.[2], even for the conditions of a hot SOL plasma the results of calculations done with the diffusion approximation deviate less than by 30% from those obtained kinetically and can be successfully applied for first estimates.

For diverse purposes numerous openings will be made in the reactor wall, e.g., for ducts leading to optical diagnostic installations collecting the radiation from the plasma. Such installations will be protected from the direct impact of charged plasma particles, moving mostly along the magnetic lines and penetrating into ducts on a Larmor radius depth of $1mm$. Hot c-x atoms, unconfined by the magnetic field, can, however, freely hit and erode these installations. Since at the position of an opening the inflow of recycling neutrals, and, thus, the source of c-x atoms are absent, the transfer of the latter has two- or even three-dimensional pattern. A

2-D situation in the vicinity of a circular opening has been modeled in Ref.[3], by solving a 2-D diffusion equation for c-x atoms. In spite of the relevance of the diffusion approximation in the 1-D geometry, considered in Ref.[2], its applicability for multi-dimensional situations is, however, still questionable. In the present paper the approach developed in Ref.[2] to accelerate 1-D kinetic calculations is elaborated further to solve a 2-D kinetic equation for modeling of c-x atoms in the vicinity of a circular opening in the wall. The results of calculations are compared with those found in Ref. [3] by applying the diffusion approximation for c-x atoms.

The remainder of the paper is organized as follows. In the next section the basic equations are formulated and reduced to the form used for the numerical treatment. A pass method applied to asses the integral in the velocity space is also discussed here. In section 3 the results of calculations both with kinetic and diffusion description of c-x atoms are presented and compared. Conclusions are drawn in section 4.

2. Basic equations

Henceforth we operate in the coordinate system shown in figure 1 and characterized by the distances x and ρ from the wall and opening axis, respectively, under the assumption of the opening axis-symmetry. The velocity distribution function of c-x atoms, $f(x, \rho, V_x, V_\rho)$, is governed by the kinetic equation:

$$V_x \partial_x f + \frac{V_\rho}{\rho} \partial_\rho (\rho f) = -\nu_a f + \frac{S_{cx}}{\pi V_{th}^2} \exp\left(-\frac{V_x^2 + V_\rho^2}{V_{th}^2}\right), \tag{1}$$

where $\nu_a = n\left(k_{ion}^a + k_{cx}^a\right)$is the total frequency of the atom disintegration in the plasma through the processes of ionization by electrons and charge-exchange with ions, characterized by the rate coefficients k_{ion}^a and k_{cx}^a, respectively; n is the plasma density, the same for electrons and ions due to quasi-neutrality; $S_{cx} = S_{cx}^0 + n k_{cx}^a n_{cx}$ is the density for c-x atoms source where the former contribution S_{cx}^0, specified in Refs. [2, 3], is due to charge-exchange of recycling neutrals, and the latter is owing to further c-x interactions of atoms generated from recycling species; the plasma ions with the mass m_i and temperature T_i have a Maxwellian velocity distribution with the thermal velocity $V_{th} = \sqrt{2T_i/m_i}$.

2.1. Reduction to integral-differential equations

Henceforth, V_ρ is discretized as $\pm V_\rho^l$, with $V_\rho^l = \Delta V (l - 1/2)$, $\Delta V = V_{\max}/l_{\max}$ and $l = 1, \cdots, l_{\max}$. The distribution functions $\varphi_l^\pm$of particles with ρ-velocity component in the range $\pm V_\rho^l - \Delta V/2 \le V_\rho \le \pm V_\rho^l + \Delta V/2$ are governed by the equations following from the integration of equation (1) over these ranges:

$$V_x \partial_x \varphi_l^\pm \pm \frac{V_\rho^l}{\rho} \partial_\rho \left(\rho \varphi_l^\pm\right) = -\nu_a \varphi_l^\pm + \frac{S_{cx,l}}{2\sqrt{\pi} V_{th}} \exp\left(-\frac{V_x^2}{V_{th}^2}\right), \tag{2}$$

where $S_{cx,l} = \left(S_{cx}^0 + n k_{cx}^a n_{cx}\right) \delta_l$, with

$$\delta_l = \left\{\text{erf}\left[-\left(\frac{V_\rho^l + \Delta V/2}{V_{th}}\right)\right] - \text{erf}\left[-\left(\frac{V_\rho^l - \Delta V/2}{V_{th}}\right)\right]\right\} / \text{erf}\left(\frac{V_{\max}}{V_{th}}\right)$$

being the fraction of ions in the the l-range.

From equations (2) one gets equations for the new variables $\varphi_l = f_l^+ + f_l^-$ and $\gamma_l = V_\rho^l \left(f_l^+ - f_l^-\right)$:

$$V_x \partial_x \varphi_l + \frac{1}{\rho} \partial_\rho (\rho \gamma_l) = -\nu_a \varphi_l + \frac{S_{cx,l}}{\sqrt{\pi} V_{th}} \exp\left(-\frac{V_x^2}{V_{th}^2}\right), \tag{3}$$

$$\partial_x \gamma_l + \frac{\left(V_\rho^l\right)^2}{V_x \rho} \partial_\rho \left(\rho \varphi_l\right) = -\frac{\nu_a}{V_x} \gamma_l, \tag{4}$$

The latter equation is straightforwardly integrated with respect to x, providing

$$\gamma_l = -\frac{\left(V_\rho^l\right)^2}{V_x \rho} \int_{x_0}^{x} \partial_\rho \left(\rho \varphi_l\right) \exp\left(\frac{U_y - U_x}{V_x}\right) dy,$$

with

$$U_{x,y} = \int_0^{x,y} \nu_a dz, \tag{5}$$

$x_0$being the position where the boundary conditions are imposed: since there is no influx of c-x atoms from the wall, $\gamma_l \left(x_0 = 0\right) = 0$ for $V_x > 0$; besides neutral species are absent deep enough in the plasma, i.e. $\gamma_l \left(x_0 \to \infty\right) \to 0$ for $V_x < 0$. By substituting the found expression for γ_l into equation (3), one gets:

$$V_x \partial_x \varphi_l = -\nu_a \varphi_l + \frac{\left(V_\rho^l\right)^2}{V_x \rho} \int_{x_0}^{x} \partial_\rho^2 \left(\rho \varphi_l\right) \exp\left(\frac{U_y - U_x}{V_x}\right) dy + \frac{S_{cx,l}}{\sqrt{\pi} V_{th}} \exp\left(-\frac{V_x^2}{V_{th}^2}\right).$$

In the second term on the right hand of the equation above both factor are expanded into the Taylor series up to linear terms, $\partial_\rho^2 \left(\rho \varphi_l\right)(y) \approx \partial_\rho^2 \left(\rho \varphi_l\right)(x) + \partial_x \partial_\rho^2 \left(\rho \varphi_l\right)(x) \cdot (y - x)$ and $U_y \approx U_x + \nu_a (x) \cdot (y - x)$. By taking into account that that the width of the region occupied by neutral atoms is of several mean free path lengths (MFPL), i.e. typically $|x - x_0| > |V_x| / \nu_a$ for most x, the term in question is estimated as $\left(V_\rho^l\right)^2 \partial_\rho^2 \left(\rho \varphi_l\right) / \left(\nu_a \rho\right)$. As a result we get the following equation:

$$V_x \partial_x \varphi_l = -\nu_a \varphi_l + \frac{\left(V_\rho^l\right)^2}{\nu_a} \frac{\partial_\rho^2 \left(\rho \varphi_l\right)}{\rho} + \frac{S_{cx,l}}{\sqrt{\pi} V_{th}} \exp\left(-\frac{V_x^2}{V_{th}^2}\right). \tag{6}$$

This can be integrated as the first order equation with respect to x, with the same boundary conditions as for γ_l, i.e. $\varphi_l \left(x = 0\right) = 0$ for $V_x > 0$ and $\varphi_l \left(x \to \infty\right) \to 0$. Finally, for the density of atoms with $V_\rho^l - \Delta V/2 \le |V_\rho| \le V_\rho^l + \Delta V/2$, $\eta_l = \int_{-\infty}^{\infty} \varphi_l dV_x$, one obtains the following integral-differential equation:

$$-\frac{D_l}{\nu_a \rho} \partial_\rho^2 \left(\rho \eta_l\right) = \int_0^{\infty} \frac{S_{cx,l} I_{xy}}{\sqrt{\pi} V_{th}} dy - \eta_l, \tag{7}$$

where $D_l = \left(V_\rho^l\right)^2 / \nu_a$ and $I_\alpha = \int_0^\infty F_\alpha du$ with $F_\alpha (u) = u^{-1} \exp\left(-u^2 - \alpha/u\right)$ and $\alpha = |U_y - U_x| / V_{th}(y)$. The total density of c-x atoms $n_{cx} = \sum_{l=1}^{l_m} \eta_l$.

Equation (7) is reducing to an integral one obtained in the 1-D case, either equation (16) in Ref.[1] or equation (11) in Ref.[2], by neglecting the ρ-derivative and considering one V_ρ^l-value only, i.e. by assuming $l_{\max} = 1$. In this case $l \equiv 1$, $\delta_l = 1$, $n_{cx} = \eta_1$ and $S_{cx,1} = S_{cx}^0 + nk_{cx}^a n_{cx}$. As it was demonstrated in Ref.[2] this integral equation can be reduced to a ordinary differential diffusion equation if (i) the electron temperature is low and the ionization rate is much smaller than that of charge-exchange, $k^a_{ion} \ll k^a_{cx}$, and (ii) the characteristic dimension for the plasma parameter change are much larger than the typical atom MFPL V_{th}/ν_a. The same procedure can be done in the case under consideration, providing the 2-D diffusion equation:

$$-\frac{D_{th}}{\rho}\partial_\rho\left(\rho\partial_\rho n_{cx}\right) - \partial_x\left[\frac{\partial_x\left(n_{cx}T_i\right)}{\nu_a m_i}\right] = S^0_{cx} - k^a_{ion} n n_{cx}, \tag{8}$$

with $D_{th} = V^2_{th}/\left(2\nu_a\right)$. The boundary conditions to equation (8) imply: (i) c-x atoms leave the plasma with the velocity close to the ion thermal one; (ii) neutral species are absent far from the wall, $n_{cx}\left(x \to \infty\right) = 0$. Additionally, $\partial_\rho \eta_l$, $\partial_\rho n_{cx} = 0$ for $\rho = 0$ and $\rho \to \infty$ for equations (7) and (8) .

2.2. Assessment of the velocity space integral I_α

By inspecting equation (7) one can see the origin of large calculation time for kinetic computations of c-x atoms. Even in the 1-D case for each grid point one has to calculate enclosed double integrals over the ion velocity space and over the whole plasma volume, $0 \le y \le r_w$, and repeat this in an iterative procedure with respect to the whole profiles of n_{cx}. For the 2-D situation the runs through the grid points in the (ρ, V_ρ)phase subspace are involved into calculations additionally; however these can be efficiently parallelized. Fortunately, the integral I_α, resulting from the generation of c-x atoms with the ion Maxwellian distribution function and the consequent attenuation in the ionization and c-x collisions, can be assessed with a high enough accuracy by an approximate pass method [8]. The function $F_\alpha(u)$ is shown in figure 2a for $\alpha = 0.3$, 1 and 3. One can distinguish four u-intervals where $F_\alpha(u)$ behaves principally differently: $u \le u_1$, $u_1 < u \le u_m$, $u_m < u \le u_2$ and $u_2 < u$, with the points $u_{1,2}$ and u_m defined by the conditions $F'_\alpha(u_m) = 0$ and $F''_\alpha(u_{1,2}) = 0$ for the derivatives $F'_\alpha = -F_\alpha g_1/u^2$, $F''_\alpha = F_\alpha g_2/u^4$ where

$$g_1(u) = 2u^3 + u - \alpha,\ g_2(u) = \left(g_1 + u\right)^2 - \left(2 + 6u^2\right)u^2,$$

u_m is a root of a cubic equation, $u_m = \sqrt[3]{b+a} - \sqrt[3]{b-a}$ with $a = \alpha/4$ and $b = \sqrt{a^2 + 1/216}$. By searching for $u_{1,2}$, one can see that for the roots of interest the equation $g_2\left(u_{1,2}\right) = 0$ reduces to the following ones:

$$h_1(u_1) = g_1(u_1) + u_1 + u_1\sqrt{2 + 6u_1^2} = 0,$$

$$h_2(u_2) = g_1(u_2) + u_2 - u_2\sqrt{2 + 6u_2^2} = 0.$$

By selecting properly the initial approximations to $u_{1,2}$, these equations are solved with the Newton method, providing accurate enough solutions after 3-5 iterations.

To assess the integrals I_α we approximate function $F_\alpha(u)$ as a linear one at $u \le u_1$ and by polynomials of the fifth order at the intervals $u_1 < u \le u_m$ and $u_m < u \le u_2$ with the coefficients selected to reproduce $F_{\alpha,m,1,2} = F_\alpha\left(u_{m,1,2}\right)$, $F'_{\alpha 1,2} = F'_\alpha\left(u_{1,2}\right)$, $F''_{\alpha m} = F''_\alpha\left(u_m\right)$ and $F'_\alpha\left(u_m\right) = F''_\alpha\left(u_{1,2}\right) = 0$; in addition $F_\alpha\left(u > u_2\right) \approx \exp\left(-u^2 - \alpha/u_2\right)/u_2$ is assumed. This results in:

$$I_\alpha(\alpha) \approx \frac{F^2_{\alpha 1}}{2F'_{\alpha 1}} + \frac{F_{\alpha 1}\delta_1 + F_{\alpha m}\left(u_2 - u_1\right) + F_{\alpha 2}\delta_2}{2} + \tag{9}$$

$$+\frac{F'_{\alpha 1}\delta_1^2 - F'_{\alpha 2}\delta_2^2}{10} + F''_{\alpha m}\frac{\delta_1^3 + \delta_2^3}{120} + \frac{\sqrt{\pi}}{2}\frac{1 - \mathrm{erf}u_2}{u_2}\exp\left(-\frac{\alpha}{u_2}\right),$$

where $\delta_{1,2} = \left|u_m - u_{1,2}\right|$. Figure 2b shows I_α versus α found with the pass method outlined above and evaluated numerically. By approximating the direct numerical results very accurately, the procedure above allows to reduce the CPU time by a factor of 30-50 at least.

As an alternative to the usage of formula (9) one can calculate in advance the integrals $I_\alpha(\alpha)$ and save this in a table. Then, for any particular α, the integral $I_\alpha(\alpha)$ can be interpolated from the values in the table for arguments larger and smaller than that in question. However, the searching of the correct α-interval in this table requires also some time which grows up with $\alpha \to 0$ since $I_\alpha \to \infty$ as $E_1(\alpha)$ and the density of data in the table has to be increasing. Thus, it is not obvious that the usage of tables is more time saving than formula (9).

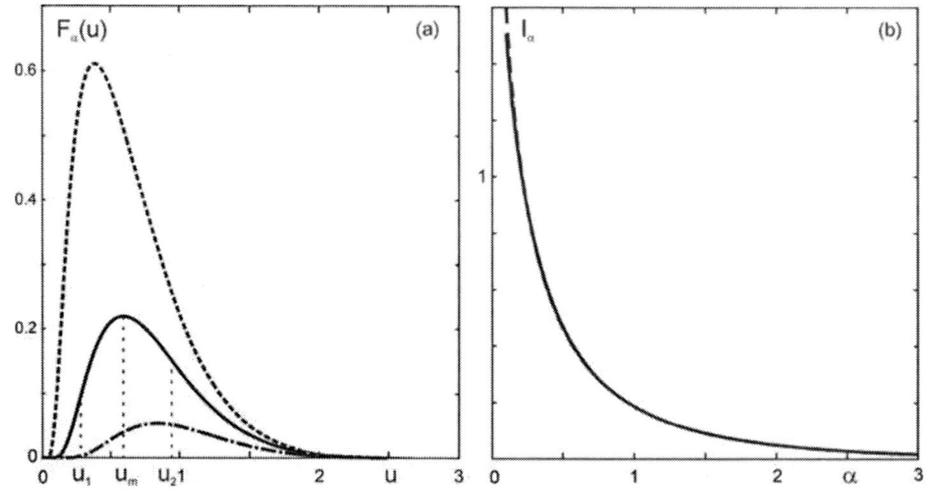

Figure 2. The functions $f_\alpha(u) = u^{-1}\exp(-u^2 - \alpha/u)$ for $\alpha = 1$ (solid curve), $\alpha = 0.3$ (dashed curve) and $\alpha = 3$ (dashed-dotted curve) (a); the integral $I_\alpha(\alpha) = \int_0^\infty F_\alpha(u)\,du$ estimated by the pass method (solid curve) and computed numerically (dashed curve) (b).

3. Results of calculations

The radial profiles of the plasma parameters used for calculations of c-x atom parameters are presented in figure 3 and were computed in Ref.[2] with the input parameters from the European DEMO project [5, 6].

In figure 4 we demonstrate the variation of the c-x atom density with the distance from the wall, far from a circular opening of the radius $\rho_0 = 0.05m$ and at the opening axis. There is a noticeable difference between the profiles calculated kinetically and in the diffusion approximation.

To comprehend the importance of this difference we assess the rate of erosion by c-x atoms of a Mo mirror imposed in the duct at the distance h from its opening. In a strict diffusion approximation the energy distribution function of c-x atoms coincides in any point in the plasma with that of the ions. Thus, the averaged energy of atoms entering the opening in the wall is close to the ion temperature at this position, $T_i(0)$. However, the ion temperature increases by several times at the penetration depth of c-x atoms, compare figures 3 and 4. C-x atoms generated here partly escape without further collisions with electrons and ions. Therefore, the distribution over the energies E of atoms, hitting the opening, can significantly differ from the local Maxwellian one of magnetized ions:

$$\varphi_i(x=0,E) = \frac{2\sqrt{E}}{\sqrt{\pi}T_i^{1.5}(0)}\exp\left[-\frac{E}{T_i(0)}\right]. \tag{10}$$

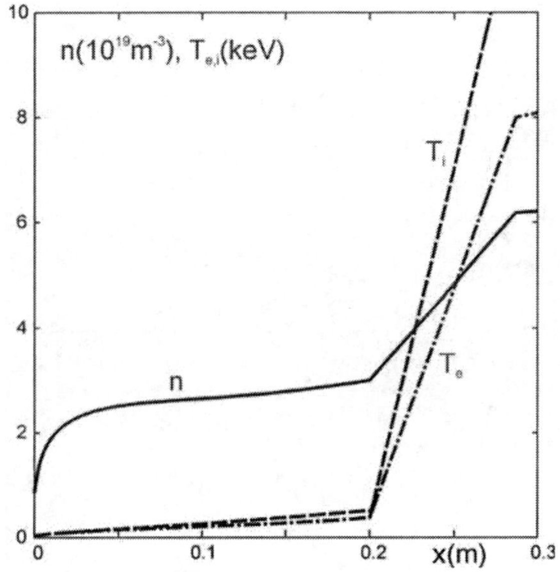

Figure 3. The profiles of the plasma density (solid curve), the ion (dashed curve) and electron (dashed-dotted curve) temperatures used for calculation of c-x atoms.

Consider an infinitesimally thin ring of the width $d\rho$, thickness dx, with the radius ρ and situated in the plasma at the distance x from the wall. C-x atoms of the energy within the range $E, E+dE$ are generated in this ring at the rate:

$$dR_{cx}(x,\rho,E) = S_{cx}(x,\rho)\,\varphi_i(x,E)\,2\pi\rho d\rho dx dE. \tag{11}$$

Since c-x atoms move in all directions some of them hit the mirror inside the duct, see figure 1. Henceforth we consider the surface position at the duct axis only. A simple geometric analysis shows that c-x atoms from the rings with $\rho \leq \rho_{\max} = \rho_0(1+x/h)$ can hit this surface point. The contribution of the plasma volume in question to the density of the c-x atom flux perpendicular to the installation surface is:

$$dj_{cx}(x,\rho,E) = dR_{cx}\frac{s\cdot\exp(-\lambda/s)}{4\pi\left[(x+h)^2+\rho^2\right]}, \tag{12}$$

where $s = \left[1+\rho^2/(x+h)^2\right]^{-1/2}$ is the cosine of the atom incidence angle with respect to the port axis. The exponential factor in equation (12), with $\lambda = \int_0^x \nu_a dx'/\sqrt{2E/m}$, takes into account the destruction, due to ionization and charge-exchange collisions, of the atoms in question on their way through the plasma. By ionization the atom completely disappears. By charge-exchange a new atom is generated. The latter process is taken into account by the second term in the source density, $S_{cx} = S_{cx}^0 + nk_{cx}^a n_{cx}$, and by the summation/integration of dj_{cx} contributions from rings situated in the plasma at all possible x.

The energy spectrum of atoms hitting the mirror is characterized by their flux density in the energy range dE, $\gamma = \int dj_{cx}/dE$, where the integration is performed over the ring radius ρ and distance from the wall x. By proceeding from ρ to s, according to the relation

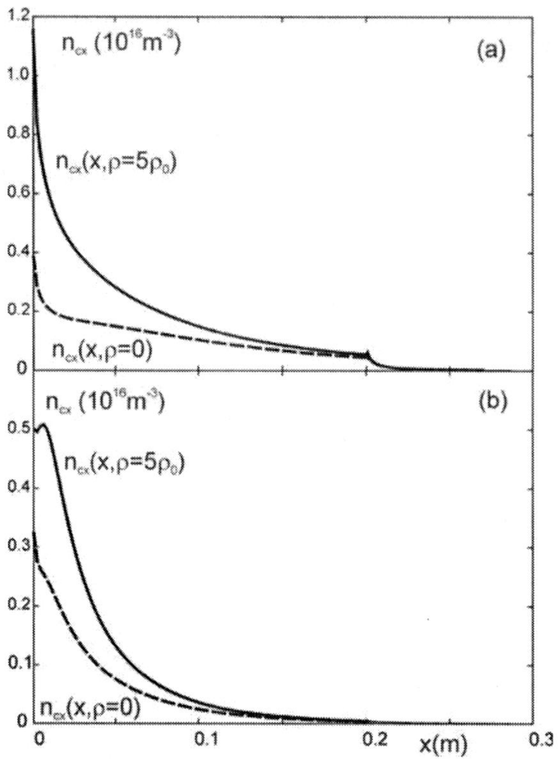

Figure 4. The profiles of the c-x atom densities far from the opening (solid curve) and at its axis (dashed curve) computed with diffusion approximation (a) and kinetically (b).

$\rho = (x + h)\sqrt{1/s^2 - 1}$, one gets:

$$\gamma(h, \rho_0, E) = \int_0^{r_w} \varphi_i(x, E)\, dx \int_{h/\sqrt{h^2+\rho_0^2}}^{1} \frac{S_{cx}}{2} \exp\left(-\frac{\lambda}{s}\right) ds. \tag{13}$$

The density of the outflow of the mirror material particles eroded by physical sputtering with c-x fuel atoms can be assessed as follows:

$$\Gamma_{sp}(h, \rho_0) = \int_0^{\infty} dE \int_0^{r_w} \varphi_i(x, E)\, dx \int_{h/\sqrt{h^2+\rho_0^2}}^{1} \frac{S_{cx}}{2} \exp\left(-\frac{\lambda}{s}\right) Y_{sp}(E, s)\, ds \tag{14}$$

Here Y_{sp} is the sputtering yield, whose dependence on the projectile energy E and the cosine s of the incidence angle is calculated by applying semi-empirical formulas from Ref. [10]. The erosion rate $h_{sp} = \Gamma_{sp}/n_{sp}$, with n_{sp} being the particle density of a mirror of Mo and measured in $nm/year$, is demonstrated in figure 5 as a function of the installation surface position h inside the port, for $n_{cx}(x, \rho)$ involved into S_{cx} calculated both kinetically from equations (7) and in the diffusion approximation, by integrating equation (8). One can see that the sputtering rate found with kinetically assessed $n_{cx}(x, \rho)$ is noticeably below that determined by using the diffusive approximation for c-x atoms in the plasma.

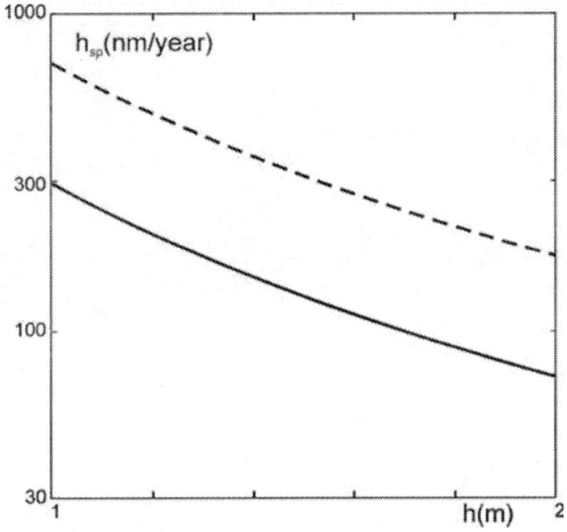

Figure 5. The erosion rate of a Mo mirror versus the surface mirror position inside the duct computed kinetically (solid curve) and with diffusion approximation (dashed curve).

4. Conclusions

The iteration approach elaborated in Ref.[1] to solve 1-D kinetic equation for c-x atoms has been developed further to describe the transport of these species in 2-D situation, in particular, in the vicinity of a circular opening in the wall of a fusion reactor. Unlike to Monte Carlo methods the iteration one does not generate statistical errors. To perform a thorough comparison with the results obtained by using a diffusion approximation for c-x atoms [3], the solving procedure is accelerated by a factor of 50, by applying pass methods to assess integrals in the velocity space [2]. Calculations done by both approaches demonstrate that the diffusion approximation provides significantly overestimated estimates for the erosion rate of a first mirror in a fusion reactor like DEMO compared to the kinetic calculations. Thus, in spite of the relevance of the diffusion approach for c-x atoms in 1-D calculations of the plasma parameters in the SOL, see Ref. [2], the kinetic one seems to be unavoidable for calculations of characteristics being very sensitive to the c-x atom energy like the erosion rate of the reactor wall and diagnostic installations.

References

[1] Rehker S and Wobig H *Plasma Phys.* **15** 1083
[2] Tokar M Z 2016 *Phys. Plasmas* **23** 122512
[3] Tokar M Z, Beckers M, Biel W "Erosion of installations in ports of a fusion reactor by hot fuel atoms" in press *Nucl. Mater. Energy* https://doi.org/10.1016/j.nme.2016.12.015.
[4] Stangeby P C 2000 *The Plasma Boundary of Magnetic Fusion Devices* (Bristol and Philadelphia: Institute of Physics Publishing)
[5] Federici G, Kemp R, Ward D, Bachmann C, Franke T, Gonzalez S, Lowry C, Gadomska M, Harman J, Meszaros B, Morlock C, Romanelli F and Wenninger R 2014 *Fusion Eng. Des.* **89** 882
[6] Wenninger R, Arbeiter F, Aubert J, Aho-Mantila L, Albanese R, Ambrosino R, Angioni C, Artaud J-F, Bernert M, Fable E, Fasoli A, Federici G, Garcia J, Giruzzi G, Jenko F, Maget P, Mattei M, Maviglia F,

IOP Conf. Series: Journal of Physics: Conf. Series **936** (2017) 012009 doi:10.1088/1742-6596/936/1/012009

Poli E, Ramogida G, Reux C, Schneider M, Sieglin B, Villone F, Wischmeier M and Zohm H 2015 *Nucl. Fusion* **55** 063003

[7] Landau D P and Binder K 2015 *A Guide to Monte Carlo Simulations in Statistical Physics* (Cambridge: University Printing House)

[8] Mathews J and Walker R L 1970 *Mathematical Methods of Physics* 2nd ed. (Menlo Park: Benjamin).

[9] Janev R K, Langer W D, Evans K J, and Post D E J 1987 *Elementary processes in hydrogen-helium plasmas* (Hamburg: Springer)

[10] Eckstein W, Garca-Rosales C, Roth J and Ottenberger W 1993 *Sputtering data* Report IPP 9/82 (Garching: Institute of plasma Physics)

IOP Conf. Series: Journal of Physics: Conf. Series **936** (2017) 012010 doi:10.1088/1742-6596/936/1/012010

Numerical Procedure for Modeling of Light Emitting Diode with Mesh-Like Electrode

Yohei Nishidate and Irina Khmyrova

University of Aizu, Aizu-Wakamatsu 965-8580, Japan

E-mail: yh.nsdt@gmail.com

Abstract. A computational procedure is presented for numerical modeling of the light emitting diode (LED) with top p-electrode designed as a mesh with the strips of rectangular cross section. Isotropic light emission in the LED's active region and light reflection from the bottom electrode are considered. Three-dimensional Laplace equation for electric potential is solved by finite element method. The numerical model incorporates mapped infinite element to account for potential decay far away from the LED structure and finite element model developed for boundary condition at semiconductor-air interface in the mesh opening. Simulation results demonstrate the effect of the mesh's geometrical parameters on the total output power.

1. Structure and physical model of the LED

Patterning of the top metal electrode of light-emitting diodes (LEDs) was proposed to enhance their output optical performance. To assess the effect of mesh-like electrode one needs to invoke numerical modeling. In the present study, we consider an LED with vertical structure schematically shown in Fig. 1. It consists of the narrow-gap active layer sandwiched between bottom n- and top p-doped wide-gap layers with thicknesses d_1 and d_2, respectively. Metal electrode attached to the bottom n-layer is solid while that one attached to the top of p-layer is designed as a mesh with pitch a. Mesh strips have rectangular cross section of width w and height h. The periodicity of the mesh allows to consider a portion of the LED structure beneath a single mesh cell which will be referred as a unit cell. Directions of coordinate axis are shown in Fig. 1. The planes of the bottom electrode and active region have coordinates $z = 0$ and $z = z_{act}$, respectively.

To find distribution of electric potential along active region one needs to solve three-dimensional Laplace equation. The boundary conditions at the bottom surface and the mesh strips are set to be constant, i.e. $\varphi(x, y, 0) = 0$ for bottom surface and $\varphi(x, y, z)|_{\text{node}} = \text{const.}$ for mesh strips, respectively. Also, the potential should decay far away from the top surface, i.e. $\varphi(x, y, \infty) = 0$ and continuity of normal component of electric flux at the semiconductor-air interface in the mesh opening is taken into account:

$$\frac{\partial\varphi}{\partial z}|_{int_+} = \varepsilon\frac{\partial\varphi}{\partial z}|_{int_-}, \tag{1}$$

where ε is the dielectric constant of semiconductor, $int_\pm$ indicate top (air) and bottom (semiconductor) sides of the interface. Due to the periodicity of the mesh the electric field

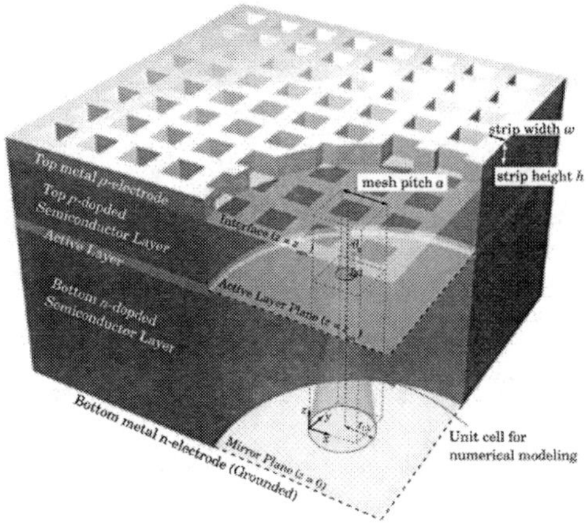

Figure 1. Schematic structure of the LED with top p-electrode patterned as a mesh.

normal to the side walls of unit cell is equal to zero, or in differential form: $\partial\varphi/\partial x|_{y=0,a} = \partial\varphi/\partial y|_{x=0,a} = 0$.

Current injected into each elementary volume ΔV is determined by the electric potential $\varphi(x, y, z_{act})$ at its location and can be expressed as [1]

$$J(x, y, z_{act}) = J_0 \exp\left[\frac{q}{a_f kT}\varphi(x, y, z_{act})\right], \tag{2}$$

where k and T are the Boltzmann constant and the temperature in Kelvin, respectively, q is electron charge. For simplicity we set an ideality factor $a_f = 1$. Saturation current density J_0 can be evaluated from the total current injected into the LED and measured as the current flowing in the external circuit [2]. Each elementary volume is treated as a point source with isotropic light emission. Power of light generated in a volume with coordinates $\{x, y, z_{act}\}$ can be expressed as [1]

$$dP(x, y, z_{act}) = \eta\frac{h\nu}{q}J(x, y, z_{act})dxdy, \tag{3}$$

where h and ν are the Planck's constant and the frequency of light photons, η is the internal quantum efficiency. Due to total internal reflection, only light coming within the escape or acceptance cone determined by the critical angle θ_c can be extracted via semiconductor-air interface. We assume that at any point of the mesh opening $\{x, y, z_{opn}\}$, the extracted light consists of a primary and secondary contributions collected from the encircled portions of the active region and bottom reflecting plane subtended by acceptance (or escape) cone (Fig. 1). The radii of these acceptance circles are $r_{c1} = d_1 \tan\theta_c$ and $r_{c2} = (d_1 + d_2)\tan\theta_c$, respectively. Output optical power density for primary contribution can be expressed as

$$P_{prm}(x, y, z_{opn}) = \int_{circl} \frac{\cos(\theta_i)}{4\pi r_1^2} T(\theta_i)\, dP(x_{circl}, y_{circl}, z_{act}), \tag{4}$$

where r_1 is the distance between a point $\{x_{circl}, y_{circl}, z_{act}\}$ within an acceptance circle in the active region plane and a point $\{x, y, z_{opn}\}$ in the mesh opening, i.e., $r_1 = [(x - x_{circl})^2 + (y - y_{circl})^2 + (z_{act} - z_{opn})^2]^{1/2}$, θ_i is the angle of incidence, $\cos(\theta_i)$ is a geometric factor which scales properly the inclined photon flux, $T(\theta_i)$ is polarization dependent transmission coefficient for the semiconductor-air interface. To evaluate secondary contribution P_{scn} we use Eq. (4) with proper substitution of coordinates at the bottom plane and $r_2 = [(x - x_{circ2})^2 + (y - y_{circ2})^2 + (z_{opn})^2]^{1/2}$ instead of r_1.

To evaluate total output optical power P_{tot}, we integrate power densities of both primary and secondary contributions obtained using Eq. (4) over the mesh opening and multiply by the number of mesh cells N_{opn}

$$P_{tot} = N_{opn} \int_{opn} (P_{prm} + P_{scn})\, dxdy. \tag{5}$$

where $N_{opn} = S_{LED}/a^2$, S_{LED} is total surface area of the LED.

2. Numerical results

The developed procedure is applied to evaluate the performance of the LED. In the computations the following parameters are used: bias voltage $V = 2$ V, internal quantum efficiency $\eta = 1$, $T = 300$ K, refractive index $n = 2.5$, dielectric constant of semiconductor layers $\varepsilon = 8.9$, and layer thicknesses and boundary conditions are indicated in Fig. 2. Total area of the LED was set $S_{LED} = 1296$ μm^2.

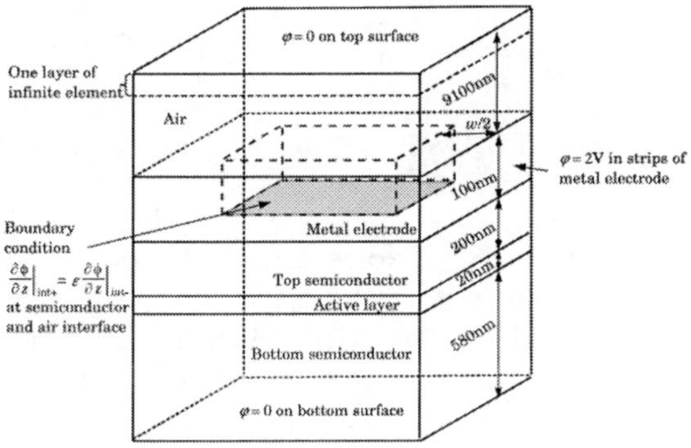

Figure 2. Computation domain of the LED's unit cell with geometric parameters and boundary conditions.

Fig. 3 shows distributions of the potential along the active region plane and the normalized output power density at mesh opening plane for mesh pitch $a = 900$ at different width of the strips: $w = 100$, 200, and 300 nm. One can see in Fig. 3 that an increase in strip width at a fixed mesh pitch boosts the potential and causes modification in the shape of distribution. It also boosts the average value of power density but smaller peak values are observed with decreasing area of mesh opening.

Fig. 4 shows total normalized output optical power P_{tnor} versus mesh pitch at different strip widths. Solid curves with filled symbols correspond to the situation when light emitted towards the top surface as well as reflected from the bottom plane was taken into account (primary plus secondary contributions) and only primary contribution was taken into account in the case of dashed curves with open symbols. It should be pointed out that at all strip widths w total output power drops with increasing mesh pitch a and approaches some saturation value when only regions in the vicinity of the strip edges contribute to the optical output, as expected. It should be noted also that at $w = 300$ nm the total output power with contribution of the reflected light reveals saturation in the range of the mesh pitches $a = 300 - 600$ nm. It can be attributed to modification of potential distribution while effect of the reduced size of mesh opening is less pronounced. However, at strip width $w = 300$ nm reduction in the size of mesh opening becomes crucial at mesh pitch $a = 600$ nm. Thus, the interplay between modification of potential distribution and reduction in size of the mesh opening at a fixed mesh pitch and increasing strip width should be addressed properly when we choose the geometrical parameters of the mesh-like electrode.

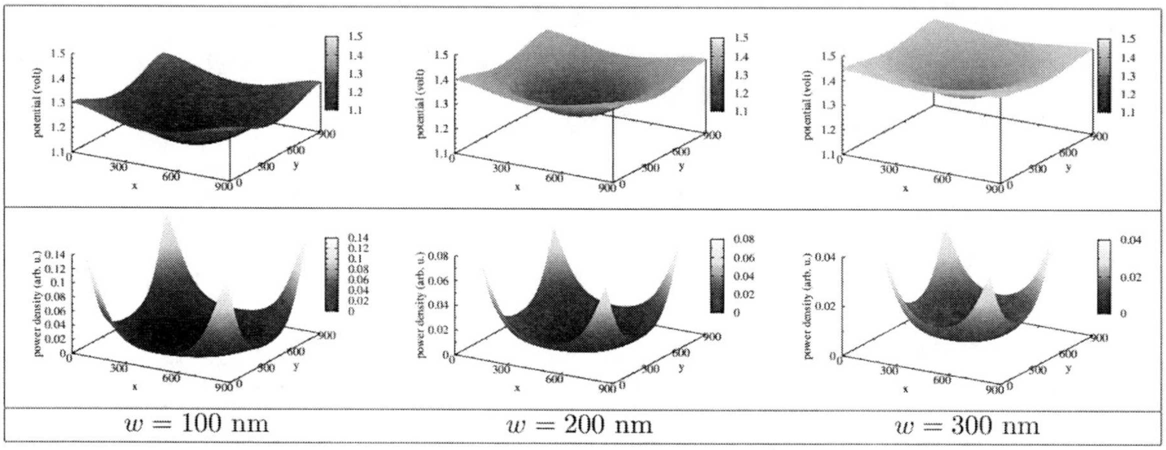

Figure 3. Potential distributions along the active region plane (top panel) and normalized output optical power density along the mesh opening (bottom panel) at mesh pitch $a = 900$. The distributions are shown for width of strips $w = 100$, 200, and 300 nm.

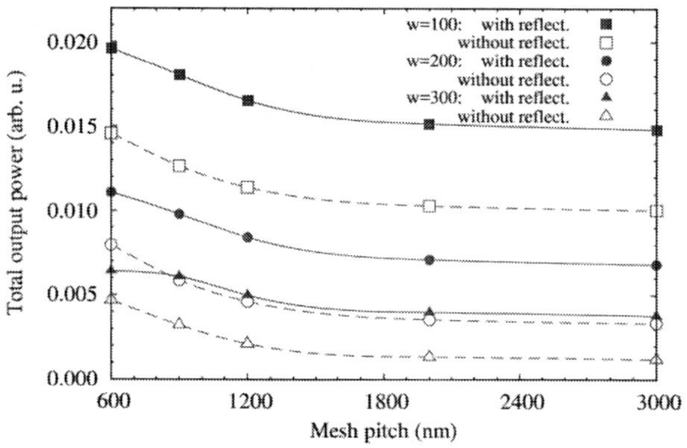

Figure 4. Normalized total output optical powe versus mesh pitches at different width of the strips Solid curves with filled symbols correspond t the situation when light emitted towards the to surface as well as reflected from the bottom plan was taken into account (primary plus secondar contributions); dashed curves with open symbols only primary contribution was taken into account

3. Conclusion

Numerical model and procedure are developed and used to study the output optical performance of the light-emitting diode with top metal electrode designed as a mesh with the strips of rectangular cross-section as in realistic LEDs. The developed model also accounts for the contribution of light reflection from the bottom electrode. Three-dimensional Laplace equation is solved by finite element method. The mapped infinite element is introduced into numerical model to account for potential decay far away from the LED structure and finite element model developed for boundary condition at semiconductor-air interface in the mesh opening. Simulation results demonstrate the effect of the mesh pitch and width of the strips on the distribution of electric potential and total output optical power. The saturation of the total output optical power is demonstrated at significant increase in mesh pitch at all strip widths. It is also shown that the interplay between increase in potential values and reduction in size of the mesh opening should be taken into consideration during design of the mesh-like patterned electrodes.

References

[1] Fukuda M 1999 *Optical Semiconductor Devices* (New York: Wiley)
[2] Khmyrova I, Watanabe N, Kholopova J, Kovalchuk A and Shapoval S 2014 *Applied Optics* **53** 4734–4741

IOP Conf. Series: Journal of Physics: Conf. Series **936** (2017) 012011 doi:10.1088/1742-6596/936/1/012011

Identification of environment symmetry for iron centers in aluminosilicates by EPR

E Apushkinskiy[a], B Popov[a], V N Romanov[a], V Saveliev[a], V Sobolevskiy[a]
[a] Peter the Great St.Petersburg Polytechnic University, 195251 St. Petersburg, Russia

Email: apushkinsky@hotmail.com

Abstract. We study an angular dependence of EPR spectra of the own defects in topaz ($Al_2[SiO_4][F,OH]_2$). The topaz crystal structure is built of silicon-oxygen tetrahedral (SiO_4) and octahedral aluminum surrounded by four oxygen ions and fluorine ions or a hydroxyl group. There are two types of substitution defects in topaz, such as octahedral and tetrahedral, depending on the location of the impurity ion. The point defects with octahedral symmetry arise on local substitution of aluminum ($Fe^{3+} \rightarrow Al^{3+}$), while tetrahedral coordinated centers are formed by substitution of silicon ($Fe^{3+} \rightarrow Si^{4+}$) in the silicon-oxygen tetrahedral (SiO_4). To determine the symmetry of the impurity centers, the angular dependence of the EPR spectrum is calculated. The excited energy states for defects are determined by the magnitude of the g-factor shift. These states equal 1.6 eV and 3.2 eV, respectively. A super-hyperthin structure (SHTS) of the EPR lines of iron has been observed. This SHTS is determined by the magnetic moments of the fluorine nuclei F^{19} located in the first coordination sphere of the paramagnetic iron center. The models of the three new paramagnetic centers in topazes are suggested, one of them being of orthorhombic symmetry for aluminum substitution and the two others being of tetragonal symmetry for silicon substitution with oxygen vacancies.

1. Introduction

The research into natural minerals shows that the local defects occurring in the course of growth are stable [1,2]. The impact of radiation on the mineral leads to the rupture of chemical bonds, to the redistribution of free electrons between the lattice ions and impurities, and to the formation of short-lived isotopes of chemical elements. The irradiation of minerals disturbs the energy balance of the crystal structure and, thereby, contributes to the formation of defects within the structure. The defects of this type are not stable and mainly determine the degradation of electrical, optical and other properties of silicates, which in turn limit their practical use.

Polyvalent elements available in aluminosilicates, in particular iron-group transition elements and rare earth elements, prevent the formation of the centers under the action of radiation. Thus, fluorine aluminosilicates can be considered as promising materials for radiation dosimetry and protective coating of solar elements against radiation.

The interest in the research of aluminosilicates doped with transition of the iron group is also associated with the search for new materials for laser physics. They are promising also for the development on their basis of light emissive devices. The compounds of fluorine aluminosilicates have good luminescent properties. The optical properties have been partially studied in [1,2,3]. However, the characteristics of the centers responsible for radiation are not available in the literature. There have not been any reliable criteria for identification of defects with natural and radioactive origins. To a large extent, this is due to the insufficient knowledge about the nature of point defects in the aluminosilicates.

It is well-known that a brilliant representative of aluminosilicates is topaz $Al_2[SiO_4],[F,OH]_2$. The topaz crystal structure is built of silicon-oxygen tetrahedral (SiO_4) and octahedral aluminum surrounded by four oxygen ions and fluorine ions or a hydroxyl group. The topaz crystal lattice is shown in Fig. 1. In our view, the defects formed by substitution ions of the iron transition group are of great practical interest. They have an unfilled 3d-shell. It means that they can play the role of cations with variable valence. A significant magnetic moment of these centers allows to use them as spin labels [4,5] and investigate the electronic configuration and symmetry of the defects.

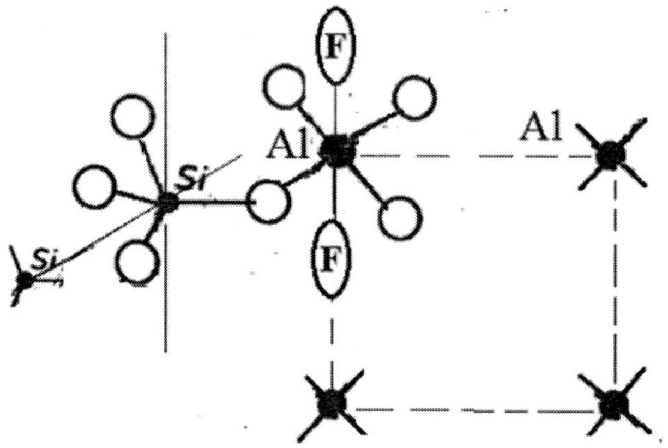

Figure 1 Structure of topaz projected on the plane (001).

For example, Fe ions in ^6S ($3d^5$) state are paramagnetic. These Fe ions, embedded in the crystalline lattice of the mineral, can provide a rather important information on the lattice structure. There are two types of substitution defects in topaz, such as octahedral and tetrahedral, depending on the location of the impurity ion. The point defects with octahedral symmetry arise on local substitution of aluminum ($Fe^{3+} \rightarrow Al^{3+}$), while tetrahedrally coordinated centers are formed by substitution of silicon ($Fe^{3+} \rightarrow Si^{4+}$) in the silicon-oxygen tetrahedral (SiO_4). The topaz composition can vary from $Al_2[SiO_4],F_2$ to $Al_2[SiO_4],[F_{0.7},(OH)_{0.3}]_2$. In accordance with this change, its optical properties are changed, too. According to [2,5], the luminescence attributed to the defects formed by the ions of tetrahedrally coordinated centers is observed in the spectra of luminescence of natural topaz. For example, the appearance of brown-orange coloration is associated with the oxygen vacancies, which formed by substitution of silicon chromium by Cr^{3+} (640 nm wavelenght) or Fe^{3+} (710 nm wavelength).

In studies of centers formed by 3d- elements, the electron paramagnetic resonance method (EPR) plays a significant role. It is an indispensable tool for investigation the electronic structure of the paramagnetic center, that is, for determining the symmetry of the center, its charge state, the total magnetic moment, the interaction between the defects and the host lattice ions. Our aim is to investigate by the EPR method the new Fe centers with local cubic symmetry embedded in $Al_2[SiO_4],[F,OH]_2$ (this was announced in [6].

2. Experiment / Calculation / Theory

An important aspect of the interpretation of EPR spectra of ions in ^6S- state is the explanation of the difference between the values of g-factor and the pure spin state ($g_0 = 2{,}0023$). Unlike to other magnetic centers with the number of d-electrons $n_d = 5$, the energy difference $\Delta = E_{t2} - E_e$, can not be considered as the excited energy state of the local center. This is due to the fact that any electron exchange between the energy levels of e and t_2 states leads to a change in total spin of the center.

The levels of unpaired 3d-electrons located in the band gap are not pure d-levels in view of a delocalisation of the corresponding wave functions. The "resonances" in the valence band, which do not affect the total magnetic moment of the center, are of explicit character. In this case, a state formed as a result of the transition of a hole from the bottom of d-levels in the forbidden zone to d_B (the level in the valence band) should be considered as the first excited state. Indeed, the neutral iron atom has the electron configuration $3d^6\ 4s^2$. Substituting a silicon atom with the electron configuration $3s^2 3p^2$ gives two 4s-electrons and one 3d-electron for the formation of chemical bonds with neighboring atoms (the electron configuration $3d^5$).

The structure of the levels in the one-electron approximation is shown in Figure 2. A characteristic feature of this structure is the presence of filled d-resonance states in the valence band. The excited state is a state $|d_B>$ in the valence band, which corresponds to the occupied states $|d_0>$ in the forbidden zone. Hence, the difference between the energies of the states $|d_0>$ and $|d_B>$ is to be taken as the energy difference Δ' for ground and excited states, which determines the *g*-factor shift (see Fig. 2). The *g*-factor shift may be calculated as follows [7]:

$$\Delta g = 8K_{\sigma\pi}\lambda\frac{H_{ij}}{\Delta'\cdot H_0}, \quad (1)$$

where $K_{\sigma\pi}$ is the factor determining the change in the spin-orbit interaction due to the d-state delocalization, λ is the constant of spin-orbit coupling to the atomic state, H_{ij} and H_0 are the one-electron matrix elements of spin-orbit coupling for the states involved in the formation of covalent bonds and purely atomic ones, respectively. Since the values of the matrix elements of spin-orbit interaction are never precisely known, the relationships H_{ij}/H_0 are considered as parameters.

Assuming a change of spin-orbital communication center with respect to a free atom insignificant, we set the parameter H_{ij}/H_0 equal to 1. We do not take into account the electron delocalisation of local centres as in the ground state as in the excited one. Therefore, we assume $K_{\sigma\pi} = 1$. The constant of spin-orbit coupling to the Fe atom is 0.008 eV [7]. Thus, the expression for the *g*-factor shift ($\Delta g = g - g_0$) can be rewritten as: $\Delta g = 8\lambda/\Delta'$. From the last equality one can find the value of the excited state energy of the center: $\Delta' = 8\lambda/\Delta g$. For Fe(III), the *g*-factor shift is $\Delta g = 0.038$, which corresponds to the energy $\Delta' = 1{,}6$ eV. This electronic transition should define a certain luminescence of topaz in 710 nm bands [1,2]. For the center of Fe (II) we have $\Delta g = 0.020$, while the corresponding excitation energy is equal to 3.2 eV.

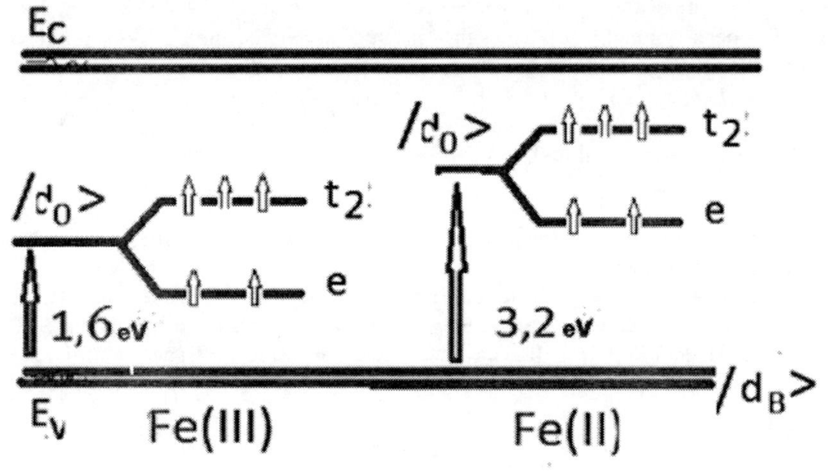

Figure 2 Scheme of energy levels of Fe^{3+} centers with tetrahedral symmetry.

Taking into account the proximity of the used model crystal field, we can observe the reasonable agreement between the estimated value of the excited energy state and the experimental one [4,6] .

The calculation of the EPR line width, measured between the inflection points, is also quite important for the interpretation of the EPR spectrum of $Al_2[SiO_4],[F,OH]_2$. Note that the transfer of the charge density of d-electrons in ligands and the associated interaction of unpaired electrons of the center with the magnetic moment of ligands, i. e. the so-called super-hyperthin interaction (SHTI), should lead to a broadening of the EPR line. In the EPR spectrum of the center of Fe(II), shown in [6], we distinctively see a group of three lines of the super-hyperthin structure (SHTS). The EPR lines for SHTS are clearly resolved for transitions (-3/2→-1/2) and (5/2→3/2). The SHTS lines for centers such as Fe(III) are actually not observed.

To determine the EPR line widths, it is sufficient to consider the interaction of the d-electron of paramagnetic center with the nuclear magnetic moments of ligands. In our case these ligands are the nuclei of F^{19} located in the vicinity of a 3d-ion. The distance between the adjacent lines of the EPR spectrum is equal to a constant A_{SHTS}, which is determined from the EPR spectrum. For Fe(II) center we have $A_{SHTS} = 5$ Oe [6]. Now one can determine the width of the central transition line between the points of maximum curvature (ΔH_{pp}) as follows. Due to SHTS, each electronic transition is spited into $2In + 1$ lines of equal intensity. Here I is the nucleum ligand spin which equals to 1/2 for fluorine, while n is the number of ligands with spin I. Thus, the width of the EPR line taken between the inflection points of the absorption line is $\Delta H_{pp}= A_{SHTS}(2In+1)$.

Since the width of the EPR line of center Fe(II) is equal to 14 Oe [6], we claim that in a vicinity of Fe^{3+} there are two fluorine ions ($n = 2$). The width of the EPR line of center Fe(III) is $\Delta H_{pp} = 17$ Oe [6]. It should be assumed that both the SHTI and the spin-lattice interaction give the contribution to the line width of the EPR of center [8]. In this case $K_{\sigma\pi}$ factor, which is responsible for the change in the spin-orbit interaction due to delocalization of d- state, is different from 1. It should be taken into account in the corresponding calculation of the EPR line width.

3. Discussion

It is shown in [6] that the constants a of the fine structure (FS) of centers Fe(II) and Fe(III) differ from the constant of FS for Fe(I). The latter indicates a significant difference in the magnitude of low-symmetric component of crystal field acting on Fe ions. This conclusion is also supported by the absence of the SHTS lines for centre Fe(III).

The super-hyperthin structure is usually permitted only for certain orientations of the magnetic field, when the ligands (in our case fluoride) form a group of magnetically equivalent centers and the magnetic field is directed along one of the crystal axes. However, the availability of the vacancy leads to a significant distortion of the center's symmetry [6]. The substitution of silicon by the Fe^{3+} ions implies the displacement of the latter from the center of tetrahedron. The replacement ion is displaced from its equilibrium position by a distance $d = 0,544 \cdot L \cdot tg\varphi$ in the direction <110>. According to the angular dependence for the EPR spectra for Fe(II) and Fe(III) [6], the angle φ is equal to 6° and 10°, respectively. The corresponding displacements from the center of tetrahedron equal to 0,17Å for Fe(II) and 0,27Å for Fe(III).

4. Conclusions

The structure of the new local centers based on the trivalent Fe has been discovered and studied for the first time. It has been shown that the center of Fe(II) is formed through the substitution of silicon by Fe accompanied by the oxygen vacancy formation in the presence of two fluorine ions. This centre of Fe(II) is responsible for the luminescence in the region 710-720 nm. The resolved hyper-thin EPR line structure is not observed near Fe(I) and Fe(III) . It should be noted that in the EPR spectra of the studied crystals one can observe the signals of Cr^{3+}, Ti^{3+} and, probably, other impurities of the low intensity.

The above-mentioned centers Fe^{3+} have been observed for the first time in the non-irradiated crystals of topaz [6]. The electronic state and structure defects are stable. Magnetic interactions between paramagnetic impurities have not been detected.

References

[1] Gatta G, Nestola F, Bromiley G, Loose A. (2006) New insight into crystal chemistry of Topaz: A multi-methodological study. American Mineralogist 91 1839-1846
[2] Skvortsova V, Mironova-Ulmane N, Trinkler L (2015) Optical properties of irradiated topaz crystals. Materials Science and Engeneering: Conference Series 80 012008
[3] Lazarenko E K, Pavlishin V I, Lett V T and Sorokin Y G (1973) Mineralogy and Genesis Chamber Pegmatites of Volyn. Lviv
[4] Petrov M P, Paugurt A P, Ivanov AV, Pleshakov IV (1986) Using the phase relations in the registration response of the spin system. Technical Phys. Letters 12 833-837
[5] Dias L, Pinheiro M, Krambrock K. (2009) Radiftion-induced defects in euclase: formation of O – hole and Ti^{3+} electron centers. Phys. Chem. Minerals 36 514-525
[6] Apushkinskaya D, Apushkinskiy E, Popov B, Romanov V N, Saveliev V, Sobolevskiy V (2015) Analysis of paramagnetic centers for theevalent iron in aluminosilicates. J.of Phys.: Conference Series 633 012115
[7] Abragam A and Bleaney B (1970) Electron Paramagnetic Resonance of Transition Ions. Oxford
[8] Chizhik V, Chernyshev Y, Donets A, Frolov V et all. (2014) Magnetic Resonance and Its Applications. Springer Cham Heidelberg New York Dordrecht London

Modeling of termokinetic oscillations at partial oxidation of methane

A V Arutyunov[1,2*], A A Belyaev[2] , I N Inovenkov[1] , V V Nefedov[1]
[1]Lomonosov Moscow State University, Faculty of Computational Mathematics and Cybernetics (CMC), Russia, 119991, Moscow, GSP-1, 1-52, Leninskiye Gory
[2] Semenov Institute of Chemical Physics of Russian Academy of Sciences (ICP RAS), 119991, Kosygina 4, Moscow, Russia
*E-mail: aarutyunovv@gmail.com

Abstract. Partial oxidation of natural gas at moderate temperatures below 1500 K has significant interest for a number of industrial applications. But such processes can proceed at different unstable regimes including oscillating modes. Nonlinear phenomena at partial oxidation of methane were observed at different conditions. The investigation of the complex nonlinear system of equations that describes this process is a real method to insure its stability at industrial conditions and, at the same time, is an effective tool for its further enhancement. Numerical analysis of methane oxidation kinetics in the continuous stirred-tank reactor, with the use of detailed kinetic model has shown the possibility of the appearance of oscillating modes in the appropriate range of reaction parameters that characterize the composition, pressure, reagents flow, thermophysical features of the system, and geometry of the reactor. The appearance of oscillating modes is connected both with the reaction kinetics, heat release and sink and reagents introduction and removing. At that, oscillations appear only at a limited range of parameters, but can be accompanied by significant change in the yield of products. We have determined the range of initial temperature and pressure at which oscillations can be observed, if all other parameters remained fixed. The boundaries of existence of oscillations on the phase plane were calculated. It was shown that depending on the position inside the oscillation region the oscillations have different frequency and amplitude. It was reviled the role of heat exchange with the environment: at the absence of heat exchange the oscillating modes are impossible. In the vicinity of the boundary of phase range, where oscillations exist, significant change of concentration of some products were observed, for example, that of CO_2, which in this case one of the principal products is. At that, insignificant increase in pressure not only change the character of CO_2 behaving with time, but as well lead to significant increase of its mole fraction simultaneously twice decreasing the mole fraction of CO.

1. Introduction

The partial oxidation of methane — the basic component of natural gas — at moderate temperatures below 1500 K is described by a complex set of kinetic equations, which should include hundreds of elementary chemical steps for an adequate description of the process. The most important special feature of this chemical system, which is a branched-chain radical process, is its nonlinearity because the main chain propagation, branching, and termination steps occur with the participation of unstable highly reactive intermediate compounds. That is, the set of kinetic equations that describes the behavior of this system and, consequently, the corresponding set of differential equations are deliberately nonlinear with respect to the initial reagents [1]. This creates prerequisites for the complex behavior of the system, including the existence of different regimes under almost the same conditions with a sharp transition between them: for example, the appearance of the region of a negative temperature coefficient of reaction rate, cold flames, a temperature hysteresis of reaction rate, and also oscillating modes and other nonlinear processes. The majority of these phenomena were observed experimentally in studies of the partial oxidation of methane under various conditions [2]. However, with consideration for the enormous number of kinetic parameters that describe this system, it is impossible to study all its special features in a potentially interesting parameter region experimentally.

At the same time, at the industrial implementation of such processes, it is necessary to insure their stability: that is, a random jump of any parameter beyond the specified limits will not cause unstable process conditions with an uncontrollable course of events. On the other hand, it is not inconceivable that the highest yield of target products can be obtained at the values of parameters that cannot be reached under laboratory conditions; that is, the process efficiency can be considerably increased. Therefore, the study of this complex nonlinear system by mathematical simulation methods is of paramount importance for its stable industrial implementation; simultaneously, this is an effective tool for its further improvement. Here, we report on a study of the appearance of unsteadystate conditions in the partial oxidation of natural gas based on the analysis of a detailed kinetic model of the process.

2. Calculation model

The possibility of the appearance of oscillating modes in the oxidation and combustion of methane was studied by a numerical calculation method. The onset of oscillations in chemical systems belongs to the critical phenomena of chemical kinetics. Their appearance is related to the accumulation of active products or heat in the reacting system [3]. In simple model systems, it is possible to find a range of parameters at which oscillations occur based on a parametric study of the problem (for example, see [4]). However, the real kinetic mechanism of the oxidation and combustion of methane contains hundreds of elementary steps with the participation of tens of components. Therefore, an analytical study is almost impossible in this case, and only a computer simulation makes it possible to characterize this system in sufficient detail and to predict at which values of the parameters oscillatory modes can appear.

This study was performed within the framework of the model of a chemical continuous stirred tank reactor because steady-state periodic oscillations in the oxidation of methane were found previously [5, 6] under these conditions. Note that the possibility of the onset of chemical oscillations is raised in flow systems, as compared with that in closed systems, because the parent substances consumed in reactors of this type are continuously supplied to the reactor. Figure 1 shows a schematic diagram of the continuous stirred tank reactor with the main parameters of input and output flows. The mathematical formulation of problem for this model is written in the form [7].

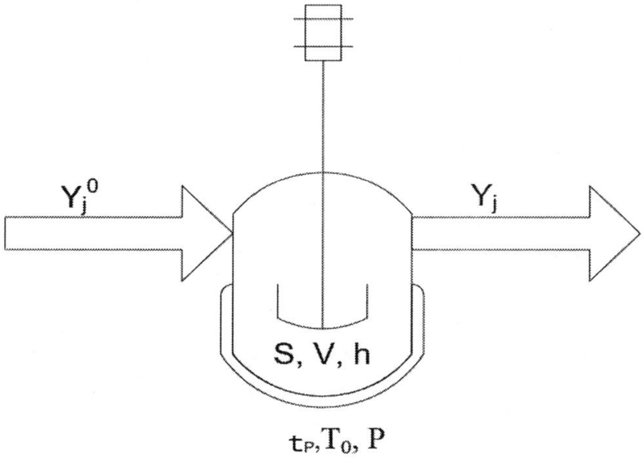

Figure 1. Schematic diagram of a continuous stirred tank reactor and the main parameters of the inlet and outlet flows.

IOP Conf. Series: Journal of Physics: Conf. Series **936** (2017) 012012 doi:10.1088/1742-6596/936/1/012012

$$\begin{cases} \rho \dfrac{\partial Y_j}{\partial t} = \omega_j + \dfrac{\rho}{\tau_r}\left(Y_j^0 - Y_j\right), \; j = 1,2,\ldots,M \\ \rho c_p \dfrac{\partial T}{\partial t} = \Phi + \dfrac{\rho c_p}{\tau_r}(T_0 - T) + h\dfrac{S}{V}(T_0 - T) \end{cases} \tag{1}$$

Initial condition:

$T = 0$, $T = T_0$, $Y_j = Y_j^0$ (2)

Accepted notations are: t is time; M is the number of mixture components; Y_j is the weight fraction of the j_{th} component; T is the temperature of the mixture; ρ is the density of the mixture; c_p is the heat capacity at constant pressure; ω_j is the rate of formation or consumption of the j_{th} component in the course of chemical reaction; τ_r is the average turnover time of the reactor (the residence time of a reaction mixture in the reactor); Φ is the heat release rate; h is the coefficient of heat transfer into the environment through the reactor walls (the reactor is spherically shaped); V and S are the reactor volume and surface area, respectively; and the subscript "0" refers to the conditions at the reactor inlet. We assumed that the reactor wall temperature was kept constant and equal to the reactor inlet temperature T_0.

It is necessary to add a number of relationships to the Eqs. (1). These are expressions for the rates of reaction and heat emission, a perfect gas equation, and a relation between the weight and mole, fractions of a component [8]:

$$\omega_j = W_j \sum_{k=1}^{N}\left(\upsilon_{j,k}^{//} - \upsilon_{j,k}^{/}\right) A_k T^{b_k} \exp\left(-\frac{E_k}{R^0 T}\right) \prod_{i=1}^{M}\left(\frac{X_i p}{R^0 T}\right)^{\upsilon_{i,k}^{/}} \tag{3}$$,

$$\Phi = \sum_{k=1}^{N} H_k A_k T^{b_k} \exp\left(-\frac{E_k}{R^0 T}\right) \prod_{i=1}^{M}\left(\frac{X_i p}{R^0 T}\right)^{\upsilon_{i,k}^{/}} \tag{4}$$

$$p = \rho R^0 T \sum_{j=1}^{M} \frac{Y_j}{W_j} \tag{5}$$

$$X_j = \left(\frac{Y_j}{W_j}\right)\left(\sum_{j=1}^{M} \frac{Y_j}{W_j}\right)^{-1} \tag{6}$$

The following notations were used: N is the number of elementary steps of a chemical reaction; W_j is the molecular weight of the j_{th} component; X_j is the mole fraction of the j_{th} component; p is the pressure; R_0 is the gas constant; $v_{j,k}^{\prime}$ and $v_{j,k}^{\prime\prime}$ are the stoichiometric coefficients of the j_{th} component as a reactant and a product in the kth elementary reaction, respectively; A_k, b_k, and E_k are the preexponential factor, the exponent for temperature, and the activation energy of the k_{th} reaction, respectively; and H_k is the heat effect of the k_{th} reaction.

The heat capacity of a gas mixture is found from the equation [8]

$$c_p = \sum_{j=1}^{M} Y_j c_{pj} \quad (7)$$

The heat capacity of each mixture component, c_{pj}, is related to temperature by the polynomial equation

$$c_{pj} = \frac{R_0}{W_j}\left(a_1 + a_2 T + a_3 T^2 + a_4 T^3 + a_5 T^4\right) \quad (8),$$

where the polynomial's coefficients (a_1, a_2, a_3, a_4, and a_5) of each particular component are taken from a thermodynamic data base [9].

For describing a chemical reaction of methane with oxygen we have used the well-known detailed kinetic mechanism of the oxidation and combustion of methane [10], which includes 261 elementary steps with the participation of 30 components. We assumed that the mixture supplied to the reactor consisted of methane (CH_4) and oxygen (O_2) strongly diluted with nitrogen. The ratio between methane and oxygen was varied and characterized by the fuel–air equivalence ratio

$$\varphi = \left(\frac{X_{CH_4}}{X_{O_2}}\right)\left(\frac{X_{CH_4}}{X_{O_2}}\right)_{st}^{-1} \quad (9),$$

where the subscript *st* refers to a stoichiometric ratio.

The real test process usually occurs in a reactor under specific conditions (thermophysical properties, the composition and feed rate of a mixture, and the geometric parameters of the reactor). Therefore, the values of the parameters that characterize these properties should be specified. In our case these parameters are T_0, p, φ, X_{N2}, h, τ_r, and V (S is related to V), the values of which were varied in the course of a numerical study.

The calculations were performed with the aid of the Chemkin program package [11]. Note that de Joannon et al. [6] also studied such a system at X_{N2}= 0.9 with the use of numerical calculations (together with experiments), but the aim of this work was essentially different from the problems solved-in [6]. First, in [6] experiments were performed at a fixed atmospheric pressure, a fixed reactor volume, a fixed residence time of a reaction mixture in the reactor, the constant conditions of heat emission into the environment, and X_{N2}= 0.9. As noted above, we studied the dependence of the domain of existence of oscillating modes on these parameters, whose values were varied. Furthermore, unlike [6], we studied the possibility of minimizing the number of elementary steps in the kinetic mechanism at which the occurrence of the oscillating mode would be retained. Finally, the analysis of the behavior of the concentrations of gas mixture components at the boundary of existence of the oscillating mode is a fundamentally new issue in this work.

3. Results of the calculations

In the course of computations, it was established that the periodic sustained oscillations of process temperature were observed in a specific range of varied parameters. The amplitude and frequency of these oscillations depend on the parameters that characterize the properties of a gas mixture and a reactor. In certain cases, the amplitude of temperature oscillations can reach hundreds of degrees (Figure 2). Note that analogous oscillations were also observed for the concentrations of many components of this chemical mixture. The range of variation of the parameters T_0, p, φ, τ_r, h and V

in which oscillations were observed is relatively small. In all of the other cases, a steady state process was established with time; that is, the temperature and concentrations became time-constant.

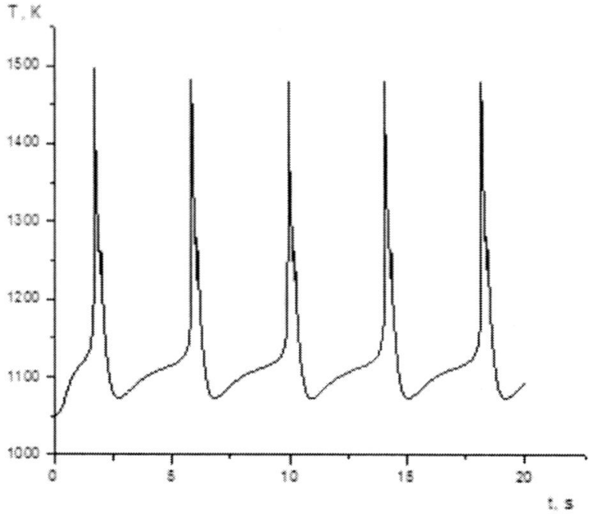

Figure 2. Process temperature oscillations at T_0 = 1050 K, p = 1 atm, φ = 0.8, X_{N2} = 0.9, h = 0.0042 cal /(см^2·s·K), τ_r = 0.5 s, and V = 0.1 м^3.

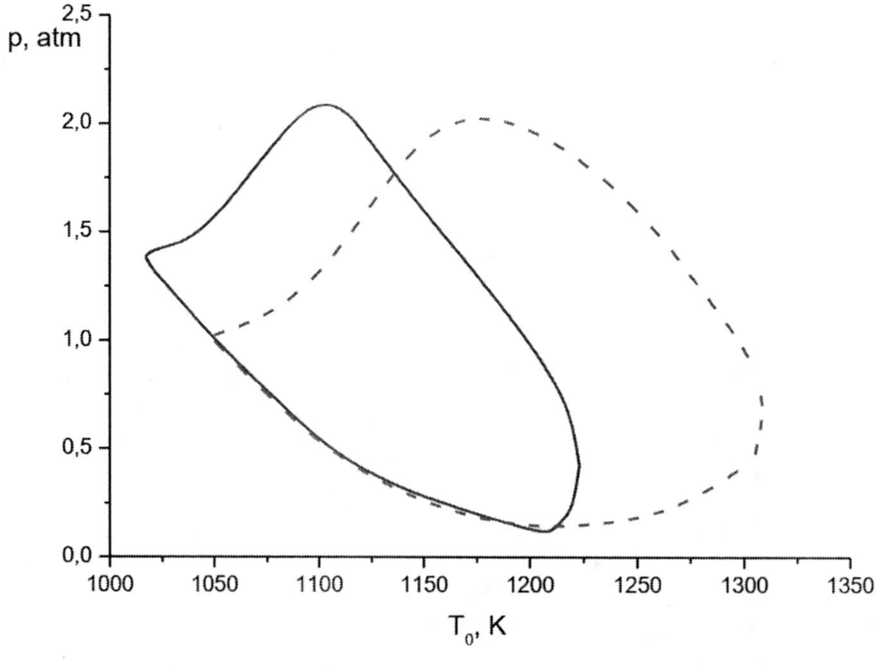

Figure 3. Domain of existence of the oscillating mode on the initial temperature–pressure phase plane at φ= 0.8, X_{N2} = 0.9, h = 0.0042 cal/(см^2·s·K), τ_r= 0.5 s, and V = 0.1 м^3. Solid and broken lines refer to the complete kinetic mechanism [10] and a reduced kinetic mechanism (57 reactions and 20 components), respectively.

We have determined the ranges of the initial temperature and pressure in which oscillations were observed at the fixed values of other parameters ($\varphi = 0.8$, $X_{N2} = 0.9$, $h = 0.0042$ cal(см^2·c·K), τ_r = 0.5 s, and $V = 0.1$ м^3). The boundary of a region in which oscillations were observed is shown by a solid line on the (T_0, p) phase plane (Figure 3). The oscillating mode has different frequency and amplitude depending on the position of a point inside the region. As the upper boundary of the region is approached (with increasing pressure), the oscillation frequency increases, but a decrease in the amplitude is simultaneously observed and oscillations disappear at the boundary. On the contrary, oscillation frequency decreases and its amplitude increase as the lower boundary approaches (with decreasing pressure). At the boundary, the oscillation period approaches to infinity. At the left boundary, changes in the amplitude and frequency are not so pronounced. At the right boundary, there is a special feature related to the fact that oscillations appear after a time rather than immediately.

We studied the possibility of reducing the number of elementary steps in a kinetic mechanism. The number of elementary steps was decreased so that, in this case, not only the main properties of the process of methane oxidation but also the occurrence of the oscillating mode were retained. As a result of a regular decrease, we managed to reduce the number of elementary steps from 261 to 57 and the number of chemical components from 30 to 20. The domain of existence of the oscillating mode for this reduced mechanism is shown by a broken line in the (T_0, p) phase plane (Figure 3). It is evident that the domain of existence of the oscillating mode for the reduced kinetic mechanism is shifted toward higher temperatures.

For other varied parameters, it is also possible to find a range of their changes within which oscillations are observed. However, to construct the region of oscillations on the reference plane based on all of the parameters is a very labor-consuming process; therefore, we have restricted ourselves to the demonstration of the behaviors of the system under variations of other parameters. These calculations were performed using a published complete kinetic mechanism [10].

Figure 4 shows different reactor conditions depending on the fuel–air equivalence ratio φ the other parameters being fixed. It is evident that periodic oscillations were observed at $\varphi = 0.8$ and 0.9; however, there were no fluctuations in both lean and stoichiometric mixtures. In the two last-named cases, a swift temperature rise was observed with the subsequent decrease to a steady-state value.

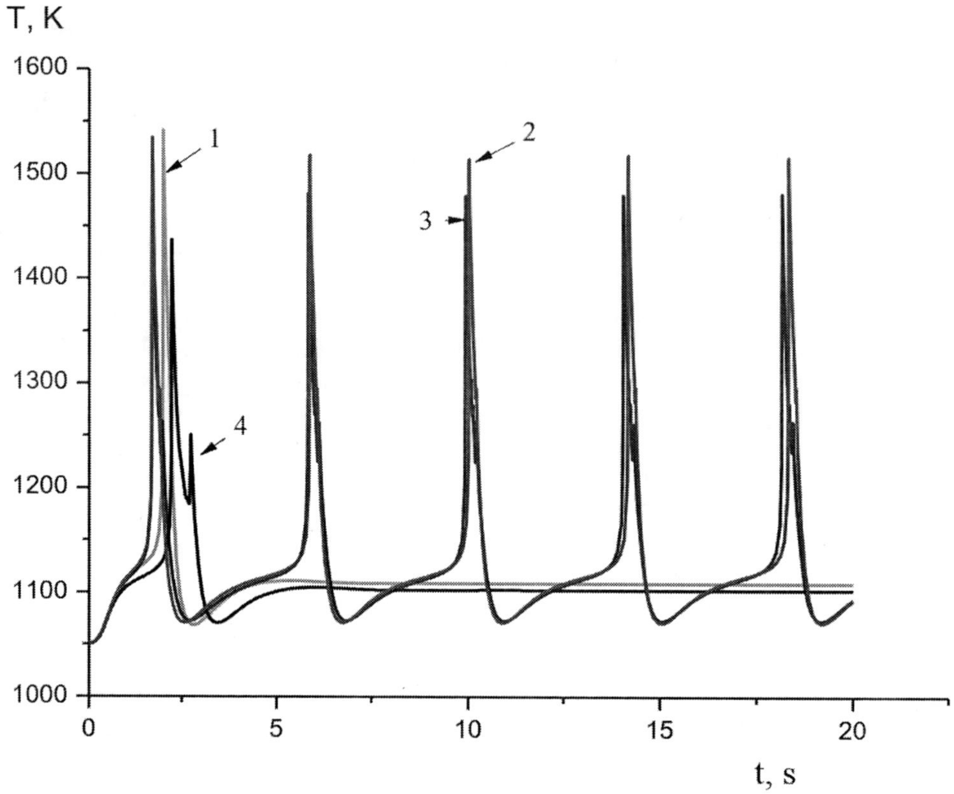

Figure 4. Time dependence of temperature at different fuel–air equivalence ratios φ: (*1*) φ = 1 (stoichiometry), (*2*) $\varphi = 0.9$, (*3*) $\varphi = 0.8$, and (*4*) $\varphi = 0.7$; $T_0 = 1050$ K, $p = 1$ atm, $\varphi = 0.8$, $\varphi = 0.9$, $h = 0.0042$ cal/(см^2·s·K), $\tau_r = 0.5$ s, and $V = 0.1$ м^3.

We also studied the dependence of the behavior of steady-state reactor conditions on parameters such as the heat-transfer coefficient h, the reactor volume V, and the parameter (which is inversely proportional to the mass feed rate to the reactor). For each of these parameters, there is also a specific range of changes in which oscillations are observed. As an example, Figure 5 shows the dependence of temperature on the heat transfer coefficient h. In the absence of heat emission ($h = 0$) or at its low value, there were no oscillations. But they appeared with increasing h and their amplitude initially increased; however, the oscillations disappeared with a further increase in the heat emission. It is obvious that the steady-state temperature (or its equilibrium value about which oscillations occur) decreases with h. Thus, oscillations do not appear at either strong or weak heat emission.

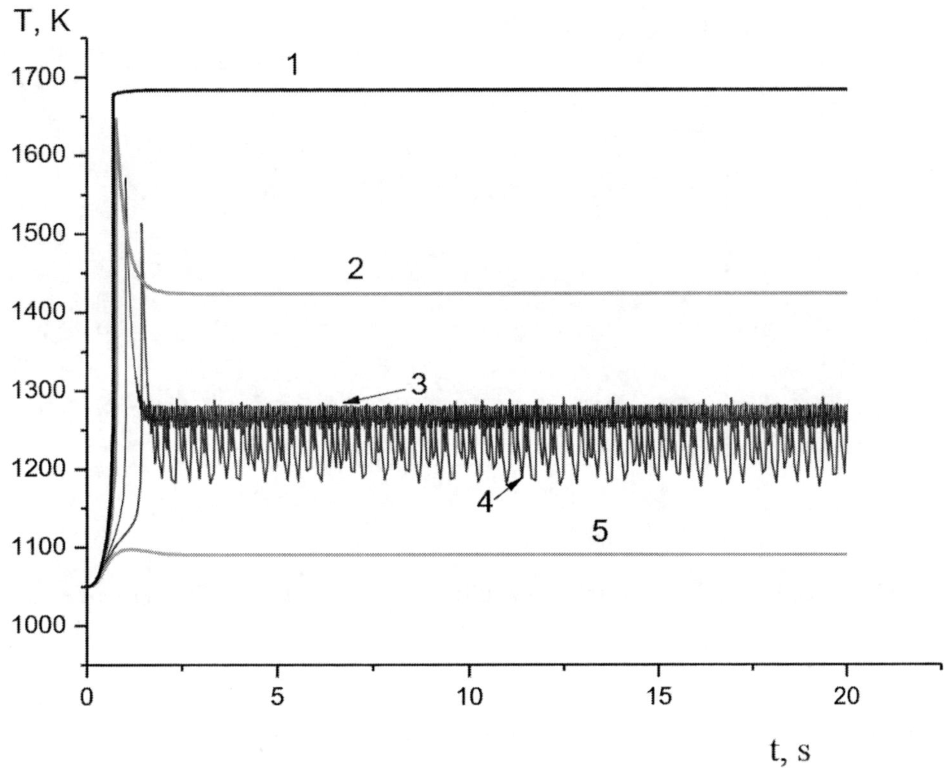

Figure 5. Time dependence of temperature at different heat transfer coefficients h, cal/(см^2·s·K): (*1*) h = 0, (*2*) h = 0.001, (*3*) h = 0.003, (*4*) h = 0.004, and (*5*) h = 0.005; *T*0 = 1050 K, p = 1 atm, φ = 0.8, = 0.9, τ_r = 0.5 s, and V = 0.1 м^3.

The nitrogen content of the mixture should be specially noted. We studied the influence of the fraction of nitrogen in the mixture on the range of pressures in which the oscillating mode occurred. Figure 6 shows this dependence. Two lines indicate the upper and lower boundaries, respectively, of a pressure range in which the oscillating mode occurs at the fixed other parameters. It is evident that the pressure range initially somewhat enlarged but then gradually decreased with decreasing the mole fraction of nitrogen. The above dependence was obtained at T_0 = 1050 K, but an analogous behavior was also observed at other initial temperatures. This behavior may be explained by two tendencies: on the one hand, a decrease in the fraction of nitrogen facilitates the complete occurrence of the reaction; however, at the same time, the rate of consumption of the initial products (methane and oxygen) is not compensated by their supply into the reactor as the value of is decreased. Note that both the amplitude and the frequency of oscillations increased with decreasing; this fact is indicative of the more rapid occurrence of the reaction and an increase in the effect of the supply of a fresh mixture on the process.

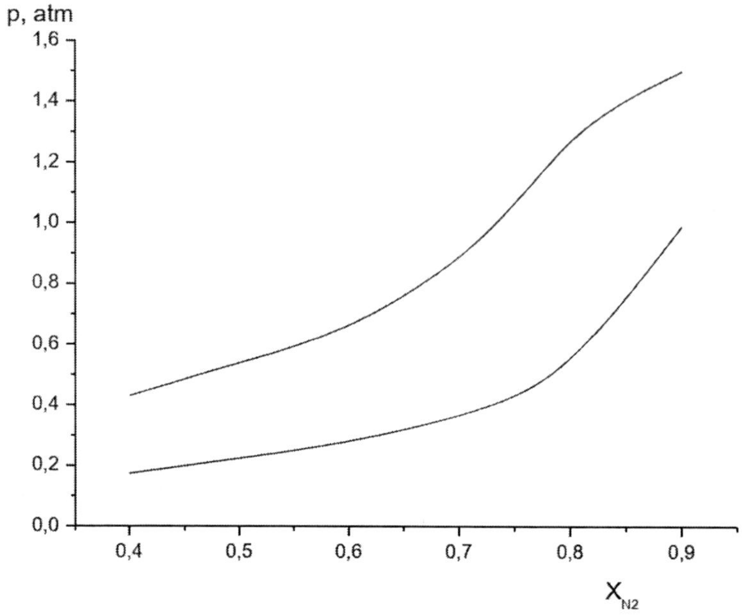

Figure 6. Dependence of the domain of existence of the oscillating mode in terms of pressure on the nitrogen content of the mixture at T_0 = 1050 K, φ = 0.8, h = 0.0042 cal/(cм^2·s·K), τ_r = 0.5 s, and V = 0.1 м^3.

As noted above, the detailed kinetic mechanism can be strongly reduced with the retention of basic regularities in the behavior of the system at the specific values of parameters. Therefore, at the given values of the parameters, many elementary steps weakly influence (or, under the given conditions, some of them do not actually influence) the occurrence of the chemical reaction as a whole. However, there are elementary steps that have a very strong effect on the process. De Joannon et al. [6] found that the following recombination reaction of methyl radicals ($CH_3^\bullet$) exerts a very strong effect on the occurrence and characteristics of the oscillations:

$$CH_3^\bullet + CH_3^\bullet \Leftrightarrow C_2H_6$$

Our studies confirmed a dramatic effect of this reaction on the process as a whole. In addition to this, we found that the chain branching reaction also has a very strong effect:

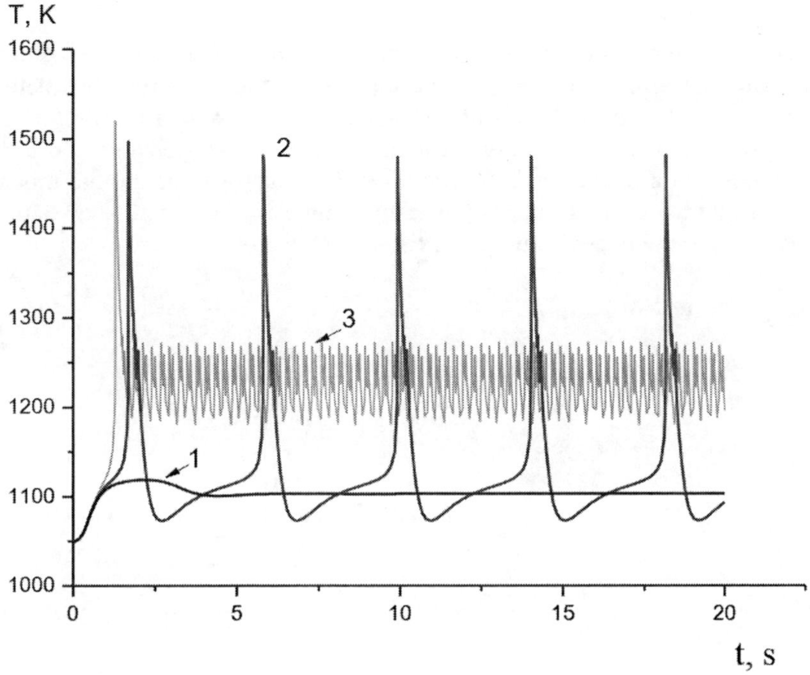

Figure 7. Time dependence of temperature at different values of the pre-exponential factor A_{br}: (*1*) $A_2 = 0.95A_2^0$, (*2*) $A_2 = A_2^0$, and (*3*) $A_2 = 1.1A_2^0$; $T_0 = 1050$ K, $p = 1$ atm, $\varphi = 0.8$, $\varphi = 0.9$, $h = 0.0042$ cal/(см^2·s·K), $\tau_r = 0.5$ s, and $V = 0.1$ м^3.

$H^{\cdot} + O_2 <=> OH^{\cdot} + O^{\cdot\cdot}$

Let us consider this reaction in more detail. We designate the pre-exponential factor of forward reaction as A_{br} (the parameters of reverse reaction are calculated through the parameters of forward reaction based on an equilibrium constant [10]) and the base value of the pre-exponential factor as A_{br}^0 ($A_{br}^0 = 2{\cdot}10^{14}$ cm^3 mol^{-1} s^{-1}). Figure 7 demonstrates that even a small change in the pre-exponential factor of this reaction has a strong effect on the course of the process. Thus, as the pre-exponential factor decreases by only 5%, the oscillating mode disappears, and a steady-state solution is reached. It is evident that, as the pre-exponential factor of this elementary step increases by a factor of 1.1, the oscillation frequency dramatically increases but the amplitude considerably decreases.

Figures 2, 4, 5, and 7 show the behavior of temperature with time because temperature oscillations the most illustrative. However, as mentioned above, the oscillations of concentrations of chemical mixture components were also observed. In certain cases, a strong change in the concentration of reaction products was observed near the boundary of the phase domain of existence of temperature oscillations (Figure 3). The behavior of the concentration of carbon dioxide (CO_2), which is one of the main products of the reaction of methane with oxygen under the given conditions, serves as a very clear example. Figure 8a shows this behavior at the pressure $p = 1.4$ atm. A small, by only 5 K, increase in the initial temperature not only changed the behavior of carbon dioxide concentration with time (oscillations appeared) but also led to a several-fold increase in the mole fraction of CO_2. This was related to a significant decrease in the mole fraction of unburned methane, that is, with an increase in the degree of its conversion (Figure 8b). In this case, the mole fraction of carbon monoxide (CO) decreased by a factor of more than 2. The analysis of sensitivity showed that, under the given conditions, the interaction of CO with the hydroxyl radical is the main step of the formation of CO_2:

$$CO + OH^{\cdot} <=> CO_2 + H^{\cdot}$$

It is likely that an increase in the temperature by 5 K shifts the equilibrium of this reaction to the side of CO_2 formation accompanied by a notable increase in the conversion of methane (Figure 8b). Thus, a transition to the region of the oscillating mode due to a redistribution of rates between the processes of the partial and deep oxidation of methane dramatically changes both the course of the process and the composition of the resulting products. We note once again that this abrupt change in the behavior of concentrations with an insignificant change in the initial temperature was observed only near the boundary of the phase domain of existence of oscillations.

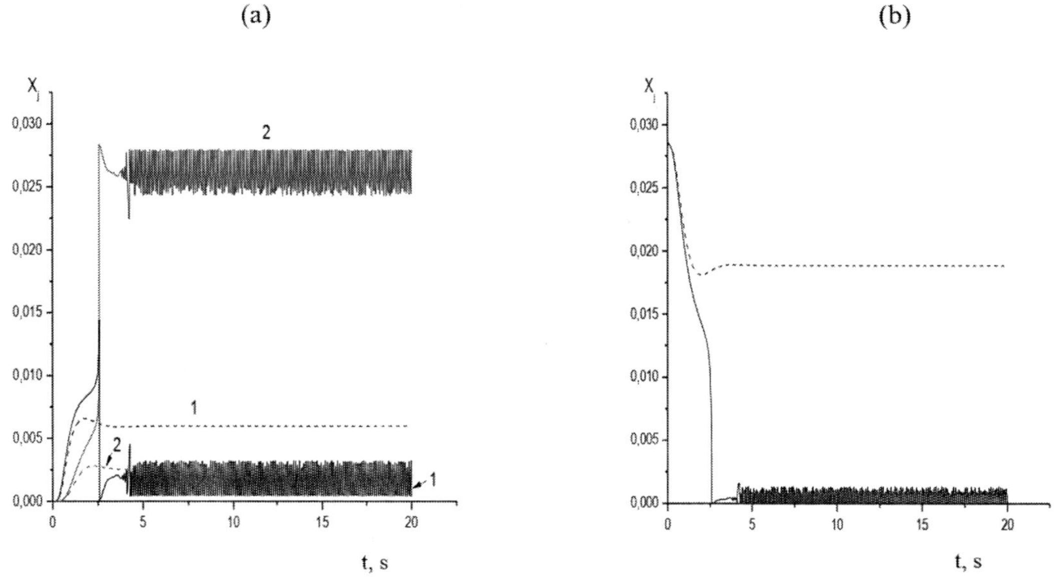

Figure 8. Time dependence of the mole fractions of (a) (1) CO_2, (2) CO_2, and (b) CH_4 at two initial temperatures: dashed and solid lines refer to T_0= 1010 and 1015 K, respectively; p = 1.4 atm, φ = 0.8, φ = 0.9, h = 0.0042 cal/(см^2·s·K), τ_r = 0.5 s, and V = 0.1 м^3.

4. Conclusions

The study of the system of kinetic equations that describes the oxidation of methane under the conditions of a continuous stirred tank reactor showed the possibility of the appearance of oscillating modes. These modes appear at the specific values of the parameters that characterize the thermophysical properties of a mixture, its composition, the feed rates of reagents, and the reactor geometry. The oscillating process conditions are related to the chemical mechanism (reaction kinetics) and the release and removal of heat and reactants. The role of heat exchange between the reactor and the environment is great: under the conditions of this work, it was impossible to obtain oscillating modes in the absence of heat exchange. Oscillations are observed only in a comparatively small range of changes in the parameters that characterize the properties of the system; however, they can be accompanied by considerable changes in the course of the process and in the yields of products.

5. References

[1] V. S. Arutyunov, *Oxidative Conversion of Methane* (KRASAND, Moscow, 2011) [in Russian].

[2] V. Arutyunov, *Direct Methane to Methanol: Foundations and Prospects of the Process* (Elsevier, Amsterdam, 2014).

[3] D. A. Frank-Kamenetskii, *Diffusion and Heat Exchange in Chemical Kinetics* (Nauka, Moscow, 1987; Springer,Heidelberg, 1995).

[4] V. I. Bykov and S. B. Tsybenova, *Nonlinear Models in Chemical Kinetics* (KomKniga, Moscow, 2011) [in Russian].

[5] B. F. Gray, J. F. Griffiths, G. A. Foulds, B. G. Charlton, and G. S. Walker, Ind. Eng. Chem. Res. **33**, 1126 (1994).

[6] M. de Joannon, P. Sabia, A. Tregrossi, and A. Cavaliere, Combust. Sci. Technol. **176**, 769 (2004).

[7] Theory Manual. Chemkin Software, RD01400-C01-004-001A (June 2004). https://ay14-15.moodle.wisc.edu/prod/pluginfile.php/119333/mod_resource/content/1/CHEMKIN_Theory.pdf.

[8] F. A. Williams, *Combustion Theory* (Benjamin, Menlo Park, CA, 1985).

[9] Thermodynamic Data at the Web Site of the Laboratory for Chemical Kinetics. Alexander Burcat's. Ideal Gas Thermodynamic Data in Polinomial Form for Combustion and Air Pollution Use. http://garfield.chem.elte.hu/Burcat/burcat.html.

[10] J. Warnatz, U. Maas, and R. W. Dibble, *Combustion:Physical and Chemical Fundamentals, Modeling and Simulation, Experiments, Pollutant Formation* (Springer, Berlin, 2006).

[11] CHEMKIN-PRO Release 15083 (2009).

Lorenz curves in a new science-funding model

Ding-wei Huang

Department of Physics, Chung Yuan Christian University, Chung-li, Taiwan

E-mail: dwhuang@cycu.edu.tw

Abstract. We propose an agent-based model to theoretically and systematically explore the implications of a new approach to fund science, which has been suggested recently by J. Bollen et al.[1] We introduce various parameters and examine their effects. The concentration of funding is shown by the Lorenz curve and the Gini coefficient. In this model, all scientists are treated equally and follow the well-intended regulations. All scientists give a fixed ratio of their funding to others. The fixed ratio becomes an upper bound for the Gini coefficient. We observe two distinct regimes in the parameter space: valley and plateau. In the valley regime, the fluidity of funding is significant. The Lorenz curve is smooth. The Gini coefficient is well below the upper bound. The funding distribution is the desired result. In the plateau regime, the cumulative advantage is significant. The Lorenz curve has a sharp turn. The Gini coefficient saturates to the upper bound. The undue concentration of funding happens swiftly. The funding distribution is the undesired results, where a minority of scientists take the majority of funding. Phase transitions between these two regimes are discussed.

1. Introduction

The funding distribution is relevant to both research policy and sociophysics, where social interactions are investigated by mathematical tools inspired from physics. Recently, Johan Bollen et al. have suggested a new approach to fund science [1, 2]. The scientific community's responses are favorable. Small-scale experiments are in progress [3, 4]. The idea is to fund every scientist equally and to ask all scientists to give a fraction of their funding to other scientists. The more one receives funding from others, the more one gives away funding to others. Every scientist serves as a funding agency to distribute a small amount of funding. As the funding circulates fluidly in the community, the system can potentially be fairer and more efficient. The funding distribution is expected to converge toward an optimal result favored by the entire community of scientists.

Along this line, we propose an agent-based model to theoretically and systematically explore the implications of this approach. The model is presented in Section 2. The phase transition is addressed in Section 3. Discussions are in Section 4.

2. Model

Consider a community consists of 10^3 scientists. Assume that the annual research budget is also 10^3, i.e., the average budget per scientist is taken as the arbitrary unit. The traditional system is to set up a small committee to allocation 10^3 budget to 10^3 scientists. Quite a few scientists would expect to receive nothing. In the new system, all scientists receive an equal share of 1, in the arbitrary unit. All scientists are also required to give a fixed ratio F to support others, where

$0 \leq F \leq 1$. More specifically, the funding of each scientist is denoted as u_i, where $i = 1, 2 \cdots 10^3$. Every scientist is asked to donate $(F \cdot u_i)$ to five other scientists. We introduce two parameters α and β to specify the random process of how to choose these five scientists and how to share the donation among the chosen. The scientist i is chosen with a probability proportional to $(u_i)^\alpha$; once chosen, the share of donation is proportional to $(u_i)^\beta$. When $\alpha = \beta = 0$, the scientists are chosen randomly and the donation are shared equally. When $\alpha > 0$ and $\beta > 0$, the cumulative advantage is introduced into the model. The Matthew effect becomes prominent when α and β are large.

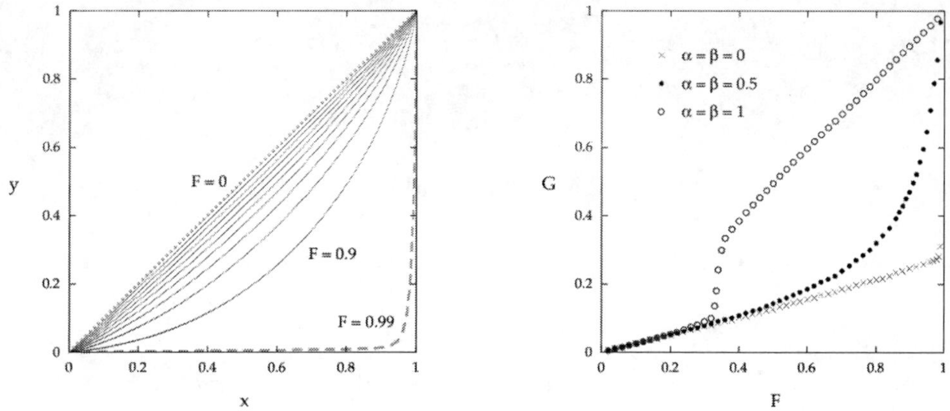

Figure 1. **(LEFT)** Lorenz curves at various F, where $\alpha = \beta = 0.5$. Solid lines show $F = 0.1, 0.2, \cdots 0.9$, where $\Delta F = 0.1$. Grey dotted line shows $F = 0$. Grey dashed line shows $F = 0.99$. **(RIGHT)** Gini coefficient G as a function of F at various α and β.

The concentration of funding is mainly controlled by the parameter F. Typical results of the Lorenz curves are shown in Fig. 1 (LEFT), where the cumulative share of funding (y) is plotted on the cumulative share of scientists (x) from lowest to highest funding. Line of equality is $y = x$, where all scientists receive the same amount of funding. The inequality increases with the increase of F. The inequality can be conveniently measured by the Gini coefficient, which is defined as the ratio between two areas in the plot, i.e., $G \equiv A/B$. The area A is between the Lorenz curve and line of equality; the area B is between x-axis and the line of equality and $B = 1/2$. Fig. 1 (RIGHT) shows the typical results of Gini coefficient as F varies. The Gini coefficient increases monotonically with the increase of F. In this model, every scientist keeps the minimum funding of $(1 - F)$. The Lorenz curve always lies above the line $y = (1 - F)x$. As a result, the area A is less than $F/2$ and the Gini coefficient is less than F. The parameter F becomes the upper bound of the Gini coefficient. When α and β are small, this upper bound is reached only when F approaches 1. When α and β are large, this upper bound can be reached by a small F.

3. Phase Transition

The two parameters α and β control the effect of cumulative advantage. The typical 3D profiles are shown in Fig. 2. Two distinct regimes can be observed as valley and plateau. In the valley regime, the Gini coefficient is well below the upper bound. When α and β increase, the Gini coefficient has a gentle rise. In the plateau regime, the Gini coefficient saturates to the upper bound. The parameter F sets the height of plateau. As F increases, the plateau is expanding and the valley is shrinking. In general, parameter α has more influence than parameter β. The valley regime can be associated with a small α. The plateau regime can be associated with a large α. When β is large, the transition between valley and plateau is smooth and along the line $\alpha = const$. The Matthew effect is controlled by α alone. When β is small, the transition is sharp and along the line $\alpha + \beta = const$. Parameters α and β are additive to the Matthew effect. As β becomes smaller, the transition appears at a larger α.

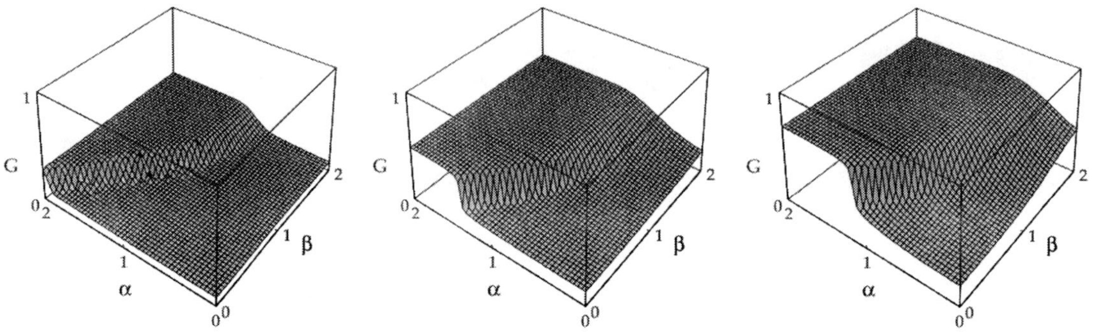

Figure 2. Gini coefficient G as a function of α and β. **(LEFT)** $F = 0.3$. **(MIDDLE)** $F = 0.5$. **(RIGHT)** $F = 0.7$.

The phase transition can also be revealed by the variation of Lorenz curves. Fig. 3 (LEFT) shows the sharp transition at $\beta = 0$ ($F = 0.5$), where eleven curves are plotted at an equal-distanced $\Delta\alpha = 0.2$. Basically the curves are indistinguishable in the range $0 \leq \alpha \leq 1.2$. When $\beta \geq 1.6$, the curves collapse to the asymptotic line of $y = (1-F)x$. In contrast, Fig. 3 (MIDDLE) shows the smooth transition at $\beta = 2$. The curves collapse to the asymptotic result when $\alpha \geq 1$. Gradual changes can be observed in the range $0 \leq \alpha \leq 0.8$.

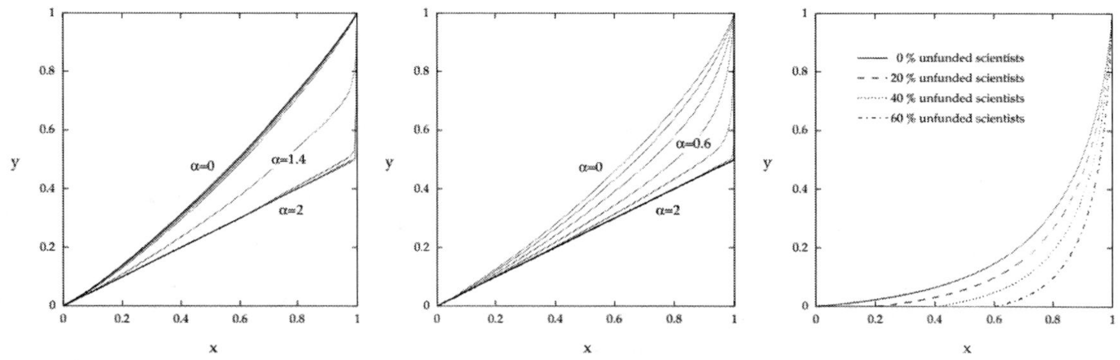

Figure 3. **(LEFT)** Sharp transition as α increases and fixed $\beta = 0$, where $\Delta\alpha = 0.2$ and $F = 0.5$. **(MIDDLE)** Smooth transition as α increases and fixed $\beta = 2$, where $\Delta\alpha = 0.2$ and $F = 0.5$. **(RIGHT)** Lorenz curves with various levels of unfunded scientists, where $F = 0.9$ and $\alpha = \beta = 0.55$.

4. Discussions

The new model has three parameters: $0 \leq F \leq 1$ (ratio of donation), $\alpha \geq 0$ (how to choose), and $\beta \geq 0$ (how to share). Parameter F controls the scale of redistribution and sets an upper

IOP Conf. Series: Journal of Physics: Conf. Series **936** (2017) 012013 doi:10.1088/1742-6596/936/1/012013

bound to the Gini coefficient. Parameters α and β control the strength of Matthew effect and move the system between valley and plateau regimes. It is expected that the concentration of funding can be effectively suppressed. The Gini coefficient can be easily controlled by adjusting the value of F. However, the desired results only present in the valley regime. The plateau regime is to be avoided. In the valley regime, the fluidity of funding is significant. The Lorenz curve is smooth. The funding distribution is continuous. All scientists are equal in regards to acquire different amount of funding. In contrast, in the plateau regime, the cumulative advantage is significant. The Lorenz curve collapses to the asymptotic result and has a sharp turn. The undue concentration of funding still happens swiftly, even though all scientists are treated equally and follow the well-intended regulations. The seemingly perfect regulations cannot prevent the undesired results from happening, where a minority of scientists take the majority of funding.

We introduce a stochastic process for the redistribution of funding. The more one has, the more one is asked to give away. With naive expectation, the accumulation of funding is totally discouraged. However, it is well known that funding is positively correlated to the research output. The positive feedback is controlled by the two parameters α and β. An unselfish scientist will give funding to support the best scientists. And it is also reasonable to expect that the donation is not shared equally, where a larger portion should be given to the better scientist. When the positive feedback is too strong, the undesired result appears. To control the concentration of funding, choosing randomly (small α) has more effect than sharing equally (small β). Sharing biasly (large β) alone is not enough to trigger the transition. But choosing biasly (large α) alone will trigger the transition.

As a final remark, we expect that this model can be used to describe the funding distribution in the current system. The mechanism of current funding system is very different from the presented model, especially in the details of operations. We believe that the model captures the two essential features of funding allocation: cumulative advantage and competition with each other. Either it's a direction interaction among the scientists (as in the model) or an indirect interaction through a small committee (as in the current system), the parameterization of the model should be able to describe the data. The model assumes that all scientists in the community are funded. In the current system, not all scientists are funded. A recent study indicated that 18%-45% of researchers are unfunded [5]. Varied in disciplines, engineering is the lowest 18% and basic medical sciences is the highest 45%. To compare with empirical data, the model can be further modified to include a finite ratio of scientists without funding. The typical results are shown in Fig. 3 (RIGHT). By adjusting the ratio of unfunded scientists, the results are conform to the proverbial 80-20 rule, i.e., 80% of funds are concentrated in 20% of scientists, which implies a Gini coefficient $G \sim 0.6$. By assuming a large F and locating (α, β) in the valley regime but near the transition, the model is expected to give a fair description of empirical data.

5. References

[1] Bollen J, Crandall D, Junk D, Ding Y, and Börner K 2017 *Scientometrics* **110** 521.
[2] Bollen J, Crandall D, Junk D, Ding Y, and Börner K 2014 *EMBO Reports* **15** 131.
[3] Gewin V 2014 *Nature* **507** 265.
[4] http://retractionwatch.com/2016/09/20/what-if-scientists-funded-each-other/
[5] Larivière, V, Macaluso, B, Archambault, É, and Gingras Y 2010 *Research Evaluation* **19** 45.

IOP Conf. Series: Journal of Physics: Conf. Series **936** (2017) 012014 doi:10.1088/1742-6596/936/1/012014

Restoration of valence density of states from XPS spectra

A V Krasavin, V A Kashurnikov, Ya V Zhumagulov

National Research Nuclear University MEPhI, 31 Kashirskoye shosse, 115409, Moscow, Russia

E-mail: avkrasavin@mephi.ru

Abstract. The technique is presented of restoring the electronic density of states of the valence band from data of X-ray photoelectron spectroscopy. The originality of the technique consists in using a stochastic procedure to solve an integral equation relating the density of states and X-ray photoelectron spectra. The results are presented for bulk metals of gold, silver, and copper. The method can be applied to materials with unknown crystal structure, such as ensembles of nanoclusters.

1. Introduction

Low-dimensional structures, being at intermediate position between single atoms and bulk materials, possess unique physical properties, which separate them in the special area of condensed matter physics. Studying the properties of such structures, and nanoscale systems in particular, attracts now an increased interest both from a fundamental point of view, and in connection with numerous applications. The practical use of low-dimensional structures is based mainly on the features of their electronic properties due to strong electron correlations. The necessity of the accounting for electron correlations leads to considerable complexity of the theoretical description of such systems. For this reason, the development of methods which combine both experimental and numerical techniques for the study of the electronic properties of nanoscale structures is an urgent task.

X-ray photoelectron spectroscopy (XPS) allows studying the evolution of the electronic properties of low-dimensional structures depending their size. This fact makes the XPS technique sensitive to the surface density of electronic states, and gives the opportunity to study physical and chemical properties of low-dimensional samples. The advantage of this technique is also the possibility of direct measurement of the density of electronic states (DOS) of the valence band. The smaller is the instrumentation broadening, which is determined mainly by nonmonochromaticity of the X-ray source and by the operation mode of the energy analyser, the more precise is the determination of DOS. Modern X-ray photoelectron spectrometers equipped with a monochromator have the value of the broadening ranged from 0.4 to 0.6 eV. This broadening leads to smearing of DOS and usually prevents direct study of spectra; therefore, a more sophisticated approach is needed to restore narrow features of DOS.

The purpose of the paper is to propose the technique of restoring the electronic density of states of X-ray photoelectron spectra with the use of the numerical stochastic procedure, which accurately accounts for the instrumentation broadening. XPS spectra from the regions of valence states and core levels allow calculation of the broadening function of the spectrometer; for this purpose, the core level line is used. Further, the stochastic procedure is applied to restore DOS by solving an integral equation. This

IOP Conf. Series: Journal of Physics: Conf. Series **936** (2017) 012014 doi:10.1088/1742-6596/936/1/012014

technique was used to calculate DOS of bulk gold, silver, and copper. To confirm the accuracy, the obtained DOS were compared with DOS calculated from density functional theory (DFT).

2. Method

The XPS signal can be described by the following convolution [1]:

$$I(\epsilon) = S(\epsilon) * I_\tau(\epsilon) * I_\alpha(\epsilon) * I_\sigma(\epsilon), \tag{1}$$

where energy ϵ is measured with respect to the Fermi level,

$$S(\epsilon) \sim \delta(\epsilon - \epsilon_0) \tag{2}$$

is the spectrum of a level with the binding energy ϵ_0, free of any broadening factors;

$$I_\tau(\epsilon) \sim \frac{\gamma}{\sqrt{(\epsilon^2 + \gamma^2)}} \tag{3}$$

describes the broadening of the line caused by the finite lifetime τ of a hole on the level; $\gamma \sim \hbar/\tau$ is the natural broadening;

$$I_\alpha(\epsilon) \sim \theta(\epsilon)\epsilon^{\alpha-1} \tag{4}$$

defines the broadening of the line due to many-electron effects; $0 \leq \alpha < 1$ is Anderson parameter of singularity [2];

$$I_\sigma(\epsilon) \sim e^{-\epsilon^2/2\sigma^2} \tag{5}$$

has the form of a Gaussian and corresponds to the instrumentation broadening.

For valence electrons, the lifetime τ is negligibly small, and (1) takes the form

$$I_{val}(\epsilon) = \rho_{val}(\epsilon) * I_\alpha(\epsilon) * I_\sigma(\epsilon), \tag{6}$$

where $\rho_{val}(\epsilon)$ is the density of valence states. Therefore, for XPS spectra of the core line and valence band, we have the following system:

$$\begin{cases} I_{core}(\epsilon) = \int_{-\infty}^{+\infty} H(\epsilon - \epsilon')\, I_\tau(\epsilon')\, d\epsilon'; \\ I_{val}(\epsilon) = \int_{-\infty}^{+\infty} H(\epsilon - \epsilon')\, \rho_{val}(\epsilon')\, d\epsilon'. \end{cases} \tag{7}$$

Here $H(\epsilon) = I_\alpha(\epsilon) * I_\sigma(\epsilon)$ is the broadening function, which takes into account technical characteristics of a spectrometer and many-electron processes. It follows from (7), that, if the parameters γ, α, σ are someway known, one can restore the density of states in the valence band $\rho_{val}(\epsilon)$ from the second equation. This equation belongs to the class of Fredholm integral equations of the first kind, and its solution is the ill-posed problem.

One of the very effective approaches which numerically address ill-posed problems of such kind is the "algorithm of random rectangles", proposed in [3]. The solution $\rho_{val}(\epsilon)$ is approximated with a piecewise constant function

$$\rho_{val}(\epsilon) \longleftarrow \tilde{\rho}_{val}(\epsilon) = \sum \chi_{c,w,h}(\epsilon), \quad (8)$$

which is the sum of rectangles determined by the centre position c, the width w and the height h (Fig. 1). The algorithm is based on minimizing the deviation between the input spectrum $I_{val}(\epsilon)$ and the approximated one $\tilde{I}_{val}(\epsilon) = \int_{-\infty}^{+\infty} H(\epsilon - \epsilon')\, \tilde{\rho}_{val}(\epsilon')\, d\epsilon'$ by generating stochastic configurations of various numbers of rectangles $\chi_{c,w,h}$ with the use of Monte Carlo and gradient descent schemes.

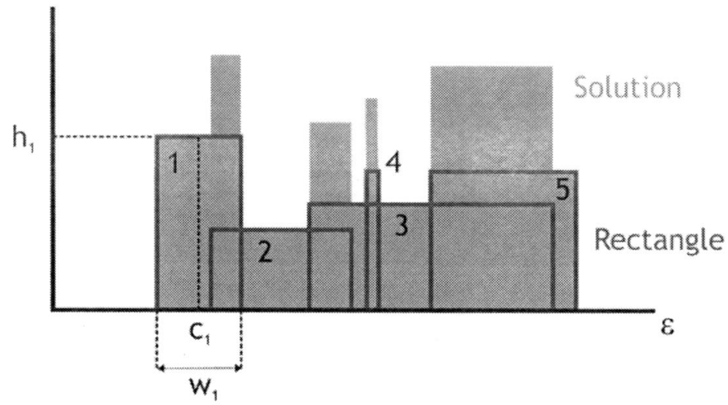

Figure 1. Approximation of the solution $\rho_{val}(\epsilon)$ (green) with the sum of random rectangles (red).

3. Results

The XPS spectra of gold, silver, and copper were taken from Research Data Australia [4]. The broadening function $H(\epsilon)$ was restored by fitting the XPS spectra of the core levels by the least square method in accordance with (1). For example, for gold the spectrum is the doublet $Au4f_{7/2,5/2}$ formed by the spin-orbit splitting. Its approximation gives the following values: $\gamma = 0.23$ eV; $\alpha = 0.04$ eV; $\sigma = 0.23$ eV (Fig. 2); the corresponding broadening function $H(\epsilon)$ is shown in Fig. 3. The same procedure was applied for calculation of the broadening functions of bulk silver and copper.

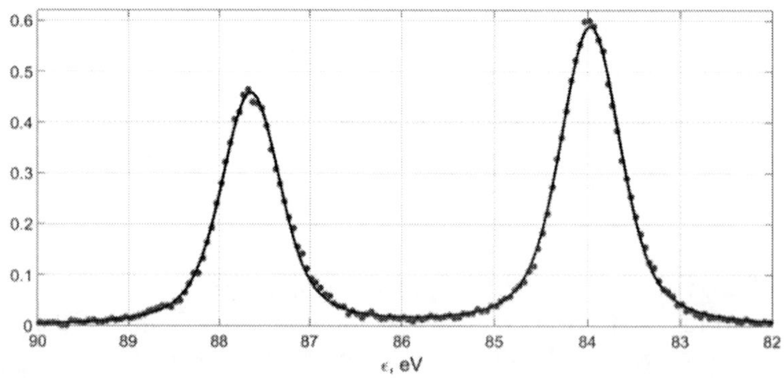

Figure 2. XPS spectrum of $Au4f_{7/2,5/2}$ doublet (black line) and its approximation (red dots).

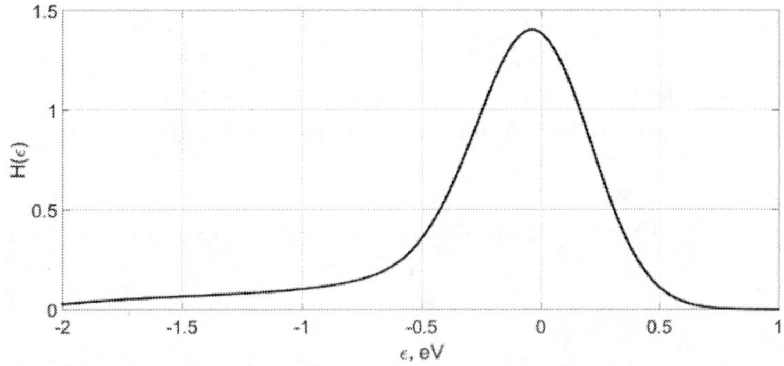

Figure 3. The restored broadening function $H(\epsilon)$ for bulk gold.

The obtained broadening functions $H(\epsilon)$ were used for the calculation of DOS in accordance with the system of integral equations (7). Figure 4 shows the calculated DOS in comparison with DOS obtained by DFT method, which was taken from Computational Electronic Structure Database (CompES-X) of National Institute for Materials Science, Japan [5]. It can be seen that the main features of DOS were correctly reproduced except sharp δ-function peaks in DOS of bulk gold, which are apparently caused by approximations used in DFT technique.

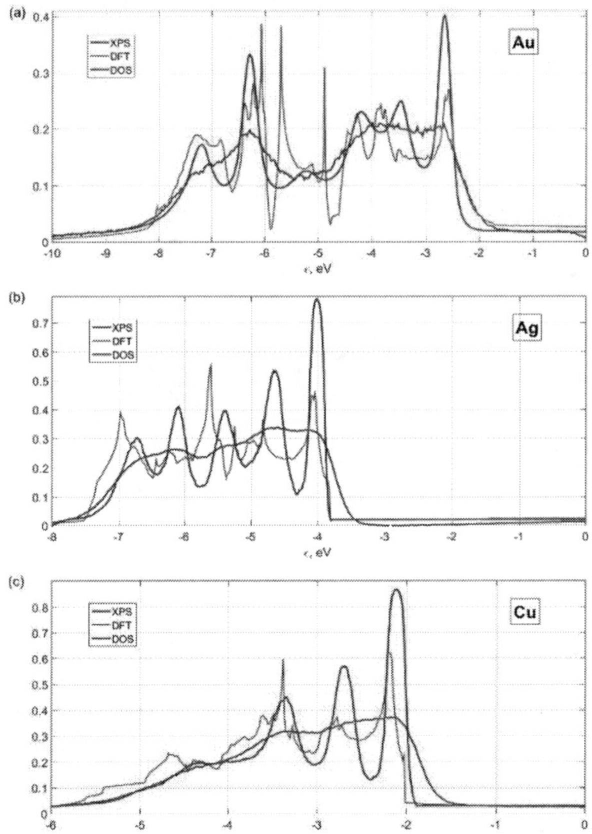

Figure 4. The XPS spectra of the valence band from the database [5] for bulk gold (a), silver (b), and copper (c), the restored DOS, and DOS calculated by DFT method.

4. Conclusions

In this work, we proposed the technique of restoring the electronic density of states from experimental XPS spectra. The method accurately takes into account the instrumentation broadening, and allows obtaining acceptable results even for low-quality XPS spectra. This technique may be applied to materials, which are complicated enough for DFT calculation, such as large low-symmetry systems, or to objects with unknown crystal structure, such as ensembles of nanoclusters.

Acknowledgements
The work was supported by Russian Found for Basic Research (project # 17-08-01021).

References

[1] Borman V D, Pushkin M A, Tronin V N, Troyan V I 2010 *J. Exp. Theor. Phys.* **110** 1005
[2] Anderson P W 1967 *Phys. Rev. Lett.* **18** 1049
[3] Mishenko A S, Prokof'ev N V, Sakamoto A, Svistunov B V 2000 *Phys. Rev. B.* **62** 6317
[4] researchdata.ands.org.au
[5] http://compes-x.nims.go.jp

IOP Conf. Series: Journal of Physics: Conf. Series **936** (2017) 012015 doi:10.1088/1742-6596/936/1/012015

Calculation of electron density of states for ensemble of gold nanoclusters

Ya V Zhumagulov, V A Kashurnikov, A V Krasavin

National Research Nuclear University MEPhI, 31 Kashirskoye shosse, 115409, Moscow, Russia

E-mail: yaroslav.zhumagulov@gmail.com

Abstract. The method was developed for calculation of electronic properties of an ensemble of nanoclusters with the use of cluster perturbation theory. This method was applied to the ensemble with various geometry of gold nanoclusters with octahedral symmetry. The dependence of the density of electronic states on the mutual spatial arrangement of nanoclusters in the ensemble is obtained. It is shown that the metallic behavior of the ensemble maintains even when strong disorder is present.

1. Introduction

Metal nanoclusters possess unique physical and chemical properties, which differ from the properties of both bulk metals and individual atoms. In recent years, nanoclusters have attracted increased attention due to numerous possible applications in various fields from biomedicine [1] to nanoelectronics [2]. Existing widely accepted numerical methods of studying the electronic structure of many-body systems face serious difficulties when considering ensembles of metal nanoclusters, such as the need to take into account a large number of elements of the system when applying the density functional theory (DFT), or the sign problem in quantum Monte Carlo (QMC) algorithms. Cluster perturbation theory (CPT), introduced in [3] and developed in [4], uses the wave function of a single nanocluster to model the wave function of the ensemble, and accounts for the interaction between nanoclusters as single-particle hopping terms. CPT was successfully used to calculate the band structure of NiO [5], to obtain the ground-state phase diagram of $t - t' - t'' - U$ Hubbard model describing high-temperature cuprate superconductors [6], to estimate hopping integrals in fullerenes [7].

The purpose of the paper is to apply CPT for calculating electronic properties of a finite system, namely, an ensemble of gold nanoclusters. Analogously to [4], the entire system (an ensemble of nanoclusters) is considered to be composed of a finite number of structure elements, i.e. single nanoclusters. Each nanocluster is big enough so that the exact diagonalization method cannot solve the problem for one cluster because of the large size of the Hilbert space, therefore, the Green's function for each nanocluster is calculated with the use of DFT technique. With the use of CPT, it became possible to model and investigate a realistic ensemble. The proposed technique is much less resource and time consuming compared to DFT and QMC.

2. Method

The ensemble of identical metal nanoclusters can be described by the following Hamiltonian,

$$H = H_0 + V, \tag{1}$$

where

$$H_0 = \sum_{R} H_{R}^{0}, \quad V = \sum_{\substack{R,R' \\ r,r'}} V_{rr'}^{RR'} \, a_{Rr}^{+} a_{R'r'}. \tag{2}$$

Here H_{R}^{0} is the Hamiltonian of a single nanocluster located at a radius-vector $\boldsymbol{R}$; V is the hopping term between nanoclusters; r is the combined index which counts the sites within a nanocluster and different orbitals on these sites.

The single nanocluster should be chosen so that it is possible to calculate its Green's function within DFT approach; the obtained Green's function has the form

$$G_{\nu\nu'}^{DFT}(\boldsymbol{k}, \omega) = \frac{\delta_{\nu\nu'}}{\omega + \mu - \varepsilon_{\nu}(\boldsymbol{k})}, \tag{3}$$

where $\varepsilon_{\nu}(\boldsymbol{k})$ is the band structure of the nanocluster; ν defines the bands; μ is the chemical potential.

To apply CPT, one should switch to the basis of localized orbitals,

$$G_{rr'}^{0}(\omega) = \sum_{k} \sum_{\nu\nu'} P_{r\nu}(\boldsymbol{k}, \omega) G_{\nu\nu'}^{DFT}(\boldsymbol{k}, \omega) P_{\nu' r'}^{*}(\boldsymbol{k}, \omega), \tag{4}$$

where $P_{n\nu}(\boldsymbol{k}, \omega)$ is the transition matrix; $G_{rr'}^{0}(\omega)$ is the Green's function of the single nanocluster.

At the first order of strong-coupling perturbation theory, the one-electron Green's function of the whole system can be written in matrix form as [4]

$$\boldsymbol{G}(\omega) = \frac{\boldsymbol{G}^{0}(\omega)}{1 - \boldsymbol{V}\boldsymbol{G}^{0}(\omega)}. \tag{5}$$

Here $\boldsymbol{G}(\omega) \equiv G_{rr'}^{RR'}(\omega)$; $\boldsymbol{V}$ is the hopping matrix between clusters, which can be expressed in quite simplified form with the use of only two parameters, the amplitude t_0 and characteristic length l_0, as

$$\boldsymbol{V} = V_{rr'}^{RR'}(\omega) = -t_0 \exp\left(-\frac{\boldsymbol{r}_r - \boldsymbol{r}_{r'}}{l_0}\right), \tag{6}$$

$\boldsymbol{r}_r \equiv \{\boldsymbol{R}, r\}$. The density of states of the ensemble is

$$A(\omega) = -\frac{1}{\pi} \mathrm{Im}\big(\boldsymbol{G}(\omega)\big). \tag{7}$$

3. Results

To apply the formalism introduced in the previous section, consider an ensemble of identical gold nanoclusters. Each nanocluster is consisted of 13 gold atoms, has the diameter of 5.66 Å, and possess octahedral symmetry (Fig. 1 (a)). Ab initio electronic structure calculations were performed within the DFT using very accurate full-potential linearized augmented planewave (FP-LAPW) approach as implemented in Wien2k code [8]. The Perdew, Burke, Ernzerhof (PBE) gradient corrected local density approximation (GGA) for the exchange correlation potential [9, 10] was used with 64 k-points in the first Brillouin zone. Then, $G_{rr'}^{0}(\omega)$ was obtained by dmftproj TRIQS package [11], which appends Wien2k code and builds the transition matrix elements from electronic bands obtained by DFT, to the localized orbitals. Since the properties of an electron system are determined by the density of states near the Fermi level, for gold we can confine ourselves to taking into account only the s- and d-orbitals, which give the main contribution. When the Green's function in localized orbitals basis is known, (7) allows us to obtain DOS of the single nanocluster, which is shown in Fig. 1 (b). Isolated bands and sharp peaks are clearly visible; these features are caused by finite and small size of the nanocluster, which lead to dimensional quantization.

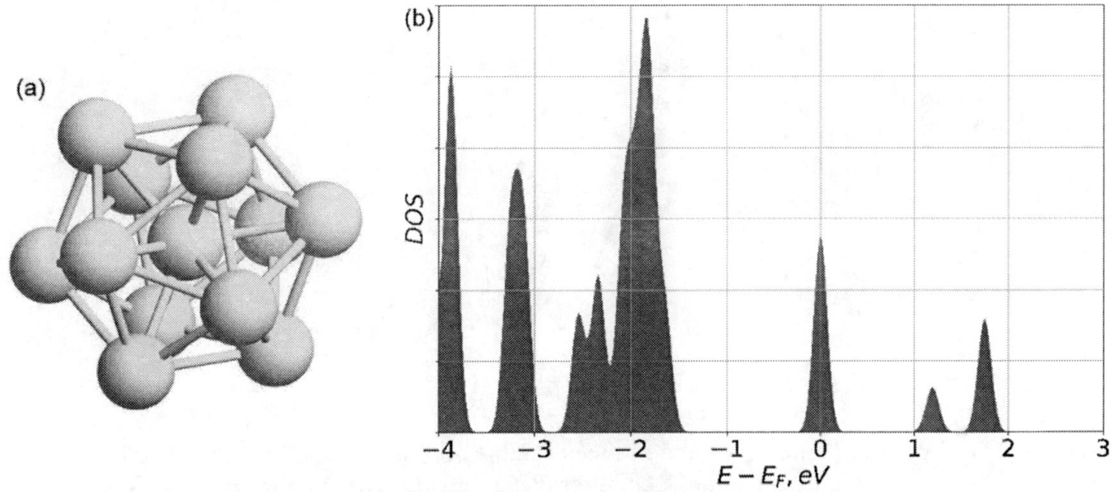

Figure 1. (a) Single nanocluster consisting of 13 gold atoms; (b) corresponding DOS calculated by DFT with the use of Wien2k code. Contributions of d-orbital at energies below −1 eV, and s-orbital at the Fermi level and upper, can be identified. This orbital structure of DOS is similar to the one of bulk gold, where deep levels are formed by d-orbital, and the band structure near the Fermi level is determined by s-orbital.

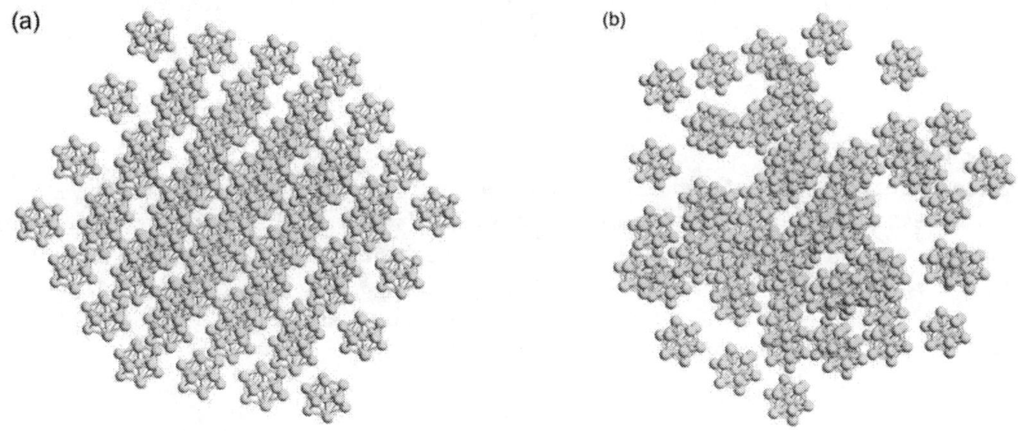

Figure 2. Ensemble of 64 nanoclusters located at equal (a) and random (b) distances from each other.

To model an ensemble, we first have constructed a regular cubic lattice consisting of 64 Au_{13} nanoclusters (Fig. 2 (a)). The distance between the centers of the clusters was equal to 10 Å. The density of states of such regular ensemble should have the same features as the DOS of the single nanocluster, except for the connections between the bands of dimensional quantization will appear. This can be clearly seen in Fig. 3 (blue line), as the DOS of the ensemble resembles the DOS of a bulk metal. To go further to more realistic situation, we have added the lattice disorder: the displacement of each nanocluster was produced by normal random generator with variance σ in the range from 0 (regular lattice) to 2Å. The configuration of the ensemble for σ = 2Å is shown at Fig. 2 (b).

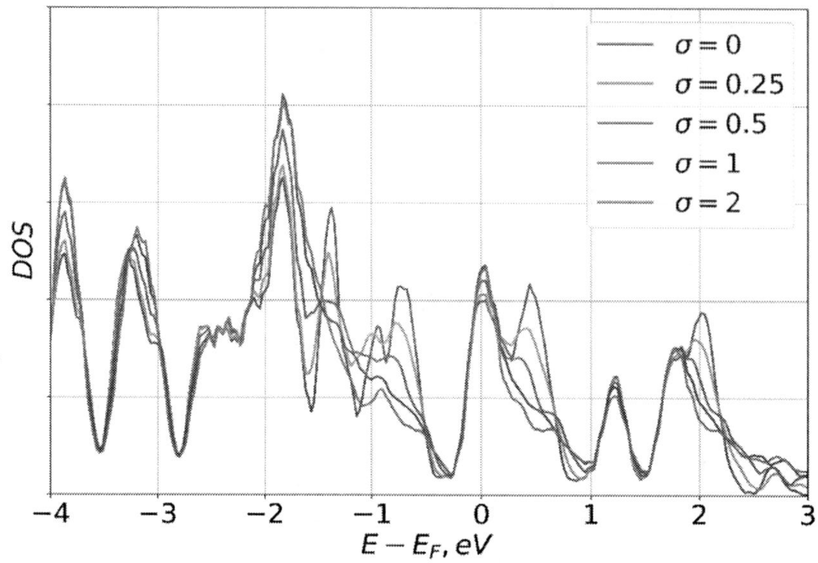

Figure 3. DOS of s-d-hybridized orbital of the ensemble of gold nanoclusters calculated by DFT with the use of Wien2k code for different values of variance σ.

Figure 3 shows the dependence of DOS of the ensemble on the variance σ. It can be seen that the value of disorder affects only those regions of DOS that have no contribution from dimensional quantization of single nanocluster. At the same time, with the increase of σ, the peaks in the region of Fermi energy disappear, and the tendency is observed of convergence of the DOS of the disordered ensemble to the DOS of single cluster. However, the initial density of the ensemble is sufficient to maintain the metallic behavior even when strong disorder is present.

4. Conclusions

In this paper, we have presented the method of calculating the density of electronic states of the ensemble of nanoclusters, which is based on the cluster perturbation theory. The method is computationally efficient since it required DFT calculation of only one nanocluster, and can be applied to numerous systems which are characterized by low symmetry and large sizes. This method will encounter difficulties when treating systems with complex hybridization, however, even in this case it will be able to obtain qualitatively satisfactory results.

Acknowledgements

The work was supported by Russian Found for Basic Research (projects # 14-08-00509 and # 15-02-02764).

References

[1] Xiao-Rong Song, Nirmal Goswami, Huang-Hao Yang and Jianping Xie 2016 *Analyst* **141** 3126

IOP Conf. Series: Journal of Physics: Conf. Series **936** (2017) 012015 doi:10.1088/1742-6596/936/1/012015

[2] Eva Koplin and Ulrich Simon 2008 Application of Metal Nanoclusters in Nanoelectronics *Metal Nanoclusters in Catalysis and Materials Science: The Issue of Size Control* Ed B Corain, G Schmid and N Toshima (Amsterdam: Elsevier) chapter 5 pp 107-128

[3] Sénéchal D, Perez D, and Pioro-Ladrière M 2000 *Phys. Rev. Lett.* **84** 522

[4] Sénéchal D, Perez D, and Plouffe D 2002 *Phys. Rev. B* **66** 075129

[5] Eder R, Dorneich A, and Winter H 2005 *Phys. Rev. B* **71** 045105

[6] Sénéchal D, Lavertu P-L, Marois M-A, and Tremblay A-M S 2005 *Phys. Rev. Lett.* **94** 156404

[7] Lin F, Sørensen E S, Kallin C, and Berlinsky A J 2007 *Phys. Rev. B* **75** 075112

[8] Blaha P, Schwarz G, Madsen D, Kvasnika K, and Luitz J 2001 *Wien2k* Technical University Wien, Austria

[9] Perdew J P, Burke K, and Ernzerhof M 1996 *Phys. Rev. Lett.* **77** 3865

[10] Perdew J P, Burke K, and Wang Y 1996 *Phys. Rev. B* **54** 16533

[11] Parcollet O, Ferrero M, Ayral T, Hafermann H, Krivenko I, Messio L, and Seth P 2015 *Comp. Phys. Comm.* **196** 398

IOP Conf. Series: Journal of Physics: Conf. Series **936** (2017) 012016 doi:10.1088/1742-6596/936/1/012016

First-order transition in XY model with higher-order interactions

Milan Žukovič

Institute of Physics, Faculty of Science, P.J. Šafárik University, Park Angelinum 9, 041 54 Košice, Slovakia

E-mail: milan.zukovic@upjs.sk

Abstract. The effect of inclusion of higher-order interactions in the XY model on critical properties is studied by Monte Carlo simulations. It is found that an increasing number of the higher-order terms in the Hamiltonian modifies the shape of the potential, which beyond a certain value leads to the change of the nature of the transition from continuous to first order. The evidence for the first-order transition is provided in the form of the finite-size scaling and the energy histogram analysis. A rough phase diagram is presented as a function of the number of the higher-order interaction terms.

1. Introduction

The XY spin model is well known to undergo a topological Berezinskii-Kosterlitz-Thouless (BKT) phase transition to a low-temperature quasi-long-range-order (QLRO) phase characterized by an algebraically decaying correlation function [1, 2]. The standard XY model only includes bilinear pairwise couplings between neighboring spins. Nevertheless, it has been shown that higher-degree pair interactions between magnetic ions may be comparable or even stronger than the bilinear ones (see, e.g., Refs. [3, 4], and references therein) and thus can play an important role in the system's behavior. They turned out useful in modeling of certain systems, such as liquid crystals [5, 6], superfluid A phase of ^{3}He [7], and high-temperature cuprate superconductors [8]. These findings resulted in an increased interest in the study of the model with higher-order interactions. The model involving terms up to the p-th order can be described by the Hamiltonian

$$\mathcal{H} = -\sum_{k=1}^{p} J_k \sum_{\langle i,j \rangle} (\mathbf{S_i S_j})^k, \tag{1}$$

where J_k are the k-th order exchange coupling constants, $\mathbf{S_i} = (S_i^x, S_i^y)$ is a two-dimensional unit vector at site i, and $\langle i,j \rangle$ denotes the sum over nearest-neighbor spins. In the present study we will assume that all the couplings are of equal strength and normalized, i.e., $J_k \equiv J = 1/p$. Then, the Hamiltonian can be rewritten in the form

$$\mathcal{H} = -J \sum_{\langle i,j \rangle} \frac{\cos\theta_{i,j}(1 - \cos^p \theta_{i,j})}{1 - \cos\theta_{i,j}}, \tag{2}$$

where $\theta_{i,j} = \theta_i - \theta_j$ is an angle between the nearest-neighbor spins.

The most studied was the bilinear-biquadratic model, i.e., the above model for $p = 2$ [5, 7, 9, 10, 11, 12]. Inclusion of a sufficiently large biquadratic coupling has been shown to lead to the separation of the quadrupole phase at higher temperatures from the dipole phase at lower temperatures. Nevertheless, the character of the phase transition to the paramagnetic phase, either from the dipole or quadrupole phases, was confirmed to belong to the BKT universality class, while the dipole-quadrupole phase transition was of the two-dimensional Ising universality class.

Motivated by orientational transitions in liquid crystals, the inclusion of the higher-order terms has also been taken in the form of the the k-th order Legendre polynomial of the bilinear term [13, 14, 15]. Although in models with $O(n)$ symmetry for $n \geq 2$ a crossover to the first order transition is expected for large enough values of k [16, 17], for the case of $O(2)$ and the studied values of $k = 2$ and 4 no such crossover has been confirmed.

Domany *et al.* [18] have introduced a modified XY model that enabled tuning its critical properties through the parameter p^2 between the standard XY model belonging to the BKT universality and the q-state Potts model. The latter is known to show a first-order transition for large q and a crossover to the first-order was also confirmed for large p^2 in this modified XY model [19, 20, 21, 22].

In this study we generalize the XY model by including up to the p-th order isotropic pairwise interactions to the bilinear term. We show that, in spite of belonging to the same universality class as the standard XY model, the generalized model changes the character of the transition to the paramagnetic phase to the first-order one for sufficiently large p.

2. Method

We use Monte Carlo (MC) simulations for spin systems on a square lattice of a size $L \times L$, with $L = 24 - 120$ and periodic boundary conditions. The spin update follows the Metropolis algorithm with the following parameters: the initial 4×10^4 MC sweeps are discarded for thermalization and the following 2×10^5 MC sweeps are used for the calculation of thermal averages. Temperature dependencies of various measured quantities are obtained by starting the simulations at sufficiently high temperatures T (measured in units J/k_B, where k_B is the Boltzmann constant), corresponding to the paramagnetic phase. Then the temperature is gradually decreased by $\Delta T = 0.025$ and the simulation at $T - \Delta T$ is initialized using the last configuration obtained at T.

In the vicinity of the transition to the paramagnetic phase we perform a finite-size scaling (FSS) analysis by using the reweighting techniques [23, 24], to convincingly identify the order of the transition. For the reweighting we use much longer simulation runs, including up to 10^7 MC sweeps.

Two basic quantities are measured: the internal energy per spin

$$e = \langle \mathcal{H} \rangle / L^2 \tag{3}$$

and the magnetization per spin

$$m = \langle M \rangle / L^2 = \left\langle \Big| \sum_{j=1}^{L^2} \exp(i\theta_j) \Big| \right\rangle / L^2. \tag{4}$$

To determine the transition order we also evaluate the magnetic susceptibility

$$\chi = \frac{\langle M^2 \rangle - \langle M \rangle^2}{L^2 T}, \tag{5}$$

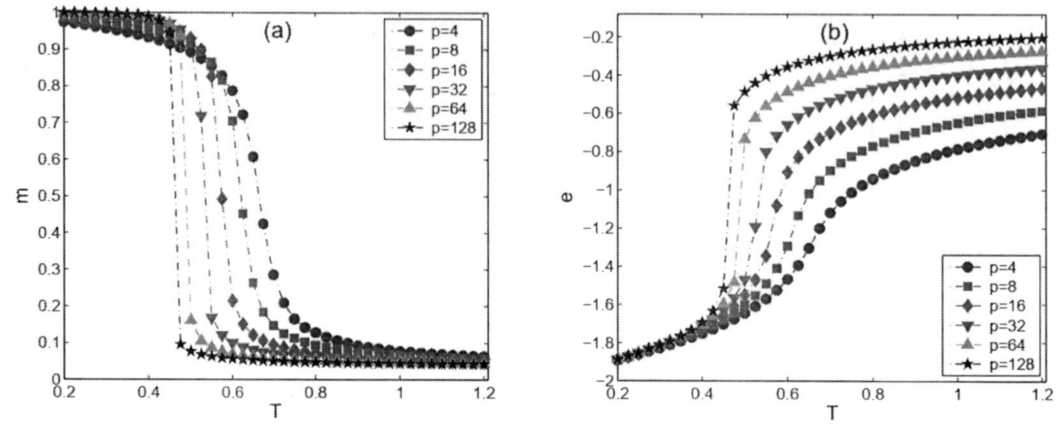

Figure 1. Temperature dependence of (a) the magnetization and (b) the internal energy, for $L = 24$ and several values of p.

and the fourth-order magnetic cumulant

$$U = 1 - \frac{\langle M^4 \rangle}{3\langle M^2 \rangle^2}. \tag{6}$$

If the transition is of first order, then at the transition point the internal energy e and the magnetization m will show a discontinuous behavior, the thermodynamic functions like the susceptibility χ will scale with volume, i.e., $\chi(L) \propto L^2$, and the fourth-order magnetic cumulant U will show an abrupt dip to negative values [25].

3. Results

Temperature dependencies of the basic quantities, i.e., the internal energy and the magnetization, are presented in Figs. 1(a) and 1(b), for $L = 24$ and the increasing number of the higher-order interactions $p = 2, 8, 16, 32, 64$ and 128. One can observe a transition between the high-temperature paramagnetic phase, characterized by $m \approx 0$ [1], and the low-temperature QLRO phase with $m > 0$. As p increases, close to the transition point the behavior of both quantities changes from a slow variation (similar to the XY model) to more abrupt changes. For $p = 64$ and 128, the variation appears clearly discontinuous, as in the case of first-order transitions.

In order to make certain that the apparent discontinuity in the magnetization and energy is not an artifact of the finite system size, we additionally performed very long runs (10^7 MC sweeps) close to the transition points and calculated energy histograms. As evidenced from Fig. 2, already for $p = 64$ the energy distributions shows a bimodal character that is characteristic for a discontinuous first-order transition. The histograms for different sizes L are reweighted to the temperatures at which both peaks are of equal height. These temperatures correspond to the pseudo-transition temperatures $T_c(L)$ and the true transition temperature can be obtained by extrapolation to the thermodynamic limit, i.e., for $L \to \infty$. The scaling properties of the energy distribution are in accordance with what we expect from the first-order transition. Namely, as L increases the heights of the peaks increase at the cost of the dip (barrier) between them. The latter tends to zero and the distance between the peaks approaches a finite value, corresponding

[1] $m = 0$ is only expected in the thermodynamic limit of $L \to \infty$.

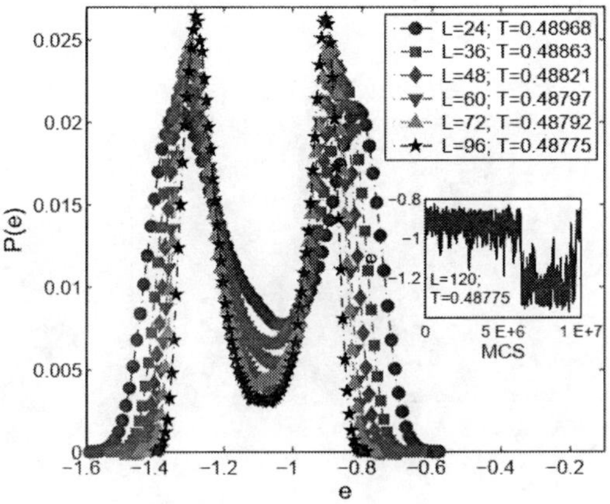

Figure 2. Energy histograms at the pseudo-transition points, for $p = 64$ and various L. The inset demonstrates enormous tunneling times between the coexisting states, for $L = 120$.

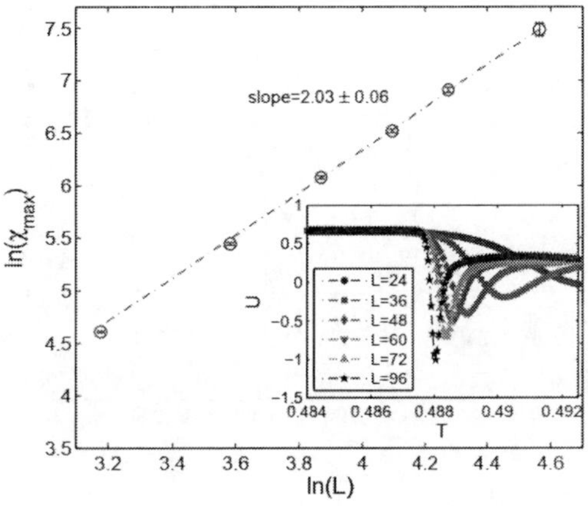

Figure 3. FSS of the magnetic susceptibility at the transition point, for $p = 64$ and various L. The inset shows the temperature dependence of the fourth-order magnetic cumulant.

to the latent heat released at the discontinuous transition. We note that the histogram for the largest considered lattice size $L = 120$ is not presented due to enormous tunneling times (see the inset), which, for the considered simulation times, prevent a reliable histogram estimation.

To obtain an additional evidence for the first-order transition scenario, we also perform FSS analysis and inspect the behavior of the fourth-order magnetic cumulant. As presented in Fig. 3, the magnetic susceptibility is confirmed to scale with volume, as it should be in the case of a first-order transition. The inset of the same figure shows the temperature variation of the fourth-order magnetic cumulant. An abrupt descent to negative values at the transition point corroborates the above findings and provides further support for the first-order transition scenario.

As can be predicted from Fig. 1, for larger p the first-order transition features become even more pronounced. The energy barrier between the two coexisting phases becomes deeper and the gap between the peak's positions becomes wider (not shown). On the other hand, the case of $p = 1$ corresponds to the standard XY model showing a continuous phase transition. Therefore, we expect the existence of a value of $1 < p_c < 64$ that would correspond to the tricritical point at which the transition changes its nature between the first and second order. A more precise location of p_c may not be trivial due to possible problems with a reliable determination of the transition order in the crossover region (near the expected tricritical point) and we leave this task for further investigation.

An approximate phase diagram in $T - p$ parameter plane is depicted in Fig. 4. Rough estimates of (pseudo)transition temperatures are obtained as positions of peaks of the magnetic susceptibility, for a fixed lattice size $L = 24$. The filled circles represent the first-order transition points (for $p \geq 64$). We note that these pseudo-transition temperatures slightly overestimate the true thermodynamic limit values, as also evidenced in the XY model (see, e.g., Ref. [15]). Overall, the increasing p shifts the transition temperature between the paramagnetic and the QLRO phases to lower values and eventually also changes its nature to the first-order one.

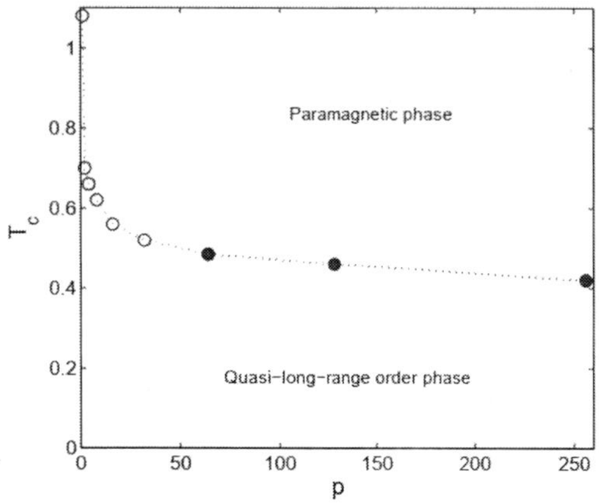

Figure 4. Phase diagram. The filled symbols show first-order transition points and the dotted line is a guide to the eye.

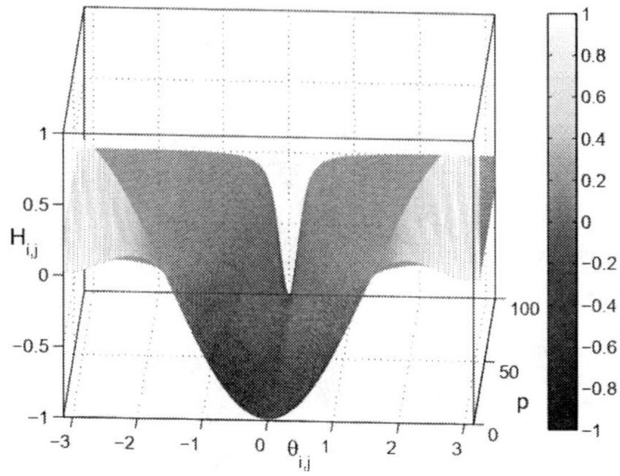

Figure 5. Potential well as a function of the neighboring spin angle $\theta_{i,j}$ and the number of higher-order interaction terms p.

4. Discussion

Finally, let us briefly discuss the origin of the crossover to the first-order transition regime. We believe that the mechanism leading to the first-order transition is similar to that observed in the nonlinear model introduced by Domany *et al.* [18]. It was concluded that the change in the nature of the transition from continuous to first order in that model resulted from an increase of the nonlinearity parameter p^2, which changed the shape of the potential. In the present model, the nonlinearity parameter p represents the number of the higher-order interaction terms, nevertheless, its effect on the potential is similar. The dependence of the potential $H_{i,j} = \cos\theta_{i,j}(1-\cos^p\theta_{i,j})/[p(1-\cos\theta_{i,j})]$ on the parameter p is depicted in Fig. 5. Like in the Domany's model, with the increasing p the cosine shape of the potential well for $p = 1$ (the standard XY model) becomes more and more narrow with a width tending to zero. The narrowing of the potential well at low temperatures can be ascribed to insufficient increase in the vertex density followed by their abrupt (discontinuous) proliferation at higher temperatures, leading to a first-order transition. A more detailed study of the vertex formation will be presented elsewhere.

Acknowledgments

This work was supported by the Scientific Grant Agency of Ministry of Education of Slovak Republic (Grant No. 1/0331/15) and the scientific grants of Slovak Research and Development Agency provided under contract No. APVV-0132-11 and No. APVV-14-0073.

References

[1] Berezinskii V I 1971 *Sov. Phys. JETP* **34** 610
[2] Kosterlitz J M and Thouless D J 1973 *J. Phys. C* **6** 1181
[3] Chiu-Tsao S T, Levy P M and Paulson C 1975 *Phys. Rev. B* **12** 1819
[4] Fedders P A and Miles C W 1979 *Phys. Rev. B* **19** 1331
[5] Lee D H and Grinstein G 1985 *Phys. Rev. Lett.* **55** 541
[6] Geng J and Selinger J V 2009 *Phys. Rev. E* **80** 011707
[7] Korshunov S E 1985 *JETP Lett.* **41** 263
[8] Hlubina R 2008 *Phys. Rev. B* **77** 094503
[9] Carpenter D B and Chalker J T 1989 *Journal of Physics: Condensed Matter* **1** 4907
[10] Shi Y, Lamacraft A and Fendley P 2011 *Phys. Rev. Lett.* **107** 240601

[11] Hübscher D M and Wessel S 2013 *Phys. Rev. E* **87** 062112
[12] Qi K, Qin M H, Jia X T and Liu J-M 2013 *J. Magn. Magn. Mater.* **340** 127
[13] Fariñas-Sánchez A I, Paredes R and Berche B 2005 *Phys. Rev. E* **72** 031711
[14] Berche B and Paredes R 2005 *Condensed Matter Physics* **8** 723
[15] Fariñas-Sánchez A I, Botet R, Berche B and Paredes R 2010 *Condensed Matter Physics* **13** 13601
[16] van Enter A C D and Shlosman S B 2002 *Phys. Rev. Lett.* **28** 285702
[17] van Enter A C D and Shlosman S B 2005 *Comm. Math. Phys.* **255** 21
[18] Domany E, Schick M and Swendsen R H 1984 *Phys. Rev. Lett.* **52** 1535
[19] Van Himbergen J E 1984 *Phys. Rev. Lett.* **53** 5
[20] Blöte H W J, Guo W and Hilhorst H J 2002 *Phys. Rev. Lett.* **88** 047203
[21] Sinha S and Roy S K 2010 *Phys. Rev. E* **81** 022102
[22] Sinha S and Roy S K 2010 *Phys. Rev. E* **81** 041120
[23] Ferrenberg A M and Swendsen R H 1988 *Phys. Rev. Lett.* **61** 2635
[24] Ferrenberg A M and Swendsen R H 1989 *Phys. Rev. Lett.* **63** 1195
[25] Tsai S and Salinas S 1998 *Brazilian journal of physics* **28** 58

Generalized logistic map and its application in chaos based cryptography

M Lawnik
Faculty of Applied Mathematics, Silesian University of Technology, ul. Kaszubska 23, 44-100 Gliwice, Poland

E-Mail: marcin.lawnik@polsl.pl

Abstract. The logistic map is commonly used in, for example, chaos based cryptography. However, its properties do not render a safe construction of encryption algorithms. Thus, the scope of the paper is a proposal of generalization of the logistic map by means of a well-recognized family of chaotic maps. In the next step, an analysis of Lyapunov exponent and the distribution of the iterative variable are studied. The obtained results confirm that the analyzed model can safely and effectively replace a classic logistic map for applications involving chaotic cryptography.

1. Introduction

Chaotic maps are widely used in many science branches, e.g. [1-4]. One of the most commonly used chaotic dynamical system in practice is the logistic map (LM) given by the following expression:

$$x_{k+1} = 4qx_k(1-x_k), \tag{1}$$

where $q \in [0,1]$ and $x \in [0,1]$. Its universal nature is certified by many applications, among others, in cryptography based on chaos theory, e.g. [5,6]. Chaos based cryptography regards chaotic maps as generators of pseudo-random numbers, and the values of their initial conditions and parameters as the secret keys. From such point of view, recurrence (1) has many disadvantages, which were indicated in professional publications [7]. They include, among others, a small space of allowable values of parameter q, not uniform distribution of the iterative variable, or unstable value of Lyapunov exponent. The Lyapunov exponent denotes the sensitivity of dynamical system $x_{k+1} = f(x_k)$ to changes in the initial conditions and is calculated in accordance with the dependence:

$$\lambda = \lim_{n\to\infty} \frac{1}{n} \sum_{k=0}^{n-1} \ln|f'(x_k)|. \tag{2}$$

In the above-mentioned applications, it is definitely recommended that the values of parameter q of (1) are close to 1. This results from its distribution, which for such set of values is flat in the middle, and determined by the density function:

$$\rho(x) = \frac{1}{\pi\sqrt{x(1-x)}} \tag{3}$$

Furthermore, for $q = 1$ the Lyapunov exponent for (1) reaches the maximum value equal to $\lambda = \ln 2$ among all allowable values of parameter q. The above results show that the admissible values of parameter q are considerably limited. This has a negative impact on the safety of algorithms and may lead to successful brutal attack. In view of this, the designation of a chaotic map that could combine the simplicity of (1) and , at the same time, render much better properties, seems relevant.

In [8] the following family of chaotic functions was presented:

$$x_{k+1} = \begin{cases} q\left(1-\left|\dfrac{p-x_k}{p}\right|^r\right), & 0 < x < p \\ q\left(1-\left|\dfrac{p-x_k}{1-p}\right|^r\right), & p \le x < 1 \end{cases}, \tag{4}$$

where p, q, r represents parameters.

If $p=0.5$, $r=2$ and $q \in [0,1]$, map (4) may be reduced to LM (1). Thus, LM is a special case of (4). In [8] only for selected values of parameters p, q and r an casual analysis of Lyapunov exponent of (4) was presented, indicating the domains for which it has a positive value. From the point of view of above mentioned applications, such analysis is insufficient.

2. The generalized logistic map

The generalized logistic map (GLM) can be given by the formula:

$$x_{k+1} = \begin{cases} \dfrac{-q}{p^2}(p-x_k)^2 + q, & 0 \le x_k \le p \\ \dfrac{-q}{(1-p)^2}(p-x_k)^2 + q, & p < x_k \le 1 \end{cases}, \tag{5}$$

where $p \in (0,1)$ and $q \in [0,1]$.

GLM (5) is a specific case of (4) for the parameter value $r = 2$.

3. Analysis

In figure1, the graph of Lyapunov exponent for the values of parameters p and q is presented. As shown in the graph, a large range of the domain $[0.89,1] \times [0,1]$ has a positive value of Lyapunov exponent. Furthermore, it may be reasoned that for the value of parameter q close to 1 and the entire variation range of parameter p, the GLM (5) has a positive Lyapunov exponent, as shown in detail in figure 2. Accordingly, the analysis of Lyapunov exponent for this map renders a considerably wider range of the allowable values in comparison with a LM (1).

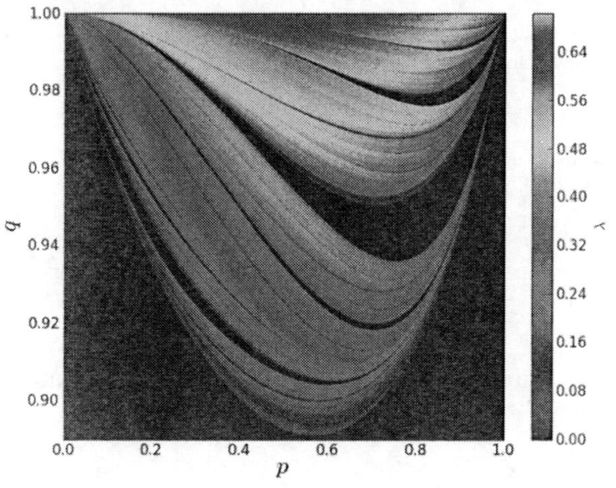

Figure 1. Graph of the Lyapunov exponent of GLM (5). The non-positive value of the exponent is marked in black.

As indicated in figure 2 the value of Lyapunov exponent is stable for parameter q close to 1. This means that Lyapunov time, which determines the number of iterations that the system requires to forget the initial condition, is also stable. In applications, Lyapunov time determines how many initial iterations of the system must be rejected, in order to obtain a chaotic solution, which has also an impact on the calculations effectiveness of the algorithms involving cryptography.

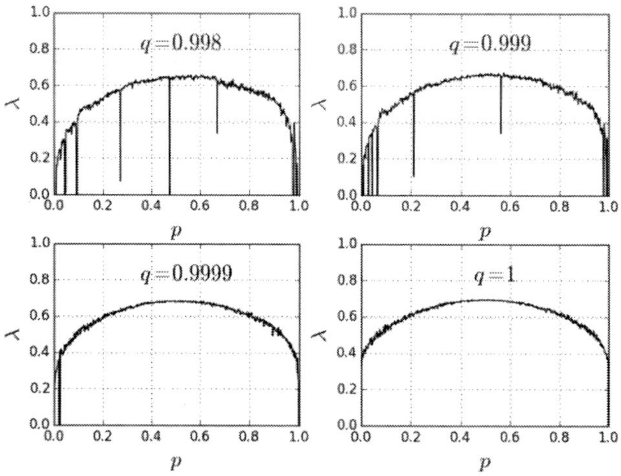

Figure 2. The graph of the Lyapunov exponent of the GLM (5) for values of parameter q close to 1.

The distribution of the iterative variable for the value of parameter q close to 1 is flat in the middle (figure 3), just as in a LM (1). However, in the discussed example, the distribution is the same for the entire variation of parameter p, and not only for one value. This significantly extends the range of the allowable parameters, and, at the same time, enhances the safety of the algorithm - see figure 4.

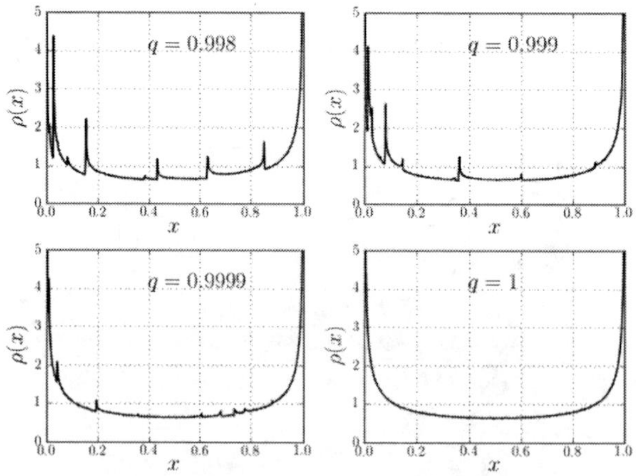

Figure 3. Distributions of the GLM (5) obtained in a numerical way for values of parameter q close to 1 and $p = 0.4$.

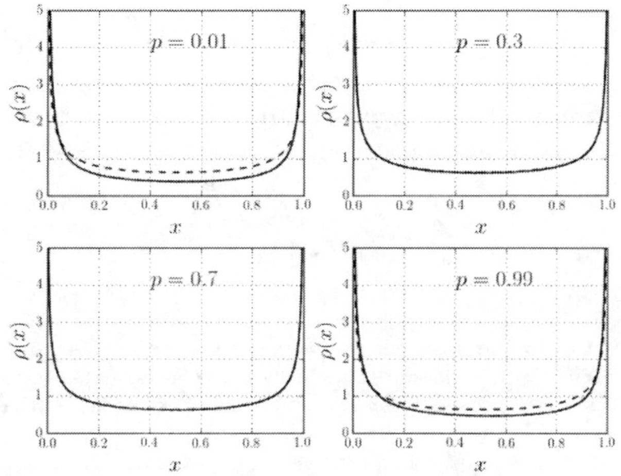

Figure 4. Distribution of GLM (5) obtained in a numerical way for set values of parameter p and $q = 1$ (continuous line). The broken line marks density function (3).

4. Conclusions

The analysis of the generalized logistic map with the use of function family (4) was conducted. The presented calculations indicated that the analyzed model has much better properties than a classic logistic map, such as wider range of the allowable parameters with a positive value of Lyapunov exponent and flat distribution of the iterative variable in the middle part. From the point of view of cryptography based on the chaos theory, the properties of the analyzed function provide its successful application. On the other hand, the analyzed model combines the simplicity of its classic correspondent. This should lead to the replacement of a logistic map by its generalized form in cryptography applications.

References

[1] Berezowski M and Lawnik M 2014 Identification of fast-changing signals by means of adaptive chaotic transformations *Nonlinear Analysis: Modelling and Control* **19(2)** 172–177

[2] Lawnik M and Berezowski M 2014 Identification of the oscillation period of chemical reactors by chaotic sampling of the conversion degree *Chem. Process Eng.* **35(3)** 387–393

[3] Lawnik M 2014 Generation of numbers with the distribution close to uniform with the use of chaotic maps *Proceedings of the 4th International Conference on Simulation and Modeling Methodologies, Technologies and Applications* pp 451–455

[4] Lawnik M 2014 The approximation of the normal distribution by means of chaotic expression *Journal of Physics: Conference Series* **490** 012072

[5] Baptista M S 1998 Cryptography with chaos *Physics Letters A* **240** 50–54

[6] Pareek N K, Vinod Patidar and Sud K K 2006 Cover image Image encryption using chaotic logistic map *Image and Vision Computing* **24(9)** 926–934

[7] Arroyo D, Alvarez G and Fernandez V 2008 *On the inadequacy of the logistic map for cryptographic applications* eds. L Hernandez A Martin X Reunin (Espa˜nola sobre Criptolog´ıa y Seguridad de la Informaci´on) pp 77–82

[8] Skrobek A 2007 *Metoda projektowania dyskretnych chaotycznych szyfrów strumieniowych oparta na kryptoanalizie* Phd thesis (in polish)

IOP Conf. Series: Journal of Physics: Conf. Series **936** (2017) 012018 doi:10.1088/1742-6596/936/1/012018

Magnetic reconnection in terms of catastrophe theory

E.Y.Echkina, I.N.Inovenkov and V.V. Nefedov

Department of Computational Mathematics and Cybernetics, Moscow State University, Moscow, Russia

E-mail: `ejane@cs.msu.su`

Abstract. Magnetic field line reconnection (magnetic reconnection) is a phenomenon that occurs in space and laboratory plasma. Magnetic reconnection allows both the change the magnetic topology and the conversion of the magnetic energy into energy of fast particles. The critical point (critical line or plane in higher dimensional cases) of the magnetic field play an important role in process of magnetic reconnection, as in its neighborhood occurs a change of its topology of a magnetic field and redistribution of magnetic field energy. A lot of literature is devoted to the analytical and numerical investigation of the reconnection process. The main result of these investigations as the result of magnetic reconnection the current sheet is formed and the magnetic topology is changed. While the studies of magnetic reconnection in 2D and 3D configurations have a led to several important results, many questions remain open, including the behavior of a magnetic field in the neighborhood of a critical point of high order. The magnetic reconnection problem is closely related to the problem of the structural stability of vector fields. Since the magnetic field topology changes during both spontaneous and induced magnetic reconnection, it is natural to expect that the magnetic field should evolve from a structurally unstable into a structurally stable configuration. Note that, in this case, the phenomenon under analysis is more complicated since, during magnetic reconnection in a highly conducting plasma, we deal with the non-linear interaction between two vector fields: the magnetic field and the field of the plasma velocities. The aim of our article is to consider the process of magnetic reconnection and transformation of the magnetic topology from the viewpoint of catastrophe theory. Bifurcations in similar configurations (2D magnetic configuration with null high order point) with varying parameters were thoroughly discussed in a monograph by Poston and Stewart.

1. Introduction

Magnetic field line reconnection (magnetic reconnection) is a phenomen in space and laboratory plasmas. Magnetic reconnection allow both the change the magnetic topology and the conservation of the magnetic energy into energy of fast particles [3], [1],[4].

The critical point (critical line or plane in higher dimensional case) of the magnetic field play a important role in process of magnetic reconnection, because in their neighborhood there is an evolution of topology of a magnetic field and redistribution of magnetic field energy.

A lot of literature is devoted to he analytical and numerical inverstigation of the reconnection process [5]-[8].The main result of these investigations is that the current sheet is formed as the result of magnetic reconnection and the magnetic topology is changed.

Althogh the studies of magnetic reconnection in 3D and 2D configurations have a led to a number of important results, many questions remain open. One of these questions is the behavior of a magnetic field in the vicinity of a critical point of a high order at the selfevolution.

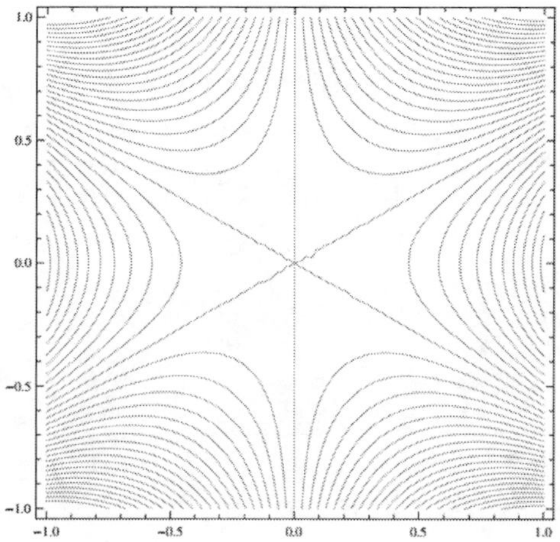

Figure 1. The magnetic configuration with third-order critical point (1)

The magnetic reconnection problem is closely related to the problem of the structural stability of vector fields. The mathematical aspects of the latter problem were discussed by Arnold [9] , who defined a structurally stable dynamic system as a system whose state remains topologically equivalent to the initial (unperturbed) state for any sufficiently small perturbation of the vector field. Since the magnetic field topology changes during both spontaneous and induced magnetic reconnection, it is natural to expect that the magnetic field should evolve from a structurally unstable into structurally stable configuration. Note that, in this case, the phenomenon under analysis is more complicated because, during magnetic reconnection in a highly conducting plasma, we deal with the nonlinear interaction between two vector fields: the magnetic field and the field of the plasma velocities. This circumstance significantly complicates the problem in comparison with the case of one vector field. The simplest example of a structuturally unstable solenoidal vector field is a magnetic field with third-order null lines in 2D space

$$\mathbf{B} = 2xy\mathbf{i} + (x^2 + y^2)\mathbf{j} + 0\mathbf{k} \tag{1}$$

The magnetic configuration described by the vector potential (1) is structurally unstable because the critical point $(x = 0, y = 0)$ of the vector field is degenerate. Small perturbations of the magnetic field described by the vector potential (1)) cause the degenerate singular point to bifurcate or disappear, in which case the perturbed magnetic field can be expressed in terms of the perturbed the magnetic filed

$$\mathbf{B} = (2xy + 2\gamma y + \delta x + \mu)\mathbf{i} + (x^2 + y^2 + 2\beta + \delta y + \epsilon)\mathbf{j} + 0\mathbf{k} \tag{2}$$

where the parameters $\beta, \gamma, \epsilon, \mu, \delta$ are assumed to be small. Bifurcations in similar configurations with var ing parameters were thoroughly discussed in a monograph by Poston and Stewart [10].

By choosing the appropriate parameters, one can make the magnetic configuration (2) physically stable but structurally unstable. Consequently, this configuration is suitable for analyzing magnetic reconnection in structurally unstable systems. In formulating the problem, we assume that the magnetic field topology changes under the action of the perturbations excited at the boundary of the region under consideration. In other words, we consider the regime of driven magnetic reconnection.

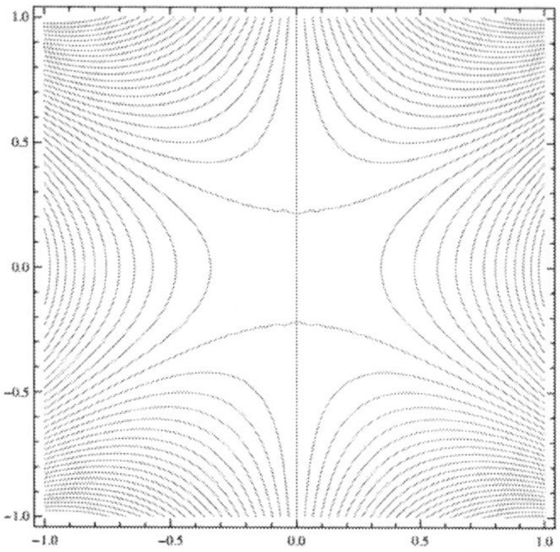

Figure 2. The magnetic configuration (2) with $\epsilon = 0.05$ another parameters is equal to zero

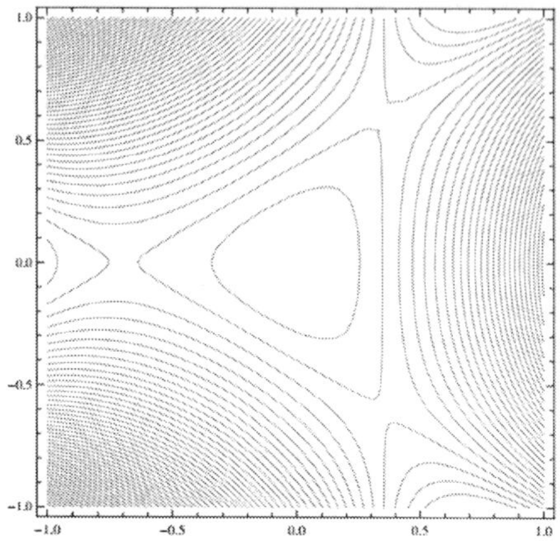

Figure 3. :The magnetic configuration (2) with $\beta = \gamma = 0.35$ another parameters is equal to zero.

In this paper we present the results of 3D and 2D computing simulation of the process of transformation of structurally unstabe into structurally one and formation of the regular current structures.

The paper is organized as follows. The mathematical model are described in Section 2. In section 3 we presented our numerical results. Comparison of results of the accidents received by means of computing experiment with catastrophe theory will be given in section 4. The numerical methods are defined in the Appendix Section.

2. Mathematical model

We inverstigate the plasma evolution using the set of MHD equations. We write the MHD equations in a form used in Ref. [8]

$$\frac{\partial \rho}{\partial t} + \nabla \cdot (\rho \mathbf{v}) = 0 \tag{3}$$

$$\rho \frac{\partial \mathbf{v}}{\partial t} = -\frac{\beta}{2} \nabla p + [(\nabla \times \mathbf{B}) \times \mathbf{B}] \tag{4}$$

$$\frac{\rho}{\gamma - 1} \frac{dT}{dt} + p \nabla \cdot \mathbf{v} = -k \Delta T + \frac{2\nu_m}{\beta} (\nabla \times \mathbf{B})^2 \tag{5}$$

$$p = \rho T \tag{6}$$

$$\frac{\partial \mathbf{B}}{\partial t} = \nabla \times (\mathbf{v} \times \mathbf{B}) - \nu_m \nabla \times \nabla \times \mathbf{B} \tag{7}$$

Here ρ, p, T correspond to density, pressure, and temperature of plasma. We probe the folllowing value of parameters abiabatic index $\gamma = 5/3$, magnetic velosity $\nu_m = c^2/4\pi s v_a \sigma$=0.006, σ - plasma conductivity, $\beta = 8\pi\rho_0/B_0^2 = 0.012$ is the ratio of the plasma pressure to the magnetic pressure on the boundary.

At the initial times the electric current density is zero and plasma is equilibrium state ($\mathbf{v}(\mathbf{0}) = \mathbf{0}$) with th constant density $\rho(0) = 1$, pressure $p(0) = 1$.

We consider the initially magnetic configuration with third-order critical point

$$\mathbf{B} = 2xy\mathbf{i} + (x^2 + y^2)\mathbf{j} + 0\mathbf{k} \tag{8}$$

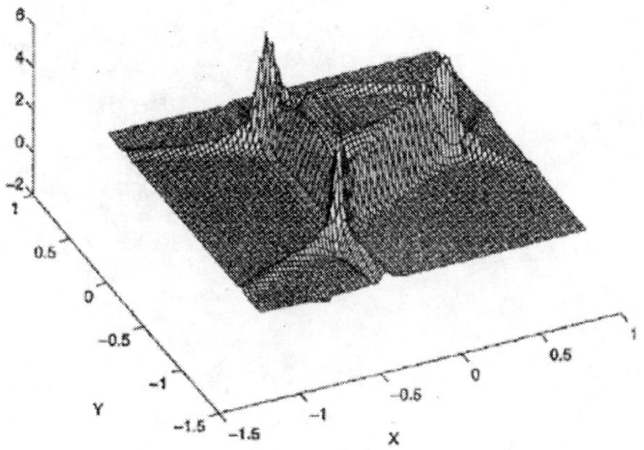

Figure 4. The distribution of current density the quasisteady regime

in 2D case and

$$\mathbf{B} = yz\mathbf{i} + zx\mathbf{j} + xy\mathbf{k} \tag{9}$$

in 3D ones.

The system of the MHD equation is solved in a cubic computation box $\{-1 \leq x \leq 1, -1 \leq y \leq 1, -1 \leq z \leq 1\}$ in case of three dimensional, and a square region $\{-1 \leq x \leq 1, -1 \leq y \leq 1\}$.

In our simulation the boundary conditions correspond to an electric current perturbation imposed at the boundary of the computational region. We considered the cylindrical MHD wave converging toward the null points $(x \pm 1, y \pm 1)$. The boundaries $z \pm 1$ correspond to the free inflow and outflow of the plasma. As for other parameters of plasma, there where plasma is entered into the computational area we set $p = 1, T = 1, \rho = 1$, in other boundary we assume free outflow conditions.

3. Simulation results

We present the result of two series of numerical calculation for the evolution of structurally unstable 2D and 3D magnetic fields with high-order critical point.

At first we consider the structurally unstable 2D magnetic field (8). In figure 4 the quasisteady regime is reached. We can see the appearance of the finite -size current region bounded by the three separatrices of the magnetic field. There are three null lines of the X-types at the corners of this region and a null line of the O-type at the center of this region. Current sheet similar to those that appear in the vicinity of null lines of the X-type are observed at the corners. The evolution of the magnetic field corresponds to a transformation of the initial magnetic field configuration to thr structurally stable one.

We will consider the process of a transformtion of this magnetic field from the point of view of the catastrophe theory in the following section.

In the second series of calculations we analyzed the structurally unstable 3D magnetic field with high-order critical point (9). To projections of power lines to the coordinate planes hyperboles are. In each octant in process of removal on infinity power lines of the field asymptotically aspire to the straight line passing through this central part. As a result in each of eight the octant forms to about one thorn.

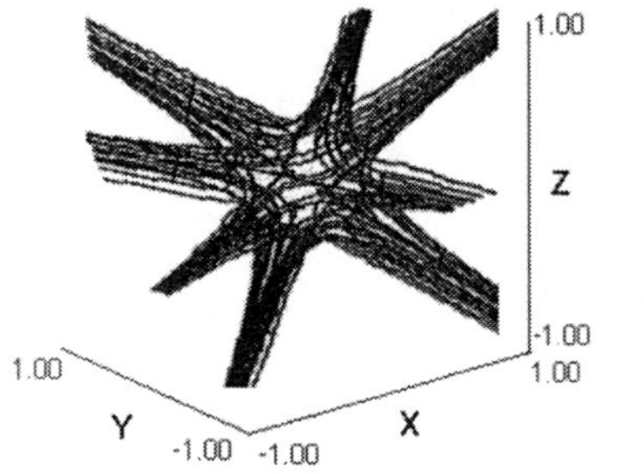

Figure 5. The lines of the 3D magnetic filed (9) at the initial time.

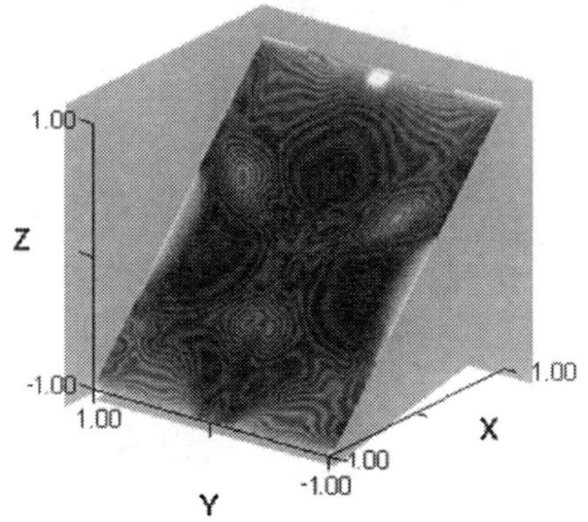

Figure 6. The distribution electric current density at the quasistationary stage.

The field has high degree of symmetry: in all octants it behaves equally. Therefore orientation of the field concerning the axis z which is an axis of symmetry of indignation isn't essential. At a quasistationary stage "disorder" of a critical point of the higher order of a magnetic field on points of the first order is clearly visible. Taking into account an initial configuration the triangular structure of disintegration is observed (fig. 6). Such thin nonlinear structures as current locks and current layers in numerical experiment in this case weren't observed.

4. Structurally unstable magnetic field and catastrophe theory

In this section we try to consider our magnetic field (8) from the view point of the catastrophe theory.

The catastrophe theory is a program. [12] The purpose of the program is to determine how the qualitative properties of solutions of equations change as the parameters that appear in the equations change. Elementary catasptohe theory is study how the critical points of a potential $V(x, c)$ move about, coalesce and annihilate each other or are created and disperse from each other in state space $x \in R^n$ as the control parameters $c \in R^k$ are varied.

The elementary catastrophe function $Cat(l, k)$ is the sum of two terms :

$$Cat(l, k) = CG(l) + Pert(l, k). \tag{10}$$

The catastrophe germ $CG(l)$ depens only on the l state variables. All its second partial derivatives vanish at the critical point. The universal perturbations depends on the k control paramenters as well as the l state variables. The dependence of $Pert(l, k)$ on the control-parameter values is linear. For "most" choices of control-parameter values (all but a set of measure zero) the function $Cat(l, k)$ has isolated critical points. The list of canonical catastrophe function is present in book [12].

In our case the magnetic configuration 8 can be described by the vector potential $\mathbf{A} = A(x, y)\mathbf{e_z}$, where

$$A(x, y) = \frac{1}{3}(x^3 - xy^2). \tag{11}$$

The magnetic configuration described by the vector potential is structurally unstable, because the zero point of the vector field $\mathbf{B} = \nabla \times A\mathbf{e_z}$ at $x = 0, y = 0$ is degenerate. This is elementary catastrophe is named the elliptic umbilic catastrophe. In this case the $k = 3, l = 2$, $CG(l) = y^2x - x^3$ $Pert(l, k) = {}_1y + a_2x + a_3y^2 + a_4x^2$.

In this paper we demonstrate the simplest case where the values of control parameters are equal to zero. It should be noted that "device" of the catastrophe theory is developed generally for two-dimensional space.

5. Conclusion

In the present paper we have studied the transformation of structurally unstable magnetic configurations into structurally stable configurations. We have investigated the transformations induced by a finite amplitude electric current excited in the plasma under the action of perturbations imposed at the boundary. The transformation into configurations with a magnetic topology that differs from the initial one is forbidden in the framework of ideal magnetohydrodynamics, but can occur during magnetic field line reconnection. We have considered two different initial magnetic configurations: one 2D structurally unstable and the other 3D structurally unstable. In the case of the initially 2D structurally unstable configuration we have observed a global redistribution of the electric current anf formation electric current "castel", and disappearance of a critical point of a high order on several ritical points of the lowest order.

In another case when we consider 3D structurally unstable configuration. Such thin nonlinear structures as current castel and current sheets weren't observed, but disappearance of the critical point was observed.

5.1. Appendices

In the present paper we use the semi-implicit method for solving the system of MHD equations. The semi-implicit method is similar to the well know predictor - corrector method. At first we made the explicit step

$$\mathbf{v}^* = \mathbf{v^n} + \alpha\Delta t\mathbf{F}(\rho, \mathbf{v}, \mathbf{B}, \mathbf{p})^\mathbf{n}/\rho^n, \tag{12}$$

$$\rho^* = \rho^n - \alpha\Delta t\nabla(\rho\mathbf{v})^n, \tag{13}$$

$$T^* = T^n + \alpha\Delta tF_1(\rho, \mathbf{B}, T)^n, \tag{14}$$

$$\mathbf{B}^* = \mathbf{B}^n + \alpha\Delta t\nabla \times (\mathbf{v}^n \times \mathbf{B}^n). \tag{15}$$

Here $0.5 \leq \alpha \leq 1.0$ is constant, $\mathbf{F}$ - presents the right hand side the motion equation, the F_1 - the function is the right hand side of the energy equation. Then the semi-implicit term is corrected the equation of velocity

$$\mathbf{v}^{n+1} - (\Delta t)^2[\nabla\times\nabla\times(\mathbf{v}^{n+1}\times\mathbf{B_0})]\times\mathbf{B_0}/\rho^* = \mathbf{v}^n + \Delta t\mathbf{F}(\rho, \mathbf{v}, \mathbf{B}, \mathbf{p})^*/\rho^* - (\Delta t)^2[\nabla\times\nabla\times(\mathbf{v}^n\times\mathbf{B_0})]\times\mathbf{B_0}/\rho^* \tag{16}$$

The $\mathbf{v^{n+1}}$ is used in calculating the magnetic field, pressure, density of plasma

$$\rho^{n+1} = \rho^n - \Delta t\nabla[\rho^*(\mathbf{v}^n + \mathbf{v}^{n+1})]/2, \tag{17}$$

$$p^{n+1} = \rho^{n+1} * T^{n+1} \tag{18}$$

$$T^{n+1} = T^n + \frac{\Delta t}{2}\frac{\gamma - 1}{\rho^*}F_1(\rho, \mathbf{B}, T, \mathbf{v}^n + \mathbf{v}^{n+1})^* - \frac{\Delta t}{2}((\mathbf{v}^{n+1} + \mathbf{v}^n)\nabla)T^*, \tag{19}$$

$$\mathbf{B}^{**} = \mathbf{B}^n + \Delta t\nabla \times [(\mathbf{v}^n + \mathbf{v}^{n+1}) \times \mathbf{B}^n]/2, \tag{20}$$

At least, we are corrected the magnetic field term

$$\mathbf{B}^{n+1} + \Delta t\nu_0\nabla \times \nabla \times \mathbf{B}^{n+1} = \mathbf{B}^{**} - \Delta t\nu_m\nabla \times \nabla \times \mathbf{B}^{**} + \Delta t\nu_0\nabla \times \nabla \times \mathbf{B}^{**} \tag{21}$$

IOP Conf. Series: Journal of Physics: Conf. Series **936** (2017) 012018 doi:10.1088/1742-6596/936/1/012018

This is a second order accuracy method capable for carrying out computations with large time step. For achievement of necessary accuracy it is necessary to use the grid containing at least $200x200x200$ points on space variables that results in need of development of the parallel computing algorithm described in paper [13].

References

[1] D. Biskamp Magnetic reconnection in Plasma. Cambridge Unoversity Press, Cambridge, 2000.

[2] E.R.Priest and T.Forbes Magnetic Reconnection : MHD Theory and Applications *Cambridge Unoversity Press, Cambridge*, 2000.

[3] S.I. Syrovatskii Formation of current sheet in a plasma with frozen in strong magnetic field *Sov. Phys. JETP*, 33: 389, 1971.

[4] E. N. Parker Nonflares and the solar X-ray corona *Astrophysics Journal* , 330:474-479, 1988.

[5] S.V.Bulanov, V.V.Pichushkin, K. Schindler Magnetic field reconnection in structurally unsable magnetic configuration *Plasma Physics Rep.* , 22:885, 1996.

[6] S.V.Bulanov, E. Echkina, I. Inovenkov, F.Pegoraro and V. Pichushkin On the structural stability of magnetic configurations with two null lines *Phys. of Plasmas*, 6: 802, 1999.

[7] S.V.Bulanov, E. Echkina, I. Inovenkov, F.Pegoraro and V. Pichushkin Formation of current sheet in structurally stable and structurally unstable magnetic configuration with two null lines *Plasma Physics Rep*, 26: 560, 2000

[8] S.V.Bulanov, E. Echkina, I. Inovenkov, F.Pegoraro Current sheet formation in three - dimansional magnetic configurations *Physics of Plasmas*,9(9):3835-3850, 2002.

[9] V. I. Arnold Geometrical methods in the theory of ordinary differential equations *Springer-Verlag*,1988.

[10] T. Poston and I.Stewart Catastrophe theory and its applications *Pitman,London*,1978.

[11] D.P.Kostomarov, E. Echkina, I. Inovenkov Application of the Catastrophe Theory in Studying the Magnetic Reconnection Process *Mathematical Models and Computer Simulations*,4(2): 135-143, 2011.

[12] R. Gilmor Catastrophe theory *Encyclopedia of applied physics*, 3, 1992.

[13] D. P. Kostomarov, I. N. Inovenkov, E.Y. Echkina, A. V. Leonenko, O. A. Pavlova Application of Parallel Programming Method for the 3D MHD Computer Simulations of Magnetic Reconnection in Plasma. *Parallel computational Fluid Dynamics. Moscow, Russia*, 258, 2003.

IOP Conf. Series: Journal of Physics: Conf. Series **936** (2017) 012019 doi:10.1088/1742-6596/936/1/012019

Applicability of the lattice Boltzmann method to determine the ohmic resistance in equivalent resistor connections

Mayken Espinoza-Andaluz[1,2*], Julio Barzola[2,3], Víctor H. Guarochico-Moreira[2,4], Martin Andersson[1]

[1]Department of Energy Sciences, Lund University, PO Box 118, 22100 Lund, Sweden

[2]Escuela Superior Politécnica del Litoral, ESPOL, PO Box 09-01-5863 Guayaquil, Ecuador

[3]Faculty of Mathematics and Physical Sciences, Universidad de Guayaquil, PO Box 090514, Guayaquil, Ecuador

[4]School of Physics and Astronomy, University of Manchester, Oxford Road, M13 9PL, Manchester UK

[*]E-mail: mayken.espinoza_andaluz@energy.lth.se, masespin@espol.edu.ec

Abstract. Knowing the ohmic resistance in the materials allow to know in advance its electrical behavior when a potential difference is applied, and therefore the prediction of the electrical performance can be achieved in a most certain manner. Although the Lattice Boltzmann method (LBM) has been applied to solve several physical phenomena in complex geometries, it has only been used to describe the fluid phase, but applicability studies of LBM on the solid-electric-conducting material have not been carried out yet. The purpose of this paper is to demonstrate the accuracy of calculating the equivalent resistor connections using LBM. Several series and parallel resistor connections are effected. All the computations are carried out with 3D models, and the domain materials are designed by the authors.

1. Introduction

During the last years, the structure of functional materials has been designed to obtain desirable responses when they are subjected to certain conditions in a system, e.g., high physical compression, flow complexity through the material, etc. Electrical and thermal transport parameters play an important role in transport processes through materials such as metal foam, porous materials, and/or semi-conductor materials.

Modeling of physical and mechanical process represents a useful tool to analyze the material properties, and it is a point of departure in the material design before the manufacturing process.

Several approaches applying the Lattice Boltzmann method (LBM) to model different transport phenomena have been carried out [1 - 3] to demonstrate the feasibility of the methodology. LBM has been applied to different physical and chemical systems. However, studies showing the applicability of LBM to determine the ohmic resistance in solid and solid/pore conducting material are scarce.

Considering the widely accepted idea of the equivalence between the electric an hydraulic system, several properties can be calculated from the hydraulic point of view, and then; its corresponding electric property being determined. Although there are some limitations when the electric-hydraulic analogy is used, it can facilitate the prediction of certain variables. The equivalences of the variables between the hydraulic and electric systems are shown in [4].

The purpose of this paper is to demonstrate the accuracy of calculating the equivalent resistor connections using LBM. Several series and parallel resistor connections are effected. All the computations are carried out with 3D models, and the domain materials are designed by the authors.

2. Methodology

The scheme applied to solve LBM to determine the ohmic resistance in solid conducting materials is D3Q19, which means a three-dimensional solution with nineteen linked velocities between the lattice elements. The NS equations, i.e., continuity and momentum equations, are recovered by solving the Lattice Boltzmann equation (LBE) and using the Bhatnagar–Gross–Krook (BGK) approximation as collision operator [5]. The discretized equation is expressed as follows:

$$\frac{\partial f_i(r,t)}{\partial t} + c_i \nabla f_i(r,t) = \frac{1}{\tau}\left[f_i^{eq}(r,t) - f_i(r,t)\right] \quad (1)$$

Where τ corresponds to the relaxation parameter, f is the denominated particle distribution function (PDF), f_i^{eq} represents the equilibrium particle distribution, and c is the bulk velocity of the fluid. The subscript i is the number of linked velocities involved in the simulations, i.e., nineteen linked velocities, r is the position vector and t is the time. Once the LBE is solved, the velocity field over the domain is computed by using the following equation:

$$u(r,t) = \frac{\sum_i f_i\, c_i(r,t)}{\rho(r,t)} \quad (2)$$

Considering the equivalences presented in [4], the velocity values are equivalent to the current density in the electrical system.

The first relationship used to analyze the applicability of the LBM to compute electrical variables is the widely known Ohm's Law:

$$j = -\sigma\, \nabla\varphi \quad (3)$$

where j is the current density; in A/m², ϕ is the potential function; in V, and σ is the material conductivity; in Ω^{-1} m^{-1}.

Since the LBM is solving the equivalent hydraulic system, the ohmic resistance, considering the hydraulic variables presented in [4], is computed as follows:

$$R_{LBM} = \frac{\Delta p}{\phi_V} \quad (4)$$

where Δp is the pressure drop applied to the ends of the material and φ, corresponds to the volumetric flow rate. Using equation (4), several measurements are carried out to demonstrate the applicability of LBM for computing the ohmic resistance of solid and solid/pore conducting materials.

3. Results and discussion

Before to carry out different proofs to demonstrate the feasibility of using LBM to compute the ohmic resistance of conducting materials, there is a condition that should be matched related to the values of the velocities obtained during the modeling process. Given the characteristics of the methodology and according to [6], LBM leads to instabilities when high velocities, i.e., high current densities in electrical systems, are reached during the simulation process.

A conducting material with constant cross-sectional area, a square of 5 *lu* by side which allow us to get excellent agreement in hydraulics phenomena [7], and for three different lengths were submitted to a range of potential difference at the ends of the material. As a result, the current-voltage curves of the material were obtained. The mentioned analysis and the I-V curve are presented in a previous study developed in [4].

According to the preliminary results, the I-V curve obtained, with no restriction on the potential difference applied, shows a similar behavior to the filament lamps [8], i.e., the current is increasing with the voltage but not linearly. However, due to the aim of this paper and taking into account that variations of temperature are not considered, a deeper analysis of the results is done to avoid such non-ohmic behaviors.

Analyzing the obtained results in detail, for a given cross-sectional area, there is a maximum value of potential difference which can give a numerical result, i.e., after this maximum value an instability is obtained. The maximum potential difference before no result is approximated 3.83 V_{LBM}, which for the given cross-sectional area, is the same independently of the material length. For each cross-sectional area, a maximum potential difference can be determined and the simulations can be effected.

3.1 Equivalent ohmic resistance for resistor connections

The validation of LBM as a tool to determine the ohmic resistance for conducting materials is obtained calculating the equivalent ohmic resistance of an arrange of resistors. As largely known, if two resistors are connected in series, the equivalent resistance is calculated adding the resistance values. On the other hand, if they are connected in parallel, the reciprocal of the equivalent resistance is the result of adding the reciprocal values of each resistance value. The calculation of an equivalent resistor of a group of resistors is based on the voltage and current distribution in the circuit. Equivalent resistors for a series and parallel circuits can be determined by using the following expressions:

$$R_{series} = \sum_{i=1}^{n} R_i \qquad (5)$$

$$\frac{1}{R_{parallel}} = \sum_{i=1}^{n} \frac{1}{R_i} \qquad (6)$$

The ohmic resistance is computed using equation (4) from modeled domains with different dimensions, i.e., different length and different cross-sectional areas. All the resistance configurations are subjected to the same potential difference along the length direction, i.e., 0.03 V_{LBM}, the LBM is applied and the resistance is computed. Four different volume dimensions, representing resistors, were mimicked to calculate the individual ohmic resistance. The obtained results are presented in Table 1.

Table 1. Computed ohmic resistance using LBM for four resistors with different dimensions (MxLxN)

Name	**Size (*lu*^{3})**	**Ohmic resistance x10^{-3} (Ω_{LBM})**
R_1	20x20x20	1.449
R_2	20x15x20	1.087
R_3	20x25x20	1.816
R_4	25x15x20	0.736

(M x N = cross-sectional area, L = length)

Various circuit configurations using the resistors presented in Table 1 were implemented. The ohmic resistance calculated by LBM, and the expected values using Eqs (5) and (6) are presented in Table 2. Last column shows the deviation error between the computed and expected values.

Table 2. Ohmic resistance configurations with their corresponding equivalent ohmic resistance computed using LBM and expected values from Eqs (5) and (6).

Configuration	Ohmic resistance x10^{-3} (Ω_{LBM})		Deviation error
	LBM	Expected	
$(R_1, R_1)^S$	2.929	2.828	+3.57%
$(R_1, R_1)^P$	0.724	0.724	--
$(R_2, R_3)^S$	2.929	2.903	+0.90%
$(R_2, R_4)^P$	4.389	4.390	-0.02%
$(R_1, R_2, R_3)^S$	4.421	4.352	+1.59%

S=Series, P=Parallel.

Deviation errors between the computed equivalent resistance using LBM and Eqs (5) and (6) show a very good agreement, i.e. lower than 4.00%. Considering the obtained results, LBM shows better accuracy when the resistors are connected in parallel than in a series circuit.

4. Conclusions

LBM has proven to be a powerful tool for computing the ohmic resistance in solid and solid/pore materials [4]. Several simulations were carried out considering the conditions in which the LBM can show ohmic behavior of the materials. LBM results show a better agreement when the parallel circuits are computed in comparison with the serial circuits. However, for both cases the deviation errors fall in an acceptable range, i.e., lower than 4%. Although this computational work shows a very good agreement, there are some limitations that should be addressed in future investigations: the relaxation parameter in LBM can be related to the material resistivity, a more realistic electron flow can be achieved if an external force is added to the LBE solution.

References

[1]. Aidun, C. K., & Clausen, J. R.: *Lattice-Boltzmann method for complex flows*. Annual review of fluid mechanics, **42**, 2010, 439-472.

[2]. Espinoza, M., Sundén, B., & Andersson, M.: *Highlights of Fuel Cell Modeling From a Lattice Boltzmann Method Point of View*. In ASME 2014 International Mechanical Engineering Congress and Exposition, 2014, pp. V06AT07A058-V06AT07A058. American Society of Mechanical Engineers.

[3]. Gebäck, T., & Heintz, A.: *A lattice Boltzmann method for the advection-diffusion equation with Neumann boundary conditions*. Communications in Computational Physics, **15**(02), 2014, pp. 487-505.

[4]. Espinoza, M., Andersson, M., & Sundén, B.: *Potential of lattice Boltzmann method to determine the ohmic resistance in porous materials*. Journal of Physics: Conference Series, **738**, 2016, pp. 1-4.

[5]. Bhatnagar, P. L., Gross, E. P., & Krook, M.: *A model for collision processes in gases*. I. Small amplitude processes in charged and neutral one-component systems. Physical review, **94**(3), 1954, 511.

[6]. Mohamad, A. A.: Lattice Boltzmann method: *fundamentals and engineering applications with computer codes*. Springer Science & Business Media, 2011.

[7]. Sukop, M. C., and Thorne Jr, D. T.: *Lattice Boltzmann Modeling: An Introduction for Geoscientists and Engineers*. Springer, 2006.

[8]. MacIsaac, D., Kanner, G., & Anderson, G.: *Basic physics of the incandescent lamp (lightbulb)*. The Physics Teacher, **37**(9), 1999, pp. 520-525

Analytic approximation for the modified Bessel function $I_{-2/3}(x)$

Pablo Martin[1], Jorge Olivares[2], and Fernando Maass[1]
[1]Department of Physics, Universidad de Antofagasta, Antofagasta, Chile
[2]Department of Mathematics, Universidad de Antofagasta, Antofagasta, Chile

E-mail: pablo.martin@uantof.cl, jorge.olivares@uantof.cl, fernando.maass@uantof.cl

Abstract. In the present work an analytic approximation to modified Bessel function of negative fractional order $I_{-2/3}(x)$ is presented. The validity of the approximation is for every positive value of the independent variable. The accuracy is high in spite of the small number (4) of parameters used. The approximation is a combination of elementary functions with rational ones. Power series and assymptotic expansions are simultaneously used to obtain the approximation.

1. Introduction

The modified Bessel functions appear in Electrodynamics and other areas of Physics [1-4]. In particular those of fractional order como $I_{-2/3}$ are also important, because they are connected to Airy functions. This kind of functions of negative and fractional order have a negative branch point at zero. To obtain a good approximation for small values of the independent variable, this singularity should appear in the approximation, which leads to some peculiar problems, which will be solved in the paper. As in a previous work [5], elementary functions will be combined with rational ones, to obtain analytic approximations valid for x≥0, and with high accuracy. The key of the present method will be the simultaneous used of power series and asymptotic expansion. The present method is an improvement of previous technique applied to 2D-Quadratic Zeemann Effect, Instantons and others [6-9].

In Section 2, the theoretical basis of the present technique will be shown, as well as, the description of the structure of the analytic approximation. The determination of the parameters and numerical results will be also treated in this Section, and the Conclusion will be the last Section.

2. Theoretical Treatment

The starting point will be the well known power series

$$I_{-2/3}(x) = \left(\frac{x}{2}\right)^{-2/3} \sum_{k=0}^{\infty} \frac{1}{k!\Gamma(\frac{1}{3}+k)} \left(\frac{x}{2}\right)^{2k} \tag{1}$$

The asymptotic expansion will be also

$$I_{-2/3}(x) \sim \left(\frac{e^x}{\sqrt{2\pi x}}\right)\left(1 - \frac{7}{72x} + ...\right) \tag{2}$$

In order to obtain analytic approximations with good accuracy for the function $I_{-2/3}(x)$, the approximate function, should have a factor $x^{-2/3}$. However this is a branch point, which are coming by pairs, thus this factor will introduce a second branch point at the infinite, which is different of the charasteristic branch point in Eq.2. In order to take care of this problem the factor to be introduce must be $(1+x)^{2/3}/x^{2/3)}$, but this is not convenient, since the adjoin power series in Eq.1, has only even powers. In order to keep this condition, the selection of the factor must be $(1+\lambda^2x^2)^{1/3}/x^{2/3}$, where λ is a free parameter, which will be very useful later, as it will be shown. This last factor introduce also a new branch point at $x^2 = -1/\lambda^2$, however this new singularity is out of the region of interest, which is the right half complex plane, where the real part of the variable x is positive.
Now, it is necessary to consider the branch point at the infinite which is $x^{-1/2}$. Similarly as in the precedent case, this factor introduces a branch point at $x = 0$, which is not in $I_{-2/3}$. The procedure to follow is to use $(1+\lambda^2x^2)^{-1/4}$, where the second branch point is out of the region of interest, and furthermore looking for the power series expansion of this function, only even powers are obtained. The same λ is used to avoid additional parameters. Fnally both factors, the one from the power series, and that from the asymptotic expansion, can be combined in a unique factor $(1+\lambda^2x^2)^{1/12}/x^{2/3}$.

The additional factor e^x at the asymptotic expansion is taken care using the auxiliary function $\cosh(x)$, since in this way the exponential factor is obtained at the infinite, and only even powers are coming in the power series expansion. Considering now, the rational function to be combined with all the previous elementary functions, and taken care of all the restrictions previously described, the structure of the analytic approximation $\tilde{I}_{-2/3}(x)$ will be considered as

$$\tilde{I}_{-2/3}(x) = \frac{\sqrt[3]{4}(1+\lambda^2x^2)^{1/12}}{\Gamma(1/3)x^{2/3}} \cosh(x) \frac{\sum_{j=1}^{n} p_j x^{2j}}{1+\sum_{k=1}^{n} q_k x^{2k}} \tag{3}$$

where the value of n, will be chosen, according with the number of parameters to use.This will lead to higher accuracy with increasing number of parameters.

The simplest approximation will be

$$\tilde{I}_{-2/3}(x) = \frac{\sqrt[3]{4}(1+\lambda^2x^2)^{1/12}}{\Gamma(1/3)x^{2/3}} \frac{p_0+p_1x^2}{1+q_1x^2} \cosh(x) \tag{4}$$

The condition with the leading term of $\tilde{I}_{-2/3}(x)$ in Eq.(2) leads to

$$\frac{p_1}{q_1} = \frac{\Gamma(1/3)}{\sqrt{\pi}\sqrt[6]{2}\lambda} \tag{5}$$

Two more equations for p_0. p_1 and q_1 are coming from the power series Eq.(1). However before to equalize coefficients between series, it is convenient to rationalize both terms of the equation, so

$$\frac{\Gamma(1/3)}{\sqrt[3]{4}} x^{2/3}(1+\lambda^2x^2)^{-1/12}(1+q_1x^2)I_{-2/3}(x) \cong \cosh(x)(p_0+p_1x^2) \tag{6}$$

$$(1-\frac{1}{12}\lambda^2x^2)(1+q_1x^2)(1+\frac{3}{4}x^2) \cong (1+\frac{x^2}{2})(p_0+p_1x^2)+... \tag{7}$$

From Eq.(7), it is obtained

$$-\frac{1}{12}\lambda^2 + q_1 + \frac{3}{4} = \frac{p_0}{2} + p_1 = \frac{1}{2} + \frac{\sqrt[6]{2}}{\sqrt{\pi}\Gamma(1/3)\lambda^{1/12}} q_1 \quad ; \quad p_0 = 1 \tag{8}$$

From Eqs. (8) and (5), q_1 and p_1 are obtained giving

$$q_1 = \frac{\lambda^2 - 3}{12(1 - \Gamma(1/3)/\sqrt[6]{2}\lambda\sqrt{\pi})} \tag{9}$$

and

$$p_1 = \frac{\Gamma(1/3)}{\sqrt{\pi}\sqrt[6]{2}\lambda} \frac{\lambda^2 - 3}{12(1 - \Gamma(1/3)/\sqrt[6]{2}\lambda\sqrt{\pi})} \tag{10}$$

A plot of q_1 as a function of λ is shown in Figure(1).

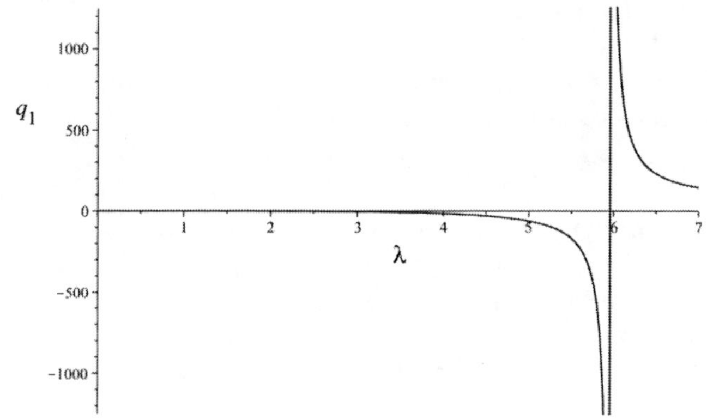

Figure 1. Value of q_1 as a function of λ

Looking the Figure(1), it is clear that q_1 is positive only for λ in the interval $(0,\sqrt{3})$ and $\lambda \geq 5.96$. The best value for λ is considered as that where there is a minimum of maximum relative error ε defined as

$$\varepsilon(x,\lambda) = \frac{\left|\tilde{I}_{-2/3}(x,\lambda) - I_{-2/3}(x)\right|}{I_{-2/3}(x)} \tag{11}$$

In this way the best value of λ is λ=0.6.
The relative error ε(x,λ=0.6) is shown in Fig (2) as a function of the independent variable x. The function $\tilde{I}_{-2/3}(x)$ is given in this case as

$$\tilde{I}_{-2/3}(x) = \frac{(1 + 0.36x^2)^{1/12}}{1.6876x^{2/3}} \frac{1 + 0.6919x^2}{1 + 0.4719x^2} \cosh(x) \tag{12}$$

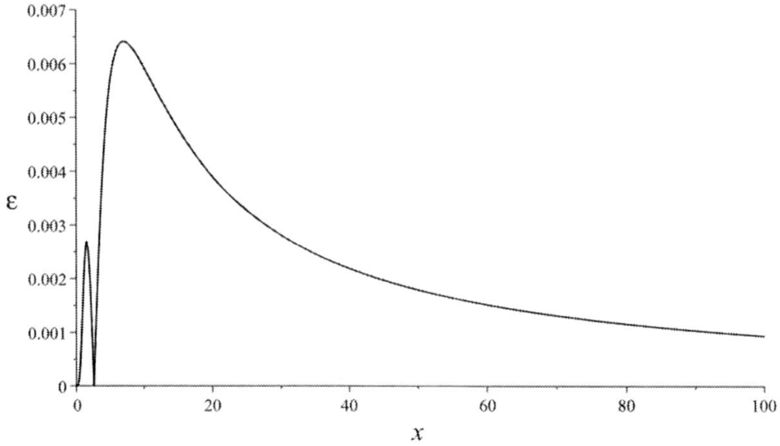

Figure 2. Relative error $\varepsilon(x, \lambda = 0.6)$ of the approximation $\tilde{I}_{-2/3}(x)$ as a function of the variable x

The maximum relative error ε_{max} is about 0.0065 for x $\sim$ 6, but taken out the region around x $\sim$ 6, the relative error becomes smaller than 0.002.

3. Conclusion

Here it is presented a method to obtain approximations to the modified Bessel functions of negative and fractional order. The case of $I_{-2/3}(x)$ is considered in detail. A general formula is shown, and the most simple approximation is considered. The form of the approximations are determined using the power series and asymptotic expansion, and they are rational functions combined with fractional power and hyperbolic functions. In the case here treated the number of parameters to be determined is four, and the maximum relative error is 0.0065, however the error is smaller than 0.002 for most of the values of x, except an small region around the values of x, where there is maximum relative error. The approximation here found is valid for any positive value of the variable x, and therefore can be integrated and differentiated. Furthermore it is easy to calculate, and this can be done with a simple pocket calculator.

Acknowledgements

We thanks to the "Vicerrector de Investigación Innovación y Postgrado" Dr. Marcelo Oliva Moreno and the "Decano de Facultad Ciencias Básicas" Dr. Guillermo Mondaca Ortiz, of the "Universidad de Antofagasta" for support to this research.

References

[1] G. N. Watson, " A treatise on the theory of Bessel functions", 2nd. ed.(Cambrigde University Press, Cambrigde,1966).
[2] M. Abramovitz and I. A. Stegun, "Handbook of mathematical Functions" (Dover Books, New York, 1972) pp. 692-697.
[3] J. D. Jackson, "Classical Electrodynamics", 3rd. ed. (John Willey and Sons, Inc., New York,1998); Ch.3.
[4] P. Martin, J. Olivares and A. Sotomayor, Rev. Mex. Fisica **63**, 130 (2017)
[5] P. Martin, J. J. Rodriguez Nuñez and J. L. Marquez, Phys. Rev. B **45**, 8359 (1992)
[6] E. Chalbaud and P. Martin, J. Math. Phys., **33**, 2483 (1992)
[7] P. Martin, E. Castro and J. L. Paz, Rev. Mexicana Fisica, **58**, 301 (2012)
[8] P. Martin, E. Castro, J. L. Paz and A. De Freitas, "Multipoint quasi-rational approximants in Quantum Chemistry", Chapter 3 of "New Developments in Quantum Chemistry" by J. L. Paz and A. J. Hernandez (Transworld Research Network, Kerala, India, 2009) pp. 55-78.

Gauge symmetries for a coupled Korteweg-de Vries system

Alvaro Restuccia[1], **Adrián Sotomayor**[2]

[1] Physics Department, University of Antofagasta, Chile
[2] Department of Mathematics, University of Antofagasta, Chile

E-mail: `arestu@usb.ve,adrian.sotomayor@uantof.cl`

Abstract. We introduce a gauge invariant Hamiltonian which generalizes one of the Hamiltonian structures of a parametric KdV system, which includes the complex KdV system. The associated canonical field equations have an infinite sequence of conserved quantities which are also gauge invariant. The gauge symmetry relates a wide class of integrable systems to the complex KdV system.

1. Introduction

In this work we introduce a gauge invariant Hamiltonian H which generalizes one of the Hamiltonian structures of a parametric KdV system. The canonical field equations $\frac{dF}{dt} = \{F, H\}$ associated to H define an integrable system, in the sense that it has an infinite sequence of conserved quantities, which reduces to the original parametric KdV system in a particular gauge. The interest of having a gauge invariant formulation arises because it allows to relate a wide class of integrable systems, obtained by different gauge fixing procedures, to the KdV system. Moreover it is well known that all physical theories describing the fundamental forces in nature are formulated in terms of gauge invariant actions. In particular, several of the properties of those theories are obtained by considering the theories in different gauges. We expect also that the other Hamiltonian structure of the KdV system can also be formulated in a gauge invariant way. If so, the Virasoro structure of the KdV system would be also compatible with the gauge symmetry we are proposing, as it occurs in String theory and Super Yang Mills theory.

2. A parametric coupled KdV system

The Korteweg-de Vries integrable system has a wide range of applications in low energy as well as in high energy physics. The Poisson structure arising from the Hamiltonian formulation via a Miura transformation coincides with the Virasoro algebra with a central extension which is directly related to the conformal symmetry in String Theory. Many integrable extensions of the KdV equation habe been considered in the literature. The most straighforward extension arises by considering the KdV equation with the field taking values on the complex algebra. Further integrable extensions arises by considering the field with values in a general Cayley-Dickson algebra [1]. Although the complex extension seems a trivial one, it has remarkable properties which are not present in the KdV equation with a real field. See [2, 3, 4] for applications to a two-layer liquid model. In [5] it was shown to have solutions developing singularities on a finite time. Also, solitonic solutions have been reported in [6, 7, 8]. The existence of an infinite sequence of Hamiltonians with a pencil of Poisson structures has been discussed in [9].

In the present work we consider a parametric coupled KdV system [2] which contains for a particular value of the parameter, $\lambda = -1$, the complex valued KdV equation. It is given by

$$u_t + uu_x + u_{xxx} + \lambda vv_x = 0 \tag{1}$$

$$v_t + u_x v + v_x u + v_{xxx} = 0 \tag{2}$$

In what follows we will assume that the fields u and v belong to the Schwartz space defined by

$$C_\downarrow^\infty = \left\{ w \in C^\infty(\mathbb{R}) / \lim_{x\to\pm\infty} x^p \frac{\partial^q}{\partial x^q} w = 0; p, q \geq 0 \right\}. \tag{3}$$

We will consider prepotential fields w, y defined as

$$\begin{aligned} u(x,t) &= w_x(x,t) \\ v(x,t) &= y_x(x,t). \end{aligned} \tag{4}$$

where the subindex x denotes derivative with respect to the spatial coordinate x. We also use a subindex t or a dot to denote derivative with respect to the time coordinate t.

3. The gauge invariant Hamiltonian

The parametric KdV system described previously can be obtained from the following Hamiltonian

$$H_1 = \int_{-\infty}^{+\infty} dx\, \mathcal{H}_1, \tag{5}$$

$$\mathcal{H}_1 = \frac{1}{6} w_x^3 - \frac{1}{2} w_{xx}^2 + \frac{\lambda}{2} w_x y_x^2 - \frac{\lambda}{2} y_{xx}^2, \tag{6}$$

subject to the following two second class constraints [9]

$$\phi_1 \equiv p + \frac{1}{2} w_x = 0, \tag{7}$$

$$\phi_2 \equiv q + \frac{\lambda}{2} y_x = 0. \tag{8}$$

The Poisson structure on the complete phase space is the canonical one given by

$$\begin{aligned} \{w(x), p(\hat{x})\}_{PB} &= \delta(x - \hat{x}) \\ \{y(x), q(\hat{x})\}_{PB} &= \delta(x - \hat{x}), \end{aligned} \tag{9}$$

with all other brackets between these variables being zero.

In order to present the Poisson structure on the constrained submanifold, we introduced in [9] the Dirac brackets associated to the second class constraints. The resulting Dirac brackets among the original independent fields are

$$\begin{aligned} &\{u(x), u(\hat{x})\}_{DB} = -\partial_x \delta(x - \hat{x}) \ , \ \{v(x), v(\hat{x})\}_{DB} = -\tfrac{1}{\lambda} \partial_x \delta(x - \hat{x}), \\ &\{u(x), v(\hat{x})\}_{DB} = 0, \end{aligned} \tag{10}$$

where we have assumed that $\lambda \neq 0$.

The canonical equations for the fileds $u(x,t)$ and $v(x,t)$ are

$$\begin{aligned} u_t &= \{u, H_1\}_{DB} = -uu_x - u_{xxx} - \lambda vv_x \\ v_t &= \{v, H_1\}_{DB} = -u_x v - v_x u - v_{xxx}, \end{aligned} \tag{11}$$

IOP Conf. Series: Journal of Physics: Conf. Series **936** (2017) 012021 doi:10.1088/1742-6596/936/1/012021

which of course are the same as the original equations describing the parametric system.

We may now extend the Hamiltonian and Lagrangian of the above formulation to a more general theory, which becomes invariant under a gauge transformation on the canonical fields. The new Hamiltonian is restricted only by a first class constraint, which is the generator of the gauge symmetry. On a particular gauge fixing the new field equations reduce to the equations of the original parametric system.

The interesting point is that we may consider different gauge fixing procedures and all of them are integrable systems, in the sense that the field equations have an infinite sequence of conserved quantities.

To obtain the gauge invariant Hamiltonian we first notice that

$$\begin{aligned} \{\phi_1(x), \phi_1(\hat{x})\} &= \partial_x \delta \\ \{\phi_1(x), \phi_2(\hat{x})\} &= 0 \\ \{\phi_2(x), \phi_2(\hat{x})\} &= \lambda \partial_x \delta, \end{aligned} \tag{12}$$

hence we can combine ϕ_1 and ϕ_2 in a way that the combination conmutes with itself,

$$\{\phi(x) + k\phi(x), \phi_1(\hat{x}) + k\phi_2(\hat{x})\} = 0. \tag{13}$$

This is so for any real parameter satisfiyng

$$k^2 = -\frac{1}{\lambda}, \lambda \neq 0. \tag{14}$$

$\phi_1(x,t) + k\phi_2(x,t)$ is thus a candidate to be a first class contraint. The other combination $\phi_1(x,t) - k\phi_2(x,t)$ does not conmute with $\phi_1(x,t) + k\phi_2(x,t)$ and is a candidate to fix the gauge of the gauge symmetry generated by $\phi_1(x,t) + k\phi_2(x,t)$.

In order to have a first class constraint $\phi_1(x,t) + k\phi_2(x,t)$ must conmute, ina a weak sense, with a new Hamiltonian. The new Hamiltonian can be constructed by adding to the original $\mathcal{H}_1$ a finite number of powers of $\phi_1(x,t) - k\phi_2(x,t)$ with suitable coefficients, which may be functions of the fields and its derivatives. The formulation is in terms of the prepotentials w and y. Although, the Hamiltonian can be expressed in terms of u and v and the canonical field equations for u and v may be obtained in terms of the Poisson brackets without reference to w and y, the quantum formulation of the system on sure has to be formulated in terms of w, y and not in terms of u, v. There are several examples of field theories where analogous formulations available, and in all case the quantization has to be performed in terms of the original prepotentials.

The Hamiltonian we introduce is given by

$$H = \int_{-\infty}^{+\infty} dx\, \mathcal{H}, \tag{15}$$

$$\mathcal{H} = \frac{1}{6}w_x^3 - \frac{1}{2}w_{xx}^2 + \frac{\lambda}{2}w_x y_x^2 - \frac{\lambda}{2}y_{xx}^2 + f(\phi_1 - k\phi_2) + g(\phi_1 - k\phi_2)^2 + \frac{1}{24}(\phi_1 - k\phi_2)^3, \tag{16}$$

$$f = -\frac{1}{2}\left(\frac{1}{2}w_x^2 + w_{xxx} + \frac{\lambda}{2}y_x^2 + k\lambda w_x y_x + k\lambda y_{xxx}\right), \tag{17}$$

$$g = \frac{1}{4}k\lambda y_x, \tag{18}$$

where ϕ_1 and ϕ_2 are defined as previously and k is a real parameter satisfiyng $k^2 = -\frac{1}{\lambda}$, we assume $\lambda \neq 0$.

The infinitesimal gauge transformation for a general field F is defined as

$$F \to F + \delta F$$
$$\delta F = \left\{ F, \int_{-\infty}^{+\infty} dx \;\; \eta(x,t)(\phi_1 + k\phi_2) \right\}_{PB} \tag{19}$$

where $\eta(x,t)$ is the infinitesimal gauge parameter.

It turns out that

$$\delta H = 0 \tag{20}$$

which means that the Hamiltonian given by (16) is invariant under the gauge symmetry. Besides we obtain

$$\begin{aligned} \delta w &= \eta(x,t) \\ \delta y &= k\eta(x,t) \\ \delta p &= \tfrac{1}{2}\partial_x \eta(x,t) \\ \delta q &= \tfrac{1}{2}k\lambda\partial_x \eta(x,t). \end{aligned} \tag{21}$$

It turns out that $u - \phi_1$ and $v - k\phi_1$ are gauge invariant fields and all conserved quantities can be formulated in terms of them. The first few conserved quantities are given by

$$\begin{aligned} h_1 &= \int_{-\infty}^{+\infty} dx\,(u - \phi_1)\,, \quad h_2 = \int_{-\infty}^{+\infty} dx\,(v - k\phi_1)\,, \\ h_3 &= \int_{-\infty}^{+\infty} dx \left[(u - \phi_1)^2 + \lambda(v - k\phi_1)^2\right], \\ h_4 &= \int_{-\infty}^{+\infty} dx\,(u - \phi_1)\,(v - k\phi_1)\,. \end{aligned} \tag{22}$$

The general proof of the existence of infinite conserved quantities will be presented elsewhere. All of them conmute with the Hamiltonian H, hence

$$\dot{h}_i = \{h_i, H\}_{PB} = 0. \tag{23}$$

Notice that, since $h_i, i = 1, \ldots,$ are gauge invariant, they conmute with the first class constraint (in the sense of the Poisson bracket), hence in the above equation one can use H or $H + \int_{-\infty}^{+\infty} dx \;\; \mu(x,t)(\phi_1 + k\phi_2)$, with $\mu(x,t)$ being an independent Lagrange multiplier.

Conclusions

We started from one of the Hamiltonian structures of a parametric KdV system, which includes the complex KdV system, and extended it to a gauge invariant Hamiltonian formulation , which has an infinite sequence of gauge invariant conserved quantities. In a particular gauge fixing they reduce to the already known conserved quantities of the original parametric KdV system. The gauge invariant Hamiltonian is explicitly expressed in terms of a canonical pair of conjugate fields, which are prepotentials of the fields defining the original system.The generator of the gauge symmetry is explicitly obtained. In particular it maps a wide class of integrable systems to the complex KdV system.

The construction we present cannot be reduced to the real KdV equation, however we expect it can be extended to the class of Cayley-Dickson integrable systems beyond the complex algebra. The extension of the second Hamiltonian structure to a gauge invariant formulation is also being under research.

Acknowledgments

A. R. and A. S. are partially supported by Project Fondecyt 1161192, Chile.

IOP Conf. Series: Journal of Physics: Conf. Series **936** (2017) 012021 doi:10.1088/1742-6596/936/1/012021

References

[1] A. Restuccia, A. Sotomayor and J. P. Veiro, arXiv: 1609.05410 v1 [math-ph] (2016).
[2] L. C. Vega, A. Restuccia and A. Sotomayor, Contribution to the Proceedings of the 2nd International Conference on Mathematical Modeling in Physical Sciences 2013, J. Phys. Conf. Series 490, 012024 (2014).
[3] J. A. Gear and R. Grimshaw, Stud. Appl. Math. 70, 235 (1984); J. A. Gear, Stud. Appl. Math. 72, 95 (1985).
[4] S. Y. Lou, B. Tong, H. C. Hu and X. Y. Tang, J. Phys. A: Math. Gen. 39, 513-527 (2006).
[5] J. L. Bona, S. Vento and F. B. Weissler, Discrete and Continuous Dynamical Systems 33, V 11-12, 4811 (2013).
[6] J. R. Yang and J. J. Mao, Commun. Theor. Phys. 49, 22-26 (2008).
[7] R. Hirota, Phys. Rev. Lett. 27, number 18, 1192-1194 (1971).
[8] L. C. Vega, A. Restuccia and A. Sotomayor, arXiv:1407.7743 v3 [math-ph].
[9] A. Restuccia and A. Sotomayor, Open Phys. 14, 95-105 (2016).

Application of the multicriterion optimization techniques and hierarchy of computational models to the research of ion acceleration due to laser-plasma interaction

I.N. Inovenkov, E.Yu. Echkina, V.V. Nefedov[1] and L.S. Ponomarenko
Lomonosov Moscow State University, Department of Computational Math & Cybernatics, Leninskie gory, bld. 1/58, Moscow, 119991, Russian Federation

E-mail: vv_nefedov@mail.ru

Abstract. In this paper we discuss how a particles-in-cell computation code can be combined with methods of multicriterion optimization (in particular the Pareto optimal solutions of the multicriterion optimization problem) and a hierarchy of computational models approach to create an efficient tool for solving a wide array of problems related to the laser-plasma interaction. In case of the computational experiment the multicriterion optimization can be applied as follows: the researcher defines the objectives of the experiment - some computable scalar values (i.e. high kinetic energy of the ions leaving the domain, least possible number of electrons leaving domain in the given direction, etc). After that the parameters of the experiment which can be varied to achieve these objectives and the constrains on these parameters are chosen (e.g. amplitude and wave-length of the laser radiation, dimensions of the plasma slab(s)). The Pareto optimality of the vector of the parameters can be seen as this: x^0 is Pareto optimal if there exists no vector which would improve some criterion without causing a simultaneous degradation in at least one other criterion. These efficient set of parameter and constrains can be selected based on the preliminary calculations in the simplified models (one or two-dimensional) either analytical or numerical. The multistage computation of the Pareto set radically reduces the number of variants which are to be evaluated to achieve the given accuracy. During the final stage we further improve the results by recomputing some of the optimal variants on the finer grids, with more particles and/or in the frame of a more detailed model. As an example we have considered the ion acceleration caused by interaction of very intense and ultra-short laser pulses with plasmas and have calculated the optimal set of experiment parameters for optimizing number and average energy of high energy ions leaving the domain in the given direction and minimizing the expulsion of electrons.

Keywords: computational experiment, mathematical modeling, multicriterion optimization problem, Pareto optimal set, laser pulse, laser-plasma interaction

Introduction

In the last twenty years laser-plasma interaction was a subject of a growing number of researches due to the theoretical and practical applicability of the results in medicine (e.g. hadronic therapy [1]), energy systems (e.g. nuclear fusion [2]) and some other applications.
In this paper we discuss how a particles-in-cell computation code can be combined with the methods of multicriterion optimization (in particular the Pareto optimal solutions of the multicriterion

[1] To whom any correspondence should be addressed.

optimization problem) and a hierarchy of computational models approach to create an efficient tool for solving a wide array of problems related to the laser-plasma interaction.

We also discuss some of the results obtain during the experimental runs of the computational system consisting of a relatavistic PIC code (based on [3]) and a hierarchical Pareto-optimality based experiment scheduler.

1. Multicriterion optimization problem

The problem of multicriterion optimization is defined as follows: a vector of decision variables which satisfy constraints and optimizes a vector function whose elements represent the objective functions is to be found [4]. These functions form a mathematical description of performance criteria and they can be in conflict with each other. The solution of such a problem can be defined as follows: a vector of decision variables $x^0 \in X$ is Pareto optimal if there does not exist another $x \in X$ such that $f_i(x) \leq f_i(x^0)$ for all $i = \overline{1,k}$ and $f_j(x) < f_j(x^0)$ for at least one index j. Here, X denotes the region of the problem where the constraints are satisfied. In most cases this definition will yield us not a single vector of decision variables but a set of Pareto optimal vectors, this set is called the Pareto optimal set.

In case of the computational experiment these definitions can be read as follows: the researcher defines the objectives of the experiment – some computable scalar values (i.e. high kinetic energy of the ions leaving the domain, least possible number of electrons leaving domain in the given direction, etc). In this paper it is assumed that all the objectives are equally significant to the researcher's cause yet this restriction can be circumvented when needed to. After that the parameters of the experiment which can be varied to achieve this objectives and the constrains on these parameters are chosen (for example, amplitude and wave-length of the laser radiation, dimensions of the plasma slab(s)). The Pareto optimality of the vector of the parameters can be seen like this: vector x^0 is Pareto optimal if there exists no vector $x \in X$ which would decrease some criterion without causing a simultaneous increase in at least one other criterion.

These efficient set of parameters and constrains can be selected based on the preliminary calculations in the simplified models (one or two-dimensional) either analytical or numerical. As it can be seen the number of possible combinations of the parameter values in most cases will exceeds capabilities of the contemporary and near-future computers even with the most strict discretization of the parameter values (if we have N parameters with 100 possible values for each we get 100^N possible variants) thus further optimization is still needed.

As a means of such an optimization the following approach can be chosen: we split the process of the problem solving into several stages. During the first stage we solve the problem with the coarse discretization of the parameters' values and compute the Pareto optimal set. Then we add finer distributed parameter values in the vicinity of the vectors in the Pareto set and solve the problem for these finer distributed decision variables. The process can be repeated several times until we get the needed accuracy for the parameters' distribution. For the N parameters and 2 stages we can achieve nearly the same accuracy as with 100^N variants with just 2×100^N calculations (or if the initial domain is wide and we expect serious deviations in results $C \times 100^N$ calculations where C is the number of unique Pareto-optimal variants on all steps). As the finishing stage we can recompute some of the variants on the finer grids and/or with more particles to verify the results.

2. Some computational results of a laser-plasma interaction

We have tested the described experiment scheduling algorithm in the problem of finding a mode of interaction of the single pulse with the plasma foil which permits us to obtain the maximal energy of ions propagating in the same direction as the laser pulse with the constraint of keeping the energy of electrons leaving the vicinity of initial plasma domain as low as possible.

This problem can be interpreted as the problem of the multicriterion optimization with two criteria: $W_1 \to \max$ and $W_2 \to \min$, where W_1 is the energy of ions blown out of the domain and W_2 is the electron energy.

In all computational experiments performed we have considered a laser with a wave length of 1.0 micron and wave period of 3.3 femtoseconds. Ratio of ion to electron mass has been set to 1836 and ratio of plasma electron frequency to pulse frequency has been set to 0.41. The domain in which we have performed our calculations had the size of $200\lambda \times 60\lambda$ and the initial plasma domain occupied a quarter of that.

We have chosen the following 3 parameters for variation – dimensionless amplitude of the pulse ($a_0 = eE_0 / m_e c\omega$) in $[0.5,5.0]$ (corresponds to intensity around of $10^{18} \div 10^{19}\ W/cm^2$), '$y$' - polarization of the impulse in Oyz ([0.1,0.9]) and the time moment when the pulse is switched off (from 5 to 55 periods).

We have made 4 iterations. During the first three iterations of the Pareto optimization procedure we have performed 322 computational experiments with the 10^6 electrons and ions in the domain and $\Delta_{x,y} = 0.1\lambda$, for each parameter we have chosen 5 values at which it's impact has been estimated by scheduler. Thus on the first iteration we have performed 125 experiments. Based on the results of these experiments a Pareto set has been formed and a single Pareto-optimal vector of parameters has been chosen – it was the following components $a_0 = 2.75$, 'y' - polarization is 0.7, switch off time is 38.75 and W_1 / W_2 is approximately 2.

For the second iteration we have repeated all the action of the first step but we have redistributed the evaluation points around the Pareto-optimal vector from the previous step, i.e. we have been selecting our new evaluation points from the region $[1.625,3.875]\times[0.5,0.9]\times[27.5,55.0]$. We do not change the computational model or accuracy of computation for this step thus we can safely omit the values computed during the previous iterations. Total number of experiments for this iteration is 98 (we evaluate each individual parameter in 5 points and some of the combinations are already computed during the previous step). After the second iteration the vector chosen for further investigation is ($a_0 = 3.03125$, 'y' - polarization = 0.7, switch off time = 41.125) and W_1 / W_2 exceeds 3.5 and gives W_1 of 1.2×10^6.

On the third iteration another 98 experiments are performed but the vector ($a_0 = 3.03125$, 'y' - polarization = 0.7, switch off time = 41.125) is still in the Pareto set and is selected for the next iteration.

During the final fourth iteration the Pareto-optimal vector has been evaluated once more with higher precision – 10^9 electron and ion pairs are distributed in the plasma domain $\Delta_{x,y} = 0.05\lambda$.

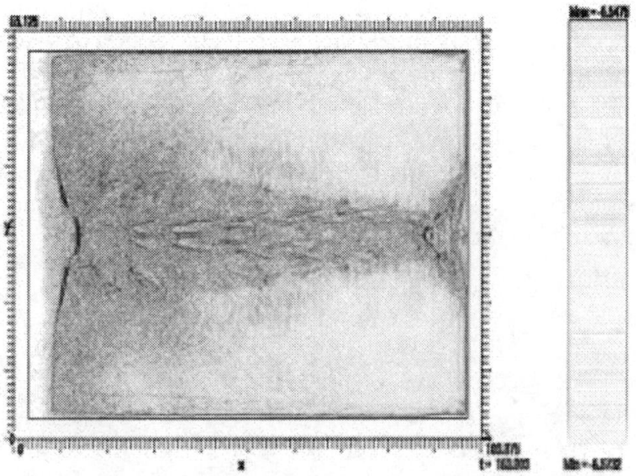

Fig.1. Electron density in the computation domain

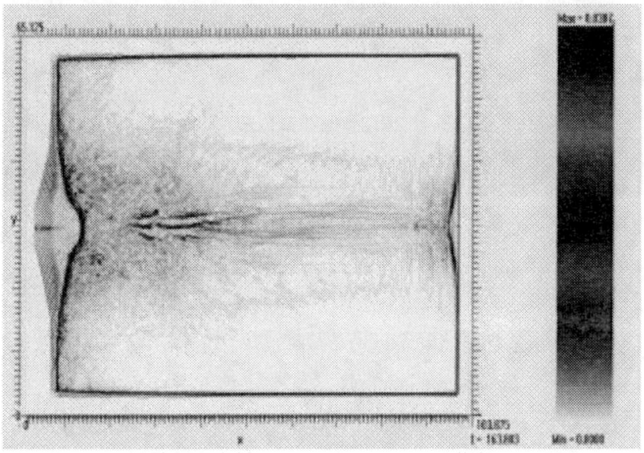

Fig.2 Ion density in the computation domain

The results of numerical computations are shown on the Figures 1 and 2. These figures demonstrate us the distribution of the electron and ion densities obtained for one of the Pareto optimal vectors and also the formation of jet streams of accelerated heavy ions is clearly visible. It can be seen on the Figure 2 that ion density is maximal along the axis of laser pulse propagation in the plasma with the slight sinusoidal variation in distance from the strait line through the domain. This propagating zone of high ion density is what leads to the high energy of ion beams which guarantee the high value of W_1.

3. Conclusions

The use of multicriteria optimization methods in selecting the conditions of a computational experiment makes it possible to effectively use the hierarchy of models and make the selection of the best values of the parameters using the Pareto optimal solution methodology.

The experimental application of our computational approach confirmed its effectiveness with respect to the problem of optimizing a computational experiment on the interaction of ultra-short laser pulses with matter.

References

[1] Bulanov S.V., Esirkepov T.Zh., Khoroshkov V.S., Kuznetsov A.V. and Pegoraro F. Oncological hadrontherapy with laser ion accelerators // Physics Letters A. 2002. Vol. 299. Issues 2-3. P.P. 240-247.

IOP Conf. Series: Journal of Physics: Conf. Series **936** (2017) 012022 doi:10.1088/1742-6596/936/1/012022

[2] Terranova F., Bulanov S.V., Collier J.L., Kiriyama H., Pegoraro F. Enabling pulse compression and proton acceleration in a modular ICF driver for nuclear and particle physics applications // Nuclear Instruments and Methods in Physics Research Section A: Accelerators, Spectrometers, Detectors and Associated Equipment. 2006. Vol. 558. Issue 2. P.P. 430-436.

[3] Esirkepov T.Zh. Exact charge conservation scheme for particle-in-cell simulation with an arbitrary form-factor // Computer Physics Communications. 2001. Vol. 135. N 2. P.P. 144–153.

[4] Zhukovskiy V.I., Salukvadze M.E. The Vector-Valued Maximin (In Series: Mathematics in Science and Engineering (Vol. 193), Edited by William F. Ames, Georgia Institute of Technology). New York: Academic Press, Inc. 1994. 404 P.

IOP Conf. Series: Journal of Physics: Conf. Series **936** (2017) 012023 doi:10.1088/1742-6596/936/1/012023

Computer simulation of population dynamics inside the urban environment

A.S. Andreev, I.N. Inovenkov, E.Yu. Echkina, V.V. Nefedov[1], L.S. Ponomarenko and V.V. Tikhomirov
Lomonosov Moscow State University, Department of Computational Math & Cybernatics, Leninskie gory, bld. 1/58, Moscow, 119991, Russian Federation

E-mail: vv_nefedov@mail.ru

Abstract. In this paper using a mathematical model of the so-called "space-dynamic" approach we investigate the problem of development and temporal dynamics of different urban population groups. For simplicity we consider an interaction of only two population groups inside a single urban area with axial symmetry. This problem can be described qualitatively by a system of two non-stationary nonlinear differential equations of the diffusion type with boundary conditions of the third type. The results of numerical simulations show that with a suitable choice of the diffusion coefficients and interaction functions between different population groups we can receive different scenarios of population dynamics: from complete displacement of one population group by another (originally more "aggressive") to the "peaceful" situation of co-existence of them together.

Keywords: mathematical modeling, the urban population, the non-linear differential equations of the diffusion type, spatiotemporal dynamics of urban population, stable and unstable regimes

1. Introduction

In recent years models of spatial economy have been widely used for describing different urban processes.

Processes of urbanization along with urban consolidation have been observed for a long time in both developed and developing countries, but modern cities are more and more characterized by decentralization tendencies. At present we can frequently observe heterogeneous rather than homogeneous urban formations.

The examples of urban formations of both established and still establishing forms may be such metropolises as Tokyo, New York, Mexico City, Moscow, Shanghai, Singapore etc.

It is well known that distribution of different population groups according to living areas represents quite complicated function of spatial variables and time period.

The given work presents computerized results which were obtained by means of "soft" mathematical model ([1]; Arnold) of interaction of two groups of urban population with non-linear spatial diffusion taken into account.

[1] To whom any correspondence should be addressed.

2. Mathematical model

The described mathematical model presents only two groups of urban population according to their social and economic indicators.

For example, urban population may be classified by their income level (the "rich" and the "poor"), according to their level of education ("more educated" and "less educated") and by a number of other factors.

In a number of economically and socially developed countries coexistence of population groups belonging to different social layers brings about serious social problems. Therefore, at present such processes are intensively studied from different viewpoints including mathematical models within spatial economy.

Interaction of similar population groups within a single urban agglomeration may be described at qualitative level using non-standard equations in partial derivatives of the diffusion type.

For simplicity and visibility the given work considers only two groups of population characterized by density of distribution $u = u(x, y, t)$ and $v = v(x, y, t)$ in point (x, y) of the fixed dimensional area D at a given period of time t.

The system of non-linear differential equations describing time dynamics of these population groups looks formal as:

$$\begin{cases} \dfrac{\partial u}{\partial t} = L_1[u] + F_1(u,v), \\ \dfrac{\partial v}{\partial t} = L_2[v] + F_2(u,v), \end{cases} \quad (1)$$

where elliptical differential operator of the second order $L_s[\omega] = \nabla \cdot (K_s \nabla \omega)$, $s = \overline{1,2}$, describes spatial diffusion of density functions of u and v population groups. In general, coefficients of diffusion K_s may have non-linear dependence on the unknown functions u and v.

Functions $F_s(u,v)$, $s = \overline{1,2}$, describe "interaction" of different population groups living in the same area. Quite realistic seems the following type of these functions:

$$\begin{aligned} F_1(u,v) &= u(a_1 - b_1 u - c_1 v) - d_1\, u v, \\ F_2(u,v) &= v(a_2 - b_2 v - c_2 u) - d_2\, u v. \end{aligned} \quad (2)$$

In this work we consider the simple area as a two dimensional area D, though this is not a necessary restriction, but it leads to a better understanding of the processes under study, namely rectangle $D = \{(x,y) : (0 \le x \le l_1) \times (0 \le y \le l_2)\}$ on the plane of variables (x, y). At the boundary $\Gamma = \partial D$ are given heterogeneous boundary conditions (boundary conditions of the 3rd type):

$$\left. h_s \frac{\partial w}{\partial n} + p_s w \right|_{\Gamma} = \varphi_s(x,y,t), \; s = \overline{1,2}, \quad (3)$$

where $\dfrac{\partial .}{\partial n}$ shows a derivative along the single outer normal at the points of boundary Γ.

It is necessary to add original conditions (the initial Cauchy conditions for equations (1) and boundary conditions (3)) and, as a result, we have the following boundary value problem consisting of two non-linear differential diffusion (parabolic) equations:

$$\begin{cases} \frac{\partial u}{\partial t} = \frac{\partial}{\partial x}\left(K_1(u,v)\frac{\partial u}{\partial x}\right) + \frac{\partial}{\partial y}\left(K_1(u,v)\frac{\partial u}{\partial y}\right) + F_1(u,v), \\ \frac{\partial v}{\partial t} = \frac{\partial}{\partial x}\left(K_2(u,v)\frac{\partial v}{\partial x}\right) + \frac{\partial}{\partial y}\left(K_2(u,v)\frac{\partial v}{\partial y}\right) + F_2(u,v), \\ h_1\frac{\partial u}{\partial n} + p_1 u\Big|_\Gamma = \varphi_1(x,y,t),\ (x,y)\in\Gamma, \\ h_2\frac{\partial v}{\partial n} + p_2 v\Big|_\Gamma = \varphi_2(x,y,t),\ (x,y)\in\Gamma, \\ u(x,y,t)|_{t=0} = u_0(x,y),\ (x,y)\in D, \\ v(x,y,t)|_{t=0} = v_0(x,y),\ (x,y)\in D. \end{cases} \tag{4}$$

This system of equations (4) was solved numerically by finite difference scheme [2; Samarskii] using the well-known Peaceman & Rachford method (or Alternating direction implicit (ADI) method) [3; Peaceman & Rachford]. The program code was realized in the MatLab development environment.

While conducting computational experiments both the initial conditions (functions) $u_0(x,y)$ and $v_0(x,y)$ and the parameters for diffusion coefficients $K_s(u,v)$ as well as functions of the so called interaction $F_s(u,v)$, $s=\overline{1,2}$, were varied.

3. Computer simulation results

Suppose that the relations between two groups of population, chosen by us, may be both "friendly" and "neutral" or even "unfriendly". A similar approach was proposed earlier in [4; Zhang].

In Zhang model numerical coefficients are $a_1 = a_2 = a$, $b_1 = b_2 = b$ and $c_1 = c_2 = c$, the summand of the form $u(a-bu-cv)$serves to describe the reaction of the population to existing economic conditions of living within urban environment.

It is possible to interpret coefficient a as physical capacity of urban space at a point (x_1,x_2). When parameter a is constant, physical capacity is homogeneous in space. If we suppose that value $(bu+cv)$ is a quantitative measure of space occupied by both groups, the value $(a-bu-cv)$ may be considered as excessive space for physical capacity.

When this value at some point exceeds zero, the living place becomes more attractive for population. It is evident when it equals zero and notion $(-d_1uv)$ and diffusion effects are negligibly small, population migration stops. Notion $(-d_1uv)$ serves to measure interaction of groups.

Coefficient d_1 may be both positive and negative (or zero). If it is positive, population "group 1" does not feel like living with population "group 2". If $d_1 = 0$, there are no prejudices between groups. If coefficient d_1 is negative high density of "group 2" attracts population of "group 1".

In the given model boundary conditions (3) are homogeneous and look the following way:

$$\begin{cases} h_1\frac{\partial u}{\partial n} + p_1 u\Big|_\Gamma = 0,\ (x,y)\in\Gamma, \\ h_2\frac{\partial v}{\partial n} + p_2 v\Big|_\Gamma = 0,\ (x,y)\in\Gamma. \end{cases} \tag{5}$$

As initial distribution density for both interacting population groups let's take "symmetrical" functions, so that the whole urban area was partially occupied by this or that population group (Fig. 1). It may be shown how urban structure for different parameters of the modeled system will change. For example, when relations between two groups of the population are of "aggressive" character and diffusion coefficients depend on the opposite functions, i.e. $K_1(u,v) = v$ and, accordingly $K_2(u,v) = u$, the obtained calculation results (Fig. 2, a-d) show that both population groups equally fill the given area, but due to "hostility" one group (the more "aggressive" one) dislodges the other one as a result, and, therefore, the urban structure of the population becomes practically homogeneous.

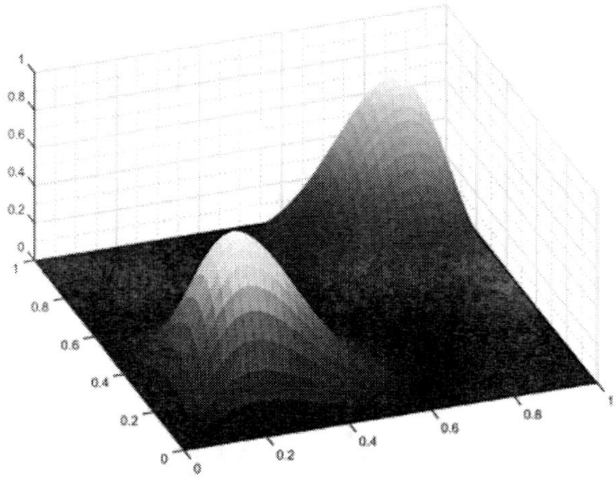

Figure 1. Initial density of distribution for functions u and v.

If coefficients $d_1 < 0$ and $d_2 = 0$, diffusion coefficients K_1 and K_2 depend only on function u, one can see the following picture of spatial-time diffusion of the population (Fig. 3, a-f).

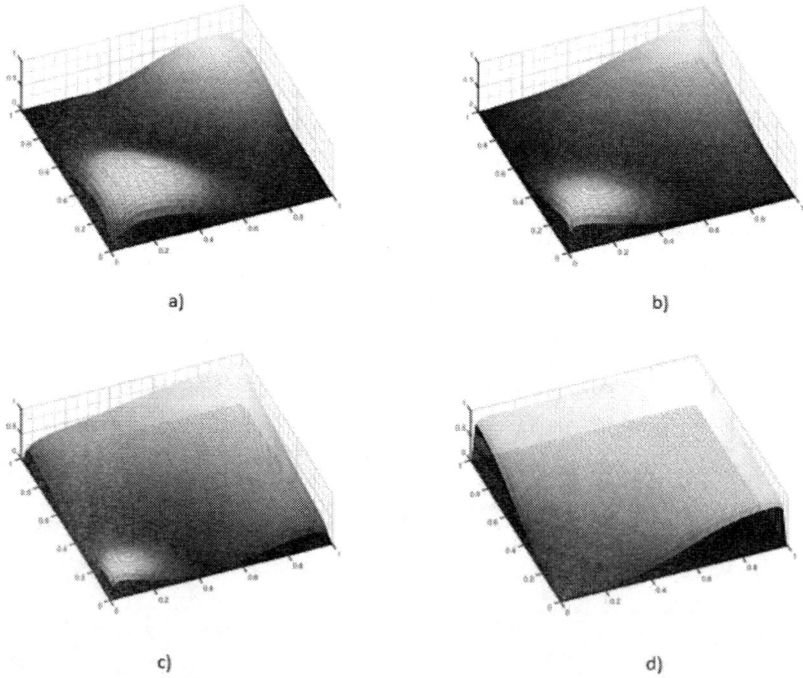

Figure 2. Effect of "geographical" diffusion of the population under "hostile" relations between population groups.

This example can be classified according to educational level when less educated people tend to live in areas with more educated population. It can also be said that areas with more favourable facilities are preferable for life which defines evident population migration from one area to another.

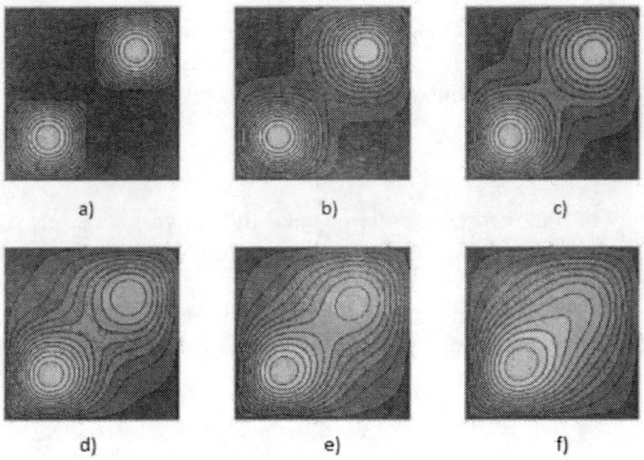

Figure 3. Redistribution of population with neutral or friendly relations ($K_1 = u, K_2 = u$).

As an example, suppose that interaction between population groups can be classified as purely competitive. Diffusion coefficients in equations are considered to have sedate dependence. Under such supposition one population group dominates over the other, dislodging it outside the city borders (Fig. 4, a-f).

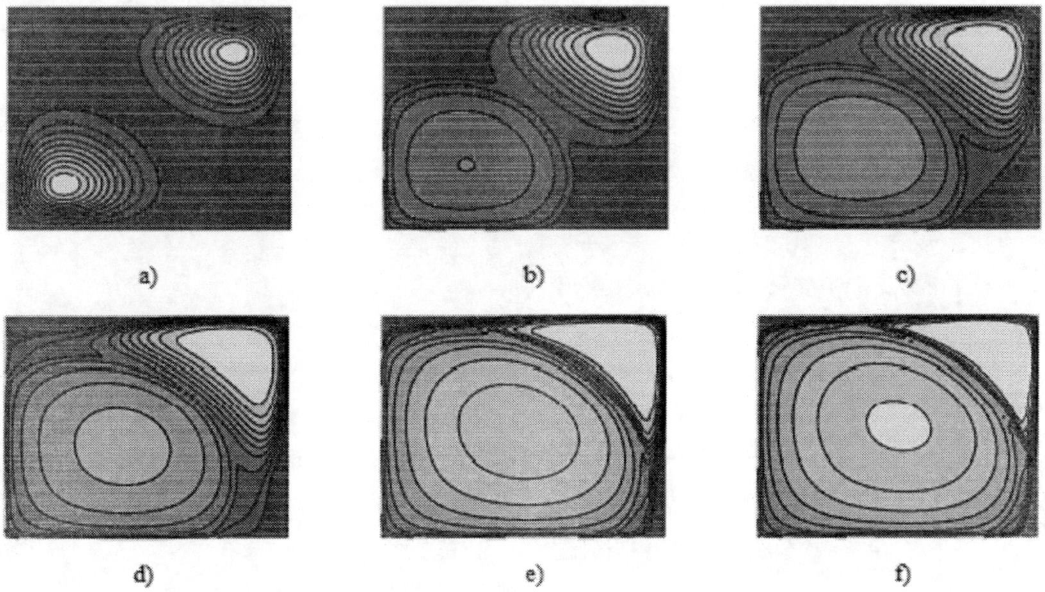

Figure 4. Redistribution of population under competitive relations ($K_1 = u^k, K_2 = u^\sigma, k < \sigma$).

4. Conclusions

The present work considers the mathematical model aimed at describing the mechanism of interaction between different population groups leaving in the same urban area.

This problem is formulated in terms of non-linear differential equations of the diffusion type. It is no use applying the so-called "hard" model in such kind of mathematical simulation. At present we can speak only about "soft" mathematical modeling.

The term "soft" mathematical model implies only the parameters and factors that are essential or available. Having a confirmed and verified mathematical model one can proceed to solve purely applied problems.

In his time (to be more exactly in 1997) an outstanding Russian mathematician, a member of the Russian Academy of Sciences Vladimir I. Arnold noted that "... success comes not so much with application of ready-made recipes ("hard" models), but rather with mathematical approach to the problems of the real world phenomena" [1; Arnold].

5. Appendix

To confirm the accuracy of the difference schemes and the convergence rate of the numerical solutions we made the corresponding analysis using precise analytical solutions of non-linear differential equations of the diffusion type.

The similar analysis was made on the example of the system of non-linear equations of "reaction-diffusion" type [5; Polyanin & Zaitsev]:

$$\begin{cases} \dfrac{\partial u}{\partial t} = \dfrac{a}{x^n}\dfrac{\partial}{\partial x}\left(x^n \dfrac{\partial u}{\partial x}\right) + u\,f(bu - cw) + g(bu \quad cw), \\ \dfrac{\partial w}{\partial t} = \dfrac{a}{x^n}\dfrac{\partial}{\partial x}\left(x^n \dfrac{\partial w}{\partial x}\right) + w\,f(bu - cw) + h(bu \quad cw), \end{cases} \tag{A.1}$$

where parameter $n = 1$ defines problem with the axial symmetry and parameter $n = 2$ defines problem with the central symmetry.

In literature function u it is customary to call 'activator' and the second function w, which leads to slowing down the described diffusion process, is called 'inhibitor'.

The system of equations (A.1) has precise solution of the type:

$$\begin{cases} u(x,t) = \phi(t) + c \cdot exp\left(\int f(b\phi - c\psi)dt\right)\theta(x,t), \\ w(x,t) = \psi(t) + b \cdot exp\left(\int f(b\phi - c\psi)dt\right)\theta(x,t), \end{cases}$$

where functions $\phi = \phi(t)$ and $\psi = \psi(t)$ are determined by the system of standard differential equations:

$$\begin{cases} \dfrac{d\varphi}{dt} = \varphi\, f(b\phi - c\psi) + g(b\phi \quad e\psi), \\ \dfrac{d\psi}{dt} = \psi\, f(b\phi - c\psi) + h(b\phi \quad c\psi), \end{cases}$$

and function $\theta = \theta(x,t)$ satisfies the linear partial derivatives equation:

$$\frac{\partial \theta}{\partial t} = \frac{a}{x^n}\frac{\partial}{\partial x}\left(x^n \frac{\partial \theta}{\partial x}\right). \tag{A.2}$$

Let's multiply the first equation of system (A.1) by 'b' and add it to the second one, multiplied by '$-c$'. Then we get the following equation:

$$\frac{\partial \zeta}{\partial t} = \frac{a}{x^n}\frac{\partial}{\partial x}\left(x^n \frac{\partial \zeta}{\partial x}\right) + \zeta\, f(\zeta) + b\,g(\zeta) \quad c\,h(\zeta), \tag{A.3}$$

where $\zeta = bu - cw$.

This equation may be obtained by transforming the first equation of the initial system of the type:

$$\frac{\partial u}{\partial t} = \frac{a}{x^n}\frac{\partial}{\partial x}\left(x^n \frac{\partial u}{\partial x}\right) + u\, f(\zeta) + g(\zeta). \tag{A.4}$$

Equation (A.3) may be transformed by division method. If solution $\zeta = \zeta(x,t)$ of the equation (A.3) is given, function $u = u(x,t)$ may be defined by solving linear equation (A.4) and function $w = w(x,t)$ according to formula $w = (bu - \zeta)/c$.

Let`s note two more important solutions of equation (A.3):

- in general case, equation (A.2) implies stationary solution $\zeta = \zeta(x)$, while the corresponding precise solution of equation (A.4) has the form $u = u_0(x) + \sum e^{\beta_n t} u_n(x)$;
- if the condition $\zeta \cdot f(\zeta) + b \cdot g(\zeta) - c \cdot h(\zeta) = k_1 \cdot \zeta \quad k_0$ takes place, the equation (A.2) is linear and takes the form $\frac{\partial \zeta}{\partial t} = \frac{a}{x^n}\frac{\partial}{\partial x}\left(x^n \frac{\partial \zeta}{\partial x}\right) + k_1\zeta + k_0$ or is transformed to the linear equation (A.2) by the change of variables $\zeta = e^{k_1 t}\bar{\zeta} - k_0 k_1^{-1}$.

Of particular, interest is the stationary case, when system (A.1) is brought to the system of non-linear elliptic equations:

$$\begin{cases} \Delta u = u\, f(bu - cw) + g(bu \quad cw), \\ \Delta w = w\, f(bu - cw) + h(bu \quad cw), \end{cases} \tag{A.5}$$

where $\Delta . = \frac{\partial^2 .}{\partial x^2} + \frac{\partial^2 .}{\partial y^2}$ is Laplace operator.

Precise solution of system (A.5) may be given as:

$$\begin{cases} u = \phi(x) + c\,\theta(x,y), \\ w = \psi(x) + b\,\theta(x,y), \end{cases}$$

with function $\phi = \phi(x)$ and $\psi = \psi(x)$ are defined from the system of usual differential equations:

$$\begin{cases} \frac{d^2\phi}{dx^2} = \phi\, f(b\phi - c\psi) + g(b\phi \quad e\psi), \\ \frac{d^2\psi}{dx^2} = \psi\, f(a\phi - b\psi) + h(a\phi \quad b\psi), \end{cases}$$

where function $\theta = \theta(x,y)$ satisfies Schrödinger linear differential equation of special type:

$$\frac{\partial^2\theta}{\partial x^2} + \frac{\partial^2\theta}{\partial y^2} = F(x)\theta(x,y)$$

with function $F(x) = f(bu - cw)$.

It can be solved by method of separation of variables. If we multiply the first equation of system (A.4) by 'b' and add the second equation, multiplied by '$-c$', we get the following equation:

$$\frac{\partial^2\zeta}{\partial x^2} + \frac{\partial^2\zeta}{\partial y^2} = \zeta \cdot f(\zeta) + b \cdot g(\zeta) - c \cdot h(\zeta), \tag{A.6}$$

where $\zeta = bu - cw$.

This equation may be transformed by means of the first equation of the initial system (A.1). As a result we obtain the equation:

$$\frac{\partial^2 u}{\partial x^2} + \frac{\partial^2 u}{\partial y^2} = u\, f(\zeta) + g(\zeta). \tag{A.7}$$

Equation (A.7) can be resolved consistently by separation technique. Existence of exact solutions for such kinetic functions as $F(\zeta) = \zeta \cdot f(\zeta) + b \cdot g(\zeta) - c \cdot h(\zeta)$ can be found in [5].

In conclusion with respect to this equation we mention two important special cases:

- in the first case, equation (A.7) gives solution of the running wave $\zeta = \zeta(z)$, where $z = k_1 x + k_2 y$ and k_1, k_2 – arbitrary constants;

IOP Conf. Series: Journal of Physics: Conf. Series **936** (2017) 012023 doi:10.1088/1742-6596/936/1/012023

- in the second case, if condition $\zeta \cdot f(\zeta)+b\cdot g(\zeta)-c\cdot h(\zeta)=c_1\cdot\zeta \quad c_0$ is fulfilled, equation (A.6) is a well-known Helmholtz linear differential equation.

Thus, for a wide class of non-linear differential systems of the form (A.1), exact solutions can be found, and so the study of which makes it possible to investigate the general properties of such systems.

6. References

[1] Arnold V.I. "Hard" and "soft" mathematical models // Published by Moscow Center for Continuous Mathematical Education (MCCME, Moscow). 2004. 32 P. (in Russian).

[2] Alexannder A. Samarskii. The theory of difference schemes // New York – Basel. Marcel Dekker, Inc. 2001. 761 P.

[3] Peaceman D.W., Rachford Jr. H.H. The numerical solution of parabolic and elliptic differential equations // Journal of the Society for industrial and Applied Mathematics. 1955. Vol. 3. No 1. P.P. 28-41.

[4] Zhang W.-B. Synergetic economics: Time and change in nonlinear economics (Springer series in Synergetics) // Berlin, Germany: Springer-Verlag. 1991. 246 P.

[5] Polyanin A.D., Zaisev V.F. Handbook of Nonlinear Partial Differential Equations (second edition). Published by Chapman & Hall/CRC Press. 2011. 1912 P.

A mathematical model for the occurrence of historical events

Teruaki Ohnishi
Institute of Science and Technology for Society,
Irifune 3, Urayasu, Chiba 2790012, Japan

E-mail: ohnishiteruaki@yahoo.co.jp

Abstract. A mathematical model was proposed for the frequency distribution of historical inter-event time τ. A basic ingredient was constructed by assuming the significance of a newly occurring historical event depending on the magnitude of a preceding event, the decrease of its significance by oblivion during the successive events, and an independent Poisson process for the occurrence of the event. The frequency distribution of τ was derived by integrating the basic ingredient with respect to all social fields and to all stake holders. The function of such a distribution was revealed as the forms of an exponential type, a power law type or an exponential-with-a-tail type depending on the values of constants appearing in the ingredient. The validity of this model was studied by applying it to the two cases of Modern China and Northern Ireland Troubles, where the τ-distribution varies depending on the different countries interacting with China and on the different stage of history of the Troubles, respectively. This indicates that history is consisted from many components with such different types of τ-distribution, which are the similar situation to the cases of other general human activities.

1. Introduction

A historical event in itself does not occur independently. Its occurrence is originated from some cause and the event thus appeared becomes a cause of another successive event. Historical events are therefore continuous in time so that they make the cause and effect to each other. If it really is the case, two events that successively occur must have a certain rule, but what a concrete rule? When the event occurs independently to others, its occurrence obeys the Poisson distribution where the time interval between two events obeys an exponential function. According to the statistics of the wars in the past by Richardson [1], although their occurrence rates approximately obey the Poisson distribution, there exist some periods when the statistics do not seem the case. Moreover the probability for the occurrence of wars seems to have varied from time to time [1]. For the general historical events, however, a statistical treatment as such has not been done so that their distinctive features in statistics are unclear. An approach to such a problem as the statistical treatment of historical events will be made here by introducing the chronological table as a data base.

In that table chronological events are listed in the order of occurrence. The following matters must be satisfied as it is used as the data base; (1) important facts and their relevant events must be objectively took up without any bias, (2) the extent of the significance of each event is indicated or it can be derived by some means, and (3) the selection of the event must not be so precise nor so coarse for the convenience of their statistical treatment. As for examples, we adopt here the following two chronological tables in modern histories where the events and their importance seem to have been sufficiently and objectively documented: (1) History of Modern China in 1941-2008 [2] and (2) Northern Ireland Troubles in 1968-1999 [3].

In the following section statistical traits regarding those histories derived from the tables are

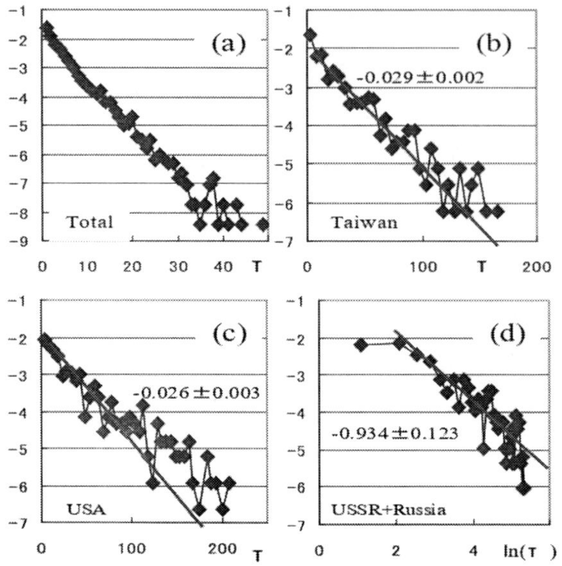

Figure 1. Logarithmic distribution of the inter-event time τ in Modern China (a) as a whole; only for the foreign relations with (b)Taiwan, (c) USA, and (d) USSR and Russia.

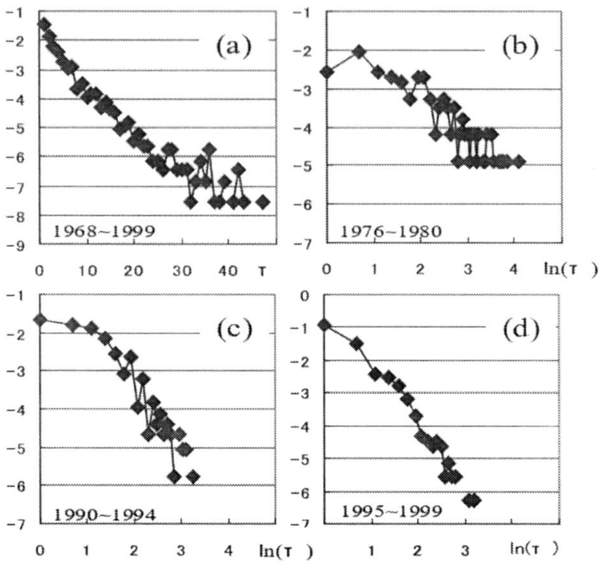

Figure 2. The same as figure 1 but in Northern Ireland Troubles (a) as a whole; only for the period of (b) 1976~1980, (c) 1990~1994, and (d) 1995~1999.

described, and in section 3 a model is proposed for the explanation of those traits. Section 4 is a mathematical description of the model, followed by a conclusion in section 5.

2. Chronological tables as historical material

The material [2] comprehensively collects the events occurred in Modern China for all genres of politics, economics, cultures, foreign affairs, military affairs, and social affairs, whereas the material [3] is for the collection of events restricted only to the Troubles. In these tables description is made for each event from one line to some tens of lines to explain the event, which is usable as a measure of the importance of the event, so that we adopt the number of explanation lines as the magnitude of its significance. The collected numbers of events in [2] and [3] are 4542 and 1913, respectively, which correspond to the average inter-event times 5.46 and 5.82 days/event, respectively.

The normalized distributions of the inter-event time τ between two successive events in the materials are respectively shown in a logarithmic scale in figures 1(a) and 2(a), while figure 3 shows the magnitude distribution of the events. In the case of [2], the τ (in days) exponentially distributes in a wide range of $\tau<40$, whereas it is also exponential in $7<\tau<30$ in the case of [3]. Those events, therefore, seem to have randomly occurred with time at a first glance. The magnitude w (in the number of lines) of the events also seems to exponentially distribute in $w<20$ if we take the segregation effect into account in the low significance region in $w<3\sim4$. When we treat the histories not comprehensively as a whole but partially within each genre of histories or within a limited time period, however, the τ differently distributes from the entirety. Figures 1(b), (c) and (d) show the τ-distributions of the foreign affairs of China with Taiwan, USA, and USSR and Russia, respectively, where we see the subtle difference of the distribution with the different partner. Figures 2(b)~(d) show, on the other hand, the cases during different stages of the Troubles, where we can see the different features of distribution from each other with different time periods.

Further to see the situation of the Troubles, the power of the τ-distribution in each year is shown in figure 4, where a power-law distribution of τ is assumed for every year in $\tau_{min}<\tau$. Such a power is determined as the value to minimize Kolmogorov-Smilnov statistic together with the value of lower limit τ_{min}, by using the method of Goldstein *et al.*[4]. The τ_{min} was determined as 3~6 days for every

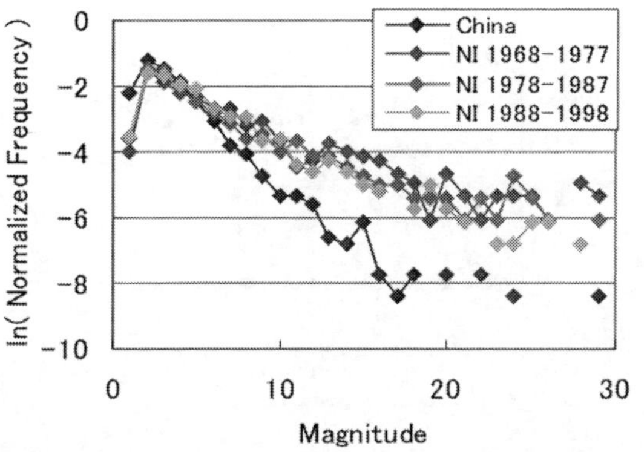

Figure 3. Magnitude distribution of events. NI is the Northern Ireland Troubles for which the distributions are separately shown for three periods.

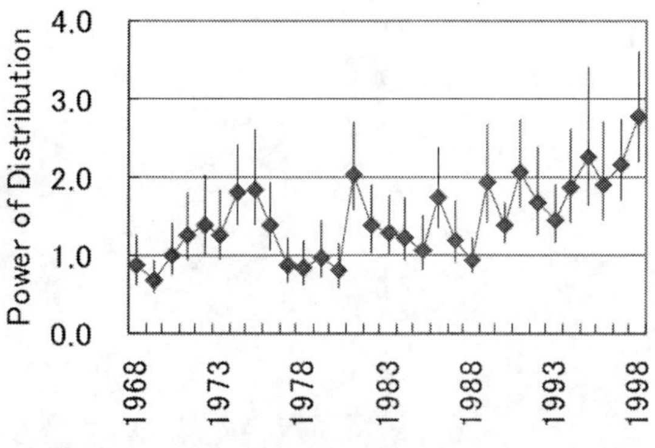

Figure 4. Powers of the distribution of τ in $\tau>\tau_m$ in the Northern Ireland Troubles. Vertical lines indicate the 95% confidence levels.

year. The upper and lower limits shown in this figure represent the 95% confidence levels, which are determined by deriving 2500 model distributions by the Monte Carlo [4]. This figure also indicates the subtle change of the τ-distribution with the progress of the situation.

To investigate what factors are acting on the mechanism that brings successive events into existence, we next introduce a model for the occurrence of historical events.

3. A model for the occurrence of historical events

3.1. Model

When an event occurs in the society, it makes the various stake holders (SHs) relevant to its problem to be in excited states. Here the SH is a general name of constituents who have an interest in the problem and make some type of reactions to thr action regarding to the problem exerted from the outside. When the action the SH receives is of a certain current and common topic for the society, the reaction that the SH makes to the action may become a new action for the other SHs to result in the continuous propagation of such an action-reaction network in the society as schematically shown in figure 5(a). In those actions the event that induces many significant reactions becomes to be cited as a historical event in the chronological table afterward.

Figure 5(b) shows a basic ingredient of such a process, where the magnitude w indicating the significance of the action. In this figure the action originating from the SH_1 stimulates or excites the SH_2 that makes a reaction after a time τ. Here the society is closed so that the information on the action is supposed to propagate within a short time period throughout the society. This seems to be an appropriate prerequisite for the modern society where there exist many types of media that make information spread in the society almost instantaneously as compared with the τ. After the occurrence of a significant event, some events relevant to it occur in a cascade manner, their magnitudes depending only on the magnitude of an event just before the event without depending on the magnitude of the original event itself. Such an assumption seems to be the case at least in the exchange of terrorism [5].

We set E as the discipline or genre such as politics or economics to which the SH_2 belongs, $\varphi(w)\mathrm{d}w$ as the probability of an action with the magnitude $[w, w+\mathrm{d}w]$ to occur within a unit time, w' as the magnitude of the influence on SH_2, $\Phi_{E,SH2}(w'';w)\mathrm{d}w''$ as the probability for the SH_2 stimulated with $[w'', w''+\mathrm{d}w'']$ that arises from the action with w, and $\psi(\tau;w')\mathrm{d}\tau$ as the probability of a reaction to be made by the SH_2 after a time τ. With the elapse of time, the amazement to an occurred event becomes dim in general and the necessity of a prompt reaction decreases since the significance and the meaning of an initial event change with time. Between the variables w'' and w', therefore, there appears an

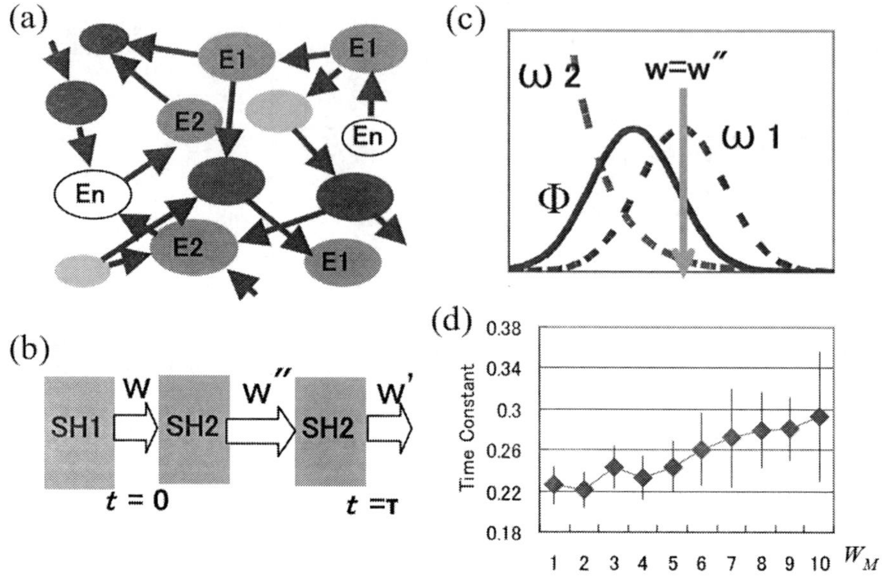

Figure 5. (a) Schematics of action-reaction network. Arrows indicate the direction of the influence which successively comes into existence, (b) basic ingredients of (a), (c) schematics for the relation of Φ, ω_1 and ω_2, (d) time constant $\lambda(w_M)$ for the exponential decrease with τ for the number of event $N(w\geq w_M, \tau)$.

effect of the oblivion of memory and the decrease of impression, so that

$$w' = \eta(\tau)\cdot w'' \tag{1}$$

We call $\eta(\tau)$ as the forgetting function in the followings. The number of events which occur during $[\tau, \tau+d\tau]$, $dN(\tau)$, is given by

$$\frac{dN(\tau)}{d\tau} = \sum_{E}\sum_{SH2}\int_{w_m}^{\infty}\int_{w_m}^{\infty}\psi_{E,SH2}(\tau;w')\,\Phi_{E,SH2}(w'';w)\,\varphi(w)\,dwdw' \tag{2}$$

$$\cong K\int_{0}^{\infty}\int_{0}^{\infty}\overline{\psi}(\tau;w')\,\overline{\Phi}(w',\tau;w)\,\overline{\varphi}(w)\,dwdw' \tag{3}$$

where K is a constant and w_m is a minimum magnitude of the event that occurs historically ($w_m \simeq 0$). The quantities $\overline{\Psi}$, $\overline{\Phi}$ and $\overline{\varphi}$ are the average values throughout the E and SH, all of which are given without the upper bars for simplicity hereafter.

3.2. Mathematical formulation

Here we derive the concrete form of equation (3) by using the data of Modern China and Northern Ireland Troubles and study under what condition of parameters it can reproduce the real τ-distribution.

According to figure 3, the distributions of the magnitude of events that occurred during three different periods of Troubles are substantially the same as each other within a margin of error. Hence an assumption will be accepted such that the distribution of the magnitude of events does not depend on time when they occur during a short time period of history, for instance for several tens of years. In this case, the quantity $\varphi(w)dw$ is given by

$$\varphi(w)dw = a\exp(-bw)dw \tag{4}$$

where a and b are parameters. On the other hand the SH_2 that perceives the action of magnitude w is supposed to make a reaction of a similar magnitude to w as it is imagined from the battles between the Catholic and Protestant camps in the Troubles [6] and from the exchange of arms under some ethnic disputes [7] although its quantitative feature is unclear. Hence we assume the probability $\Phi(w''; w)dw''$ as the product of a normal distribution function $\omega_1(w'';w)dw''$ and a compensation factor $\omega_2(w'')$ to revise it as

$$\Phi(w''; w)dw'' \equiv \omega_1(w''; w) \cdot \omega_2(w'')dw'' \tag{5}$$

$$\omega_1(w''; w) = (c/\pi)^{1/2} \exp\{-c(w-w'')^2\} \tag{6}$$

$$\omega_2(w'') = d\exp(-ew'') \tag{7}$$

where *c*, *d* and *e* are constants. Such a situation is schematically shown in figure 5(c).

Since the process for the SH_2 to be first stimulated to the magnitude w'' to the time when it give rise to a reaction is restricted only to the inside of the SH_2, not depending on the other SHs, it may be a Poisson process. Hence in this case the function $\psi(\tau;w')d\tau$ is given by

$$\psi(\tau; w')d\tau = f\lambda\exp(-\lambda\tau)d\tau \tag{8}$$

where $\lambda \equiv \lambda(w')$ and f is a constant. When the number of the event with the magnitude $w \geq w_M$, $N(w \geq w_M; \tau)$ is plotted as a function of τ in the case of the Troubles, it shows an exponentially decreasing function with τ. Here w_M ia a constant parameter (figure 3 just corresponds to the case of $w_M=1$). From such a figure we can derive the time constant λ with respect to τ for a fixed value of w_M. Then by varying the value of w_M from 1 to 10, we obtain the $\lambda(w_M)$ as a function of w_M as shown in figure 5(d), according to which the $\lambda(w_M)$ is approximately proportional to w_M, and therefore the $\lambda(w')$ is also expected to be proportional to w'. Namely

$$\lambda \equiv \bar{\lambda}_{E,SH2}(w') = g + hw' \tag{9}$$

where *g* and *h* are constants. Since the magnitude w'' becomes large when the w' is large, equation (9) means that the larger becomes the magnitude or the shock SH_2 receives by an action, the faster appears the reaction by the SH_2.

Substituting the above functions into equation (2) and performing the integration, we obtain

$$\frac{dN}{d\tau} = \frac{Kadfg}{2}\exp\left(\frac{b^2}{4c}\right) erfc\left(\frac{b}{2c^{1/2}}\right) \frac{\eta\exp(-g\tau)}{b+e+h\eta\tau}\left(G + \frac{bhH\eta}{2cg}\right) \tag{10}$$

where

$$G \equiv 1 - \exp\left(-\frac{b(b+e)}{2c}\right) \cdot \left(erfc\left(\frac{b}{2c^{1/2}}\right)\right)^{-1} \cdot G' \tag{11}$$

$$H \equiv 1 - \frac{b\alpha - 2c^{1/2}}{b\alpha}\exp\left(\frac{b-\alpha}{2c^{1/2}} - \beta\right) - \exp\left(\frac{\alpha^2}{4} - \beta\right) \cdot \left(erfc\left(\frac{b}{2c^{1/2}}\right)\right)^{-1} \cdot H' \tag{12}$$

Here $\alpha \equiv c^{-1/2}(b+e+h\eta\tau)$, $\beta \equiv bc^{-1/2}\alpha/2$, G' and H' are variables (which are given in the Appendix) including constants *a*~*f*, and α and β, and $erfc(x)$ is the conjugate error function. Equation (10) can be rewritten as

$$\frac{dN}{d\tau} \propto \frac{\eta\exp(-g\tau)}{b+e+h\eta\tau} F(\eta, \tau) \tag{13}$$

where $F(\eta,\tau) \equiv G + bhH\eta/(2cg)$.

As seen in equation (13), the distribution function of τ is mainly determined by the constants *b,c,e,g* and *h* and the form of the forgetting function η. Although we can not make clear the form of η for long-term memory as a function of time even now, it is known that there does not appear any serious inconsistency when it is assumed as a power-law function of time [8-11]. Hence we also assume as

$$\eta = \gamma\left(\frac{\tau_0}{\tau + \tau_0}\right)^k + (1-\gamma) \tag{14}$$

where γ is the oblivion fraction when $\tau \rightarrow \infty$ (hence $1-\gamma$ is the fraction that remains in the memory), *k* and τ_0 are constant.

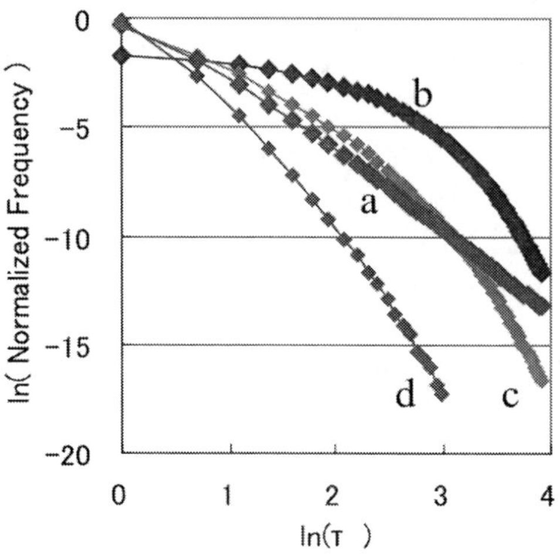

Figure 6. Model calculation for the distribution of τ with a=28, d=0.25, c=(d/π)1/2, f=1.0 and (a) b=e=0.2, g=h=0.0001, γ=1.0, τ_0=1.0, k=2.0, (b) b=e=0.2, g=0.2, h=0.01, γ=0.0, τ_0=10.0, k=2.0, (c) b=e=0.002, g=0.2, h=0.01, γ=0.0, τ_0=1.0, k=2.0, and (d) b=e=0.2, g=0.2, h=0.0001, γ=1.0, τ_0=1.0, k=3.0.

When we restrict the values of parameters within each probable range, the factor G becomes a small value whereas the H becomes a factor almost independent on τ. Hence equation (13) approximately becomes

$$\frac{dN}{d\tau} \propto \frac{\eta^2 \exp(-g\tau)}{b+e+h\eta\tau} \tag{15}$$

Depending on the relative magnitude of constants, equation (15) has the following forms of distribution

$$\frac{dN}{d\tau} \propto \exp(-g\tau) \quad \textit{when } h \textit{ is small and } (\gamma \textit{ or } \tau_0^{-1}) \textit{ are small} \tag{16}$$

$$\propto \eta\tau^{-1} = \tau^{-k-1} \quad \textit{when } g \textit{ is small and } \tau \textit{ is large} \tag{17}$$

$$\propto \tau^{-2k} \exp(-g\tau) \quad \textit{when } h \textit{ is small} \tag{18}$$

$$\propto \tau^{-1} \exp(-g\tau) \quad \textit{when } \tau_0 \textit{ is large or } \gamma \textit{ is small} \tag{19}$$

To ascertain the behavior of the distribution, figure 6 shows some normalized patterns of dN/dτ which were directly calculated by using equation (10). According to our model, the integrated result becomes several forms of an exponential type, a power-law type or an exponential-with-a-tail type depending on the various conditions even when the individual SH has a reaction that follows the Poission process.

3.3. *Parameters of the histories of Modern China and the Troubles and their interpretation*

The τ-distribution in our model is mainly determined by the extent of oblivion, namely the values of k, τ_0 and γ , and the probability for an event to occur under the influence of the preceding event, g and h. In the case of Modern China (figure 1), the τ-distribution in the foreign affairs differs with interacting countries. The dN/dτ for Taiwan shows an exponential behavior that corresponds to the curve b in figure 6, so that the h and γ must be small. Hence in this case both countries of Modern China and Taiwan have passed time without forgetting their foreign policies and diplomatic relations in the past irrespective of their significance. The dN/dτ for USSR and Russia seems to obey a power-law in τ>9 days, which corresponds to the case of small g and h (hence a small λ), and a large γ. This is the situation such that the memory scarcely remains between two countries though the occurrence of reactive event is retarded. On the other hand the dN/dτ for USA is an exponential-with-a-tail type so that the values of b, e and γ are all small, or h is small and γ is large. In the former case a reaction occurs with almost similar magnitude as the action with scarcely forgetting it, whereas in the latter case the probability for occurring the reaction is not so dependent on the magnitude of the shock SH receives by an action, with scarcely retaining the memory.

In the case of Northern Ireland Troubles, the power of dN/dτ differs with the situation of the Troubles as seen in figure 4, the value of power being $-(k+1)$. Since the power is small in 1970s, peoples slowly forgot the events. The extent of oblivion became large although it fluctuated with time in 1980s to result in a much larger value in 1990s, indicating faster oblivion of the event with the longer continuation of the Troubles.

4. Conclusion

In the history of Modern China, the distributional feature of inter-event time has subtly differed with the different countries of foreign interaction, and in Northern Ireland the distribution has also changed with the stage of the Troubles. The τ-distribution as a whole, therefore, is just a superposition of many components with subtly different distributions from each other.

The occurrence of historical events is well noticed as the resultant that is participated from many stake holders. When we consider the cases of foreign affairs between two countries and the cases like the Troubles where friends and enemies are definitely divided, however, history can be interpreted as the repetition of the exchange of action and reaction between two substances that have different values from each other. Extending such an image to a general case, the occurrence of a historical event can be interpreted as the similar phenomenon to the exchange of information between two participants as the correspondence between two persons [12,13], the replying to a received e-mail [13,14], or the conflict around the articles in Wikipedia [15,16]. Since the historical event is just the resultant of the human activities, it is not so strange that its characteristics are similar to the other activities as such.

Appendix: Forms of the factors *G'* and *H'*

The mathematical expressions of G' and H' which appeared in equations (11) and (12) are respectively given by

$$G' = \left[2 - erfc\left\{\frac{b}{2c^{1/2}} - \frac{b+e+h\eta\tau}{2c^{1/2}}\right\}\right] \cdot \exp\left\{-\frac{bh\eta\tau}{2c} + \frac{(b+e+h\eta\tau)^2}{4c}\right\} \qquad (A1)$$

$$H' = 2 - erfc\left(\frac{b}{2c^{1/2}} - \frac{\alpha}{2}\right) + \frac{c^{1/2}(\alpha^2-2)}{b\alpha}\left\{2 - erfc\left(\frac{b-c^{1/2}\alpha}{2c^{1/2}}\right)\right\} - \frac{2c^{1/2}}{\pi^{1/2}b}\exp\left(\frac{(b-c^{1/2}\alpha)^2}{4c}\right) \qquad (A2)$$

References

[1] Richardson L F 1960 *Statistics of Deadly Quarrels* (California, Pacific Grove)
[2] Ando M 2010 *Modern China Chronology 1941-2008* (Tokyo, Iwanami Shoten)
[3] Bew P and Gillespie G 1999 *Northern Ireland: A Chronology of the Troubles 1968-1999*, 2nd ed (Dublin, Gill & Macmillan)
[4] Goldstein M L, Morris S A and Yen G G 2004 *Europ.Phys.J.* **B 41** 255-8
[5] Jhonson N, Carran S, Botner J *et al.* 2011 *Science* **333** 81-3
[6] Arther P and Jeffery K 1996 *Northern Ireland since 1968*, 2nd ed (Oxford, Blackwell pub.)
[7] Ryan S 1995 *Ethnic Conflict and International Relations*, 2nd ed (Dartmouth, Adershot)
[8] Johansen A 2001 *Physica* **A 296** 539-46
[9] Harder U and Paczuski M 2006 *Physica* **A 361** 329-36
[10] Crane R and Sornette D 2008 *PNAS* **105** 15649-53
[11] Sano Y, Yamada K, Watanabe H *et al.* 2013 *Phys.Rev.* **E 87** 012805
[12] Malmgren R D, Stouffer D B, Campanharo A and Nunes Amaral L A 2009 *Science* **325** 1696-700
[13] Barabasi A L 2005 *Nature* **435** 207-211
[14] Eckmann J P, Moses E and Sergi D 2004 *PNAS* **101** 14333-7
[15] Yasseri T, Sumi R, Rung A *et al.* 2012 *Plos ONE* **7** e28869
[16] Yun J, Lee S H and Jeong H 2016 *Phys.Rev.* **E 93** 012307

IOP Conf. Series: Journal of Physics: Conf. Series **936** (2017) 012025 doi:10.1088/1742-6596/936/1/012025

Modeling of multidimensional light bullets in Fermi liquid and ADS/CFT correspondence

NN Konobeeva[1], MB Belonenko[1,2]
[1] Volgograd State University, Volgograd, Russia, 400062
[2] Volgograd Institute of Business, Volgograd, Russia, 400048

E-mail: yana_nn@volsu.ru

Abstract. This paper discusses two-dimensional and three-dimensional solitary electromagnetic waves propagation in Fermi liquid in the framework of ADS/CFT correspondence. The electromagnetic field is classically considered within the framework of Maxwell's equations. The obtained effective equation is numerically analyzed and the state of the electromagnetic field that is localized in two / three spatial dimensions is revealed.

1. Introduction

The study of the Fermi liquid properties led to investigate various kinds of nonlinear optical phenomena in these media. So recently, the dynamics of few cycle optical pulses (FCP) with a plane wave front in the Fermi liquid was studied [1]. However, there are a number of questions connected with going beyond the approximation of a plane wave front and studying the dynamics of propagation with taking into account the transverse dispersion.

Another interesting challenge is investigation of the dynamics of three-dimensional optical pulse localized in all spatial directions. This confirms the fact that the study of such pulses, called "light bullets"[2,3], has recently become very popular due to the many practical applications [5,6].

In the last few years, there has been a growing interest in ideas from both superstring theory and quantum gravitation [7,8], and they find application in the physics of condensed matter. And, namely, ADS/CFT correspondence allows us to get the Green's function [13,14] for both Fermi liquid and non-Fermi liquid, and for marginal Fermi liquid. The ideas of ADS/CFT correspondence (the anti-de Sitter/conformal field theory correspondence), which is closely related to the conformal field theory, which has the applications in various fields of solid state theory and works particularly fruitfully in the region of phase transitions.

The following approach corresponds to the ideology of the so-called "semi-holography" [15], where the dispersion law of strongly interacting subsystems is taken based on the ADS/CFT correspondence, and further consideration is "classically" based on the developed approaches based on the Maxwell and Boltzmann equations.

2. Basic equations

Let us begin with the dispersion law (here k is measured from the Fermi level):

$$\varepsilon = \varepsilon\left(\left|\vec{k}\right|\right), \left|\vec{k}\right| = \left(k_x^2 + k_\perp^2\right)^{0.5} \qquad (1)$$

Or it is more convenient to present in the form:

IOP Conf. Series: Journal of Physics: Conf. Series **936** (2017) 012025 doi:10.1088/1742-6596/936/1/012025

$$\varepsilon = \varepsilon\left(k_x, k_\perp^2\right) \tag{2}$$

We have the following expression for the density of current [16]:

$$j = n_0 \int dk\, v_x\left(k_x - A, k_\perp^2\right) f\left(k_x, k_\perp^2\right) \tag{3}$$

where f is the distribution Fermi function, $v_x = \partial\varepsilon\left(k_x, k_\perp^2\right)/\partial k_x$. Here we consider that electric field has only one component: $E_x=-\partial A/c\partial t$, then the speed of light c is assumed equal to unity.

In the case of low temperatures, we replace f with 1 in a layer near the Fermi surface. The depth of the layer will select for reasons of particles conservation:

$$\int_{-\Delta}^{\Delta} dk = \int_{ZB} dk\, f \tag{4}$$

Then

$$j = n_0 \int_{-\Delta}^{\Delta} dk_x \int_0^{\sqrt{2}\Delta} k_\perp dk_\perp \int_0^{2\pi} dk_\varphi \frac{\partial\varepsilon\left(k_x - A, k_\perp^2\right)}{\partial k_x} = \pi n_0 \int_0^{2\Delta^2} dx\left(\varepsilon\left(\Delta - A, x\right) - \varepsilon\left(-\Delta - A, x\right)\right) \tag{5}$$

We specify the dispersion law using ADS/CFT correspondence ideas and following the work [17].

Let us consider two- and one- dimensional boundary theories, corresponding to a gravity theory in asymptotically ADS4. Lagrangian can be chosen in standard type for both gravitational field and gauge field with U(1) symmetry [17]:

$$L = \frac{1}{2g^2}\left(R + \frac{6}{\Lambda^2}\right) - \frac{1}{4e^2} F_{\mu\nu} F^{\mu\nu} \tag{6}$$

where g is gravitational interaction constant, e is a charge, $F_{\mu\nu}$ is stress tensor for the gauge field. The given choice of the Lagrangian is in accordance with the solution as the Reissner-Nordstrom-AdS black hole. As it is shown in [17] the Green's function has the form:

$$G_R\left(w, \vec{k}\right) = \frac{h}{k - k_F - w/w_f - \sum\left(w, k_F\right)} \tag{7}$$

where $k = \left|\vec{k}\right|, k_F$ is Fermi momentum, v_F is Fermi velocity. The main difference of the given Green's function which poles determine the dispersion law of quasiparticles lies in the fact that the quasiparticles dissipation highly depends on parameter: $v_k = \left(\dfrac{m^2L^2\mu^2 + 3k^2 - \gamma^2\mu^2}{6}\right)$.

We note that v_k (hereinafter simply) is closely related to the critical exponent.

The dispersion law in the case of low temperatures is given by the following expression:

$$\varepsilon(k) = c\left(k_x^2 + k_\perp^2\right)^{v/2} \tag{8}$$

Then, we obtain the current density in the following form:

$$j(A) = \frac{\pi n_0 c}{0.5v + 1}\left(\left(\Delta + A\right)^{v+2} + \left(\Delta - A\right)^{v+2} + \left(\left(\Delta - A\right)^2 + 2\Delta^2\right)^{0.5v+1} - \left(\left(\Delta + A\right)^2 + 2\Delta^2\right)^{0.5v+1}\right) \tag{9}$$

Let us write a wave equation in 2D (10a) and 3D (10b) cases:

$$\frac{\partial^2 A}{\partial y^2} + \frac{\partial^2 A}{\partial z^2} - \frac{1}{c^2} A_{tt} + \frac{4\pi}{c} j(A) = 0 \tag{10a}$$

$$A_{tt} = \frac{1}{r}\frac{\partial}{\partial r}\left(r\frac{\partial A}{\partial r}\right) + \frac{\partial^2 A}{\partial z^2} + \frac{1}{r^2}\frac{\partial^2 A}{\partial \varphi^2} + 4\pi j(A) \tag{10b}$$

Further, we consider in the cylindrical symmetry:

$$\frac{\partial}{\partial\varphi} \to 0$$

An important point that needs to pay special attention is associated with a cylindrically symmetric case. Due to the inhomogeneity of the field along an axis (for definiteness, let field is to be specific along the x axis), and the current is not uniform. In consequence of the current inhomogeneity we observe accumulation of a charge at some region. The easiest way to assess this accumulation from the law of the charge conservation:

$$\frac{d\rho}{dt}+\frac{dj}{dx}=0 \tag{11a}$$

$$\rho\approx\tau\frac{j}{l_x} \tag{11b}$$

ρ is the charge density, j is the density of current along x axis, τ is the electromagnetic field duration, l_x is the characteristic length on which the electric field changes.
(11b) is shown, that accumulated charge, which generates an additional electric field interferes with the field of few cycle pulses, essentially depends on the FCP duration. Estimates show that the FCP duration is $\tau\sim10^{-15}$s, and the current density is $j\sim10^{12}$ A/sm^2, ε~2. Then we have for additional field the following value: $E\sim10^{-3}$ V/sm. It should be noted, that the FCP field is about 10^6 V/sm, therefore we can neglect with accumulation of a charge for femtosecond pulses. According to calculation results in Ref. [18] for carbon nanotubes case and pulse with $\tau\sim10\text{-}20^{-15}$s, the accumulated charge is about 1-2% of the charge contributing to the current (10).

3. Results and conclusion

The effective equation (12) was solved numerically by using direct difference scheme of the "cross" type [19]. The time and coordinate steps were determined from the standard stability conditions. Difference scheme steps where iteratively decreased twice until the solution became unchanged in the eighth decimal place. The initial pulse profile is supposed to have a following form:

$$A(z,0)=Q\exp\left(-\left(\frac{z-z_0}{\gamma_z}\right)^2\right)\exp\left(-\left(\frac{y-y_0}{\gamma_y}\right)^2\right),$$
$$\frac{dA(z,0)}{dt}=\frac{2Q(z-z_0)v_z}{\gamma_z^2}\exp\left(-\left(\frac{z-z_0}{\gamma_z}\right)^2\right)\exp\left(-\left(\frac{y-y_0}{\gamma_y}\right)^2\right) \tag{12}$$

here Q is the initial amplitude, its initial velocity along z-axis, $\gamma_{z,y}$ are the pulse width along z and y, z_0, y_0 are the initial coordinate of the pulse center.
Figure 1 demonstrates an example of electromagnetic wave propagation in Fermi liquid. It shows the dependence of the electric field strength $E(z,y,t)$ from the coordinate at different times (v=0.8).
We can observe a spreading of the light bullet with a loss of amplitude at the propagation in the sample (it is less by a factor of 5 than for figure d in comparison with figure a). Moreover, the pulse is not destroyed. It only changes the form, which indicates the possibility of fairly stable pulse propagation in Fermi liquid.

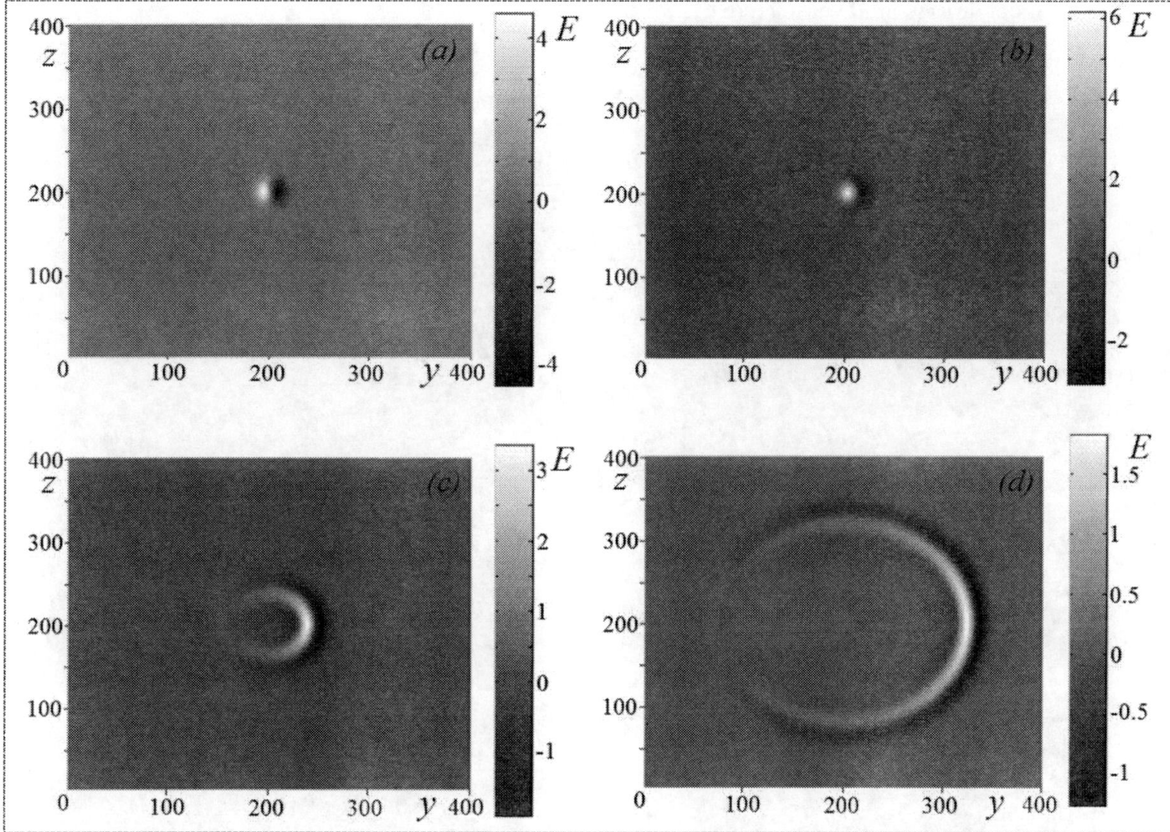

Figure 1. Dependence of the strength $E(z,y,t)$ for two-dimensional electromagnetic pulse (1 r.u.=10^6 V/m) on coordinates: a) the initial profile of the pulse; b) the profile of the pulse at t=5.0·10^{-13}s; c) 1.5·10^{-12}s (ν=0.8). Coordinate y corresponds to 30 nm and z – 20 nm.

Figure 2 shows the pulse intensity at the different time points in the case when ν=0.6.

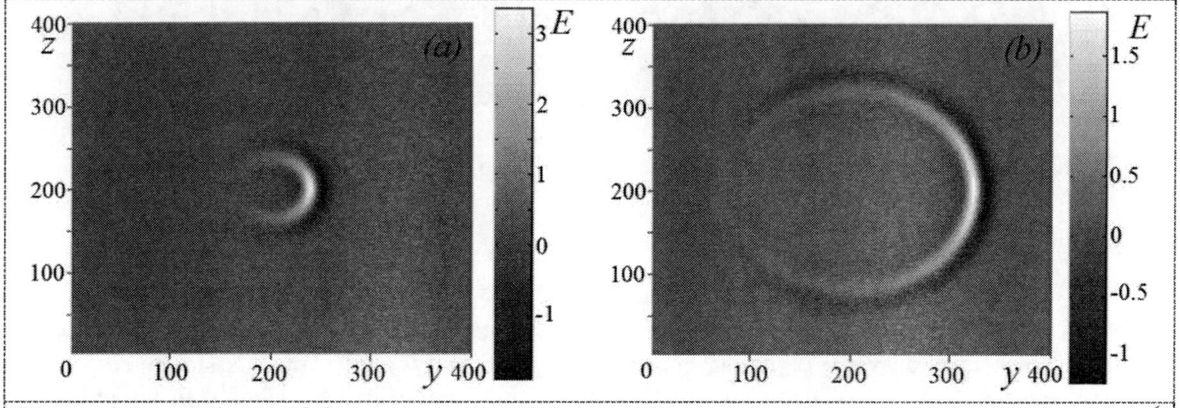

Figure 2. Dependence of the strength $E(z,y,t)$ for two-dimensional electromagnetic pulse (1 r.u.=10^6 V/m) on coordinates: a) t=5.0·10^{-13}s; b) 1.5·10^{-12}s (ν=0.6). Coordinate y corresponds to 30 nm and z – 20 nm

As can be seen from Figures 1 and 2, an influence of the parameter ν is manifested in the following: as the value of ν decreases, an increase in the pulse amplitude is observed, and the smaller spreading in the transverse direction does not change the character of the propagation of the pulse.

Further we consider the pulse in the cylindrically symmetric coordinate system at the ν=0.8 (figure 3). The initial conditions in 3D case have the following form:

$$A(z,0)=Qexp\left(-\left(\frac{z-z_0}{\gamma_z}\right)^2\right)exp\left(-\left(\frac{r}{\gamma_r}\right)^2\right),$$
$$\frac{dA(z,0)}{dt}=\frac{2Q(z-z_0)v_z}{\gamma_z^2}exp\left(-\left(\frac{z-z_0}{\gamma_z}\right)^2\right)exp\left(-\left(\frac{r}{\gamma_r}\right)^2\right) \quad (13)$$

here r is the radius, γ_r determines the pulse width in the radial direction.

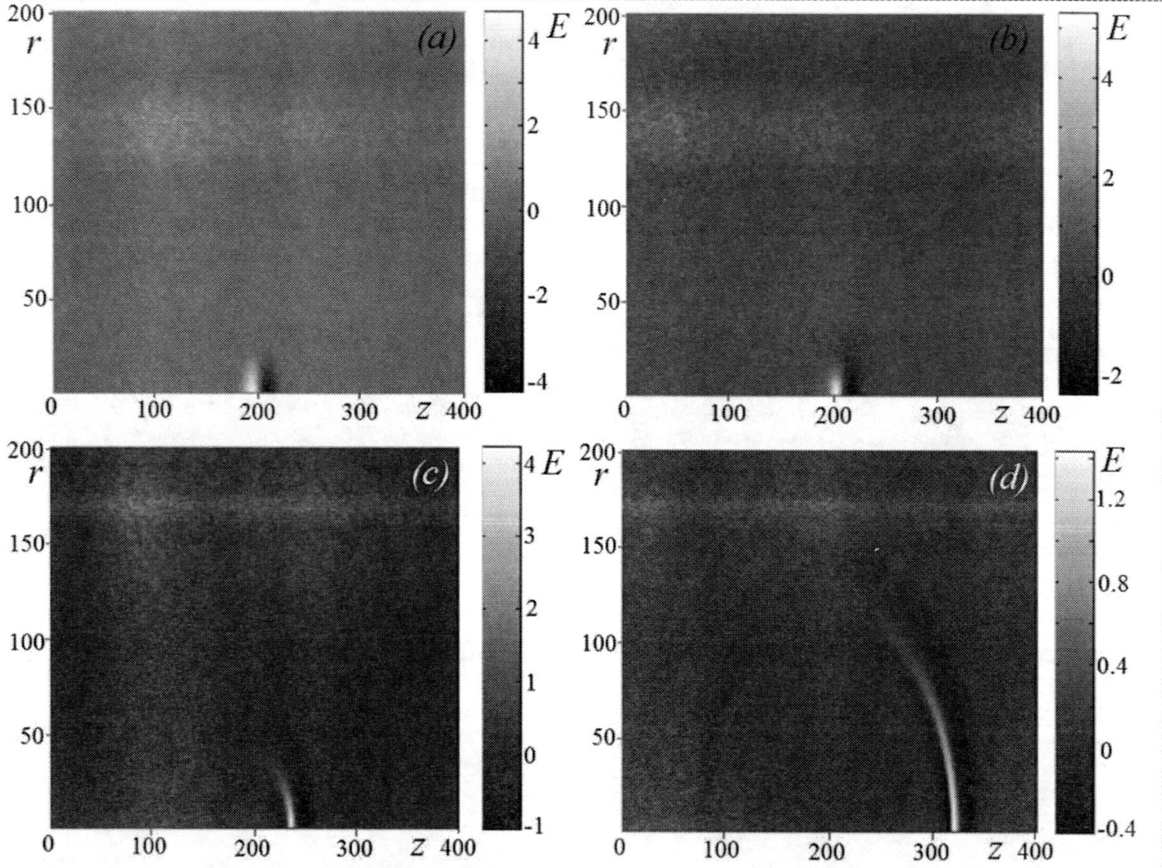

Figure 3. Dependence of the strength $E(z,r,t)$ for 3D electromagnetic pulse (1 r.u.=10^6 V/m) on coordinates: a) the initial profile of the pulse; b) t=10^{-13}s; c) t=5.0·10^{-13}s; d) 1.5·10^{-12}s (ν=0.8). Coordinate r corresponds to 30 nm and z – 20 nm.

In 3D case, there is a greater spreading of the pulse compared to the 2D consideration. Note that the pulse structure is changed. The pulse is divided into several peaks along the propagation axis, that breaking into the transverse direction, continue to spread with conservation its structure. Also, we investigate the influence of parameter ν at the pulse propagation in Fermi liquid in the cylindrical coordinate system. We obtain that a decrease in the parameter ν also leads, as in the previous case, to an increase in the pulse amplitude, while it has a stabilizing effect, manifested in a smaller diffraction spreading. That is, using the value of ν, one can influence the process of the pulse propagation in the Fermi liquid.

An important question is also the stability of the obtained solutions with respect to perturbations that depend on the angle φ. The stability analysis can be carried out on the basis of the linearized equation (10b), which for small perturbations δA will have the form:

$$\delta A_{tt} = \frac{1}{r}\left(r\frac{\partial \delta A}{\partial r} \right) + \frac{\partial^2 \delta A}{\partial z^2} + \frac{1}{r^2}\frac{\partial^2 \delta A}{\partial \phi^2} + 4\pi \frac{\partial j(A)}{\partial A}\delta A \tag{14}$$

We note that the last term in (14) is calculated on the solutions $A(z,r,t)$ of equation (10b). Because of the linearity of (14) δA, one can search as:

$$\delta A \propto \delta A(z,r,t)exp(im\phi) \tag{15}$$

Next, we can calculate the corresponding corrections to the electric field: $\delta E = -c^{-1}\partial \delta A/\partial t$

The corresponding results for the electric field are presented in Figure 4 and Figure 5. These figures show the maximum magnitude modulus δE (throughout the calculation area) as a function of time and number m.

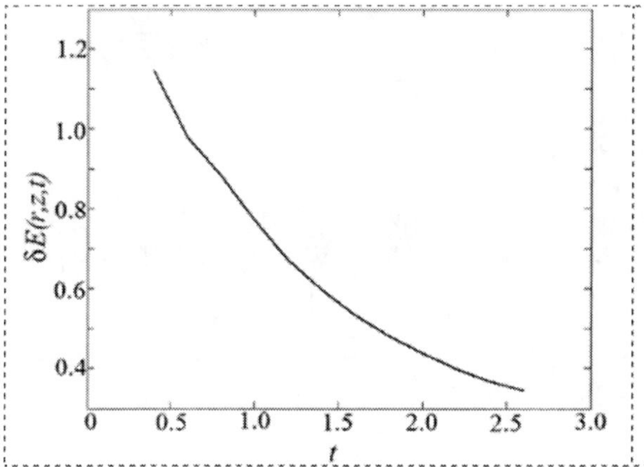

Figure 4. Dependence of maximum modulus $E(r,z,t)$ on time t (1 r.u.=10^{-12} s).

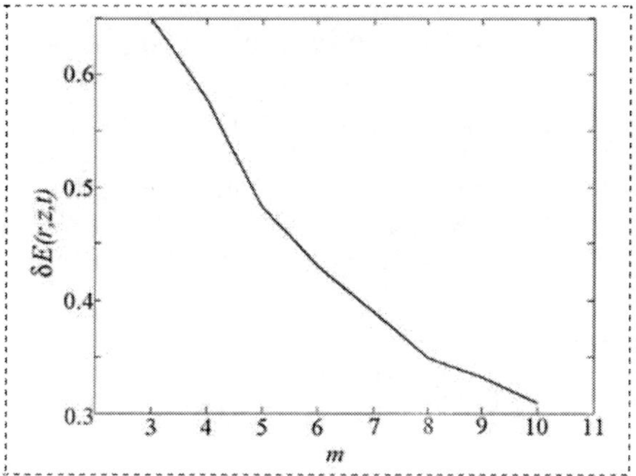

Figure 5. Dependence of maximum $\delta E(r,z,t)$ on m

Figure 5 (for m=5) shows that the perturbations decrease with time. According to Figure 6 (calculations are made for one instant of time), the perturbation decreases the faster, the larger the number m. These results allows us to conclude that the obtained solutions are stable to the angular perturbations.

The above dependences are typical, and it follows that in Fermi liquid a stable propagation of pulses localized in two/three dimensions is possible, which are often called "light bullets" in the literature. Calculations showed that, although there is a diffraction spreading of the pulse in the transverse direction to the direction of propagation, the pulse as a whole retains its shape. It should also be noted that there is a partial curvature of the pulse front, which arises from diffraction.

Key results of this work may be summarized as follows:

1. We proposed a model describing the dynamics of 2D and 3D laser pulse in a Fermi liquid in the framework of the ADS/CFT correspondence. The approximations used in constructing the model are indicated.
2. Numerical calculations have shown that in the 2D case stable nonlinear waves, light pulses localized in two directions, which are analogs of "light bullets" are possible.
3. In the propagation of a "bullet" in a Fermi liquid, its spreading in the transverse direction is rather weak, and the energy is mainly concentrated in the central region of the pulse.
4. It is shown that the value of the parameter ν has a significant effect on the pulse form.

IOP Conf. Series: Journal of Physics: Conf. Series **936** (2017) 012025 doi:10.1088/1742-6596/936/1/012025

5. The results of the calculations allow us to conclude that the obtained solutions are stable with respect to the perturbations in the angle φ.

References

[1] Belonenko M B, Konobeeva N N, Galkina E N 2015 *Mod. Phys. Lett.* B **29** 1550096
[2] Vlasov V N, Petrischev I A, Talaniv V I 1971 *Izv. Vuzov. Radiof.* **14** 1353
[3] Silberberg Y 1990 *Opt. Lett.* **15** 1282
[4] Popov A S, Belonenko M B, Lebedev N G, Zhukov A V, Paliy M 2011 *Eur. Phys. J.* D **65** 635
[5] Panagiotopoulos P, Papazoglou D G, Couairon A, Tzortzakis S 2013 *Nat. Commun.* **4** art. 2622
[6] Minardi S et al. 2010 *Phys. Rev. Lett.* **105** 263901
[7] Chowdhury D, Raju S, Sachdev S, Singh A, Strack Ph 213 *Phys. Rev.* B **87** 085138
[8] Charmousis C, Gouteraux B, Kim B S, Kiritsis E, Mayer R 2010 *J. High En. Phys.* **1011** 151
[9] Maldacena J M 1998 *Adv. Theor. Vath. Phys.* **22** 231
[10] Maldacena J M 1999 *Int. J. Theor. Phys.* **38** 1113
[11] Policastro G, Son D T, Strinets A O 2002 *J. High En. Phys.* **9** 43
[12] Pal Sh S 2011 *Phys. Rev.* D **84** 126009
[13] Lee S-S 2009 *Phys. Rev.* B **80** 165102
[14] Herzog C P 2009 *J. Phys.* A **42** 343001
[15] Faulkner T, Polchinski J 2011 *J. High En. Phys.* **2011** 12
[16] Konobeeva N N, Belonenko M B 2013 *Tech. Phys. Lett.* **39** 87
[17] Benini F 2012 *Fortschr. Phys.* **60** 810
[18] Zhukov A V, Bouffanais R, Fedorov E G, Belonenko M B 2013 *J. Appl. Phys.* **114** 143106
[19] Bahvalov N S, Zhidkov N P, Kobelkov G M 2008 Numerical Methods (Moscow: BINOM)

Acknowledgments
This work was supported by the Ministry of Education and Science of the Russian Federation, project No. MK-4562.2016.2 and the government task No.2.852.2017/4.6.

IOP Conf. Series: Journal of Physics: Conf. Series **936** (2017) 012026 doi:10.1088/1742-6596/936/1/012026

A model of jam formation in congested traffic

N Zh Bunzarova[1,2], N C Pesheva[2], V B Priezzhev[1] and J G Brankov[1,2]

[1] Bogoliubov Laboratory of Theoretical Physics, Joint Institute for Nuclear Research, 141980 Dubna, Russia
[2] Institute of Mechanics, Bulgarian Academy of Sciences, 1113 Sofia, Bulgaria

E-mail: nadezhda@imbm.bas.bg

Abstract.
We study a model of irreversible jam formation in congested vehicular traffic on an open segment of a single-lane road. The vehicles obey a stochastic discrete-time dynamics which is a limiting case of the generalized Totally Asymmetric Simple Exclusion Process. Its characteristic features are: (a) the existing clusters of jammed cars cannot break into parts; (b) when the leading vehicle of a cluster hops to the right, the whole cluster follows it deterministically, and (c) any two clusters of vehicles, occupying consecutive positions on the chain, may become nearest-neighbors and merge irreversibly into a single cluster. The above dynamics was used in a one-dimensional model of irreversible aggregation by Bunzarova and Pesheva [Phys. Rev. E 95, 052105 (2017)]. The model has three stationary non-equilibrium phases, depending on the probabilities of injection (α), ejection (β), and hopping (p) of particles: a many-particle one, MP, a completely jammed phase CF, and a mixed MP+CF phase. An exact expression for the stationary probability P(1) of a completely jammed configuration in the mixed MP+CF phase is obtained. The gap distribution between neighboring clusters of jammed cars at large lengths L of the road is studied. Three regimes of evolution of the width of a single gap are found: (i) growing gaps with length of the order $O(L)$ when $\beta > p$; (ii) shrinking gaps with length of the order $O(1)$ when $\beta < p$; and (iii) critical gaps at $\beta = p$, of the order $O(L^{1/2})$. These results are supported by extensive Monte Carlo calculations.

1. Introduction

We study a discrete-time discrete-space model of congested traffic on finite segments of a single-lane road, which allows a detailed description of the growth of clusters of jammed vehicles under different boundary conditions. The stochastic dynamics is a particular case of the generalized Totally Asymmetric Simple Exclusion Process (gTASEP), which has been used in a one-dimensional model of irreversible aggregation [1].

Various generalizations of the TASEP have been developed and used to model traffic flow, with a special focus on the formation and spreading of traffic jams, see, e.g., [2, 3]. The problem is of particular interest also for various transport processes taking place in the living cells. From the viewpoint of physics, molecular motors are proteins or macromolecular complexes that utilize chemical energy to move collectively on a single track in a way resembling vehicular traffic [4].

In Sec. II we formulate the model, present the phase diagram and some basic properties which were obtained in [1]. In Sec. III an exact (in the thermodynamic limit) analytic expression for the stationary probability P(1) of finding completely jammed configurations is derived. In

Sec. IV we consider the evolution of the gaps between consecutive clusters under different boundary conditions: three regimes of asymptotic behavior of the gaps lifetimes are found and the corresponding asymptotic form of P(1) is evaluated close to the threshold of complete jamming. Perspectives for further developments are given in the concluding Sec. V.

2. The Model

We consider an open chain of L sites, labeled consecutively by the index $i = 1, 2, \dots, L$. Each site of the lattice can be empty or occupied by just one particle. The bulk dynamics of the generalized totally asymmetric simple exclusion process (gTASEP) corresponds to a discrete-time backward-ordered update with two hopping probabilities p and $\tilde{p}$. The update of the whole configuration starts with the last site: if site L is occupied, the particle is removed from it with probability β. In the bulk, consecutive updates of all pairs of nearest-neighbor sites are performed in the backward order $(L-1, L), \dots, (i, i+1), \dots, (1, 2)$. Single particles, as well as the leading particle of a cluster of close packed particles, hop one site to the right with probability p, when that site is vacant. If the leading particle of a cluster has hopped, each of the remaining particles of that cluster hop to the empty site on the right with probability $\tilde{p}$. Finally, the first site of the chain is updated: a particle is injected with probability $\alpha > 0$, if the site was vacant at the beginning of the configuration update, and with probability $\tilde{\alpha} = \min\{\alpha\tilde{p}/p, 1\}$, if it was initially occupied by a particle, but became vacant after the update of the bond $(1, 2)$. Note that when $\tilde{p} = p$ one has the TASEP with backward-sequential update, and when $\tilde{p} = 0$ - the TASEP with parallel update. In the general case of $0 < \tilde{p} < 1$, the model was studied under periodic boundary conditions in [5, 6].

We adopt the above stochastic dynamics with $\tilde{p} = 1$, which implies the following properties: (a) Existing clusters of jammed vehicles do not break into parts; (b) Whenever the leading vehicle of a cluster hops with probability p to an empty site on the right, the whole cluster follows it as a whole entity. When a cluster reaches the last site of the chain, it starts hopping forward with probability β, rather than p. (c) Any two clusters of vehicles, occupying consecutive positions on the chain, may become nearest-neighbors and merge irreversibly into a single cluster.

As shown in [1], this model has three non-equilibrium stationary phases: a many-particle one, MP, a phase with completely filled configuration, CF, and a mixed MP+CF phase, see Fig. 1.

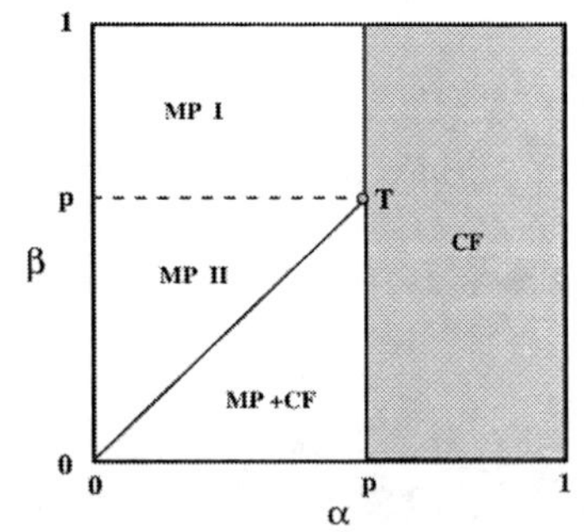

Figure 1. Phase diagram in the plane of injection (α) - ejection (β) probabilities. The small circle with the letter 'T' beside denotes the triple point.

The many-particle phase MP occupies two regions, MP I and MP II; it contains a macroscopic number of clusters of jammed cars of size $O(1)$ as $L \to \infty$; MP I and MP II differ by the shape of the local density profile and the type of evolution of the rightmost gaps. The phase CF consists of completely jammed configurations carrying the current $J = \beta$. The phase MP+CF is mixed: the completely filled configurations are perturbed by short living gaps entering the chain from the first site.

IOP Conf. Series: Journal of Physics: Conf. Series **936** (2017) 012026 doi:10.1088/1742-6596/936/1/012026

3. Probability of complete jamming

An expression for the probability P(1) of a completely jammed configuration in the mixed MP+CF phase can be derived from the global stationarity condition $J_{\rm in} = J_{\rm out}$, where $J_{\rm in}$ ($J_{\rm out}$) is the input (output) current. If the system is in a completely jammed configuration, which happens with probability P(1), the first site becomes empty with probability β, as a result a particle ejection from the last site of the chain and a deterministic shift of the whole cluster one site to the right; then a particle may enter the system with probability $\tilde{\alpha} = \alpha/p$. The total probability of this event is $\mathrm{P}(1)(\alpha/p)\beta$. When the system is not completely jammed, which takes place with probability 1-P(1), a particle enters at the first site with probability α, as shown in [1]. By equating the input current $J_{\rm in} = P(1)\alpha\beta/p + [1 - P(1)]\alpha$ to the output current $J_{\rm out} = \beta$ (since $\rho_L^{\rm MP+CF} = 1$), one solves for P(1) and obtains: $P(1) = p(\alpha - \beta)/[\alpha(p - \beta)]$, $\beta \le \alpha < p$. This result is in excellent agreement with the Monte Carlo simulation data.

4. Evolution of the system configurations

In particle hopping discrete-space traffic models, one often uses the dual representation of configurations in terms of empty sites positions, instead of particle coordinates. In the case of our model, such a representation leads to a very peculiar dynamics of the inter-cluster gaps. First, because existing clusters cannot break up, gaps may appear only at the first site of the chain. Second, gaps disappear when two consecutive clusters merge or when the rightmost cluster leaves the system and the next one reaches the last site of the chain. Third, and most interesting feature of the gap dynamics is that, as long as two consecutive gaps exist, the distance between them remains constant due to the conservation of particles in the cluster between them. Thus, in the space-time picture, the neighboring edges of two coexisting gaps are forced to move in parallel and gaps may not cross or merge, see a computer simulation in Fig. 2. The evolution of a mixed configuration in MP+CF, close to the boundary with MP II, is shown in Fig. 3. When $(1 - \tilde{\alpha}) \ll 1$, a vacancy at the first site of a completely filled configuration appears, on the average, $\bar{N} = [(1 - \tilde{\alpha})\beta]^{-1} \gg 1$ updates (discrete time steps). The width of each gap performs a random walk, which begins with an initial state containing one or several neighboring empty sites, and ends up when the random walk reaches the origin. If the average lifetime $\bar{n}$ of the gap is $\bar{n} < \bar{N}$, then an estimate of the probability P(1) is given by $\mathrm{P}(1) \simeq 1 - \bar{n}/\bar{N}$.

In general, when $\beta \neq p$, the size of the rightmost gap performs an *asymmetric* random walk: its right edge belongs to the cluster which hops one site to the right with the ejection probability β; on the other hand, its left edge moves one site to the right with probability p. Therefore, after each update, the gap width increases by one site with probability $p_g = \beta(1 - p)$, decreases by one site with probability $q_g = p(1 - \beta)$, and remains the same with probability $r = 1 - q_g - p_g$. Three regimes of evolution of the gap width follow:

(1) *Growing gaps.* In the region MPI $\beta > p$ and $p_g > q_g$. Hence, the gap width asymptotically grows with the number $n \gg 1$ of time steps as $W(n) \propto (p_g - q_g)n$. The gap will exist until its lower edge reaches the end of the chain, that is for L/p time steps on the average, while its two edges coexist until the time moment L/β, when the upper edge reaches the end of the chain. According to the random walk theory, the probability that a gap with unit initial width collapses, before the time moment L/β, tends to $q_g/p_g < 1$, as $L \to \infty$. Therefore, the probability that the growing gap survives until the end of the chain is $P_{surv} = 1 - q_g/p_g = (\beta - p)/[\beta(1-p)]$. On the average, such a gap will appear after $\bar{N}_c$ updates, hence the estimate $\mathrm{P}(1) \simeq 1 - \bar{n}\bar{N}_c$ follows.

(2) *Short-living gaps.* When $\beta < p$, one has $p_g < q_g$ and the gap between the newly growing cluster and the cluster which is leaving the system from its right boundary closes in a finite number of time steps. An upper estimate can be given by the result for a random walk on an infinite chain with initial state at site $i = 1$ and one absorbing state at the origin $i = 0$: $\bar{n} < (q_g - p_g)^{-1} = (p - \beta)^{-1}$.

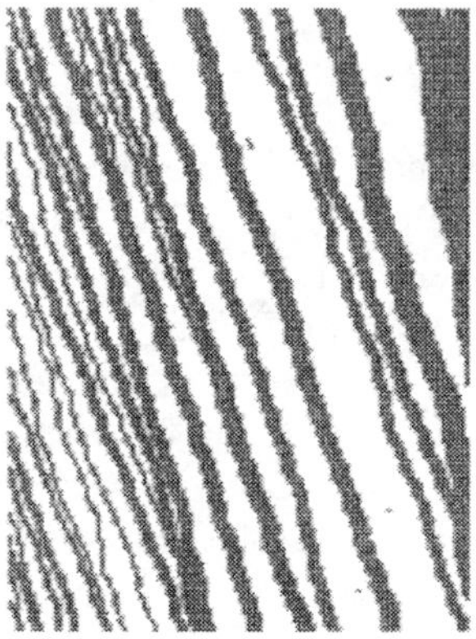

Figure 2. Simulated time evolution of clusters and gaps (white strips) in the MPII phase, at $L = 200$, $\alpha = 0.2$, $\beta = 0.3$, and $p = 0.6$.

Figure 3. Simulated time evolution of clusters and gaps (white strips), in the MP+CF phase, at $L = 200$, $\alpha = 0.31$, $\beta = 0.3$, and $p = 0.6$.

(3) *Critical gaps.* On the line segment $\beta = p$, $\alpha \leq p$, the gap width performs a symmetric random walk with $p_g = q_g = p(1-p)$, $r = 1-2p(1-p)$. The calculation of $\bar{n}$ in this case becomes involved, since the average length of a symmetric random walk on the infinite chain, with initial state at $i = 1$ and one absorbing state at $i = 0$, diverges. Therefore, we have to take into account that the lifetime of the random walk on a finite chain is limited, on the average, by L/p. We introduce the average lifetime $\bar{n}_M$ of a gap during $M = L/p$ updates, $\bar{n}_M = \sum_{m=1}^{M} m f_{1,0}^{(m)}$, where $f_{1,0}^{(m)}$ is the probability that the initial gap of unit size will vanish at update m. By using the generating function of $f_{1,0}^{(m)}$, see Eq. (53) in Ref. [7], and applying a chain of approximations, we obtain for $M \gg 1$: $\bar{n}_M \simeq \sqrt{L}/[\sqrt{2\pi}p^{3/2}(1-p)]$. This implies the finite-size scaling result $P(1) \simeq 1 - b_1 L^{1/2}(p - \alpha)$, verified by our computer simulations.

5. Conclusions

The gap representation seems promising for calculation of the density profile over the whole chain segment. The generalized TASEP with $\tilde{p} \in (0, 1)$ can be applied to take into account processes of cluster fragmentation.

Acknowledgment

VBP acknowledges the support of the RFBR, grant 16-02-00252.

References

[1] Bunzarova N Zh and Pesheva N C 2017 *Phys. Rev.* E **95** 052105
[2] Chowdhury D, Santen L and Schadschneider A 2000 *Phys. Rep.* **329** 199
[3] Helbing D 2001 *Rev. Mod. Phys.* **73** 1067
[4] Tipathi T and Chowdhury D 2008 *Phys. Rev.* E **77** 011921
[5] Derbyshev A E, Povolotsky A M and Priezzhev V B 2015 *Phys. Rev.* E **91** 022125
[6] Aneva B L and Brankov J G 2016 *Phys. Rev.* E **94** 022138
[7] Cox D R and Miller H D 2001 *The Theory of Stochastic Processes* (Boca Raton: Chapman and Hall/CRC) chapter 2 pp 22-75

Simple method to improve comb frequency characteristic in passband and folding bands

G. Jovanovic Dolecek[1]
Department of Electronics,
Institute INAOE, Enrique Erro 1, 72740 Tonantzintla, Puebla, MEXICO

E-mail: gordana@ieee.org

Abstract. The most simple decimation filter is comb filter, which has all coefficients equal to unity. The filter must attenuate aliasing which occur in the bands around the comb zeros, called folding bands. Additionally, it must have flat magnitude characteristic in the passband in order to avoid the deterioration of the decimated signal. However, comb filter has pure attenuation in the folding bands and droop in the passband of interest. The attenuation in the folding bands can be increased by cascading comb filters. Nevertheless, this process increases comb passband droop. We present here a simple method to improve simultaneously comb magnitude characteristic in both: passband and folding bands. The cascade of combs is used to improve the attenuation in the folding bands, and the compensator, cascaded with combs, improve the passband characteristic. Two cases are elaborated depending if the passband of interest is a wideband or narrowband. The comparisons with some methods from the literature confirm the benefit of the proposed method.

1. Introduction

Decimation is decreasing of sample rate by an integer [1]. This process generates aliasing, which must be eliminated by a filter, called decimation filter. The most simple decimation filter is comb filter which does not require multipliers because all its coefficients are equal to unity. The system function of comb is given as:

$$H(z) = \left[\frac{1}{M}\frac{1-z^{-M}}{1-z^{-1}}\right]^K, \qquad (1)$$

where M is decimation factor and K is the number of cascaded combs, also known as comb order.
Comb filter must have high attenuation in folding bands, where aliasing occur, and a flat passband characteristic to avoid the deterioration of decimated signal.
Simplest way to increase attenuation in folding bands is achieved by increasing the number of cascaded combs. However, the passband droop increases with the increase of the comb parameter K. As a result, the cascade of combs is usually limited to five. The comb passband droop is corrected using compensator filters proposed in [2-6]. Similarly, different methods have been proposed to improve aliasing rejection, like [7-10]. In [11] is introduced sharpening technique to simultaneously improve comb magnitude characteristic in passband and folding bands. The five comb corrector filters

[1] Any correspondence address to: gordana@ieee.org

are introduced in [12], each for a given value of *K*, *K*=1,...,5, to simultaneously improve passband and folding band comb magnitude characteristic.
We propose here to use a high value of parameter *K* to improve alias rejection in folding bands, and to use compensator to compensate for a high passband droop. We consider two cases depending if the passband of interest is wideband or narrowband. Te rest of the paper is organized in the following way. Next Section describes wideband case, while the Section 3 is dedicated to a narrowband case. The comparisons are presented in Section 4.

2. Wideband case

In this section we discuss the comb magnitude characteristic improvement for wide passband of interest, which is defined by the passband edge frequency [11]:

$$\omega_p = \frac{\pi}{2M}. \tag{2}$$

First, the choice of compensator is discussed and followed by the proposed method, which is illustrated with two examples and compared with the original comb filter.

2.1. Choosing the compensator

The magnitude response of the comb compensator has to approximate the inverse of the comb magnitude response in the comb passband. As a result, the cascade of combs and compensator should have the magnitude characteristic approximately equal to 1, in the passband [6]. Our goal here is that the compensated comb has an maximum absolute value of passband deviation equal or less than 0.1 dB. To this end, we consider here the compensator [6], recently proposed in literature with the following magnitude response, at low rate:

$$\left|G(e^{j\omega})\right| = [1 + A\sin^4(\omega/2)][1 + B\sin^2(\omega/2)], \tag{3}$$

where *A* and *B* are the amplitude of sinewave functions which depend on the comb parameter *K*.
The system function of the compensator is given as:

$$G(z) = \{2^{-4}A[1 + z^{-4} - 2^2(z^{-1} + z^{-3}) + (2^2 + 2)z^{-2}] + z^{-2}\} \times \{2^{-2}B[-1 + 2z^{-1} - z^{-2}] + z^{-1}\}. \tag{4}$$

2.2. Proposed method

We proposed here using the comb parameter *K*=6 and the compensator from [6], requiring 9 adders. As a difference to method in [6], using MATLAB simulation we find the parameters *A* and *B* both equal to unity. This choice requires 9 adders keeping the maximum of the absolute value of the passband deviation lesser than 0.1dB. As a result, the proposed compensator has the following system function:

$$G_p(z) = \{2^{-4}[1 + z^{-4} - 2^2(z^{-1} + z^{-3}) + (2^2 + 2)z^{-2}] + z^{-2}\} \times \{2^{-2}[-1 + 2z^{-1} - z^{-2}] + z^{-1}\}. \quad (5.$$

The method is illustrated in two examples.
Example 1:
Consider *M*=20 and *K*=6. Fig.1 compares magnitude responses of comb and the proposed filter. The zooms in the passband and the first folding bands are also shown. Note that the passband requirement is satisfied, and that the worst case attenuation is about 60dB.

Example 2:
In this example we chose *M*=32 and *K*=6, to show that the choice of *A*=*B*=1 does not depend on decimation factor *M*. Fig.2 confirms that the maximum of absolute value of passband deviation and the worst case of attenuation are similar to those, obtained in Example 1.

3. Narowband case

In this section we discuss the comb magnitude characteristic improvement taking narrow passband, which is defined by the passband edge frequency [11]:

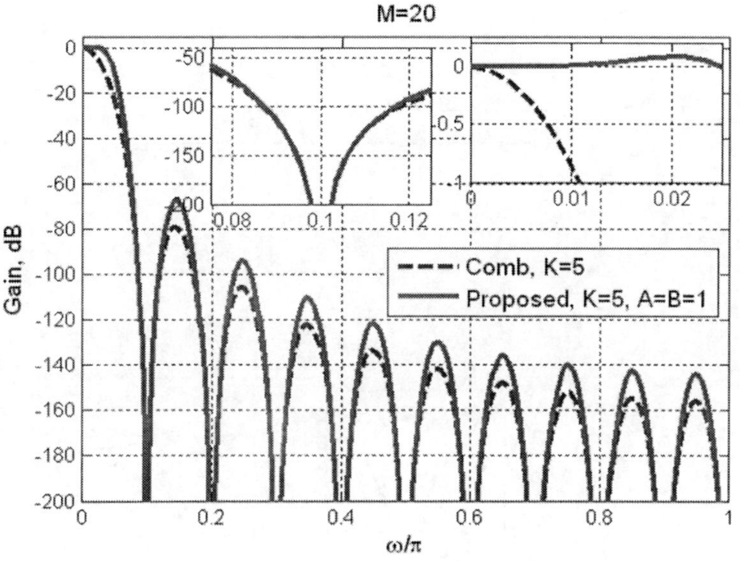

Figure 1. Magnitude responses in Example 1.

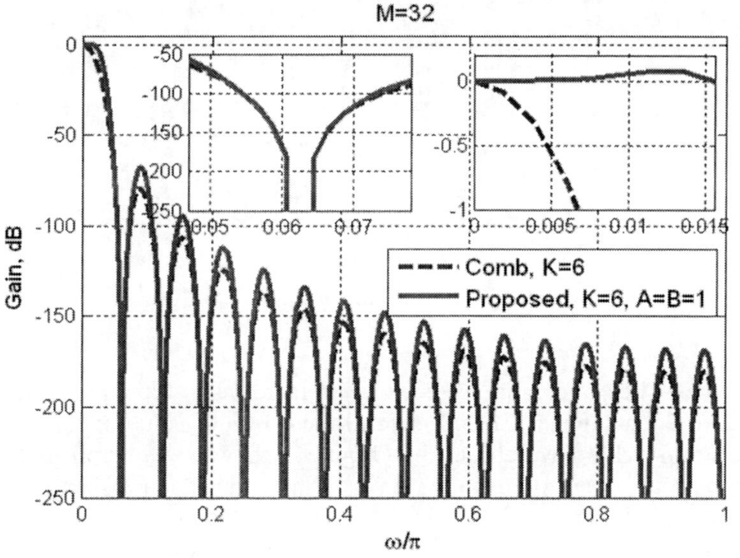

Figure 2. Magnitude responses in Example 2.

$$\omega_p = \frac{\pi}{8M}. \tag{6}$$

Lake in wideband case, we use the compensator from [6] with parameters A=B=1. However, in this case we consider K=6,…,8, to get more choices for the worst case attenuation, keeping the maximum absolute value of passband deviation lesser than 0.1dB. We also consider the value of K=9. Yet, in this case the maximum absolute value of the passband deviation is about 1.6 dB. The parameters are shown in Table 1.The method is illustrated in the following two examples.

Example 3:

We chose M=15 and the minimum attenuation in the first folding band of 150 dB. From Table 1 it follows K=7. Fig. 3 contrasts magnitude responses of comb and the proposed filter, including the corresponding first folding bands and the passbands.

Table 1. The worst case attenuations for different values of *K*.

K	*A* [dB]
6	142
7	165
8	190
9	213

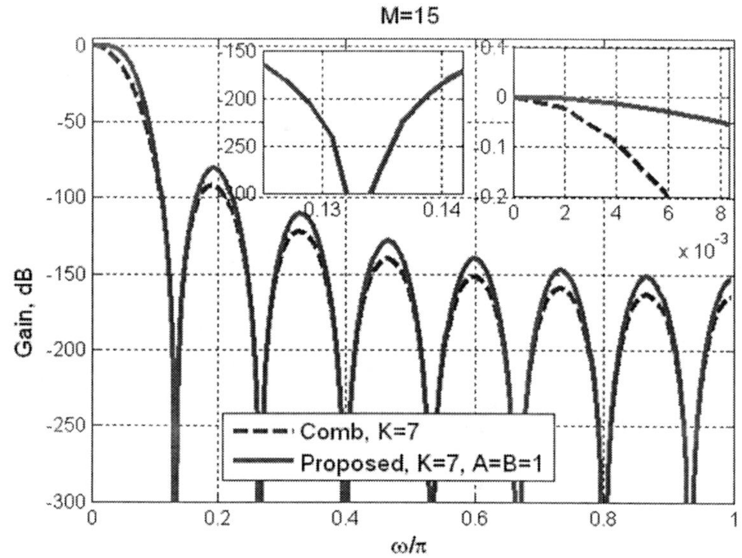

Figure 3. Magnitude responses in Example 3.

Example 4:
In this example we chose the same value for decimation factor as in Example 3, M=15. However, here the minimum required attenuation is of 200 dB and the absolute value of the passband deviation is lesser than 0.2 dB. From Table I we chose K=9. Figure 4 shows the overall magnitude responses of comb and the proposed filter. The zooms in the first folding banda and passbands are also shown.

4. Some comparisons

In this section we compare the proposed method with two methods, proposed so far, for simultaneously improvement of passband and folding bands of comb filter.

4.1. Comparison with sharpening method

The sharpening method to simultaneously improve passband and folding bands in comb filter, (narrowband case), using the sharpening polynomial in the form: $H_{sh}=3H^2-2H^3$, where H is the prototype filter, is introduced in [11]. For a sake of comparison we take M=16, and the comb filter with the parameter K_1=3, as a prototype filter. In the proposed method the parameter K=9. The corresponding magnitude responses are contrasted in Fig.4. Note that the folding bands are similar, while the proposed method provides better passband compensation. However, the sharpening method is more complex.

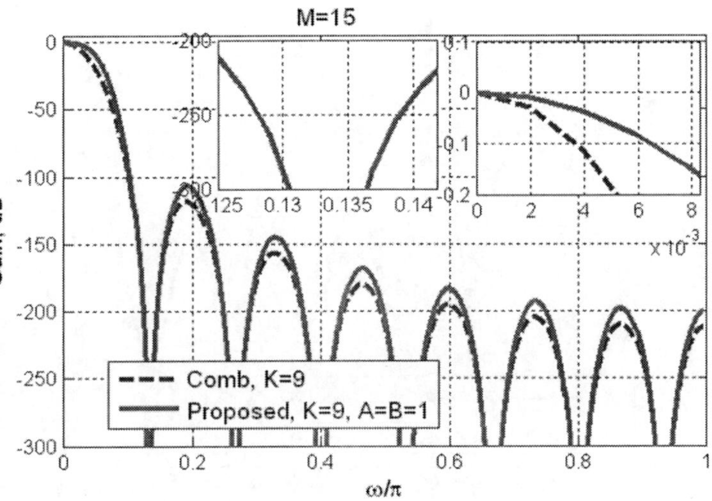

Figure 4. Magnitude responses in Example 4.

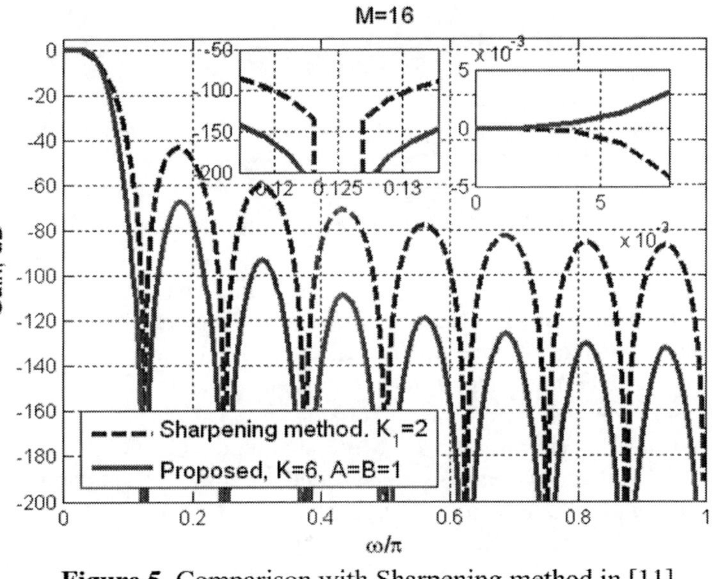

Figure 5. Comparison with Sharpening method in [11].

4.2. Comparison with Corrector filter method

The five corrector filters for simultaneous improvement of comb magnitude responses in wide passband and folding bands, for values of K_c=1,…,5, are introduced in [12].

For the sake of comparison with method in [12] we choose K_c=4. The corrector filter has coefficients [12]:

$$[1,1,-2,-8,1,24,24,1,-8,-2,1,1].$$

In the proposed method K=6, A=B=1. The magnitude responses are compared in Fig.6. The proposed method has better attenuation in even folding bands and also has better passband compensation. The attenuations in the first folding bands are similar. The method in [12] requires 11 adders, (two more than in proposed method. Additionally, the method in [12] is limited to even decimation factors, while the proposed method has limited value of K.

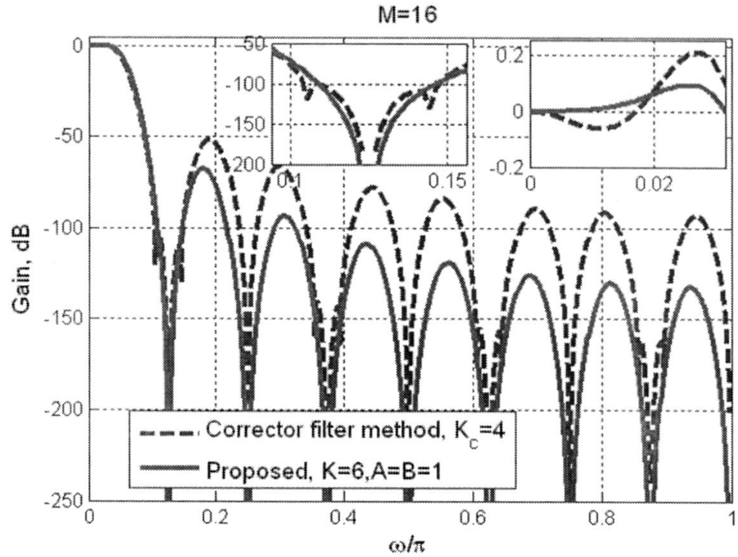

Figure 6. Comparison with Corrector filter method.

Acknowledgement
This work is supported by Conacyt Mexico.

References

[1] Jovanovic Dolecek G 2002 *Multirate Signal Processing: Design and Application* (Hershey: IGP).
[2] Jovanovic Dolecek G and Mitra S K 2008 *Electronics Letters* Simple method for compensation of CIC decimation filter **44** 1162-1163.
[3] Jovanovic Dolecek G 2009 *Electronics Letters* Simple wideband CIC compensator **45**, 1270-1272.
[4] Molnar G Vucic M 2011 *IEEE Transactions on Circuits and Systems II: Express Brief* Closed-form design of CIC compensators based on maximally flat error criterion **58** 926-930.
[5] Jovanovic Dolecek G and Fernandez-Vazquez A 2014 Trigonometrical approach to design a simple wideband comb compensator *International Journal for Electronics and Communications (AEUE)* **68** 437–441.
[6] Jovanovic Dolecek G, Garcia Baez R, Molina Salgado G and de la Rosa JM 2016 Novel multiplierless wideband comb compensator with high compensation capability. *Circuits, Systems and Signal Processing* Published on line August 2016, 1-19
[7] Presti L 2000 *IEEE Transactions on Circuits and Systems II: Express Brief* Efficient modified sinc filters for sigma-delta A/D converters **47** 1204–1213.
[8] Laddomada M 2007 *IEEE Trans on Circ. and Syst.-I* Generalized comb decimation filters for sigma-delta A/D converters **54**, 994–1005.
[9] Baez R and Jovanovic Dolecek G 2015 *International Journal of Electronics* **4** 569-582,
[10] Jovanovic Dolecek G, Garcia Baez R, and Laddomada M 2017 Design of efficient multiplierless modified cosine-based comb decimation filters: Analysis and implementation. *IEEE Transactions on Circuits and Systems I: Regular Papers* **5** 2031–2049.
[11] Kwentus A and Wilson Jr A 1997 *IEEE Trans. on Signal Processing* Application of filter sharpening to cascaded integrator-comb decimation filters **45**, 457-467.
[12] Jovanovic Dolecek G Feranandez-Vazquez A Novel droop-compensated comb decimation filter with improved alias rejections *International Journal for Electronics and Communication (AEUE)* **67**, 387-396.

IOP Conf. Series: Journal of Physics: Conf. Series **936** (2017) 012028 doi:10.1088/1742-6596/936/1/012028

Finding exact constants in a Markov model of Zipfs law generation

V.V. Bochkarev, E.Yu. Lerner, A.A. Nikiforov, A.A. Pismenskiy

Kazan Federal University, Kremlevskaya str.18, Kazan 420018, Russia

E-mail: vbockarev@mail.ru, eduard.lerner@gmail.com, antnikiphorov@yandex.ru, aa.pismenskiy@gmail.com

Abstract. According to the classical Zipfs law, the word frequency is a power function of the word rank with an exponent −1. The objective of this work is to find multiplicative constant in a Markov model of word generation. Previously, the case of independent letters was mathematically strictly investigated in [Bochkarev V V and Lerner E Yu 2017 International Journal of Mathematics and Mathematical Sciences Article ID 914374]. Unfortunately, the methods used in this paper cannot be generalized in case of Markov chains. The search of the correct formulation of the Markov generalization of this results was performed using experiments with different ergodic matrices of transition probability P. Combinatory technique allowed taking into account all the words with probability of more than e^{-300} in case of 2 by 2 matrices. It was experimentally proved that the required constant in the limit is equal to the value reciprocal to conditional entropy of matrix row P with weights presenting the elements of the vector π of the stationary distribution of the Markov chain.

1. Introduction. Formulation of the main result.

The Zipf law [1] determines dependence between frequency f of usage of words and rank r of these words (the number in the frequency list sorted by lack of growth). According to the classical Zipf law (empirically established by him for the most frequently used words) $f(r) \approx cr^{-1}$.

This law was justified in [6, 3] for random texts with independent letters (the so-called monkey model) with arbitrary letter probabilities. Let $p_1, p_2, \ldots, p_n, p_0$ be the probabilities of pressing letters of some alphabet $L = \{A_1, A_2, \ldots, A_n\}$ and a space (the symbol of the end of the word) on the keyboard, respectively, $\sum_{i=0}^{n} p_i = 1$. In this article, we will consider the case when the probability p_0 of pressing a space is fixed and sufficiently small. The classical Zipf law arises in the limiting case $p_0 \to 0$. First, let's formulate the known results for this case.

The word w is any finite sequence of letters:

$$w = (A_{i_1}, \ldots, A_{i_N}). \tag{1}$$

The word weight (in the case of a model with independent letters) is determined by the formula:

$$\mathbb{P}(w) = p_{i_1} \times p_{i_2} \times \ldots \times p_{i_N}. \tag{2}$$

Let's consider all kinds of words ordered by lack of growth of weights and matched with a rank r (the position of a word in the sorted list). Let us denote the weight of a word of rank r by $p(r)$. The statement was proved in [6, 3]:

Theorem 1 *For any positive vector* $\mathbf{p} = (p_1, \ldots, p_n)$, $n \geqslant 2$ *satisfying the condition* $\sum_{i=1}^{n} p_i = 1$, *for some positive* c_1, c_2, *the following inequality occurs:*

$$c_1 r^{-1} < p(r) < c_2 r^{-1} \tag{3}$$

It should be noted that when determining (2), the sum of the weights of all possible words is equal to infinity, and the sum of the weights of words of the same fixed length is equal to one. The infinite sum problem of the weights of all words was originally contained in the formulation of the Zipf law with -1 exponent of the power law, because of the divergence of the harmonic series. The papers [6, 3] that actually consider a model with symbols of letters and space does not have such problem. It leads to the same assertion with the exponent less than -1. To specify the probability of a word with a space, it is necessary to use not the formula (1), but a formula in which the weight $\mathbb{P}(w)$ is multiplied by the constant p_0. The basic inequality proved in the works [6, 3] has the following form:

$$c_1 r^{-\alpha} < p_0 p(r) < c_2 r^{-\alpha},$$

where $\alpha = 1/\gamma$, and γ is the root of the equation $\sum_{i=1}^{n} p_i^{\gamma} = 1$ (in the case of $p_0 > 0$). Inequality (3) is an equivalent mathematical result. Also, it should be noted that $\alpha \to 1$ for $p_0 \to 0$.

Further, in sections 1 and 2 (as well as in the previous case with Theorem 1), we will work with models without a space ($p_0 = 0$), write about word weights (the sum of all weights is not equal to one) and correspondingly formulate the results of all the works cited below. To apply these results to the probabilities of the words normalized into the range (0, 1], a modification of the case $p_0 > 0$ is required, which we consider in section 3.

The theorem 1 was extended for the case of Markov dependence in the papers [2, 4] (a somewhat weaker result was independently proved in [7]). Let there be some initial probability distribution $\mathbf{q} = (q_1, \ldots, q_n)$, $q_i \geqslant 0$, $\sum_{i=1}^{n} q_i = 1$ and a stochastic matrix P of transient probabilities (with the size $n \times n$) with the following property: some power m of this matrix exists so that all the elements of P^m are not equal to zero (the matrix P has ergodic property). Let us redefine the weight of the word (1) by the formula:

$$\mathbb{P}(w) = q_{i_1} \times p_{i_1,i_2} \times \ldots \times p_{i_{N-1},i_N}, \tag{4}$$

where p_{ij} are the corresponding elements of the matrix P. Using the formula (4) instead of the formula (2), we correspondingly redefine $p(r)$ — the weight of the r-th word in the sorted word list. In [2, 4] the following theorem was proved.

Theorem 2 *Suppose that the matrix P has the ergodicity property and the weight of the word w is given by the formula (4). Thus, the inequality (3) with some positive constants c_1, c_2 is satisfied for any initial distribution* $\mathbf{q}$.

The natural question connected with the inequality (3) is the existence of a limit

$$\lim_{r\to\infty} \frac{p(r)}{r^{-1}}. \tag{5}$$

It is not difficult to verify that in general case the limit may not exist. For example, this is true for independent letters having equal probabilities $p_1 = p_2 = \ldots = p_n = 1/n$. Nevertheless, the following theorem was proved [5, Theorem 2]:

Theorem 3 *Suppose that the conditions of Theorem 1 are satisfied and the relation* $\frac{\log p_i}{\log p_j}$ *for some pair of letters A_i, A_j is irrational. Thus, the limit (5) exists and is equal to $H^{-l}(\mathbf{p})$, where $H(\mathbf{p}) = -\sum_{i=1}^{n} p_i \log p_i$ is entropy of the vector* $\mathbf{p}$.

Unfortunately, the proof methods of Theorem 3 are not applicable in the Markov situation. The aim of this paper was to find the correct formulation of the Markov analogue of this theorem. We have established experimentally the validity of the following assertion:

Statement 1 *Let the elements of the stochastic matrix of transition probabilities P from Theorem 2 be generated using any multidimensional continuous distribution which guarantees with a single probability that the conditions of this Theorem are satisfied and the vector $\mathbf{q}$ is chosen arbitrarily. Then the limit (5) exists and is equal to $H^{-1}_{cond}(P)$, where $H_{cond}(P)$ is the conditional entropy of rows of the matrix P with probabilities equal to the corresponding elements of the stationary distributed vector $\pi = (\pi_1, \dots, \pi_n)$, that is*

$$H_{cond}(P) = -\sum_{i=1}^{n} \pi_i \sum_{j=1}^{n} p_{ij} \log p_{ij}, \tag{6}$$

where $\pi P = \pi$ and $\sum_{i=1}^{n} \pi_i = 1$.

Obviously, Theorem 3 is a special case of Statement 1.

2. The procedure of the experimental proof of the Statement 1.

The graph of the function $p(r)$ is a set of steps of natural length and positive heights arranged in a descending order. The steps have a height $\leqslant 1$ defined by the formula (4)), but they are usually relatively "long" (their length is equal to the number of words having the same weight). We compute the length of each step from the combinatorial considerations for all different values of $\mathbb{P}(w)$, $\mathbb{P}(w) > p'$. Then we sort these steps in a descending order $\mathbb{P}(w)$ and as a result find all the pivot points of the graph of the function $p(r)$ for $p(r) > p'$. The value of p' is chosen from taking into account that the number of steps allows us to effectively perform all of the above operations on the computer. Thus, the main idea was to consider classes of words, having the same probability instead of considering all the words.

Besides, the proof of the Statement 1 for any initial distribution $\mathbf{q}$ is reduced to the initial distribution, concentrated in one of the letters. In fact, any stochastic vector $\mathbf{q}$ can be represented as a convex combination of unitary vectors. Correspondingly, the weight of the word (4) can be represented as a convex combination of weights for the cases of initial distributions $\mathbf{q}$ concentrated in one of the letters. Assume that the Statement 1 will be proved for these particular cases. This means that for each of these initial distributions, the weights of all words, except for a finite number, lie in the ε-neighborhood of the number $H^{-1}_{cond}(P)$. Then the number of exceptional words will be finite for the convex combinations under consideration.

Thus, we will consider only the case of an alphabet containing two symbols $L = \{A, B\}$ and an initial distribution concentrated in state A. In Section 3, we discuss the reasons why we are sure of the validity of the Statement 1 for an alphabet containing more than 2 symbols.

Let us describe in more detail the considered steps and the method of calculating their lengths and heights. There are 3 kinds of words starting with the letter A, i.e. corresponding to our initial distribution.

1. Words consisting only of the letters A. Let A occur k times in a word w. For each fixed $k \geqslant 0$ such word is unique (the length of the step is equal to one) and its weight is given by the formula: $\mathbb{P}_1(w) = p_{11}^{k-1}$.

2. Words starting with the letter A and ending with it, in which the letter B occurs a certain number of times. Let's denote the number of transitions in such a word from the state A to B and back by x ($x \geqslant 1$), the number of transitions from A to A ($y \geqslant 0$) by Y, and the number of transitions from B to B ($z \geqslant 0$) by z. Then, the weight of such a word w is given by the formula: $\mathbb{P}_2(w) = (p_{12}p_{21})^x p_{11}^y p_{22}^z$.

The number of such words is the number of ways to insert y transitions $A \to A$ among $x + y$ transitions from A, multiplied by the number of ways to insert z transitions $B \to B$ among $x + z - 1$ transitions from B (the last transition from B must be in A). Thus, the length of the corresponding step is equal to $\binom{x+y}{y}\binom{x+z-1}{z}$.

3. Words beginning with the letter A and ending with B. The weight of such a word w in the preceding paragraph is given by the formula: $\mathbb{P}_3(w) = (p_{12}p_{21})^x p_{11}^y p_{22}^z p_{12}$, $x, y, z \geqslant 0$. The number of such words (the length of the corresponding step) is $\binom{x+y}{y}\binom{x+z}{z}$.

Note that it is convenient to plot not $p(r)$, but $-\log p(r)$ ("minus" since $0 < p(r) \leqslant 1$) on the ordinate axis. Let us introduce the following notations: $a = -\log(p_{12}p_{21})$, $b = -\log p_{11}$, $c = -\log p_{22}$, $d = -\log p_{21}$. Thus, the logarithms of the weights of the words from Sections 1 to 3 are given by formulas

$$-\log \mathbb{P}_1(w) = (k-1)b \tag{7}$$
$$-\log \mathbb{P}_2(w) = xa + yb + zc \tag{8}$$
$$-\log \mathbb{P}_3(w) = xa + yb + zc + d \tag{9}$$

If $p' = e^{-300}$, then to find the number of classes of words satisfying the condition $\mathbb{P}(w) > p'$ we should sort out all possible values $k \geqslant 1$ so that the right-hand side of the formula (7) is less than 300.

Besides, we need to sort out all possible values $x \geqslant 1$, $y, z \geqslant 0$ so that the right side of the formula (8) is less than 300. Finally, we should sort out all possible values of $x, y, z \geqslant 0$ so that the right-hand side of the formula (8) is less than 300. In the case the values p_{11}, p_{21} have uniform distribution within the interval, $[0, 1]$ (in this case, the values of the remaining elements of the matrix P are uniquely determined: $p_{12} = 1 - p_{11}$, $p_{22} = 1 - p_{12}$) it turns out that the number of different classes of words (the number of different steps) has the order of 10^7.

To specify the graph of the function $-\log p(r)$, we need to know the abscissas and ordinates of the pivot points of this graph. To calculate them, we sorted the steps in the order of increasing values given by formulas (7,8,9) (the ordinate of the pivot points). In this case, the lengths of the steps are given by combinatorial formulas from sections 1–3. Thus, as a result of sorting, we obtained a series of natural numbers — lengths of the steps. The abscissas of the pivot points are obviously partial sums of this series. Note that the Statement 1 is equivalent to the assertion

$$\lim_{r\to\infty} (\log r + \log p(r)) = -\log H_{cond}(P). \tag{10}$$

Thus, to verify the Statement 1 experimentally, it is convenient to introduce a logarithmic scale along both the ordinate and the abscissa axes. Therefore, we took the logarithm of the abscissas of the pivot points found by the method described in the previous paragraph.

As a result, we obtained a step-by-step graph of the dependence of $-\log p(r)$ on $\log r$, which lies "along the bisector of the first coordinate angle". The differences between the abscissa and ordinate values for the pivot points of this graph (that is, the values of $\log r + \log p(r)$) are shown in Fig. 1. It can be seen that for large values of $\log r$, the oscillations of the graph are damped. Having calculated the average value $(\log r + \log p(r))$ by the last thousand of the pivot points, we experimentally obtained the required constant from the formula (10). Later we managed to formulate Statement 1 and the result coincided with the experimental one up to 3–4 digits after the decimal point (note that the values $\log H_{cond}(P)$ for randomly generated matrices P are of the order of 1).

3. Discussion of the obtained results.

We have experimentally proved Statement 1 for $n = 2$. We considered approximately $e^{300} \approx 10^{130}$ number of words (in fact, about 10^7 classes of words). We checked the validity of the main

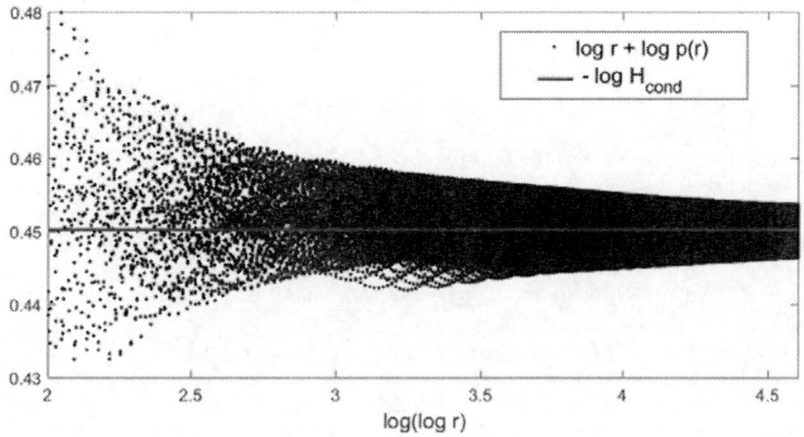

Figure 1. The angular points of the graph of the function $p(r)$ as $-\log p(r) < 100$ for the matrix $P = \begin{pmatrix} 0.7 & 0.3 \\ 0.4 & 0.6 \end{pmatrix}$ the initial distribution $\mathbf{q} = \begin{pmatrix} 1 \\ 0 \end{pmatrix}$.

statement both for completely non-zero matrices P of dimension 2×2 and for matrices which have $p_{11} = 0$ or $p_{22} = 0$.

Inequalities $-\log \mathbb{P}_2 < 300$, $-\log \mathbb{P}_3 < 300$ (taking into account the relations (8, 9)) are analogous to the restrictions on the summation domain of combinations in the Pascal triangle, which were used in [5] for the proof of Theorem 3. This suggests that the Statement 1 is valid not only for randomly generated matrices P, but in general for all matrices in which there is a pair of elements in one row with an irrational ratio of their logarithms.

M.D. Ward drew our attention that the Theorem 3 has an analog in the form of some statement for digital trees (also known as "tries"). In this case, the mean for which the statement is formulated satisfies a completely different type of recurrence relation, than $p(r)$ (see [9, p.404]). After the experimental proof of Statement 1, we have found the article [8] with the same formula (6) for digital trees in the case of Markov dependence. We are going to apply the complex analysis technique used there, for mathematical proof of the Statement 1. We are also going to investigate the nature of fluctuations $\frac{p(r)}{r-1}$ near its limit value (see Figure 1).

Finally, let us discuss the application of Statement 1 to the case when there is a certain small probability p_0 to get from each state into the space (the symbol of the end of the word). Then, the probability of each word is $\mathbb{P}(w)p_0$, where the weight $\mathbb{P}(w)$ is given by the formula (4). Thus, the weight $\mathbb{P}(w)$ of the word can be restored by its probability, and the value $p(r)$ of the weight of a word of rank r can be restored with the sorted list of probabilities. It was actually proved in [4] that in this case for some positive c_1, c_2 the following inequality is correct

$$c_1 r^{-1/\gamma} < p(r) < c_2 r^{-1/\gamma},$$

where γ is the power to which all elements of the matrix P must be raised, so that its spectral radius is equal to one. Note that for sufficiently small values p_0, the value γ is arbitrarily close to one. Assuming that the limit $\lim_{r\to\infty} \frac{r^{-1/\gamma}}{p(r)}$ exists and depends continuously on P and p_0, then, according to Statement 1 this limit will be arbitrarily close to $-\sum_{i=1}^{n} \pi_i \sum_{j=1}^{n} p_{ij} \log p_{ij}$, In the case of sufficiently small values of p_0, where π is the eigenvector of the matrix P corresponding to its largest eigenvalue, $\sum_{i=1}^{n} \pi_i = 1$.

Acknowledgements
This work was supported by the Russian Foundation for Basic Research, Grant no. 15-06-07402. The research of the first author was supported by the Russian Government Program of Competitive Growth of Kazan Federal University.

References
[1] Baayen R H 2001 *Word Frequency Distributions* (Dordrecht: Kluwer Academic Publishers)
[2] Bochkarev V V and Lerner E Yu 2012 Zipf and non-Zipf Laws for Homogeneous Markov Chain *Preprint* arXiv:1207.1872
[3] Bochkarev V V and Lerner E Yu 2012 *Russian Mathematics* **50** 25
[4] Bochkarev V V and Lerner E Yu 2014 *Electron. J. Linear Algebra* **27** 534
[5] Bochkarev V V and Lerner E Yu 2017 *International Journal of Mathematics and Mathematical Sciences* **2017** Article ID 914374
[6] Conrad B and Mitzenmacher M 2004 *IEEE Trans. Inform. Theory* **50** no 7 1403
[7] Edwards R, Foxall E and Perkins T J 2012 *Electron. J. Linear Algebra* **23** 966
[8] Jacquet Ph and Szpankowski W 1991 *IEEE Trans. Information Theory* **37** 1470
[9] Sedgewick R and Flajolet P *An Introduction to the Analysis of Algorithms* 1995 Addison–Wesley Boston Mass

IOP Conf. Series: Journal of Physics: Conf. Series **936** (2017) 012029 doi:10.1088/1742-6596/936/1/012029

Structural characterization of nano and micro-fractals using lacunarity analysis and small-angle scattering

A. Slyamov

Joint Institute for Nuclear Research, Dubna 141980, Russian Federation
Institute of Nuclear Physics, Almaty, Kazakhstan

E-mail: slyamov.azat@gmail.com

Abstract. The paper presents structural characterization of deterministic nano and micro-fractals using the lacunarity analysis and small-angle scattering technique. We show that for the considered mass-fractal models, the lacunarity spectrum reveals the main structural parameters of the fractal, such as overall size of the system, iteration number, scaling factor and the size of basic units.

1. Introduction

Fractals are known as the infinitively complex systems with self-similar spatial patterns across different scales [1]. Having such properties, fractals became very efficient mathematical models for describing a large number of natural formations from macro to micro scales [2] (galaxy clusters, clouds, trees, micro-molecules, nanostructures, etc).

Modern fractal research techniques usually focus on the calculation of the fractal dimension which, in the case of mass fractals, can be defined as the exponent that describes the distribution of mass $M(r)$ inside a spherical surface of radius r centered on the fractal, through the relation $M(r) \propto r^{\mathrm{D_m}}$, where D_m is the fractal dimension [3, 4].

Despite that fractal dimension is considered as the main fractal property, it does not provide complete information about the fractal. The ambiguity of characterizing the spatial fractal pattern only by a fractal dimension is such that two fractals may share the same fractal dimension, but may differ in a distribution of mass. To overcome such uncertainty B. Mandelbrot [1] proposed an additional property, called lacunarity (from Latin "lacuna" meaning "gap") that shows the inhomogeneity (gappiness) of the fractal structure, describing the spatial distribution of elements (mass) inside the fractal.

In this paper we apply lacunarity analysis to digital images of the deterministic fractal sets for revealing their structural and fractal properties. The approach we use to determine lacunarity is the differrential box counting method defined by Voss [5]. We show that the main morphological parameters of a fractal, such as the overall size, the size of basic units, iteration number and scaling factor can be extracted from the lacunarity spectrum. We compare our results with the ones obtained from the well-known small-angle scattering technique.

2. Theoretical background

2.1. Lacunarity

Several definitions and algorithms for estimating lacunarity have been proposed since B. Mandelbrot introduced the concept. For practical purposes, in our calculations we restrict ourself with definition introduced by R. Voss based on the differential box-counting (DBC) algorithm [5]. The process starts with covering a fractal image with non-overlapping square boxes of the size r, then the number of occupied boxes, with M number of pixels (mass) inside, is determined as $n(M,r)$. Dividing $n(M,r)$ by the total number of boxes N that cover a fractal image, one obtains the probability function $P(M,r)$ that a box of size r contains M number of pixels. By introducing the statistical moments of $P(M,r)$, as

$$Z^{(q)}(r) = \sum_{M=1}^{N} M^q P(M,r), \tag{1}$$

the lacunarity can be defined as the fluctuations of mass-distributions over its mean

$$\Lambda(r) = \frac{Z^{(2)}(r) - (Z^{(1)}(r))^2}{(Z^{(1)}(r))^2}. \tag{2}$$

Obviously, the lacunarity has dependence on the size of boxes, the larger a box, the lesser the $\Lambda(r)$ due to the smaller variations of $M(r)$. On the scales of small r, lacunarity will depend on the distribution of mass inside the fractal. Clustered or inhomogeneous distribution of mass leads to higher values of lacunarity.

2.2. Small-angle scattering

The common used technique for structural analysis of nano and micro fractals is the small-angle scattering (SAS), which yields the differential elastic cross-section per unit solid angle as a function of the momentum transfer, and describes the spatial density-density correlations in the investigated sample [6, 7]. The main indicator of the fractal structure is the fractal dimension, and it can be obtained from the power-law behavior of the scattering intensity $I(q)$ given by [8, 9]

$$I(q) \sim q^{-\alpha} \tag{3}$$

where $q = (4\pi \sin 2\theta)/\lambda$, θ is the scattering angle and λ is the wavelength of incident radiation. The exponent α shows the fractal dimension [10] of the fractal, $\alpha = D_m$ for mass fractals [8] and $\alpha = 6 - D_s$ for surface fractals [3], where D_m is mass fractal dimension and D_s is surface fractal dimension. Besides the fractal dimension, SAS can reveal additional structural parameters as: scaling factor, number of units composing the fractal, and the iteration number of regular mass and surface fractals [11, 12, 13, 14] and fat fractals [15, 16, 17, 18].

3. Deterministic mass-fractal models

In order to calculate lacunarity of the fractals of the same fractal dimension, but with different mass distribution we consider images (300x300 pixels) of four deterministic mass-fractal models. In Fig. 1 are represented fractals up to the third iteration.

One of the well-known algorithms for constructing the deterministic fractal is the following [13, 19]:

(i) Start with the disk with the diameter $a = 300\, pixels$ (initiator), represented by the iteration number $m = 0$.

(ii) Scale down the initiator by the factor of three ($\beta_s = 1/3$), and make four copies of it.

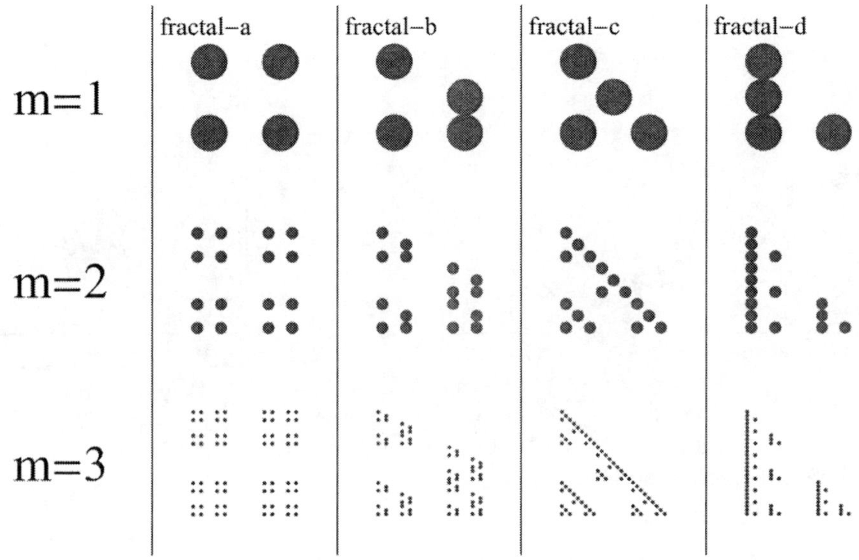

Figure 1. (Color online) Construction of the deterministic mass-fractals up to third iteration $m = 3$.

(iii) Shift each scaled copy of side length $a_m = \beta_s a$ to four different vertices of a rectangle (for Fractal-a) into which the initiator is inscribed at $m = 0$ and obtain the fractal at $m = 1$.

(iv) Repeat the same procedure for each new disk in order to obtain the next iteration.

For an arbitrary m-*th* iteration, the number of units that compose the fractals is given by

$$N_m = 4^m. \tag{4}$$

For the fractals -b, -c and -d from Fig. 1, Step (iii) is modified, and the translation of one of the copies is performed arbitrarily. Please note that according to the proposed algorithm all fractals have the same fractal dimension

$$D = \lim_{m\to\infty} \frac{\log N_m}{\log (a/a_m)} \approx 1.26, \tag{5}$$

4. Results and discussion

For the considered deterministic mass-fractal models we calculated the lacunarity spectra according to Eq. 2 for iteration number $m = 2, 3$. The results are shown in Fig. 2(left side for $m = 2$ and right side for $m = 3$).

From the construction of the models shown in the Fig. 1, it can be seen that fractals -b, -c and -d have a more clustered structure and the corresponding lacunarity curves from Fig. 2 have higher value in comparison with fractal-a. For iteration of higher order such correlation become more significant, allowing one to quantitatively differentiate a distribution of mass between fractals.

Spectra of lacunarity reveal main morphological features of fractals. Non-zero value of lacunarity appears when the size of boxes covering the image is less than half of the image size due to the mass-distribution fluctuations. In the case of the considered mass-fractal models, non-zero lacunarity starts from 150 pixels and shows that the overall size of the fractal is twice

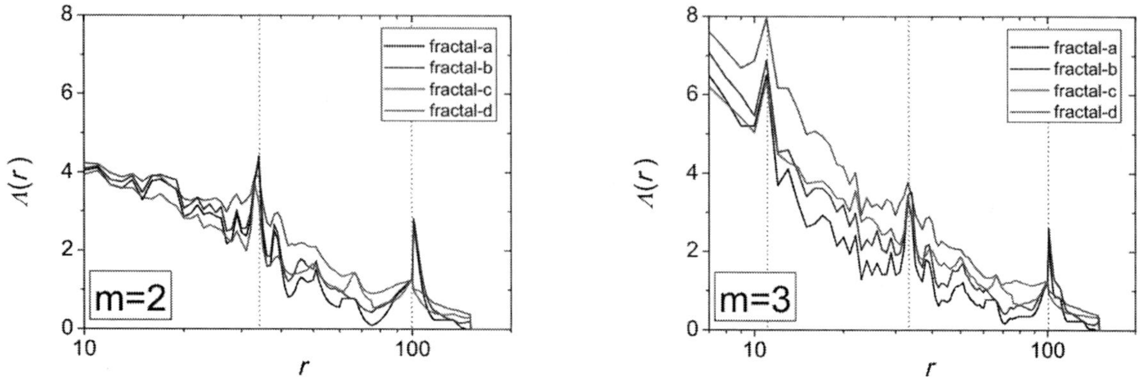

Figure 2. (Color online) Lacunarity spectra for the iteration number $m = 2$ and 3 of the deterministic mass-fractal models.

this value, as expected. When we cover the fractal with boxes of exact size as the size of its elements of the particular iteration m, the number of empty boxes will take maximum value, increasing the lacunarity in this region. The number of the most pronounced local maxima shows the iteration number of the fractal, positions of this maxima reveal the size of the units at particular iteration and by taking the ratio between two consecutive maxima we can obtain the scaling factor.

Characterizing the structure of the fractal using the lacunarity is performed in a real space. On the contrary, small-angle scattering spectrum gives information in the reciprocal space. Fig. 3 represents the scattering structure factors of deterministic mass-fractals for $m = 2$ and $m = 3$.

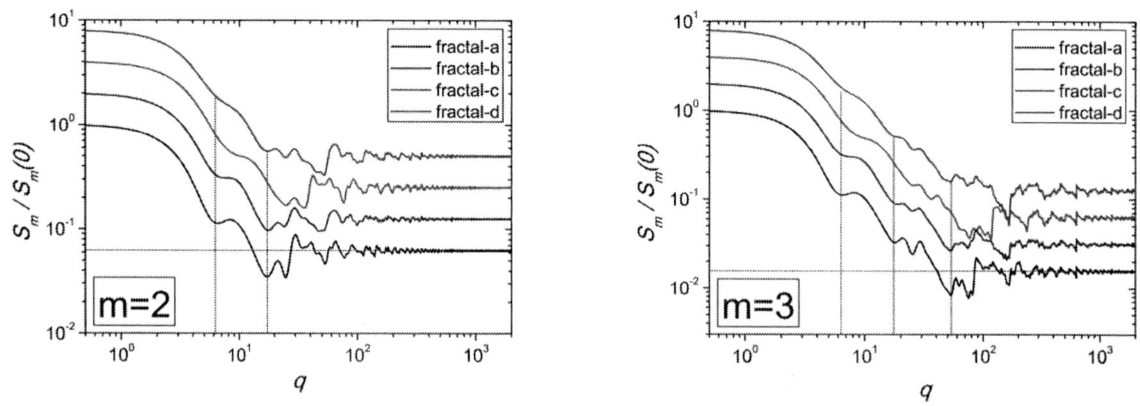

Figure 3. (Color online) Scattering intensities for the iteration number $m = 2$ (left side) and $m = 3$ (right side) of the deterministic mass-fractal models. The values of the scattering intensity for the fractals -b, -c and -d are scaled up for clarity by the factor 2, 4 and respectively 8.

In the SAS data, the region with a constant intensity at small values of q is called Guinier region. The rightmost part of the region shows the overall size of the fractal as $q = 2\pi/a$, where a is the side length of the fractal. The number and positions of the most pronounced minima (denoted by vertical dotted lines) indicates the iteration number, and respectively the sizes of units at corresponding iteration.

The scaling factor of the fractal can be calculated by the ratio of the values of two consecutive minima. Additionally, the asymptotic behavior of SAS spectrum at high values of q provides the information about number of basic units N_m at particular iteration.

5. Conclusions

We considered the deterministic mass-fractal models that are characterized by the same fractal dimension, but different spatial distribution of mass. Lacunarity spectrum of each model is calculated using differential box-counting algorithm proposed by Voss. The results obtained give the information about main morphological and fractal properties as overall size of fractal, iteration number and the scaling factor.

We confirm our results by comparing them with data from the simulation of the small-angle scattering from the considered mass fractal models. Corresponding structure factors are calculated using the algorithm developed by Pantos [21].

References

[1] Mandelbrot B. B. *The Fractal Geometry of Nature* (W. H. Freeman, 1983).
[2] Barnsley M. F. *Fractals Everywhere* 2nd ed. (Morgan Kaufmann, 2000).
[3] H. D. Bale and P. W. Schmidt 1984, *Phys. Rev. Lett.* **53**, 596.
[4] J. E. Martin and B. J. Ackerson 1985, *Phys. Rev. A* **31**, 1180.
[5] R. F. Voss 1986, *Phys. Scr.* **13**, 27.
[6] L. A. Feigin, D. I. Svergun *Structure Analysis by Small-Angle X-Ray and Neutron Scattering* (Springer 1987).
[7] Brumberger, H. *Modern Aspects of Small-Angle Scattering* (Springer Netherlands 1995).
[8] Teixeira J. 1988, *J. Appl. Cryst.* **21**, 781.
[9] J. E. Martin 1985, *J. Appl. Cryst.* **19**, 25.
[10] Hausdorf F. 1918, *Math. Ann.* **79**, 157.
[11] A. Yu. Cherny, E. M. Anitas, V. A. Osipov and A. I. Kuklin 2017, *Phys. Chem. Chem. Phys.* **19**, 2261.
[12] A. Y. Cherny, E. M. Anitas, V. A. Osipov, A. I. Kuklin 2017, *J. Appl. Cryst.* **50**, 919.
[13] A. Yu. Cherny, E. M. Anitas, V. A. Osipov, and A. I. Kuklin 2011, *Phys. Rev. E* **84**, 036203.
[14] A. Y. Cherny, E. M. Anitas, V. A. Osipov, A. I. Kuklin 2014, *J. Appl. Cryst.* **47**, 198.
[15] E. M. Anitas 2014, *Eur. Phys. J. B* **87**, 139.
[16] E. M. Anitas 2015, *J. Opt. Adv. Mat.* **17**, 1122.
[17] E. M. Anitas 2015, *Rom. Journ. Phys.* **60**, 647.
[18] E. M. Anitas, and A. M. Slyamov, and R. Todoran, and Z. Szakacs 2017, *Nanoscale Res. Lett.* **12**, 389.
[19] P. W. Schmidt and X. Dacai 1986, *Phys. Rev. A* **33**, 560.
[20] Debye P. 1915, *Ann. Phys* **46**, 809.
[21] Pantos E., van Gardener H. F., Hilbers A. J., Beelen T. P. M. and van Santen R. A. 1996, *J Mol Struct.* **383**, 303.

IOP Conf. Series: Journal of Physics: Conf. Series **936** (2017) 012030 doi:10.1088/1742-6596/936/1/012030

Simulation of small-angle scattering patterns using a CPU-efficient algorithm

E M Anitas[1,2]

[1]Joint Institute for Nuclear Research, Dubna 141980, Moscow region, Russia
[2]Horia Hulubei National Institute of Physics and Nuclear Engineering, RO-077125 Bucharest, Romania

E-mail: `anitas<at>jinr.ru`

Abstract. Small-angle scattering (of neutrons, x-ray or light; SAS) is a well-established experimental technique for structural analysis of disordered systems at nano and micro scales. For complex systems, such as super-molecular assemblies or protein molecules, analytic solutions of SAS intensity are generally not available. Thus, a frequent approach to simulate the corresponding patterns is to use a CPU-efficient version of the Debye formula. For this purpose, in this paper we implement the well-known DALAI algorithm in Mathematica software. We present calculations for a series of 2D Sierpinski gaskets and respectively of pentaflakes, obtained from chaos game representation.

1. Introduction

Small-angle neutron/x-ray/light scattering (SANS/SAXS/SALS) is widely used to study the structure of disordered systems at nano and micro scales, with applications to metallic alloys, fibers, viruses, proteins, colloidal suspensions, etc. [1–10]. The technique yields the scattering intensity (i.e. elastic cross section per unit solid angle, normalized per unit volume) $I(q)$ vs. q, where $q = (4\pi \sin\theta)/\lambda$ is the magnitude of the scattering vector, θ is half the scattering angle, and λ is the wavelength of the radiation. Provided the scattering intensity is measured in relative units and depending on the structure being analyzed, various geometrical and/or fractal characteristics about scattering systems can be obtained, such as: radius of gyration, correlation length, specific surface, the largest dimension, fractal dimension, scaling factor or the number of scattering units [11–18]. If the scattering curve is measured in absolute units, then weight parameters can also be determined.

For simple geometrical structures, analytic expressions for the form factor are well-known [19] and for several types of exact self-similar (deterministic) fractals they have been recently developed [18, 20–27]. However, for more complex systems which don't have the exact self-similarity property, such as super-molecular assemblies or protein molecules, analytic solutions are generally not available. Therefore, in such cases one has to resort to Debye formula [2, 28].

Although Debye formula is quite straightforward to implement into a computer program, it may become very time consuming when the number of scattering units exceeds few thousands. An elegant solution to this problem has been suggested in Refs. [29, 30] by discretizing the pair distances in a histogram. In Ref. [29] a pseudo code, which is the precursor of the DALAI program, has been provided also.

In this paper, Mathematica symbolic language is used to implement the DALAI's pseudo code and to efficiently simulate small-angle scattering (SAS) patterns of large structures, containing thousands of scattering centers. The obtained program is used to calculate the scattering curves for different structures

IOP Conf. Series: Journal of Physics: Conf. Series **936** (2017) 012030 doi:10.1088/1742-6596/936/1/012030

generated using chaos game representation such as 2D Sierpinski gasket and pentaflakes. The computing time is greatly reduced as compared to the classical Debye model, and thus the main regions of interest (Guinier, intermediate/fractal and Porod) are obtained within minutes.

2. Theoretical background

In this work we consider a two-phase approximation where the objects have scattering length b_j and scattering length density (SLD) $\rho_s(\boldsymbol{r}) = \sum_j b_j \delta(\boldsymbol{r} - \boldsymbol{h}_j)$. Here, $\boldsymbol{h}_j$ are the position vectors of the scattering objects. Then, the differential elastic cross section is $\mathrm{d}\sigma/\mathrm{d}\Omega = |A_t(\boldsymbol{q})|^2$, where $A_t(\boldsymbol{q}) = \int_{V^{tot}} \rho_s(\boldsymbol{r}) \exp(i\boldsymbol{q}\cdot\boldsymbol{r})\mathrm{d}\boldsymbol{r}$ is the total scattering amplitude and V^{tot} is the total volume irradiated by the radiation used. Considering that the scattering objects are immersed in a solid matrix of SLD ρ_0, the total scattering intensity is

$$I(q) \equiv \frac{1}{V^{tot}} \frac{\mathrm{d}\sigma}{\mathrm{d}\Omega} = c\,|\Delta\rho|^2\, V^2 \left\langle |F(\boldsymbol{q})|^2 \right\rangle, \tag{1}$$

where $\Delta\rho = \rho - \rho_0$ is the scattering contrast, c is the concentration of fractal, V is the volume of each fractal, and $F(\boldsymbol{q}) \equiv (1/V) \int_V \exp(-i\boldsymbol{q}\cdot\boldsymbol{r})\mathrm{d}\boldsymbol{r}$ is the normalized form factor, with $F(0) = 1$. Averaging over all orientations in 3D space is performed according to $\langle f(\boldsymbol{q})\rangle = (1/4\pi)\int_0^{\pi} \mathrm{d}\theta \sin\theta \int_0^{2\pi} \mathrm{d}\phi f(q,\theta,\phi)$.

It is known that the total scattering intensity can be written as [22]

$$I(q) = I(0)S(q)\,|F_0(qR)|^2 / N, \tag{2}$$

where $S(q)$ is the structure factor, $F_0(qR)$ is the form factor of the basic scattering 'unit' composing the object, and N is the number of the units. The classical Debye formula [28] gives

$$I^D(q) = N I_s(q) + 2F_s(q)^2 \sum_{i=1}^{k-1} \sum_{j=i+1}^{k} \frac{\sin qr_{ij}}{qr_{ij}}, \tag{3}$$

where $I_s(q)$ is the intensity scattered by each scattering unit, and r_{ij} is the distance between units i and j. When the number of units exceeds few thousands, the computation of the term $\sin(qr_{ij})/(qr_{ij})$ is very time consuming, and thus it is handled via a pair-distance histogram $g(r)$, with a bin-width commensurate with the experimental resolution [29,30]. Therefore Eq. (3) can be rewritten as

$$I^D(q) = N I_s(q) + 2F_s^2(q) \sum_{i=1}^{N_{\mathrm{bins}}} g(r_i) \frac{\sin qr_i}{qr_i}, \tag{4}$$

where $g(r_i)$ is the pair-distance histogram at pair distance r_i. The latter quantity is calculated from the positions of scattering units. Since we are interested in Guinier, Porod and asymptotic regions, we can neglect the form factor, and consider $I_s(q) = F_s^2(q) = F_0(qR) = 1$. Thus, the intensity given by Eq. (4) becomes

$$I^D(q) \equiv S^D(q) = N + 2 \sum_{i=1}^{N_{\mathrm{bins}}} g(r_i) \frac{\sin qr_i}{qr_i}, \tag{5}$$

and gives the structure factor. Taking into account the normalization used in Eq. (2), subsequently, the final expression of the scattering structure factor in Eq. (5), is represented as $S^D(q)/N^2$.

3. Mathematica algorithm for calculating SAS intensity

Using the direct calculation form of the structure factor equation (3), one can develop an efficient implementation of the Debye formula, similar to that used for DALAI's program [29, 30]. In both versions, the distances r_{ij} between all pairs of scattering 'units' have to be known. With the Debye formula, one must calculate the terms $\sin(qr_{ij})/(qr_{ij})$ which is very time consuming, when the number N of scattering units exceeds few thousands. In the efficient version of the algorithm this process is accelerated by discretizing the pair distances in a histogram of bin size commensurate with the required spatial resolution in the pattern [29,30]. In what follows, a simple explanation of the method is provided, along with key Mathematica code implementation and steps.

Given a data set $X : \{(x_i, y_i) \in \mathbb{R}^2, i = 1, 2, \cdots, n\}$ with n data points representing the positions of the scattering units, and assuming they are stored in an ASCI *file*, then we import data from the ***file***, and store them in the list **pos** by using the command

```
pos = Import["file"];
```

The following code calculates the distances between all the pairs $(x_i, y_i), i = 1, 2, \cdots, n$ and store them in the list **dist**:

```
d = {};
For[s = 1, s ≤ Length[pos] − 1, ++s,
AppendTo[d, EuclideanDistance[pos[[s]], pos[[1 + #]]]&/@Drop[Range[Length[pos] − 1], s − 1]];
];
dist = Flatten[d];
```

The next command tallies the elements in **dist**, storing all distinct distances together with their multiplicity in **tallyDistances**:

```
tallyDistances = Tally[dist];
```

Then, the values of the distances are sorted from the lowest to the highest value by the following command:

```
val = Sort[tallyDistances][[All, 1]];
```

The corresponding multiplicities of the distances are obtained in a similar way:

```
freq = Sort[tallyDistances][[All, 2]];
```

The list of all distances, with each individual distance appearing as many times as given by its corresponding multiplicity, is constructed as:

```
hist = Flatten@Array[Table[val[[#]], {freq[[#]]}]&, Length[val]];
```

The list of bins and the histogram heights with bins specified by **w** and commensurate with the required spatial resolution in the pattern (see next section) can be obtained from:

```
hl = HistogramList[hist, {w}];
```

Then, the pair-distance histogram at pair distance r_i (see Eq. (4)) is obtained by:

```
pd = Drop[Table[{hl[[1, i]], hl[[2, i]]}, {i, 1, Length[hl[[2]]]}], 1];
```

By defining the **sinc** function involving the the pair-distances **pd**, such as:

$$\mathbf{sincpd[q\_]:=\frac{Sin[\mathit{q}\ pd[[All,1]]]}{\mathit{q}\ pd[[All,1]]};}$$

then, finally, the expression of scattering intensity can be written as:

$$\mathbf{S[q\_]:=\frac{1}{Length[pos]^2}\left(Length[pos]+2\sum_{\mathit{i}=1}^{Length[pd]}pd[[\mathit{i},2]]sincpd[\mathit{q}][[\mathit{i}]]\right);}$$

4. Applications

We make use of the algorithm above to calculate the scattering intensity for $2D$ Sierpinski gaskets and pentaflakes.

4.1. Sierpinski gaskets

We consider a Sierpinski triangle of side length $a = 1$, and centered in the origin. SG is generated by playing chaos game on three vertices. The points of the SG are obtained starting with an initial point chosen at random, and calculating each subsequent point as one half of the distance between the previous point and one of the vertices (randomly selected). The results for few sets of points are shown in Fig. 1.

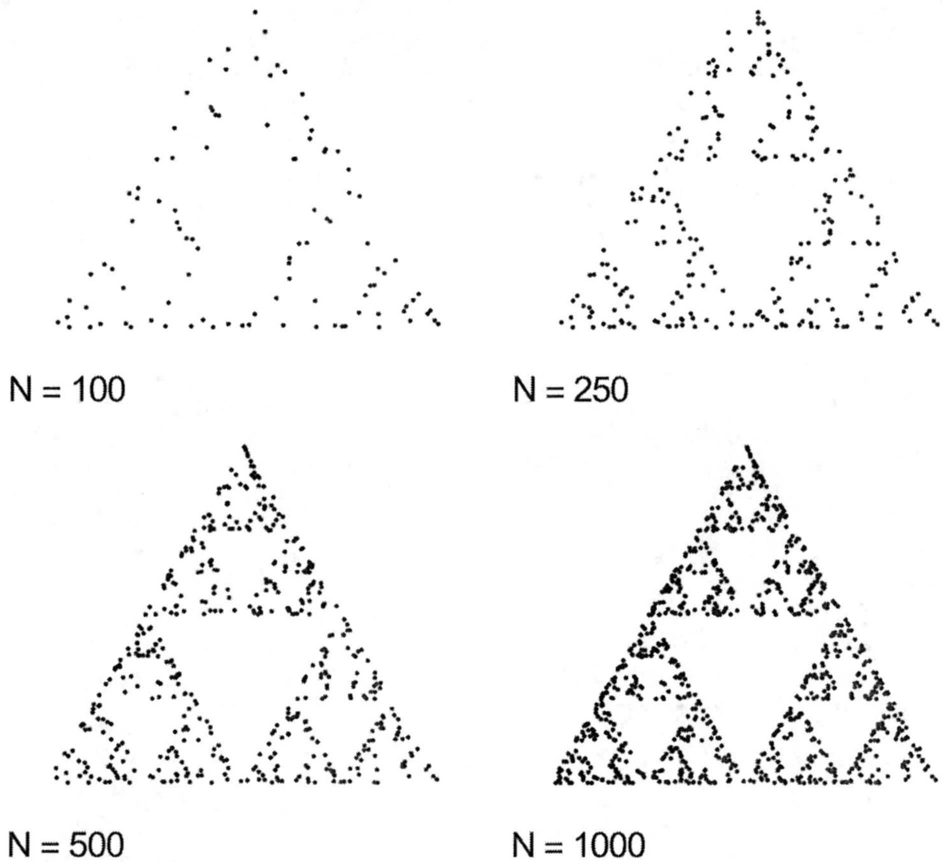

Figure 1. The structure of two-dimensional SG from chaos game representation, for $N = 100, 250, 500$ and respectively $N = 1000$ points.

In Fig. 2 is shown the structure factor of SG generated from the CGR, for different values of the number of units composing the fractal. All the scattering curves are characterized by the presence of four main regions. At $q \lesssim 1/a$ a plateau is observed (Guinier region, where $I(q) \propto q^0$), and from the end of this plateau, information about the overall size of the fractal can be extracted. At $1/a \lesssim q \lesssim 1/d$ (where d is of the order of the distances between the scattering points) we have the fractal regime (where $I(q) \propto q^{-1.585}$), from which the fractal dimension ($D = 1.585$) can be extracted. Note that by increasing the number of points, the distances between them decreases and thus, the length of the fractal regime increase also. For $1/d \lesssim q$ a transition region appears between fractal and asymptotic regions. For $N = 1000$ this region is at $10^2 \leq qa \leq 8 \cdot 10^2$ (Fig. 2). Finally, when $1/d \ll q$, the asymptotic region is attained.

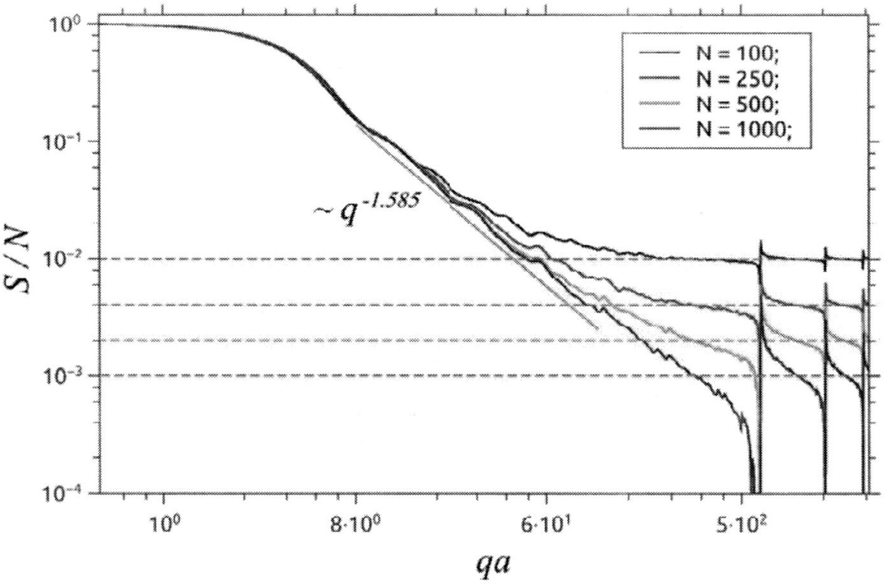

Figure 2. (Color online) The structure factor of SG using CGR. The horizontal lines indicate the value of the asymptotes in the limit of high q values.

4.2. Pentaflakes

We generate the fractal pentaflake starting from the center of a five-sided polygon of side length a. A point is drawn at a fraction $\beta_s = 0.35$ of the distance between a randomly chosen vertex and the polygon center. Fig. 3 show the fractal pentaflake when the process is repeated for $N = 4000$ points. Thus, the corresponding fractal dimension is given by

$$D \approx -\log 5/\log 0.35 \simeq 1.533. \tag{6}$$

The structure factor corresponding to the pentaflake is shown in Fig. 4. The numerical value of the fractal dimension obtained from the slope of the scatering curve in the fractal region coincides with the one given by theoretically by Eq. (6). The numerical value $\beta_s = 0.35$ of the scaling factor can be recovered from the periodicity of minima in the fractal region, and this is a specific feature of scattering from mass fractals with a single scale.

IOP Conf. Series: Journal of Physics: Conf. Series **936** (2017) 012030 doi:10.1088/1742-6596/936/1/012030

Figure 3. The structure factor of two-dimensional pentaflake using CGR for $N = 4000$ points and scaling factor 0.35.

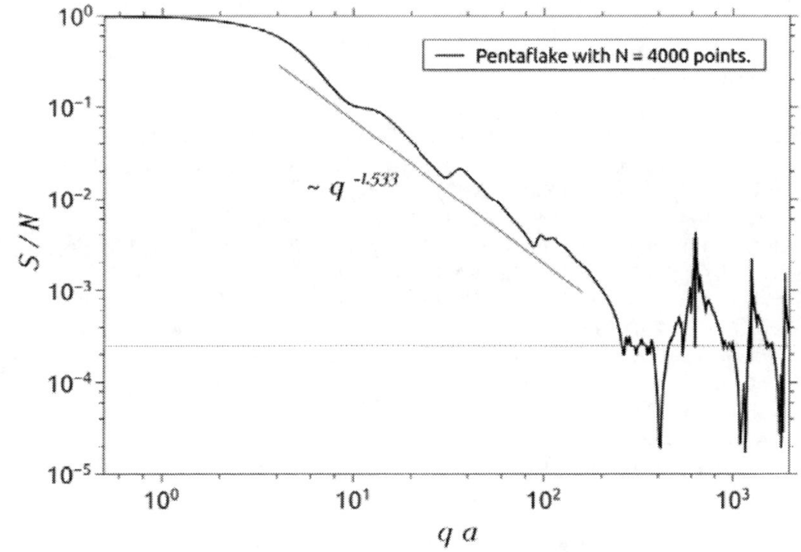

Figure 4. (Color online) The structure factor of pentaflake using CGR. The horizontal line indicates the value of the asymptote in the limit of high q values.

5. Conclusion

We implement in Mathematica software a CPU-efficient version of the Debye formula to calculate the small-angle scattering intensity from complex structures. We use the algorithm to reveal the main structural properties (fractal dimension, the overall fractal size, scaling factor and the number of particles composing the fractal) of fractal structures generated from chaos game representation.

The computational time is few orders of magnitude lower that in the traditional Debye approach, and this can be used to analyze various structures containing thousands of scattering centers (particles) when analytic solutions are not available. The efficiency of the algorithm can be further increased by using the **Compile** function available in Mathematica.

References

[1] Feigin L A and Svergun D I 1987 *Structure Analysis by Small-Angle X-Ray and Neutron Scattering* (NY: Plenum press)
[2] Glatter O and Kratky O 1982 *Small-angle X-ray Scattering* (London: Academic Press)
[3] Stevenson W D, Ahmed Z, Zeng X B, Welch C, Ungar G and Mehl G H 2017 *Phys. Chem. Chem. Phys.* **19**(21) 13449–13454
[4] Eicher B, Heberle F A, Marquardt D, Rechberger G N, Katsaras J and Pabst G 2017 *J. Appl. Cryst.* **50** 419
[5] Demchenko V, Riabov S and Shtompel' V 2017 *Nanoscale Research Letters* **12** 235
[6] Lynchak O V, Prylutskyy Y I, Rybalchenko V K, Kyzyma O A, Soloviov D, Kostjukov V V, Evstigneev M P, Ritter U and Scharff P 2017 *Nanoscale Research Letters* **12** 8
[7] Wisotzki E I, Tempesti P, Fratini E and Mayr S G 2017 *Phys. Chem. Chem. Phys.* **19**(19) 12064–12074
[8] Pandey P K, Rawat K, Aswal V K, Kohlbrecher J and Bohidar H B 2017 *Phys. Chem. Chem. Phys.* **19**(1) 804
[9] Trinh V and Xu J 2017 *Nanoscale Research Letters* **12** 216
[10] Singh A, Vihinen J, Frankberg E, Hyvärinen L, Honkanen M and Levänen E 2016 *Nanoscale Research Letters* **11** 447
[11] Bahadur J, Medina C R, He L, Melnichenko Y B, Rupp J A, Blach T P and Mildner D F R 2016 *J. Appl. Cryst.* **49** 2021–2030
[12] von Gundlach A R, Garamus V M, Willey T M, Ilavsky J, Hilpert K and Rosenhahn A 2016 *J. Appl. Cryst.* **49** 2210–2216
[13] Benavent V, Frizon F and Poulesquen A 2016 *J. Appl. Cryst.* **49** 2116–2128
[14] Hirosawa K, Fujii K, Ueki T, Kitazawa Y, Littrell K C, Watanabe M and Shibayama M 2016 *Phys. Chem. Chem. Phys.* **18**(27) 17881–17889
[15] Sanchez-Fernandez A, Edler K J, Arnold T, Alba Venero D and Jackson A J 2017 *Phys. Chem. Chem. Phys.* **19**(13) 8667–8670
[16] Rossi B, Paciaroni A, Venuti V, Fadda G C, Melone L, Punta C, Crupi V, Majolino D and Mele A 2017 *Phys. Chem. Chem. Phys.* **19**(8) 6022–6029
[17] White C E, Olds D P, Hartl M, Hjelm R P and Page K 2017 *J. Appl. Cryst.* **50** 61–75
[18] Cherny A Y, Anitas E M, Osipov V A and Kuklin A I 2017 *Phys. Chem. Chem. Phys.* **19**(3) 2261–2268
[19] SASFit `https://kur.web.psi.ch/sans1/SANSSoft/sasfit.pdf` accessed: 9 June 2017
[20] Cherny A Y, Anitas E M, Osipov V A and Kuklin A I 2017 *J. Appl. Cryst.* **50** 919
[21] Anitas E M, Slyamov A, Todoran R and Szakacs Z 2017 *Nanoscale Research Letters* **12** 389
[22] Cherny A Y, Anitas E M, Kuklin A I and Osipov V A 2011 *Phys. Rev. E* **84** 036203-1–036203-11
[23] Cherny A Y, Anitas E M, Osipov V A and Kuklin A I 2015 *Rom. Journ. Phys* **60** 658–663
[24] Cherny A Y, Anitas E M, Osipov V A and Kuklin A I 2014 *J. Appl. Cryst.* **47** 198–206
[25] Anitas E M 2014 *Eur. Phys. J. B* **87** 139
[26] Anitas E M 2015 *Advances in Condensed Matter Physics* **2015** Article ID 501281
[27] Anitas E M 2015 *Rom. Journ. Phys* **60** 647–652
[28] Debye P 1915 *Ann. Phys.* **46** 2927
[29] Pantos E and Bordas J 1994 *Pure & Appl. Chem.* **66** 77
[30] van Garderen H F, Pantos E, Dokter W H, Beelen T P M and van Santen R A 1994 *Modeling Simul. Mater. Sci. Eng.* **2** 295

IOP Conf. Series: Journal of Physics: Conf. Series **936** (2017) 012031 doi:10.1088/1742-6596/936/1/012031

Small-angle scattering from 3D Sierpinski tetrahedron generated using chaos game

Azat Slyamov
Joint Institute for Nuclear Research, Dubna 141980, Russian Federation
Institute of Nuclear Physics, Almaty, Kazakhstan
E-mail: slyamov.azat@gmail.com

Abstract. We approximate a three dimensional version of deterministic Sierpinski gasket (SG), also known as Sierpinski tetrahedron (ST), by using the chaos game representation (CGR). Structural properties of the fractal, generated by both deterministic and CGR algorithms are determined using small-angle scattering (SAS) technique. We calculate the corresponding monodisperse structure factor of ST, using an optimized Debye formula. We show that scattering from CGR of ST recovers basic fractal properties, such as fractal dimension, iteration number, scaling factor, overall size of the system and the number of units composing the fractal.

1. Introduction

A large class of natural objects from nano to macroscales has the property of self-similarity, meaning that they preserve geometrical characteristics under the scale transformations. Such objects are known as fractals and depending on exact or statistical self-similarity they can be classified as deterministic, and respectively stochastic (random) [1]. The presence of fractality in nano and microscale systems has influence on chemical [2, 3], electromagnetic [4], dynamical [5] and other features, therefore the study of relation between structure and shape from one hand, and physical properties from another hand is a topic of significant research interest [6].

One of the common used techniques in investigating nano and microscale fractals is small-angle scattering (SAS; neutron, X-ray, light) which yields the differential elastic cross-section per unit solid angle as a function of momentum transfer [7, 8]. The main indicator of the fractal structure is the fractal dimension, and it can be obtained from the power-law behavior of the scattering intensity $I(q)$ given by [9, 10, 11]

$$I(q) \sim q^{-\alpha} \tag{1}$$

where $q = (4\pi \sin 2\theta)/\lambda$, θ is the scattering angle and λ is the wavelength of incident radiation. The exponent α shows the fractal dimension [12] of the fractal, $\alpha = D_m$ for mass fractals [9] and $\alpha = 6 - D_s$ for surface fractals [13, 14, 15], where D_m is mass fractal dimension and D_s is surface fractal dimension.

It has been shown recently that besides the fractal dimension, SAS can reveal additional structural parameters, such as: scaling factor, number of units composing the fractal, and the iteration number for both, regular [16, 17] and fat fractals [18, 19, 20, 21], thus making this technique an indispensable tool in studying structural properties of self-similar nano and micro scale objects.

In this paper we calculate numerically, using optimized version of Debye [22] formula developed by Pantos *et.al* [23], the monodisperse SAS structure factor from three-dimensional ST generated by both deterministic and CGR algorithms, and show that SAS from CGR of ST recovers the main fractal properties, such as fractal dimension, iteration number, scaling factor, overall size of the system and number of fractal units.

2. Small-angle scattering and Debye formula

For monodisperse fractal, the scattering intensity $I(q)$ is defined as [9]

$$I_0(q) = F^2(q)S(q), \tag{2}$$

where $F(q)$ is the form factor and $S(q)$ is the structure factor of the fractal. Since we use a CGR algorithm, defined in Section 3, that generates directly coordinates of positions of scatterers, we can compute intensity spectrum using Debye formula [22]

$$I(q) = NI_s(q) + 2F_s(q)^2 \sum_{i=1}^{N-1} \sum_{j=i+1}^{N} \frac{\sin qr_{ij}}{qr_{ij}}, \tag{3}$$

where $I_s(q)$ is the intensity scattered by each fractal unit, and r_{ij} is the distance between units i and j. Computational time of the term $\sin(qr_{ij})/(qr_{ij})$ is increasing proportional to number of units and when it exceeds few hundred this approach become very time consuming. The problem can be resolved by introducing a pair-distance histogram $g(r)$ with a bin-width commensurate with the experimental resolution [23]. Therefore Eq. (3) can be rewritten as

$$I(q) = NI_s(q) + 2F_s^2(q) \sum_{i=1}^{N_{\text{bins}}} g(r_i)\frac{\sin qr_i}{qr_i}, \tag{4}$$

where $g(r_i)$ is the pair-distance histogram. Since we are interested only in fractal properties of the sample given by structure factor, we can consider $I_s(q) = F_s^2(q) = 1$, and in order to be consistent with previous works [14, 16, 17, 18] we use normalization $I_0(q) = I(q)/N$ and using Eq. 2, we obtain expression for structure factor

$$S(q) = 1 + \frac{2}{N} \sum_{i=1}^{N_{\text{bins}}} g(r_i)\frac{\sin qr_i}{qr_i}. \tag{5}$$

3. Sierpinski tetrahedron and Iterated Function Systems

The Iterated Function System (IFS) of a three-dimensional fractal is given by the set of linear equations $x = a_i x + d_i$, $y = b_i y + e_i$ and $z = c_i z + f_i$ that generates values of x, y, z coordinates. ST centered at the origin, with a side length a can be described by IFS of the affine map, written in a matrix form as

$$w_i \begin{bmatrix} x \\ y \\ z \end{bmatrix} = \begin{bmatrix} \mathrm{a}_i & 0 & 0 \\ 0 & \mathrm{b}_i & 0 \\ 0 & 0 & \mathrm{c}_i \end{bmatrix} \begin{bmatrix} x \\ y \\ z \end{bmatrix} + \begin{bmatrix} \mathrm{d}_i \\ \mathrm{e}_i \\ \mathrm{f}_i \end{bmatrix}, \tag{6}$$

where the coefficients $\mathrm{a}_i, \mathrm{b}_i, \mathrm{c}_i, \mathrm{d}_i, \mathrm{e}_i, \mathrm{f}_i$ with $i = 1, 2, 3, 4$ are IFS parameters and given in Tab. 1; x, y, z represent three-dimensional coordinates of generated points.

By definition, chaos game is described by an IFS, and simply the process of CGR is the following:

Table 1. IFS parameters $a_i, b_i, c_i, d_i, e_i, f_i$ of the affine maps w_i (first column) for deterministic algorithm and the probabilities p_i (last column) for random algorithm.

w	a	b	c	d	e	f	p
1	1/2	1/2	1/2	0	0	$1/(2\sqrt{3})$	1/4
2	1/2	1/2	1/2	-1/4	$-1/(4\sqrt{3})$	$-1/(4\sqrt{3})$	1/4
3	1/2	1/2	1/2	1/4	$-1/(4\sqrt{3})$	$-1/(4\sqrt{3})$	1/4
4	1/2	1/2	1/2	0	$1/(2\sqrt{3})$	$-1/(4\sqrt{3})$	1/4

(i) Determine a number of subparts into which fractals will be divided at each iteration (four for ST).

(ii) Find coordinates of the vertices of fractal and label them (1, 2, 3, 4).

(iii) Choose a random initial point inside the fractal.

(iv) Randomly (with equal probability) pick one of the labeled vertices and depending on scaling factor β_s put a new point between (on a halfway for ST, since $\beta_s = 1/2$) previous point and chosen vertex.

(v) To obtain the fractal continue the same procedure for each generated point.

The set of points that fill ST is generated starting with a random initial point and for every consecutive one, algorithm chooses one of mappings w_i, with a corresponding probability p_i, to act on the previous point.

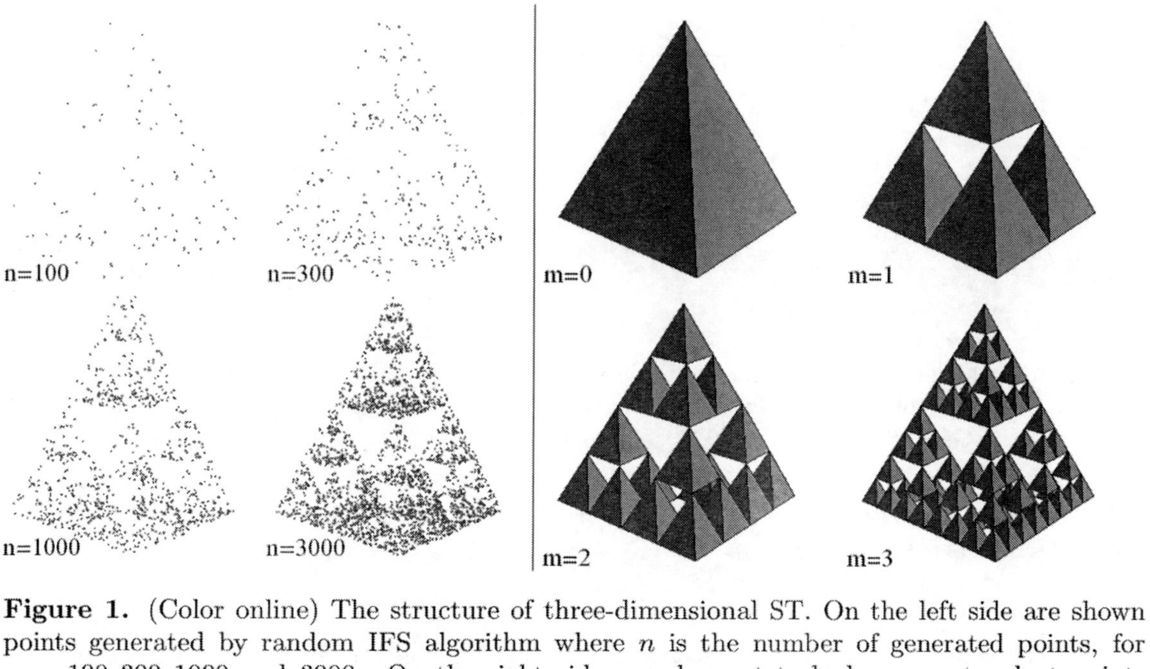

Figure 1. (Color online) The structure of three-dimensional ST. On the left side are shown points generated by random IFS algorithm where n is the number of generated points, for $n = 100, 300, 1000$ and 3000. On the right side are shown tetrahedrons, centered at points generated by deterministic IFS, with side length $a_m = \beta_s^m a$, where β_s is the scaling factor and m is the iteration number, for $m = 0, 1, 2, 3$.

The way of constructing the deterministic ST is following:

(i) Start with the tetrahedron of the side length a (initiator), represented by the iteration number $m = 0$.

(ii) Scale down the initiator by the factor of two (i.e. $\beta_s = 1/2$), and make four copies of it.

(iii) Shift each scaled copy to four different vertices of tetrahedron at $m = 0$ and obtain the ST at $m = 1$.

(iv) Repeat the same procedure for each new tetrahedron in order to obtain the next iteration.

For an arbitrary m-*th* iteration, the number of units that compose the ST is given by

$$N_m = 4^m. \tag{7}$$

We calculate the structure factors using the function of the relative positions of the points considered as the centers of the tetrahedrons, neglecting the shape of fractal units. Fig. 1 (left side) clearly shows that with increasing the number of generated points n, CGR tends to approximate ST constructed using deterministic approach, when $m \to \infty$ (Fig. 1 (right side)).

4. Results and discussion

For the considered model of ST we calculate numerically the structure factor for deterministic case up to the 5-*th* iteration. The results are shown in Fig. 2 (left side). The main structural and fractal properties of the sample can be extracted from the intensity spectrum [16]. We clearly see the presence of Guinier region [7], as a plateau at small values of q, and the rightmost part of this plateau indicates the overall size of the system as $q = 2\pi/a$, where a is the side length of ST at $m = 0$.

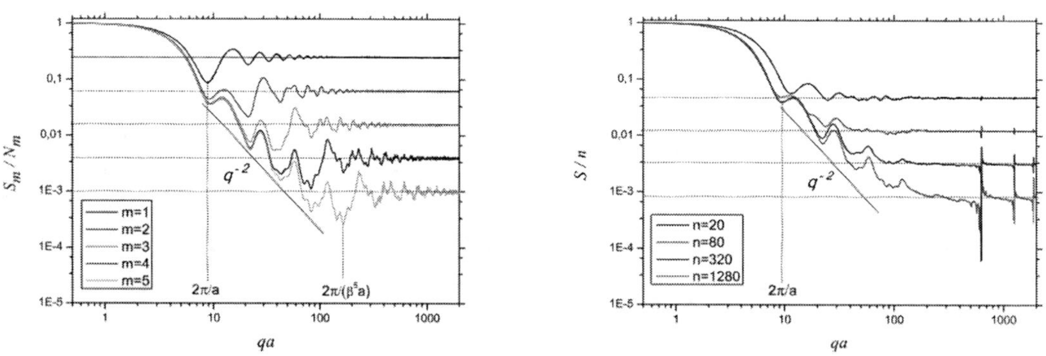

Figure 2. (Color online) On the left side is represented the structure factor for the first five iterations of deterministic ST with side length a. On the right side is represented the structure factor of the GCR of ST for $n = 20, 80, 320$, and respectively 1280, generated points.

The exponent of scattering vector between the first and the last minima reveals the fractal dimension, which is given by:

$$D = \lim_{m \to \infty} \frac{\log N_m}{\log (a/a_m)} = 2, \tag{8}$$

where m is the iteration number, $a_m = \beta_s^m a$ is the side length of fractal units centered in the points generated by the deterministic IFS of the ST at m-*th* iteration. For the deterministic case, the number of such minima indicates the iteration number of the fractal. The asymptotic behavior of the structure factor at high q region tends to the value $1/N_m$ and gives the information about the number of units composing the fractal at a given iteration [16].

The right side of Fig. 2 represents the intensity spectrum from the CGR of ST, and it shows how the scattering curve is changing with increasing the number of points n. The Guinier region depends on the number of points as expected and such dependence becomes negligible as $n \to \infty$ [18]. The fractal region, that reveals the fractal dimension by the slope of the curve, is increasing with n, the slope tends to theoretical value given by Eq. 8, and the asymptotes define the number of constituents, by ratio $1/n$.

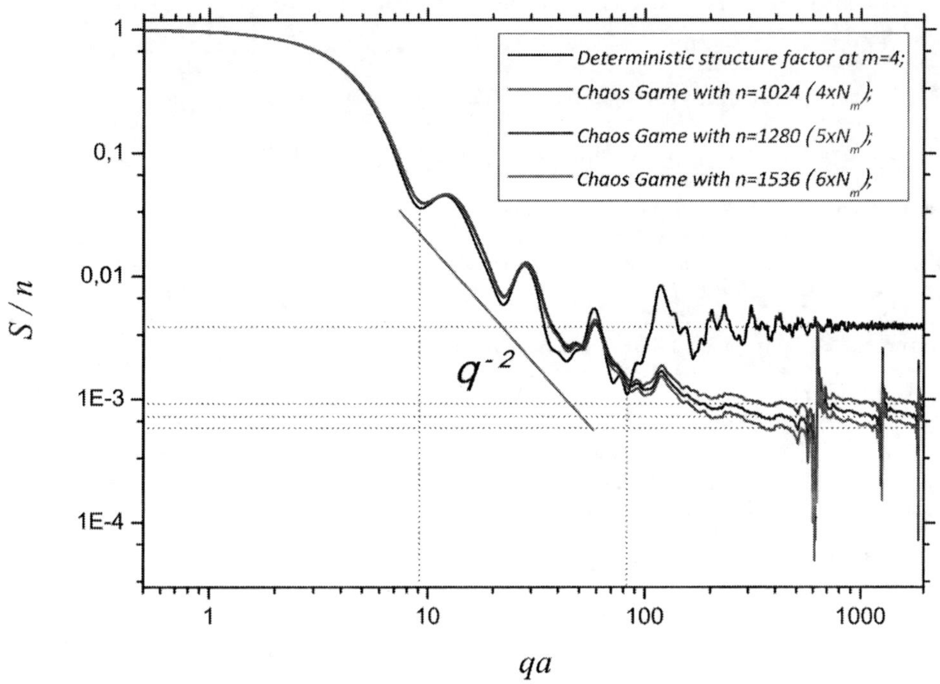

Figure 3. (Color online) The comparison of structure factors of the deterministic ST at iteration number $m = 4$, with the CGR of ST generated for $n = 1024(4\text{x}N_m), 1280(5\text{x}N_m), 1536(6\text{x}N_m)$.

To understand under which condition (i.e. the number of generated points) the CGR algorithm gives a better approximation of the structure of ST, it is necessary to compare directly the structure factor of the deterministic ST at particular iteration with the structure factor using CGR generated for different number of points. Fig. 3 shows the approximation of structure factor of CGR of ST to the deterministic one, for the 4-*th* iteration. The main characteristic regions (Guinier, fractal) and features (positions of the minima and their number) are in agreement for both algorithms, when the number of points in the CGR is at least 4 or 5 times N_m. The asymptotic behavior, as expected tend to the value $1/(number\ of\ units)$, that is in agreement with previous theoretical results.

5. Conclusions

We considered the model of the 3D Sierpinski tetrahedron, which is characterized by fractal dimension $D = 2$ and generated using both, deterministic and CGR constructing algorithms. The structure factor of the deterministic ST is calculated using optimized version of Debye formula [23]. The results give the information about main structural and fractal properties, such as the fractal dimension (from the slope of the curve in the fractal region), the overall size of the fractal (from the Guinier region), the iteration number (from the number of minima), and the number of units composing the fractal (from the asymptotes of spectrum).

The structure factor of the CGR of ST is calculated for different number of generated points, and we show that the fractal properties appear when that number starts to exceed few hundreds. We compared deterministic structure factor with CGR in order to determine the number of generated points that are needed to approximate ST at particular iteration number and we conclude that the number of CGR points should be 4 or 5 times N_m, the number of of units of fractal at m-*th* iteration. The results can be used in obtaining the structural properties of fractals with known CGR, using the small-angle scattering technique.

References

[1] Mandelbrot B. B. *The Fractal Geometry of Nature* (W. H. Freeman, 1983).
[2] L. Spanhel 2006, *J Sol-Gel Sci Techn.* **39**, 7.
[3] Li H, Kim J, O'Keeffe M and O. M. Yaghi 2003, *Angew. Chem. Int. Ed.* **42**, 1819.
[4] Jun He et.al, *Inorg. Chem.* **47**, 7948.
[5] Cheminiak P. *Phys. Lett. A* **376**, 2917
[6] Q. Zhao, J. Qian, Z. Gui, Q. An and M. Zhua 2010, *Soft Matter* **6**, 1129.
[7] L. A. Feigin, D. I. Svergun *Structure Analysis by Small-Angle X-Ray and Neutron Scattering* (Springer 1987).
[8] Brumberger, H. *Modern Aspects of Small-Angle Scattering* (Springer Netherlands 1995).
[9] Teixeira J. 1988, *J. Appl. Cryst.* **21**, 781.
[10] J. E. Martin and B. J. Ackerson 1985, *Phys. Rev. A* **31**, 1180.
[11] J. E. Martin 1985, *J. Appl. Cryst.* **19**, 25.
[12] Hausdorf F. 1918, *Math. Ann.* **79**, 157.
[13] H. D. Bale and P. W. Schmidt 1984, *Phys. Rev. Lett.* **53**, 596.
[14] A. Yu. Cherny, E. M. Anitas, V. A. Osipov and A. I. Kuklin 2017, *Phys. Chem. Chem. Phys.* **19**, 2261.
[15] A. Y. Cherny, E. M. Anitas, V. A. Osipov, A. I. Kuklin 2017, *J. Appl. Cryst.* **50**, 919.
[16] A. Yu. Cherny, E. M. Anitas, V. A. Osipov, and A. I. Kuklin. *Phys. Rev. E* **84**, 036203.
[17] A. Y. Cherny, E. M. Anitas, V. A. Osipov, A. I. Kuklin 2014, *J. Appl. Cryst.* **47**, 198.
[18] E. M. Anitas 2014, *Eur. Phys. J. B* **87**, 139.
[19] E. M. Anitas 2015, *J. Opt. Adv. Mat.* **17**, 1122.
[20] E. M. Anitas 2015, *Rom. Journ. Phys.* **60**, 647.
[21] E. M. Anitas, and A. M. Slyamov, and R. Todoran, and Z. Szakacs 2017, *Nanoscale Res. Lett.* **12**, 389.
[22] Debye P. 1915, *Ann. Phys* **46**, 809.
[23] Pantos E., van Gardener H. F., Hilbers A. J., Beelen T. P. M. and van Santen R. A. 1996, *J Mol Struct.* **383**, 303.
[24] Barnsley M. F. *Fractals Everywhere* 2nd ed. (Morgan Kaufmann, 2000).

Mathematical model of highways network optimization

R L Sakhapov[1,2,3], R V Nikolaeva[2], M H Gatiyatullin[2] and M M Makhmutov[2]
[2]Kazan State University of Architecture and Engineering, 420043, Kazan, Zelenaya str., 1, Russia

[3]E-mail: rusakhapov@gmail.com

Abstract. The article deals with the issue of highways network design. Studies show that the main requirement from road transport for the road network is to ensure the realization of all the transport links served by it, with the least possible cost. The goal of optimizing the network of highways is to increase the efficiency of transport. It is necessary to take into account a large number of factors that make it difficult to quantify and qualify their impact on the road network. In this paper, we propose building an optimal variant for locating the road network on the basis of a mathematical model. The article defines the criteria for optimality and objective functions that reflect the requirements for the road network. The most fully satisfying condition for optimality is the minimization of road and transport costs. We adopted this indicator as a criterion of optimality in the economic-mathematical model of a network of highways. Studies have shown that each offset point in the optimal binding road network is associated with all other corresponding points in the directions providing the least financial costs necessary to move passengers and cargo from this point to the other corresponding points. The article presents general principles for constructing an optimal network of roads.

Keywords: road network, road transport, correspondence points, mathematical model, route, optimization

1. Introduction

The availability of a reliable and efficient transport system is an important factor in the development of the economy of any country. Today, transport is an important part of the world economy, as it is a material carrier between states. Presence of a well-developed transport system is one of the important factors of attracting people and production and is an important advantage for locating productive forces, which gives an integration effect.

The dominant type of transport around the world is automobile, which is rapidly growing every year. The importance of the road network can also be judged by the share of the total volume of passenger and freight traffic carried out by automobiles. Studies of foreign scientists have shown that road transport is one of the largest contributions to budget revenues. For example, road user taxes and fees in the US amounted to $ 78 billion in 1994 (6.2 percent of federal revenues of the state budget) and $ 33 billion in the United Kingdom in 1995-96 (of which only $ 10 billion was spent on roads). The share of the transport system in the main production assets of Russia is 27%, and a significant share of transportation services in the gross domestic product is 8%. Net financial flows from the road sector tend to be positively correlated with economic development [10].

[1] To whom any correspondence should be addressed.

The road network is a complex, large-scale system. The road network is the basis for the functioning of the economy and society through the movement of people, goods and services. At present, in most developed countries of the world, there is a serious imbalance in supply and demand in the transport sector: traffic volumes, especially in large megacities, are approaching the limits of the capacity of existing highways, and sometimes surpass them in rush hour [11,15].

From the economic point of view, the construction of the road network should be based on the basic topology of the network, the demand of the users of the road network and the cost of its construction. Equilibrium occurs when the number of trips between the corresponding departure points (for example, the place of residence, the place of work) and the points of arrival (place of work, residence) are equal to the demand for travel taking into account the cost policy [9].

The study of the road network has been developed at the intersection of various fields of knowledge - physics, mathematics, economics, operating systems research, transportation processes, etc. Measurements, observations and modeling of individual processes in the transport sector and the totality of demographic and socioeconomic relations are important for the development of a transport infrastructure development strategy [4,8].

A consequence of the difficulty of optimizing the road network is a certain imbalance between the results of scientific research and mathematical calculations, on the one hand, and the practically observed results, on the other. That's why today we have various approaches to mathematical modeling and a sufficiently large number of software products modeling transport flows.

In recent years, successful mathematical and computational methods have been developed to formulate the model of a dynamic transport system. Many of them generalize traditional statistical methods; advanced and realistic models use a traffic simulator, traffic flow dynamics. The number of software products for the dynamic transport systems, which are sufficiently tested in the developed countries of the world, includes: DYNASMART [2], Dynameq [7], AIMSUN [1]. Creating models of transport systems in these programs is a rather cumbersome task that involves the presence of a large number of elements and requires a long process of calibrating hundreds or thousands of coefficients.

The purpose of this article is to present a network of highways in the form of a mathematical model and to select the necessary conditions for optimizing the road network. The road network is represented as a set of corresponding points (departure and arrival points) connected by a set of routes. As a result of its use, it is possible to reduce the task of optimizing the road network to a multitude of similar tasks of distributing the transport flow between two nodes along the routes and obtaining the equilibrium distribution of the transport stream in an explicit form. In turn, the explicit form of such distributions will make it possible to substantially reduce the time spent on computations.

2. Fundamental solutions, concepts and models of road networks

The road network model is the basis for solving many transport tasks, such as optimal network planning, improvement of traffic organization, optimization of public transport routes system, etc.

In 1952, Wardrop suggested that any transport system after a certain time comes to an equilibrium state and formulated two principles of the equilibrium distribution of traffic flows [12]. According to the first principle, "the travel time for all used routes is the same for all participants of the traffic, and less time that any participant will spend changing their route," and according to the second, "the average travel time is minimal."

Beckmann, McGuire, and Winsten (1956) were first who mathematically formulated these conditions [13]. Specifically, Beckmann, McGuire, and Winsten (1956) established an equivalence between transport network equilibrium conditions that argue that all paths connecting a pair of correspondent points will have equal and minimum travel times (or costs) [13]. Consequently, in this case, the equilibrium of the transport network can be obtained by solving the problem of mathematical programming [4]. This approach allowed to formulate, analyze and then solve the transport problem on the basis of actual transport networks.

Dafermos and Sparrow (1969) came up with terms "optimization-user" and "optimization-system" for the transport network [14]. These terms distinguish two situations, in the first situation, users act unilaterally in their own interests when choosing routes, in the second situation, users choose the optimal routes from a public point of view, taking into account the minimum cost in the transport system. Fig. 1 presents the two behavioral principles underlying transport networks.

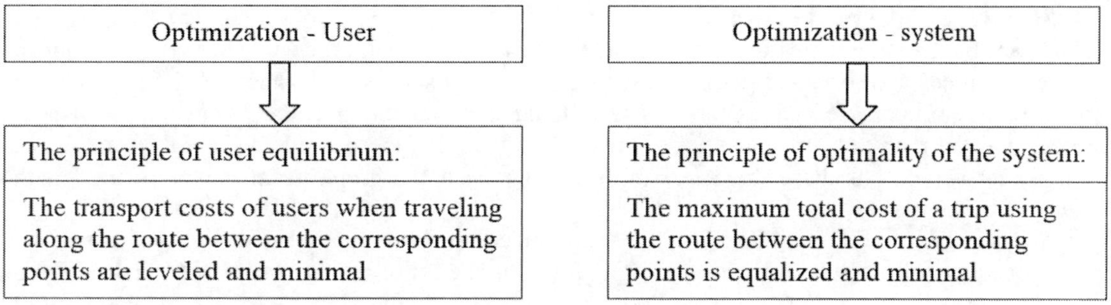

Figure 1. Behavioral principles of transport networks

Traditional methods of transport planning, including methods of network design, focus on improving mobility based on parameters such as speed and travel time. However, a mobile-oriented approach to transportation planning can stimulate the demand for travel and lead to congestion [3]. While mobility planning can increase the availability of the network, the availability of planning extends the range of possible solutions compared to the mobile-oriented strategy [6]. One of the obstacles is the lack of methods to solve the problem of accessibility directly, rather than mobility in transporting the network design.

In this situation, the goal of optimizing the transport network is to determine the shortest distance between the corresponding points. The road network should be planned in such a way as to minimize travel time and financial costs in the transport system.

3. Requirements for the optimal location of the road network

The location, density and composition of the road network form under the influence of numerous factors, which can be conditionally divided into three groups: economic, social and natural. When designing roads, taking these factors into account leads to the need for a quantitative assessment of their impact. The variety of these factors makes this assessment difficult. The question arises of identifying the most significant factors that determine the development of the road network. For the construction of the optimal version of the road network, it is necessary to investigate the influence of each factor on the result of the solution. It is impossible to conduct such studies in real conditions, because for this it is necessary to firstly build roads and only then to make a comparison. In such cases, studies can be carried out on mathematical models. The mathematical model will allow describing the process under study in the form of equations and inequalities.

When creating a mathematical model of the road network, it is necessary to first of all determine the criterion of optimality and to choose an objective function that reflects the requirements for the road network in the design.

To model a transport network it's necessary to have:

• cartographic material; Usually these are large-scale maps, since they allow you to measure with great accuracy the distances between points;

• information on the location of the main objects of the transport system and its environment (depending on the task to be solved: load-forming and load-absorbing enterprises, residential areas, places of employment, etc.);

• additional information from municipal and road organizations in the form of a list of streets with a characteristic of their carriageway;

• Information on the organization of traffic, i.e., traffic management schemes at intersections, squares and traffic intersections, as well as information on various traffic restrictions associated with established road signs.

4. Structure of the road network

To describe the structure of the transport network we apply graph theory, where the transport network is represented as an oriented graph consisting of a set of successively numbered vertices and a set of consecutively numbered edges (arcs) symbolizing the elements and their connections. The graph G modeling the road network must necessarily be tied, so that there is always a path from any vertex to any other vertex. Numbers characterizing the links of such a graph usually express the length of the route, time or cost of travel.

Let's introduce the notation: m is the set of successively numbered nodes of the graph G; Z is the set of sequentially numbered arcs of the graph G; I is the set of nodes that are the departure points, $i \subseteq m$; J is the set of nodes that are arrival points, $j \subseteq m$; It is assumed that $i \cap j = \emptyset$; M_{ij} is the set of routes between the departure points $i \in I$ and the arrival points $j \in J$.

The graph F modeling the road network is determined by the set of vertices and the set of pairs of vertices, between which there is a link (Fig. 2).

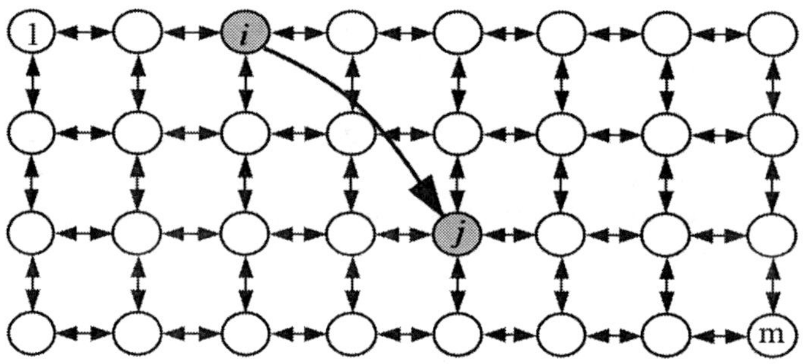

Figure 2. Model of road network graph

The task in this case is to decide: to include the k -th link in the projected network or not to include. The number of links of the k -th type is indivisible and can take only two values: 0 and 1. The definition of the unknown links of the road network (Z_k) must satisfy the condition:

$$Z_k = 0;1 \ (k = 1,2,3,...), \tag{1}$$

or

$$Z_k = \begin{cases} 1, \text{if arc} \in Z \text{ is on the route Mij} \\ 0, \text{ in the other case} \end{cases}. \tag{2}$$

Let's suppose, to take the minimum of road costs as a criterion of optimality (costs for construction, maintenance and repair of the road network). In this case, the optimal option will be met by the network with the least total length, for which the amount of travel costs is minimal:

$$\sum_k D_k l_k = \min, \tag{3}$$

where:

D_k – road costs associated with the construction, maintenance and repair of the road network;
l_k – length of road network;

k – number of sections (links of the road network).

If the road network is required to reduce the cost of transport work or the travel time of passengers or cargo, the only solution that meets these requirements is to ensure all transport links by independent roads along the shortest path, i.e. direct connection between each of the corresponding points (departure and arrival points). The solution of the problem is to determine the unknown routes between the points i and j (M_{ij}) satisfying the following condition:

$$M_{ij} \geq (i = 1,2,3,...,m;\ j = 1,2,3,...,m), \tag{4}$$

for which transport operation on the route ij (P_{ij}) or connection time between i and j (t_{ij}) will be minimal:

$$\sum_i \sum_j P_{ij} = \min, \tag{5}$$

$$\sum_i \sum_j t_{ij} = \min. \tag{6}$$

Considering the complexity and practical limitations of numerical programming, combinatorial analysis methods can be used in solving such problems, viewing different combinations of variable values, but only a reasonably selected part of possible combinations.

The realization of transportation with the lowest possible financial costs corresponds to the reduction in the costs of the means of production consumed in transport, and the criterion for assessing the optimality in the design of the road network corresponds to the general criterion of optimality for the economy of the country. This most fully satisfies the condition of optimality of the minimum financial costs, changing with the change in the outline of the network:

$$\sum_i \sum_j E_{ij} = \min, \tag{7}$$

where $\sum_i \sum_j E_{ij}$ - total financial costs, i. e. costs associated with the construction, maintenance and repair of roads and with the movement of vehicles in the directions ij.

To meet the requirement (7), it is necessary and sufficient that at each of the directions of the projected network, a possible minimum financial cost is provided, i.e. to:

$$\left.\begin{aligned} E_{1-2} &= \min; \\ E_{1-3} &= \min; \\ &\ldots \\ E_{1-n} &= \min; \\ &\ldots \\ E_{ij} &= \min. \end{aligned}\right\} \tag{8}$$

The validity of condition (8) can be proved in the following way. Let's say that there is an optimal road network for which:

$$E_{1\text{-}2} = \min$$
$$E_{1\text{-}3} = \min$$
$$\ldots$$
$$E_{1\text{-}n} = \min$$
$$\ldots$$

but $E_{ij} \neq \min.$

If we replace the route for which $E_{ij} \neq min$, by such a route that $E'_{ij} = min$, then the sum of all financial costs on the network will decrease by the amount:

$$\Delta = E_{ij} - E'_{ij}, \tag{9}$$

and it appears that:

$$\sum_i \sum_j E'_{ij} < \sum_i \sum_j E_{ij}. \tag{10}$$

It follows that for the case under consideration $\sum_i \sum_j E'_{ij} \neq \min$ the original proposal on the optimality of the road network, for which at least one route $E_{ij} \neq min$, is incorrect.

Thus, unlike the original proposal, the road network can not be optimal if inequality (8) is not respected.

5. Economic and mathematical model of road network optimization

The economic-mathematical model of a network of highways should be considered as a special case of the model of optimization of a unified transport network.

On the basis of the above considerations, it seems possible to draw up a system of equations characterizing the main regular relations between the formulated requirements of road transport and the deployment of a network of highways:

$$\left.\begin{array}{c} \sum_i \sum_j E_{ij} = \min; \quad E_{ij} = \min; \\ F\left(l_{ij}, Q_{ij}, D_{ij}, v_{ij}, E_{ij}\right) = 0; \\ M_{ij} \geq \left(i = 1,2,3,\ldots,m; \ j = 1,2,3,\ldots,m\right) \end{array}\right\}, \tag{11}$$

where:

l_{ij} – route length between the corresponding points i and j;
Q_{ij} – traffic between the corresponding points i and j;
D_{ij} – road costs that ensure the movement of vehicles in the direction;
A_{ij} – transport costs in the same direction;
v_{ij} – average vehicle speed.

The system (11) reflects the basic necessary condition for optimal road networks and the general principle of their construction in the mathematical formulation. Given mathematical model instructs on ways of the decision of a problem contains: definition of road and transport expenses; Selection of routes on which these costs are minimal, which leads to the fulfillment of the requirement expressed by the first ratio. The penultimate ratio in the system indicates that there is a functional dependence on a certain route between the objects of transportation, traffic speed, road and transport costs. The solution of this dependence with respect to E_{ij}, i.e. determination of total road and transport costs, is of independent interest and is an essential element of the task as a whole.

The principal relationship between total costs and their road and transport components can be represented by the equation:

$$E_{ij} = l_{ij}\left(D_{ij} + A_{ij}\right). \tag{12}$$

Considering the issue of optimal allocation of the road network, it is legitimate to assume that the most loaded links of the network should first of all be selected in the optimal network. If, during the selection of links, it becomes necessary to identify passenger transportations and consider them separately, this will require additional calculations and new principles for constructing optimal binding networks of roads.

The general principle of constructing the optimal binding network of roads is that any corresponding point in the optimal connecting road network is connected to all other corresponding points of the links, which ensure the least specific road and transport costs for moving goods from this point to all other corresponding network points.

Drawing up a mathematical model of the road network, one should start from the following limiting conditions:

• the optimal network must be a binder, i.e. must connect all without exception the corresponding points (there must be at least one route between each pair of points);

• for various reasons, some link z_k must be excluded from the construction of the optimal network or, conversely, necessarily included in it.

This condition is expressed by equality (1), and the transport-operational level of the projected network must be such that the speed $\bar{v}_{ij}$ between points i and j can not be less than a certain predetermined speed, the traffic intensity N_{ij} is not more than $\overline{N}_{ij}$, the safety factor of traffic $K_{бij}$ is not less than $\overline{K}_{бij}$.

As a result, the economic-mathematical model for optimizing the road network can be represented by the following system of equations and inequalities:

$$\left.\begin{array}{c} M_{ij} \geq (i = 1,2,3,\ldots,m;\ j = 1,2,3,\ldots,m); \\ z_k = 0;1; \\ L_{ij} = \dfrac{l_{ij}}{Q_{ij}}\left(D_{ij} + A_{ij}\right); \\ L_{ij} = \min; \\ \sum_i \sum_j L_{ij} = \min; \\ v_{ij} \geq \bar{v}_{ij};\ N_{ij} \geq \overline{N}_{ij};\ K_{бij} \geq \overline{K}_{бij}; \end{array}\right\}. \quad (13)$$

On the basis of the general principle, the task is reduced to the selection of such links, in which routes would be formed, providing the lowest specific road and transport costs between any two corresponding points.

6. Conclusion

The goal of optimizing the road network is to improve the efficiency of transport. The most fully satisfying condition for optimality is the minimum financial costs associated with the construction, maintenance and repair of the road network and the movement of vehicles along them. In the study, this indicator was adopted as a criterion of optimality in the economic and mathematical model of the road network. Therefore, the main requirement for the road network from the road transport side is to ensure the realization of all the transport links served by it with the least cost required for this.

From the example considered, it can be concluded that the main condition for the optimal binding networks of roads is the following: each corresponding point in the optimal connecting road network is connected with all other corresponding points in the directions providing the least financial costs necessary for moving passengers and cargo from this point to other corresponding points.

This also implies the general principle of constructing an optimal connecting road network, which consists in the fact that any isolated point in the optimal connecting road network is connected to all other corresponding points, which ensure the least financial costs for the movement of passengers and goods from this point to all other corresponding Network point.

On the basis of the general principle, the task is to select such links, under which routes would be created that would ensure road and transport costs between two corresponding points.

References

[1] Barcelo J and Casas J 2002 Dynamic Network Simulation with AIMSUN. *International Symposium Proceedings on Transport Simulation*, Yokohama. Kluwer Academic Publishers

[2] Florian M, Mahut M and Tremblay N 2005 Application of a simulation-based dynamic traffic assignment model. In: Kitamura, R., Kuwahara, M. (Eds.), *Simulation Approaches in Transportation Analysis: Recent Advances and Challenges*. Springer, New York, pp. 1–22
[3] Handy S 2005 Planning for accessibility: In theory and in practice. Access to destinations, pp. 131–147
[4] Krilatov A Yu Optimal strategies for traffic flow management on the transportation network of parallel links *Bulletin of St. Petersburg University*. 2014, Vol. 10. № 2, pp. 120–129
[5] Ligthill M J, Whitham F R On kinetic waves. A theory of traffic flow on crowded roads. *Proc. of the Royal Society Ser. A*. 1995, Vol. 229, No. 1178
[6] Litman T 2003 Measuring transportation: traffic, mobility and accessibility. *ITE Journal*, **73** (10), pp 28–32
[7] Mahmassani H S 2001 Dynamic network traffic assignment and simulation methodology for advanced system management applications. *Networks and Spatial Economics*. **1** (3/4), pp. 267–292
[8] Nagel K, Wagner R, Woesler R Still flowing: Approaches to traffic flow and traffic jam modeling, January 2, 2003
[9] Nagurney A 2004 Spatial Equilibration in Transport Networks, *in Handbook of Transport Geography and Spatial Systems, D. A. Hensher, K. J. Button, K. E. Haynes, and P. R. Stopher, editors, Elsevier, Amsterdam, The Netherlands*, pp. 583–608
[10] Sakhapov R L, Nikolaeva R V, Gatiyatullin M H and Makhmutov M.M. Risk management model in road transport systems. *Journal of Physics: Conference Series*. 2016, Vol. 738. № 1, pp. 1–10
[11] Volkov D O, Garichev S N, Gorbachev R A, Moroz N N Mathematic simulation of transport network load with a view to assessing viability of building new types of network systems. *PROCEEDINGS MIPT*. 2015, Vol. 7, № 3, pp. 69–77
[12] Wardrop J G Some theoretical aspects of road traffic research. *Proc. Institution of Civil Engineers*. 1952, Vol. 2, pp. 325–378
[13] Beckmann, Martin, C. B. McGuire, and Christopher B. Winsten. *Studies in the Economics of Transportation*. No. 226 pp. 1956.
[14] Dafermos, Stella C., and Frederick T. Sparrow. "The traffic assignment problem for a general network." *Journal of Research of the National Bureau of Standards B* 73.2 (1969): 91-118.
[15] Sakhapov R.L., Absalyamova S.G. Integration of universities and business as the condition of formation of the innovative economy. International Conference on Interactive Collaborative Learning, ICL 2013 2013. pp. 189-191

Modeling of traction-coupling properties of wheel propulsor

R L Sakhapov[1,2,3], R V Nikolaeva[2], M H Gatiyatullin[2] and M M Makhmutov[2]
[2]Kazan State University of Architecture and Engineering, 420043, Kazan, Zelenaya str., 1, Russia

[3]E-mail: rusakhapov@gmail.com

Abstract. In conditions of operation of aggregates on soils with low bearing capacity, the main performance indicators of their operation are determined by the properties of retaining the functional qualities of the propulsor. Therefore, the parameters of the anti-skid device can not be calculated by only one criterion. The equipment of propellers with anti-skid devices, which allow to reduce the compaction effect of the propulsion device on the soil, seems to be a rational solution to the problem of increasing traction and coupling properties of the driving wheels. The mathematical model is based on the study of the interaction of the driving wheel with anti-skid devices and a deformable bearing surface, which takes into account the wheel diameter, skid coefficient, the parameters of the anti-skid device, the physical and mechanical properties of the soil. As a basic mathematical model that determines the dependence of the coupling properties on the wheel parameters, the model obtained as a result of integration and reflecting the process of soil deformation from the shear stress is adopted. The total value of the resistance forces will determine the force of the hitch pressure on the horizontal soil layers, and the value of its deformation is the degree of wheel slippage. When the anti-skid devices interact with the soil, the traction capacity of the wheel is composed of shear forces, soil shear and soil deformation forces with detachable hooks. As a result of the interaction of the hook with the soil, the latter presses against the walls of the hook with the force equal to the sum of the hook load and the resistance to movement. During operation, the linear dimensions of the hook will decrease, which is not taken into account by the safety factor. Abrasive wear of the thickness of the hook is approximately proportional to the work of friction caused by the movement of the hook when inserted into the soil and slipping the wheel.

Keywords: anti-skid device, drag force, hook load, deformation, wheel diameter, tangential traction force

1. Introduction

When a wheel operates with removable devices, there are forces of resistance to movement from the wheel and removable hooks. We made a scheme for a calculation the resistance to the movement of the wheel and the process of interaction with the soil. When the anti-skid devices interact with the soil, the traction capacity of the wheel is composed of shear forces, soil shear and soil deformation forces with detachable hooks. We took V.V. Guskov model as a basic model, which determines the dependence of traction-coupling properties on the wheel parameters.

[1] To whom any correspondence should be addressed.

IOP Conf. Series: Journal of Physics: Conf. Series **936** (2017) 012033 doi:10.1088/1742-6596/936/1/012033

2. Research results and their discussion.

As skidding decreases, the drag force increases [1,2,3]. Let's consider a driving wheel equipped with anti-skid devices. When the torque is applied, forces of resistance arise, due to the movement of the unit and the hook load. The total value of the resistance forces will determine the force of the hitch pressure on the horizontal soil layers, and the value of its deformation is the degree of wheel slippage.

We developed a removable device (Fig. 1) fixed to the rim disk. Fig. 2 demonstrates the scan of the bracket.

The device consists of a bracket (1), a stand (2) fixed to the bracket with bolts (3), a lug (4) connected to the bottom of the stand by a bolt connection (5). In order to increase the resistance of the bracket to the bending and torque, the stiffening rib is welded (6). The rubber gasket (7) serves to soften the impact of dynamic loads on the wheel disc. Bending the sweep of the bracket along the line AC, the triangle's legs are welded, and the cut out triangle of the CBD later serves as a stiffener [4, 5, 6, 7].

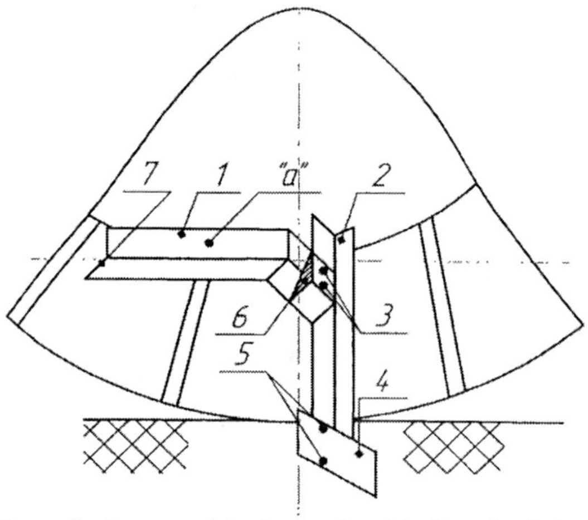

Figure 1. A removable device fixed to the wheel rim

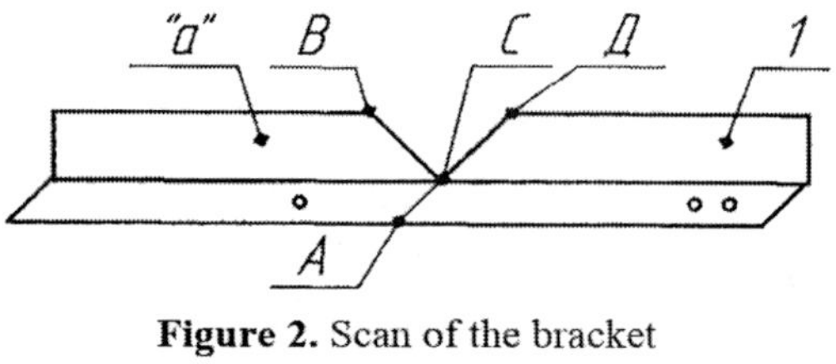

Figure 2. Scan of the bracket

The principle of operation is as follows:

To fix the device to the rim of the wheel, it is enough to unscrew one of the nuts of the disc and, in combination with the hole "a", fix the bracket 1. Then, using the bolts 3 to the bracket 1, fasten the rack 2 with the lug 4.

When the wheel rolls on a soil with a low bearing capacity, the hook 4 is set to the maximum height of its interaction with the soil. With increasing soil density, the height, due to the bolts 3, is reduced. When the unit leaves the road, the anti-skid device, by turning the bracket 4 by 180 °, is moved to the transport position [8].

The advantage of this device lies in the simplicity of the design, which allows, due to one connection, fastening to the rim wheel rim, to mount and dismantle it on the propulsion device of the vehicle. The bracket at one end has a rubber gasket, and on the other the rack is fixed [9,10,11,12].

The bracket being mounted to the wheel rim with a single bolt connection, made in the form of a L-shaped form containing a stiffener.

Let's consider a driving wheel equipped with anti-skid devices (Figure 3). When the torque is applied, forces of resistance arise, due to the movement of the unit and the hook load. The total value of the resistance forces will determine the force of the hitch pressure on the horizontal soil layers, and the value of its deformation is the degree of wheel slippage.

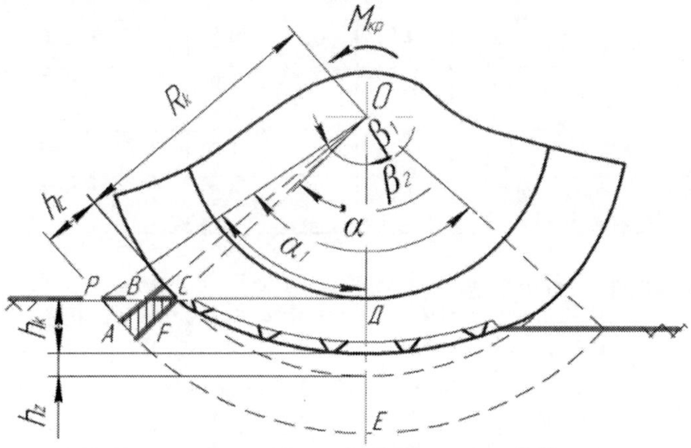

Figure 3. Scheme for calculating wheel slippage

The amount of crushing of the soil is determined by the product of the area ABCF by the length of the hook, then:

$$S_{ABCF}=S_{AOF}-S_{OBC}=S_{AOF}-(S_{OBD}-S_{OCD}) \quad (1)$$

$$S_{AOF}=[\pi\cdot(R_k+h_c)^2(\beta_1-\beta_2)]/360=0{,}0087[(R_k+h_c)^2(\beta_1-\beta_2)],$$

$$S_{OBD}=[/OD/\times/DB/]/2=[((R_k-h_k-h_z)^2/2)]\cdot tg\,\beta_1,$$

$$S_{OCD}=[/OD/\times/CD/]/2=[((R_k-h_k-h_z)^2/2)]\cdot tg\,\beta_2,$$

$$/OD/=R_k-h_k-h_z;\ /DB/=(R_k-h_k-h_z)\cdot tg\,\beta_1,$$

$$/CD/=(R_k-h_k-h_z)\cdot tg\,\beta_2.$$

Therefore,

$$S_{ABCF}=0{,}0087\cdot(R_k+h_c)^2\cdot(\beta_1-\beta_2)-[(R_k-h_k-h_z)^2/2)]\cdot(tg\,\beta_1-tg\,\beta_2) \quad (2)$$

$$\beta_1-\beta_2=\delta_{cu}\cdot\alpha,$$

where R_k – the radius of the whee, m; δ_{cu} – coefficient of wheel skidding at the moment of interaction of removable and tire hooks; h_c– height of anti-skid devices, m; h_k , h_z – tire deflection and gauge depth, m; α – the angle of interaction of the hook with the soil, deg.

Expressing the difference of angles β_1 and β_2 from formula (1), substituting in expression (2), we obtain the model of the slip process. Bearing in mind that the tangential force of the thrust P_C, realized by a detachable hook, is the product of the volume of the crushing of the soil by the coefficient of the volume crushing $K_Г$, we obtain:

$$cos\alpha=2\cdot(R_k-h_k-h_z)/(R_k+h_c);\quad \beta_1=(\delta_{cu}\cdot\alpha)+\beta_2;$$

$$\beta_1=\delta_{cu}\cdot(2\cdot arccos\,(R_k-h_k-h_z)/(R_k+h_c))+\beta_2.$$

Taking into account that $P_C=l_c\cdot K_Г\cdot S_{ABCF}$, we obtain:

$$P_C=l_c\cdot K_Г\cdot 0{,}0087\,(R_k+h_c)^2\cdot\delta_{cu}\cdot arccos\,(R_k/(R_k+h_c))-(R_k-h_k-h_z)^2/2)\cdot tg\,(\delta_{cu}\cdot(2\cdot arccos\,(R_k-h_k-h_z)/(R_k+h_c))+\beta_2)-tg\beta_2 \quad (3)$$

Analysis of the obtained model in the range of variable factors (Table 1) shows that with an increase in the height of anti-skid devices, regardless of their length the P_C rises along a curve with negative acceleration at a central level of factor variation (Fig. 3). For l_c = 0.1 m, in the interval h_c from 0.1 to 0.2 m, the P_C increases by 198%, at l_c = 0.2 m - by 177%, at l_c = 0.3 m - by 192% . With an increase in h_c from 0.2 to 0.3 m, the P_C rises, depending on its length by 66; 58 and 14% respectively.

Table 1. Levels and intervals of factor variation

Levels of variation	Interval	l_C, m	h_C, m	D_k, m	$\delta_{сш}$	β_2,^{0}	W_a, %	$K_П$	h_z, m
Lower	- 1	0,1	0,1	0,5	0,15	20	17	1,0	0,02
Central	0	0,2	0,2	1,0	0,20	30	20	1,5	0,04
Upper	+ 1	0,3	0,3	1,5	0,25	40	23	2,0	0,06
Variation interval	ΔX	0,1	0,1	0,5	0,05	10	3	0,5	0,02

With increasing wheel diameter, regardless of the height of anti-skid devices, at a central level of factor variation P_C rises along a curve with negative acceleration (Figure 4). For D_k = 0.5 m, in the interval h_c from 0.1 to 0.2 m, the P_C increases by 175%, at D_k = 1.0 m - by 91%, at D_k = 1.5 m - by 68% . With an increase in h_c from 0.2 to 0.3 m, the P_C rises, depending on the diameter of the wheel by 90, 86 and 46%, respectively.

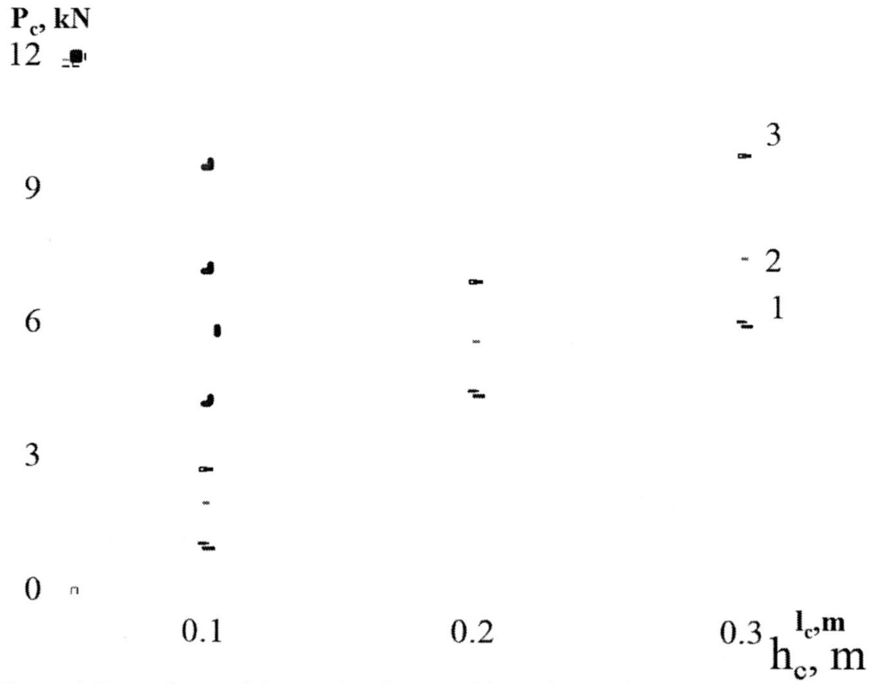

Figure 4. Dependence of the traction force on the height of the hook and its length:
1 – l_c= 0,1 m; 2 – l_c= 0,2 m; 3 – l_c= 0,3 m

With the increase in the coefficient of wheel slippage, at the moment of interaction of the removable and tire hooks, regardless of the height of the anti-skid devices, at the central level of factor variation, P_C rises along the curve with positive acceleration (Fig. 5). For $\delta_{сш}$ = 0.15, in the interval h_c from 0.1 to 0.2 m, the P_C increases by 80%, at $\delta_{сш}$ = 0.20 - by 95%, at $\delta_{сш}$ = 0.25 - by 112%. With an increase in h_c from 0.2 to 0.3 m, the P_C rises, depending on the skidding ratio of 67; 59 and 50% respectively.

With an increase in the angle β_2, regardless of the height of the anti-skid devices, at the central level of factor variation, P_C decreases along the curve with negative acceleration (Fig. 4). For β_2 = 200, in the interval h_c from 0.1 to 0.2 m, P_C decreases by 34%, with β_2 = 300, 400 - by 54% and 71%, respectively. With an increase in h_c from 0.2 to 0.3 m, the P_C decreases, depending on the angle β_2 by 35; 34 and 33% respectively.

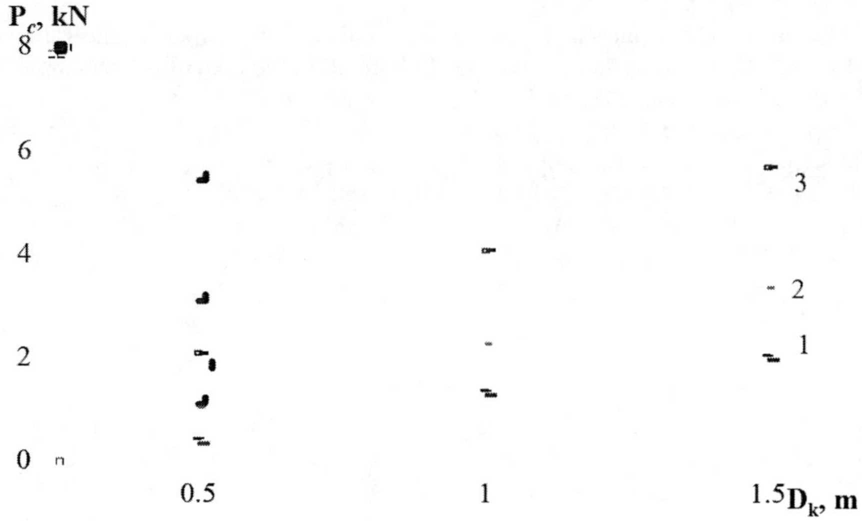

Figure 5. Dependence of traction force on wheel diameter and hitch height
1 – h_c = 0,1 m; 2 – h_c = 0,2 m; 3 – h_c = 0,3 m

With an increase in the amount of air in the soil, regardless of the height of the anti-skid devices, at a central level of factor variation, P_C decreases along the curve with negative acceleration (Figure 6). For $K_П$ = 1.0, in the interval h_c from 0.1 to 0.2 m, the P_C decreases by 19%, at $K_П$ = 1.5, by 25%, at $K_П$ = 2.0, by 29%. With an increase in h_c from 0.2 to 0.3 m, the P_C decreases, depending on the amount of air in the soil at 28; 30 and 33% respectively.

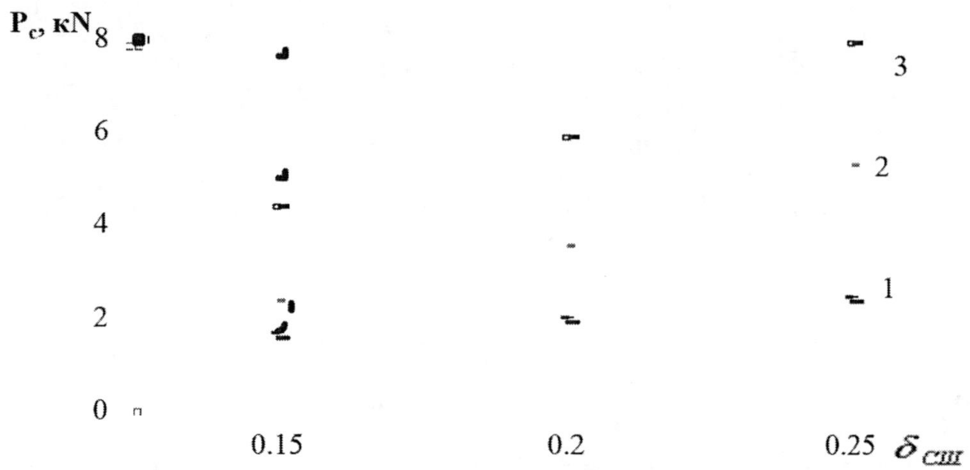

Figure 6. Dependence of the traction force on the wheel slip and hitch height: : 1 – h_c = 0,1 m; 2 – h_c = 0,2 m; 3 – h_c = 0,3 m

Graphically, analysis of the model (3) shows that in practice it is difficult to apply it because of the need to determine the angles of the hooks in the period of their interaction with the soil [13]. The

IOP Conf. Series: Journal of Physics: Conf. Series **936** (2017) 012033 doi:10.1088/1742-6596/936/1/012033

height of interaction of the hooks with the soil, depending on the angle of rotation of the wheel, changes, so in order to eliminate these drawbacks, we will determine the reduced height of the anti-skid devices h_p, m:

$$h_p = \frac{\sum_{1}^{\alpha_{max}} (R_K + h_C)\cdot(\cos(0{,}0175\cdot\alpha)) - R_K}{\arccos(R_K / (R_K + h_C))}, \qquad (4)$$

$0 < \alpha < \arccos(R_k / (R_k + h_c))$;

The reduced altitude of anti-skid devices can also be determined by the ratio of the area of the segment formed as a result of the interaction of the hook to the area of the rectangle (Fig. 7). Then, the area of ANBK S_{cez} segment is defined:

$S_{cez} = ((\pi \cdot R_k^2 \alpha)/360) - (R_k - h_c)\cdot a;$

Given, that $\alpha = 2\times \arccos((R_k - h_k - h_z)/(R_k + h_c))$;

$$a = \sqrt{(R_K + h_C)^2 - (R_K - h_K - h_Z)^2}$$

We have: $S_{cez} = ((\pi R_k^2 \cdot \arccos((R_k - h_k - h_z)/(R_k + h_c))/180) - (R_k - h_c)\cdot \sqrt{(R_K + h_C)^2 - (R_K - h_K - h_Z)^2}$;

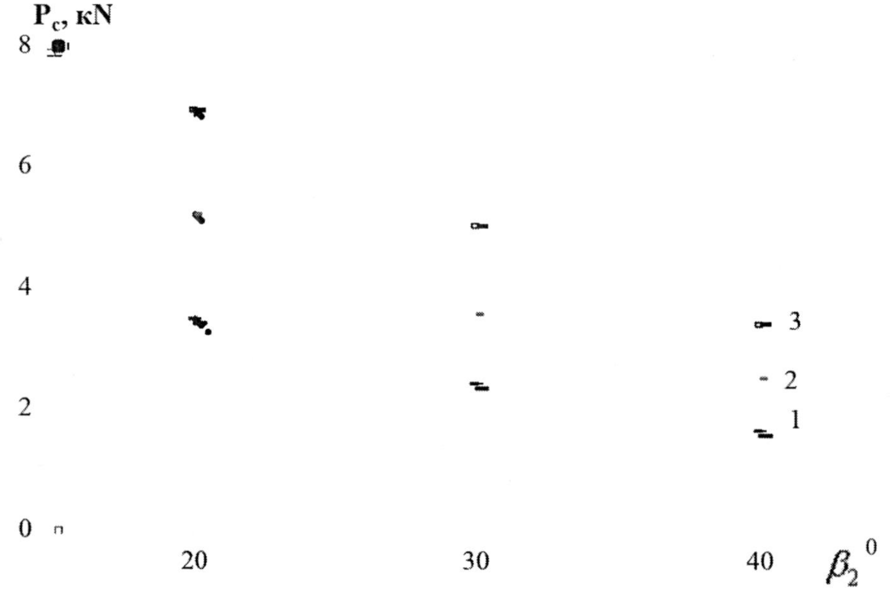

Figure 7. Influence of the angle β_2 on the thrust force on the height of the hooks: 1 – h_c = 0,1 m; 2 – h_c = 0,2 m; 3 – h_c = 0,3 m

The area of the rectangle ABCD will be determined as:

$S_{np} = 2\, a\cdot h_c = 2\cdot h_c\cdot \sqrt{(R_K + h_C)^2 - (R_K - h_K - h_Z)^2}$.

Then $$h_p = \frac{S_{cez}\cdot h_C}{S_{np}}. \qquad (5)$$

Analysis of the obtained model, in the range of variable factors, shows that with increasing height of anti-skid devices, regardless of the diameter of the wheel, at the central level of factor variation, h_p increases linearly (Fig. 8). In the interval h_c from 0.1 to 0.2 m, h_p increases by 102%, and with increasing h_c from 0.2 to 0.3 m, h_p increases by 51%.

IOP Conf. Series: Journal of Physics: Conf. Series **936** (2017) 012033 doi:10.1088/1742-6596/936/1/012033

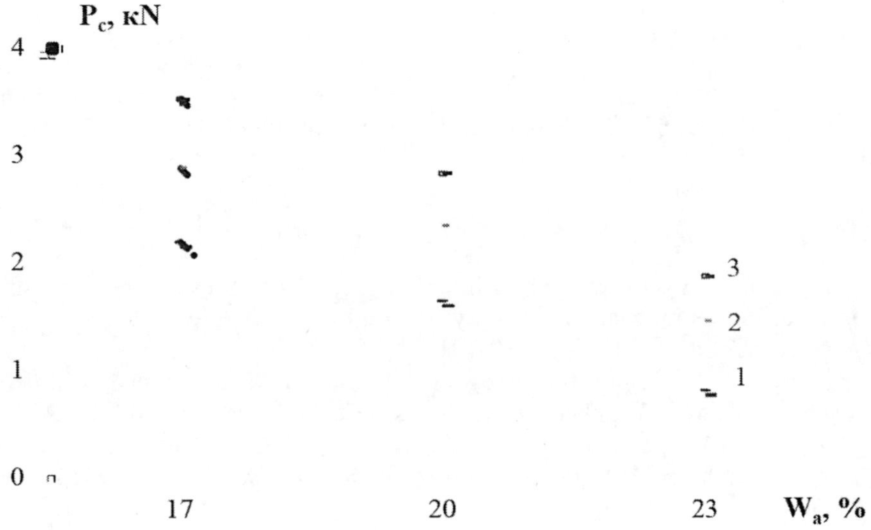

Figure 8. Change in traction force from soil moisture and height of anti-skid devices:
1 – h_c = 0,1 m; 2 – h_c = 0,2 m; 3 – h_c = 0,3 m

With the increase in the diameter of the wheel, regardless of the depth of the track, at a central level of factor variation, the reduced altitude is reduced along a curve with negative acceleration. For h_k = 0.05 m, 0.10 m, 0.15 m in D_k interval from 0.5 to 1.0 m, h_p decreases by 3.4%, 3.7%, 5.3%, respectively. With an increase in D_k from 1.0 to 1.5 m, h_p decreases, depending on the depth of the track, by 1.4, 2.6 and 2.9%, respectively.

In general, in analyzing these formulas, it should be noted that the values of the reduced height of the hooks [14] according to model (4), depending on α_{max}, are obtained 2-4% lower than in the model (5).

Thus, applying the parameter of the reduced height of the anti-skid devices, the tangential force of the thrust of the P_C, realized by a detachable hook, is determined:

$$P_c = l_c \cdot h_p \cdot K_Г \cdot \delta_{сш} \cdot S_{T_2}, \quad (6)$$

where S_{T_2} – The theoretical path of wheel movement, respectively, in the joint interaction of tire and anti-skid devices with soil, m:

$$S_{T_2} = 0{,}035 \cdot R_k \cdot arccos(R_k / (R_k + h_c)).$$

Analysis of the model for calculating the tangential force of traction, which appears from anti-skid devices (6), shows the identical influence of the studied factors on the response of the function; however, with respect to the model (3), the values for traction forces are obtained lower by 4-5% due to the average value of the hook height.

3. Conclusion

When the anti-skid devices interact with the soil, the traction capacity of the wheel is composed of the forces of shear and shear of soil "bricks" sandwiched between the tire hooks and the deformation forces of the soil by removable hooks.

With the increase in the coefficient of wheel slippage, at the moment of interaction of the removable and tire hooks, regardless of the height of the anti-skid devices, at the central level of factor variation, P_C rises along the curve with positive acceleration (Fig. 5). For $\delta_{сш}$ = 0.15, in the interval h_c from 0.1 to 0.2 m, the P_C increases by 80%, at $\delta_{сш}$ = 0.20 - by 95%, at $\delta_{сш}$ = 0.25 - by 112%. With an increase in h_c from 0.2 to 0.3 m, the P_C rises, depending on the skidding ratio of 67; 59 and 50% respectively.

With a decrease in skidding, the drag force increases. Consequently, the process is extreme. To find the conditional extremum, the optimization of the process was carried out by jointly solving the regression equations to determine the slipping and the resistance to movement from the device. For this purpose, the problem of process optimization was considered and solved taking into account the quantitative and qualitative parameters.

References

[1] Sakhapov R L, Absalyamova S G Integration of universities and business as the condition of formation of the innovative economy. *International Conference on Interactive Collaborative Learning, ICL 2013*, 2013. pp. 189-191

[2] Belyakov V V Transport-technological all-terrain machine. *Fundamentals of the theory of movement*. ed. by V V Belyakov and A P Kuleshova. N. Novgorod: TALAM, 2010. 960 P.

[3] Sakhapov R L, Zagretdinov R V, Gabdullin T R, The Modeling of control systems of road-building equipment. *Izvestiya Samarskogo nauchnogo centra RAN*. 2014. vol. 16. № 1(2). pp. 394-396.

[4] Kanygin V V, Orda A N The impact of running systems on soil. *Mechanization and electrification of agriculture*. 2008. № 4. pp. 17-18.

[5] Kaczynski N A Physics of soils. M.: Vishaya schola, 2001. 323 P.

[6] Kotlyarenko, V I The Main directions of improving the patency of wheeled vehicles. M.: Moscow state Industrial University, 2008. 285 P.

[7] Makhmutov M M Optimization of parameters of taper of the strut device chains. *Tractorii chelhozmachinye*. 2013. № 6. pp. 31-33.

[8] Patent RUS № 2453445 Anti-skid device for the wheels of the vehicle, the application 2010150215/11 / Makhmutov M M, Makhmutov M M, Patent holder FGBOU VPO "Russian state agrarian correspondence University", publ. 07.12.2010, publ. 20.06.2012, bull. № 17, 2 p.

[9] Sakhapov R L Mechanical-technological justification of resource-saving cultivators: Thesis. Doctor of Technical Sciences: 05.20.01. Kazan, 2002. 381 P.

[10] Bekker M G Wheels and track. *Journal Automobile engineers*. 2001. № 6. pp. 15–17.

[11] Bekker M G Introduction to Terrain-Vehicle Systems. M. G. Bekker, 2002. 520 P.

[12] Reece A R, Adams G. One Aspect of Tranchayer Performanse. *Transactions of the ASAE*, 1966. pp. 6-9.

[13] Sakhapov, R L, Nikolaeva, R V, Gatiyatullin, M H, & Makhmutov, M M (2016, August). Risk management model in road transport systems. In Journal of Physics: Conference Series (Vol. 738, No. 1, p. 012008). IOP Publishing.

[14] Sakhapov, R L, Nikolaeva, R V, Gatiyatullin, M H, & Makhmutov, M M (2016, August). Modeling the dynamics of the chassis of construction machines. In Journal of Physics: Conference Series (Vol. 738, No. 1, p. 012119). IOP Publishing.

IOP Conf. Series: Journal of Physics: Conf. Series **936** (2017) 012034 doi:10.1088/1742-6596/936/1/012034

The Wang Landau parallel algorithm for the simple grids. Optimizing OpenMPI parallel implementation

A S Kussainov[1,2]

[1] Physics and Technology Department, al-Farabi Kazakh National University, Almaty, Republic of Kazakhstan

[2] National Nanotechnology Laboratory of Open Type, al-Farabi Kazakh National University, Almaty, Republic of Kazakhstan

E-mail: arman.kussainov@gmail.com

Abstract. The Wang Landau Monte Carlo algorithm to calculate density of states for the different simple spin lattices was implemented. The energy space was split between the individual threads and balanced according to the expected runtime for the individual processes. Custom spin clustering mechanism, necessary for overcoming of the critical slowdown in the certain energy subspaces, was devised. Stable reconstruction of the density of states was of primary importance. Some data post-processing techniques were involved to produce the expected smooth density of states.

1. Introduction

Wang-Landau algorithm became a universal tool to study complex energy landscapes of the multidimensional quantum systems with multiple degrees of freedom and complex bonding mechanism between its members [1]. Numerous publications demonstrate successful application of this method to the wide range of phenomena. It guides itself through the plethora of accessible states by the explicit Hamiltonian of the system and by keeping track of the visited energy states and their accessibility. It could easily adopt or be a part of almost every effective modern algorithm simulating the behavior of the spin glasses. Complex optimization, graph theory and other state-of-the-art numerical methods greatly benefit from this algorithm. Wang-Landau algorithm is based on the Monte Carlo method and Metropolis algorithm [2] and provides an indispensable tool to study the complex systems. It could include in its fabric a cluster flip scheme to overcome critical slowdown around the transition points.

2. Methods

We implemented the all-inclusive universal version of the Wang-Landau Monte Carlo method. We were motivated by the need to get reasonable performance for a systems bigger than 8 by 8 points and also expected to provide a tool against the possible critical slowdown not only around the transition points but also during the regular thermalization stage at low temperatures. The single spin flip algorithm is straightforward and easily substantiated theoretically but generally inefficient. Convergence to the equilibrium values is usually slow and changes introduced by a single local spin flip propagate diffusively. Various algorithms have been proposed to speed up the process by flipping the clusters of spins at once. Swendsen-Wang [3] and Wolff [4] Monte

IOP Conf. Series: Journal of Physics: Conf. Series **936** (2017) 012034 doi:10.1088/1742-6596/936/1/012034

Carlo methods are among them. The Wolff algorithm is an improvement over the SwendsenWang algorithm since it has a larger probability of flipping the bigger clusters. It is not always possible to provide the exact value of the temperature to the algorithm and we have implemented the method of invasion cluster procedure elaborated by Machta *et al* [5].

Nowadays everyone should always consider and implement the code in parallel [6], [7] by distributing the tasks between the multiple walkers [8], [9] or split the energy range into pieces. In all cases the unique and custom made procedure to combine the results from a multiple sources is required and was built in our work.

Versatile Open MPI library [10] for the multithread coding in the multicore CPU (central processing unit) environment or in a network of computers connected in the computational cluster was used. OpenMPI is extensively documented in electronic resources and much easier to deploy on the Linux systems. Another, OpenMP project, is among the available alternatives though it works at its full power on the paid operational system platform and software [11].

The types of $2D$ grids we considered for our density of states simulation were originally studied by Domb [12] and represent a two-dimensional Euclidian $m \times n$ lattices hosting a single spin in every node of its discreet structure. The spin is taking q possible values distributed around the circle with the following steps

$$\theta_n = 2\pi n/q, \tag{1}$$

where n goes from 1 to q and interaction Hamiltonian is defined as

$$\mathcal{H} = -J\sum_{i,j}\cos(\theta_{\mathrm{s_i}} - \theta_{\mathrm{s_j}}) \tag{2}$$

This model is known as vector Pott model or clock model [13]. Here J is some spin-spin coupling constant and θ_{s_i} and θ_{s_j} are the spins' positions around the clock dial. The simpler standard Pott model has the Hamiltonian

$$\mathcal{H} = -J\sum_{i,j}\delta(\mathrm{s_i}, \mathrm{s_j}), \tag{3}$$

where δ is the Kroneker's delta. When, in eq.(2), spin orientation takes only two values $\mathrm{s_i}=\{+1,-1\}$ it is known as the Ising model [14].

To draw an energy landscape for the simple Hamiltonian, see eq.2-3, we need to consider interactions only between the closest neighbors which is a valid assumption considering the magnetic interaction strength is dropping as fast as $1/r^3$ with the distance. Periodic boundary conditions, known as toroidal, were used.

If the external magnetic field $\boldsymbol{B} \neq 0$ is superimposed, the total interaction energy is modified by the presence of individual magnetic dipoles into the following expression

$$E = -J\sum_{i,j}\mathrm{s_i}, \mathrm{s_j} - \mathbf{B}\sum_{i=1}^{m\times n}\mathrm{s_i}, \tag{4}$$

where J is again the spin-spin interaction strength and $\boldsymbol{B}$ is the external magnetic field strength. If $J > 0$ the system is ferromagnetic, otherwise, if $J < 0$, it is paramagnetic. Indices i and j are combined into all accessible pairs of adjacent spins on the grid, excluding the double counts of ij and ji pairs. The upper bound for the first sum in eq.(4) is determined by the range of interaction, which is a simple pairwise interaction, and by the number of spins on the simulation grid.

The complete thermodynamical and statistical description of the system requires the knowledge of thermodynamic partition function and density of states. For canonical ensemble the partition function Z is given by the following expression

$$Z = \sum_{conf.space} e^{-E_r/k_B T}, \tag{5}$$

where summation is run over all configuration space, k_B is the Boltzmann constant, E_r and T are the enumerated energy values and the temperature for this particular configuration [15].

High degree of the system's degeneracy, that is the same values of E_r are attained by the different system's spin configurations, allows us to describe all possible configurations only in terms of energy, that is

$$Z = \sum_{E} g(E) e^{-E/k_B T}, \tag{6}$$

where $g(E)$ is the density of states the Wang Landau algorithm is aimed to estimate.

For our analysis we need to keep track of the two values, that is of the $H(E)$ - which is the number of times this particular energy value is measured at the single Monte Carlo run of the algorithm and $\widetilde{g}(E)$ which is the accumulated relative density of states. Proper normalization will give the true value of $g(E)$. Probability of the system transition to the next configuration is given by an expression

$$p(E_1 \to E_2) = min\left[\frac{\widetilde{g}(E_1)}{\widetilde{g}(E_2)}, 1\right] \tag{7}$$

Process starts with the array of $H(E)$ values filled with zeros everywhere. The unknown density of states $\widetilde{g}(E)$ is also uniform and taken to be 0 for any value of E. Both values are modified every time any configuration with this particular value of E is visited. Due to the overwhelming multitude of the Monte Carlo steps it is better to keep track of $g(E)$ changes through the following expression $ln\left[\widetilde{g}(E)\right] \to ln\left[\widetilde{g}(E)\right] + ln\left[f\right]$. For $ln[\widetilde{g}(E)]$ the initial value of modification factor f is chosen to be bigger than 1 and equal $e \approx 2.71828$ in this particular implementation.

As soon as the condition of flatness for $H(E)$ is met we change modification factor to a new value $f_{i+1} = \sqrt{f_i}$ and repeat the procedure starting with $H(E)$ equals zero everywhere again. Measure of flatness is chosen to be mean$(H(E))$/max$(H(E)) * 100\%$. $H(E)$ is incremented by 1 every time the state with particular E is visited or the system stays with the old value of E. After multiple iterations, $\widetilde{g}(E)$ should converge to the unnormalized density of states for the simulated system.

Thus, to find the density of states the flat histogram random walk method of Wang and Landau goes through following steps:

1. Choose at random any spin $\mathbf{s_i}$ and flip its sign so $\mathbf{s_i}' = -\mathbf{s_i}$
2. Calculate the total energy according to eq.(2) or eq.(3).
3. Calculate the transition probability according to eq.(7) and keep the updated configuration if the criterion is met. If not, reverse the spin flip and keep the current configuration.
4. Repeat the previous steps to achieve the proper flatness of the histogram.

Every instance of visiting the different spin configuration with certain value of energy E or keeping the old one, when the transition rules in eq.(7) are not met, is tracked by an increase $ln\left[\widetilde{g}(E)\right] \to ln\left[\widetilde{g}(E)\right] + ln\left[f\right]$ and $H(E) \to H(E) + 1$.

These steps, repeated multiple times, which is usually on the order of number of sites in the simulation grid or bigger, represent a single Monte Carlo step. To calculate any observable value for a given temperature or energy range, Monte Carlo steps have to be repeated numerous times.

Different physical properties, including magnetization, specific heat, density of states etc, could be calculated via similar tedious algorithms or directly from the partition function, see the next three expressions for the thermodynamic total energy, heat capacity at constant volume and entropy

$$\begin{aligned} \langle E\rangle &= -\frac{\partial \ln Z}{\partial \beta}, \\ C_v &= \frac{1}{k_B T^2}\frac{\partial^2 \ln Z}{\partial \beta^2}, \\ S &= \frac{\partial}{\partial T}\left(k_B T \ln Z\right) \end{aligned} \tag{8}$$

Splitting the energy range between the processes require some additional thoughts and design. The different energy range pieces are meant to overlap to fit the latest development of the algorithm, see [16]. Each particular E value is enumerated and has its own slot in $\widetilde{g}(E)$ array to accumulate the number of visits $H(E)$ from one Monte Carlo step to another. It is easy to show that the number of these slots, for the $m \times n$ $2D$ grid, is $2nm + 1$.

3. Results and Discussions

The eight-core desktop PC which was used for simulations runs on Intel Core i7 4790K, 4.0GHz/LGA-1150/22nm/Haswell/8Mb L3 Cache, DDR-3 16Gb/1866MHz PC14900. *gcc* C++ compiler and Octave packages were deployed on Debian Linux OS .

On Figure 1 we plotted the sample output data from our implementation of the algorithm for the Ising model. The grid size is 16^2 and the flatness criterion is 85%. For simplicity, magnetic field $\boldsymbol{B}$ is taken to be zero everywhere.

First subplot (a) on Figure 1 represents the $H(E)$ - which is the number of visits per energy value accumulated for each separate energy range after the last modification of $g(E)$, before flashing the data to the hard drive. The separate energy ranges and their overlap are clearly visible on the subplot. The second subplot (b) shows the densities of states $g(E)$ obtained from the parallel processes before stitching them to a single function. For a better visual analysis the every data piece head is normalized. The thick line of zeros at the bottom represents the states included into $2mn + 1$ range but not accessible due to the system configuration and its interaction Hamiltonian. The third subplot (c) is the final reconstructed density function.

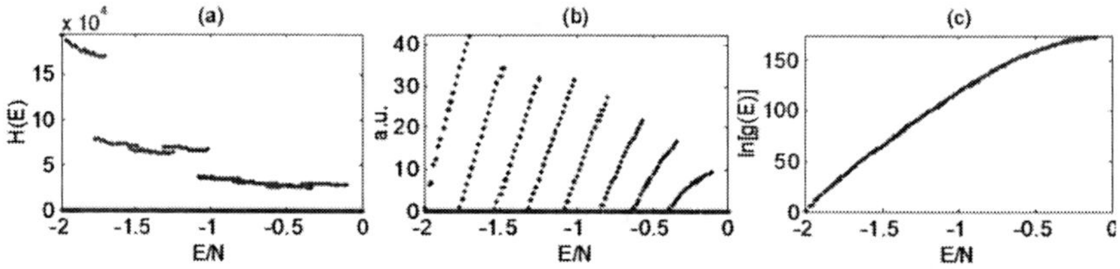

Figure 1. From left to right: (a) Accumulated histogram of the $H(E)$ visits, (b) intermediate density function data and (c) final reconstructed density of state $g(E)$ function for the Ising model.

The final graphical data postprocessing and display is done in Octave package. At this stage, depending on the number of artifacts in the data, each head of the consecutive data piece is aligned with the tail of the previous. The head of the first data piece is normalized to 2, because there are only two states when the spins are aligned in the same direction.If head-to-tail stitching at the particular point is unsuccessful we are automatically switching to the different but overlapping in the energy space points.

For the sake of simplicity, we studied only the region from -2 to 0 $[E/N]$. In case of the Ising model the positive part is easily reconstructed by the mirror reflection. In all other cases the energy range could be expanded as far as it necessary.

One can see that the difference between the number of visits may be significant for the different pieces of the same model, see subplot (a), nonetheless the first derivative (the factor determining the quality of stitching) for the two adjacent $g(E)$ pieces at the connection points, as could be understood from the process, is mainly dependent on the smoothness of $H(E)$.

On Figure 2 we have plotted the data for the three various spin models, that is the Ising model on the subplot (a), the Pott model on the subplot (b) and the clock model on the subplot (c). For the Ising model we increment the grid size from 16 to 128 points. For the Pott model the q takes the value of $8, 9, 19$ and for the clock model the q takes the value of $7, 9$ and 11 correspondingly.

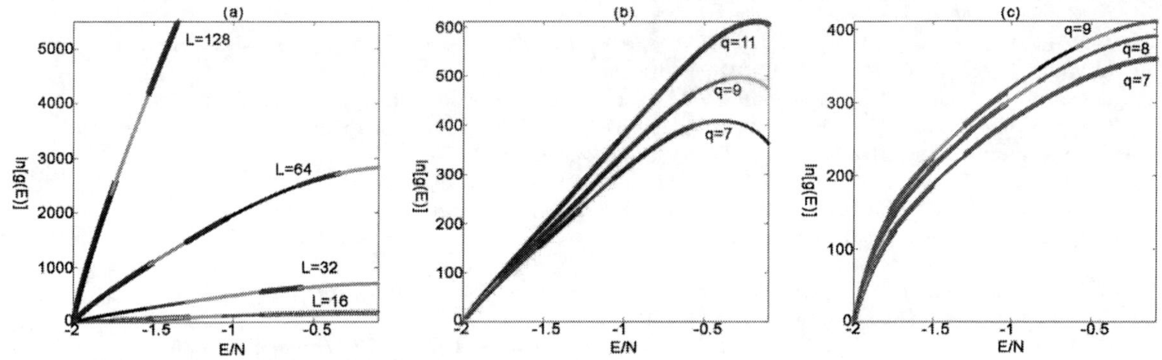

Figure 2. (a) the Ising, (b) the Pott , and (c) the clock model simulations for the different parameters sets, see text for the complete description.

We also implemented the cluster flip algorithm with the set of stopping rules effectively determined by the energy range of the particular thread in the parallel simulations, see Figure 3.

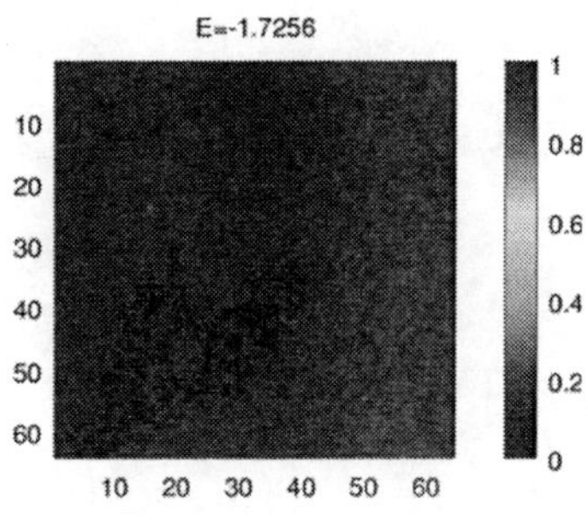

Figure 3. Invaded cluster algorithm implementation data with corresponding grid energy value $E = -1.7258$.

Test run and analysis have shown that this algorithm is still more suitable for the transition points but not for keeping the walker at the low energy range and requires a further study.

4. Conclusions

We have implemented multitool Wang-Landau algorithm to calculate density of states based on the Monte Carlo method with importance sampling in $2D$ spin glasses for the Ising, Domb and Pott models as an example of multithreading and performance optimization in scientific computing in general and in density of states calculations in particular. We reported results produced by a generalized set of rules represented in terms of parallel algorithms and programs to construct a density of states. The algorithm works fine for all type of simple models and only requires correct definition of the interaction Hamiltonian in C++ script. Significant speedup compare to the linear model on the multicore computer has been observed. The model and its development represent the computational basis for the whole generation of the quantum algorithms for approximating partition function. Matlab/Octave scripts are supplied to the code package to produce the final graphical output density of states. Knowing, that it is not always possible to provide the exact value of the temperature to the algorithm we have implemented the invaded cluster procedure to supplement our single spin flip code.

Acknowledgments
This research was supported by grant $3824/GF4$ provided by the Science Committee at the Ministry of Science and Education of Republic of Kazakhstan to the principal investigator at the National Nanotechnology Laboratory of Open Type, Physics and Technology Department, al-Farabi Kazakh National University.

References
[1] Wang F, Landau D P 2011 *Phys. Rev. Lett.* **86** 2050
[2] Metropolis N, Rosenbluth A W 1953 *J. Chem. Phys.* **21** 1087
[3] Wang J, Swendsen R H 1990 *Physica A: Statistical Mechanics and its Applications* **167(3)** 565
[4] Wolff U 1989 *Phys. Rev. Lett.* **62(4)** pp 361-364
[5] Machta J, Choi Y S, Lucke A and Schweizer T and Chayes L M 1996 *Phys. Rev. E* **54(2)** 1332
[6] Kussainov A S and Kussainov S G 2015 *J. Phys.: Conf. Series* **633** 012072
[7] Kussainov A S and Pya N 2016 *J. Phys.: Conf. Series* **738** 012076
[8] Altevogt P, Linke A 1993 *Parallel Computing* **19(9)** 1041
[9] Heermann D W Burkitt A N 1991 *Parallel Algorithms in Computational Science* (New York:Springer-Verlag)
[10] Pješivac-Grbović Jelena, Angskun T *et al.* 2004 *Proceedings, 11th European PVM/MPI Users' Group Meeting* 97
[11] Chapman B 2007 *Using OpenMP: Portable Shared Memory Parallel Programming (Scientific and Engineering Computation)* (Massachusetts:The MIT Press) 384
[12] Domb C, Green M S 1974 *Ising Model. Phase Transitions and Critical Phenomena* **V3** 357
[13] Potts R B 1952 *Proc. Cambridge Philos. Soc.* **48(1)** 106
[14] Ising E 1925 *Zeitschrift fr Physik* **31** 253
[15] Zhang C, Ma J 2007 *Phys. Rev. E.* **76** 036708
[16] Vogel T *et al. Phys. Rev.E* **90** 023302

IOP Conf. Series: Journal of Physics: Conf. Series **936** (2017) 012035 doi:10.1088/1742-6596/936/1/012035

On the natural small vibrations of dislocation in an isotropic medium

I. L. Bataronov and V V Dezhin

Voronezh State Technical University, Moskovsky prospect 14, Voronezh 394026, Russian Federation

E-mail: viktor.dezhin@mail.ru

Abstract. The equation for the natural bending vibrations of an infinite dislocation is written. The long-wavelength limit is considered. The orientation dependence of the vibrational spectrum has been studied. Solutions for two cases (the bending wave velocity along the dislocation line is not equal to the speed of sound waves and the bending wave velocity along the dislocation line is close to the speed of sound waves) are obtained. Local and quasilocal branches of edge and screw dislocations vibrations are found.

1. Introduction

The important role of dislocations (linear crystal defects) in the formation of the physical properties of crystals is known. This explains the interest in the problem of a dislocation crystal oscillations. The dislocation oscillations were considered earlier in a number of papers. Laub and Eshelby [1] considered the dislocation lying along the axis Ox and oscillating in the plane xOz in accordance with the law $u = u_0 \cos k_x x \cos \omega t$, where k_x is wave vector component along dislocation line, ω is frequency, t is time. Lifshitz and Kosevich [2] analyzed long-wave oscillations, and the perturbation introduced by dislocation was reduced to a change in the elastic modules of the medium. In the works of Ninomiya [3-5] the problem of crystal vibrations with a dislocation is solved by application of self-consistent dynamic theory of dislocations. Lifshitz and Pushkarov [6] found a spectrum of short-wave phonons near the dislocation. Localized oscillations near a screw dislocation based on the Green's function method were considered by Tewary [7]. In subsequent years, individual problems of crystal vibrations with dislocations were studied. However, these studies do not show a general relationship between the crystal response function and the linear dislocation response function on the external force. In paper [8], using the lagrangian formalism, the equation of small oscillations of a crystal with a dislocation was obtained, and the generalized susceptibility (linear response function) of an infinite dislocation was found. Based on the results of [8], a generalized susceptibility of a dislocation in a dissipative crystal is found [9-11]. The oscillations of the dislocation segment [12, 13] are also investigated on the basis of paper [8]. In [14], a generalized susceptibility of a dislocation that interacts elastically with point defects is found. In the present paper, on the basis of work [8], we obtain the equation for the natural small vibrations of a dislocation and carry out its analysis.

2. Equation of dislocation natural small vibrations

Using the result of [8], we write the expression for the inverse generalized susceptibility for an infinite dislocation lying along the axis Oz:

$$\alpha^{-1}(k_z,\omega) = -\frac{\mu}{8\pi}\left\{-(b^2+b_e^2)k_z^2 + \left(b^2\frac{\omega^2}{c_t^2} - 3b_s^2k_z^2\right)\ln\frac{k_m^2}{k_z^2-\omega^2/c_t^2}\right.$$

$$\left.+\xi\left[2k_z^2(2b_s^2-b_e^2)+b_e^2\frac{\omega^2}{c_l^2}\right]\ln\frac{k_m^2}{k_z^2-\omega^2/c_l^2} + (4b_s^2-b_e^2)\frac{k_z^4c_t^2}{\omega^2}\ln\frac{k_z^2-\omega^2/c_l^2}{k_z^2-\omega^2/c_t^2}\right\}, \quad (1)$$

where k_z is wave vector component along dislocation line, μ is shear modulus of the crystal, $\boldsymbol{b}$ is Burgers vector of the dislocation, b_e and b_s are edge and screw components of the Burgers vector, c_t and c_l are the transverse and longitudinal sound velocities, k_m is maximum wave number, $\xi = c_t^2/c_l^2$. The dispersion law for the natural oscillations of a dislocation can be found from equation $\mathrm{Re}\,\alpha^{-1}(k_z,\omega)=0$. Taking into account the designations

$$L_t = \ln\left|\frac{k_z^2-\omega^2/c_t^2}{k_m^2}\right|, \quad L_l = \ln\left|\frac{k_z^2-\omega^2/c_l^2}{k_m^2}\right|,$$

from (1), we obtain equation of dislocation natural small vibrations:

$$\frac{\omega^2 2}{c_t^2k_z^2}k_z^2\left[b_e^2(L_t+\xi L_l)+b_s^2L_t\right]+b_e^2\left[-1-\xi-2\xi L_l-\frac{c_t^2k_z^2}{\omega^2}(L_l-L_t)\right]$$

$$+b_s^2\left[-1-3L_t+4\xi L_l+4\frac{c_t^2k_z^2}{\omega^2}(L_t-L_l)\right]=0. \quad (2)$$

To pass to the long-wave limit, the solution of equation (2) is sought directly for $k_z \to 0$, taking into account that $\omega \approx c_d(k_z)k_z$, where c_d is the propagation of waves velocity along the dislocation. Then the quantity $c_t^2k_z^2/\omega^2$ tends to the constant c_t^2/c_d^2, and for the equality (2), it is necessary that the coefficients before the infinite functions L_t and L_l tend to zero for $k_z \to 0$. In accordance with the nature of the functions behavior noted above, two different solutions are possible.

3. Analysis of the dislocation oscillations equation

We consider separately two cases: $c_d \neq c_{t,l}$ and $c_d \approx c_{t,l}$.

1) If $c_d \neq c_{t,l}$ for $k_z \to 0$, then $L_t \approx L_l \approx 2\ln(k_z/k_m)$ and from (2) we obtain a solution for c_d:

$$c_d^2 = \frac{2\xi\beta^2+(3-4\xi)(1-\beta^2)}{1+\xi^2\beta^2}c_t^2 < c_l^2. \quad (3)$$

Here $\beta = \sin\varphi$, φ is the angle between the dislocation line and the Burgers vector. Expression (3) reaches its maximum as a function of φ for a screw dislocation ($\beta=0$):

$$c_d^2 = (3-4\xi)c_t^2 = \frac{1+\nu}{1-\nu}c_t^2 > c_t^2 \quad (4)$$

and the minimum for an edge dislocation ($\beta=1$), for which (3) coincides with the result of [4]:

$$c_d^2 = \frac{2\xi}{1+\xi^2} c_t^2 = \frac{4-12\nu+8\nu^2}{5-12\nu+8\nu^2} c_t^2 < c_t^2, \tag{5}$$

where ν is Poisson's ratio. Thus, for some value of $\beta = \beta_c$:

$$\beta_c^2 = \sin^2 \varphi_c = \frac{2(1-2\xi)}{3-6\xi+\xi^2} = \frac{8(1-\nu)\nu}{1+8\nu-8\nu^2} \tag{6}$$

in (3) $c_d = c_t$, which contradicts the original assumption, that is, the expression (3) becomes incorrect. Therefore, (3) actually defines two branches of oscillations: local for $\beta > \beta_c$ with the limiting value (5) and quasilocal for $\beta < \beta_c$ with the limiting value (4).

2) If $c_d \approx c_{t,l}$ for $k_z \to 0$, then we write

$$c_d^2(k_z) = c_{t,l}^2 \pm \delta c_{t,l}^2(k_z), \tag{7}$$

where the small additive to $c_{t,l}^2$ should have the form $\delta c_{t,l}^2 = c_{t,l}^2 (k_z / k_m)^a$ ($a > 0$).We consider separately the solution of the dispersion equation (2) near c_l and c_t.

a) For $c_d \to c_l$ we have $L_t \approx 2\ln(k_z/k_m)$, $L_l \approx (2+a)\ln(k_z/k_m)$ and after substituting these values in (2) it becomes clear that in the case under consideration equation (2) does not have solutions and, consequently, quasilocal frequencies of oscillations are absent for $c_d \to c_l$.

b) For $c_d \to c_t$ we have $L_t \approx (2+a)\ln(k_z/k_m)$, $L_l \approx 2\ln(k_z/k_m)$ and then in the long-wave limit equation (2) reduces to

$$\frac{\omega^4}{c_t^4 k_z^4}\left(1+\frac{a}{2}+\beta^2\xi^2\right) = \frac{\omega^2}{c_t^2 k_z^2}\left[3\left(2\xi-1-\frac{a}{2}\right)\beta^2 + 3 + \frac{3}{2}a - 4\xi\right] + \frac{a}{2}(5\beta^2-4).$$

From this equation, taking into account that in the case under consideration $\omega^2/(c_t^2 k_z^2) \to 1$, we find

$$a = \frac{(1-\xi)[4-\beta^2(5-\xi)]}{1-\beta^2} - 2, \quad (\beta < \beta_c), \tag{8}$$

where the inequality corresponds to the values $a > 0$. Then from the relation $\omega = c_d(k_z)k_z$, taking into account (7), we find the dispersion equation of frequencies

$$\omega^2 = c_t^2 k_z^2 \pm c_t^2 k_z^2 (k_z / k_m)^a. \tag{9}$$

For $k_z \to 0$, from (9) we have $\omega \approx c_t k_z \pm \frac{1}{2} c_t k_z (k_z / k_m)^a$. Consequently, the solution of (7), (8) makes sense only for $a > 0$, and for $a \to 0$ the corresponding frequency "departs" from $c_t k_z$, where the condition $a = 0$ corresponds to (6). Thus, equation (9) also defines two branches of oscillations, one of which is local, and the second - quasilocal.

All of the branches found become "indeterminate" at $\beta \to \beta_c$, but it is clear that they are, first, continuous, and secondly, they can not cross the frequency $\omega = c_t k_z$.

Equation (2) is also solved by numerical methods at a value of $\nu = 1/3$, which corresponds to $\xi = 1/4$. The results of the solution are shown in figures 1a,b. Here, the values of the angle between the dislocation line and the Burgers vector are plotted on the horizontal axis, and the values of the ratio of the wave velocity along the dislocation to the transverse sound waves velocity are plotted on the vertical axis.

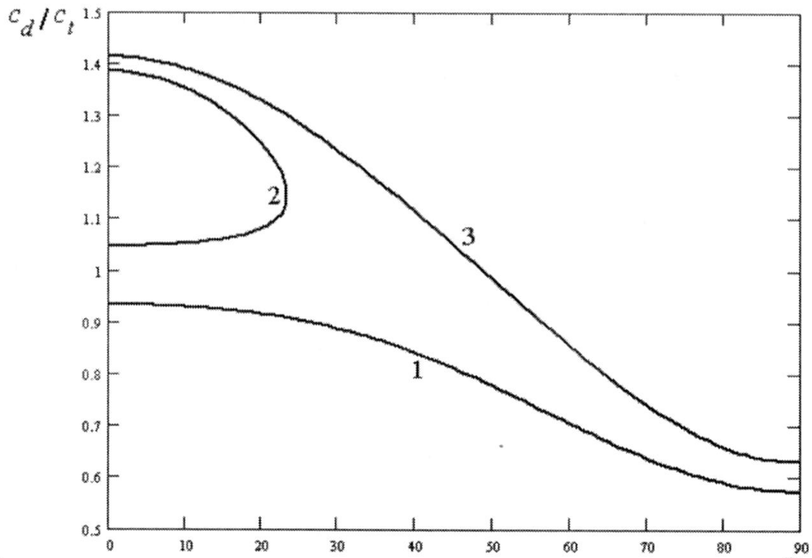

a)
screw dislocation *edge dislocation*

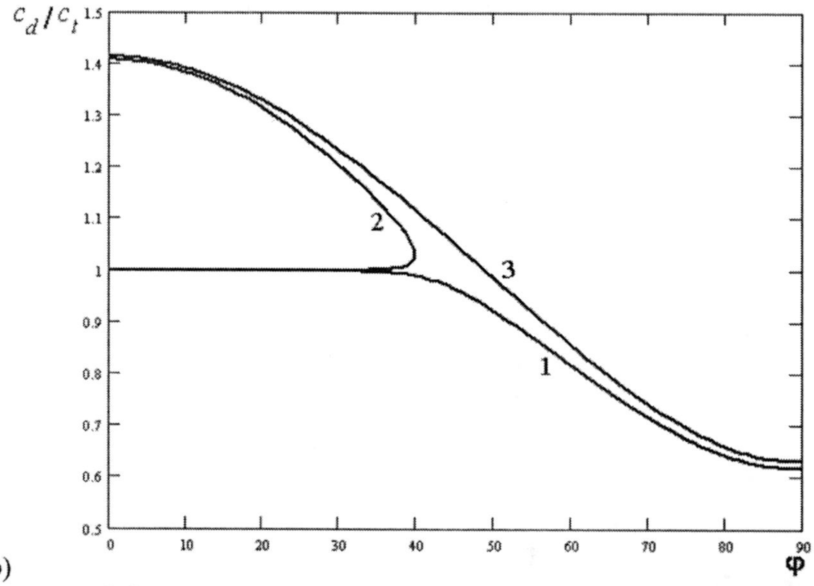

b)
screw dislocation *edge dislocation*

Figure 1. Dependence of the propagation waves velocity along a dislocation on the angle between the dislocation line and the Burgers vector. 1 and 2 are the branches of local and quasilocal oscillations, 3 is the dependence determined by formula (3).
a) Short-wave case ($k_m/k_z = 10$); b) long-wave case ($k_m/k_z = 10^4$).

It can be seen from the figures that with an increase of the ratio k_m/k_z, the local (line 1) and quasilocal (line 2) branches converge, approaching the value of $c_d/c_t = 1$, and tend to the limiting dependence (line 3).

4. Conclusion
The results obtained in this paper allow one to interpret the data of a number of experiments (internal friction and scattering of ultrasound in dislocation crystals, the effect of variable electric and magnetic fields on the dynamics of dislocations, etc.) and stimulate further experimental studies in this field.

References
[1] Laub T and Eshelby J D 1966 *Phil. Mag.* **14** 1285
[2] Lifshitz I M and Kosevich A M 1966 *Reports Progr. Phys.* **29** 217
[3] Ninomiya T and Ishioka Sh 1967 *J. Phys. Soc. Japan* **23** 361
[4] Ninomiya T 1968 *J. Phys. Soc. Japan* **25** 830
[5] Ninomiya T 1970 *Fundamental Aspects of Dislocations Theory*, ed J A Simmons, R de Wit and R Bullough (Nat. Bur. Stand. (U.S.) Spec. Publ. 317) p 315
[6] Lifshitz I M and Pushkarov Kh I 1970 *JETP Letters* **11** 310
[7] Tewary V K 1974 *J. Phys. C. Solid State Phys.* **7** 261
[8] Bataronov I L, Dezhin V V and Roshchupkin A M 1993 *Bulletin of the Russian Academy of Sciences: Physics* **57** 1947
[9] Roshchupkin A M, Bataronov I L and Dezhin V V 1995 *Bulletin of the Russian Academy of Sciences: Physics* **59** 1648
[10] Dezhin V V 2016 *Vestnik Tambovskogo Universiteta. Ser.: Yestestv. i Tekhnich. Nauki* **21** 956 (In Russian)
[11] Dezhin V V 2016 *Vestnik Tambovskogo Universiteta. Ser.: Yestestv. i Tekhnich. Nauki* **21** 959 (In Russian)
[12] Bataronov I L and Dezhin V V 2013 *Vestnik Tambovskogo Universiteta. Ser.: Yestestv. i Tekhnich. Nauki* **18** 1566 (In Russian)
[13] Bataronov I L and Dezhin V V 2016 *J. Phys.: Conf. Ser.* **738** 012108
[14] Bataronov I L, Dezhin V V and Roshchupkin A M 1995 *Bulletin of the Russian Academy of Sciences: Physics* **59** 1690

IOP Conf. Series: Journal of Physics: Conf. Series **936** (2017) 012036 doi:10.1088/1742-6596/936/1/012036

Numerical and approximate calculation of the generalized susceptibility matrix elements of dislocation segment in nondissipative crystal

I. L. Bataronov and V V Dezhin

Voronezh State Technical University, Moskovsky prospect 14, Voronezh 394026, Russian Federation

E-mail: viktor.dezhin@mail.ru

Abstract. Numerical calculation of the generalized susceptibility and the inverse generalized susceptibility matrix elements of the dislocation segment for edge and screw dislocations, different frequencies and different values of the dislocation segment length is performed. Certain regularities have been established. Based on the graphical results of numerical calculations, an approximate calculation of the generalized susceptibility matrix elements was carried out. Expressions for the diagonal and off-diagonal generalized susceptibility matrix elements of the dislocation segment are obtained.

1. Introduction

Earlier, to describe the dynamics of the dislocation segment, Koehler [1] proposed a string model. Further development of Kohler's theory was made by Granato and Lücke [2]. Garcia-Moliner and Thomson [3] used the Fourier transform of the response function to calculate the Granato-Lücke model. In [4], the equation of motion of a dislocation segment was obtained by means of the Lagrange function. The most consistent solution to the problem of a dislocation crystal vibrations using a self-consistent dynamic dislocation theory was carried out in the works of Ninomiya [5-7]. In [9], using the results of [8], a string model with nonlocal dynamic dislocation characteristics was constructed. The generalized susceptibility (the response function) of the dislocation segment, which is a matrix of dislocation oscillators generalized susceptibilities, is found in [10-12] based on the results of [8]. Analytic expressions for the matrix elements B_{mn} of the dislocation oscillators generalized susceptibilities inverse matrix are obtained. Later on, studies of the two adjacent dislocation segments oscillations were started [13, 14] on the basis of papers [8, 10-12]. In this paper, a numerical and approximate calculation of the dislocation oscillators generalized susceptibilities matrix elements $\alpha_{mn} = (B^{-1})_{mn}$ is carried out.

2. Numerical calculation of the matrix elements

In this section, we numerically compute the elements of the matrix α and the inverse matrix B for different types of dislocations (edge and screw), different frequencies ω, and different values of the dislocation segment length L. The results of calculations were made in the form of graphs. We use the notation: k_m is the maximum wave number, c_t is the transverse elastic waves velocity, $q_t = \omega L/c_t$ is the normalized frequency.

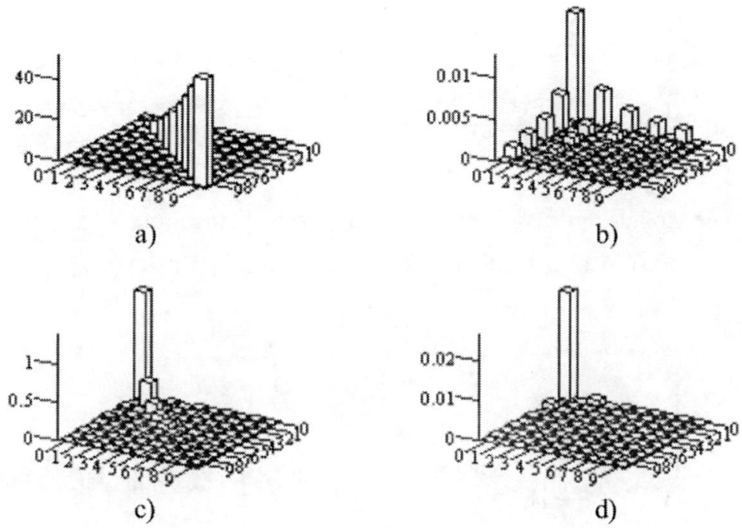

Figure 1. Modules of matrix elements for edge dislocation, $k_m L = 100$, $q_t = 1$.
a) $|\mathrm{Re}\,B_{mn}|$, b) $|\mathrm{Im}\,B_{mn}|$, c) $|\mathrm{Re}\,\alpha_{mn}|$, d) $|\mathrm{Im}\,\alpha_{mn}|$.

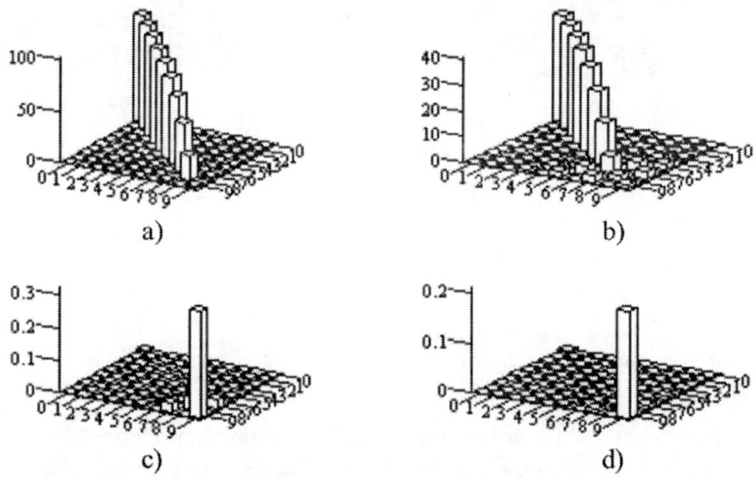

Figure 2. Modules of matrix elements for edge dislocation, $k_m L = 1000$, $q_t = 25$.
a) $|\mathrm{Re}\,B_{mn}|$, b) $|\mathrm{Im}\,B_{mn}|$, c) $|\mathrm{Re}\,\alpha_{mn}|$, d) $|\mathrm{Im}\,\alpha_{mn}|$.

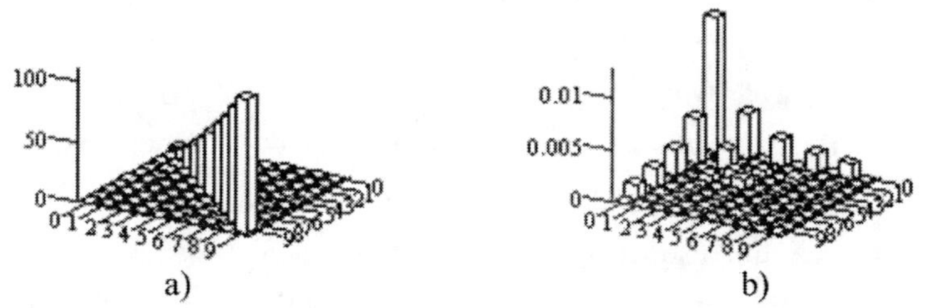

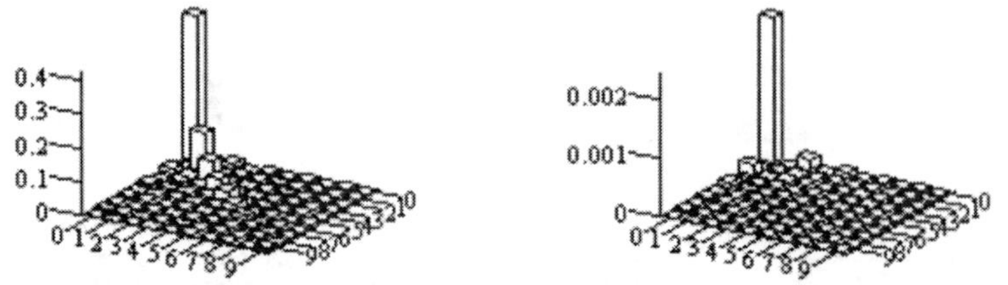

c) d)

Figure 3. Modules of matrix elements for screw dislocation, $k_m L = 100$, $q_t = 1$.
a) $|\mathrm{Re}\, B_{mn}|$, b) $|\mathrm{Im}\, B_{mn}|$, c) $|\mathrm{Re}\,\alpha_{mn}|$, d) $|\mathrm{Im}\,\alpha_{mn}|$.

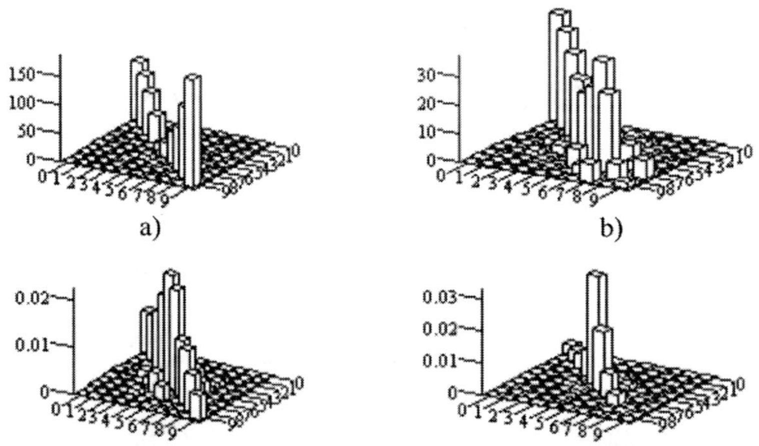

Figure 4. Modules of matrix elements for screw dislocation, $k_m L = 1000$, $q_t = 25$.
a) $|\mathrm{Re}\, B_{mn}|$, b) $|\mathrm{Im}\, B_{mn}|$, c) $|\mathrm{Re}\,\alpha_{mn}|$, d) $|\mathrm{Im}\,\alpha_{mn}|$.

Based on the constructed graphs, we do the following conclusions.

1. The diagonal elements of all matrices are predominant in absolute value over off-diagonal elements.

2. The minimum value of the inverse matrix element B_{kk}, corresponds to the maximum value of the corresponding element α_{kk}.

3. As the frequency increases, the minimum value of the diagonal elements of the inverse matrix B is moves from B_{11} to B_{NN}. Simultaneously, the minimum value of the diagonal elements of the matrix α moves from α_{NN} to α_{11}.

3. Approximate calculation of the matrix elements

We will carry out an approximate calculation of the matrix elements α, using the results of the previous section. Consider the matrix elements B_{mn}, which form a matrix of size $N \times N$. This matrix is symmetric, i.e. $B_{mn} = B_{nm}$ [10-12]. We define the vector $b = (B_{1k}/\beta \;\; B_{2k}/\beta \;\; \ldots \;\; b_k = \beta \;\; B_{Nk}/\beta)^{\mathrm{T}}$ and express the matrix elements in terms of this vector components: $B_{mn} = b_m b_n + B'_{mn}$ or $B = bb^{\mathrm{T}} + B'$. Here, the number k is chosen from the minimum modulus condition of the matrix B diagonal element, i.e. from the resonance condition, the number β is defined as $\beta = \sqrt{B_{kk} - \eta}$, η is an arbitrary number under the condition $B_{kk} - \eta > 0$, the auxiliary matrix B' has the form

$$B' = B - bb^{\mathrm{T}} = \begin{pmatrix} B_{11} - \frac{B_{1k}^2}{\beta^2} & B_{12} - \frac{B_{1k}B_{2k}}{\beta^2} & \dots & 0 & \dots & B_{1N} - \frac{B_{1k}B_{Nk}}{\beta^2} \\ B_{12} - \frac{B_{2k}B_{1k}}{\beta^2} & B_{22} - \frac{B_{2k}^2}{\beta^2} & \dots & 0 & \dots & B_{2N} - \frac{B_{2k}B_{Nk}}{\beta^2} \\ \vdots & \vdots & \ddots & \vdots & \ddots & \vdots \\ 0 & 0 & \dots & B_{kk} - \beta^2 & \dots & 0 \\ \vdots & \vdots & \ddots & \vdots & \ddots & \vdots \\ B_{1N} - \frac{B_{Nk}B_{1k}}{\beta^2} & B_{2N} - \frac{B_{Nk}B_{2k}}{\beta^2} & \dots & 0 & \dots & B_{NN} - \frac{B_{Nk}^2}{\beta^2} \end{pmatrix}.$$

We write the matrix inverse to the matrix B in the form of a sum of a matrix inverse to B' and an additive summand $B^{-1} = (B')^{-1} + \gamma CC^{\mathrm{T}}$. We divide the B' into two matrices $B' = B'^{(d)} + B'^{(nd)}$. In this case, the matrix $B'^{(d)}$ consists of the diagonal elements of the matrix B', and the matrix $B'^{(nd)}$ consists of off-diagonal elements. It can be seen from the numerical calculations of the previous section that the diagonal elements of the matrix B' substantially predominate over the off-diagonal elements in absolute value, so we seek the inverse to it in the form of the sum of the diagonal matrix with the elements inverse to the corresponding elements of the matrix B' and the matrix D with elements that are significantly smaller in absolute value for the elements of the diagonal matrix, i.e. $(B'^{-1})_{ij} = \delta_{ij} / B'_{ii} + D_{ij}$. After further cumbersome calculations, setting $\eta \to -\infty$, we obtain an elements of the matrix $\alpha = B^{-1}$:

$$B_{kk}^{-1} = \frac{1}{B_{kk} - \sum\limits_{\substack{i \\ (i \neq k)}} \frac{(B_{ki})^2}{B_{ii}} + \sum\limits_{\substack{i,j \\ (i,j \neq k)}} \frac{B_{ki} B_{ij}^{(nd)} B_{jk}}{B_{ii} B_{jj}}},$$

$$B_{ik}^{-1} \approx \frac{\frac{1}{B_{ii}} \left(-B_{ik} + \sum\limits_{\substack{j \\ (j \neq k)}} \frac{B_{kj} B_{ji}^{(nd)}}{B_{jj}} \right)}{B_{kk} - \sum\limits_{\substack{m \\ (m \neq k)}} \frac{(B_{km})^2}{B_{mm}} + \sum\limits_{\substack{m,j \\ (m,j \neq k)}} \frac{B_{km} B_{mj}^{(nd)} B_{jk}}{B_{mm} B_{jj}}}, \quad (i \neq k),$$

$$B_{ii}^{-1} \approx \frac{1}{B_{ii}} \left(1 + \frac{\frac{1}{B_{ii}} \left(B_{ik} - \sum\limits_{\substack{j \\ (j \neq k)}} \frac{B_{kj} B_{ji}^{(nd)}}{B_{jj}} \right)^2}{B_{kk} - \sum\limits_{\substack{j \\ (j \neq k)}} \frac{B_{jk}^2}{B_{jj}} + \sum\limits_{\substack{m,j \\ (m,j \neq k)}} \frac{B_{km} B_{mj}^{(nd)} B_{jk}}{B_{mm} B_{jj}}} \right), \quad (i \neq k),$$

IOP Conf. Series: Journal of Physics: Conf. Series **936** (2017) 012036 doi:10.1088/1742-6596/936/1/012036

$$B_{ij}^{-1} \approx \frac{1}{B_{ii}B_{jj}}\left(-B_{ij}^{(nd)} + \frac{\left(B_{ik} - \sum\limits_{\substack{m \\ (m\neq k)}} \frac{B_{km}B_{mi}^{(nd)}}{B_{mm}} \right)\left(B_{jk} - \sum\limits_{\substack{l \\ (l\neq k)}} \frac{B_{kl}B_{lj}^{(nd)}}{B_{ll}} \right)}{B_{kk} - \sum\limits_{\substack{m \\ (m\neq k)}} \frac{B_{mk}^2}{B_{mm}} + \sum\limits_{\substack{m,l \\ (m,l\neq k)}} \frac{B_{km}B_{ml}^{(nd)}B_{lk}}{B_{mm}B_{ll}}} \right), \quad (i, j \neq k\,,\, i \neq j\,).$$

Calculations based on these formulas showed good agreement with numerical calculations of the previous section.

4. Conclusion

In this paper, the elements of the generalized susceptibility matrix of dislocation oscillators are found. The results obtained can be used to determine the orientation and size dependence of the dislocation segment vibrational spectrum, the dependence of the dislocation segment vibrational spectrum on the Poisson's ratio.

References

[1] Koehler J S 1952 *Imperfections in Nearly Perfect Crystals*, eds. W Shockley et al. (Wiley, New York) p 197
[2] Granato A and Lücke K 1956 *J. Appl. Phys.* **27** 583
[3] Garcia-Moliner F and Thomson R 1966 *J. Appl. Phys.* **3** 83
[4] Bross H and Stenzel G 1967 *Phys. Lett. A* **25** 236
[5] Ninomiya T and Ishioka Sh 1967 *J. Phys. Soc. Japan* **23** 361
[6] Ninomiya T 1968 *J. Phys. Soc. Japan* **25** 830
[7] Ninomiya T 1970 *Fundamental Aspects of Dislocations Theory*, ed J A Simmons, R de Wit and R Bullough (Nat. Bur. Stand. (U.S.) Spec. Publ. 317) p 315
[8] Bataronov I L, Dezhin V V and Roshchupkin A M 1993 *Bulletin of the Russian Academy of Sciences: Physics* **57** 1947
[9] Bataronov I L and Dezhin V V 2006 *Vestnik Voronezhskogo gosudarstvennogo tekhnicheskogo universiteta* **2** (8) 15 (In Russian)
[10] Bataronov I L and Dezhin V V 2013 *Vestnik Tambovskogo universiteta. Ser.: Yestestv. i Tekhnich. Nauki* **18** 1566 (In Russian)
[11] Bataronov I L and Dezhin V V 2014 *Aktual'nyye napravleniya nauchnykh issledovaniy XXI veka: teoriya i praktika* **2** (4-2) 65 (In Russian)
[12] Bataronov I L and Dezhin V V 2016 *J. Phys.: Conf. Ser.* **738** 012108
[13] Dezhin V V 2016 *Matematicheskiye metody v tekhnike i tekhnologiyakh*: Proc. XXIX Int. Sci. Conf. (Saratov State Tech. University, Saratov) **3** p 5 (In Russian)
[14] Dezhin V V 2016 *Mekhanika deformiruyemogo tverdogo tela*: Proc. IX All-Russian Conf. (Voronezh State University, Voronezh) p 4 (In Russian)

IOP Conf. Series: Journal of Physics: Conf. Series **936** (2017) 012037 doi:10.1088/1742-6596/936/1/012037

Models of evaluation of public joint-stock property management

N M Yakupova[1][2*] , S Levachkova[2], S G Absalyamova[2] and G Kvon[3]
[2]Institute of Management, Economics and Finance, Kazan Federal University, Kazan, Russia;
[3]Department of Economic Theory, Kazan National Research Technical University named after A.N.Tupolev, Kazan, Russia.

E-mail: yakupova.nm@mail.ru

Abstract. The paper deals with the models of evaluation of performance of both the management company and the individual subsidiaries on the basis of a combination of elements and multi-parameter and target approaches. The article shows that due to the power of multi-dimensional and multi-directional indicators of financial and economic activity it is necessary to assess the degree of achievement of the objectives with the use of multivariate ordinal model as a set of indicators, ordered by growth so that the maintenance of this order on a long interval of time will ensure the effective functioning of the enterprise in the long term. It is shown that these models can be regarded as the monitoring tools of implementation of strategies and guide the justification effectiveness of implementation of management decisions.

1. Introduction

A topical issue of the current stage of economy development is forming an efficient mechanism of public stock ownership management. By 01.01.2015, the register of federal property listed 2008 legal entities (Public Corporations, Ltd., Close Corporations), 1928 holdings of stocks and shares, 79 federal public enterprises, 1557 federal public unitary enterprises, 17288 federal public organizations. The imperfection of the management mechanism led to the situation, when the enterprises with a public share sometimes serve not the interests of the state-owner, but of the top management. The state has actually dissociated itself from the process of forming an efficient owner, by relying on the automatism of market relations, which has led to the low efficiency of the mechanisms of public stock portfolios management.

Many aspects of efficient property management were researched in the works by T. Keller, T. Copeland, M. Porter, M. Scott, N Antill, G Cokins et al.

Under the current conditions, the topical task is to form the efficient mechanism of public stock ownership management. One of the tools to solve this problem within the integrated structures – holdings, in the Russian and foreign practice [6, 7] is a system of the public stock trust managing. Management companies may serve as the basic element of such a system. Management companies are commercial organizations, which have the right to manage certain property (shares), and fulfill the managerial functions on behalf of the owners on the contract repayable basis.

[1] To whom any correspondence should be addressed.

Another essential aspect of forming an efficient mechanism of public stock ownership management is elaboration of the efficiency assessment tools, of performance of both the management company as a whole and individual subsidiaries.

2. Methods

The research has been based on the normative-legal acts of the Russian Federation and Tatarstan Republic, which regulate the issues of public stock management procedures; academic works of the Russian and foreign authors on the issues of public stock ownership management; results of fundamental and applied research of the modern scholars who studied the specific issues of improving the institution of public stock ownership management; reviews of periodicals on the researched issue, as well as the actual data of the researched enterprises.

During the research, general scientific methods (systemic analysis, synthesis, induction, deduction) and special economic-mathematical methods (rank correlation) were used. Each method was used in conformance with its functionality and resolution to solve the appropriate research tasks at each stage.

To construct and identify the economic-mathematical model, we studied the example of one of the largest holdings of Tatarstan Republic, with the predominant share belonging to the state. To observe privacy, notation conventions were attributed to the enterprises.

In order to build a performance evaluation model, both for the management company and the managed companies of the holding, we proceeded from the fact that firstly, the main objective of the management company is to ensure the interests of the state in making and implementing decisions on the management of these companies.

Following the research results, we propose to use management companies in the system of public stock management, by transferring into their trust managing the public holdings stocks of the companies grouped on the territorial-sector basis. The main task of the management company should be securing the public interests when making and implementing the managerial decision [7].

In our opinion, a management company should function by employing independent directors. The independent directors, as representative of the state, must not be officials of any other entity, apart from the management company; must not own, personally or via affiliated persons, shares in the managed joint-stock companies; must not obtain remuneration for consultations and other services to the managed establishments, apart from the remuneration for the activities in the management company; must have good business reputation, etc. [5].

To assess the performance efficiency of both the management company as a whole and individual subsidiaries, we propose the approach, which combines elements of multivariate and target approaches. It implies that the performance efficiency of the management company should be assessed basing on the degree to which its targets are achieved.

Traditionally the performance efficiency of enterprises is assessed basing on individual indicators of financial-economic activity. However, they provide only indirect measurement of the results and costs, and each individual indicator cannot serve as an indicator of the performance efficiency of public stock companies in general. Besides, a large amount of diverse indicators, used in assessment and management of public stock ownership in the form of holdings, requires summarizing the initial data in the form of integral indicators before using them for decision making.

Due to the multivariate and diverse character of the indicators of financial-economic activity, we propose to use the multifactor model in the form of an ordinal compression of indicators. The proposed model is based on formulating certain strategic conceptions or requirements of the enterprises' functioning and development, which are a complex of indicators, ordered by growth rates, so that this order maintenance during a lengthy period will provide efficient functioning of the enterprise long-term.

To construct the model, we have selected the most significant indicators of the productive-economic activity of enterprises: net profit (NP); income (I); prime cost (PC); monetary funds (MF);

short-term and long-term investments (ShLI); own current assets (OCA); current assets (CA); own capital (OC); total assets (TA); noncurrent assets (NCA); payroll fund (PRF); long-term liabilities (LL); short-term liabilities (ShL). In our research, basing on the coefficient analysis, we have found the normative ordering of indicators by the growth rates, which serves as the standard for integral assessment of the level of economic efficiency of the public stock ownership management and is a normative dynamic model of the enterprise management system (DMEM):

$$NP > I > PC > MF > ShLI > OCA > CA > OC > TA > NCA > PRF > LL > ShL \qquad (1)$$

The degree of closeness of the two orderings – the actual and the normative – is determined by the indicator of the structural-dynamic assessment of the enterprise efficiency (ADAE). This indicator is calculated in compliance with the modified formula of the Kendall rank coefficient correlation for two rank series [3]

$$\text{ADAE} = 1 - \frac{\sum_{i=1}^{n} m_i}{n(n-1)} = 1 - \frac{M(A,N)}{n(n-1)} \qquad (2)$$

where n – number of indicators in the dynamic normative model of the enterprise management (DMEM);

M(A,N) – sum of inversions in the actual order of indicators (A) related to DMEM (N);

mi – number of inversions in the actual order for the indicator having the i-th rank in DMEM:

$$m_i = \sum_{j=1}^{n} a_{ij} \qquad (3)$$

where a_{ij} – variable reflecting the presence or absence in the actual order of indicators of the relation "faster" between the i-th and j-th indicators, set by DMEM of the management company (i = 1,…,n; j = 1,…,n):

$$a = \begin{cases} 1, \text{ if } r_i > r_j \text{ at } i < j \\ \text{н}, \text{ if } r_i < r_j \text{ at } i > j \\ 0, \text{ otherwise,} \end{cases} \qquad (4)$$

where r_i and r_j – ranks of the i-th and j-th indicators in the actual order;

We consider ADAE to be a tool for monitoring the enterprise' strategies implementation and a basis to determine the integral indicator of the performance efficiency, of both the management company as a whole and individual subsidiaries.

We built a management company performance evaluation model (assuming state as an owner) on the basis of transformation of regulatory and dynamic model of enterprise management system (DMEM) and structural and dynamic performance evaluation of their activities (ADAE). The modification is based on the premise that it is important for management companies to comply with all the dynamic relationships between indicators of the norm (DMEM) for all companies that it controls, whereby each of the indicators DMEM dynamic management model can be represented as a vector:

$$\begin{bmatrix} NP_1 \\ NP_2 \\ \cdots \\ NP_k \\ \cdots \\ NP_d \end{bmatrix} \gg \begin{bmatrix} I_1 \\ I_2 \\ \cdots \\ I_k \\ \cdots \\ I_d \end{bmatrix} \gg \cdots \gg \begin{bmatrix} ShL_1 \\ ShL_2 \\ \cdots \\ ShL_k \\ \cdots \\ ShL_d \end{bmatrix} \qquad (5)$$

where NP_k – net profit of the *k*-th enterprise;

I_k – income of the *k*-th enterprise;

k ϵ [1;d].

IOP Conf. Series: Journal of Physics: Conf. Series **936** (2017) 012037 doi:10.1088/1742-6596/936/1/012037

However, the significance of carrying the same proportions between indicators for various companies may be different for management company. This can be explained by the difference of the significance of the contribution of enterprises to the final result of the activities of the management company, which eventually will be evaluated by the state. We have defined the main factors that determine the level of significance of the contribution of each enterprise: the value added of each enterprise and the share of state ownership in the authorized capital of these enterprises. The level of significance of the contribution of each enterprise in the total activities of the management company is proposed to be measured using the indicator "degree of significance" (DS). The algorithm of calculation of this indicator for each enterprise consists of the following steps:

1. Calculation of the added value of enterprise with a public share (statistical form 1-AV).
2. Determination of the public property share in the authorized capital of enterprises within the management company system.
3. Calculation of the size of added value, created by the enterprises, which is contributed by the public holdings stock (product of indicators calculated at stages 1 and 2).
4. Calculation of the degree of significance (DS) of each enterprise is determined as the size of contribution of each enterprise into the total size of the added value contributed by the public holdings stock (ratio of the indicator calculated at stage 3 to the sum of similar indicators or all managed enterprises).
5. Taking into account the introduced indicator of the degree of significance, the methodology of calculating the structural-dynamic assessment of the performance efficiency of the management company (ADAEmc) will take the following form:

$$\mathrm{ADAEmc} = 1 - \frac{\sum_{i=1}^{n}\sum_{k=1}^{n} DS_k \times \sum_{j=1}^{n} \mathrm{a}_{ij}^{k}}{\mathrm{n(n-1)} \times \sum_{k=1}^{d} \mathrm{DS}_k} \quad (6)$$

where: n – number of indicators in the dynamic normative model;

d – number of enterprises, whose holding stocks belong to the management company;

DS_k– degree of significance of the k-th enterprise;

a^k_{ij}– variable reflecting the presence or absence in the actual order of indicators of the relation "faster" between the i-th and j-th indicators, set by the dynamic normative model (i = 1,...,n; j = 1,...,n) for the k-th enterprise.

3. Results

The practical realization of the offered model of estimation of efficiency of functioning of enterprises with state share was made by us on the example of one of the largest holdings of the Republic of Tatarstan, the prevailing blocks of shares in which belong to the state. In order to maintain the confidentiality of this holding has a designation of JSC "Alpha". The name of the other companies and the estimated values of ADAE are presented in table 1.

In compliance with ADAE calculation technique, we have calculated the rate of changing the indicators of the public stock ownership management efficiency of the researched enterprises (Table 1). As can be seen from the results of calculations, enterprises PJSC "Beta", PJSC "Gamma", Public Corporation "Epsilon", PJSC "Zeta", PJSC "Eta" и Joint-stock Company "Iota" tend to decrease the level of the performance efficiency in the analyzed period, which testifies to the negative influence on the development of the public holding in general. The slight change in the level of the performance and management efficiency can be seen in PJSC "Delta". The trend to increase the level of the performance and management efficiency can be seen in PJSC "Theta" and Joint-stock Company "Kappa".

Table 1

Values of ADAE indicators of the JSC "Alfa" enterprises

NN	Enterprises with the public share	ADAE indicator				
		Periods				
		2006	2007	2008	2009	2010
1	PJSC "Beta"	0.6667	0.5128	0.6410	0.564103	0.423077
2	PJSC "Gamma"	0.3846	0.3718	0.4872	0.6410	0.2179
3	PJSC "Delta"	0.3974	0.4487	0.5897	0.4744	0.4872
4	PC "Epsilon"	0.5256	0.6923	0.5897	0.5769	0.3974
5	PJSC "Zeta"	0.5128	0.6154	0.6282	0.6026	0.3846
6	PJSC "Eta"	0.7308	0.5769	0.4359	0.6282	0.6026
7	PJSC "Theta"	0.4872	0.5256	0.4615	0.6410	0.7308
8	JSC "Iota"	0.5513	0.2821	0.2949	0.6410	0.5897
9	JSC "Kappa"	0.5128	0.6667	0.5513	0.4615	0.5128

Graphic illustration of ADAE indicators' dynamics is shown in the figure 1.

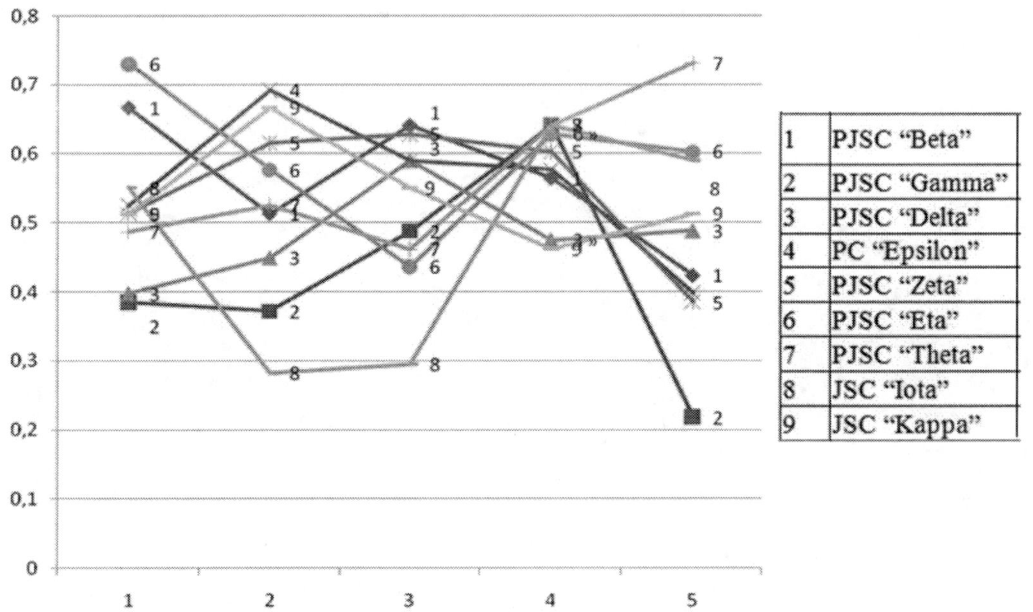

Fig. 1. ADAE indicators dynamics for each enterprise

It should be noted that the values of ADAE integral indicators are an adequate informative base to determine the level of the management efficiency of each individual enterprise, but do not allow to make a full-fledged and unambiguous assessment of the management company performance in general.

An example of calculating the integral ADAEmc in 2013 by the above methodology is given in Table 2.

Table 2
Structural-dynamic assessment of the "Alfa" joint-stock management company's efficiency in 2013

	Enterprises	Share of the management company in the authorized capital, %	2013			
			Added value (AV), mln rubles	degree of significance (DS)	Σa^k_{ii}	$DS \times \sum_{j=1}^{m} a^k_{ij}$
1	PJSC "Beta"	26.64	203	0.001	66	0.12
2	PJSC "Gamma"	25.21	7 897	0.07	34	2.32
3	PJSC "Eta"	25.00	93	0.001	94	0.07
4	PJSC "Delta"	36.00	59 280	0.73	76	55.62
5	PJSC "Theta"	33.60	7 796	0.09	114	10.24
6	JSC "Kappa"	48.98	34	0.001	80	0.05
7	PC "Epsillon"	100.00	652	0.02	62	1.39
8	PJSC "Zeta"	87.21	2 739	0.08	60	4.91
9	JSC "Iota"	100.00	77	0.001	92	0.24
	Total		78 771	1		74.96
	$n(n-1) \times \sum_{k=1}^{d} DS_k$					156
	ADAEmc					0.4805

The dynamics of structural-dynamic assessment of the "Alfa" joint-stock management company's efficiency in the analyzed period is shown in Figure 2.

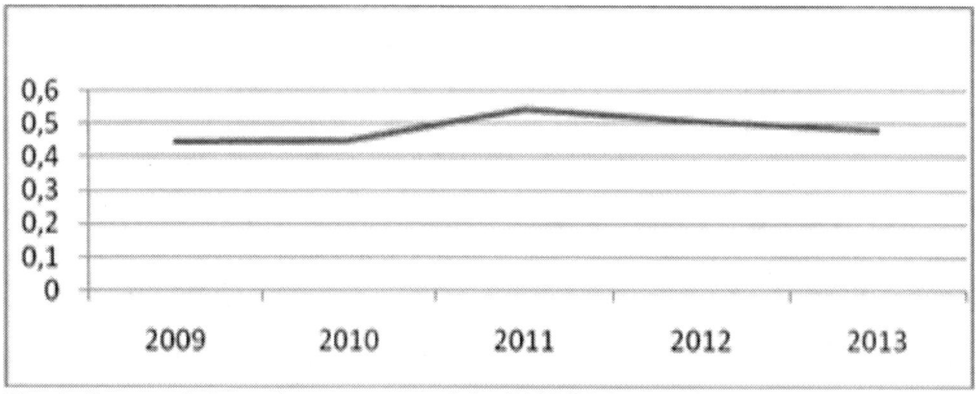

Fig. 2. Structural-dynamic assessment of the "Alfa" joint-stock management company's efficiency

It should be noted that the ADAEmc level is in general very far from the maximally possible value – 1. To make adequate managerial decisions, the management company should timely reveal the existing and emerging negative trends in the individual enterprise's functioning.

4. Conclusion

Thus, we consider it appropriate to carry out the public stock ownership management basing on the transfer of state holdings of stock for the trust managing by management companies. Using the normative-dynamic models of enterprise's management and models of the performance efficiency assessment of both the individual enterprises and the management company. It is possible not only to assess the efficiency of their functioning but serve as a monitoring tool of their strategies implementation, as well as the guidelines to ground the efficiency of managerial decisions.

We can list the advantages of the proposed methods of forming DMEM and calculation integral indicators of ADAE and ADAEmc:

First, the proposed method is based on systematic, comprehensive (multidimensional) approach to the assessment of such complex phenomena as the performance of activities of holding companies; DMEM, in contrast to the existing approaches to modeling regulatory ratios measures the movement of indices, gives not a piecemeal consideration of their relations, but linkages between many of the indicators of effectiveness of management of state join-stock properties.

Second, the distinguishing feature of DMEM (normative dynamic model of enterprise management) is that it represents the "ideal" (normative) model of functioning of the object and answers the question "how it should be". The main purpose of its construction is not so much in the reflection of reality, but in the definition of rational behavior.

Third, integral ADAE gives you the opportunity to make comparisons in space with other businesses, and in time for a series of periods;

Fourth, using DMEM can be retrospective (if you are using actual values of indicators) and planning (plan, forecast, prior) if using the expected values of indicators, analysis of changes of efficiency of enterprise activity;

Fifth, to obtain an integral evaluation a fairly simple computational algorithm is implemented on a PC;

Sixth, this methodology takes into account the characteristics of enterprises, the overall economic situation, the interests of users, by modifying the composition and arrangement of the indicators in the DMEM;

Seventh, if necessary, the methodology allows to update the model with new strategic directions, or, conversely, to exclude from the model the least informative ratios to the growth rate, the change in the number and nomenclature used in the model indicators does not require changing the computational algorithm.

References

[1] Copeland T., Koller T, Murrin J. Valuation Measuring & managing the value of companies. McKinsey & Company, Inc. John wiley & Sons, INC.

[2] Antill N., Lee K. Company valuation under ifrs. Interpreting and Forecasting Accounts Using. International financial reporting standards. Harriman House Publishing

[3] Cokins G.. Performance management. Finding the missing pieces (to close the intelligence gap). John Wiley & Sons, Inc.

[4] Du M., Boateng A. State ownership, institutional effects and value creation in cross-border mergers & acquisitions by Chinese firms // International Business Review. Volume 24, Issue 3, 1 June 2015, pp. 430-442

[5] Song J., Wang R., Cavusgil S.T. State ownership and market orientation in China's public firms: An agency theory perspective. // International Business Review. Volume 24, Issue 4, 1 August 2015, pp. 690-699

[6] Adithipyangkul P, Leung T.Y. State Ownership, Legal Institution, and Independent Director Compensation: An Exploratory Study in China. // Chinese Economy. Volume 48, Issue 6, 2 November 2015, pp. 430-448

[7] Song Z., Nahm A., Zhang Z. The value of partial state ownership in publicly listed private sector enterprises: evidence from China. // Post-Communist Economies. Volume 27, Issue 3, 3 July 2015, pp. 336-353

[8] Nhung L.T.P., Okuda H. Effects of state ownership on companies' capital structure and profitability: 2015, pp. 64-78

[9] Pan X., Tian G.G. Does banks' dual holding affect bank lending and firms' investment decisions? Evidence from China. // Journal of Banking and Finance. Volume 55, June 01, 2015, pp. 406-424

[10] Wu Y., Yao H. Income Inequality, State Ownership, and the Pattern of Economic Growth – A Tale of the Kuznets Curve for China since 1978. // Atlantic Economic Journal. Volume 43, Issue 2, 28 April

2015, pp. 165-180
[11] Zhang L., Wu W., Wei Y., Pan R. Stock holdings over the life cycle: Who hesitates to join the market? // Economic Systems. Volume 39, Issue 3, September 2015, pp. 423-438
[12] Dasilas A., Leventis S. Corporate Governance, Dividend Status, Ownership Structure, and the Performance of Greek Seasoned Equity Offerings. // International Journal of the Economics of Business. Volume 20, Issue 3, November 2013, pp. 387-419
[13] Garaeva E.K. International holding company as an optimal structure for optimizing tax liabilities of taxpayers. // Asian Social Science. Volume 11, Issue 11, 30 April 2015, pp. 137-141

IOP Conf. Series: Journal of Physics: Conf. Series **936** (2017) 012038 doi:10.1088/1742-6596/936/1/012038

Finsler Geometry Modeling of an Orientation-Asymmetric Surface Model for Membranes

Evgenii Proutorov

Department of Physics, Cherepovets State University, Pr. Lunacharskii 5, Cherepovets 162600, Russia

Hiroshi Koibuchi

Department of Industrial Engineering, National Institute of Technology, Ibaraki College, Nakane 866, Hitachinaka, Ibaraki 312-8508, Japan

E-mail: koibuchi@mech.ibaraki-ct.ac.jp; koibuchih@gmail.com

Abstract. In this paper, a triangulated surface model is studied in the context of Finsler geometry (FG) modeling. This FG model is an extended version of a recently reported model for two-component membranes, and it is asymmetric under surface inversion. We show that the definition of the model is independent of how the Finsler length of a bond is defined. This leads us to understand that the canonical (or Euclidean) surface model is obtained from the FG model such that it is uniquely determined as a trivial model from the viewpoint of well definedness.

Keywords: Triangulated surface model, Non-Euclidean metric, Finsler geometry, Membranes, Surface inversion

1. Introduction

The Finsler geometry (FG) modeling technique expands the range of application of the surface model of Helfrich and Polyakov for membranes [1, 2]. Indeed, the so-called soft elasticity and elongation under temperature changes observed in liquid crystal elastomers are understood via the 3D FG modeling technique [3]. It has been shown that the J-shaped stress-strain curve of biological materials, such as human skin and muscle, can be calculated using a 2D FG model [4]. Moreover, the origin of the line tension energy, which plays an important role in the model for two-component membranes [5], is explained by another 2D FG model.

In Ref. [6], it was mentioned that this FG model is independent of how the Finsler length of a bond is defined. In this paper, we demonstrate that the FG model in Ref. [6] remains unchanged if the definition of the Finsler length of a bond is altered. The Finsler length of a bond is in general an essentially new and important ingredient in an FG model. In this paper, we discuss the details of this interesting property and its consequence.

2. Model

2.1. FG modeling

A surface model is defined by a mapping $\mathbf{r}: M \ni (x_1, x_2) \mapsto \mathbf{r} \in \mathbf{R}^3$, where M is a two-dimensional parameter space. For a discrete model, both M and $\mathbf{r}(M)$ should be triangulated. We should note that the Finsler length of a bond refers to the edge length of triangles in M. This edge length can be direction dependent in the FG modeling technique, in contrast with the standard modeling, where the bond length is always direction independent on the triangulated surfaces.

The discrete partition function and Hamiltonian are given by [6]

$$\begin{aligned}
&Z(\zeta,\kappa) = \sum_{\chi}\sum_{\mathcal{T}} \int' \prod_{i=1}^{N} d\mathbf{r}_i \exp\left[-S(\mathbf{r},\chi)\right], \quad S = S_1 + \kappa S_2 + \zeta S_3, \\
&S_1 = \sum_{ij} \gamma_{ij}\ell_{ij}^2, \quad S_2 = \sum_{ij} \kappa_{ij}\left(1 - \mathbf{n}^+ \cdot \mathbf{n}^-\right), \\
&S_3 = \sum_{\pm}\left(1 - \chi^+ \cdot \chi^-\right), \quad \left(\chi^{\pm} \in \{1,-1\}\right), \\
&\gamma_{ij} = \frac{1}{4}\left(\gamma_{ij}^+ + \gamma_{ij}^-\right), \quad \kappa_{ij} = \frac{1}{4}\left(\kappa_{ij}^+ + \kappa_{ij}^-\right), \\
&\gamma_{ij}^+ = \begin{cases} \rho_i^+ + 1/\rho_j^+ & (\chi(\Delta^+) = 1) \\ 1/\rho_i^+ + \rho_j^+ & (\chi(\Delta^+) = -1) \end{cases}, \quad \gamma_{ij}^- = \begin{cases} \rho_j^- + 1/\rho_i^- & (\chi(\Delta^-) = 1) \\ 1/\rho_j^- + \rho_i^- & (\chi(\Delta^-) = -1) \end{cases}, \\
&\kappa_{ij}^+ = \begin{cases} \rho_j^+ + 1/\rho_i^+ & (\chi(\Delta^+) = 1) \\ 1/\rho_j^+ + \rho_i^+ & (\chi(\Delta^+) = -1) \end{cases}, \quad \kappa_{ij}^- = \begin{cases} \rho_i^- + 1/\rho_j^- & (\chi(\Delta^-) = 1) \\ 1/\rho_i^- + \rho_j^- & (\chi(\Delta^-) = -1) \end{cases},
\end{aligned} \tag{1}$$

where $\mathbf{r}$, $\mathcal{T}$, and χ denote the vertex position, the triangulation, and the variable corresponding to the orientation of triangle Δ, respectively (Fig. 1). S_1, S_2 and S_3 are the Gaussian bond potential, the bending energy, and the energy for the triangle orientation, respectively. The symbols ℓ_{ij} and $\mathbf{n}^{\pm}$ represent the length of bond ij and unit normal vectors of triangles $+$ and $-$, respectively. The Finsler metric assumed here is

$$g_{ab} = \begin{pmatrix} 1/\rho & 0 \\ 0 & \rho \end{pmatrix}, \tag{2}$$

where ρ is a function on triangle Δ.

In Ref. [6], the Finsler lengths of bonds 12 and 21 are defined by (see Fig. 1)

$$\begin{aligned}
&L_{12}(\Delta_L^+) = (1/2)\left(1/\rho_1^+ + \rho_2^+\right), \quad L_{21}(\Delta_L^-) = (1/2)\left(1/\rho_2^- + \rho_1^-\right), \\
&\bar{L}_{12}(\Delta_R^-) = (1/2)\left(1/\rho_1^- + \rho_2^-\right), \quad \bar{L}_{21}(\Delta_R^+) = (1/2)\left(1/\rho_2^+ + \rho_1^+\right).
\end{aligned} \tag{3}$$

In contrast, in this paper, we define the lengths of bonds 12 and 21 as follows:

$$\begin{aligned}
&L_{12}(\Delta_L^+, \Delta_L^-) = (1/2)\left(1/\rho_1^+ + \rho_1^-\right), \quad L_{21}(\Delta_L^+, \Delta_L^-) = (1/2)\left(\rho_2^+ + 1/\rho_2^-\right), \\
&\bar{L}_{12}(\Delta_R^+, \Delta_R^-) = (1/2)\left(\rho_1^+ + 1/\rho_1^-\right), \quad \bar{L}_{21}(\Delta_R^+, \Delta_R^-) = (1/2)\left(1/\rho_2^+ + \rho_2^-\right).
\end{aligned} \tag{4}$$

The expressions in Eq. (4) are different from those in Eq. (3). In this paper, we show that the canonical (Euclidean) model is obtained from the model defined in Eq. (1) as a well-defined model even when the lengths of bonds are given by Eq. (4). We should note that there are two other expressions for L_{12} and L_{21} for the combination of two neighboring triangles (Δ_L^+, Δ_R^-) and (Δ_R^+, Δ_L^-). For the expressions of $\bar{L}_{12}$ and $\bar{L}_{21}$, we use the inverted triangles (Δ_R^+, Δ_L^-) and (Δ_L^+, Δ_R^-). However, the following discussions remain unchanged although the expressions for several conditions become dependent on the definition of the bond length. In this paper, we confine ourselves to the lengths described in Eq. (4).

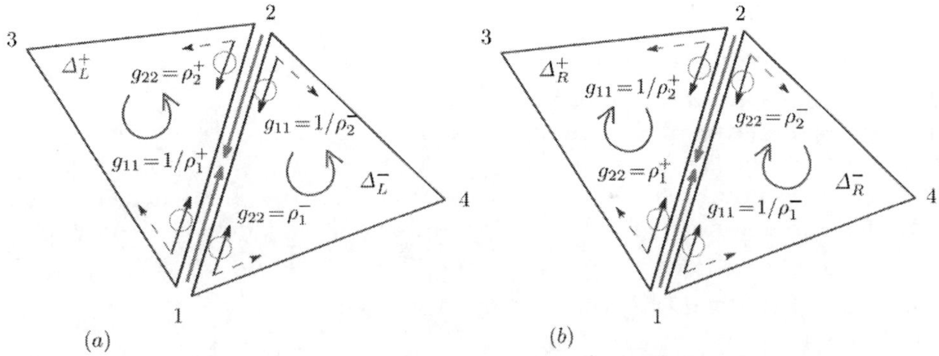

Figure 1. (a) Two neighboring triangles $\Delta^{\pm}$ and elements of g_{ab} for the direction-dependent length of bond 12 between the vertices 1 and 2 and (b) the inverted triangles $\Delta_R^{\mp}$ (inside view) of $\Delta_L^{\pm}$ in (a). The direction-dependent lengths of bonds 12 and 21 are indicated by the long arrows in both (a) and (b).

2.2. Discussions

In this subsection, we follow the arguments in Ref. [6]. We call a discrete surface model *trivial* (*non-trivial*) if the following conditions are (not) satisfied:

$$\gamma_{ij} = \text{constant}, \quad \kappa_{ij} = \text{constant}. \tag{5}$$

It is well-known that the trivial models are equivalent to the canonical (or Euclidean) model, which is given by $\gamma_{ij} = \kappa_{ij} = 1$. This is understood from the scale invariance of the partition function [7].

A discrete surface model is called *well defined* (*ill defined*) if the following conditions are (not) satisfied:

(D1) *The bond length is independent of its direction.*

(D2) *The bond length is independent of the surface orientation.*

Note that the bond length in (D1) and (D2) is the bond length of triangles in the parameter space M. Thus, from the expressions in Eq. (4), we have the following equations:

$$1/\rho_1^+ + \rho_1^- = 1/\rho_2^- + \rho_2^+, \qquad (\Leftrightarrow \text{(D1)}), \tag{6}$$

$$1/\rho_1^+ + \rho_1^- = 1/\rho_1^- + \rho_1^+, \quad 1/\rho_2^- + \rho_2^+ = 1/\rho_2^+ + \rho_2^- \quad (\Leftrightarrow \text{(D2)}). \tag{7}$$

These equations are different from those in Ref. [6] because these come from Eq. (4). We also have the following condition for the model to be orientation symmetric:

$$1/\rho_1^- + \rho_2^- + 1/\rho_2^+ + \rho_1^+ = 1/\rho_2^- + \rho_1^- + 1/\rho_1^+ + \rho_2^+. \tag{8}$$

This condition comes from the Hamiltonian in Eq. (1) and the following definition:

(D3) *A discrete surface model is called orientation symmetric if the Hamiltonian is orientation symmetric.*

We should note that the condition in Eq. (8) is identical to the one given in Ref. [6]. This is because the Hamiltonian is independent of how the length of a bond (in M) is defined.

The condition in Eq. (8) is also proved independently of the Hamiltonian. We make the following remarks:

(R1) *Any well-defined model is orientation symmetric.*

(R2) *For any well-defined model, we have*

$$1/\rho_i + \rho_i = a (= \text{const}). \tag{9}$$

(R3) *If Eq. (9) is satisfied, then the model is trivial.*

(R1) is a direct consequence of Eqs. (6) and (7). (R2) is proved as follows: multiplying $\rho_1^+\rho_1^-$ by the first equation in Eq. (7), we have $\rho_1^+ = \rho_1^-$. We also have $\rho_2^+ = \rho_2^-$ from the second equation in Eq. (7). Then, we obtain $1/\rho_1 + \rho_1 = 1/\rho_2 + \rho_2$ from Eq. (6). This equality is satisfied for all bonds; therefore, we have Eq. (9). (R3) is proved as follows: since $1/\rho_i + \rho_i$ is independent of vertices and triangles, we have $\gamma_{ij} = (1/4)\left(\gamma_{ij}^+ + \gamma_{ij}^-\right) = a/2$ and $\kappa_{ij} = (1/4)\left(\kappa_{ij}^+ + \kappa_{ij}^-\right) = a/2$.

We obtain from (R2) and (R3) the following theorem:

(Th) *All well-defined models are trivial.*

This means that *all non-trivial models are ill defined.* However, this ill definedness turns out to be well defined in the context of FG modeling because the Finsler length depends on the direction in general, where the bond length is called the *Finsler length* of a bond. The Finsler length is allowed to be dependent on the surface orientation in the FG model of Eq. (1). This is the main result reported in Ref. [6].

Now, we consider the implication of the theorem described above. Let us assume that the model given by Eq. (1) is well defined in the sense of Eqs. (6) and (7). In this case, the Finsler length of the bond is given by Eq. (4). Then, we find from the theorem that this model is trivial, and this trivial model is equivalent to the Euclidean model, which is given by $\gamma_{ij} = \kappa_{ij} = 1$. It is also possible to obtain the Euclidean model by using another Finsler length in Eq. (3) as a trivial model. This implies that the Euclidean model is determined independently of the definition of the Finsler length of the bond, and therefore, the Euclidean model is uniquely determined from the FG model in Eq. (1).

3. Summary

We have shown that all well-defined models are trivial. This indicates that a well-defined model is uniquely determined because the well-defined model, which is trivial, is determined independently of how the Finsler length of a bond is defined. This uniqueness is also supported by the physically understood fact that all trivial models are equivalent to the canonical (or Euclidean) model in the sense that the differences in the two models (except for a scale difference) can be absorbed in the bending rigidity. Thus, the result in this paper implies that the canonical model is uniquely determined as a trivial model in the framework of FG modeling from the viewpoint of well definedness.

Acknowledgment This work is supported in part by JSPS KAKENHI grant No. 17K05149.

References

[1] D. Nelson, in Statistical Mechanics of Membranes and Surfaces, Second Edition, edited by D. Nelson, T. Piran, and S. Weinberg, (World Scientific, 2004) p.1.

[2] H. Koibuchi and H.Sekino, Monte Carlo studies of a Finsler geometric surface model, Physica A **393**, 37-50 (2014).

[3] K. Osari and H. Koibuchi, Finsler geometry modeling and Monte Carlo study of 3D liquid crystal elastomer, Polymer **114**, 335-369 (2017).

[4] Y. Takano and H. Koibuchi, J-shaped stress-strain diagram of collagen fibers: Frame tension of triangulated surfaces with fixed boundaries, Phys. Rev. E **95**, 042411(1-11) (2017).

[5] S. Usui and H. Koibuchi, Finsler geometry modelling of phase separation in multi-component membranes, Polymers **8**, 284(1-18) (2016).

[6] E. Proutorov and H. Koibuchi, Orientation Asymmetric Surface Model for Membranes : Finsler Geometry Modeling, Axioms **6**, 10 (2017).

[7] J.F. Wheater, Random surfaces: From polymer membranes to strings, J. Phys. A Math. Gen. **27**, 3323-3353 (1994).

IOP Conf. Series: Journal of Physics: Conf. Series **936** (2017) 012039 doi:10.1088/1742-6596/936/1/012039

Network structures sustained by internal links and distributed lifetime of old nodes in stationary state of number of nodes

Nobutoshi Ikeda
Tohoku Seikatsu Bunka Junior College,
1-18-2 Niji-no-Oka, Izumi-ku, Sendai 981-8585, Japan

E-mail: ikeda.nobu@nifty.com

Abstract. In network models that take into account growth properties, deletion of old nodes has a serious impact on degree distributions, because old nodes tend to become hub nodes. In this study, we aim to provide a simple explanation for why hubs can exist even in conditions where the number of nodes is stationary due to the deletion of old nodes. We show that an exponential increase in the degree of nodes is a natural consequence of the balance between the deletion and addition of nodes as long as a preferential attachment mechanism holds. As a result, the largest degree is determined by the magnitude relationship between the time scale of the exponential growth of degrees and lifetime of old nodes. The degree distribution exhibits a power-law form $\sim k^{-\gamma}$ with exponent $\gamma = 1$ when the lifetime of nodes is constant. However, various values of γ can be realized by introducing distributed lifetime of nodes.

1. Introduction

The knowledge of network structures provides a very fundamental framework for studying complex systems constructed by many constituents interacting with each other [1]. One of the most interesting properties commonly seen in real networks must be the existence of hub nodes whose degree (number of links incident to the node) is typically a power function of the network size. One simple explanation for the existence of hub nodes is based on the growth property of networks [2]. However, there are some real networks including social networks, in which network growth is attenuated by the continuous deletion of nodes [3]. In these networks, the deletion of old nodes is inevitable for reviewing how their structures developed, for deletion of old nodes has a serious impact on degree distributions of growing networks. Network structures in the stationary state of number of nodes have not been completely clarified even in the explanation for the existence of hub nodes.

In the growth model of networks where a new node attaches to pre-existing nodes with the probability proportional to the node at each time step (BA model [2]), the random deletion of a node at each time step leads to the divergence of the power-law exponent of degree distribution as the growth rate vanishes, and hub nodes disappear in the stationary state of number of nodes [4]. To maintain the power-law behavior in the degree distribution in the stationary state of number of nodes, other mechanism besides BA rule is needed. For example, Miura et.al [5] showed that the coagulation of nodes can maintain the power-law form of degree distribution even in the stationary state of number of nodes.

In the present study, we consider the selective deletion of old nodes (hub nodes) instead of the random deletion of nodes. In this case, it is obvious that the hub nodes disappear immediately by the deletion of nodes if only the BA rule is employed, because old nodes tend to become hub nodes in this rule. Here, we aim to provide a simple mechanism based on internal links [6] by which hubs can exist even in conditions where old nodes are being deleted. First, we will show that an exponential increase in the degree of nodes is a natural consequence of the constancy of the internal factor which describes the contribution of internal links to the rate of increase in degrees. Secondly, we will show that the values of the largest degree and the power-law exponent describing the degree distribution are determined by the magnitude relationship between the time scale of the exponential growth of degrees and lifetime of old nodes.

2. Effect of constant lifetime of nodes

To carry out a concrete examination of the effect of the deletion of old nodes on the degree distribution of networks, let us consider the following network model in which all nodes have the same lifetime.

(i) A complete graph with m nodes is prepared initially ($t = 1$).

(ii) At each discrete time step t, a vertex with m edges newly attaches to the network with a preferential attachment rule (each edge attaches to the pre-existing node with a probability proportional to the degree of the node [2].).

(iii) At each time step t, two vertices are selected with a probability proportional to the product of degrees of the selected vertices [6], and if the selected vertices have not been linked, these two vertices are linked. If the two vertices have already been linked, nothing happens. This process repeats rN_t times, where N_t is the number of nodes at time t, and $0 \le r \le 1$ is a constant parameter.

(iv) Let t_i be the time at which node i attached to the network ($i = 1, 2, \cdots, t+m$). When $t - t_i > L_0$ is satisfied, node i and the incident links are deleted, where L_0 is the constant life time of all nodes.

(v) t is increased by 1, and goes to step (ii).

Owing to step (iii), the network is accelerated in the time interval $0 < t \le L_0$, that is, the mean vertex degree increases with time [8]. Owing to (iv), however, N_t and the sum of all vertex degrees D_t take constant values $N_{\rm eq} = L_0 + m \simeq L_0$ and $D_{\rm eq}$ when $t \gg L_0$. (Isolated nodes will be counted in $N_{\rm eq}$ in later sections.) In this model, the increment of degree k_i of node i per unit time is evaluated by the following equation in terms of D_t,

$$\frac{dk_i}{dt} = \begin{cases} \frac{Bk_i}{t} + \frac{mk_i}{D_t}, & (t \le L_0) \\ \tilde{B}k_i + \frac{mk_i}{D_{\rm eq}} - f_{k_i}(t), & (t > L_0) \end{cases} \tag{1}$$

where B and $\tilde{B}$ are internal factors corresponding to $t \le L_0$ and $t > L_0$, which are defined by

$$Bt^{-1} = 2rN_t(1 - \sum_{j \in {\rm n}_i} k_j/D_t)/D_t, \tag{2}$$

$$\tilde{B} = 2rN_{\rm eq}(1 - \sum_{j \in {\rm n}_i} k_j/D_{\rm eq})/D_{\rm eq}, \tag{3}$$

where ${\rm n}_i$ is the set of nodes directly connected to node i [7]. In contrast to (2), the right hand side of (3) is independent of time, because $D_{\rm eq}$ is nearly constant. The second term in (1) is the contribution of external links, and $f_k(t)$ represents the effect of the deletion of nodes on dk_i/dt.

To ensure exact treatment of (1), there is a need to know to which nodes deleted nodes are linked. However, as a first approximation, we can evaluate the behavior of degrees k_i for

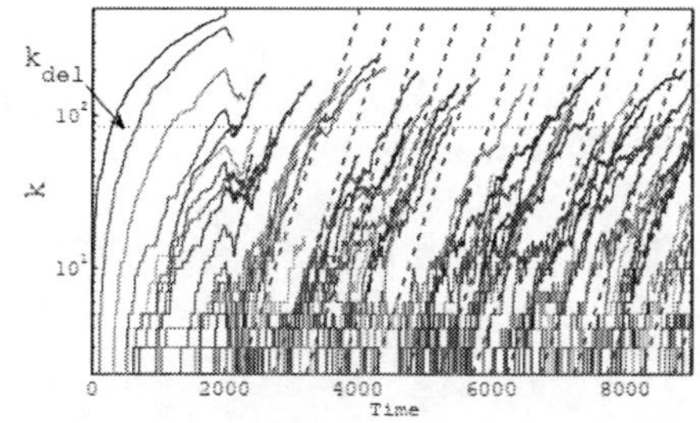

Figure 1. Numerical results for the time-dependence of degrees when $r = 0.05$, $m = 2$ and $L_0 = 2000$. Dashed lines indicate exponential increases $\sim \exp\left\{(\tilde{B} + m/D_{\text{eq}})(t - t_i)\right\}$ in which the numerical value of (3), $(\tilde{B}+m/D_{\text{eq}}) = 1/327.2$, is substituted. The horizontal dotted line indicates a numerical value of $\bar{k}_{\text{del}}$ obtained by (6).

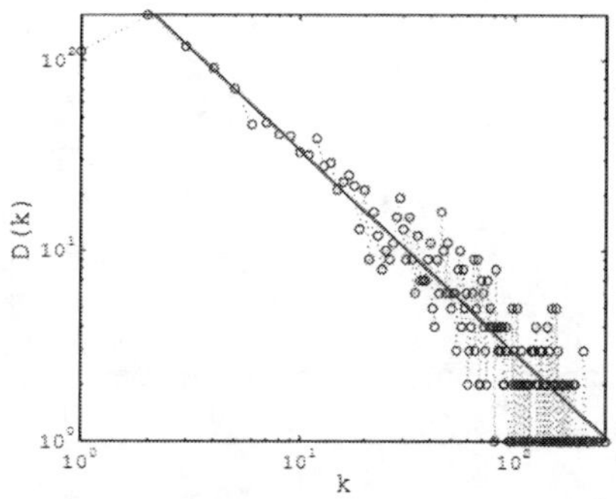

Figure 2. Numerical result for degree distribution at $t = 10000$ using the same parameters with Fig. 1. The bold line indicates a fitting curve with slope -1.06.

$t > t_i > L_0$ using the assumption that the first term in (1) $\tilde{B}$ is constant and dominant enough to ignore $f_k(t)$,

$$\log k_i \simeq (\tilde{B} + m/D_{\text{eq}})(t - t_i). \quad (\text{for } \ t > t_i > L_0, \ \ t - t_i < L_0) \tag{4}$$

Specifically speaking, $\tilde{B}$ is a decreasing function with respect to k with a very small slope. However, approximated behaviors of k can be described by the value of $\tilde{B}$ at $k \simeq 0$ even if considering k-dependence of $\tilde{B}$ [7]. According to (4), the density of number of nodes with degree k is proportional to $d \log k = dk/k$, because t_i is uniformly distributed in the time axis (see Fig. 1). As the result, the degree distribution is obtained as

$$P(k)dk \sim \frac{dk}{k}. \tag{5}$$

Fig. 1 presents the numerical result for the time-dependence of degrees when condition $e^{\tilde{B}L_0} \gg 1$ is satisfied. (Condition $e^{\tilde{B}L_0} \gg 1$ must be satisfied for degree to survive to a certain size.) The deviation from the ideal behavior (4) in high degrees found in Fig. 1 can be understood by the following equation on the balance between the increase and decrease in the number of edges,

$$\alpha r N_{\text{eq}} = \bar{k}_{\text{del}}, \tag{6}$$

where α represents the probability that two nodes selected in step (iii) have not linked until time t,

$$\alpha = 1 - \sum_i \sum_{j \in \text{n}_i} k_i k_j / D_{\text{eq}}^2, \tag{7}$$

and $\bar{k}_{\text{del}}$ is the mean degree of deleted nodes. The value of $\bar{k}_{\text{del}}$ is smaller than $e^{\tilde{B}L_0}$, because $e^{\tilde{B}L_0}$ is an ideal maximum degree which can be realized by ignoring the effect of $f_k(t)$. Power-law exponent 1 indicated in (5) is not adequate for describing the result in range $\bar{k}_{\text{del}} < k < e^{\tilde{B}L_0}$, because most of the nodes with degree $\bar{k}_{\text{del}} < k < e^{\tilde{B}L_0}$ are deleted. Fig. 2 shows that (5) describes the numerical result with considerable accuracy in range $k < \bar{k}_{\text{del}}$, although the increase tendency of degrees described by (4) is perturbed by $f_{k_i}(t)$ in Fig. 1. At least, condition $1 \ll \bar{k}_{\text{del}}$ must be satisfied for the power-law form to appear clearly in the degree distribution.

3. Effect of distribution of lifetime of nodes

By introducing the distribution of lifetime of nodes, the impact of deletion of nodes on the degree distribution will decrease, because, for the same value of internal factor, the maximum degree can become larger as the lifetime of nodes tends to be longer. To carry out concrete calculation, we employed the following lifetime distribution that expresses that node i is not deleted until time $t(> t_i)$,

$$P(t-t_i) = \begin{cases} 1 & (\text{for} \quad t-t_i \le L_0) \\ \exp\{-(t-t_i-L_0)/\tau\} & (\text{for} \quad t-t_i > L_0) \end{cases} \tag{8}$$

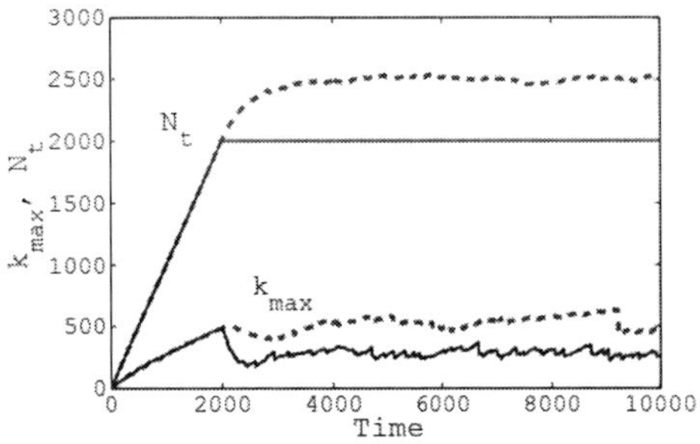

Figure 3. Numerical results for the time dependences of the number of nodes and the maximum degree when $r = 0.05$, $m = 2$ and $L_0 = 2000$. The bold lines and dashed lines indicate cases when $\tau = 0$ and when $\tau = 500$, respectively.

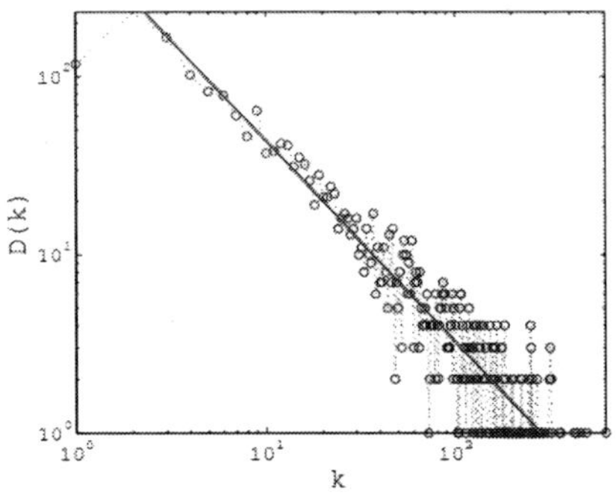

Figure 4. Numerical result for degree distribution at $t = 15000$ when $r = 0.05$, $\tau = 500$, $m = 2$ and $L_0 = 2000$. The bold line indicates a fitting curve with slope -1.13.

In Fig. 3, the time dependences of the number of nodes and maximum degrees for case $\tau = 0$ are compared with that for case $\tau = 500$. As shown in the figure, the distributed lifetime of nodes raises the maximum degree. As a result, the range of k where the power law is found widens (Fig. 4). For node i in condition $t - t_i > L_0$, the degree distribution is generalized by introducing probability (8) to,

$$P(k)dk = \frac{P(t-t_i)dk}{k}, \tag{9}$$

$$\sim k^{-1/(\tilde{B}+m/D_{\text{eq}})\tau-1}, \tag{10}$$

where we have used (4), $t - t_i = \log k_i/(\tilde{B} + m/D_{\text{eq}})$. As the result, we obtain,

$$\gamma = 1/(\tilde{B} + m/D_{\text{eq}})\tau + 1. \tag{11}$$

The relation (11) shows that the value of γ is determined by the balance between the magnitude of the internal factor and range of distribution of lifetime of nodes. The power-law behavior disappears as $\tau \to 0$ (in $k > \bar{k}_{\text{del}}$), and γ approaches to 1 as $\tilde{B}+m/D_{\text{eq}}$ becomes large. Fig. 5 and Fig 6 illustrate a typical result for the time-dependence of degrees and the degree distribution when $\tau \neq 0$ and $\bar{k}_{\text{del}}$ is not that large. The validity of relation (10) is confirmed for parameters used in the numerical calculation. Note that the relation (10) is valid for $k > \bar{k}_{\text{del}}$, but not for $k < \bar{k}_{\text{del}}$. The value of $\gamma(= 1)$ remains the same as the case $\tau = 0$ for small k.

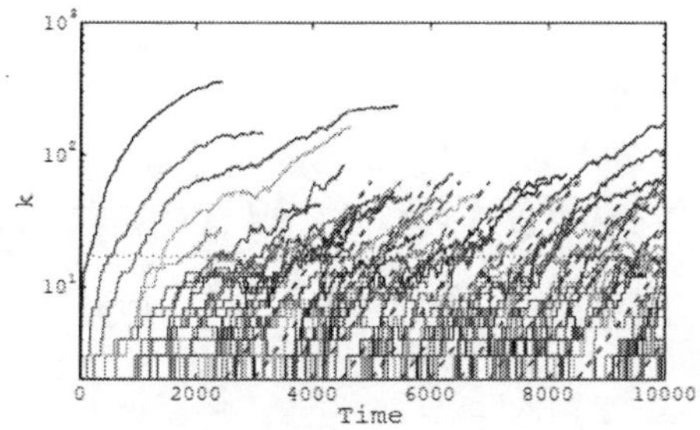

Figure 5. The numerical results for the time-dependence of degrees when $r = 0.01$, $\tau = 1000$, $m = 2$ and $L_0 = 2000$. The dashed lines indicate exponential increases $\sim \exp\left\{(\tilde{B} + m/D_{\rm eq})(t - t_i)\right\})$ in which $(\tilde{B} + m/D_{\rm eq}) = 1/722.8$ numerically obtained by (3) is substituted. The horizontal dotted line indicates the numerical value of $\bar{k}_{\rm del}$ numerically obtained by (6).

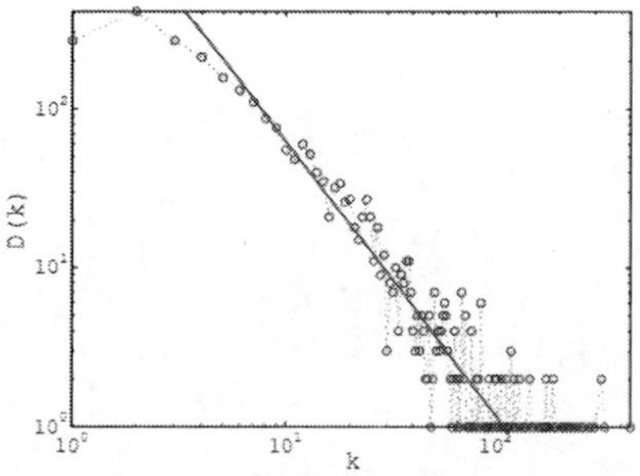

Figure 6. Numerical result for degree distribution at $t = 20000$ using the same parameters with Fig. 5. The guide line indicates a power-law form with (11) in which the numerical result, $\gamma = 722.8/1000 + 1 \simeq 1.72$, is substituted.

4. Summary

We have examined network structures in the stationary state of number of nodes where the rate of addition of new nodes and deletion of old nodes is balanced. We explained the exponential increase in vertex degrees by the constancy of the internal factor which describes the contribution of internal links to the rate of increase in degrees. If the internal factor $\tilde{B}$ is large enough to develop the value of degrees before the node is deleted, that is, if condition $\alpha r N_{\rm eq} = \bar{k}_{\rm del} \gg 1$ is satisfied, hub nodes with order $\bar{k}_{\rm del}$ can exist, and the degree distribution can have a power-law form with exponent $\gamma = 1$, even though the deletion of old nodes causes deletion of hub nodes. When the lifetime of nodes is governed by an exponential decayed distribution, the rate of deletion of hub nodes decreases, and the power-law form of degree distribution with exponent $\gamma > 1$ is maintained by the balance between the magnitude of the internal factor and range of distribution of lifetime of nodes.

In this paper, we employed global links to describe internal links that connect pre-existing nodes. However, there are other mechanisms that emerge as an internal factor that maintains the power-law behavior in accelerating networks, such as intermediary process [7]. Local links may provide different feature from networks examined here, which we plan to investigate in future studies.

References

[1] Boccaletti S, Latora V, Moreno Y, Chavez M and Hwang D-U 2006 *Physics Reports* **424** 175
[2] Barabási A-L and Albert R 1999 *Science* **286** 509
[3] Saavedra S, Reed-Tsochas F, and Uzzi B 2008 *PNAS* **105** 16466
[4] Moore C, Ghoshal G, and Newman M E J 2006 *Phys. Rev.* E **74** 036121
[5] Miura W, Takayasu H, and Takayasu M 2012 *Phys. Rev. Lett* **108**168701
[6] Dorogovtsev S N and Mendes J F F 2000 *EPL* **52** 33
[7] Ikeda N 2017 *Physica* A **484** 378
[8] Smith D M D, Onnela J-P and Jones N S 2009 *Phys. Rev* E **79** 056101

IOP Conf. Series: Journal of Physics: Conf. Series **936** (2017) 012040 doi:10.1088/1742-6596/936/1/012040

Propagation of Ultra High Energy Cosmic Rays from Galactic Sources in a Fractal Interstellar Medium and Origin Studies

Pantea Davoudifar
Research Institute for Astronomy and Astrophysics of Maragha (RIAAM)/P.O.B:55134-441
E-mail: dfpantea@riaam.ac.ir

Keihanak Rowshan Tabari

Saeed Doostmohammadi
Shahid Bahonar University of Kerman, Kerman, Iran

Abstract. Cosmic rays propagation within the Galaxy is simulated considering a fractal-like distribution of matter and magnetic fields. In such an anomalous diffusion process, the sources of ultra-high energy particles were considered to be galactic pulsars. The coordinate and birth times of sources were chosen from SN and pulsar catalogues. From these sources the primary spectrum of cosmic rays (H, He, CNO, Ne-Si, Fe) in energy range EeV and above were reproduced. The resulted spectrum is used to discuss about the change in cosmic ray sources form Galactic to Extragalactic.

1. Introduction

Different astronomical sources and sites are considered to be responsible of particle acceleration in the Galaxy. In a usual picture, acceleration mechanism are considered to be second and first order Fermi and diffusive shock acceleration mechanisms, all of which are resulting in power law spectrums. Another fact about source fluxes comes from analyzing abundance ratios of secondary to primary particles. For example, analyzing boron-to-carbon abundance ratio, suggests a soft power law energy spectrum at source, $E^{-\gamma}$, with $\gamma \sim 2.4$ and $\delta \sim 0.5$ for the propagation path length, $\Lambda(E)$ (i.e. $\Lambda(E) = E^{\delta}$) [1] (Considering other studies, $\delta \sim 0.6$). Thus in a diffusive propagation with negligible energy loss processes (i.e. Leaky Box Model), the energy spectrum of Cosmic Rays (CR) at source is predicted to be $\propto E^{-2.1}$. This flux index it produced in different acceleration models, for example the first order fermi mechanism is very efficient in regions with small scale turbulences [2] and in this situation the spectral indexes are nearly ~ 2.

Diffusive shock acceleration mechanisms were first studied using simple first order Fermi mechanisms by different groups [3, 4, 5, 6], later more complicated situation were analyzed; for example it is showed that ultra-relativistic shocks result power law spectrums with index values of $\approx 2.2 - 2.3$ [7]. Today acceleration mechanism are studied in more complicated situation

[8, 9, 10]. At the highest energies the striking problem is that the first order Fermi mechanism and diffusive shock acceleration does not work at highly relativistic shocks, unless in the presence of very weak magnetic fields [11].

When studying the propagation of CR in media with sharp density variations the diffusion approach is inappropriate and an alternative case is propagation in a fractal medium. The anisotropy studies also gives different results as the matter is not uniformly distributed.

2. Fractal Mediums

Diffusion in fractal mediums is governed by different formalisms. In fact, in the frame of diffusion approach where the inhomogeneities of matter and field within the Galaxy are small scale, the evolution of particle number density is formulated by[12]:

$$\frac{\partial N}{\partial t} = D(E)\triangle N(\mathbf{r}, t, E) + S(\mathbf{r}, t, E) \tag{1}$$

in which, D is diffusivity.

But the super-diffusion without energy losses and nuclear interactions has the form of [13]:

$$\frac{\partial N}{\partial t} = -D(R, \alpha, \beta) D_{0+}^{1-\beta}(-\triangle)^{\frac{\alpha}{2}} N(\mathbf{r}, t, R) + S(\mathbf{r}, t, R) \tag{2}$$

with $D(R, \alpha, \beta) = D_0(\alpha, \beta)R^{\delta}$, anomalous diffusivity; and D_{0+}^{μ}, Riemann-Liouville fractional derivative and $(-\Delta)^{\frac{\alpha}{2}}$ Riss'operator [14]. In steady state case $\frac{\partial N}{\partial t} = 0$ and the equation is possible to be solved using Green's functions [15].

2.1. Estimations

To numerically produce the expected values of fluxes and compare it with the observed particle fluxes at different energies, some estimated values were used. The key difference of this work is considering the millisecond pulsars as the point source.

More experimental data on proton fluxes makes it a good subject in case studies. One should consider that in the study of Galactic sources for high energy protons, we can not consider a galactic origin for protons with energies higher than $\sim 10^{19}\ eV$. Because theoretically, the most attainable energy reported for a proton primary in a pulsar wind is around $\sim 10^{19}\ eV$ [16, 17, 18] and just in the case of a young pulsar. Anyhow, we considered the millisecond pulsars of ATNF catalogue [22], and considered their observed positions to estimate their spatial distribution in a cylindrical volume. We used [23]:

$$F(r, z) = (\frac{r}{r_0})^a \exp\Big(-b(\frac{r}{r_0} - 1) - \frac{z}{H_z}\Big) \tag{3}$$

The parameters were calculated using millisecond pulsars of ATNF catalogue, for which $a = 1.2$, $b = 2.8$, $r_0 = 8.5Kpc$, and $H_z = 0.2$. This function were used later to calculate radial gradient of galactic CR intensity (as described in [24]).

At this assumption it is possible to also produce the flux of other nuclei (figure 2 and 3).

3. Results

At this stage, we used our previous studies [19, 20, 21] in which the propagation of CR is turbulent magnetic fields were studied. We considered the same pulsar distribution function as in equation (3) and compared our results for propagation of only protons in our Galactic magnetic field model. for the simple case of $\beta = 1$ (see equation 2) our Kolmogorov Magnetic Field (MF) produces a flux which is showed in figure 4. Different values of α also are considered to compare the result.

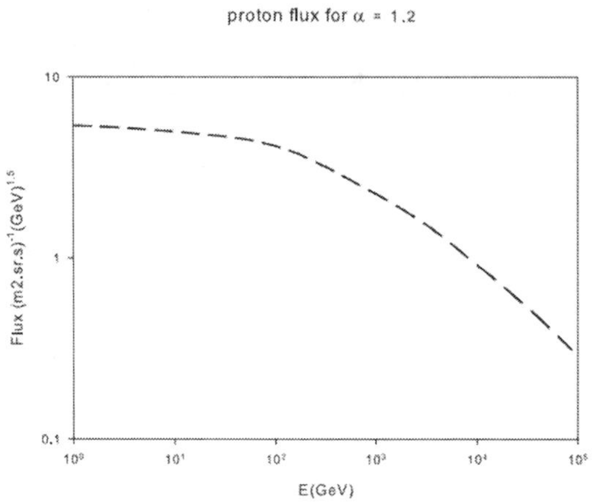

Figure 1. CNO flux.

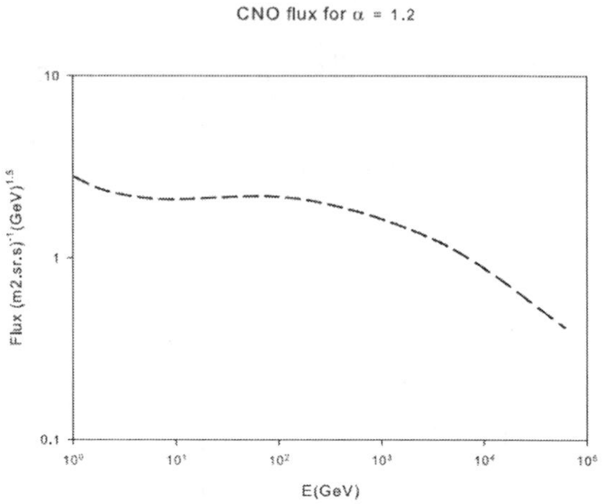

Figure 2. Fe Flux.

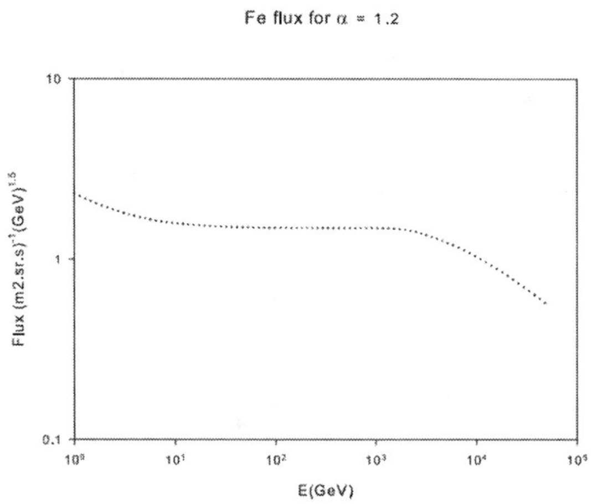

Figure 3. CNO flux.

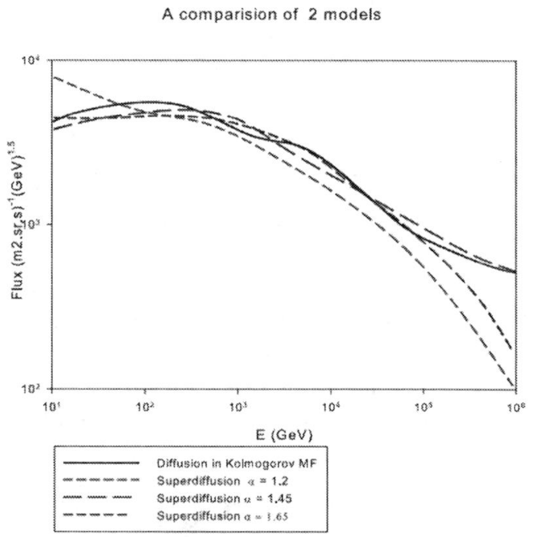

Figure 4. Fe Flux.

It is shown that the Kolmogorov MF of our study produces the results comparable with superdiffusion with $\alpha = 1.45$, considering pulsar origin of CR up to $10^6\ GeV$. In fact continuing this flux up to higher energies (i.e. above a few EeV) the results of to model are not fitted for $\alpha = 1.45$. It may lead to this conclusion that one source model at higher energies fails. It is a logical result because at the highest energies, one source models cause large anisotropies in directions of detected ultra high energy events. And such anisotropy is not seen yet.

References

[1] Obermeier A., Boyle P., Hörandel J., and Müller D., **2012** Astrophysical Journal, *The Boron-to-Carbon Abundance Ration and Galactic Propagation of Cosmic Radiation*, **752**, 69(7pp)

[2] Spurio M., **2015** Springer, *Particles and Astrophysics: A Multi-Messenger Approach*

[3] Axford W.I., Leer E., Skadron G., **1977** in Proceedings of the 15th International Cosmic Ray Conference, *The acceleration of cosmic rays by shock waves*, **11**, 132135

[4] Jokipii J.R., **1987** Astrophysical Jornal, *Rate of energy gain and maximum energy in diffusive shock acceleration.*, **313**, 842846
[5] Blandford R., and Eichler D., **1987** Physics Reports (Review Section of Physics Letters), *Particle Acceleration at Astrophysical Shocks: A Theory of Cosmic Ray Origin*, **154**, 1-75
[6] Jones F.C., Ellison D.C., **1991** Space Sciense Reviews, *The plasma physics of shock acceleration*, **58**, 259
[7] Achterberg A., Gallant Y.A., Kirk J.G., Guthmann A.W., **2001** Montly Notices of Royal Astronomical Society, *Particle acceleration by ultrarelativistic shocks: theory and simulations*, **328**, 393-408
[8] Sironi L., and Spitkovsky A., **2012** International Journal of Modern Physics: Conference Series, *Particle Acceleration at the termination Shock of Striped Pulsar Winds*, **8**, 144-150
[9] Amato E., **2014** Proceedings of Sciences, *Particle acceleration and radiation in Pulsar Wind Nebulae*, **224**, 043
[10] Olmi B., Del Zanna L., Amato E., Bucciantini N., and Bandiera R., 2015 Journal of Physics: Conference Series, *Particle acceleration and non-thermal emission in Pulsar Wind Nebulae from relativistic MHD simulations*, **642**, 012019
[11] Sironi L., and Spitkovsky A., **2009** Astrophysical Journal, *Particle Acceleration in Relativistic Magnetized Collisionless Pair Shocks: Dependence of Shock Acceleration on Magnetic Obliquity*, **698**, 1523-1549
[12] Ginzburg V.L., and Syrovatskii S.I., **1964** Pergamon Press, *Origin of Cosmic Rays*
[13] Lagutin A.A., Tyumentsev A.G., and Volkov N.V., **2009** Bulletin of the Russian Academy of Sciences: Physics, *Cosmic Ray Spectrum in a Fractal-Like Galactic Medium for Different Particle Acceleration Mechanisms in a Source*, **73**, 561-563
[14] Samko G.S., Kilbas A.A., Marichev O.I., **1993** Gordon and Breach Science Publishers, *Fractional Integrals and Derivatives, Theory and Applications*
[15] Lagutin A.A., Nikulin Yu.A., and Uchaikin V.V., **2001** Nuclear Physics B, Proceedings Supplements, *The knee in the Primary Cosmic Ray Spectrum as Consequence of the Anomalous Diffusion of the Particles in the Fractal Interstellar Medium*, **97**, 267-270
[16] Berezhko E.G., **1991** in Proceedings of the 22nd International Cosmic Ray Conference, *Generation of Super-High Energy Cosmic Rays in the Vicinity of Pulsars*, **2**, 436 Pulsars
[17] Berezhko E.G., **1994** Astronomy Letters, *Generation of ultrahigh-energy cosmic rays in the surroundings of pulsars*, **20**, 75-79
[18] Wibig T., and Wolfendale A. W., **2011** Open Astronomy Journal ,*The Knee in the Cosmic Ray Energy Spectrum: a Pulsar, Supernova Origin?*, **4**, 211-217
[19] Davoudifar P., and Rowshan Tabari K., **2014** Journal of Physics: Conference Series, *Modeling the Galactic Magnetic Field and its application in verifying a Pulsar Origin of Very High Energy Cosmic Rays*, **490**, 012029
[20] Davoudifar P., and Rowshan Tabari K., **2015** Journal of Physics: Conference Series, *Development of a Galactic Magnetic Field Model and its application in identifying sources of Ultra-High-Energy Cosmic Rays in Northern Sky*, **490**, 012029
[21] Davoudifar P., **2016** Journal of Physics: Conference Series, *Modeling a Kolmogorov-Type Magnetic Field in the Galaxy and its Effect on an Extragalactic Isotropic Flux of Ultra High Energy Cosmic Rays*, **738**, 012035
[22] http://www.atnf.csiro.au/research/pulsar/psrcat
[23] Erlykin A.D., Lagutin A.A., and Wolfendale A.W., **2003** Astroparticle Physics, *Properties of the interstellar medium and the propagation of cosmic rays in the Galaxy*, **19**, 351362
[24] Doostmohammadi S., and Fatemi S.J., **2012** ISRN High Energy Physics, *The Characteristics of Cosmic Rays in a Fractal Medium*, **2012**, 673250

IOP Conf. Series: Journal of Physics: Conf. Series **936** (2017) 012041 doi:10.1088/1742-6596/936/1/012041

Modeling of light scattering by polymer coatings of rolled metal surface

A Maksimov[1], **E Maksimova, V Egorov**

Cherepovets State University, Cherepovets, Lunacharskii Pr., Russia

E-mail: a_v_maximov@mail.ru

Abstract. The statistical characteristics of light scattering on textured polymer coatings of rolled metal surface are studied on the base screen model with two-dimensional Weierstrass function. In the frame of the Kirchhoff method, the average light scattering and gloss coefficients are numerically calculated for different coating surface calibration (profiles) and falling angles. It is shown; that the scattering is symmetrical relatively to the plane of incidence and it is maximal in the direction$\theta_3 = 0$. The scattering wave is observed in the cone of lateral directions relatively the normal direction$\theta_2 = 0$. For such coatings it is nessesary to develop a special convergence experimental technique allowing identifying a correspondence between the gloss and the texture of the surface .

1.Introduction

Rolled metal with a textured polymer coating must be characterized by high mechanical and optical characteristics, which give the building structures (roofs, etc.) good decorative qualities, namely, a natural and noble appearance typical for the structure and color of natural tiles. The mechanical properties (the strength, durability and etc.) of the branded products “Steel Velvet” and “Steel Silk”of PAS “Severstal” (Cherepovets) correspond to parameters required by world standarts [1, 2]. However, photometric instruments underestimate the gloss coefficient and total deviation of the color coordinates of the sample from the corresponding optical properties of the standart. At present, the comparison of optical charactristics (the gloss, color etc.) of textured polymer coatings with the corresponding ones of standart occurs often visually (“by eye”), without the use of magnifying and special optical devices.

Therefore, it’s nessary to find out the reasons for this discrepancy. For this purpose, preliminary, we studied experimentally the sample surface of a textured polymer coating using scanning probe microscopy method. As can be seen from Fig. 1, the surface of textured coating has a rough fractal character. Therefore, we supposed that one from reasons for this discrepancy is the significant light scattering on the by randomly rough surface of the coating.

In this paper the average coefficient of light scattering and the gloss coefficient are calculated in the frame of the Kirchhoff method. A normalized Weierstrass function is used for modeling 2D fractal rough surface. The analogues calculations of scattering indicatrices diagrams for various surfaces and incidence angles were performed by others authors [3-7], but our results have some distinctive features, related with calcucatiog of the gloss coefficient.

[1] To whom any correspondence should be addressed.

Figure 1. The sample of steel sheet with 200-fold magnification with the probe attached to it.

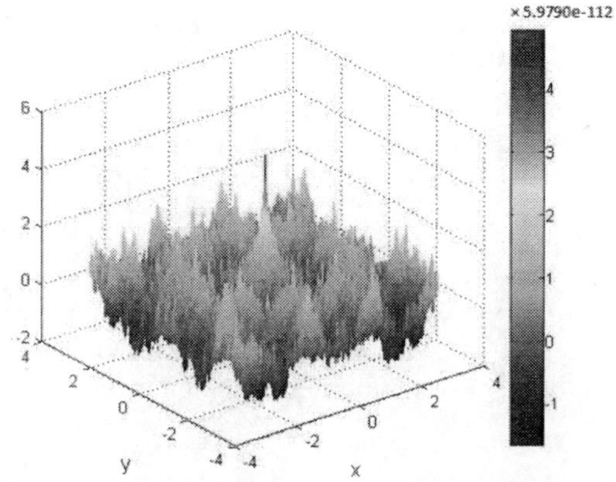

Figure 2. The example of a rough fractal surface simulated by means of the Eq.(1) at $N = 10$, $M = 10$, $D = 2.5$, $q= 2$, K=3.

2. Fractal-like model for two-dimensional rough surfaces

Usually [4,5], for the simulation of rough surfaces, the modified two-dimensional Weierstrass function z(x,y) for the height of roughness is chosen. This function has the form:

$$z(x,y) = c_W \sum_{n=0}^{N-1} \sum_{m=1}^{M} q^{(D-3)n} \sin\left\{\mathrm{K}\mathrm{q}^{\mathrm{n}}\left[\mathrm{x}\cos\frac{2\pi \mathrm{m}}{\mathrm{M}} + \mathrm{y}\sin\frac{2\pi \mathrm{m}}{\mathrm{M}}\right] + \varphi_{\mathrm{nm}}\right\}, \quad (1)$$

where c_w is a normalizing constant. The number K determines the wave length of the surface's fundamental harmonic. The value q characterizes both the amplitude and frequency of harmonics (q >1). The numbers N, M, D and q determine the surface calibration degree because of imposing additional harmonics on the fundamental wave, and N and M determine the number of overtones. The number D is the fractal dimension that determines the amplitude of harmonics (usually $2<D<3$). It was shown [4-6] that the large scale spatial heterogeneity of the surface increases, too with an increase in N, M, D and q.

The phases φ_{nm} in Eq. (1) can be chosen in appointed or casual ways, what gives deterministic or stochastic function $z(x, y)$, respectively. As the rule [3-7], the functions φ_{nm} are considered as casual values distributed uniformly over the interval $[-\pi, \pi]$. Particular realization of the function $z(x,y)$ with the parameters c_w, q, K, D, N, M chosen beforehand can be received with each particular choice of numerical values of all $N \times M$ phases φ_{nm} with the help of a random-number generator. Therefore, the function determined by Eq. (1) is a combination of determinate periodic and random constituents. This function may be anisotropic in both directions, given M and N are not too large. Since natural surfaces are generally neither purely random nor purely periodic, and often anisotropic, this function proposed is a good candidate for modeling natural surfaces [3-7]. The example of a rough surface simulated by means of the eguation (1) is shown on Figure 2.

Every possible realization of the function $z(x,y)$ forms an ensemble of surfaces. Deflection of the rough surface points from the basic plane appears to be proportional to the value c_w; therefore, this parameter is connected with the height of structure imperfections over the surface. It is found more convenient to set a rough surface by choosing the root-mean-square height of its profile σ which can be calculated using the expression:

$$\sigma \equiv \sqrt{\langle h^2 \rangle} \quad (2)$$

where $h = z(x,y)$, the notition $< \cdots \geq = \prod_{n=0}^{N-1} \prod_{m-1-\pi}^{M} \int_{-\pi}^{\pi} \frac{d\varphi_{nm}}{2\pi} (\ldots)$ means ensemble averaging over the surfaces.The relation between the parameters c_w and σ can be established taking the integrals directly[4-6]:

$$\sigma = \left[\prod_{n=0}^{N-1} \prod_{m=1}^{M} \int_{-\pi}^{\pi} \frac{d\phi_{mn}}{2\pi} z^2(x,y) \right]^{\frac{1}{2}} = c_w \left[\frac{M\left(1 - q^{2N(D-3)}\right)}{2\left(1 - q^{2(D-3)}\right)} \right]^{\frac{1}{2}}. \tag{3}$$

Thus, the rough surface in the accepted model is described by six parameters: c_w(or σ), q, K, D, N, M. In refs. [4-6], the influence of different parameters onthe profiles of surfaces was investigated detaily on the base of the numerical calculations of the Weierstrass function

3. Light scattering on surface fractal structures

The scheme of experiment on light scattering by the surface S is shown in Figure 3. Let us consider the wave falling on the rough surface S at an angle θ_1 and scattering at a polar angle θ_2 and an azimuthal angle θ_3 in the spherical coordinate system. The gloss coefficient for the coating surface is determined as the ratio

$$g = lg \frac{\langle I_s \rangle}{I_0}, \tag{4}$$

where $< I_s > = < \vec{E_s} \vec{E_s^*} >$ and I_0 are the average intensity of scattering light on the ensemble of surfaces at the angle $\theta_1 = \pi/8$ and $\theta_1 = 0$ correspondingly. Therefore, to compare with the experimental data it is necessary to calculate the average intensity of scattering ligth.

.We will find the scattered field $E_s(\vec{r})$ after its interaction with the fractal surface (1) on the base of the scalar Kirchhoff method [8]. This method allows to find the scattered field under the following conditions: (1) the incident wave is mono-chromatic and plane; (2) the scattered surface is rough inside a rectangle ($-X < x_0 < X$, $-Y < y_0 < Y$) and smooth outside the borders; (3) the size of the rough site is much larger than the length of the incident wave; (4) the reflection coefficient is the same at all points of the surface; (5) the scattering field is observed far enough from the screen surface [3-7].

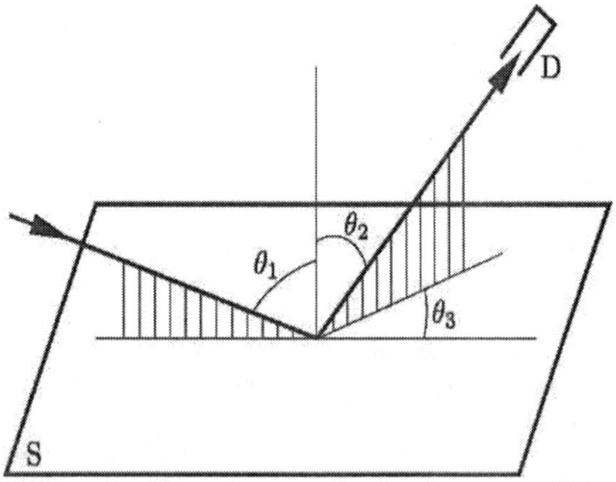

Figure 3. The scheme of experiment on light scattering by the surface S: D is detector, θ_1 is a falling light angle; θ_2 is a polar angle; θ_3 is an azimuthal angle.

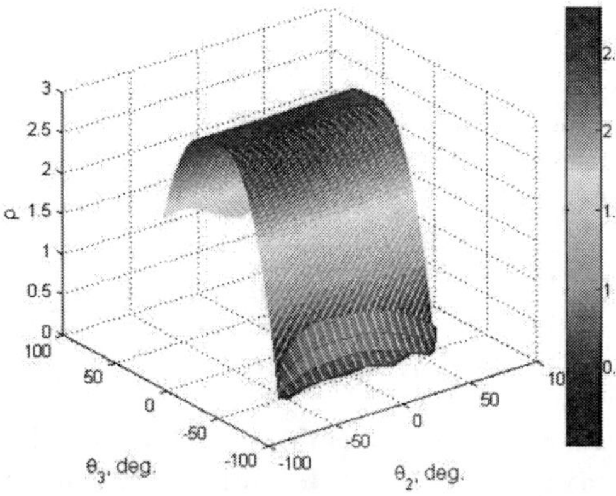

Figure 4. The gloss coefficient for the coating surface versus scattering angles 0_2 and 0_3. The parameters of Eq.(1)$N = 5$, $M = 10$, $D = 2.5$, $q = 3$.

Under these conditions, the scattered field is expressed by the following formula:

$$E_s(\vec{r}) = -ikrF(\theta_1,\theta_2,\theta_3)\frac{e^{ikr}}{2\pi r}\int_S \exp\big(ik\Phi(x_0,y_0)\big)\,dx_0dy_0 + E_e(\vec{r}), \qquad (5)$$

where k is the wave number of the incident wave; the function $F(\theta_1,\theta_2,\theta_3) = -\frac{R}{2C}(A^2 + B^2 + C^2)$ is an angle factor; R is the scattering coefficient; $\Phi(x_0,y_0) = Ax_0 + By_0 + Ch(x_0,y_0)$ is a phase function; $h(x_0,y_0) = z(x_0,y_0)$; $A = sin\theta_1 - sin\theta_2\, cos\theta_3$; $B = -sin\theta_2 sin\theta_3$; $C = -cos\theta_1 - cos\theta_2$. The expression (5) may be written in the form: $E_e(\vec{r}) = -\frac{R}{C}\frac{e^{ikr}}{4\pi r}(AI_1 + BI_2)$, where the coefficients are expressed thro the intergrals:

$$I_1 = \int_{-Y}^{Y}\left[e^{ik\Phi(X,y_0)} - e^{ik\Phi(-X,y_0)}\right]dy_0, \qquad I_2 = \int_{-X}^{X}\left[e^{ik\Phi(x_0,Y)} - e^{i\Phi(x_0,-Y)}\right]dx_0. \qquad (6)$$

Having taken integrals (6) using the formula $e^{izsin\Phi} = \sum_{l=-\infty}^{\infty} I_l\,(z)e^{il\Phi}$, where $I_l(z)$ is the Bessel function of integer order l, it is possible to obtain the expression:

$$E_s(\vec{r}) = -2ikFXY\frac{e^{ikr}}{\pi r}\sum_{l\{r\,s\}}\left\{\left[\prod_{u\,v} I_{l_{uv}}(\xi_u)\right] exp[i\sum_{nm} l_{nm}\,\varphi_{nm}]\right\}\times sinc(k_AX)sinc(k_BY) + E_e(\vec{r}) \qquad (7)$$

with the following notitons are used: $\sum_{l\{r\,s\}} \equiv \sum_{l_{0,1}=-\infty}^{\infty}\sum_{l_{0,2}=-\infty}^{\infty}\cdots\sum_{l_{(N-1),M}=-\infty}^{\infty}$, $\prod_{uv} = \prod_{u=1}^{N-1}\prod_{v=0}^{M}$, $\sum_{nm} = \sum_{n=1}^{N-1}\sum_{m=0}^{M}$, $\xi_u \equiv kc_wCq^{(D-3)u}$, $sincx \equiv \frac{sinx}{x}$, $k_A = kA + K\sum_{nm} q^n l_{nm} cos\frac{2\pi m}{M}$, $k_B = kB + K\sum_{nm} q^n l_{nm} cos\frac{2\pi m}{M}$. Then, the expression (7) may be represented as:

$$E_e(\vec{r}) = -ikXY\frac{R}{C}(A^2 + B^2)\frac{e^{ikr}}{\pi r}sinc(kAX)sinc(kBY). \qquad (8)$$

Thus, the expression (8) in the frame of the Kirchhoff method allows solving the problem of finding field scattering by the fractal surface (1) setting the parameters of this surface c_w (or σ), D, q, K, N, M, X, Y, φ_{nm}, parameter k (or $\lambda = \frac{2\pi}{k}$) of the incident wave, also geometrical parameters $\theta_1, \theta_2, \theta_3$. This intensity characterizes scattering at particular realization of surface $z(x, y)$ with the casual phase's φ_{nm}. Calculating $< I_s > = < \overrightarrow{E_s}\overrightarrow{E_s^*} >$ and using Equation (8), the exact expression can be obtained:

$$\langle I_s\rangle = \left[\frac{F(\theta_1,\theta_2,\theta_3)}{cos\theta_1}\right]^2 \sum_{l\{r\,s\}}\left\{\prod_{uv} I_{l_{uv}}^2(\xi_u)\, sinc^2(k_AX)sinc^2(k_BY)\right\} + \left[\frac{R(A^2+B^2)}{2Ccos\theta_1}\right]^2 sinc^2(kAX)sinc^2(kBY)\,. \quad (9)$$

While the expression (9) contains an infinite sum, using it for numerical calculations appears to be inconvenient. An essential simplification can be reachedin the case $\xi_n < 1$. Using the Bessel function expansion in the series and neglecting members of orders greater thanξ_{2n}, an approximate expression can be obtained for the average intensity:

$$< I_s > \approx \left[\frac{F(\theta_1,\theta_2,\theta_3)}{cos\theta_1}\right]^2 \Big\{[1 - (k\varrho C)^2]sinc^2(kAX)sinc^2(kBY) + \frac{1}{2}\sum_{nm} q^{2(D-3)n} sinc^2[\big(kA + Kq^n cos\frac{2\pi m}{M}\big)X]\sum\; sinc^2[\big(kB + Kq^n cos\frac{2\pi m}{M}\big)Y]\Big\} + \left[\frac{R(A^2+B^2)}{2Ccos\theta_1}\right]^2 sinc^2(kAX)sinc^2(kBY), \qquad (10)$$

IOP Conf. Series: Journal of Physics: Conf. Series **936** (2017) 012041 doi:10.1088/1742-6596/936/1/012041

4. Numerical results

On the base of expressions (1), (4) and (10) the numerical modeling of geometrical and optical properties of light scattering with polymer coatings is investigated, the diagrams of scattering (indicatrices for different incident angles θ_1) and the gloss coefficient versus polar (θ_2) and horizontal (θ_3) scattering angles for fractal surfaces of textured polymer coatings of rolled metal were calculated. A root-mean-square value σ, a surface's fundamental wave number K, and dimensions X, Y of the surface fragment are expressed in k-units because the wave number k of the incident wave is used in the form of combinations kσ, kX, and kY.The Frenel reflection coefficient for a surface is taken as R = 1.

An analysis of the scattering indicatrices shows the scattering is symmetrical relatively the plane of incidence and it is maxmmal intensity in the direction $\theta_3 = 0$. The scattering wave is observed in the cone of lateral directions relatively the normal direction $\theta_2 = 0$. When the incident angle increases, the incident wave as though "stops to notice" the height of imperfections, so their contribution diminishes.The registered scattering peculiarities are a consequence of a combination of both the chaotic character and self-similarity of a scattering rough fractal surface.

5. Conclusion

The gloss coefficient for the coating surface vs. scattering angles θ_2 and θ_3 is shown on Figure 4. Thus, this coefficient differs from zero in other lateral directions: over the interval $\theta_2 \in [0, \pi/3]$ ("forward" scattering) and $\theta_2 \in [0, -\pi/3]$ ("back" scattering). Therefore, the gloss detector, settting in the mirror direction $\theta_2 = -\theta_1 = -\pi/8$, measures more less light intensity from the rough surface than from the smooth one. Therefore, it's nessecary to develop a special convergence experimental technique for such metal coatings allowing identifying a correspondence between the gloss, colour and the texture of the sample surfaces.

References

[1] Varfolomeev IA, Maksimova OG, Ershov EV, Maksimov AV and Vinogradova LN 2013 *Production of rolled metal* 4 ⅋ 26.

[2] Maksimova OG. Maksimov AV 2014 Macromol. *Symp.* 346 59.

[3] Zamani M, Shafiei F, Fazeli S M, Downer M C and Jafari G R 2016 *Phys. Rev.* E 94 042809

[4] Jafari GR, Mahdavi SM, A. Iraji Zad, Kaghazchi P 2005 *Surf. Interface Anal.*37 641.

[5] LinNet al.1995*J. Mod. Opt.* 42 225.

[6] Pinchuk A Oand Gozhenko V. 8th-International Conference on Electromagnetics of Complex Media, Lisbon, Portugal, on 27-29 September 2000. Bianisotropics 2000 305

[7] Semchuk O Yu., Grechko L G, Vodopianov D L, KunitskaL Yu 2009 *Task Quarterly* 13 199.

[8] Ogilvy J A, *Theory of Wave Scattering from Random Rough Surfaces*. New York: Adam Hilger, 1991.

Effect of thickness on the magnitude of spontaneous polarization in thin ferroelectric films

O G Maksimova[1],O S Baruzdina and E A Maksimova
Department of Physics, Cherepovets State University,
5 LunacharskiiPr., Cherepovets, Russia

E-mail:og62@mail.ru

Abstract. The influence of the thickness of a ferroelectric film and the depolarizing field on its spontaneous polarization and the order parameter has been investigated by means of the Monte-Carlo method. Dependences of the polarization of thethin ferroelectric film on temperature at different values of its thickness and the potential well depth of the Lennard-Jones potential are calculated. The thickness of the "dead" layer is analyzed depending on the temperature and of the potential well depth.

1. Introduction

The introduction of ferroelectric films in modern microelectronic devices, related with limitations on the reduction in the size of the element base, has revived interest in the question of the possible existence of a critical thickness of the films necessary for the appearance of ferroelectricity[1].

The critical size of the ferroelectric film is usually understood to mean the minimal thickness of the sample, at which its ferroelectric properties disappear, and it becomes a usualdielectric material [2]. To describe the properties of the film vs. its thickness, different models are used, taking into account: a) correlation effects [3,4], b) the presence of a transient dielectric layer (“dead” layer)[5], c) the bulk charge density and the surface charge density near boundaries of ferroelectric [6], d) thermoelastic properties [7,8].

However, these effects cannot be considered isolated, because they all are related each with other.We used the Ising model for computer simulation of ferroelectric systems.This model has a discrete symmetry group. Onsager has been strictly proved that there is a spontaneous polarization in this model of monolayer [9]. However, it has been experimentally established that there are sufficient ferroelectric thin films, in which the spontaneous polarization may be absent [2]. In this paper, the Ising model was modified to take into account thermoelastic properties and a depolarizing field.

2. Model

To describe properties of the ferroelectric films and tostudy of ordering effects we use a three-dimensional lattice model (Fig.1), consisting of N_1,$N_2$and N_3 nodes along the respective axes of the Cartesian coordinate system.The position of the lattice node is characterized by the set of three numbers $\vec{n} = (n_1, n_2, n_3)$.

In this paper, the interaction energy of dipoles is described by a potential that takes into account orientational interactions of particles (as in the classical Ising model) and the additional termrepresenting the Lennard-Jones potential:

$$, \quad H = H_{or} + \sum_{\vec{n},\vec{m}} \varepsilon \left(\frac{r_0^{12}}{r_{\vec{n},\vec{m}}^{12}} - \frac{2r_0^6}{r_{\vec{n},\vec{m}}^6} \right) \quad (1)$$

where ε is well depth of the Lennard-Jones potential, $r_{\vec{i},\vec{j}}$ isthe distance between the dipoles, r_0 is average distance between the dipoles (in the absence of orientational interactions).The second term of equation (1) does not depend on temperature and polarization, in contrast to the first term.

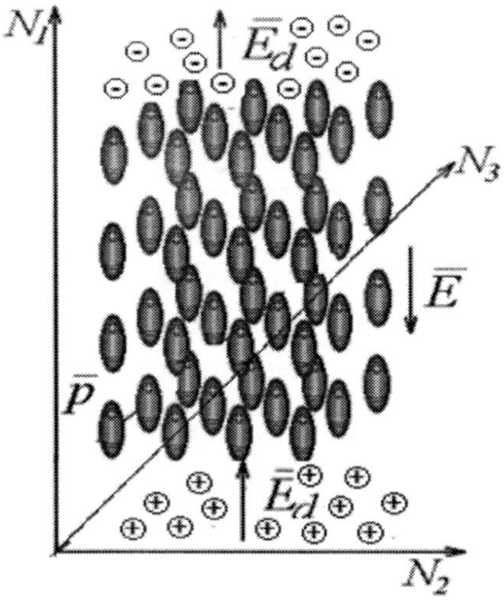

Figure 1. The three-dimensional lattice model of the thin ferroelectric film.

When the polarization reduced transverse film dimensions in the directions N_2 and N_3 , the distances between the dipoles change in these directions.Therefore, the potential of orientational interactions H_{or} may be represented by the formula:

$$H_{or} = -\sum_{\vec{n}} K_1 S_{n_1,n_2,n_3} S_{n_1-1,n_2,n_3} - \sum_{\vec{n}} K_{20} \frac{r_0^3}{r^3} S_{n_1,n_2,n_3} S_{n_1,n_2-1,n_3} -$$
$$, \quad -\sum_{\vec{n}} K_{20} \frac{r_0^3}{r^3} S_{n_1,n_2,n_3} S_{n_1,n_2,n_3-1} + p\sum_{\vec{n}} S_{\vec{n}} E_d \quad (2)$$

where the quantity $S_{\vec{n}}$ takes two values: +1 and -1, K_1 is coefficient of exchange interactions in the longitudinal direction N_1 , p is the module of dipolemoment, K_{20} is theconstant of exchange interactions between the dipoles in the transverse directions N_2 and N_3 , The quantity E_d is the projection of the vector of strength of the depolarizing field on the direction N_1 .

The short-range orientational order parameters in the longitudinal and transverse directions $\mu_{\|}$ and $\mu_{\perp}$ are determined correspondingly by the following formulas:

$$\mu_{\|} = \frac{1}{N_1 N_2 N_3} \left(\sum_{\vec{n}} S_{n_1,n_2,n_3} S_{n_1-1,n_2,n_3} \right) \quad (3a)$$

$$\mu_{\perp} = \frac{1}{2N_1N_2N_3}\left(\sum_{\bar{n}} S_{n_1,n_2,n_3} S_{n_1,n_2-1,n_3} + \sum_{\bar{n}} S_{n_1,n_2,n_3} S_{n_1,n_2,n_3-1}\right) \tag{3b}$$

To simplify the calculations the variable $x^3 = \frac{r_0}{r}$ was introduced.The values $\mu_{\parallel}$ and $\mu_{\perp}$ depend on the variable x. As an example, figure 2 shows the dependence of the short-range orientational order parameter $\mu_{\perp}$ on the variable x at different reduced temperatures.

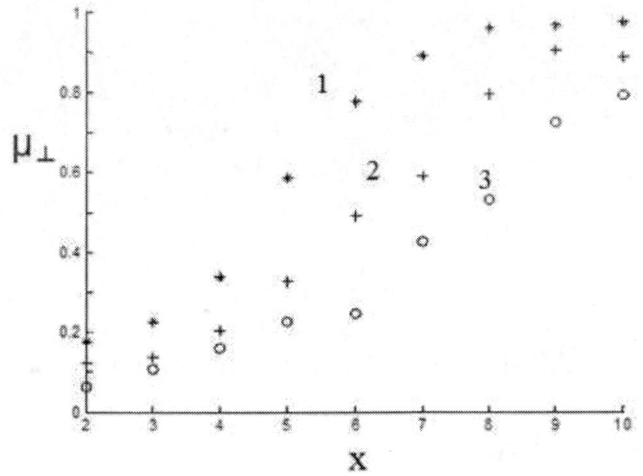

Figure 2. The short-range orientational order parameter $\mu_{\perp}$ vs. the variable x at different reduced temperatures $k_BT / K_1 = 15(1), 20(2), 25(3)$ $\varepsilon / K_1 = 0.01, K_{20} / K_1 = 1$.

In contrast to the classical Ising model, the interactions between dipoles in the model considered depend on the distance between them. Therefore, it was necessary to calculate the average value of the distance between dipoles at which the potential energy has a minimum. This problem is reduced to finding the minimum of the function:

$$h = 4\varepsilon\left(\frac{r_0^{12}}{r^{12}} - \frac{2r_0^6}{r^6}\right) - 2K_1\mu_{\parallel} - 4K_{20}\mu_{\perp}\frac{r_0^3}{r^3} \tag{4}$$

The differentiation of equation (4) with respect to r leads to the equation:

$$\varepsilon(4x^3 - 4x) - \frac{1}{2}K_1\frac{\partial\mu_{\parallel}(x)}{\partial x} - K_{20}\frac{\partial\mu_{\perp}(x)}{\partial x}x - K_{20}x\mu_{\perp}(x) = 0 \tag{5}$$

To solve equation (5) the functions $\mu_{\parallel}(x)$ and $\mu_{\perp}(x)$ were calculated by the Monte Carlo method, and they were approximated by the function $erf\left(\frac{x-m}{\sigma}\right)$, where m and σ are fitting coefficients.

3. Results of simulation

Usually, the change of properties of the system under an external action is described as its response to this action.For example, the dielectric susceptibility reflects a change in the polarization upon an electric field.

:

$$\chi = \frac{1}{k_BT}\left[\left\langle S_{\bar{n}}^2\right\rangle - \left\langle S_{\bar{n}}\right\rangle^2\right] \tag{6}$$

Figure 3 shows the dependence of the average distance between the dipoles (a) and the susceptibility of the ferroelectric system (b) on the reduced temperature for different values of the potential well depthin the absence of the depolarizing field.

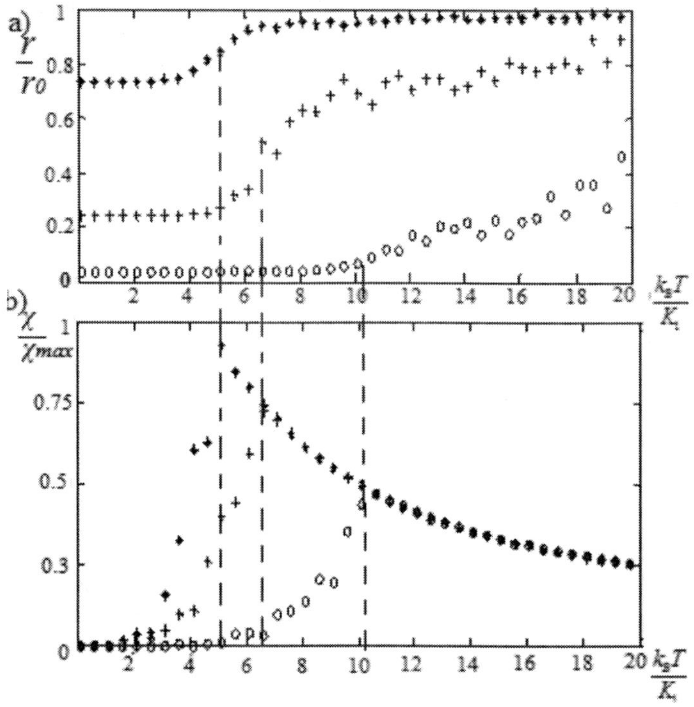

Figure 3. The renormalized average distance between the dipoles r/r_0 (a) and the susceptibility of the ferroelectric system $\chi/\chi_{\max}$ (b) vs. the reduced temperature k_BT/K_1 for values of the potential well depth $\varepsilon/\mathrm{K}_1 = 1(*), 0.1(+), 0.01(o)$ inthe absence of the depolarizing field.

As is seen from Fig.3, if the temperature is increased, the average distance between dipoles near the phase transition point increases abruptly. At high temperatures, the distance $r \to r_0$.With a decrease in the of the potential well depthof the Lennard-Jones potential, the average distance between the dipoles decreases, that leads to a shift of the phase transition point to higher temperatures.

4. Influence of depolarizing field

Under the action of internal electric field caused by spontaneous polarization, free particles move to the outer surfaces of the film and create an additional depolarizing field which depends on the value of the long-range orientational order parameter:

$$E_d(n_1) = E_0\mu\left(e^{-\lambda(N_1-n_1)}S_{n_1,n_2,n_3} + e^{-\lambda(n_1-1)}S_{n_1,n_2,n_3}\right) \tag{7}$$

where $E_0$and λ are constants determined by the number of free carriers in the film, the quantity μ is the long-range order parameter in the ferroelectric film, which value is determined by the following formula:

$$\mu = \frac{1}{N_1N_2N_3}\sum_{\vec{n}} S_{\vec{n}} \tag{8}$$

Figure 4 shows the dependence of the long-range orientational order parameter μ on the reduced temperature $k_BT/K_1$for different values of the depolarizing field. It is shown that the inclusion of the depolarizing field leads to following effects: 1) a decrease in the order parameter and 2) a shift of the phase transition point to the region of lower temperatures.

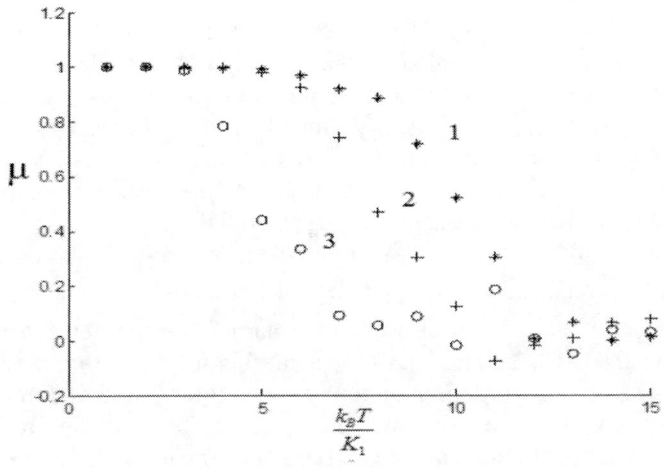

Figure 4. Thelong-range orientational order parameter μvs. The reduced temperature k_BT/K_1 without the depolarizing field (1) and in the presence ofthe field at the value $pE_d/K_1 = 5(2), 10(3)$ at the potential well depth $\varepsilon/K_1 = 0.01$.

Figures 5-6 show the dependence of the long-range order parameter on the thickness of the ferroelectric film N_1 at different reduced temperatures and the potential well depth. As the thickness of the film decreases, the polarization decreases, and decreases sharply to zero at a certain thickness.

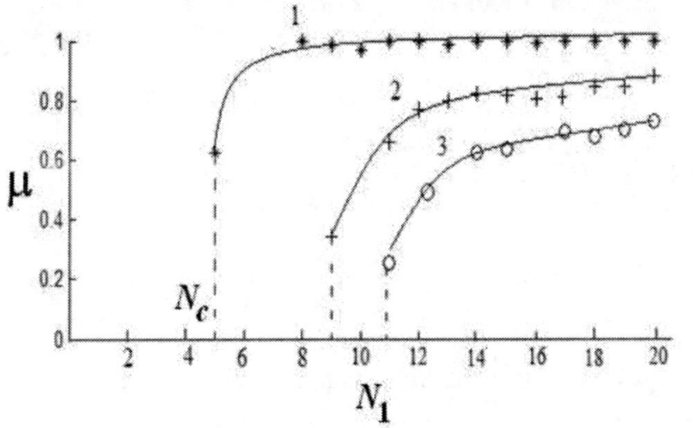

Figure 5. The long-range order parameter μ vs. thethickness oftheferroelectricfilm N_1at different reduced temperatures $k_BT/K_1 = 2(1), 7(2), 10(3)$ and the potential well depth $\varepsilon/K_1 = 0.01$.

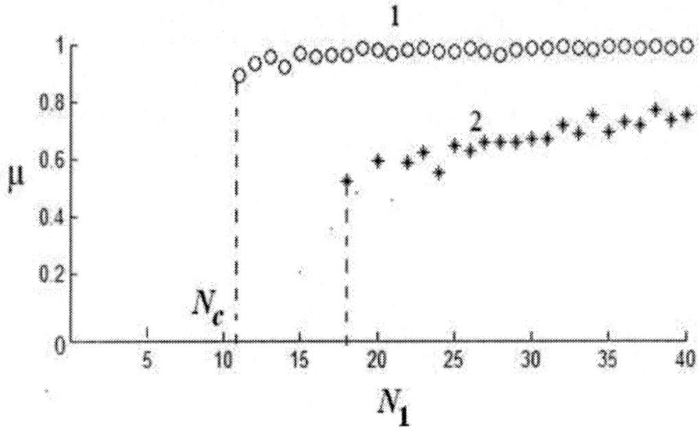

Figure 6. The long-range order parameter μ vs. the thickness of the ferroelectric film N_1 at different potential well depths $\varepsilon/K_1 = 0.01(1), 1(2)$ at the reduced temperature $k_BT/K_1 = 5$.

5. Conclusion

Therefore, correlation effects related with the ordering in thin ferroelectric films lead to the appearance of a surface charge density, the depolarizing field, and also to the presence of the "dead" layer. It is shown that the depolarizing fielddecreases the polarization of the film and shifts the phase transition point to the region of lower temperatures.In sufficiently thin films, the polarization is not observed even at low temperatures. This result can be explained by the fact that its size is less than doublethickness of the "dead" layer that is considered as the critical size of the film, at which its ferroelectric properties disappear, and the film becomes a usual dielectric material.
As shown in figures 5-6, the thickness of the "dead" layer is 2-10 unit cells. The value of the critical thickness N_C increases with decrease in the well depth of the potential of Lennard-Jones (Fig.5) and increasing temperature (Fig.6). Experimental techniques allowed to obtainperovskite ferroelectric films with a thickness of 40 Å (ten unit cells) [10]. The authors [11] identified a critical thickness of about three unit cells, below which there was no ordering at low temperatures. In the frame of approach proposed in this paper, such difference in the experimental results can be explained by the difference nthe potential welldepths and the concentrations of free carriers in the materials considered.

6. References

[1] VorotilovK A andSigovA S2012 Ferroelectric memory*Physics of the Solid State* **54:5**894-899
[2] FridkinV M 2006 Critical size in ferroelectric nanostructures *UFN***176:2** 203–212
[3] ZubkoS P 2010 A model of the size effect in a ferroelectric thin film under conditions of the phase Transition *Physics of the Solid State* **52:12** 2565-2569
[4] PirozerskiĭA L, Charnaya E V 2010 Ising model for a ferroelectric phase transition in a system of interacting small particles *Physics of the Solid State***52:3**620-624
[5] Vendik O G., Zubko S P, Medvedeva N Yu 2009 "Dead layer" characteristics based on a correlation of the ferroelectric polarization under relevant boundary conditions in a parallel plate capacitor*J. Appl. Phis.***105: 5** 053515(1-4)
[6] Sidorkin A S, Darinskii B M and SigovA S 1997 Evolution of domain structure in ferroelectrics in the presence of polarization screening by charges bound to surface states and by free charge carriers *Physics of the Solid State 39:5*823-825
[7] Davitadze S T, Strukov B A, Vysotskiĭ D V, Lemanov V V, Shul'man S G,Uesu Yand Asanuma S2008 *Physics of the Solid State***50:12** 2303-2306
[8] VendikOG, ZubkoSP 2004 *Crystallography reports* **49:6** 1046-1052
[9] Onsager L 1994 *Phys. Rev.***65**117
[10] Junquera J and. Ghosez P 2003 Critical thickness for ferroelectricity in perovskite ultrathin films *Nature* **422:3** 506-509
[11] Paul Jaita, Nishimatsu Takeshi, Kawazoe Y and Waghmare Umesh V2007 Ferroelectric phase transitions in ultrathin films of BaTiO3 *Physical Review Letters* **09**

Numerical simulation in nonlinear dynamic systems with retiming of motions of individual components

D Apushkinskaya[a], E Apushkinskiy[b],
[a] Department of Mathematics, Saarland University, P.O. Box 151150, 66041 Saarbrücken, *Germany*
[b] Peter the Great St.Petersburg Polytechnic University, *195251 St. Petersburg, Russia*

Email: *apushkinsky@hotmail.com*

Abstract. We study differential equations describing nonlinear processes in dynamical systems (macro-systems) that consist of a large number of components (micro-systems) and allow a synchronization in behavior of these components. In physics, these processes underlie the echo phenomena. An empirical solution of such equations is proposed. A solution is given in the form of power series of power series of the spectra of external of external perturbations acting on a macro-system. Numerical simulations of this solution give a good agreement with a number of experimentally observed echo phenomena.

1. Introduction

We investigate the behavior of dynamical systems (macro-systems) that contain a generous number of components (micro-systems) with nonlinear properties. In such systems, the echo phenomena can be observed, i.e., the appearance some response after a finite external influence on the time-axis.

According to Gould [1], the formation of an echo phenomenon is provided by the presence in micro-systems the nonlinearities of three types.

Nonlinearity of type I occurs only during the external influence to all micro-systems. As a result, we get a phase shift of the oscillations in every micro-system. The latter is ensured in view of the fact that the micro-systems prevent the external influence [1]. The nature of the nonlinearity of type I is different for various kind of micro-systems. However, all the cases with this nonlinearity can be described by the following equation

$$\frac{\partial^2 U_l}{\partial t^2} + 2\Gamma(\chi_{\{s>0\}} s(t))\frac{\partial U_l}{\partial t} + \omega_l^2(\chi_{\{s>0\}} s(t))U_l = \chi_{\{s>0\}} \sum_{n=0}^{N} \mu_n U_l^n(t). \qquad (1)$$

Here U_l *is an unknown function characterizing the state of the* l^{th} *micro-system,* Γ *is a given function determined by the properties of the system,* s *is an external signal acting to the system,* ω_l *denotes the oscillation frequency of the* l^{th} *micro-system,* μ_n *are given constants, and* $\chi_{\{s>0\}}$ *denotes the characteristic function of the set* $\{t : s(t) > 0\}$ *i.e.* $\chi_{\{s>0\}} = 1$ *if* $s(t) > 0$ *and* $\chi_{\{s>0\}} = 0$ *otherwise. Observe that the well-known Bloch equations describing the spin echo can be transformed into the form (1).*

Nonlinearity of type II implies to the appearance of a nonlinear dispersion. This nonlinearity manifests in the absence of an external influence, and its nature is related only to the form of a micro-system. This type of nonlinearity occurs, for example, in a phonon echo from grains of powder [2-4] and in echo in plasma [5]. Nonlinearity of type III is related to the nonlinear damping of oscillations of a micro-system, i.e., to the dependence of the damping coefficient on the amplitude of oscillations. Such a nonlinearity arises, for instance, in plasma [6-8]. We note that for some macro-systems two or even all three types of nonlinearity may occur simultaneously [9].

Observe that the cases with nonlinearities of types II and III can be described with the help of the following equation:

$$\frac{\partial^2 U_l}{\partial t^2} - \frac{\partial^2 U_l}{\partial x^2} + 2\Gamma(U_l)\frac{\partial U_l}{\partial t} = \chi_{\{s>0\}} s(t,x) + \sum_{n=0}^{N} \mu_n U_l^n(t,x). \tag{2}$$

The functions and constants involving in (2) are understood analogously to those from (1).

The behavior of nonlinear systems exposed to external influences demonstrates a variety of motions. Among the most common motions are periodic oscillations and solitary waves. The type of the prevailing motion is determined by the properties of the system and by the type and intensity of an external action. It is known that a response of a nonlinear system to a uniform (with respect to the space) external influence can be obtained by approximate analytical methods in the form of a spectral representation of an external action [10]. The expansion of this response in powers of the spectra of external action allows to identify some rules applied to formation of echo signals [11-12].

2. Experiment / Calculation / Theory

Consider a macro-system consisting of the micro-systems which are described either by Eq. (1) or by Eq. (2), and subject all the micro-systems to the same external action $s(t)$ *given by the sequence of* M *real non-overlapping pulses, i.e.,*

$$s(t) = \sum_{m=1}^{M} s_m(t - \tau_m).$$

Here τ_m *stands for the beginning of the* m *-th pulse. Let us redefine the function* $s(t)$ *along negative time-axis by the even reflection. It is easy to see that the spectrum of* $s(t)$ *can be written in the following form:*

$$\hat{S}(\omega) = \sum_{m=1}^{M} [\hat{s}_m(\omega)e^{-i\omega\tau_m} + \hat{s}_m^*(\omega)e^{i\omega\tau_m}], \tag{3}$$

where i *is the imaginary unit,* $\hat{s}_m(\omega) = \int_0^\infty s_m(t)e^{-i\omega t}dt$ *, and* $\hat{s}_m^*(\omega)$ *is the complex conjugation of* $\hat{s}_m(\omega)$.

Further, denoting the spectrum of $U_l(t)$ *by* $\hat{U}_l(\omega)$ *and applying the same arguments as in [11-12] one can show that*

$$\hat{U}_l(\omega) = \sum_{n=0}^{N} \{a_n[\hat{S}(\omega)]^n\} \tag{4}$$

where a_n *are factors determining the level of the corresponding components. Thus,* $U_l(t)$ *follows from (4) with a use of the inverse Fourier transform* F^{-1}.

It is obvious that there is no echo if we set $N \le 2$ *in formula (4). Consider, for simplicity, two cases* $N = 3, M = 1$ *and* $N = 3, M = 2$. *For* $M = 1$, *we have one external impulse* $s_1(t)$ *starting at the moment* $\tau_1 = 0$ *and ending up at the moment* $t = \Delta t_1$. *Substituting representation (3) in (4), applying the inverse Fourier transform, and taking into account the moments of the phases reversal, we conclude that the echo signal arising after* Δt_1 *is expressed only by the term*

$$F^{-1}[a_3(\hat{s}_1(\omega))^3]. \tag{5}$$

For $M = 2$, *there are two external pulses,* $s_1(t)$ *starting at the moment* $\tau_1 = 0$ *and* $s_2(t)$ *starting at the moment* τ_2. *Repeating the above arguments, we deduce that the echo signal arising during the doubled time interval between the first and second pulses is expressed by the term*

$$F^{-1}[3a_3\hat{s}_1^*(\omega)(\hat{s}_2(\omega))^2 e^{-2i\omega\tau_2}]. \tag{6}$$

For the detailed justification of formulas (5) and (6) we refer the reader to [11-12]. Observe also that the cases $N \ge 4$ *and* $M \ge 3$ *are treated in the similar manner.*

Using the above described algorithms, we perform the numeric simulations for long and complex external signals, respectively. In the first case, we simulate a single-pulse spin echo observed in thin ferromagnetic cobalt films from a pulse with harmonic filling of 10 μs duration. In the second case, we model a two-pulses echo observed in the powder high temperature superconductor (HTSC). It is easy to see that the first case corresponds to Eq. (1), while the second one relates to Eq. (2). The corresponding results are shown in figures 1 and 2. The experiments and simulations are in good agreement.

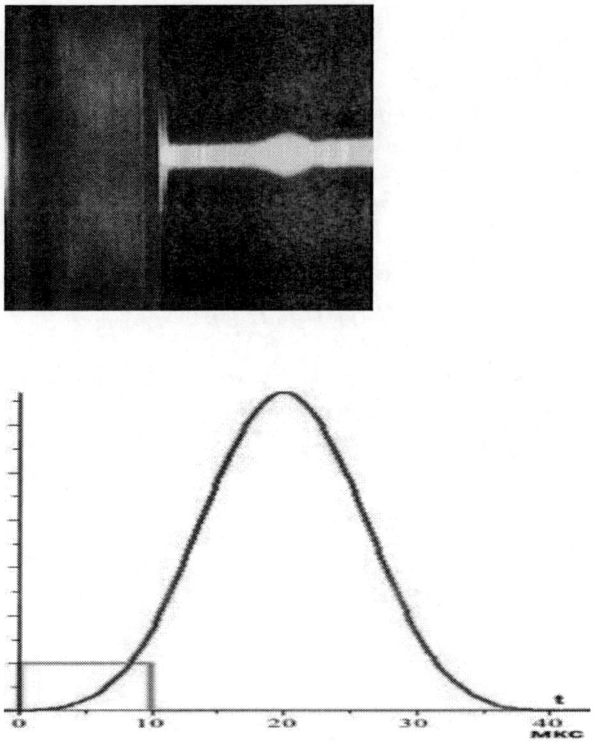

Figure 1. *Experimentally observed the single-pulse spin echo in thin ferromagnetic cobalt films from a pulse with harmonic filling of 10 μs duration.and its simulation on Maple 14.*

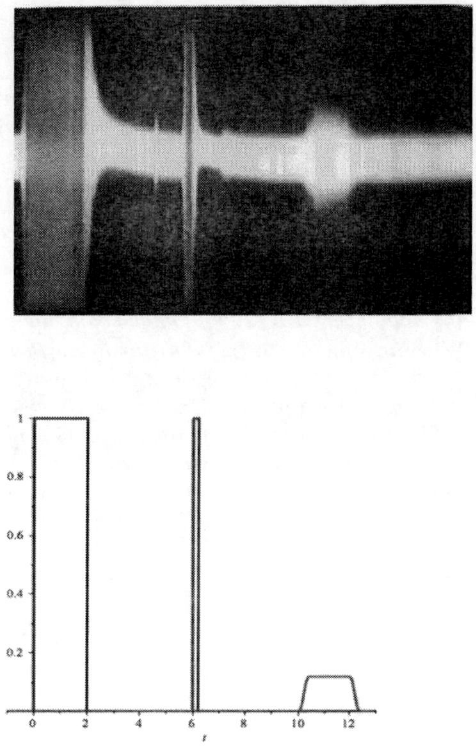

Figure 2. *Experimentally observed the two-pulses echo in the powder high temperature superconductor (HTSC) and its simulation on Maple 14. The first pulse has the duration of 2 μs and noise coverage, the second pulse has the duration of 0.2 μs and harmonic filling. The delay between pulses is equal to 6 μs.*

3. Discussion

Comparison of experimental and numerical results allows us to conclude that the proposed algorithms for representing the solutions of equations (1) and (2) provide the correct results for distinct types of external signals and echoes of a different nature. It would be of special interest to generalize the suggested method to spatially multidimensional cases as well as to provide the corresponding simulations and compare them with experimental data.

As in the case of simple external actions considered in [12], one can conclude that in the case of complex external influences with equal carrier frequencies, the minimum value of parameter N in formula (4) providing the existing of an echo signal is $N = 3$. It should be also noted that considered nonlinear phenomena are manifested only at the level of a macro-system, while nonlinearities take places in the corresponding micro-systems.

4. Conclusions

The proposed algorithms are especially attractive for analysis of echo-signals upon application of complicated and noise external actions since the obtaining of approximate solutions of nonlinear equations is extremely difficult.

Acknowledgements

The first author is partly supported by the Russian Foundation of Basic Research (RFBR) through the grant 15-01-07650.

References

[1] *Gould R W (1969) Cyclotron echo phenomena. Amer. J. Phys. 37 (6) 585-597*

[2] *Goldin B A, Kotov L N, Zarembo L K, and Karpachev S N (1991) Spin-Phonon Interactions of Crystal-Ferrites. Moscow, Nauka (in Russian)*

[3] *Kajimura K (1982) Dynamic polarization echoes in powdered materials. Phys. Acoustics XVI (4) 295-340*

[4] *Pleshakov I V, Nechitailov A A, Paugurt A P, Matveev V V, Fofanov Ya A, Ylinen E (2007) Phonon echo probing of high-temperature superconductors containing magnetic ions. Tech. Phys. Letters 33 (12) 1050-1053*

[5] *Pavlenko V N, Sitenko A G (1988) Echo Phenomena in Plasma and Plasmalike Media. Moscow, Nauka (in Russian)*

[6] *Kadomtsev B B (1968) Landau damping and echo in a plasma. Sov. Phys. Usp. 11 (3) 328*

[7] *Klimontovich Yu L (1994) Nonlinear Brownian motion. Phys.-Usp. 37 737*

[8] *Bassi G, Mais H, Bazzani A, Turcketti G (2002) Time evolution of the particle distribution in beam physics: the echo effect. Proc. of EPAC*

[9] *Richards P M, Christensen C R, Guenther B D, Daniel A C (1971) Nuclear magnetic resonance echo enhancement in an antiferromagnet. Phys. Rev. B 4 (7) 2216 -2224*

[10] *Korpel A, Chatterjee M (1981) Nonlinear echoes, phase conjugation, time reversal, and electronic holography. Proc. IEEE 69 (12) 1539-1556*

[11] *Apushkinskiy E (2012) Nonlinear transformation of signal spectra. Nauch. Techn. Vedomosti SPbSPU 153 182-190 (in Russian)*

[12] *Kosarev A, Apushkinskiy E, Maximov K, Ripchinskaya E, Shamray P (2015) Investigation of nonlinear transformations of the signal spectrum. J. Phys.: Conference Series 574 012025.*

IOP Conf. Series: Journal of Physics: Conf. Series **936** (2017) 012044 doi:10.1088/1742-6596/936/1/012044

Dual Energy Tomosynthesis breast phantom imaging

V Koukou[1], N Martini[1], G Fountos[2], G Messaris[3], C Michail[2], I Kandarakis[2] and G Nikiforidis[1]

[1] Department of Medical Physics, Faculty of Medicine, University of Patras, 265 00 Patras, Greece
[2] Radiation Physics, Materials Technology and Biomedical Imaging Laboratory, Department of Biomedical Engineering, Technological Educational Institute of Athens, Egaleo, 122 10 Athens, Greece
[3] Department of Medical Physics, University Hospital of Patras, 265 04 Rion, Greece

Email: gfoun@teiath.gr

Abstract. Dual energy (DE) imaging technique has been applied to many theoretical and experimental studies. The aim of the current study is to evaluate dual energy in breast tomosynthesis using commercial tomosynthesis system in terms of its potential to better visualize microcalcifications (μCs). The system uses a tungsten target X-ray tube and a selenium direct conversion detector. Low-energy (LE) images were acquired at different tube voltages (28, 30, 32 kV), while high-energy images at 49 kV. Fifteen projections, for the low- and high-energy respectively, were acquired without grid while tube scanned continuously. Log-subtraction algorithm was used in order to obtain the DE images with the weighting factor, w, derived empirically. The subtraction was applied to each pair of LE and HE slices after reconstruction. The TORMAM phantom was imaged with the different settings. Four regions-of-interest including μCs were identified in the inhomogeneous part of the phantom. The μCs in DE images were more clearly visible compared to the low-energy images. Initial results showed that DE tomosynthesis imaging is a promising modality, however more work is required.

1. Introduction

X-ray mammography is currently the best method to search for alterations that might indicate cancer [1]. Mammography has been demonstrated to reduce breast cancer mortality [2], however it has low positive predictive value and sensitivity [3, 4] due to anatomical noise resulting from superimposition of normal structures. Magnetic resonance imaging (MRI) may be used for screening purposes in patients with dense breasts. However, the long examination time along with the expensive equipment, makes MRI screening extremely costly compared with screening mammography and digital breast tomosynthesis (DBT) [5].

DBT is of great interest in screening and diagnostic breast imaging. The ability to provide "slices" through the breast and reduce the noise of the tissue structures is the major advantage of tomosynthesis. Due to this benefit, DBT has been shown to improve detection of abnormalities in women with dense breast tissue, reduce recall rates and improve diagnosis of benign lesions (reduced number of negative biopsies) [6].

Dual energy (DE) imaging technique can improve the conspicuity of a suspicious tissue. DE imaging suppresses the signal of one tissue material resulting in removal of anatomic background noise [7,8]. Thus, the contrast of the feature of interest is enhanced [1].

Microcalcifications (μCs) are calcium deposits in the soft tissue of the breast that can indicate the presence of cancer and appear as a spot or as group, called "clusters of microcalcifications" [1]. The detection of clustered μCs is an important component of mammographic analysis. Since tomosynthesis presents "slices" of the breast, the observer may not perceive the microcalcifications that are not encompassed in a single slice [9].

The purpose of this study was to investigate the potential to better visualize microcalcifications using dual energy in digital tomosynthesis. A commercial tomosynthesis system with tungsten target X-ray tube and selenium direct conversion detector was used. Log-subtraction algorithm was applied to each pair of low-energy (LE) and high-energy (HE) slices after reconstruction. Four regions-of-interest (ROIs) including μCs were identified in the inhomogeneous part of the TORMAM phantom. The μCs in DE images were more clearly visible compared to the LE images.

2. Materials and methods

This work is based on the images obtained with the 3D Hologic Selenia Dimensions system (Hologic Inc., Bedford, USA). The system has a tungsten anode X-ray tube and three filter options in the commercial release. The X-ray generator has a maximum tube current of 200 mA and a voltage range from 20 up to 49 kV. It has various automatic exposure control (AEC) modes. For the acquisition of the DBT images, the high transmission cellular anti-scatter grid was automatically removed out of the X-ray field. The screening mode was selected and 15 projections acquired over 15 degrees (-7.5° - +7.5°) scan angle in 3.7 seconds. The tube was scanned continuously, allowing fast scan time. Reconstruction is performed using a filtered back projection algorithm (FBP) obtaining contiguous planes with thickness of 1 mm. DBT images are acquired using the 0.7 mm thick aluminum (Al) filter. For the LE images three kVs were used: 28, 30 and 32 kV. Also, the auto-time AEC mode was used, at which the user selects the filter and the tube voltage, while the system selects the mAs. The HE kV was set to 49 kV, since it was the highest that could be used, but further optimization of the HE kV is to be done in follow-up study. For the HE, the manual AEC mode was used, at which the user selects all the parameters (kV, filter, mAs). The dose allocation between the LE and HE exposures was selected to be 0.5 in this study. For the calculation of the corresponding mAs that should be used for the HE exposure, in order to result in equal entrance dose to that of LE exposure, the Barracuda dosimeter (RTI Electronics) coupled to an ionization chamber (Magna A600) was positioned at the surface of the TORMAM phantom and the entrance dose was measured for each LE exposure.

Images of the TORMAM phantom (Leeds Test Object Ltd, Boroughbridge, UK) were obtained. The middle right of the phantom test plate simulates the appearance of breast tissue; it contains calcification clusters with sizes of 354-224, 283-180, 226-150, 177-106, 141-90, 106-93 μm, in addition to fibrous material and nodules. The exact location of the features in the structured background is unknown. The middle left with uniform background contains 18 low contrast objects, filaments and the calcification clusters as shown above. The test plate (1.6 cm) was positioned on the center of the polymethyl methacrylate (PMMA) slabs of 4 cm thickness, resulting in a total thickness of 5.6 cm.

The DE images were obtained by weighted log pixel-by-pixel subtraction method [8]. The LE and HE projection images were firstly reconstructed into LE and HE volume slices separately, and then weighted logarithmic subtraction was performed with each pair of LE and HE images of the same height, as defined in Eq. 1:

$$\mathrm{DE} = \ln(\mathrm{HE}) - w\ln(\mathrm{LE}) \qquad (1)$$

where HE and LE refer to pixel intensity of high- and low-energy reconstructed images, respectively and w is the weighting factor.

Four ROIs containing calcification clusters were identified in LE and HE reconstructed images of the realistic breast tissue part of the phantom. For each ROI of LE and HE images, a number of DE images were generated for all weighting factors in the range of 0 to 1, at 0.01 intervals. The standard deviation (σ) in the DE-DBT images was computed for two background regions (without calcifications present) using a custom-developed algorithm. The optimal locally adjusted w [10] was defined as that which minimizes σ.

3. Results and discussion

Figures 1, 2 and 3 illustrate the inhomogeneous part of the TORMAM phantom acquired at 28, 30 and 32 kV, respectively. Four ROIs with size of 50x50 pixels were selected in each LE image, containing μCs. The corresponding ROIs of the DE images, as defined in LE images, are also shown at the right side of the figures.

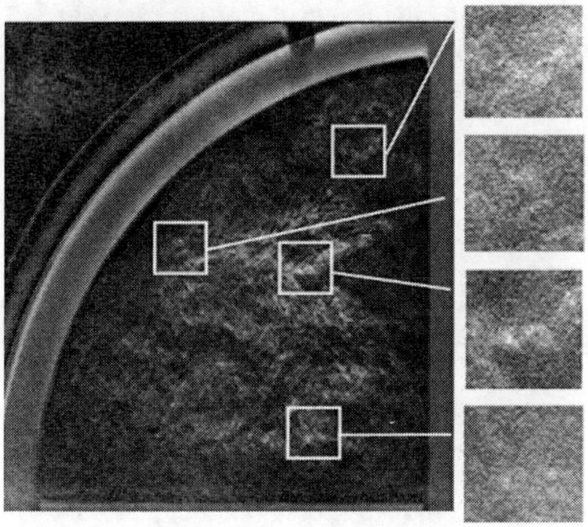

Figure 1. LE image of TORMAM phantom at 28 kV and the corresponding ROIs of the DE image.

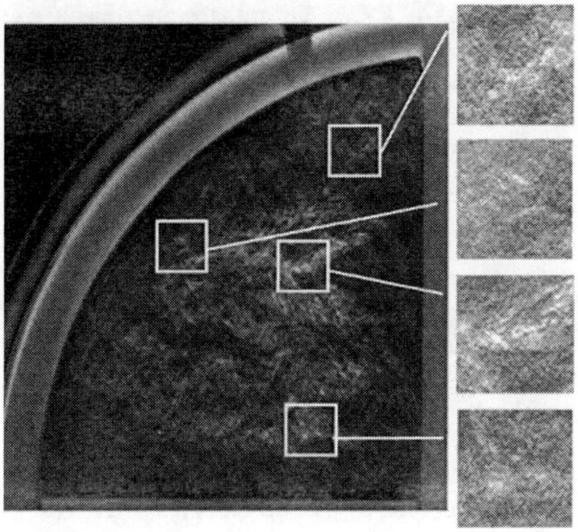

Figure 2. LE image of TORMAM phantom at 30 kV and the corresponding ROIs of the DE image.

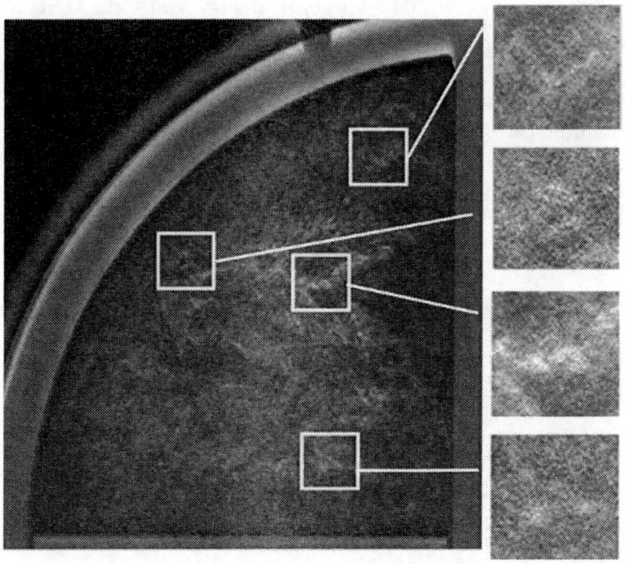

Figure 3. LE image of TORMAM phantom at 32 kV and the corresponding ROIs of the DE image.

The tomosynthesis slice was selected at the height of the test plate location. While the μCs were obscured by the surrounding tissue in LE images, in DE subtraction images the background tissue structures were well supressed. Thus, μCs are highlighted and are more clearly visible in the DE images.

The w factor was locally adjusted in this study and ranged from 0.36 to 0.53 for all examined ROIs. It has been shown that w factor varies with the breast tissue composition [11]. If a single value is selected, then the suppression of the background structure may be lower. Furthermore, w is affected by the beam quality and the average beam energy [11].

4. Conclusions

The aim of the present work was to evaluate the potential of dual energy in breast tomosynthesis in order to better visualize μCs. The LE images were acquired at different tube voltages, while HE images at 49 kV. In DE images the four μCs clusters were better defined than in the corresponding LE images. The initial results showed that DE tomosynthesis imaging could be a promising modality, however more work is required.

Acknowledgement

This work was supported by Grant E.040 from the Research Committee of the University of Patras (Programme K. Karatheodori).

5. References

[1] Gomi T and Nakajima M 2012 *Open J Med Imaging* **2012** 2
[2] Timberg P, Båth M, Andersson I, Mattsson S, Tingberg A and Ruschin M 2010 *Med. Phys.* **37** 11
[3] Carney PA, Miglioretti DL, Yankaskas BC, Kerlikowske K, Rosenberg R, Rutter CM, Geller BM, Abraham LA, Taplin SH, Dignan M, Cutter G and Ballard-Barbash R 2003 *Ann. Intern. Med.* **138** 3
[4] Poplack SP, Tosteson AT, Grove MR, Wells WA and Carney PA 2000 *Radiology* **217** 3
[5] Destounis SV, Morgan R and Arieno A 2015 *Am. J. Roentgenol.* **204** 2
[6] Good WF, Abrams GS, Catullo VJ, Chough DM, Ganott MA, Hakim CM and Gur D 2008 *Am. J. Roentgenol.* **190** 4
[7] Koukou V, Martini N, Michail C, Sotiropoulou P, Fountzoula C, Kalyvas N, Kandarakis I, Nikiforidis G and Fountos G 2015 *Comput. Math. Methods Med.* **574238**
[8] Koukou V, Martini N, Fountos G, Michail C, Sotiropoulou P, Bakas A, Kalyvas N, Kandarakis I, Speller R and Nikiforidis G 2017 *Nucl. Inst. Meth. Phys. Res. A* **848**
[9] Kopans D, Gavenonis S, Halpern E and Moore R 2011 *Breast J* **17** 6
[10] Ducote JL, Xu T and Molloi S 2006 *Med. Phys.* **33** 6
[11] Carton A, Ullberg C, Lindman K, Acciavatti R, Francke T and Maidment A 2010 *Med. Phys.* **37** 11

IOP Conf. Series: Journal of Physics: Conf. Series **936** (2017) 012045 doi:10.1088/1742-6596/936/1/012045

Darboux transformation and soliton solution for the (2+1)-dimensional complex modified Korteweg-de Vries equations

K Yesmakhanova[1], **G Shaikhova**[2], **G Bekova**[2,3], **R Myrzakulov**[2]

[1] Eurasian International Center for Theoretical Physics and Department of Higher Mathematics, Eurasian National University, Astana, 010000, Kazakhstan

[2] Eurasian International Center for Theoretical Physics and Department of General Theoretical Physics, Eurasian National University, Astana, 010000, Kazakhstan

[3] Kh. Dosmukhamedov Atyrau state university, Atyrau, Kazakhstan, 060000, Kazakhstan

E-mail: `kryesmakhanova@gmail.com`

Abstract. In this paper, we consider the (2+1)-dimensional complex modified Korteweg-de Vries equations (CMKdVE). Using the Lax pair, we construct a Darboux transformation and namely one-fold and n-fold transformations. The soliton solutions are obtained from the different "seeds" by using this Darboux transformation.

1. Introduction

The more typical and well-studied integrable evolution equations which describe nonlinear wave phenomena for a range of dispersive physical systems are the Korteweg-de Vries (KdV) equation, modified Korteweg-de Vries (mKdV) equation and the nonlinear Schrodinger (NLS) equation. A wide family of Kortweg-de Vries (KdV) equations provide for the description of the nonlinear wave propagation in hydrodynamics, plasma, transmission line theory, gas-dynamics, and in charged-particle beam dynamics as well [1-6]. Because of the real physical space-time being in higher-dimensional and more complex, research integrable equations in higher dimensions is still open and attractive [7-14].

The objective of this paper is considering the (2+1)-dimensional CMKdVE [15]. Here we use the method of Darboux transformation (DT) which has been proved to be an efficient way to find the different solutions from "seed" solutions. This transformation allows constructing non-trivial analytical solutions of nonlinear partial differential equations [16-23].

The paper is organized as follows. The Lax representation for the (2+1)-dimensional CMKdVE we consider in Section 2. In Section 3, we construct the DT for the (2+1)-dimensional CMKdVE. In Section 4, using the constructed one-fold DT, the one-soliton solutions of the (2+1)-dimensional CMKdVE are given.

2. Lax representation of the (2+1)-dimensional complex modified Korteweg de Vries equations

In this section, we concentrate on the linear eigenvalue problem of the (2+1)-dimensional complex modified Korteweg-de Vries equations [15]. The linear eigenvalue problem is expressed in the form of the Lax pair A and B as

$$\Psi_x = A\Psi, \tag{1}$$
$$\Psi_t = 4\lambda^2\Psi_y + B\Psi, \tag{2}$$

where

$$A = -\lambda J_3 + A_0,$$
$$B = \lambda B_1 + B_0,$$

with

$$J = \begin{pmatrix} i & 0 \\ 0 & -i \end{pmatrix},$$
$$B_1 = wJ + 2JA_{0y},$$
$$A_0 = \begin{pmatrix} 0 & q \\ -r & 0 \end{pmatrix},$$
$$B_0 = -\frac{v}{2}J + \begin{pmatrix} 0 & -q_{xy} - wq \\ r_{xy} + wr & 0 \end{pmatrix},$$
$$\Psi = \begin{pmatrix} \psi_1(\lambda, x, y, t) \\ \psi_2(\lambda, x, y, t) \end{pmatrix}$$

and $r = \delta q^*$, where $\delta = \pm 1$. The compatibility condition of equations (1)-(2) is still

$$A_t - 4\lambda^2 A_y - B_x + [A, B] = 0. \tag{3}$$

By direct substituting expression (1)-(2) in the equation (3) we can get the (2+1)-dimensional complex modified Korteweg-de Vries equations [15]

$$iq_t + iq_{xxy} - vq + i(wq)_x = 0, \tag{4}$$
$$v_x - 2i\delta(q^*_{xy}q - q^* q_{xy}) = 0, \tag{5}$$
$$w_x - 2\delta(|q|^2)_y = 0, \tag{6}$$

where q is complex function, v, w are real functions and δ is real constant.

3. Darboux transformation for the (2+1)-dimensional CMKdVE

In this section, we construct the Darboux transformation (DT) for the (2+1)-dimensional CMKdVE (4)-(6).

3.1. First iteration

Let Ψ and $\Psi^{[1]}$ are two solutions of the system (1)-(2) so that

$$\Psi^{[1]}_x = A^{[1]}\Psi^{[1]} = (\Lambda J + A^{[1]}_0)\Psi^{[1]}, \tag{7}$$
$$\Psi^{[1]}_t = 4\lambda^2\Psi^{[1]}_y + B^{[1]}\Psi^{[1]} = 4\Lambda^2\Psi^{[1]}_y + (\Lambda B^{[1]}_1 + B^{[1]}_0)\Psi^{[1]}. \tag{8}$$

We assume that these two solutions are related by the following transformation about linear function Ψ in the form:

$$\Psi^{[1]} = T\Psi = \Psi\Lambda - M\Psi, \tag{9}$$

where

$$M = \begin{pmatrix} m_{11} & m_{12} \\ m_{21} & m_{22} \end{pmatrix}.$$

The matrix function T obeys the following equations

$$A^{[1]} = (T_x + TA)T^{-1}, \tag{10}$$
$$B^{[1]} = (T_t + TB - 4\lambda^2 T_y)T^{-1}. \tag{11}$$

Then the relation between q, v, w and $q^{[1]}, v^{[1]}, w^{[1]}$ can be reduced from (10)-(11), which is the Darboux transformation of CMKdVE. Comparing the coefficient of λ^i $(i = 0, 1, 2)$ of the sides the equation (10)-(11), we have

$$A_0^{[1]} = A_0 + [M, J], \tag{12}$$
$$B_0^{[1]} = B_0 - MB_1 + (B_1 + 4M_y)M, \tag{13}$$
$$B_1^{[1]} = B_1 + 4M_y \tag{14}$$
$$M_x = A_0^{[1]}M - MA_0, \tag{15}$$
$$M_t = B_0^{[1]}M - MB_0. \tag{16}$$

The equation (12) gives

$$q^{[1]} = q - 2im_{12}, \tag{17}$$
$$q^{*[1]} = q^* - 2im_{21}. \tag{18}$$

Hence we get $m_{21} = -m_{12}^*$ in our attractive interaction case that is if $\delta = +1$. At the same, from the equation (13)-(14) we get

$$v^{[1]} = v + 4(m_{12}q_y^* + m_{12}^* q_y + 2im_{11}m_{11y} - 2im_{12}^* m_{12y}), \tag{19}$$
$$w^{[1]} = w - 4im_{11y} = w + 4im_{22y} \tag{20}$$

and we additionally have $m_{22} = m_{11}^*$. So the matix M has the form

$$M = \begin{pmatrix} m_{11} & m_{12} \\ -m_{12}^* & m_{11}^* \end{pmatrix}. \tag{21}$$

We now assume that

$$M = H_1 \Lambda_1 H_1^{-1}, \tag{22}$$

where

$$H_1 = \begin{pmatrix} \psi_1(\lambda_1; t, x, y) & \psi_1(\lambda_2; t, x, y) \\ \psi_2(\lambda_1; t, x, y) & \psi_2(\lambda_2; t, x, y) \end{pmatrix} := \begin{pmatrix} \psi_{1,1} & \psi_{1,2} \\ \psi_{2,1} & \psi_{2,2} \end{pmatrix}.$$

Here

$$\Lambda_1 = \begin{pmatrix} \lambda_1 & 0 \\ 0 & \lambda_2 \end{pmatrix} \tag{23}$$

and $det\ H_1 \neq 0$, where λ_1 and λ_2 are complex constants. The matrix H obeys the system

$$H_{1x} = -i\sigma_3 H_1 \Lambda_1 + A_0 H_1, \tag{24}$$
$$H_{1t} = 4H_{1y}^2 + B_1 H_1 \Lambda_1 + B_0 H_1. \tag{25}$$

In order to satisfy the constraints as mentioned above, we first notes that

$$\Psi^+ = \Psi^{-1}, \quad A_0^+ = -A_0, \tag{26}$$

$$\lambda_2 = \lambda_1^*, \quad H_1 = \begin{pmatrix} \psi_1(\lambda_1; t, x, y) & -\psi_2^*(\lambda_1; t, x, y) \\ \psi_2(\lambda_1; t, x, y) & \psi_1^*(\lambda_1; t, x, y) \end{pmatrix}, \tag{27}$$

$$H_1^{-1} = \frac{1}{\Delta} \begin{pmatrix} \psi_1^*(\lambda_1; t, x, y) & \psi_2^*(\lambda_1; t, x, y) \\ -\psi_2(\lambda_1; t, x, y) & \psi_1(\lambda_1; t, x, y) \end{pmatrix}, \tag{28}$$

where

$$\Delta = |\psi_1|^2 + |\psi_2|^2.$$

The first iterated DT is given by $\Psi^{[1]} = T\Psi = \Psi\Lambda - M\Psi$. If H_1 is the solution of $\Psi^{[1]}$ at $\Lambda = \Lambda_1$ then it should satisfy

$$\Psi_1^{[1]} = TH_1 = 0. \tag{29}$$

In other words

$$H_1\Lambda_1 - MH_1 = 0 \Rightarrow MH_1 = H_1\Lambda_1. \tag{30}$$

Expressing Eq. (30) in matrix form, we have

$$\begin{pmatrix} m_{11} & m_{12} \\ -m_{12}^* & m_{11}^* \end{pmatrix} \times \begin{pmatrix} \psi_1 & -\psi_2^* \\ \psi_2 & \psi_1^* \end{pmatrix} = \begin{pmatrix} \lambda_1\psi_1 & -\lambda_1^*\psi_2^* \\ \lambda_1\psi_2 & \lambda_1^*\psi_1^* \end{pmatrix}. \tag{31}$$

Using Cramer's rule we can determine the exact expression of m_{11} and m_{12} which in turn reads

$$m_{11} = \frac{\begin{vmatrix} \psi_1^* & -\psi_2 \\ \lambda_1^*\psi_2^* & \lambda_1\psi_1^* \end{vmatrix}}{\begin{vmatrix} \psi_1 & -\psi_2^* \\ \psi_2 & \psi_1^* \end{vmatrix}}, \quad m_{12} = \frac{\begin{vmatrix} \psi_1 & -\psi_2^* \\ \lambda_1\psi_1 & -\lambda_1^*\psi_2^* \end{vmatrix}}{\begin{vmatrix} \psi_1 & -\psi_2^* \\ \psi_2 & \psi_1^* \end{vmatrix}}. \tag{32}$$

Evaluating the above determinants, we find

$$m_{11} = \frac{\lambda_1|\psi_1|^2 + \lambda_1^*|\psi_2|^2}{\Delta}, \quad m_{12} = \frac{(\lambda_1 - \lambda_1^*)\psi_1\psi_2^*}{\Delta}. \tag{33}$$

So, the matrix M has the form

$$M = \frac{1}{\Delta} \begin{pmatrix} \lambda_1|\psi_1|^2 + \lambda_1^*|\psi_2|^2 & (\lambda_1 - \lambda_1^*)\psi_1\psi_2^* \\ (\lambda_1 - \lambda_1^*)\psi_1^*\psi_2 & \lambda_1|\psi_2|^2 + \lambda_1^*|\psi_1|^2) \end{pmatrix}. \tag{34}$$

From (33) it is evident that to determine m_{11} and m_{12} one should know the explicit expressions of ψ_1 and ψ_2 which are the solutions of the eigenvalue problem (1)-(2). They can be determined by solving the following six coupled linear first order partial differential equations, namely

$$\psi_{1x} = -i\lambda\psi_1 + q\psi_2, \tag{35}$$
$$\psi_{2x} = -r\psi_1 + i\lambda\psi_2, \tag{36}$$
$$\psi_{1t} = 4\lambda^2\psi_{1y} + \left(i\lambda w - \frac{i}{2}v\right)\psi_1 + (2i\lambda q_y - q_{xy} - wq)\psi_2, \tag{37}$$
$$\psi_{2t} = 4\lambda^2\psi_{2y} + (2i\lambda r_y - r_{xy} - wr)\psi_1 - \left(i\lambda w - \frac{i}{2}v\right)\psi_2. \tag{38}$$

Solving the system of equations given in (35)-(38) with appropriate seed solution q, v and w, one can obtain the explicit expressions of ψ_1 and ψ_2. With the known expressions of ψ_1 and ψ_2 the matrix elements m_{11} and m_{12} can now be fixed. Plugging the latter into (17), (19) and (20), we obtain the solution for the Eq. (4)-(6) in the form

$$q^{[1]} = q - 2i\frac{(\lambda_1 - \lambda_1^*)\psi_1\psi_2^*}{|\psi_1|^2 + |\psi_2|^2}, \tag{39}$$
$$v^{[1]} = v + 4(m_{12}q_y^* + m_{12}^*q_y + 2im_{11}m_{11y} - 2im_{12}^*m_{12y}), \tag{40}$$
$$w^{[1]} = w - 4im_{11y} = w + 4im_{22y}. \tag{41}$$

Through the formulas (39)-(41) one can generate a class of solutions including solitary wave solution, breather and rogue wave solution for the Eq. (4)-(6).

3.2. Second iteration

The second iterated DT reads,

$$\Psi^{[2]} = T_2\Psi^{[1]} = \Psi^{[1]}\Lambda - M_2\Psi^{[1]}, \tag{42}$$

where $\Psi^{[2]}$ and $\Psi^{[1]}$ are the second and first iterated eigenfunctions respectively and T_2 is the second iterated DT matrix. The DT (42) transforms the first iterated Lax pair of the same type

$$\Psi_x^{[2]} = A^{[2]}\Psi^{[2]} = (\Lambda J + A_0^{[2]})\Psi^{[2]}, \tag{43}$$
$$\Psi_t^{[2]} = 4\lambda^2\Psi_y^{[2]} + B^{[2]}\Psi^{[2]} = 4\Lambda^2\Psi_y^{[2]} + (\Lambda B_1^{[2]} + B_0^{[2]})\Psi^{[2]}, \tag{44}$$

where the matrices $A_0^{[2]}$, $B_0^{[2]}$ and $B_1^{[2]}$ have the same forms as that of $A_0^{[1]}$, $B_0^{[1]}$ and $B_1^{[1]}$ except that the potentials $q^{[1]}$, $v^{[1]}$ and $w^{[1]}$ have now acquired new expressions $q^{[2]}$, $v^{[2]}$ and $w^{[2]}$ in the matrices $A^{[2]}$ and $B^{[2]}$. Substituting the transformation (42) into (43)-(44) and rearranging the resultant equations, we get

$$A^{[2]} = (T_{2x} + T_2A)T_2^{-1}, \quad B^{[2]} = (T_{2t} + T_2B^{[1]} - 4\lambda^2T_{2y})T_2^{-1}. \tag{45}$$

Substituting the matrix expressions of $A^{[2]}$, $B^{[2]}$, $A^{[1]}$, $B^{[1]}$ and T_2 in Eq. (45) and equating the coefficients of various powers of Λ on both sides we get the following relations between the first iterated and second iterated potentials, namely

$$A_0^{[2]} = A_0^{[1]} + [M_2, J], \tag{46}$$
$$B_0^{[2]} = B_0^{[1]} - M_2B_1^{[1]} + (B_1^{[1]} + 4M_{2y})M_2, \tag{47}$$
$$B_1^{[2]} = B_1^{[1]} + 4M_{2y} \tag{48}$$
$$M_{2x} = A_0^{[2]}M_2 - M_2A_0^{[1]}, \tag{49}$$
$$M_{2t} = B_0^{[2]}M_2 - M_2B_0^{[1]}, \tag{50}$$

The eigenvalue problem given in (43)-(44) remains invariant under the transformation (42), provided M_2 satisfies all the Eqs. (46)-(50).

We consider the matrix M_2 is of the form

$$M_2 = \begin{pmatrix} m_{11}^{[2]} & m_{12}^{[2]} \\ m_{21}^{[2]} & m_{22}^{[2]} \end{pmatrix} \tag{51}$$

so that upon substituting (51) in Eqs. (46)-(48) we can find the following two expressions which link the new and old potentials respectively, that is

$$q^{[2]} = q^{[1]} - 2im_{12}^{[2]}, \tag{52}$$

$$v^{[2]} = v^{[1]} + 4(m_{12}^{[2]}q_y^{[1]*} + m_{12}^{[2]*}q_y^{[1]} + 2im_{11}^{[2]}m_{11y}^{[2]} - 2im_{12}^{[2]*}m_{12y}^{[2]}), \tag{53}$$

$$w^{[2]} = w^{[1]} - 4im_{11y}^{[2]} = w^{[1]} + 4im_{11y}^{[2]} \tag{54}$$

We determine the elements in the matrix M_2 by assuming the latter to be

$$M_2 = H_2 \Lambda_2 H_2^{-1} \tag{55}$$

where

$$H_2 = \begin{pmatrix} \psi_1^{[1]}(\lambda_3; t, x, y) & -\psi_2^{[1]*}(\lambda_3; t, x, y) \\ \psi_2^{[1]}(\lambda_3; t, x, y) & \psi_1^{[1]*}(\lambda_3; t, x, y) \end{pmatrix}, \quad \Lambda_2 = \begin{pmatrix} \lambda_3 & 0 \\ 0 & \lambda_3^* \end{pmatrix} \tag{56}$$

In Eq. (56) $\Psi^{[2]} = (\psi_1^{[1]}, \psi_2^{[1]})^T$ is column solution of the Lax pair (43)-(44) at $\lambda = \lambda_3$. It follows from the orthogonality condition that $(-\psi_2^{[1]*}, \psi_1^{[1]*})^T$ is also the solution of (43)-(44) at $\lambda = \lambda_3^*$. If H_2 is the solution of $\Psi^{[2]}$ at $\Lambda = \Lambda_2$ then it should satisfy

$$\Psi^{[2]} = T_2 H_2 = 0 \tag{57}$$

In other words

$$H_2\Lambda_2 - M_2H_2 = 0 \Rightarrow M_2H_1 = H_2\Lambda_2. \tag{58}$$

The explicit matrix form of Eq. (58) is given by

$$\begin{pmatrix} m_{11}^{[2]} & m_{12}^{[2]} \\ -m_{12}^{[2]*} & m_{11}^{[2]*} \end{pmatrix} \times \begin{pmatrix} \psi_1^{[1]} & -\psi_2^{[1]*} \\ \psi_2^{[1]} & \psi_1^{[1]*} \end{pmatrix} = \begin{pmatrix} \lambda_3\psi_1^{[1]} & -\lambda_3^*\psi_2^{[1]*} \\ \lambda_3\psi_2^{[1]} & \lambda_3^*\psi_1^* \end{pmatrix}. \tag{59}$$

Since we need to know the exact expression of only two matrix elements, namely $m_{11}^{[2]}$ and $m_{12}^{[2]}$, we determine them from the above equation with the help of Cramer's rule. Our result shows that

$$m_{11}^{[2]} = \frac{\lambda_3|\psi_1^{[1]}|^2 + \lambda_3^*|\psi_2^{[1]}|^2}{\Delta^{[1]}}, \quad m_{12}^{[2]} = \frac{(\lambda_3 - \lambda_3^*)\psi_1^{[1]}\psi_2^{[1]*}}{\Delta^{[1]}}. \tag{60}$$

where

$$\Delta^{[1]} = |\psi_1^{[1]}|^2 + |\psi_2^{[1]}|^2$$

From (60) it is evident that to determine $m_{11}^{[2]}$ and $m_{12}^{[2]}$ one should know the explicit expressions of $\psi_1^{[1]}$ and $\psi_2^{[1]}$ which are nothing but the solutions of the Lax pair equations,

$$\psi_{1x}^{[1]} = -i\lambda_3\psi_1^{[1]} + q^{[1]}\psi_2^{[1]}, \tag{61}$$
$$\psi_{2x}^{[1]} = -r^{[1]}\psi_1^{[1]} + i\lambda_3\psi_2^{[1]}, \tag{62}$$
$$\psi_{1t}^{[1]} = 4\lambda_3^2\psi_{1y}^{[1]} + \left(i\lambda_3 w^{[1]} - \frac{i}{2}v^{[1]}\right)\psi_1^{[1]} + (2i\lambda_3 q_y^{[1]} - q_{xy}^{[1]} - w^{[1]}q^{[1]})\psi_2^{[1]}, \tag{63}$$
$$\psi_{2t}^{[1]} = 4\lambda_3^2\psi_{2y}^{[1]} + (2i\lambda_3 r_y^{[1]} - r_{xy}^{[1]} - w^{[1]}r^{[1]})\psi_1^{[1]} - \left(i\lambda_3 w^{[1]} - \frac{i}{2}v^{[1]}\right)\psi_2^{[1]}. \tag{64}$$

Solving Eqs. (61)-(64) consistently with first iterated potentials $q^{[1]}$, $v^{[1]}$ and $w^{[1]}$ we can obtain the exact forms of $\psi_1^{[2]}$ and $\psi_2^{[2]}$. Substituting them in (60) we can get the matrix elements $m_{11}^{[2]}$ and $m_{12}^{[2]}$. Plugging the latter into (52)-(54) we arrive at the second iterated solution for the Eqs. (4)-(6) in the form

$$q^{[2]} = q^{[1]} - 2i\frac{(\lambda_3 - \lambda_3^*)\psi_1^{[1]}\psi_2^{[1]*}}{|\psi_1^{[1]}|^2 + |\psi_2^{[1]}|^2}, \tag{65}$$
$$v^{[2]} = v^{[1]} + 4(m_{12}^{[1]}q_y^{[1]*} + m_{12}^{[1]*}q_y^{[1]} + 2im_{11}^{[1]}m_{11y}^{[1]} - 2im_{12}^{[1]*}m_{12y}^{[1]}), \tag{66}$$
$$w^{[2]} = w^{[1]} - 4im_{11y}^{[1]} = w^{[1]} + 4im_{22y}^{[1]}. \tag{67}$$

Using (65)-(67) one can generate a class of solutions including 2-soliton solutions, second order breathers and so on for the Eqs. (4)-(6).

4. Soliton solutions

Having the explicit form of the DT, we are ready to construct exact solutions of the (2+1)-dimensional CMKdV equations (10)-(12). To get the one-soliton solution we take the seed solution as

$$q = v = w = 0. \tag{68}$$

Then the corresponding associated linear system takes the form

$$\psi_{1x} = -i\lambda\psi_1, \tag{69}$$
$$\psi_{2x} = i\lambda\psi_2, \tag{70}$$
$$\psi_{1t} = 4\lambda^2\psi_{1y}, \tag{71}$$
$$\psi_{2t} = 4\lambda^2\psi_{2y}. \tag{72}$$

The system of equations (69)-(72) admits the following exact solutions

$$\psi_1 = e^{-i\lambda x + i\mu y + i(4\lambda^2\mu)t + \delta_1 + i\delta_2}, \tag{73}$$
$$\psi_2 = e^{i\lambda x - i\mu y - i(4\lambda^2\mu)t - \delta_1 - i\delta_2 + i\delta_0} \tag{74}$$

or

$$\psi_1 = e^{\theta_1 + i\chi_1}, \tag{75}$$
$$\psi_2 = e^{\theta_2 + i\chi_2}. \tag{76}$$

Here $\mu = \eta_1 + i\nu_1$, $\lambda = \alpha_1 + i\beta_1$ and δ_i are real constants then

$$\theta_1 = \beta_1 x - \nu_1 y - (8\eta_1\alpha_1\beta_1 + 4\nu_1(\alpha_1^2 - \beta_1^2)t) + \delta_1, \quad (77)$$

$$\chi_1 = -\alpha_1 x + \eta_1 y + (4\eta_1(\alpha_1^2 - \beta_1^2) - 8\nu_1\alpha_1\beta_1)t + \delta_2 \quad (78)$$

and $\theta_2 = -\theta_1$, $\chi_2 = -\chi_1 + \delta_0$. Then the one-soliton solution of the (2+1)-dimensional CMKdV equations (4)-(6) takes the form

$$q^{[1]} = 2\beta_1 e^{2i\chi_1 - i\delta_0}\,\mathrm{sech}[2\theta_1], \quad (79)$$

$$v^{[1]} = 16\beta_1(\alpha_1\nu_1 + \beta_1\eta_1)\,\mathrm{sech}^2[2\theta_1], \quad (80)$$

$$w^{[1]} = -8\beta_1\nu_1\,\mathrm{sech}^2[2\theta_1]. \quad (81)$$

Dynamics of the soliton solution of the CMKdV equations are presented on Fig.1 and Fig.2

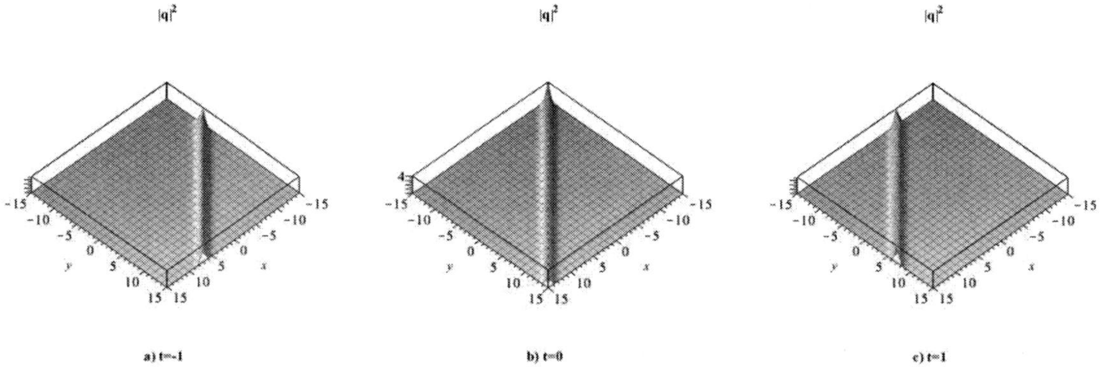

Figure 1: One-soliton solution $q^{[1]}$ of the cmKdV equations

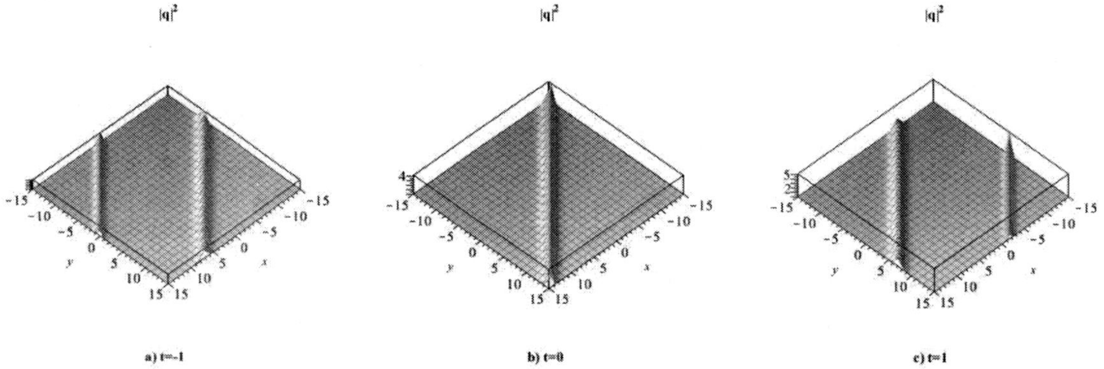

Figure 2: Two-soliton solutions $q^{[2]}$ of the cmKdV equations

5. Conclusion

In this paper, we derived the Darboux transformation of the (2+1)-dimensional complex modified Korteweg-de Vries equations. In particular, the first iteration, second iteration of DT are presented in detail. One soliton solutions of the CMKdV equations have been constructed explicitly by using Darboux transformation from trivial solutions seed solutions. In a recursive manner, one can obtain the multi-soliton and other type exact solutions of the (2+1)-dimensional CMKdV equations. .

Acknowledgement
The work of K.Y., G.Sh., G.B., and M. R. is supported by the Ministry of Education and Science of Republic of Kazakhstan (the base part of the state task No.0888/GF4)

References

[1] Kadomtsev B *Phenomenes Collectifs dansles Plasmas* Mir Moscow 1979
[2] Karpman V I *Nonlinear Waves in Dispersive Media* Oxford: Pergamon Press, 1975
[3] Lonngren K, Scott A *Solitons in Action* San Diego: Academic Press 1978
[4] Schamel H *Phys. Rev. Lett.* **79** , 2811 (1997)
[5] Ablowitz M J and Clarkson P A *Cambridge University Press, Cambridge, NY, 1991*
[6] He J, Wang L, Li L, Porseizian K and Erdely R *arXiv.org:1405.7845v1 [nlin.SI]*
[7] Toda K, Yu S J *J. Math. Phys.* **41** 4747 (2000)
[8] Peng Y Z *Commun. Theor. Phys.* **54** 863865 (2010)
[9] Lu X, Ma W X *Appl. Math. Lett.* **50** 3742 (2015)
[10] Wazwaz A M *Phys. Scr.* **84** 035010 (2011)
[11] Grses M Pekcan A *J. Math. Phys.* **52** 083516 (2011)
[12] Hereman W, Nuseir A *Math. Comput. Simul.* **43** 1327 (1997)
[13] Lou S *Commun. Theor. Phys* **28** , 4150 (1997)
[14] Biswas A *Nonlinear Dyn.* **58** 345348 (2009)
[15] Myrzakulov R, Mamyrbekova G K, Nugmanova G N, Lakshmanan M *Symmetry* **7(3)** (2015) 1352-1375
[16] Li Ch, He J *arXiv:1210.2501*
[17] Yang J, Li Ch, Li T, Cheng Zh *Chinese Physics Letters* **30** (2013) 104201
[18] Li Ch, He J, Porsezian K *Physical Review E* **87** (2013) 012913
[19] Matveev V.B. and Salle M.A. Springer-VerlagBerlin 1991
[20] Li Ch , He J *arXiv:1210.2501*
[21] Yesmakanova K R, Shaikhova G N, Bekova G T, and Myrzakulova Zh R *Advances in Intelligent Systems and Computing* **441** 183-198 (2016)
[22] Yesmakanova K R, Shaikhova G N, Bekova G T *AIP Conference Proceedings* **1759** 020147 (2016)
[23] Li Ch, He J, Porsezian K *Physical Review E* **87** (2013) 012913.

IOP Conf. Series: Journal of Physics: Conf. Series **936** (2017) 012046 doi:10.1088/1742-6596/936/1/012046

Statistical homogeneity tests applied to large data sets from high energy physics experiments

Trusina J, Franc J, and Kůs V

Department of Mathematics, Faculty of Nuclear Sciences and Physical Engineering, Czech Technical University in Prague, Trojanova 13, 120 00 Prague 2, Czech Republic

E-mail: jakub.trusina@fjfi.cvut.cz, jiri.franc@fjfi.cvut.cz, kus@fjfi.cvut.cz

Abstract. Homogeneity tests are used in high energy physics for the verification of simulated Monte Carlo samples, it means if they have the same distribution as a measured data from particle detector. Kolmogorov-Smirnov, χ^2, and Anderson-Darling tests are the most used techniques to assess the samples' homogeneity. Since MC generators produce plenty of entries from different models, each entry has to be re-weighted to obtain the same sample size as the measured data has. One way of the homogeneity testing is through the binning. If we do not want to lose any information, we can apply generalized tests based on weighted empirical distribution functions. In this paper, we propose such generalized weighted homogeneity tests and introduce some of their asymptotic properties. We present the results based on numerical analysis which focuses on estimations of the type-I error and power of the test. Finally, we present application of our homogeneity tests to data from the experiment DØ in Fermilab.

1. Introduction

Homogeneity tests are used when determining if any two samples originate from the same distribution. We use these tests to verify that data measured at certain high energy physics (HEP) experiment are homogeneous with simulated Monte Carlo (MC) sample (i.e. both samples have the same distribution). Rejection of such a test means remodeling of yield sample contribution. Since an amount of entries produced by these generators is significantly higher than the size of data sample, we have to assign an appropriate weight to every entry which simulates not only various trigger efficiencies in the detector. Classical statistical tests should be generalized in such a way that they can test the homogeneity of weighted samples. In this paper, we present χ^2 and Kolmogorov-Smirnov (KS) tests which we implemented in ROOT, the framework used by HEP community. We also compare these tests with those which are already implemented in TH1 class by estimating a probability of type-I error which is a criterion of tests correctness. Subsequently, we discuss different approaches for handling of these weights and their possible utilization in practice. Finally, we apply the tests to samples from DØ experiment.

2. Homogeneity tests

Despite the fact that we have worked with Anderson-Darling test too, in this paper, we present results of χ^2 and KS homogeneity tests only. The reason is that we can compare our proposed tests with those already implemented in ROOT. Nevertheless, it is worth mentioning that Anderson-Darling test has a higher power of test than KS test as shown in [1].

The χ^2 **test** is designed for multinomial distributed samples represented by binned data (i.e. we count the number of events belonging to the individual intervals of observed space partition). If both samples are homogeneous, they have similar numbers of events in bins. That means we test whether both samples have multinomial distribution $M(n, p_1, ..., p_k)$, where the parameter n can be different for both samples while parameters $(p_1, ..., p_k)$ are mutual. Let $\boldsymbol{X} = (X_1, ..., X_k)$ and $\boldsymbol{Y} = (Y_1, ..., Y_k)$ be random vectors with the multinomial distributions $\mathrm{M}(n, p_1^{(1)}, ..., p_k^{(1)})$ and $\mathrm{M}(m, p_1^{(2)}, ..., p_k^{(2)})$, respectively. Test statistic of χ^2 test has the form

$$\chi^2(\boldsymbol{X}, \boldsymbol{Y}) = \frac{1}{mn} \sum_{i=1}^{k} \frac{(mX_i - nY_i)^2}{X_i + Y_i}$$

and it has asymptotically χ^2 distribution with $k-1$ degrees of freedom as $n \to \infty$ and $m \to \infty$. Now, let $\boldsymbol{W} = (W_1, ..., W_n)$ and $\boldsymbol{V} = (V_1, ..., V_m)$ be vectors of weights which we perceive as random variables in current scope. Every entry which contributes to X_i by one, now contributes to $W_{i\cdot}$ by W_{ij} which is defined as $W_{i\cdot} = \sum_{j=1}^{X_i} W_{ij}$ (similarly $V_{i\cdot} = \sum_{j=1}^{Y_i} V_{ij}$). Let us define the number of events $W_{\cdot\cdot} = \sum_{i=1}^{k} W_{i\cdot}$ and $V_{\cdot\cdot} = \sum_{i=1}^{k} V_{i\cdot}$. Then we can define generalized χ^2 test statistic by

$$\widetilde{\chi}^2(\boldsymbol{W}, \boldsymbol{V}) = \frac{1}{W_{\cdot\cdot} V_{\cdot\cdot}} \sum_{i=1}^{k} \frac{(V_{\cdot\cdot} W_{i\cdot} - W_{\cdot\cdot} V_{i\cdot})^2}{W_{i\cdot} + V_{i\cdot}}. \quad (1)$$

We assume that (1) keeps the χ^2-asymptotic property as $n \to \infty$ and $m \to \infty$ (implicating $W_{\cdot\cdot} \to \infty$ and $V_{\cdot\cdot} \to \infty$) even though the vectors $(W_{1\cdot}, ..., W_{k\cdot})$ and $(V_{1\cdot}, ..., V_{k\cdot})$ do not have multinomial distributions. It will be shown that this asymptotic property seems to be true for some class of weights only. We point out that we implemented both probabilistic and equidistant binning while the referential test `TH1::Chi2Test` uses equidistant binning only.

Contrary to χ^2 test, the **Kolmogorov-Smirnov test** is applied to unbinned sample which contains more information than binned data. It uses Empirical Distribution Function (EDF) or Weighted Empirical Distribution Function (WEDF) to compute difference between samples. Let us consider $\boldsymbol{X} = (X_1, ..., X_n)$ *iid* by F and $\boldsymbol{Y} = (Y_1, ..., Y_n)$ *iid* by G. Let $\boldsymbol{W} = (W_1, ..., W_n)$ and $\boldsymbol{V} = (V_1, ..., V_n)$ be corresponding weights and $W_{\cdot} = \sum_{i=1}^{n} W_i$ and $V_{\cdot} = \sum_{i=1}^{m} V_i$ be number of events. EDF $\mathrm{F}_n(x)$ and WEDF $\mathrm{F}_{\boldsymbol{W}}(x)$ are defined as

$$\mathrm{F}_n(x) = \frac{1}{n} \sum_{i=1}^{n} \mathbf{1}_{\{X_i \le x\}} \qquad \text{and} \qquad \mathrm{F}_{\boldsymbol{W}}(x) = \frac{1}{W_{\cdot}} \sum_{i=1}^{n} W_i \mathbf{1}_{\{X_i \le x\}}. \quad (2)$$

Test statistic of KS test and its asymptotic property are the following:

$$T(\boldsymbol{X}, \boldsymbol{Y}) = \sqrt{\frac{nm}{n+m}} \sup_{x \in \mathbb{R}} |F_n(x) - G_m(x)| \xrightarrow[H_0]{\mathcal{D}} K(\lambda) = 1 - 2 \sum_{k=1}^{\infty} (-1)^{k+1} e^{-2^k \lambda^2} \quad (3)$$

under the assumption of homogeneity hypothesis H_0. Replacing $(\boldsymbol{X}, \boldsymbol{Y}, n, m, F_n, G_n)$ by $(\boldsymbol{W}, \boldsymbol{V}, W_{\cdot}, V_{\cdot}, F_{\boldsymbol{W}}, G_{\boldsymbol{V}})$ we obtain generalized test statistic. We expected that the asymptotic property would not vanish after this modification; however, similarly as for χ^2 test, such property seems to be true for some class of weights only.

3. Estimating probability of type-I error

To verify the weighted tests, we need to show that probability of the type-I error is equal or lower than the significance level α. Nevertheless, in order to maximize power of the test, the probability should be as close as possible to the significance level. Therefore, we carried out computer experiments and compared both χ^2 (under probabilistic binning) and KS tests with `TH1::Chi2Test` and `TH1::KolmogorovTest` [2]. We did not compare it with `TMath::KolmogorovTest` because it cannot be applied to weighted samples; however, it returns the same p-value as our test, if it is applied to unweighted samples. We produced two *iid* samples from N(0, 1) of the same size: $n = m \in \{100, 200, 500, 1000, 2000, 5000, 10000\}$. We generated and assigned *iid* weights to every entry of both samples individually and repeated this procedure 10 000 times. Finally, ratio of the tests that rejected null hypothesis on three significance levels $\alpha = \{0.05, 0.01, 0.001\}$ was found. If the null hypothesis (homogeneity H_0) is true then the ratios are distributed by $\mathrm{Bi}(n, \alpha)/10000$. The number of bins for χ^2 tests was chosen as $\lceil 1.88n^{2/5} \rceil$ which is proposed in [3]. The corresponding results are shown in figures 1 and 2. The subfigers 1a, 1b, and 2a show us that some correct class of weights seems to exist.

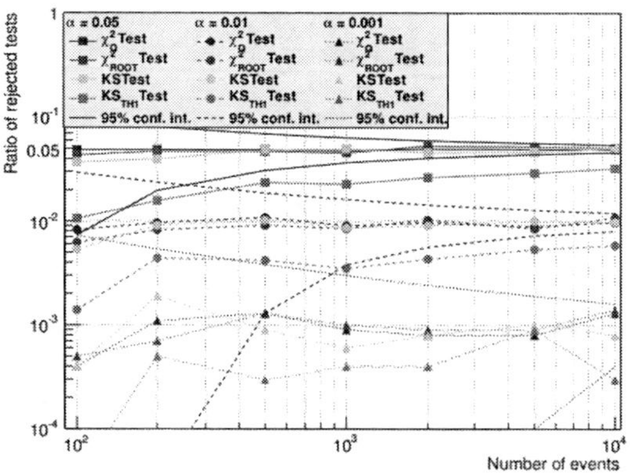

(a) All weights are equal to 1.

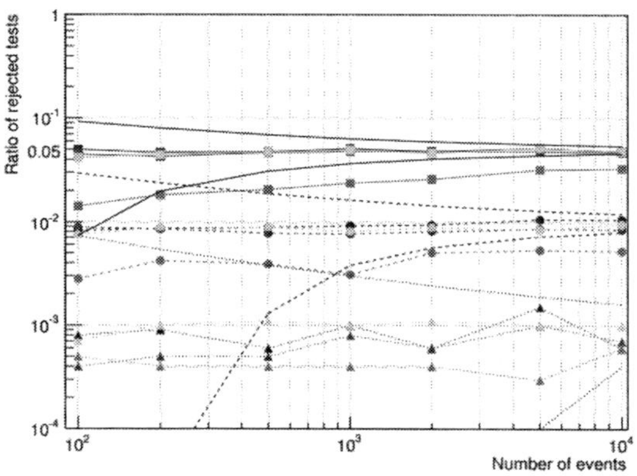

(b) All weights were generated from such U(a,b) that $\mathrm{E}[w] = 0.7$ and $\mathrm{Var}[w] = 0.7 - 0.7^2 = 0.21$.

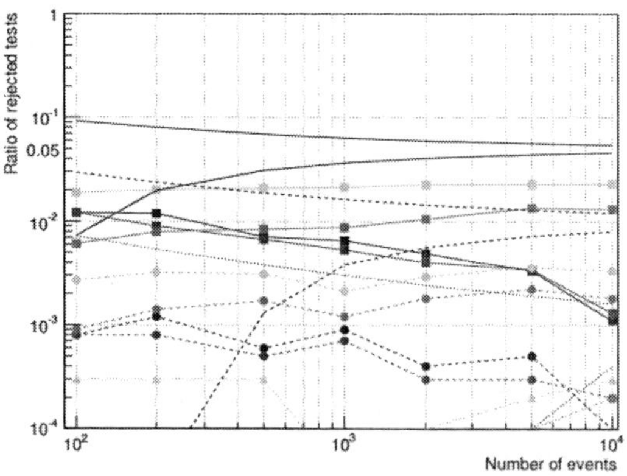

(c) All weights were generated from U(0.2,1.2).

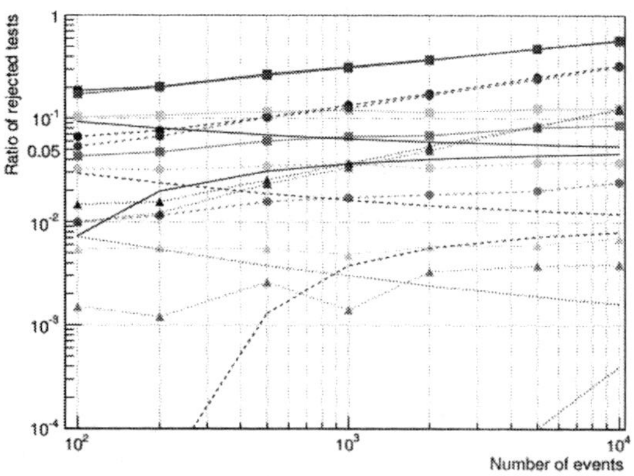

(d) All weights were generated from U(0,2).

Figure 1: Probability of type-I error estimated for various *iid* weights. χ^2_QTest and KSTest were implemented by us while χ^2_{ROOT}Test and $\mathrm{KS}_{\mathrm{TH1}}$Test are implemented in TH1 class.

From the rest subplots it is clear that the expected asymptotic property is not valid for some

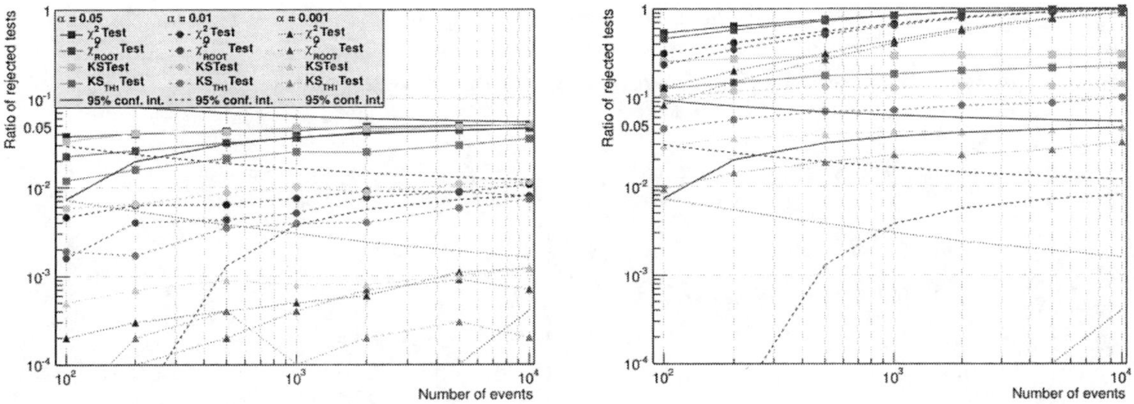

(a) All weights were generated from Exp(0.5). (b) All weights were generated from Exp(1).

Figure 2: Probability of type-I error estimated for exponential *iid* weights. χ^2_QTest and KSTest were implemented by us while χ^2_{ROOT}Test and $\mathrm{KS_{TH1}}$Test are implemented in TH1.

other classes of weights. As the mean of weights deviates from 1 the ratio deviates in the same direction. This statement is true also for variance being equal to 0 (i.e. $w = c$ almost surely, where c is some constant). Furthermore, this rejects the appropriateness of tests if the weights are not randomly generated but they are chosen as constants. If the variance increases from 0, the ratio increases too. We can also see that $\mathrm{KS_{TH1}}$Test returns higher p-value than it should produce and, as a consequence, the ratio is lower. The reason is that the binning reduces the highest difference between two WEDFs in most cases. In [4] it is shown that after transformation of weighted sample to unweighted one by summing up "near" entries to one entry with weight equal to one, the calculated p-value by non-generalized test is not significantly different from p-value computed by generalized test. However, such sample transformation does not solve the issue that the type-I error deviates from significance level.

4. Class of perfect weights

In order to find such a class of *iid* weights, where the probability of type-I error is equal to the significance level, we carried out another experiment. At first, we produced two samples from $\mathrm{N}(0, 1)$ containing 10 000 entries. Afterwards, we generated weights from $\mathrm{U}(a, b)$, where a and b were chosen in such a way that $(\mathrm{E}[W], \mathrm{Var}[W]) = (0.05, 0.075, 0.1, ..., 1) \times (0.01, 0.02, ..., 0.4)$. We applied KS test, repeated this procedure 10 000 times and counted ratio of rejected tests for significance level $\alpha = 0.05$. The result can be seen in figure 3 which indicates that the class of weights, for which the homogeneity tests have the same ratio r, preserve the parabolic behavior

$$\mathrm{Var}[W] = a(r)\mathrm{E}[W] - \mathrm{E}[W]^2, \tag{4}$$

where $a(\cdot)$ denotes a function which was manually estimated and linearly interpolated in figure 4. For $a(r) = 1$, KS test seems to have probability of type-I error equal to significance level $\alpha = 0.05$. The same phenomenon occurred when we generated weights from another distribution, for example $\mathrm{N}(p, p(1-p))$ with $p \in [0.05, 1]$. Moreover, from the figures 1b and 2a it seems that even χ^2 test's ratios of rejection lie on the same parabola; however, from figures 1c or 1d it is

IOP Conf. Series: Journal of Physics: Conf. Series **936** (2017) 012046 doi:10.1088/1742-6596/936/1/012046

clear that $a(\cdot)$ is unique for every different test (for example χ^2 test). Unfortunately, it is not clear whether $a(\cdot)$ changes with the samples produced from other distributions than N(0,1), or how it changes with different sample sizes or if both samples have different types of weights. Ideally, we estimate mean and variance of weights, we find $a(r)$ from (4), and then we know probability of type-I error of such test when rejecting it for p-value lower than significance level α. It is important to note that $a(\cdot)$ is scaled by α and it has to fulfill condition that $a(1) = \alpha$.

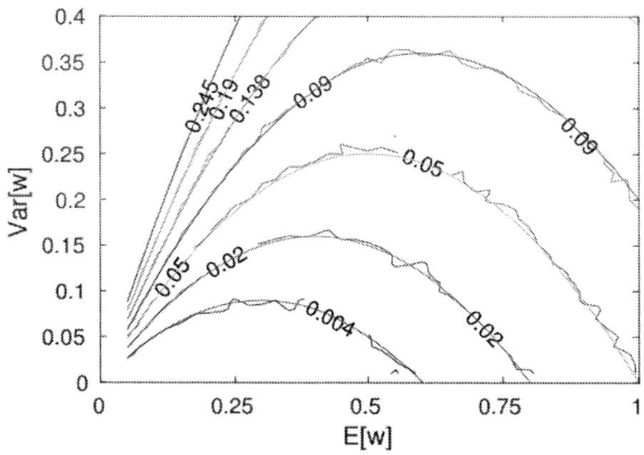

Figure 3: Contour plot of ratios of rejected KS tests for weights generated from such $U(a, b)$ that $(\mathrm{E}[W], \mathrm{Var}[W]) = (0.05, 0.075, 0.1, ..., 1) \times (0.01, 0.02, ..., 0.4)$, fitted to parabolas which are defined by (4) for $a(r) = \{0.6, 0.8, 1, 1.2, 1.4, 1.6, 1.8\}$.

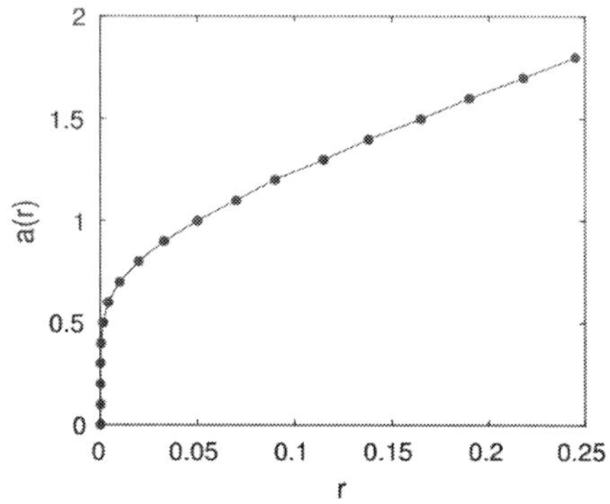

Figure 4: We estimated a(r) by visually estimating points for which contour plot's levels fits parabolas (as shown in figure 3).

5. Application of modified tests to DØ data

Proposed weighted homogeneity tests were applied to the top quark pair production search in lepton+jets channel using the data sample collected by the DØ detector at the Tevatron proton-antiproton collider at Fermilab [5]. All 6 different channels were used in the mentioned analysis, however, in this paper, we present only some particular results in $e + 3$jets channel, where the data sample contains 11905 entries and 11905 events while MC sample contains 719515 entries and 11905.5 events. All the weights of data sample are equal to 1, unlike of the MC sample, where the mean and variance are $\mathrm{E}[W] = 0.0166$ and $\mathrm{Var}[W] = 0.0036$. Histograms and WEDFs of the selected variable DRminejet can be seen in figure 5.

We already mentioned that the generalized test works well when variance of weights is equal to 0. Since in all HEP analysis the weights of final samples are not generated randomly but are assigned through the results of another analysis, we can use it without any doubts. On the other hand, if we consider our weights as *iid* random variables, calculated p-values would not be true p-values since asymptotic distribution of all four tests is violated. In addition, if we suppose the same function $a(\cdot)$ as in the previous experiment then p-values lie on parabola $\mathrm{Var}[W] = 0.23\mathrm{E}[W] - \mathrm{E}[W]^2$. Because $0.23 < 1$, probability of type-I error is smaller than significance level and the test is correct; however, we admit that power of the test can be low.

In table 1 we can see results of homogeneity tests of MC vs. data for 8 variables with highest separation power between signal and background that were used as an input to BDT in the MVA analysis to determine final discrimination. Description of each variable can be found in [5]. Notice the difference in p-values between KS tests and χ^2 tests in the Mva_max variable, which is an output from another MVA analysis determining the probability that a given event

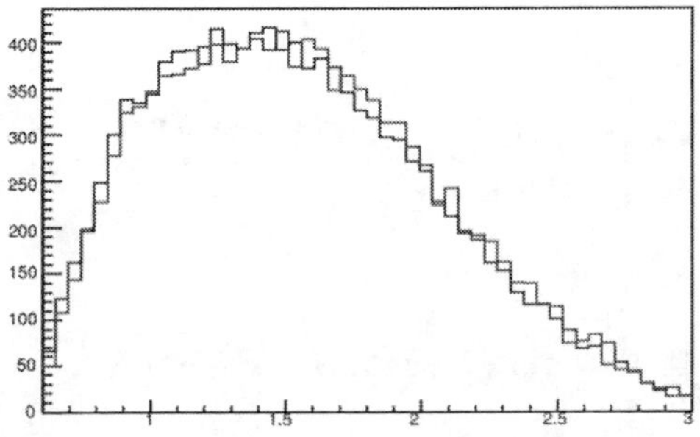

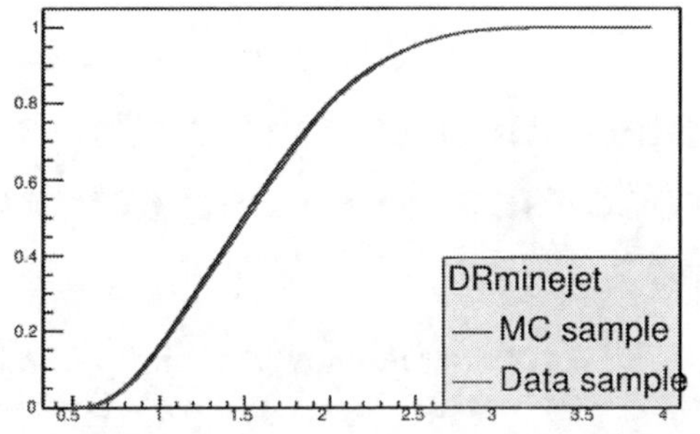

Figure 5: Histogram and WEDF of DRminejet.

Table 1: Calculated p-values for testing data vs MC sample in $e + 3$jets channel.

Test	Mva_max	DRminejet	Apla	Spher	Jtetal	LepEta	Ht	Wpt
χ^2_Qtest	0.6767	0.9372	0.9701	0.9658	0.3052	0.8234	0.8476	0.9137
χ^2_{ROOT}test	0.9483	0.9407	0.8895	0.5804	0.4409	0.7550	0.9879	0.8628
KStest	0.0001	0.0167	0.8639	0.1754	0.4309	0.1279	0.7461	0.1409
KS_{TH1}test	0.2672	0.0262	0.9817	0.2014	0.4793	0.1917	0.9552	0.1658

contains a jet originating in b-quark. Since it is artificial variable, it is expected to have worse agreement between MC and data. On the other hand, whole analysis is done with binned data stored in histograms, therefore the results of χ^2 test is more reliable.

6. Conclusion

In this paper, we showed that expected asymptotic distributions of generalized χ^2 and KS tests with random weights are not correct; however, for the class of *iid* weights which fulfills condition that $\text{Var}[W] = E[W] - E[W]^2$, the probability of type-I error seems to be equal to significance level. Moreover, our conjecture is that the relation between weights and probability of type-I error $\tilde{r}$ is given by $\text{Var}[W] = a(\tilde{r})E[W] - E[W]^2$, where $a(\cdot)$ is unknown function which can be estimated for particular test setup. Some more detailed numerical analysis is needed to be carried out. It is unknown how the probability of type-I error would change if the sample sizes are different or whether $a(\cdot)$ changes if we would produce samples from another distribution than $N(0, 1)$. Proving a statement that probability of type-I error is equal to significance level iff $\text{Var}[W] = E[W] - E[W]^2$, would be a great success.

Acknowledgments

This work was supported by the CZ grants LG15047 (MYES), LM2015068 (MYES), SGS15/214/OHK4/3T/14 (CTU), and GA16-09848S (GACR).

References

[1] Engmann S and Cousineau D 2011 *Journal of Applied Quantitative Methods* **6** 117
[2] ROOT Reference Documentation https://root.cern.ch/doc/master/index.html accessed: 2017-01-04
[3] D'Agostino R and Stephens M 1986 *Goodness-of-Fit Techniques* (New York: Dekker) p 70 iSBN 0824774876
[4] Bour P and Kůs V 2015 *SPMS 2015 proceedings* (FNSPE CTU in Prague)
[5] D0 Collaboration Abazov V M *et al* 2016 *Phys. Rev. D* **94**

IOP Conf. Series: Journal of Physics: Conf. Series **936** (2017) 012047 doi:10.1088/1742-6596/936/1/012047

Data collection system for a wide range of gas-discharge proportional neutron counters

V Oskomov, A Sedov, N Saduyev, O Kalikulov, I Kenzhina*, E Tautaev, Y Mukhamejanov, V Dyachkov, Sh Utey

Institute of Experimental and Theoretical Physics of al-Farabi Kazakh National University

*kenzhina@physics.kz

Abstract. This article describes the development and creation of a universal system of data collection to measure the intensity of pulsed signals. As a result of careful analysis of time conditions and operating conditions of software and hardware complex circuit solutions were selected that meet the required specifications: frequency response is optimized in order to obtain the maximum ratio signal/noise; methods and modes of operation of the microcontroller were worked out to implement the objectives of continuous measurement of signal amplitude at the output of amplifier and send the data to a computer; function of control of high voltage source was implemented. The preliminary program has been developed for microcontroller in its simplest form, which works on a particular algorithm.

1. Introduction

Program-technical complex for data collection is designed to measure the intensity of pulsed signals of inputs of its information channels.

The following basic requirements should be provided in developing this system:

- Stability of parameters over long periods of time (at least 8 months), such as amplification factor, voltage, detection efficiency.
- Low power consumption.
- Collection, storage and automatically transfer of information.
- Reception, processing and delivery of the received information in a convenient way.

The use of microcontroller provides an extremely low total power consumption of the entire circuit and the small size of electronic circuit board. Data and control at electronic board from the control computer are provided by the second chip that implements algorithm for information exchange via USB interface. The use of this modern interface allows you to achieve high system performance in a temporary scan mode, and enables the use of system in absence of a separate power supply for its external electronic circuits, in relation to computer.

Electronic amplifier board with microcontroller and USB connection allows you to get a full ("oscillographic") registered signals from input channels while maintaining the data on your computer and their subsequent processing. In contrast to hardware pre-processing of signals when the apparatus determines that the right signal is registered and recorded only the fact of equipment operation.

Because the software and hardware system of data collection is assumed to be used in scientific experiments, it was decided to adapt the electronic amplifying board to a wide class of gas-discharge counters (Geiger-Muller, proportional counters, ionization chambers, etc.). In this case, the bandwidth required for this class of meter is from 0.1 kHz to 50 kHz, i.e. signals registration will go in a strip of low and medium frequencies.

2. Instruments and methods

First, we note that processes in gas counters are characterized by the order of several microseconds, the time between ADC counts should be of the order of several microseconds.

As a result of precise analysis of time and operating conditions of software and hardware complex circuit solutions that meet the following specifications were selected:

- frequency characteristic was optimized in order to obtain the maximum signal/noise ratio.
- methods and modes of operation of the microcontroller were worked out to implement the objectives of continuous measurement of signal amplitude at the output of amplifier and send the data to a computer.
- function of control of high voltage source was implemented.

The amplifier consists of a filter for high-voltage, high-voltage transition capacitor (C14), charge sensitive stage with capacitor 7.5 pF in feedback, and two identical stages with a voltage gain of about 10 (Figure 1).

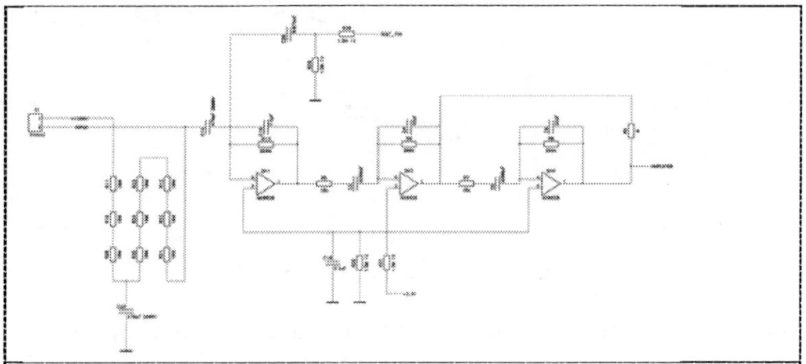

Figure 1. Signal amplifier.

To be able to accurately tie recorded events in time, a highly stable temperature-compensated FOX914B generator was used with drift about one millionth.

Amplifier and all other circuits are powered from a ADP121 stabilizer for 3.3 volts, which is powered (5V) from USB.

The signal from amplifier is supplied to microcontroller ADC, which makes 100,000 counts per second with a resolution of 10 bits, or 80,000 counts per second with a resolution of 12 bits. Couples of byte counts are sent via USB to the computer through FT245R chip (Figure 2, 3).

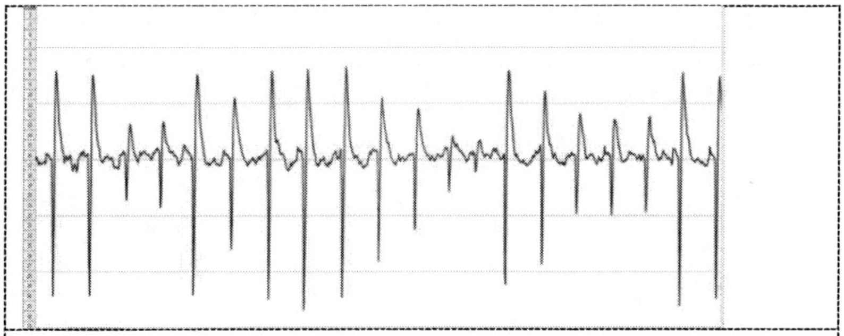

Figure 2. The pulses on output of amplifier.

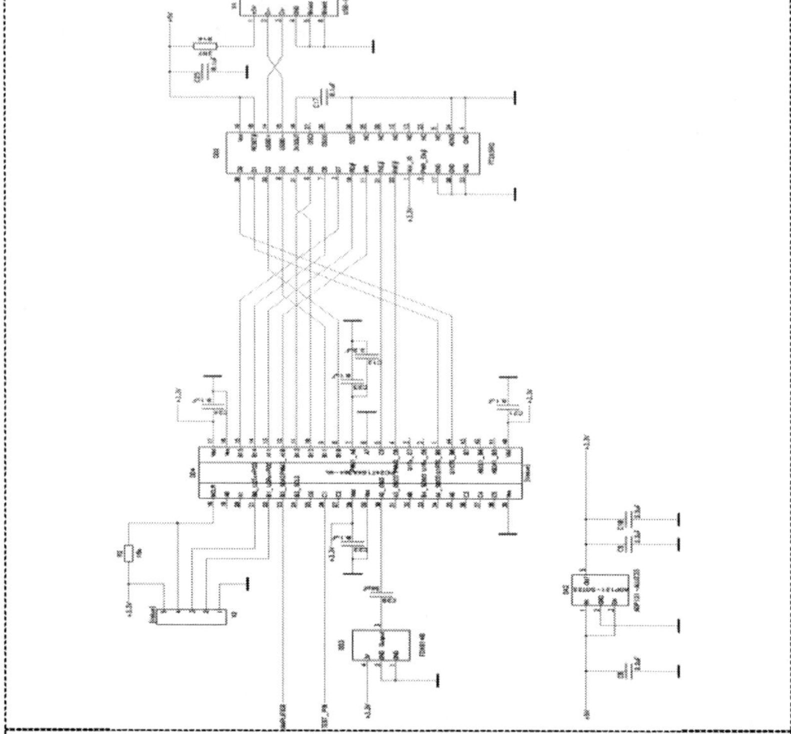

Figure 3. Microcontroller with ADC and USB interface.

High voltage unit - a transformer with rectifier and electronic multiplying circuit of analysis and control, and the high voltage source is physically consisted of transistor switch, transformer and multiplying rectifier.

Microcontroller supply an impetus to the key of certain length (straight course) via a transformer and diode, one half of rectifier capacitors is recharged.

It also accumulates the energy in inductance of transformer. Then the key is turned off, the voltage due to transformer inductance changes sign (reverse), and the other half of capacitors recharges, followed by a relatively long pause.

In pause voltage diode rectifier is distributed uniformly and is half of the voltage level, about 100 volts. This provides a small leakage current of diodes, which is strongly dependent on the voltage.

During retrace microcontroller measures the flyback voltage. It is proportional to high voltage output of rectifier.

If the measured voltage is below set one microcontroller decreases by a small amount the duration of pause between pulses, thereby increasing the transmit power in rectifier. If the voltage is higher - the microcontroller increases the length of pause between pulses, the power transmitted to rectifier is reduced.

High-voltage unit supports a pre-determined set voltage and gives information on current consumption (Figure 4).

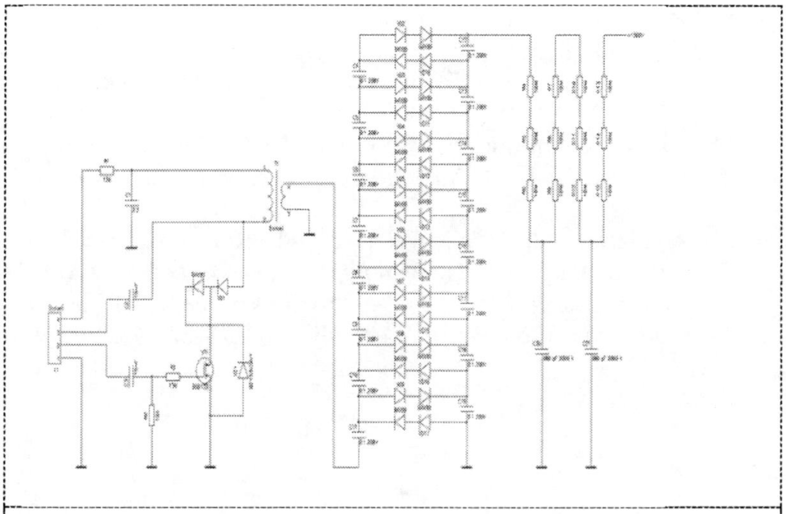

Figure 4. Schematic diagram of a high voltage source for a wide range of gas-discharge counters.

3. Results and discussion

The key is controlled by a microcontroller; the signal from primary winding of transformer is fed back to microcontroller for measurement.

Since multiplying rectifier is symmetrical, the inverter transfers energy to rectifier and on the direct course, and vice versa.

High voltage value is continuously measured with a high frequency at amplifier output, the last few counts are analyzed for the presence of a pulse. If it appears that pulse is there, that is, signal exceeds a certain threshold value, the drive is informed about another pulse, and the pulse value is used for correction of high voltage. It must be remembered that measured reverse voltage is not quite "measured high", depending on external influences the actual voltage may vary. In addition, the characteristics of gas-discharge counters may change over time, it is necessary that amplitude of pulse was approximately constant over time. Therefore, the setting value can be adjusted according to analyzing values of pulses. In addition to correcting the high measured values scatter pulses is measured that provides diagnostic information about the "health" of counter and electronics.

It is suggested the number of pulses are transmitted to the outside "during the reporting period" on demand or at preset times that number, together with diagnostic information. It is necessary to transmit the minimum and maximum values of high-voltage pauses.

4. Conclusions

The preliminary program for microcontroller in its simplest form is working on following algorithm:

- turns on ADC power and configures on a given frequency and accuracy of counts.
- Then in the next cycle is expected readiness of ADC reference. And as soon as the countdown is ready, the data is read from ADC and sent to USB chip.
- Again waiting for a new frame of reference.

In a more complex version (when setting up, for example), the program may apply test pulses to amplifier input, t may change the ADC mode.

Acknowledgments

The study was carried out under the support of grant #3838/GF4 of Ministry of Education and Science of Kazakhstan.

IOP Conf. Series: Journal of Physics: Conf. Series **936** (2017) 012047 doi:10.1088/1742-6596/936/1/012047

Referencies

[1] Nikolajchuk O I 2003 Sistemy maloj avtomatizacii *M.: SOLON-Press* 256 p
[2] Tavern'e K 2003 PIC-mikrokontrollery. Praktika primenenija *Per. s fr. M: DMK Press* 272 p.
[3] Kulakov V 2003 Programmirovanie na apparatnom urovne *Special'nyj spravochnik 2-e izdanie. – SPb.: Piter* 874 p.
[4] Bolski M I 1988 Jazyk programmirovanija Si *Spravochnik: Per. s angl. M.: Radio i svjaz'* 96 p.
[5] Pejton A, Dzh A, Bolsh V 1994 Analogovaja jelektronika na operacionnyh usiliteljah *M.: BINOM* 352 p.
[6] Vinogradova N A, Lietratov Ja I, Sviridov E V 2005 Razrabotka prikladnogo programmnogo obespechenija v srede LabVIEW Uchebnoe posobie - M.: Izdatel'stvo MJeI 51 p.
[7] Mathanov H N 1990 Osnovy analiza jelektricheskih cepej. Linejnye cepi *Ucheb. dlja jelektrotehn. i radiotehn. spec. 3-e izd. pererabot. i dop. –M.: Vyssh. shk.* 400 p.

IOP Conf. Series: Journal of Physics: Conf. Series **936** (2017) 012048 doi:10.1088/1742-6596/936/1/012048

Numerical simulation of a device with two spin crossover complexes: application for temperature and pressure sensors

Jorge Linares,[1,2] Salah Eddine Allal,[3] Pierre Richard Dahoo,[3] Yann Garcia[4]

[1]GEMAC, Université de Versailles Saint-Quentin-en-Yvelines, 45 Av. des Etats-Unis, CNRS-UMR 8635, 78035 Versailles Cedex, France

[2]Departemento de Ciencias, Sección Física, Pontificia Universidad Católica del Perú, Apartado 1761, Lima, Perú

[3]LATMOS, Université de Versailles-Saint-Quentin-en-Yvelines, CNRS-UPMC-UVSQ (UMR 8190), 11 boulevard d'Alembert, 78280 Guyancourt, France

[4]Institute of Condensed Matter and Nanosciences, Molecules, Solids and Reactivity (IMCN/MOST), Université catholique de Louvain, 1348, Louvain-la-Neuve, Belgium.

Abstract: The spin-crossover (SCO) phenomenon is related to the ability of a transition metal to change its spin state *vs.* a given perturbation. For an iron(II) SCO complexes the reversible changes involve the diamagnetic low-spin (S = 0) and the paramagnetic high-spin (HS S = 2) states [1,2,3]. In this contribution we simulate the HS Fraction (n_{HS}) for different set values of temperature and pressure for a device using two SCO complexes with weak elastic interactions. We improve the calculation given by Linares *et al.* [4], taking also into account different volume (V_{HS}, V_{LS}) changes of the SCO. We perform all the calculation in the frame work of an Ising-like model solved in the mean-field approximation. The two SCO show in the case of "weak elastic interactions", gradual spin transitions such that both temperature and pressure values can be obtained from the optical observation in the light of calculations discussed in this article.

1. Introduction

The increasing demand regarding safety and security rules require applications that involve advanced sensors with higher sensitivity, better technical specifications, lower dimensions and low energy consumption. A special class of sensors is represented by temperature and pressure solid state sensors that allow direct or remote detection. This kind of sensing devices can be realized by developing new multifunctional materials exhibiting appropriate properties that can be controlled at the atomic or molecular level. Interesting properties that could be used in these applications are exhibited by spin crossover (SCO) molecular materials [1,2,3]. The SCO phenomenon is a molecular process which combines an electronic transformation and atomic displacements in the coordination sphere of a central metal ion $3d^4$–$3d^7$. Thus, the SCO usually occurs between a diamagnetic low-spin (LS) state and a paramagnetic high-spin (HS) state, which can be triggered by various external perturbations such as temperature, pressure, light or magnetic field. In the course of the SCO phenomenon, the physical properties (magnetic, optical, electrical, vibrational, etc.) of the material change dramatically and the process can be followed by various detection techniques (magnetic, optical, electrical, etc.)[4-12]. In this work we show that SCO compounds can be used as a visual detector of strong mechanical contact pressure and temperature. This result is based on the thermal dependence of the optical

IOP Conf. Series: Journal of Physics: Conf. Series **936** (2017) 012048 doi:10.1088/1742-6596/936/1/012048

reflectivity measurements, a correlation of the optical SCO properties with the pressure contact and temperature highlighted thanks to an Ising-like model. These results provide the basis for the construction of a thermal and pressure sensor.

2. Model and principle of calculation

For the complete understanding of our phenomena, we use both experimental and theoretical results in order to determine the way SCO materials are influenced by external factors. We proceed from the knowledge of the temperature and pressure influence on the SCO behavior, as studied in the frame of the Ising-Like model. The mean field approximation is applied to solve the Hamiltonian of the interacting spins. The Ising-Like model proposed by Wajnsflasz and Pick[13,14] uses a fictitious spin operator σ which can take the value +1 or -1 to specify respectively the HS and LS state with degeneracies g_{HS} for the HS and g_{LS} for the LS state. Linares *et al.*[15,16] proposed an explicit extension of this model by introducing short- and long-range interactions, termed J and G respectively in the model. As a result, the Hamiltonian is written in the following form:

$$H = \frac{\Delta + \alpha p - Tln(g)}{2}\sum_i \sigma_i - J\sum_{ij}\sigma_i\sigma_j - G\langle\sigma\rangle\sum_i \sigma_i$$

Where Δ represents the difference between the energy of the HS and that of the LS states, $\sum(I,j)$ is the sum over nearest neighbors, $g = g_{HS}/g_{LS}$ is the degeneracy ratio and αp is the pressure operator. The term $\langle\sigma\rangle$ is the average value of σ which can also be expressed in terms of the fraction of HS species by:

$$n_{HS} = \frac{\langle\sigma\rangle + 1}{2}$$

Then we can develop our Hamiltonian by considering for each particle, the σ_j having the same value $\langle\sigma\rangle$, such that $\sum_j \sigma_j = q\langle\sigma\rangle$, where q is the number of nearest neighbors.

$$H = \frac{\Delta + \alpha p - Tln(g) - 2G\langle\sigma\rangle}{2}\sum_i \sigma_i - Jq\langle\sigma\rangle\sum_i \sigma_i$$

After collecting together the terms factorized by $\sum_i \sigma_i$ in order to simplify the equation, then the Hamiltonian is expressed as:

$$H = \frac{\Delta + \alpha p - Tln(g) - 2G\langle\sigma\rangle - 2Jq\langle\sigma\rangle}{2}\sum_i \sigma_i$$

$$H = \frac{\Delta + \alpha p - Tln(g) - 2\Gamma\langle\sigma\rangle}{2}\sum_i \sigma_i$$

Where finally $\Gamma = G + Jq$ is introduced for simplification and represents both the short- and long-range interactions.

As usual, if we have an Hamiltonian of the form $H = -h\sigma_i$, then after some transformations we can obtain $\langle\sigma\rangle$, by solving the following equation $\langle\sigma\rangle = \tanh(\beta h)$

So in our case if we consider that:

$$h = -\frac{\Delta + \alpha p - Tln(g) - 2\Gamma\langle\sigma\rangle}{2}$$

$$\beta = \frac{1}{k_B T}$$

Then our equation to solve can be written as:

$$\langle\sigma\rangle = \tanh\left(-\frac{\Delta + \alpha p - Tln(g) - 2\Gamma\langle\sigma\rangle}{2k_B T}\right)$$

$$\langle\sigma\rangle = \tanh\left(-\frac{\Delta/k_B + pc\delta V - Tln(g) - 2\Gamma\langle\sigma\rangle/k_B}{2T}\right)$$

Where c is a constant, c = 0.0724, when the external applied pressure, p, is expressed in MPa and the volume variation of the molecule, δV, is expressed in Å^3.

In order to determine the two variables (T,p) from one equation, our device must include two SCO compounds characterized by the corresponding parameters $\Delta 1, \delta V1, g1, \Gamma 1, <\sigma 1>$, and $\Delta 2, \delta V1, g2, \Gamma 2, <\sigma 2>$, respectively (we make the difference $\delta V1$, and $\delta V1$,in this contribution)
Then we have the following system:

$$\begin{cases} <\sigma 1> = \tanh\left(-\dfrac{\Delta 1 + pc\delta V1 - T\ln g1 - 2\Gamma 1 <\sigma 1>}{2T}\right) \\ <\sigma 2> = \tanh\left(-\dfrac{\Delta 2 + pc\delta V2 - T\ln g2 - 2\Gamma 2 <\sigma 2>}{2T}\right) \end{cases}$$

From these equations we can derive the expression of p and T:

$$\begin{cases} T = \dfrac{\beta_2 - \beta_1 \dfrac{\delta V2}{\delta V1}}{\lambda_2 - \lambda_1 \dfrac{\delta V2}{\delta V1}} \\ p = \dfrac{\lambda_1 . \left[\beta_2 - \beta_1 \dfrac{\delta V2}{\delta V1}\right] - \beta_1 . \left[\lambda_2 - \lambda_1 \dfrac{\delta V2}{\delta V1}\right]}{c\delta V1 . \left[\lambda_2 - \lambda_1 \dfrac{\delta V2}{\delta V1}\right]} \end{cases}$$

In order to simplify the lecture of the equations we introduce the following terms:

$$\begin{cases} \lambda_1 = \ln g1 - 2\tanh^{-1} <\sigma 1> \\ \lambda_2 = \ln g2 - 2\tanh^{-1} <\sigma 2> \end{cases} \qquad \begin{cases} \beta_1 = \Delta 1 - 2\Gamma 1 <\sigma 1> \\ \beta_2 = \Delta 2 - 2\Gamma 2 <\sigma 2> \end{cases}$$

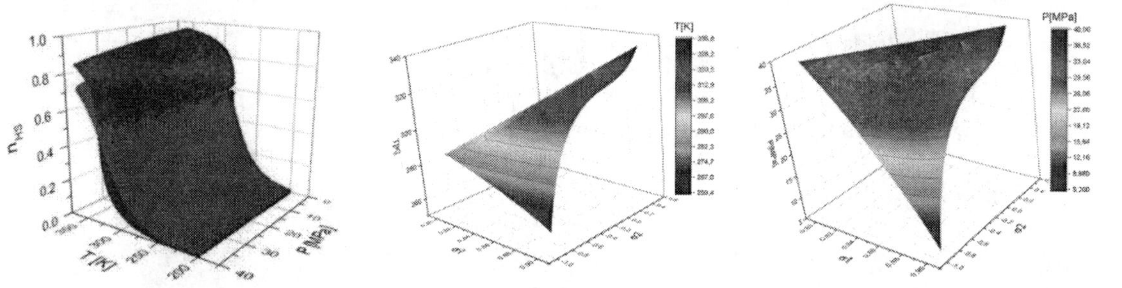

(a) (b) (c)

Figure 1. (a) Simulated HS fraction as a function of temperature and pressure variation for two values of the interaction parameter (red-Γ/k_B = 160 K and blue-Γ/k_B = 300 K). The calculation parameters are: Δ/k_B = 1978.6 K, ln(g) = 6.906, δV = 100 Å^3 [4]. (b) Simulated temperature as function of two mean spin values $<\sigma_1>$ and $<\sigma_2>$, using two SCO compounds with the same interaction parameter Γ/k_B = 300 K and different parameters' values Δ_1/k_B = 1300 k, δV_1 = 93.8 Å3, ln(g_1) = 6.01, and Δ_2/k_B = 1978.6 k, δV_2 = 100 Å^3, ln(g_2) = 6.906, respectively. (c) Simulated pressure as function of two mean spin values $<\sigma_1>$ and $<\sigma_2>$, using two SCO compounds with the same interaction parameter Γ/k_B = 300 K and different parameters' values Δ_1/k_B = 1300 k, δV_1 = 93.8 Å^3, ln(g_1) = 6.01, and Δ_2/k_B = 1978.6 k, δV_2 = 100 Å^3, ln(g_2) = 6.906, respectively.

Figure 1a gives the HS fraction *vs.* the variation of temperature and pressure for each complex, and shows clearly the overlapping between the two HS fraction 2D surfaces, the overlap range representing the solutions of the combined system from which the simultaneous graphical values of temperature and pressure can be obtained. Fig. 1b and Fig. 1c show respectively the temperature and pressure variation *vs.* $<\sigma 1>$ and $<\sigma 2>$. As we can see, we can have several pairs of values of $<\sigma 1>$ and $<\sigma 2>$ for one value of temperature or pressure. But the number of pairs decreases for high temperatures in the thermal case, and for low pressure in the pressure case.

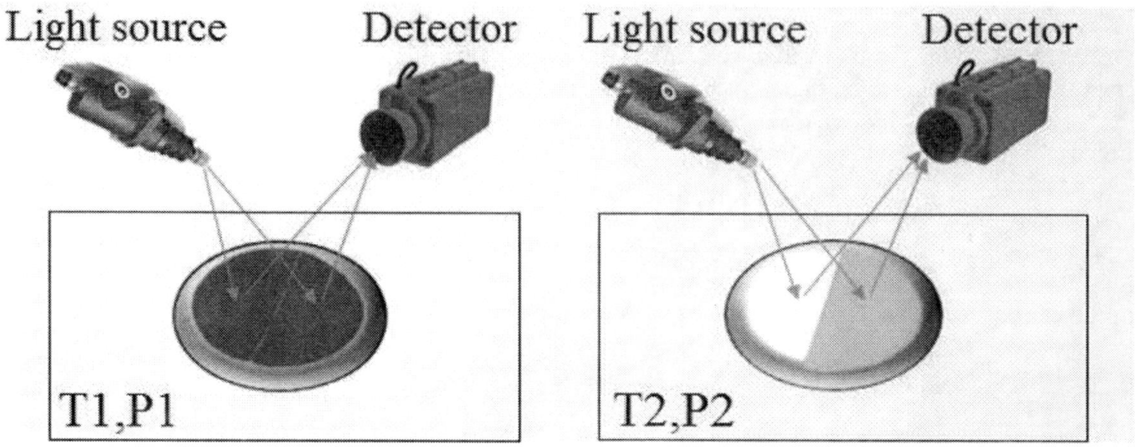

Figure 2. Principle to measure simultaneously temperature and pressure using two SCO compounds.

3. Conclusion

In this work we have presented a novel concept for spin crossover-based sensors that allows the concomitant detection of both temperature and pressure. This new approach is based on the use of two complexes exhibiting gradual spin transitions that leads to the identification of a single set of (P-T) values for a single set of optical densities of each complex. Thus, a simple colour identifier is sufficient to detect and measure the variation of both temperature and pressure, opening new perspectives in the multifunctional sensors field. As far as the necessary apparatus needed to monitor temperature and pressure changes is concerned, the one proposed here is comparatively much simpler to implement than the existing ones using sophisticated X-Ray diffractometers and apparatuses needed for Brillouin spectroscopy. It consists of a standard laser source, a detector and an associated computer for data analysis. Signal to noise ratio is much better with the detection of a reflected laser beam than that of laser induced fluorescence or luminescence.

Acknowledgments

We thank the CHAIR Materials Simulation and Engineering, the UVSQ and Université Paris Saclay. The Fonds de la Recherche Scientifique-FNRS (PDR T.0102.15), and COST actions CM1305 and CA15128 are gratefully acknowledged.

References

[1] Gütlich P, Gaspar AB and Garcia Y 2013 *Beilstein J. Org. Chem.* **9** 342.

[2] Nasser J., Boukheddaden K., Linares J., 2004, *Eur. Phys. J. B.* 39, 219-227

[3] Varret F., Bleuzen A., Boukheddaden K., Boussekou A., Codjovi E., Enachescu C., Goujon A., Linares J., N. Menendez N. and M. Verdaguer M., 2002, *Pure Appl. Chem.*, 74, 2159-2168

[4] Jureschi C., Linares J., Boulmaali A., Dahoo P.R., Rotaru A. and Garcia Y., 2016, *Sensors*,16, 187

[4] Ksenofontov V, Levchenko G, Spiering H, Gütlich P, Létard J F, Bouhedja Y, Kahn O 1998 *Chem. Phys. Lett.* **294** 545.

[5] Gaspar AB, Ksenofontov V, Real JA, Gütlich P 2003 *Chem. Phys. Lett.* **373**, 385.

[6] Seredyuk M, Gaspar AB, Ksenofontov V, Reiman S, Galyametdinov Y, Haase W, Rentschler E, Gütlich P 2006 *Chem. Mater.* **18** 2513.

[7] Jeftic J, Menéndez N, Wack A, Codjovi E, Linarès J, Goujon A, Hamel G, Klotz S, Syfosse G, Varret F 1999, *Meas. Sci. Techn.*, **10**, 1059-1064

[8] Mishra V, Mukherjee R, Linares J, Balde C, Desplanches C, Létard J.F, Collet E, Toupet L, Castro M, Varret F 2008, *Inorg. Chemistry*, **47**, 7577-7587

[9] Stassen AF, Roubeau O, Ferrero-Gramage I, Linares J, Varret F, Mutikainen I, Turpeinen U, Haasnoot JG, Reedijk J 2001, *Polyhedron*, **20**, 1699-1707
[10] Morscheidt W, Jeftic J, Codjovi E, Linares J, Bousseksou A, Constant-Machado H, Varret F 1998, *Meas. Sci.Techn.*, **9**, 1311-1315
[11] Klokishner S, Linares J, Varret F 2000, *Chem. Phys.*, **255**, 317-323
[12] Rotaru A, Varret F, Gîndulescu A, Linares J, Stancu A, Létard JF, Forestier T, Etrillard C 2011, *Eur. Phys. J. B* **84**, 439-449
[13] Wajnflasz J, Pick R 1971, *J. Phys.Colloques*, **32**, 91-92
[14] Bousseksou A, Nasser J, Linares J, Boukheddaden K, Varret F 1992 *J. Physique* I **2** 1381
[15] Linares J, Jureschi C, Boukheddaden K 2016, *Magnetochemistry*, **2**, 24
[16] Linares J, Jureschi C, Boulmaali A, Boukheddaden K 2016, *Physica B*, **486**, 164–168

Mathematical modeling of elementary trapping-reduction processes in positron annihilation lifetime spectroscopy: methodology of Ps-to-positron trapping conversion

Ya Shpotyuk[1,2], J Cebulski[1], A Ingram[3] and O Shpotyuk[4,5]

[1] Centre for Innovation and Transfer of Natural Sciences and Engineering Knowledge, Faculty of Mathematics and Natural Sciences, University of Rzeszow, 1, Pigonia str., 35-959 Rzeszow, Poland

[2] Department of Sensor and Semiconductor Electronics, Ivan Franko National University of Lviv, 107, Tarnavskogo str., 79017 Lviv, Ukraine

[3] Opole University of Technology, 75, Ozimska str., 45370 Opole, Poland

[4] Institute of Physics of Jan Dlugosz University in Czestochowa, 13/15, Armii Krajowej str., 42200 Czestochowa, Poland

[5] O.G. Vlokh Institute of Physical Optics, 23, Dragomanov str., 79005 Lviv, Ukraine

E-mail: yashpotyuk@gmail.com

Abstract. Methodological possibilities of positron annihilation lifetime (PAL) spectroscopy in application to nanostructurized substances treated within three-term fitting procedure are reconsidered to parameterize their atomic-deficient structural arrangement. In contrast to conventional three-term fitting analysis of the detected PAL spectra based on admixed positron trapping and positronium (Ps) decaying, the nanostructurization due to *guest* nanoparticles embedded in *host* matrix is considered as producing modified trapping, which involves conversion between these channels. The developed approach referred to as x3-x2-coupling decomposition algorithm allows estimation free volumes of interfacial voids responsible for positron trapping and bulk lifetimes in nanoparticle-embedded substances. This methodology is validated using experimental data of Chakraverty et al. [Phys. Rev. B71 (2005) 024115] on PAL study of composites formed by *guest* $NiFe_2O_4$ nanocrystals grown in *host* SiO_2 matrix.

1. Introduction

To be successful in nanomaterial engineering, it is necessary to possess an advanced instrumentation tool that can efficiently explore a nanospace at both atomic-specific (inner space occupied by atoms, atomic groups and molecules) and atomic-deficient (free of electron and atomic densities) levels [1]. The latter is of high importance in view of essential deficit in reliable void-testing probes sensitive to length scales of atomic and subatomic ranges. One of such probes that can fit the gap in methodological resolutions is based on positronics [2-6], i.e. analytical space-time continuum determination for electron interaction with its antiparticle, positron. This phenomenon realized as positron annihilation lifetime (PAL) spectroscopy, can be applied to study free-volume imperfections (such as vacancies, vacancy-like clusters, voids, pores, void agglomerates, and even macroscopic cracks) in crystals, glasses, coarse-grained or fine-milled powders, ceramics, bulk alloys, etc.

This work is aimed to analyze possibilities of positronics in application to free-volume defects caused by nanostructurization in nanoparticle (NP) embedded solids, where *guest* NPs can affect both positron-trapping and Ps-decaying channels in an overall balance of annihilation events in *host* matrix. Nanostructurization is considered as modified trapping, which involves three elementary components, these being positron/Ps trapping reduction/enhancement and Ps-to-positron trapping conversion.

2. Model

2.1. Methodological background.

In nanomaterials, the PAL spectra typically yield three distinct lifetimes $\tau_{1,2,3}$ with $I_{1,2,3}$ intensities reflecting competitive input from mixed positron-Ps trapping channels under normalization [3,6]:

$$I_1 + I_2 + I_3 = 1. \tag{1}$$

The longest PAL component (τ_3,I_3) originates from decaying of spin-triplet ortho-positronium o-Ps states in free-volume holes. This process stretched intrinsically in a vacuum ending by emitting three γ-rays with theoretical lifetime of 142 ns, whereas in a matter it is quenched in shorten lifetimes of a few ns because of Ps interaction with electrons from surrounding. As a result, the o-Ps annihilates extrinsically in a medium emitting two γ-rays, the process known as *pick-off annihilation* [3-5].

The intermediate PAL component (τ_2,I_2) results from positron annihilation from free-volume defect sites such as atomic and sub-atomic voids (vacancies, vacancy-like clusters, etc.). Being trapped by such defect, the positron annihilate with lifetimes τ_2=τ_d ranging from 0.2-0.3 to 0.5 ns [3,5].

In case of three-component fitting of the detected PAL spectra, the resolution of all inputs is essentially disturbed by additional contribution in the first shortest component (τ_1,I_1) arising from spin-singlet para-positronium p-Ps states, giving τ_p=0.125 ns with relative population of I_p=$I_3/3$ for a vacuum. Thus, the annihilation from defect-free bulk states are currently admixed to p-Ps decaying, thus complicated the physical meaning of this component. So direct correlations are not allowed for this (τ_1,I_1) component and material-related parameters of a medium.

2.2. Methodology of Ps-to-positron trapping conversion within x3-x2-CDA.

In our model, we distinguish all inputs in the first PAL component to extract the "pure" feedback caused by NP-related free-volume positron traps themselves. Assuming an additive two-state positron trapping model for these defects, we obtain a possibility of their quantitative parameterization. Like [7,8], we use the generalization procedure allowing transformation of the measured PAL spectra from x3-term to x2-term (most generalized) decomposition forms.

Firstly, this analysis is applied to *host* matrix without embedded *quest* NPs having (τ_1^{host},I_1^{host}), (τ_2^{host},I_2^{host}), (τ_3^{host},I_3^{host}) inputs in three-term decomposed row PAL spectrum ($I_1^{host} + I_2^{host} + I_3^{host} = 1$). This model can be easily transferred to the known x2-component trapping model by removing I_p=$I_3/3$ input of p-Ps annihilation with τ_p=0.125 ns lifetime from the first channel and (τ_3^{host},I_3^{host}) input from the third channel to the generalized trapping channel (within the generalized x2-term approach). Thus, we can estimate the contribution (τ_a,I_a) to the first channel other than p-Ps:

$$\tau_a \cdot I_a = \tau_1^{host} \cdot I_1^{host} - \tau_p \cdot I_p; \tag{2}$$

$$I_a = I_1^{host} - I_p. \tag{3}$$

Then, returning to NP-embedded substance having (τ_1^*,I_1^*), (τ_2^*,I_2^*) and (τ_3^*,I_3^*) inputs in the x3-term decomposed PAL spectrum ($I_1^* + I_2^* + I_3^* = 1$), we find input from additional channel (τ_{int},I_{int}) assuming its relation to *int*-sites and remainder of *o-Ps*-sites taken like in *host* matrix:

$$I_2^* = I_{int} + I_3^* \cdot (I_2^{host}/I_3^{host}), \tag{4}$$

$$\tau_2^* \cdot I_2^* = \tau_{int} \cdot I_{int} + \tau_2^{host} \cdot (I_2^* - I_{int}). \tag{5}$$

Under this condition, it seems reasonably to equilibrate ($\tau_n \cdot I_n$) input in (τ_a^*,I_a^*) component (i.e. without p-Ps) with the same as ($\tau_{int} \cdot I_{int}$) input in (τ_2^*, I_2^*) component (the full inter-channel balance):

$$\tau_n \cdot I_n / \tau_a^* \cdot I_a^* = \tau_{int} \cdot I_{int} / \tau_2^* \cdot I_2^*. \tag{6}$$

Transferring x3-term PAL spectrum into x2-term one with (τ_a^*,I_a^*) and (τ_t^*,I_t^*) components like in *host* matrix, we find compensating (τ_n,I_n) input arising from this additional trapping channel (τ_{int},I_{int}).

Parameterization of NP-sites can be performed accepting (τ_n,I_n) and (τ_{int},I_{int}) as components of the generalized x2-term PAL spectrum of nanostructurized substance. The second component with τ_{int} lifetime reflects positron trapping sites appeared due to embedded NP. Under accepted prerequisites, these extended free-volume defects can be associated with pseudogap holes at the interface between the outer surface layer of agglomerated NPs and innermost layer of surrounding host matrix [2,7,8]. The bulk lifetime recalculated respectively to these components in according to eq. (4) can be accepted as bulk positron lifetime related to NPs. For monolith NPs, this value tends to bulk lifetime of corresponding substance, while in looser media it is higher. The positron trapping rate of NP-related traps κ_d can be also estimated within this scheme in terms of two-state trapping model [3,6].

Thus, this approach allows description of nanostructurization in substances in terms of *Ps-to-positron trapping conversion* within the same *host* matrix, i.e. the process occurring as transformation of o-Ps-sites in a *host* matrix to positron-trapping sites in the NP-embedded *host-guest* matrix. In contrast to *conventional x3-term decomposition*, describing PAL spectra in terms of mixed positron-Ps trapping, this approach can be defined as *x3-x2-coupling decomposition algorithm (x3-x2-CDA)*. It becomes meaningless, provided nanostructurization concerns defect-free annihilation channel, thus changing PAL parameters of annihilation from delocalized Bloch states in a *host* matrix. Therefore, this algorithm can be considered as *test-indicator* for nanostructurization, separating the processes of host-matrix modification from "pure" interplay between positron- and Ps-trapping.

3. Mathematical modelling and PAL-data interpretation algorithm

As an example, let's consider the experimental results of S. Chakraverty et al. [7] on PAL response produced by *guest* $NiFe_2O_4$ nanocrystals grown in *host* SiO_2 glassy matrix. The x3-term decomposed PAL spectra of these $NiFe_2O_4$-SiO_2 nanocomposites evidently demonstrate gradual decrease in the intensity of o-Ps lifetime component I_3 under nanostructurization, this process being inter-balanced by essential increase in the second component intensity I_2 and defect-related lifetime τ_2 (see Table 1).

Table 1. Fitting parameters describing x3-term decomposed PAL spectra of $NiFe_2O_4$ nanocrystals of different grain sizes grown in SiO_2 matrix after Chakraverty et al. [7].

$NiFe_2O_4$ grain size	Fitting parameters					
	τ_1, ns	I_1, a.u.	τ_2, ns	I_2, a.u.	τ_3, ns	I_3, a.u.
Pure SiO_2	0.116	0.501	0.370	0.272	1.62	0.227
40.0 nm	0.133	0.515	0.400	0.389	1.64	0.096
25.0 nm	0.137	0.558	0.400	0.385	1.76	0.057
15.0 nm	0.139	0.540	0.416	0.433	1.50	0.027
6.8 nm	0.143	0.561	0.420	0.417	1.43	0.022
5.6 nm	0.131	0.508	0.392	0.462	1.13	0.030
4.8 nm	0.133	0.506	0.406	0.452	1.08	0.042
3.5 nm	0.142	0.520	0.432	0.451	1.29	0.029

Using analogy with inert gas bubbles in a metal matrix, the authors [7] explain such behavior due to positron-trapping free volumes at the interfaces of agglomerated NPs appeared instead of Ps-sites. The PAL modes were calculated from a condition of balanced intensities (assuming that percentage contribution to I_1 by intensity of positrons annihilating within NPs I_n is the same as that of I_{int} to I_2). This allows authors [7] to ascribe their τ_n (close to 0.140 ns) to specific positron lifetime in NPs themselves, which seems somewhat underestimated as for typical non-defect spinels [9,10].

Within x3-x2-CDA, we assume that full conversion between positron-trapping and Ps-decaying channels is alone process under nanostructurization in $NiFe_2O_4$-SiO_2 composites, so that no intensities, but rather component inputs are inter-balanced in respect to eq. (9). So, we accept (τ_n,I_n) and (τ_{int},I_{int}) as first and second components of the x2-term PAL spectrum ascribed to hypothetical medium formed under nanostructurization. The defect-free bulk lifetime τ_b^{NP} within 0.195÷0.203 ns domain (Table 2) recalculated in respect to formalism of the x2-component trapping model given by eq. (4) [5-9], can be treated as specific lifetime in NPs themselves, i.e. in $NiFe_2O_4$ spinel. The defect-specific lifetime τ_{int} is a signature of positron-trapping volumes related to embedded NPs. In general, this value shows growing tendency with decrease in $NiFe_2O_4$ grain size towards 3.5 nm as it follows from Table 2. The same is character for positron trapping rate in these defects at the interfaces of NPs κ_d, which is a result of NP-related positron-trapping input in the PAL spectrum. Slight deviations in (τ_{int}-τ_b^{NP}) and τ_2/τ_b values, which can be accepted in x2-component model as signatures of defect size and type [2-6], testify in a favor of rather unchanged nature of these defects irrespectively to $NiFe_2O_4$ grain size.

Table 2. PAL trapping modes describing nanostructurization in $NiFe_2O_4$-SiO_2 nanocomposites [7] for *guest* $NiFe_2O_4$ nanocrystals in respect to *host* pure SiO_2 within x3-x2-CDA

$NiFe_2O_4$	PAL trapping modes within x3-x2-CDA							
grain size	τ_n, ns	I_n, a.u.	τ_{int}, ns	I_{int}, a.u.	τ_b^{NP}, ns	κ_d, ns^{-1}	τ_{int}-τ_b^{NP}, ns	τ_{int}/τ_b^{NP}, a.u.
3.5 nm	0.145	0.468	0.437	0.416	0.211	2.17	0.226	2.07
4.8 nm	0.136	0.434	0.411	0.402	0.200	2.37	0.210	2.05
5.6 nm	0.133	0.456	0.394	0.426	0.195	2.42	0.199	2.02
6.8 nm	0.145	0.515	0.423	0.391	0.203	1.95	0.221	2.09
15.0 nm	0.141	0.488	0.420	0.401	0.202	2.11	0.218	2.08
25.0 nm	0.143	0.433	0.406	0.317	0.197	1.91	0.209	2.06
40.0 nm	0.143	0.329	0.413	0.274	0.203	2.09	0.210	2.03

Principally, the reverse process, tending the positron trapping towards Ps-decaying, can be also well described within this approach due to simultaneous negative values of I_n and I_{int} intensities, which was demonstrated previously at the example of other nanostructurized substances [2,11].

4. Conclusions

Nanostructurization in *host* matrix substances due to embedded *guest* nanoparticles is considered as a modified process in PAL spectroscopy, which involves conversion between Ps-decaying and positron-trapping channels. The developed algorithm allows estimation free volumes of interfacial voids responsible for positron trapping and bulk lifetimes in nanoparticle-embedded substances.

5. References

[1] Vollath D 2013 *Nanomaterials: An Introduction to Synthesis, Properties and Applications. Second Edition* (Weinheim: Wiley-VCH Verlag GmbH)
[2] Shpotyuk O, Filipecki J, Ingram A, Golovchak R, Vakiv M, Klym H, Balitska V, Shpotyuk M and Kozdras A 2015 *Nanoscale Res. Lett.* **10** 77-1
[3] Krause-Rehberg R and Leipner H 1999 *Positron annihilation in semiconductors: defect studies* (Heidelber: Springer)
[4] Jean Y C, Mallon P E and Schrader D M 2003 *Principles and application of positron and positronium chemistry* (New Jersy-London-Singapore-Hong Kong: World Sci. Publ.)
[5] Shpotyuk O and Filipecki J 2003 *Free volume in vitreous chalcogenide semiconductors: possibilities of positron annihilation lifetime study* (Czestochowa: Ed. WSP)
[6] Tuomisto F and Makkonen I 2013 *Rev. Mod. Phys.* **85** 1583
[7] Chakraverty S, Mitra S, Mandal K, Nambissan P M G and Chattopadhyay S 2005 *Phys. Rev. B* **71** 024115-1
[8] Mitra S, Mandal K, Sinha S, Nambissan P M G and Kumar S 2006 *J. Phys. D* **39** 4228
[9] Jensen K O and Nieminen R M 1987 *Phys. Rev. B* **35** 2087
[10] Nambissan P M G, Mondal O, Chakrabarty S and Pal M 2013 *Mater. Sci. Forum* **733** 219.
[11] Shpotyuk O, Ingram A, Shpotyuk Ya, Bujňáková Z, Baláž P 2017 *Polym. Eng. Sci.* **57** 502.

IOP Conf. Series: Journal of Physics: Conf. Series **936** (2017) 012050 doi:10.1088/1742-6596/936/1/012050

Kinetics models describing degradation-relaxation effects in nanoinhomogeneous substances

O Shpotyuk[1,2,*], V Balitska[3] and M Brunner[4]

[1] Institute of Physics, Jan Dlugosz University in Czestochowa, 13/15 al. Armii Krajowej, 42200, Czestochowa, Poland
[2] Vlokh Institute of Physical Optics, 23 Dragomanov str., 70005 Lviv, Ukraine
[3] Lviv State University of Life Safety, 35 Kleparivska str., 79040 Lviv, Ukraine
[4] Technische Hochschule Köln, University of Technology, Arts, Sciences, 2 Betzdorfer Strasse, 50679 Köln, Germany

E-mail: olehshpotyuk@yahoo.com

Abstract. The mathematical models of degradation-relaxation kinetics are considered for jammed systems composed of screen-printed spinel $Cu_{0.1}Ni_{0.1}Co_{1.6}Mn_{1.2}O_4$ and conductive Ag or Ag-Pd alloys. Structurally-intrinsic nanoinhomogeneities due to Ag and Ag-Pd diffusants embedded in spinel phase environment are shown to define governing kinetics of thermally-induced degradation obeying an obvious non-exponential behaviour in the resistance drift. The stretched-to-compressed exponential crossover is detected for degradation-relaxation kinetics in these systems with conductive contacts made of Ag-Pd and Ag alloys. Under essential migration of conductive phase, the resulting kinetics is though to be considerable two-step diffusing process originated from Ag penetration deep into spinel ceramics.

1. Introduction

The jammed systems prepared from high-temperature state by so-called fluid-to-solid transition compose an important class of substances for practical using [1]. Being quenched in far-from-equilibrium configuration, these materials demonstrate response on different degradation-relaxation inputs, tending towards equilibrium in the controlled parameter *N(t)* through compressed-exponential relaxation (CER) kinetics (super-exponential, i.e. faster than simple exponential decay):

$$N_\eta(t) = exp\left[\left(-\frac{t}{\tau}\right)^\beta\right], \tag{1}$$

with scaling (viz. compressing) exponent $\beta>1$ and characteristic decay time τ.

Despite recent progress in the functional characterization of such substances [1-6], their phenomenological behaviour has been remained mysterious in many aspects, especially from a point of mathematical modelling and prediction.

* To whom any correspondence should be addressed.

This work deals with specificity of thermally-induced electric degradation-relaxation phenomena revealed in heterogeneous jammed solid-state system composed of the same ceramics matrix with different types of diffusion agents (diffusants), exemplified by thick-film structures of spinel $(Cu,Ni,Co,Mn)_3O_4$ manganites with screen-printed conductive layers composed of "pure" Ag and Ag-based (made of Ag-Pd conductive paste C1216) alloys [7-11].

2. Experimental

2.1. Nanoinhomogeneous ceramics preparation
The starting ceramics were synthesized by sintering route (1040 °C, 4 h duration) using reagent grade Cu carbonate hydroxide, and Ni, Co, Mn carbonate hydroxide hydrates [12-14]. The paste was prepared by mixing powder of basic $Cu_{0.1}Ni_{0.1}Co_{1.6}Mn_{1.2}O_4$ ceramics with МБ-60 glass, Bi_2O_3 (used as inorganic binder) and some organic vehicle. Then, the paste was printed on alumina substrates (Rubalit 708S) with conductive Ag-Pd layer screen printed from C1216 paste using a manual screen-printing device equipped with a steel screen. Finally, the thick-film structures were fired at 850 °C.

The prepared thick-film spinel-conductor systems were subjected to degradation under storage at 170 °C within (24-360 h) [7-10]. The results of aging tests were controlled by relative resistance drift (RRD), e.g. changes in electrical resistance $\Delta R/R_0$ detected under normal conditions, the confidence interval in the RRD measuring error-bar being no worse than ±0.2 %. Additional deviations in some points were allowed due to faults in an exact reproduction of degradation cycles (cooling regime from aging temperature, influence of environment and humidity, etc.). Statistical analysis testified these factors introduced additional error of about ±0.2 % in $\Delta R/R_0$ values. So, the overall uncertainties in electrical measurements within this ageing protocol did not exceed ±(0.4-0.5) %.

2.2. Mathematical simulation procedure
With a purpose of adequate mathematical description of degradation-relaxation kinetics in $\Delta R/R_0$ values, the numerical values of fitting parameters in typical relaxation functions (RF) were calculated in such a way to minimize the mean-square deviation *err* of experimentally measured points from chosen RF, taken as adequate solution of general degradation equation in a power-like form [15-17]:

$$\frac{d\eta}{dt} = -\lambda \eta^{\alpha} t^{\gamma}, \tag{2}$$

where power indexes α and γ as well as λ coefficient are some material-related constants.

All RFs can be determined from this eq. (2) in dependence on controlled relaxation parameter $\eta(t)$. As was pointed out in [15-17], there are five typical kinds of RFs, which can be considered as partial solutions of the equation (2) in dependence on numerical α and γ variables.

In case of $\alpha=1$ and $\gamma=0$, we have a well known monomolecular relaxation kinetics with RF 1, expressed by simple exponential dependence on time t with $\tau=1/\lambda$ ($\lambda\neq0$):

$$\eta = a\left(e^{-t/\tau} - 1\right). \tag{3}$$

If the degradation is caused by recombination of specific types of defect pairs, the governing kinetics is defined by bimolecular RF 2 at $\alpha=2$ and $\gamma=0$ ($\tau=1/\lambda$, $\lambda\neq0$):

$$\eta = a\left(\left(1 + t/\tau\right)^{-1} - 1\right). \tag{4}$$

The exact solution of equation (1) at $\gamma=0$ gives RF 3, which exhibits character "stretched" behaviour owing to standard α^{th}-order kinetics of degradation ($\tau=1/\lambda(\alpha-1)$, $\beta=1/(\alpha-1)$, $\alpha\neq1$, $\lambda\neq0$):

$$\eta = a\left(\left(1+ {t}/{\tau}\right)^{-\beta} - 1\right). \tag{5}$$

In case of $\alpha=1$ and $\gamma\neq 0$, the relaxation process is described by stretched exponential RF 4 with $\tau=(1+\gamma)/\lambda$, $\beta=1+\gamma$, $\beta\neq$-1 and $\lambda\neq 0$:

$$\eta = a\left(e^{\left(-{t}/{\tau}\right)^{\beta}} - 1\right). \tag{6}$$

Noteworthy, in dependences of β values, this RF 4 can be presented in stretched-exponential (β<1) or compressed-exponential (β>1) form, thus corresponding to above eq. (1). The exact solution of eq. (2) with arbitrary values of α and γ (both different from 0 or 1) can be presented by RF 5 with $\tau=(1+\gamma)/\lambda(\alpha-1)$, $\beta=1+\gamma$, $r=1/(\alpha-1)$, $\alpha\neq 1$, $\beta\neq -1$ and $\lambda\neq 0$:

$$\eta = a\left(\left[\left(1+ {t}/{\tau}\right)^{\beta}\right]^{-r} - 1\right). \tag{7}$$

This formalism of eqs. (4)-(8) is applied to select the best RF describing degradation-relaxation kinetics in terms of minimal mean-square deviations between experimental points and modeling curve.

3. Results and discussion

The both types of the studied nanoinhomogeneous substances demonstrate monotonically growing tendency in the RRD (in electrical resistance $\Delta R/R_0$), this behavior having more threshold-like character in case of screen-printed $Cu_{0.1}Ni_{0.1}Co_{1.6}Mn_{1.2}O_4$ spinel ceramics with conductive Ag contacts as it shown previously in [7-11]. These dependences were examined through above RFs 1-5 to determine the best one describing this kinetics (the results for $Cu_{0.1}Ni_{0.1}Co_{1.6}Mn_{1.2}O_4$ thick films prepared with screen-printed Ag-Pd and Ag contacts are given in Tables 1 and 2, respectively).

Table 1. Parameters of the RFs, describing $\Delta R/R_0$ kinetics in thick-film $Cu_{0.1}Ni_{0.1}Co_{1.6}Mn_{1.2}O_4$ with Ag-Pd contacts.

	err	*a*	τ	β	*r*
RF 1	0.22	6.9	22.5	-	-
RF 2	0.06	7.8	17.4	-	-
RF 3	0.04	10.6	5.2	0.32	-
RF 4	0.05	7.8	32.3	0.58	-
RF 5	0.05	9.3	12.5	0.81	0.66

Table 2. Parameters of the RFs, describing $\Delta R/R_0$ kinetics in thick-film $Cu_{0.1}Ni_{0.1}Co_{1.6}Mn_{1.2}O_4$ ceramics prepared with conductive screen-printed Ag contacts.

	err	*a*	τ	β	*r*
RF 1	0.06	6.9	279	-	-
RF 2	0.07	11.3	444	-	-
RF 3	0.08	8.3	1014	3.07	-
RF 4	0.001	4.8	154	1.68	-
RF 5	0.002	5.1	251	1.82	2.73

The best results of modelling for thick-films $Cu_{0.1}Ni_{0.1}Co_{1.6}Mn_{1.2}O_4$ ceramics with Ag-Pd contacts (see table 2) can be presented by a well-known stretched-exponential RF 4 given by equation (7) with a character non-exponentionality index β approaching ~0.58 and effective time constant τ tending towards ~32 h. In contrast, with changing in the conductive material of electrical contacts (under transition to the same thick-films $Cu_{0.1}Ni_{0.1}Co_{1.6}Mn_{1.2}O_4$ ceramics with Ag contacts), the governing degradation-relaxation kinetics is drastically changed attaining an obvious CER character with over-unity compressibility index β = 1.68 and time constant τ reaching ~154 h.

The crossover of the detected degradation-relaxation kinetics from stretched-exponential (β = 0.58) to CER (β = 1.68) in the studied jammed system (caused by change in the diffusant type in a contact with $Cu_{0.1}Ni_{0.1}Co_{1.6}Mn_{1.2}O_4$ spinel ceramics) is well depicted in the linearized form as it shown on figure 1. Within this presentation in *y-x* variables, the stretched-exponential behaviour (with β = 0.58) is defined by experimental points grouped along red-coloured dotted straight line in the 4-8 octants, while the CER behaviour (with β = 1.68) is defined by experimental points grouped along green-coloured dash-double-dotted line penetrating the 3-7-th octants.

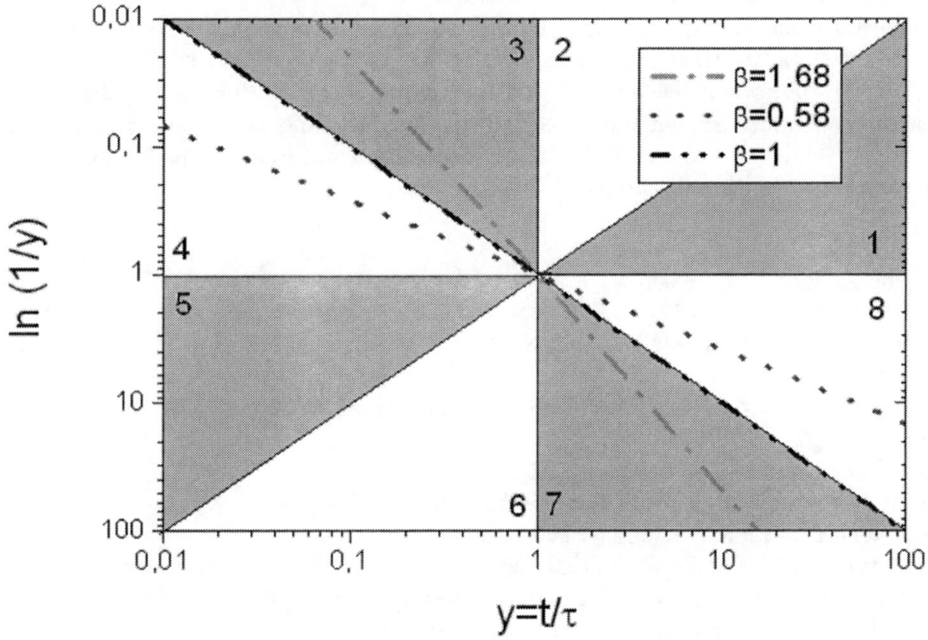

Figure 1. The linearized presentation of stretched- (β=0.58) and compressed-exponential (β=1.68) relaxation kinetics, determined by $y=\eta(t)=exp[(-x)^{\beta}]$ functional in dependence on $x=t/\tau$.

The EDX study of Ag distribution profiles taken on fresh cut-sections of thick-film ceramics-Ag structures before and after degradation testing shows essential metallic penetration into ceramics [11].

The metallic Ag appears in the regions of spinel ceramics adjusted to conductor contact, as well as in deep intrinsic region in ceramics bulk. In case of conductive Ag-Pd contact, the metallic Ag penetrates only in short pre-contact region. It is known that conductive material penetrating spinel ceramics reduces electrical resistivity, thus resulting in negative RRD for such thick-film systems [10,11]. This diffusive-related process is strongly thermally-activated, in contrast to own structural changes occurring in ceramics body, which produces rather positive feedback in the RRD due to increased defectiveness of spinel matrix [14].

The negative RRD quickly saturates with conductive material penetrating into ceramics as it character for thick films of $Cu_{0.1}Ni_{0.1}Co_{1.6}Mn_{1.2}O_4$ spinel supplied by conductive Ag-Pd alloy contacts. Under these conditions, the whole Ag-Pd alloy behaves as one "cumulative" diffusing agent with a significantly suppressed possibility for Ag atoms migration. The resulting kinetics of such diffusive degradation-relaxation process attains strong tendency to yield stretched-exponential scenarios.

However, if Ag penetration is not sufficiently inhibited in a spinel ceramics by Pd addition, as for $Cu_{0.1}Ni_{0.1}Co_{1.6}Mn_{1.2}O_4$ thick films with contacts made of pure screen-printed Ag, the kinetics of the resulting diffusive process is principally changed. The overall diffusion-limited relaxation in the studied thick-film system occurs as two-step penetration of conductive agent (Ag) into spinel body.

4. Conclusions

Nanoinhomogeneous due to metallic diffusants (Ag and Ag-Pd alloys) in $Cu_{0.1}Ni_{0.1}Co_{1.6}Mn_{1.2}O_4$ spinel ceramics are shown to be decisive factors governing kinetics of thermally-induced degradation tests under 170 °C obeying an obvious non-exponential behaviour in relative resistance drift. Two different degradation-relaxation kinetics are detected for the same thick-film $Cu_{0.1}Ni_{0.1}Co_{1.6}Mn_{1.2}O_4$ spinel ceramics in dependence on supplied metallic contacts. If Ag migration is inhibited in ceramics by Pd addition due to conductive Ag-Pd alloy, the governing kinetics attains a stretched exponential form with stretching exponent $\beta = \sim 0.58$ typical for one-stage diffusion in structurally dispersive media. Under Ag penetration into ceramics, as for thick-films supplied by Ag contacts, the degradation-relaxation kinetics drastically changes, attaining a compressed-exponential form with compressibility index $\beta = 1.68$. The resulting kinetics in this case is though to be attributed to two-step diffusing process originated from Ag penetration deep into spinel $Cu_{0.1}Ni_{0.1}Co_{1.6}Mn_{1.2}O_4$ ceramics body.

5. References

[1] Cipelletti L, Ramos L, Manley S, Pitard E, Weitz D A, Pashkovski E E, Jahnsson M 2003 *Faraday Discuss* **123** 237
[2] Narayanan R A, Thiyagarajan P, Lewis S, Bansal A, Schadler L S, Lurio L B 2005 *Phys. Rev. Lett.* **97** 075505
[3] Srivastava S, Kandar A K, Basu J K, Mukhopathyay M K, Lurio L B, Narayanan S, Sinha S K 2009 *Phys. Rev. E* **79** 021408
[4] Hashimoto C, Ushiki H 2000 *Polym J* **32** 807
[5] Hamm P, Helbing J, Bredenbeck J 2006 *Chem Phys* **323** 54
[6] Nakamura H K, Sasai M, Takano M 2004 *Chem Phys* **307** 259
[7] Klym H, Balitska V, Shpotyuk O, Vakiv M 2005 *Chemia Ochrona Srodowiska (Pol.)* **10** 33
[8] Hadzaman I, Klym H, Shpotyuk O 2014 *Int J. Nanotechnol* **11** 843
[9] Klym H, Hadzaman I, Shpotyuk O 2014 *Can J. Phys* **92** 822
[10] Klym H, Balitska V, Shpotyuk O, Hadzaman I 2014 *Microelectronics Reliability* **54** 2843
[11] Shpotyuk O, Brunner M, Hadzaman I., Balitska V, Klym H 2016 *Nanoscale Res Lett* **11** 499
[12] Shpotyuk O, Balitska V, Brunner M, Hadzaman I, Klym H 2015 *Phys. B* **459** 116
[13] Altenburg H, Plewa J, Plesch G, Shpotyuk O 2002 *Pure Appl Chem* **74** 2021
[14] Vakiv M, Shpotyuk O, Balitska V, Butkiewicz B, Shpotyuk L 2004 *J Europ Cer Soc* **24** 1243
[15] Balitska V, Shpotyuk O 2006 Archives of Mater **4** 189.
[16] Balitska V, Butkiewich B, Shpotyuk O, Vakiv M 2002 *Microelectronics Reliability* **42** 2003
[17] Vakiv M, Shpotyuk O, Mrooz O, Hadzaman I 2001 *J Europ Ceram Soc* **21** 1783

Investigation of humidity using the muon component of cosmic rays

V Oskomov, A Sedov , N Saduyev , O Kalikulov*, I Kenzhina, A Naurzbayeva, N Alimgazinova , A Zhumabaev , S Shinbulatov , N Erezhep

Institute of Experimental and Theoretical Physics of al-Farabi Kazakh National University

*orazaly_82@mail.ru

Abstract. Determination of humidity is one of the most important types of hydrometeorological and glaciological observations performed in agriculture, hydropower and water supply. The work is devoted to the development of physical basis of moisture determination method, based on attenuation of the flux of cosmic-ray muons. The relationship between the intensity of muons registered in the underground room of the Tien Shan mountain research station (Almaty) and relative humidity was studied. The results of studies show that the values of the normalized mutual correlation function between the rows of muon intensity and relative humidity vary from 0.3 to 0.7, depending on the coincidence scheme. The data obtained from the muon telescope located at the the Tien Shan mountain research station was used in the work.

1. Introduction

The possibility of using muon fluxes to estimate the amount of water and humidity is considered. This problem is solved using muons generated by cosmic rays in the atmosphere. The properties of muons are considered with reference to the problem of determining the amount of water and moisture using cosmic rays.

2. Instruments and methods

During studying of meson fluxes with energies within 100 GeV, one can neglect the recording of π- and K mesons in pion-hadron and kaon-hadron collisions, and also in the production of pions in K decays. In this case, for the energy spectra of vertically incident muons at sea level, the difference does not exceed 3%. In these approximations, the transport of mesons in the atmosphere is described by the following equation:

$$\left\{\frac{\partial}{\partial h} + \lambda_\mu^{-1} + m_\mu\left[p\tau_\mu\rho(h,\theta)\right]^{-1}\right\}\mu(E,h,\theta) = G_\mu(E,h),$$

where λ_μ - mean meson mean free path to inelastic interaction; m_μ, τ_μ, p –meson mass, its lifetime and momentum, respectively; $\rho(h,\ \theta)$ – air density at depth h and in direction θ; $G_\mu(E,h) = G_\mu(E,h,R_i)$ - function of meson generation by μ-nucleons and nuclei; $\mu\ (E,\ h,\ \theta)\ dE = \mu\ (E,\ h,\ \theta,\ R_i)\ dE$ – intensity of mesons of μ type ($\pi^\pm$, $K^\pm$) with energies from E to $E+dE$ at atmosphere depth h, propagating in a single solid angle under the zenith angle θ, R_i - rigidity of geomagnetic cutoff.

IOP Conf. Series: Journal of Physics: Conf. Series **936** (2017) 012051 doi:10.1088/1742-6596/936/1/012051

The solution of above equation has the following form:

$$\mu(\mathrm{E},\mathrm{h},\theta)=\int_0^h dh'\exp\left[-\frac{h-h'}{\lambda_\mu(E)}-\frac{m_\mu}{\tau_\mu p}\int_{h'}^{h}\frac{dh''}{\rho(h'',\theta)}\right]G\,(E,h')$$

The relationship between the temperature and pressure of the atmosphere h_0 has the form:

$$T(h_0)=T_s\left[1+\alpha\theta(h_0-h_t)\left(\frac{h_0}{h_t}-1\right)\right]$$

where T_s - stratospheric temperature; h_t - pressure over the tropopause; α - parameter characterizing the temperature gradient in troposphere; $\theta(x)=1$ at $x>0$ and $\theta(x)=0$ at $x<0$. During calculations, it was used $T_s=250\,K$, $h_t=200\,g/cm^2$, $\alpha=0{,}073$. Relation ρch and θ has the following form:

$$\rho(h,\theta)=\rho\,(h,cos\theta^*)\approx ghcos\theta^*/[R_0T(cos\theta^*h)]$$

where θ^* - effective angle, coinciding with θ in the plane atmosphere limit. For a spherical atmosphere with $cos\theta\gg\sqrt{2H/R_i}$:

$$\cos\theta^*=\frac{\left(\frac{2H_0}{\pi R_E}\right)^{\frac{1}{2}}\exp\left(-\frac{cos^2\theta R_E}{2H_0}\right)}{erfc\cdot cos\theta(\frac{R_E}{2H_0})};\qquad H_0=\frac{R_0T_S}{g},$$

where R_E - Earth's radius, R_0 - specific gas constant; H – height above sea level.

Estimates show that taking nonisothermicity into account is necessary for low-energy mesons at sea level:

One-dimensional kinetic equation describing the transport of muons in the atmosphere without taking into account Coulomb scattering and the motion of muons in the geomagnetic field:

$$\left\{\frac{\partial}{\partial h}+m_\mu[p\tau_\mu\rho(h,\ \theta)]^{-1}\right\}\mu\,(E,h,\theta)=\frac{\partial[\beta_\mu(E)\mu\,(E,\ h,\ \theta)]}{\partial E}+G_\mu(E,h,\theta),$$

Where is the function of generating muons in π_{μ_2} and K_{μ_2} – decays:

$$G_\mu(E,h,\theta)=\sum_\mu\frac{m_\mu B_\mu}{\tau_\mu\rho\,(h,\theta)}\left(1-\frac{m_\mu^2}{m_\pi^2}\right)^{-1}\int_{E_\mu^-}^{E_\mu^+}\frac{dE_\mu}{p_\mu^2}\,\mu\,(E_\mu,h,\theta),$$

B_μ - relative decay probability; B_μ(E)= $-dE/dh$ - continuous losses of muon energy. Limits of integration are the following:

$$E_\mu^\pm=\frac{1}{2}\,[(\frac{m_\mu^2}{m_\pi^2}+1)E\pm(\frac{m_\mu^2}{m_\pi^2}-1)p].$$

at β_μ=β= const up to $E\approx 10^2$ the solution will be the following:

$$\mu(E,h,\theta)=\int_0^h dh'\,\omega_\mu(E,h',h,\theta)G_\mu(E_{h-h'},h',\theta)$$

where $\omega_\mu(E,h',h,\theta)$ - probability that the muon does not decay as the layer passes from $h'$to h, is equal:

$$\omega_\mu(E,h',h,\theta) =$$
$$\tilde{\psi}\ (E,h'_m,h)\ ^{-\alpha\theta(hcos\theta^*-h_t)\frac{p_{\mu c}}{\beta h_t}} \times \left\{\psi(E,h'_m,h)\ ^{-\alpha\theta(h\,cos\theta^*-h_t)}\psi(E,h',h)\right\}\frac{p_{\mu c}\sec\theta^*}{p_h}($$

Also, there are the following denotions:

$$\psi(E,h',h) = \left[1-\frac{m_\mu^2}{\omega^2}\frac{\beta(h-h')}{E}\right]\left[1+\frac{\beta(h-h')}{E}\right]\ ^{-1}\left(\frac{h'}{h}\right);$$
$$\tilde{\psi}\ (E,h',h) = \left[1-\frac{m_\mu^2}{\omega^2}\frac{\beta(h-h')}{E}\right]\left[1+\frac{\beta(h-h')}{E}\right];$$
$$\omega^2(E,h',h) = \frac{(Ep_{h-h'}+pE_{h-h'})(E_hE_{h-h'}+p_hp'_h-m_\mu^2)}{E(p_h+p_{h-h'})+E_{h-h'}(p_h+p)};$$
$$\tilde{\omega}\ (E,h',h) = \frac{(Ep_{h-h'}+pE_{h-h'})(Ep_{h-h'}+pE_{h-h'})}{E+E_{h-h'}}$$
$$E_p = E+\beta h;\ p_h = (E_h^2-m_\mu^2)\ ^{1/2};\ p_{\mu c} = \frac{H_{0}m_\mu}{\tau_\mu},$$
$$h'_m = \max\,(h',h_t sec\theta^*)$$

at energies much larger βh, ω_μ has the following form:

$$\omega_\mu(E,h',h,\theta) \simeq \exp\left[-\frac{\alpha\theta(hcos\theta^*-h_t)(h-h'_m)}{h_t\left(\frac{p_{\mu c}}{p}\right)}\right] \times \left\{\frac{h}{h'}\left(\frac{h'_m}{h}\right)\ ^{-\alpha\theta(hcos\theta^*-h_t)}\right\}\frac{p_{\mu c}}{p}cos\theta^*.[1\text{-}4]$$

3. Results and discussion

A relation was obtained between the intensity of muons (Figure 1) recorded in the underground room of the Tien Shan mountain research station for the coincidence circuit 1 (Figure 2) and the relative humidity of atmosphere. 2-minute data of the intensity of muons was used of Tien Shan mountain research station of Ionosphere Institute of NCITC.

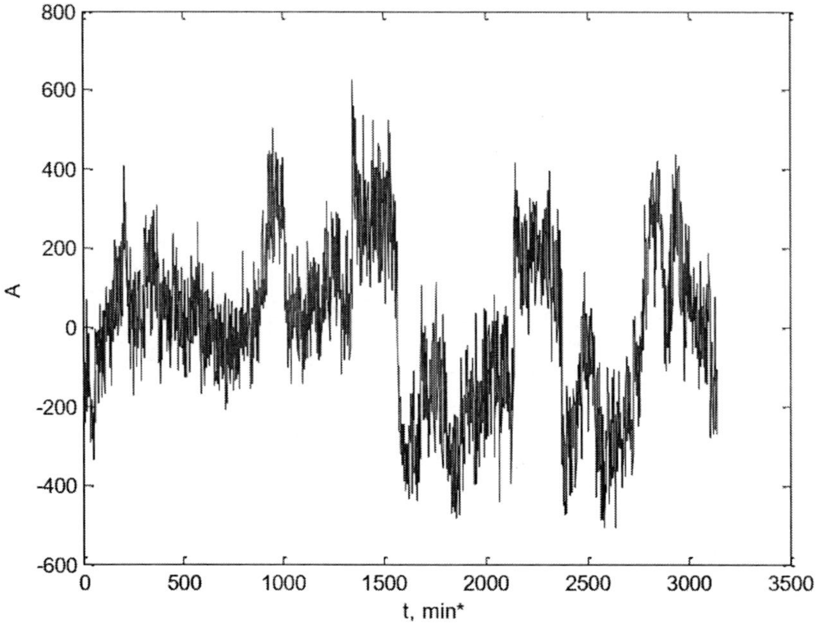

Figure 1. Uncorrected for 2-minutes data muon intensity in underground station for October 2016.

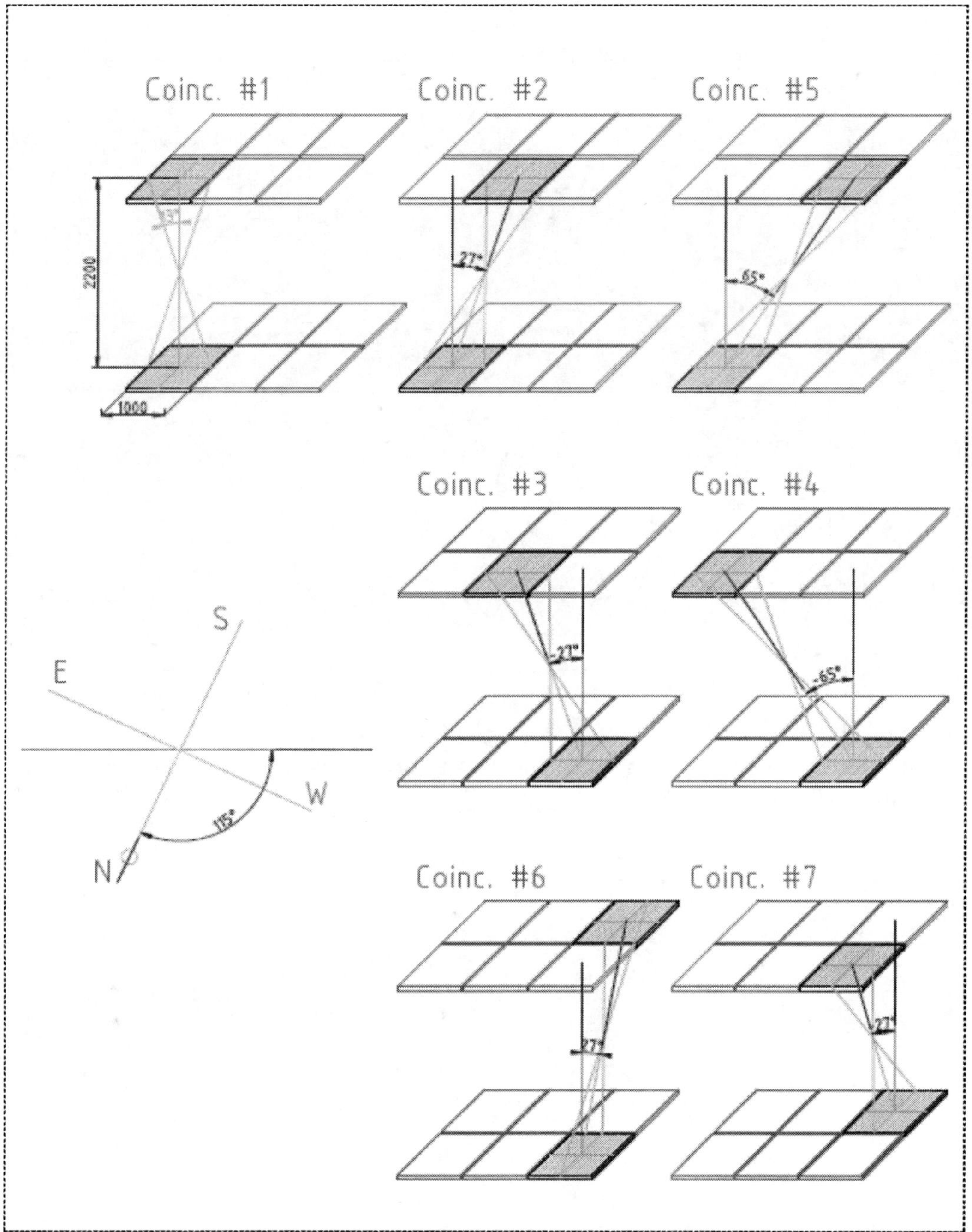

Figure 2. The scheme of muon coincidences.

4. Conclusions

Correlation analysis was carried out for those periods of time when an average correlation was observed between time series of minute values of counting rate and relative humidity. The results of the studies show that the values of normalized mutual correlation function between the rows of muon intensity and relative humidity vary from 0.3 to 0.7, depending on the coincidence scheme under consideration. So, for example, for the time period from October 1 to October 4, 2016, we see the strongest link in the scheme of matches according to scheme 5, as well as autocorrelation in scheme 3. (See Figures 3,4)

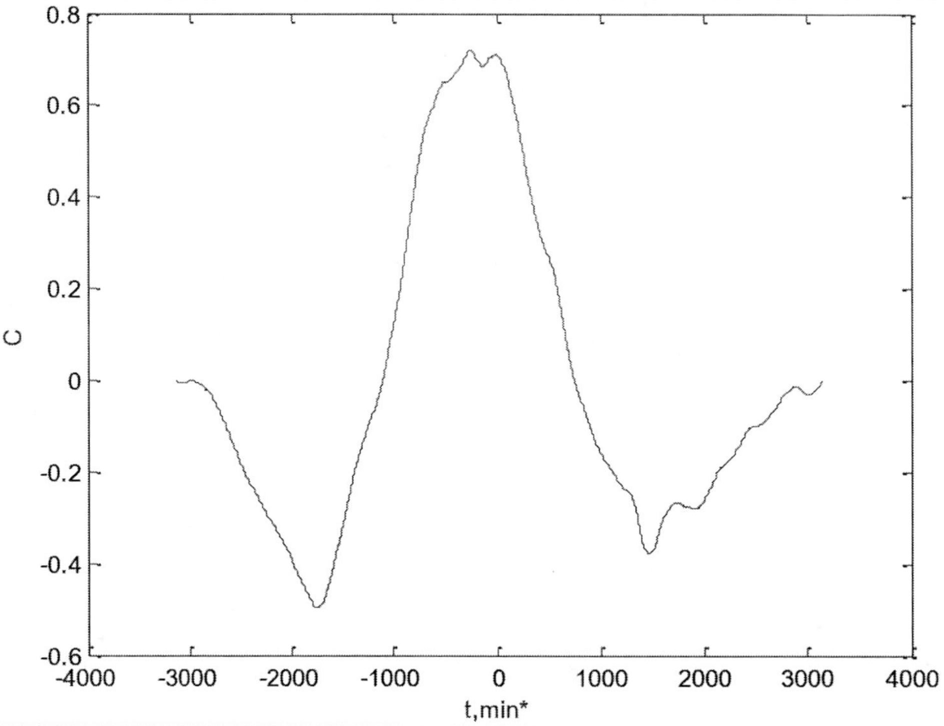

Figure 3. The normalized cross-correlation function calculated for time series of muon intensity (Scheme 5) and relative humidity for the period from October 1 to October 4, 2016

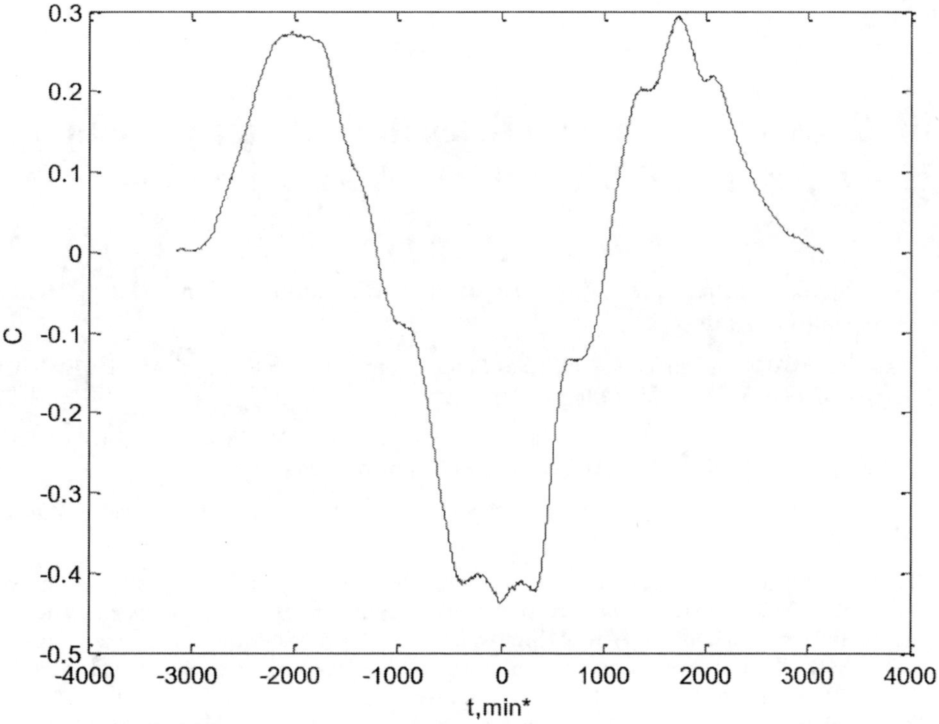

Figure 4. The normalized cross-correlation function calculated for time series of muon intensity (Scheme 3) and relative humidity for the period from October 1 to October 4, 2016

Acknowledgments
The study was carried out under the support of grant #3842/GF4 of Ministry of Education and Science of Kazakhstan.

References

[1] Pierog T, Werner K 2008 Muon production in extended air shower simulations *Phys. Rev. Lett.* 101 171101.
[2] Basarov E V, Beisembaev R U, Besshapov S P 1975 The observation of muon groups at Tien-Shan station *Proceedings of the 14th ICRC* 6 2067-2071
[3] Basarov E V, Beisembaev R U, Besshapov S P 1977 The search for directly produced muon pairs in extensive air showers *Proceedings of the 15th ICRC* 6 179-184.
[4] Stamenov J N, Georgiev N K, Kabanova N V 1979 Phenomenological characteristics of the muon component of extensive air shower sat mountain level *Trudy FIAN Nauka,Moscow* 109 132-151.
[5] Britvich I, Chernichenko S K, Chubenko A P, Gilitsky Y V 2006 The largest cintillation charged particles detector of theTien- Shan complex 'ATHLET' *Nucl.Instrum.Methods* A5 64(1) 225-234.

IOP Conf. Series: Journal of Physics: Conf. Series **936** (2017) 012052 doi:10.1088/1742-6596/936/1/012052

2D Spin Crossover Nanoparticles described by the Ising-like model solved in Local Mean-Field Approximation

Salah Eddine Allal,[1,2] Jorge Linares,[1,2] K. Boukheddaden,[1] Pierre Richard Dahoo,[2] F. de Zela[3]

[1]GEMAC, Université de Versailles Saint-Quentin-en-Yvelines, 45 Av. des Etats-Unis, CNRS-UMR 8635, 78035 Versailles Cedex, France

[2]LATMOS, Université de Versailles-Saint-Quentin-en-Yvelines, CNRS-UPMC-UVSQ (UMR 8190), 11 boulevard d'Alembert, 78280 Guyancourt, France

[3] Departemento de Ciencias, Sección Física, Pontificia Universidad Católica del Perú, Apartado 1761, Lima, Perú.

Abstract: Some six-coordinate iron (II) coordination compounds exhibit thermal-, optical-, electrical-, magnetic- and pressure-induced switching between the diamagnetic low-spin (LS, S=0) and the paramagnetic high-spin (HS; S=2) states [1]. This may lead to potential application of these complexes in molecular devices such as temperature and pressure sensors [2]. An Ising-like model has been proposed to explain the occurrence of the thermal hysteresis behaviour [3,4] of this switchable solids. In this contribution, the local mean field approximation is applied to solve the Hamiltonian modelling interactions pertaining to 2D nanoparticles embedded in a magnetically-inactive matrix.

1. Introduction

It is very well known that spin-transition phenomena are related to the thermodynamic competition between high-spin (HS) and low-spin (LS) states of spin-crossover (SCO) molecules. Spin-transition phenomena can be triggered by various external perturbations such as temperature, pressure, light and magnetic fields[1-7]. During the occurrence of the SCO phenomenon, several physical properties of the material (magnetic, optical, electrical, vibrational, etc.) can change dramatically, making it possible to track the process by using various detection techniques (magnetometry, optical spectroscopy, Mössbauer, calorimetry, dielectric constant, etc.).

In this work, we implement an approach based on the mean field approximation, which is commonly employed as a first step when trying to construct a more accurate theory. Actually, there are many possible mean field models, but they all share the same spirit. Here we deal with a local mean field approximation (LMFA), which takes into account for the dependence of the number of neighbors on whether the molecules are in the bulk or at the border (or on the surface) of the lattice. By this LMFA, we study both shape and size effects on the SCO compounds.

2. Model and principle of calculation

Various physical techniques and methods, such as magnetic, optical and vibration spectroscopy, X-ray diffraction and heat capacity measurements are used to study the behaviour of switching SCO materials. To reproduce the different results obtained by these techniques, researchers have proposed different models such as atom-phonon coupling[8-9], Ising-like[10-11-12-13] or mechano-elastic and electro-elastic[14] coupling. One can predict the way SCO materials react to external factors, depending on their lattice architecture and the short- and long-range interactions among the particles of the materials. In this work, we have applied a local mean field approximation of the Ising-like model using realistic parameters to calculate and obtain the bi-stability property of the spin state. This will help to get more information about the switching properties of the complex under study. The Ising-Like model proposed by Wajnsflasz and Pick uses a fictitious spin operator σ, whose eigenvalues, ±1 correspond to the high-spin (HS) state and to the low-spin (LS) state, respectively, whose degeneracies are denoted by g_{HS} for the high spin and g_{LS} for the low spin. The Hamiltonian is written in this form:

$$H = \frac{\Delta - Tln(g)}{2}\sum_i \sigma_i - J\sum_{ij} \sigma_i\sigma_j$$

Here, Δ represents the difference between the energy of the HS and the energy of the LS state, ∑(i,j) is the sum over nearest neighbors and $g=g_{HS}/g_{LS}$ is the degeneracy ratio. We can express the fraction of high spin, n_{HS}, in terms of the average value $\langle\sigma\rangle$, as $n_{HS} = \frac{\langle\sigma\rangle+1}{2}$. In the usual homogeneous mean-field approximation, the total Hamiltonian writes:

$$H = \frac{\Delta - Tln(g)}{2}\sum_i \sigma_i - Jq\langle\sigma\rangle\sum_i \sigma_i$$

and then:

$$H = \frac{\Delta - Tln(g) - 2Jq\langle\sigma\rangle}{2}\sum_i \sigma_i$$

Where q is the number of neighboring sites considered here as independent on the site i. Because our Hamiltonian is of the form $H = -h\sigma_i$, it readily follows that $\langle\sigma\rangle$ is given by the equation $\langle\sigma\rangle = \tanh(\beta h)$ where,

$$h = -\frac{\Delta - Tln(g) - 2Jq\langle\sigma\rangle}{2} \text{ and } \beta = \frac{1}{k_B T}.$$

To determine the thermal-dependence of the average magnetization, one must solve the following equation:

$$\langle\sigma\rangle = tanh(-\frac{\Delta - Tln(g) - 2Jq\langle\sigma\rangle}{2k_B T})$$

For a "classical" homogeneous mean field in the case of a 2D square lattice, the used coordination number is q = 4. This approximation remains valid as long as we consider an infinite lattice, which means that the surface effects are negligible. In this work, we study nanoparticles, for which surface and border effects play a significant role on their properties. To include these effects, we take into account that the number of neighbors for each molecule depends on its position (see Fig. 1).

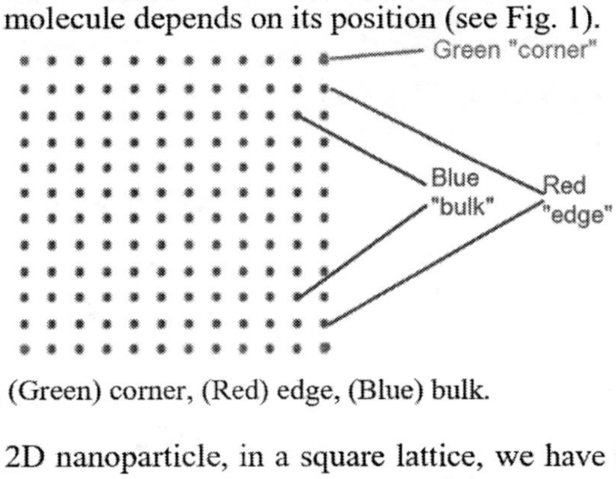

Figure 1. Matrix of particles, (Green) corner, (Red) edge, (Blue) bulk.

As we see in Fig. 1, for a 2D nanoparticle, in a square lattice, we have three types of sites or atoms: corner, edge and bulk atoms. Thus, atoms at a "corner" (green dots in Fig. 1) have two neighboring sites while particles located on the "edge" (red dots)have three neighbors and those of the bulk (in blue) connect to four neighbors. If we take into account the interactions between the molecules with the environment, we must include the term $-L\sum_i \sigma_i$ for "corner" and "edge" molecules.

In the following, we consider that the bulk, edge and corner magnetizations are different, and that the mean-field approximation can be applied to each of these sites, which means that we consider a local mean-field. So, to obtain the average value of the spin state we need to write the corresponding expression for each type of atom, which gives:

- for the corner $$\langle\sigma_C\rangle = \tanh(-\frac{\Delta-4\times L-T\ln(g)-2\times J\times 2\langle\sigma\rangle}{2k_BT})$$
- for the edge $$\langle\sigma_E\rangle = \tanh(-\frac{\Delta-2\times L-T\ln(g)-2\times J\times 3\langle\sigma\rangle}{2k_BT})$$
- And for the bulk $$\langle\sigma_B\rangle = \tanh(-\frac{\Delta-T\ln(g)-2J\times 4\langle\sigma\rangle}{2k_BT})$$

The average magnetization over the lattice is easily obtained as from the expression of each spin type, as follows

$$\langle\sigma\rangle = \frac{N_C\langle\sigma_C\rangle + N_E\langle\sigma_E\rangle + N_B\langle\sigma_B\rangle}{N_T}$$

Where N_C (=4) N_E (= 4N-8) and N_B (=$(N-2)^2$) are, respectively, the number of corner, edge and bulk particles and N_T (=N^2) is the total number of particles.
To fully understand the evolution of the hysteresis, it is worth mentioning that in our case, the first-order phase transition takes place only when the condition $T_{OD}>T_{eq}$ is satisfied. Here, $T_{OD.}$ is the average order-disorder (or Curie) temperature of the corresponding pure Ising model solved in the local mean-field approximation, by considering $\Delta/k_B = 0$ K, ln(g)= 0, $L/k_B = 0$ K. One can easily find:

$$T_{OD} = \frac{N_CT_{ODC} + N_ET_{ODE} + N_BT_{ODB}}{N_T}$$

where $T_{ODC} = 2.J$ and $T_{ODE} = 3.J$ and $T_{ODB} = 4.J$

The analytical expression of the order-disorder transition is obtained as

$$T_{OD} = \mathrm{J}\frac{8 + 3(4N-8) + 4(N-2)^2}{N^2} = \mathrm{J}\frac{4(3N-4) + 4(N-2)^2}{N^2}$$

Which indeed gives $4J$ for large systems (N >>).

Similarly, T_{eq} is the average value of the transition temperature. It reads

$$T_{eq} = \frac{N_CT_{eqC} + N_ET_{eqE} + N_BT_{eqB}}{N_T}$$

with : $T_{eqC} = \frac{\Delta - 4\times L}{\mathrm{k_B}\ln(g)}$ and $T_{eqE} = \frac{\Delta - 2\times L}{\mathrm{k_B}\ln(g)}$ and $T_{eqB} = \frac{\Delta}{\mathrm{k_B}\ln(g)}$

Similarly, the analytical expression of T_{eq} writes

$$T_{eq} = \frac{\Delta}{k_B \ln(g)} - \frac{2 \times L}{k_B \ln(g)} \frac{4 + 2(4N - 8) + (N-2)^2}{N^2} = T_{eq}(L = 0) - \frac{2 \times L}{k_B \ln(g)} \frac{4(2N-3) + (N-2)^2}{N^2}$$

Which gives for an infinite system, $T_{eq} = \frac{\Delta}{k_B \ln(g)} - \frac{2 \times L}{k_B \ln(g)}$.

The calculation of the previous two temperatures, $T_{OD}(N)$ and and $T_{eq}(N)$, help us to predict the presence or not of a thermal hysteresis loop according to the lattice size and model parameters. Table 1 shows these temperatures as a function of the principal model parameters.

Table 1. Order-desorder and equilibrium temperatures calculated for different sizes and shapes.

Nx	Ny	J	Delta	L	TOD (LMFA)	Teq (LMFA)	TOD (MFA)	Teq (MFA)
20	20	70	1300	500	266	183,0283	280	216,3062
15	15	70	1300	500	261,3333	171,9357	280	216,3062
10	10	70	1300	500	252	149,7504	280	216,3062
6	6	75	1300	0	250	216,3062	300	216,3062
2	18	75	1300	0	216,6667	216,3062	300	216,3062
5	5	70	1300	500	224	83,19468	280	216,3062
5	5	70	1300	0	224	216,3062	280	216,3062
5	5	70	1300	100	224	189,6839	280	216,3062

3. Results and discussion

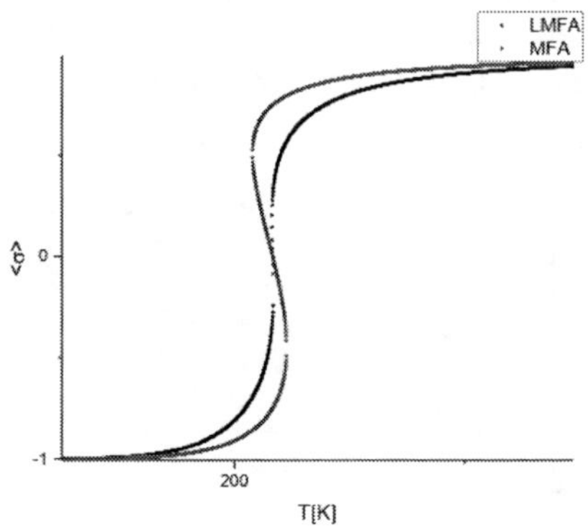

Figure 2. Difference between the homogeneous mean field approximation (red curve) and the local MFA (black curve) with border effects for a 5x5 lattice; parameter values: $\Delta/k_B = 1300$ K, ln(g)=6.01, $J/k_B = 70$ K, $L/k_B = 0$ K.

Fig.2 shows results obtained by the classical MF (CMFA) and the "local" mean field approximations proposed in this contribution. We observe the presence of a thermal hysteresis of the average magnetization in the CMFA while it disappears in LMFA for a 5x5 lattice, due to a decrease of the effective coordination number. Indeed the latter is equal to 4 in the CMFA while it becomes equal to

$$< q > = \frac{8+3(4N-8)+4(N-2)^2}{N^2}$$

Which gives the effective value 3.2 for $N = 5$, which then reduces the effective short-range interaction energy, and so the order-disorder transition. The advantage of the LMFA is the possibility to study the effect of the shape of the compound on the the thermal properties, through the change of the number of "edge" molecules, which increases for a rectangular case, for example compared to the square with the same total number of sites. Fig.3 shows the thermal behavior of the high spin fraction for a square (Fig.3.a) and for a rectangular (Fig.3.b) shaped-lattice. We see that the hysteresis obtained in a 6x6 lattice disappeared for the 2x18 one, due to the increase of the number of edge particles with respect to the number of "bulk" particles, something which reflects the reduction of the effective number of neighbors.

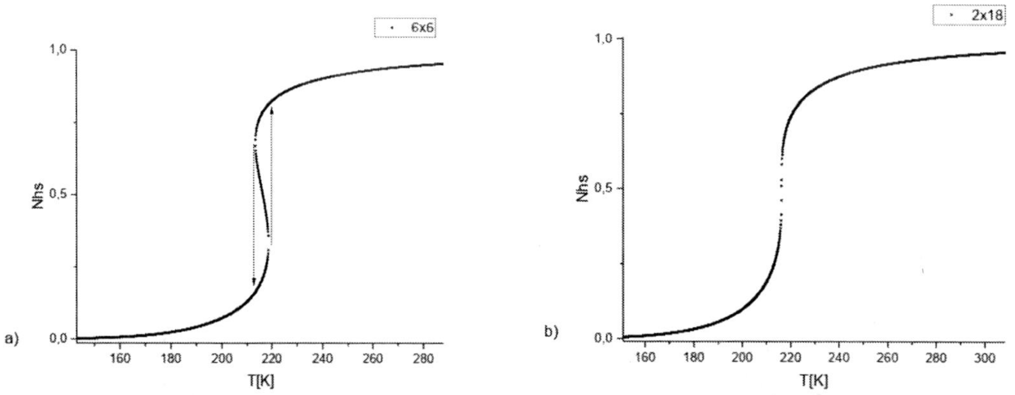

Figure 3. Thermal behavior of the high spin fraction for two lattice shapes using the Local Mean Field Approximation. (a) square-shaped of 6x6 lattice, (b) rectangular-shaped 2x18 lattice. Parameter values are Δ/k_B = 1300 K, ln(g)=6.01, J/k_B = 75 K, L/k_B = 0 K.

We have also studied the environmental effect by introducing the term $-L\sum_i \sigma_i$ in the Hamiltonian for the surface molecules. In this case, we start by studying a 5x5 lattice with a low value of L/k_B = 100 K (Fig.4) and we observe the recurrence of hysteresis compared with the result of LMFA in Fig. 1. This behavior is explained by the reduction of the transition temperature T_{eq} due to environmental effects.

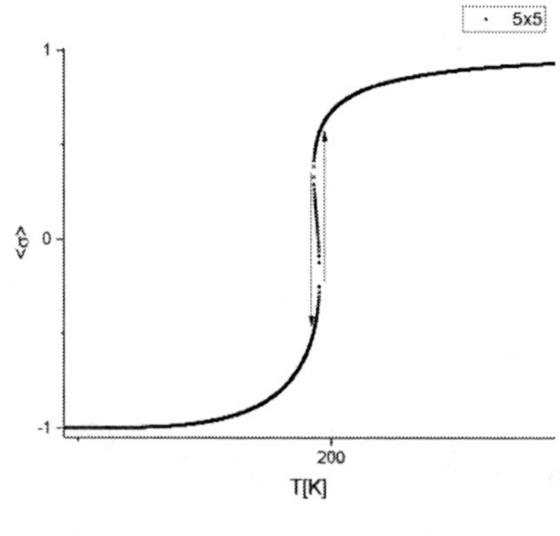

Figure 4. Thermal behavior of Sigma for a 5x5 lattice using the Local Mean Field Approximation, parameter values are: Δ/k_B = 1300 K, ln(g)=6.01, J/k_B = 70 K, L/k_B = 100 K.

By rising the strength of the environment interaction to L/k_B = 500 K, the result we obtain is a double step transition with a remarkable hysteresis in the first step (Fig.5.a), this behavior being due to the interplay between the three contributions corner and edge effects which lead to the lower hysteresis and the bulk which leads to the gradual transition with a higher equilibrium temperature, as we can clearly see in Fig.

5b where we represented separately the corresponding thermal dependence of all these sites. Interestingly for the chosen values of Δ and L, each corner site experiences a negative effective ligand field $\Delta - 4L = -700$ K, which prevents any spin transition and stabilizes these sites in the HS state as seen in Fig. 5b. So finally, the double step transition in this case is the result of the competition between the bulk and the edge thermal transitions.

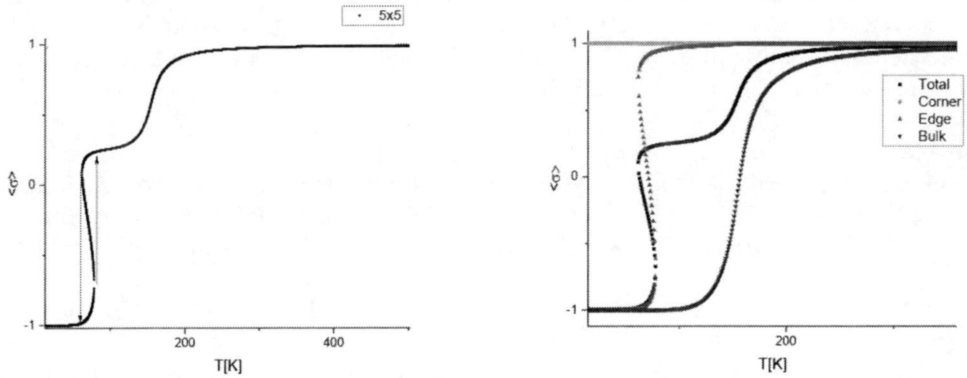

Figure 5. Thermal behavior of <σ> for a 5x5 lattice using the Local Mean Field Approximation. (a) The total variation of the spin state, (b) the variation of corners "green circles", edge "red up triangles" and bulk's "blue down triangles spin state regarding the total variation. Parameter values are Δ/kB = 1300 K, ln(g)=6.01, J/kB = 70 K, L/kB = 500 K.

The local mean field approximation gives us the possibility to analyze the size effect for different cases, on the transition phenomena, as shown in Fig.6, where we have summarized the thermal behavior of the total HS fraction for different square lattices. It is worth noticing that for small sizes (5x5) the first order transition is driven by the surface (edge) atoms while the bulk atoms undergo a gradual transition. This situation is reversed for big sizes (20x20) where the surface effects become less significant and the first-order transition turns to the bulk material, while the transition of edge sites becomes gradual. This interplay between edge and bulk atoms on the characteristics of the thermal transition is very interesting and deserves to be investigated more deeply in future works.

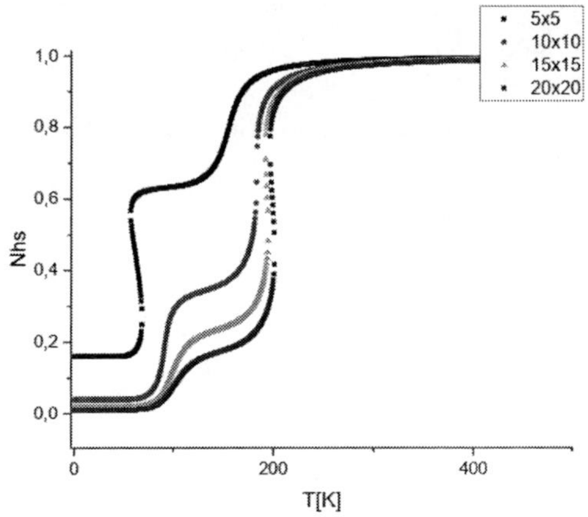

Figure 6. The simulated thermal behavior of the fraction of high spin for different system sizes, 5x5 (black square), 10x10 (red circle), 15x15 (green triangle), 20x20 (blue star); Parameter values are Δ/kB = 1300 K, ln(g)=6.01, J/kB = 70 K, L/kB = 500 K.

4. Conclusion

We have shown that the mean-field approximation (MFA) is successful in qualitatively explaining the critical behavior of SCO macroscopic compounds, and that the Local Mean Field Approximation (LMFA) can complement the last one when performing calculations on SCO nanoparticles, in which the number of molecules at the surface or corners with different number of neighbors becomes important and it could not be ignored. In particular, we found that combined with the matrix effects, the LMFA brought to light the existence of double step transitions, resulting from the thermodynamic interplay between the core and the edge molecules. We have also demonstrated that by increasing the size effects, the LMFA allows to find out the usual MFA and the role of the bulk atoms in driving the first-order spin transition in macroscopic lattice sizes.

Acknowledgments
The present work has been supported by the French "ministère de la recherché", University of Versailles, CNRS and ANR BISTA-MAT: ANR-12-BS07-0030-01, whose support is highly appreciated. CHAIR Materials Simulation and Engineering, UVSQ, University Paris Saclay is gratefully acknowledged.

References

[1] P. Gütlich, H.A. Goodwin, Topics in Current Chemistry, Spin Crossover in transitions Metal Compounds I-III, edited by Springer, Heidelberg, Berlin 2004.
[2] Gütlich P, Gaspar AB and Garcia Y *Beilstein J. Org. Chem.* 9 342 (2013).
[3] Hauser A. ,*Coord. Chem. Rev.* 111,(1991) 275-290.
[4] Varret F., Bleuzen A., Boukheddaden K., Bousseksou A., Coldjovi E., Enachescu C., Goujon A., Linares J., N.Menendez N. And M. Verdaguer M., *Pure Appl. Chem.*, 74, (2002) 2159-2168.
[5] Enachescu C., Linares J., Varret F., *J. Phys.: Condens, Matter*, 13,(2001) 2481-2495.
[6] Seredyuk M. Gaspar AB, Ksenofontov V, Reiman S, Galyametdinov Y, Haase W, Rentschler E, Gütlich P *Chem. Mater.* 18 2513 (2006).
[7] Mishra V., Mukherjee R., Linares J., Balde C., Desplanches C., Létard J.F., Collet E., Toupet L., Castro M., Varret F., *Inorg Chemistry,* 47, (2008) 7577-7587.
[8] Nasser J A *European Physical Journal* B 21 (2001) 3-10
[9] Nasser J A, Boukheddaden K and Linares J *European Physical Journal* B 39 (2004) 219-27.
[10] Wajnflasz J, Pick R, *J. Phys.Colloques*, 32, (1971) 91-92.
[11] Bousseksou A. Nasser J, Linares J, Boukheddaden K, Varret F *j. Physique* I 2 1381 (1992).
[12] Linares J, Jureschi C, Boukheddaden K, *Magnetochemistry,* 2, 24 (2016).
[13] Linares J, Jureschi C, Boulmaali A, Boukheddaden K, *Physica B,* 486, (2016) 164-168.
[14] Slimani A., Boukheddaden K., Varret F., Oubouchou H., Nishino M., and Miyashita S. , *Phys. Rev.* B 87, 014111 (2013).

Edge Effect on Nanoparticles of an Interconnect Alloy from the ABV Model

Chemseddine Saada,[1] Pierre Richard Dahoo,[1] Jorge Linares,[2] Tarek Merzouki [3], Philippe Pougnet,[4] Abdelkhalak El hami [5]

[1]LATMOS, Université de Versailles-Saint-Quentin-en-Yvelines, CNRS-UPMC-UVSQ (UMR 8190), 11 boulevard d'Alembert, 78280 Guyncourt, France

[2]GEMAC, Université de Versailles Saint-Quentin-en-Yvelines, 45 Av. des Etats-Unis, CNRS-UMR 8635, 78035 Versailles Cedex, France

[3]LISV, Université de Versailles-Saint-Quentin-en-Yvelines, 78035 Versailles, France

[4]VALEO SIEMENS eAutomotive, BP 68532 Cergy, 95892 Cergy Pontoise Cedex, France

[5]INSA de ROUEN Laboratoire de Mécanique de Normandie (LMN), Pôle Technologique du Madrillet, BP 8-76801 Saint Etienne du Rouvray, France.

E-mail: pierre.dahoo@uvsq.fr

Abstract: The physical phenomena underlying crack initiation and hence failures in interconnection alloy is investigated using the ABV model [1] (Metals A and B and void V) focusing on boundary effects at the interface with the device. The Hamiltonian which is expressed as the sum of the interaction energies between A, B and V with interaction parameters E_{AA}, E_{BB}, E_{AB} , E_{AV} and E_{BV} and reformulated in terms of fictitious 3 states spins (-1, 0, +1). And new parameters J, K, and U function of the interaction energy parameters between the metal atoms A, B and void V are defined and associated to the different spin combinations of the transformed Hamiltonian. A Monte Carlo (MC) simulation of a 2D microscopic 3 states Ising model taking into account edge effects [2] at the boundary between an active chip in a photovoltaic device or a sensor and nanoparticles of an interconnect alloy is performed. The results are discussed in terms of realistic values of interaction parameters and different algorithms for fixed compositions of A, B and V.

1. Introduction

Embedded electronic systems designed for smart technology implemented in mechatronic, robotics or internet of things devices can be studied by simulation and modelling before any experimentation is performed to test its reliability. This approach makes it possible to reduce time to market and the development cost of a correct alloy. The physics of failure of the interconnecting alloys (ICA) used in mechatronics modules are related to the degradation observed as a result of the presence of intermetallics, voids, delamination at adjacent boundaries. Ageing due to current density cycling through interconnections (solder, wire bonding ...) and bulk materials (copper busbar, silicon chip, insulating ceramic …) occurs principally because of thermal effect and the presence of voids. Figure 1 a) and b) shows the defects that are present in interconnect alloys after thermal shocks [2].

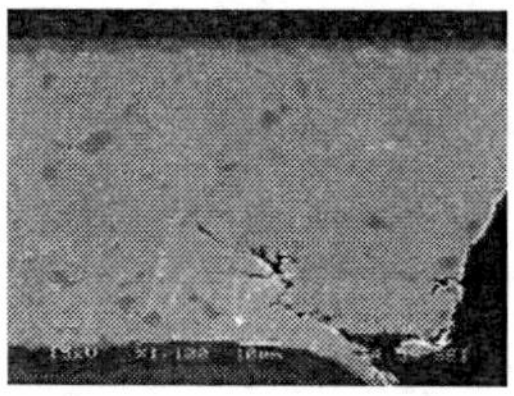

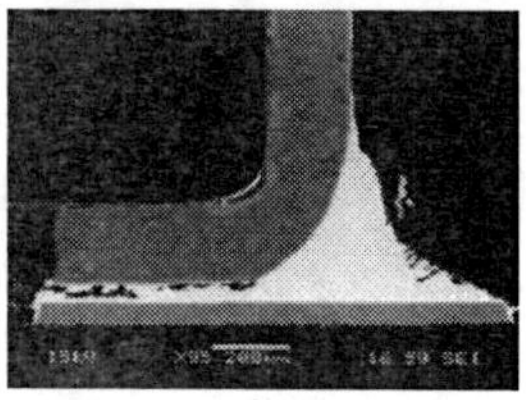

a) **b)**

Figure 1: a) Cracks and voids after thermal shocks test 1h/1h a) ASIC (683 cycles, -40/+105°C) b) Melf Resistor (1000 cycles, -40/+125°C)

The Ising model which has been developed to study phase transition physics is applied to a variety of systems not necessarily involving interacting magnetic spins in 1D, 2D or 3D configurations. A variety of models can be found: for example, the Bragg-Williams model on Binary alloys [3], the Potts model applied to Grain boundary energy [4], the Lee and Yang Lattice gas model [5] or the Magnetic Spin Cross-Over (SCO) compounds [6]. In the frame of the Ising ABV model [1], Monte Carlo simulation is performed to optimize the reliability of mechatronic devices using in their assembly innovative materials complying with regulations about hazardous substances.

2. Model For Simulation

The ABV model developed by Yaldram and Binder [1], where A and B stand for two different metals and V for void or a vacancy is based on the equivalence between the Ising-lattice gas and a binary alloy. Its application in connection to several other models, in particular the Blume-Emery-Griffths [7] has been discussed in details by Tafa, Puri and Kumar [8] with a focus on phase separation in ternary mixtures and characteristic features of the domain growth process. This model has been applied to study temperature effects and void segregation in interconnect materials using a Monte Carlo simulation (MC) in the Metropolis scheme [9]. This simulation highlighted the importance of specific temperature rise in the region of voids clusters initiating cracks and leading to delamination. The simulation work presented here in this paper, is focused on the failures observed at the interfaces of the interconnecting material of a mechatronic device. This approach is based on an ABV model which takes into account edge effects, and applies standard Metropolis Monte Carlo (MC) algorithm. The study of the edge effect is performed on the homogeneous structure of a binary alloy of constant composition containing voids. The simulation is performed on a 2D microscopic Ising model by considering a non-periodic MC application using the same three states Hamiltonian as in references [8][9] as:

$$H = H_0 - J\sum_{\langle ij\rangle}^{NN}\sigma_i\sigma_j - K\sum_{\langle ij\rangle}^{NN}\sigma_i^2\sigma_j^2 + U\sum_{\langle ij\rangle}^{NN}(\sigma_i^2\sigma_j + \sigma_i\sigma_j^2) \quad (1)$$

where σ can be set to -1, 0, 1. The different parameters K, J and U are defined as :

$$J = \frac{1}{4}(2\varepsilon_{AB} - \varepsilon_{AA} - \varepsilon_{BB})$$

$$K = \varepsilon_{AV} + \varepsilon_{BV} - \varepsilon_{VV} - \frac{1}{4}(\varepsilon_{AA} + \varepsilon_{BB} + 2\varepsilon_{AB}) \quad (2)$$

$$U = \frac{1}{4}(\varepsilon_{AA} - \varepsilon_{BB}) + \frac{1}{2}(\varepsilon_{BV} - \varepsilon_{AV})$$

IOP Conf. Series: Journal of Physics: Conf. Series **936** (2017) 012053 doi:10.1088/1742-6596/936/1/012053

Where ε_{ik} ($i = A, B$ or V and $k = A, B$ or V) are the interaction energy parameters between the metal atoms A, B and void V. Note that phase separation is mediated by vacancies although there are no pairwise interaction terms involving voids and that segregation are due to vacancy atom exchanges rather than exchanges between A and B. In the absence of vacancies, the systems tends to mix if J < 0 (antiferromagnetic) and become homogeneous (complete disorder) if J > 0, then the system tends to segregate in A and B atoms (ferromagnetic). Parameters K/J and U/J account for the order or disorder that result from one of these two possible configurations.
In order to simulate the segregation of voids in the interconnecting material of a mechatronic module at interfaces, the interaction parameters ε_{AA}, ε_{BB} and ε_{AB} are adjusted and the spatial pseudo-spin configuration is determined. The results are discussed in terms of modifications on K and very slightly on U as a function of interaction parameters and the pair distribution function with fixed void percentage.

3. Results

To simulate the defects encountered in a mechatronics device, and precisely at interfaces as for example the active semi-conductor chip and the interconnect material, a new parameter is added to the algorithm developed for a closed system to address the problem of void clustering in an interconnect material [9]. This parameter serves to control the change in the interaction energies between atoms (A or B) and voids (ε_{AV} et ε_{BV}) at the surface at the nanometer scale under the hypothesis these values are different that at surfaces and in the bulk because the ratio of surface to bulk atoms are different. The effect of this modification is then determined by applying the standard Metropolis sampling procedure to a 2D square configuration. Two different algorithms are applied to simulate edge effects on the segregation and diffusion of voids through thermal cycling, either a clockwise search (CS) of void neighbors to mimick temperature rise effect when void clusters are formed or a search carried at random (RS), in an open system with only one external edge interface or with all the four borders. This situation is realized by breaking the symmetry in the periodic condition at the corresponding edge in the nanometer range, and replacing it by an edge of voids, which doesn't affect the total energy of the system. In the first instance, a study of the closed system i.e. when the periodic condition is verified, is performed for different values of the coefficient K.

In the results presented below, a square lattice of size 50×50, composed of atoms A and B, and voids in fixed proportion, $N_A = 40\%$; $N_B = 40\%$; $N_V = 20\%$ is used for the simulation. The interaction energies are set as: $\varepsilon_{AA} = \varepsilon_{AV} = 3{,}32\ eV$; $\varepsilon_{BB} = \varepsilon_{BV} = 3{,}19\ eV$; $\varepsilon_{AB} = 3{,}25$ eV. The initial configurations are randomly distributed in the matrix during the thermal initialization process carried at a temperature of T =300K. With U set to zero, the parameter K controls the void segregation and the results are given for different values of K. A plot of the probability of finding a pair of void as a function of the distance from the border or edge (termed deepness), allows the topology and the distribution of voids in the square lattice to be evaluated.

3.1 Without edge effect

a) b) c)

Figure 2: 2D distribution without edge effect : Topology a) and b) and c) Probability of pairs for different values of K (eV)

Topology : A White, B Dark, V Grey : (a) left : K = 122.16 eV ; (b) right : K = 322.27 eV.

As can be seen in Figure 2, there are no marked differences in the topology between the left a) and the right b) distributions for the two values of K: 122.16 eV and 322.27 eV. The only difference observed is in the random distribution of void clusters of different sizes in the lattice. In figure 2 c), the probability to find a pair of voids as a function of distance from an edge (deepness) show that voids are distributed at random whatever the deepness, such that the parameter K has no effect on this distribution when the system is closed with periodic boundaries.

3.2 With edge effect and clockwise search

3.2.1 All free edges

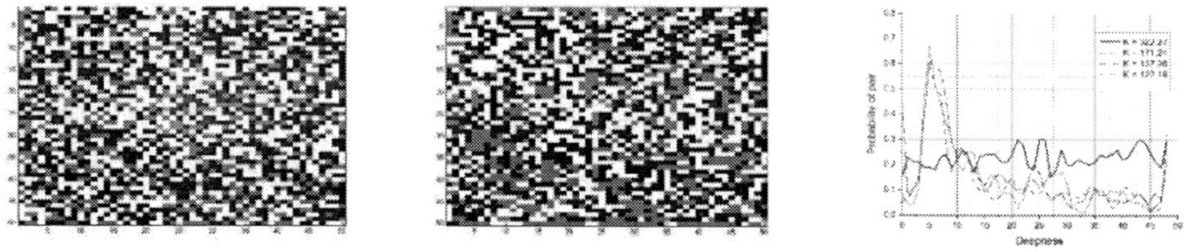

a) b) c)

Figure 3: 2D distribution with 4 free edges (CS): Topology a) and b) and c) Probability of pairs for different values of K (eV)

Topology : A White, B Dark, V Grey : (a) left : K = 122.16 eV ; (b) right : K = 171.24 eV.

Figures 3 a) and 3 b) present several differences in the void distribution, and precisely with a lower number of clusters and more voids at the borders at lower K values. At high values of K (322.17eV), void clusters increase while voids at borders decrease significantly. The pair distribution probability correlates very nicely to the analysis of the topology of voids with different values of K as shown in Figure 3 c), with a maximum in the probability of pairs for range 5 from the border and a second peak at range 25 except for K=322.27. This apparently unusual trend with no pronounced peak is attributed to the clockwise search of void neighbor algorithm as compared with the results obtained with the random search algorithm of void neighbors. This effect may be attributed to the temperature effect incorporated in the CS algorithm and which shows that voids distributes more randomly with temperature rise when K is high.

3.2.2 One free edge

a) b) c)

Figure 4: 2D distribution with 1 free edge (CS): Topology a) and b) and c) Probability of pairs for different values of K (eV)

Topology : A White, B Dark, V Grey : (a) left : K = 137.26 eV ; (b) right : K = 322.27 eV.

With only one free edge, the simulation leads to more pronounced results as shown in Figure 4. When all the edges are free, the voids tends to redistribute on the four border because they are attracted by the voids present at the boundary of the square. For the one edge case, the size of the clusters accumulating at the border are larger because there is only one free edge which tends to attract voids. The pair distribution function of Figure 4 c) correlates to the topology presented in Figures 4 a) and 4 b) with two close maxima at range 4 and 7, the peak corresponding to K= 322.27 eV being highest. No apparent discrepancy due to the search algorithm is found in this case.

3.3 With edge effect and random search

3.3.1 All free edges

a) b) c)

Figure 5: 2D distribution with 4 free edges (RS): Topology a) and b) and c) Probability of pairs for different values of K (eV)

Topology : A White, B Dark, V Grey : (a) left : K = 137.26 eV ; (b) right : K = 322.27 eV.

With the random search algorithm, the accumulation of voids at the surface of the four borders is roughly the same as shown in Figures 5a) and 5b). The parameter K controls the distribution of voids, with less clusters in the bulk for higher K values. The size of clusters is enhanced with algorithm RS which limits accumulation of voids in a preferential direction. The peaks are closer to the borders with clusters in the bulk for low K values. The pair distribution curve corresponding to K=322.27 shows that voids are denser at the borders with limited clusters in the bulk.

3.3.2 One free edge

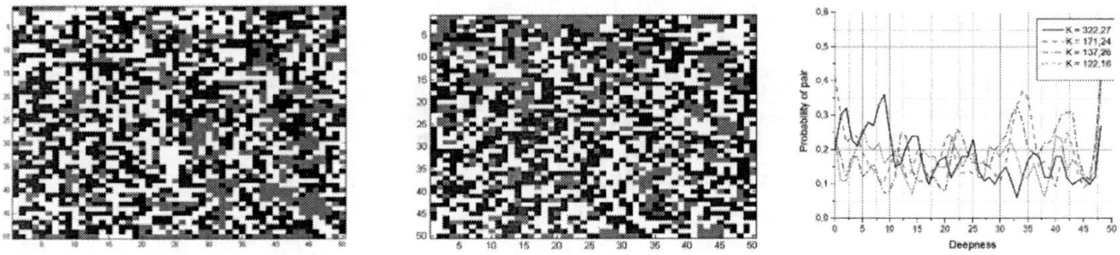

a) b) c)

Figure 6: 2D distribution with 1 free edge (RS): Topology a) and b) and c) Probability of pairs for different values of K (eV)

Topology : A White, B Dark, V Grey : (a) left : K = 137.26 eV ; (b) right : K = 322.27 eV.

With one free edge, the RS algorithm leads to results different from those obtained with the CS algorithm and looks more like results obtained with a closed system as presented in Figure 2. The voids density at the edge is not peaked as with the CS algorithm as shown in Figure 6 c). Remembering that the CS algorithm mimicks the effect of temperature rise when clusters of voids form, then the distribution of voids depends on the competition between the K effect and the temperature effect.

4. Conclusion

Two algorithms based on the ABV model are applied to simulate by standard metropolis Monte Carlo method, edge effects on the distribution of voids in a 2D configuration of a binary alloy interconnect material. The algorithms termed CS and RS are respectively closely linked to temperature rise effect with void clustering and K effect on the void distribution. Results show that temperature effect tends to favour preferential directional spread of voids where as K effect tend to favour cluster formation in the bulk for low K values and at the border for high K values. Then, the distribution of voids resulting from temperature effect compete with the tendency of void clustering at the surface for the high K, and homogeneous spread at the lowest K value.

Acknowledgments
We thank the CHAIR Materials Simulation and Engineering, of UVSQ and Université Paris Saclay for funding.

References
[1] Yaldram, K. and Binder, K. Monte-Carlo simulation of phase-separation and clustering in the ABV model. 1991, *Journal of Statistical Physics*, **62**, 161. 4
[2] Dahoo P R, Alayli N, Girard A, Pougnet P, Tan K L and Morelle J M, 2010 *Mater. Res. Soc. Symp. Proc. Warrendale, PA,* **1195** 183
[3]W L Bragg and E J Williams, The effect of thermal agitation on atomic arrangement in alloys, 1934, 1934, *Proc. Roy. Soc. London*, **145A**, 699-730 (1934).
[4] Potts R.B., Some Generalized Order-Disorder Transformations, 1952, *Math. Proc. Camb. Phil. Soc.* **48**, 106-109
[5] Lee T.D. and Yang C.N., Statistical theory of Equation of States and Phase Transitions: II Lattice gas and Ising Model, 1952, *Phys. Rev. E*, **87**, 3, 410-419
[6] Linares J., Chiruta D., Jureschi C-M., Alayli Y., Turcu C.O, Dahoo P.R., "Simulations of Edge Effect in 1D Spin Crossover Compounds by Atom-Phonon Coupling Model, 2016 *Journal of Physics : Conference Series, IOP*, **738**, pp 012052
[7] Blume, M., Emery, V. J., Griffiths, R. B., Ising Model for the λ Transition and Phase Separation in He_3-He_4 Mixtures,1971, *Phys.Rev.A,* **4**(3), 1071-1077
[8] Tafa K, Puri S and Kumar D, Kinetics of phase separation in ternary mixtures, 2001, *Phys. Rev. E.*, **64**, 056139-1
[9] Dahoo P.R., Linares J, Chiruta D., Chong C., Pougnet P., Meis C. and El Hami A., Monte Carlo Method Applied to the ABV Model of an Interconnect Alloy , 2016 *Journal of Physics : Conference Series, IOP*, **738**, pp 012073

IOP Conf. Series: Journal of Physics: Conf. Series **936** (2017) 012054 doi:10.1088/1742-6596/936/1/012054

On efficiency of fire simulation realization: parallelization with greater number of computational meshes

Lukas Valasek, Jan Glasa
Slovak Academy of Sciences/Institute of Informatics, Dubravska cesta 9, 845 07 Bratislava, Slovakia

E-mail: lukas.valasek@savba.sk

Abstract. Current fire simulation systems are capable to utilize advantages of high-performance computer (HPC) platforms available and to model fires efficiently in parallel. In this paper, efficiency of a corridor fire simulation on a HPC computer cluster is discussed. The parallel MPI version of Fire Dynamics Simulator is used for testing efficiency of selected strategies of allocation of computational resources of the cluster using a greater number of computational cores. Simulation results indicate that if the number of cores used is not equal to a multiple of the total number of cluster node cores there are allocation strategies which provide more efficient calculations.

1. Introduction

Simulation of fires belongs to computationally demanding tasks requiring efficient realization on a HPC (high-performance computer) system. Contemporary fire simulation systems are capable to incorporate many physical and chemical processes related to fire such as combustion, pyrolysis, thermal radiation, heat transfer, turbulence, gas spread dynamics, etc. Fire Dynamics Simulator (FDS) [1] is an advanced fire simulator utilizing the knowledge of CFD (computational fluid dynamics). It numerically solves a form of Navier-Stokes equations for low-speed fire induced flows with emphasis on transport of smoke and heat from the fire. FDS has been tested and validated for various types of fires and environments. It supports four programming models: sequential model designed for sequential computers, parallel MPI (Message Passing Interface) model designed for distributed memory systems, multithreading OpenMP (Open Multi-Processing) model designed for shared memory systems and hybrid MPI&OpenMP model designed for distributed shared memory systems. In our previous research we studied the impact of parallelization on efficiency and accuracy of a corridor fire simulation [2, 3] using a single computational node of the HPC cluster at the Institute of Informatics of Slovak Academy of Sciences in Bratislava (Slovakia). The research focused on the parallel MPI model of FDS and on testing the main strategies of allocation of available computational resources to given parallel calculations. In this paper we study efficiency of the parallel MPI model for simulation of the same type of fire using greater numbers of computational meshes and nodes.

2. Corridor fire scenario

Similarly to [2, 3] we use a rectangular 100 kW fire source in a corridor with dimensions of 7.2 m x 1.2 m x 2.7 m. The fire source with dimensions of 0.1 m x 1.0 m is placed on the floor at the distance of 3.64 m from the left corridor end and about 20 cm away from the corridor walls. We consider initial 10 s of fire (Figure 1).

Figure 1. Smoke spread at the 2nd and 6th s to the fire (left and middle), and the fire HRR (heat release rate) curve (right)

3. Sequential simulation

For sequential simulation a regular cube computational mesh with the 2 cm resolution (360 cells in the x-direction, 60 cells in the y-direction and 135 in the z-direction) was used. The total number of cells equals to 2916000. For simplicity, the INERT and OPEN boundary conditions were used on corridor walls and ends, respectively. The fire source is represented by a VENT with the 1000 kW/m^2 HRRPUA (heat release rate per unit area) boundary condition. The default values of ambient and initial temperatures (20°C) and the default dominant chemical reaction (ETHYLENE) were considered. In regard of the mesh resolution, LES (Large Eddy Simulation) was used to resolve turbulence.

The mesh sensitivity study performed in [2, 3] indicates that the chosen mesh can be considered to be fine enough already since the 2nd s to the fire. The total execution time of the sequential simulation (denoted by 1M) performed on a single computational core of the HPC cluster was about 17.45 hours.

The fire course is illustrated in Figure 1. At first, the smoke spread upwards from the fire source. Then it hit the ceiling at the 3rd s to the fire and began to spread under the ceiling in all directions. At the 6th s to the fire, the smoke reached both corridor ends. The HRR of the fire increased to the value of 100 kW at the 2nd s to the fire and then fluctuated around this value till the end of the simulation.

4. Parallel simulations

In the parallel MPI model of FDS, the domain is divided into several computational meshes, the computation on each mesh is considered as a single MPI process assigned to one computational core. Communication between MPI processes is provided by MPI. We used the 64-bit MPI version 6.3.2 of FDS for Linux and the open-source Open MPI version 1.10.0 of MPI. The process of computational resources allocation for a given simulation represented by m-MPI processes consists of three phases: mapping, ranking and binding [4]. The exact choice on how MPI processes will be mapped by particular computational nodes and subsequently bound to particular sockets or cores is provided by the `mpirun` script which is a part of Open MPI.

In [2, 3], we studied the sequential simulation described in the previous section as well as 9 parallel simulations parallelized by the parallel MPI model using a single computational node of the HPC cluster. We illustrated efficiency of two strategies for allocation of computational resources: `--map-by core, --bind-to core` and `--map-by socket, --bind-to socket` for parallel simulations with m≤12 [3].

In this paper we illustrate the efficiency of two allocation strategies: `--map-by node, --bind-to core` (denoted by NC) and `--map-by node, --bind-to socket` (denoted by NS). The NC strategy binds particular MPI processes to particular computational cores of a cluster such that it binds particular MPI processes alternately node-by-node to corresponding particular cores of the socket 0 and consequently to corresponding particular cores of the socket 1. The NS strategy binds particular MPI processes to groups of cores of a socket such that it binds particular MPI processes alternately node-by-node to corresponding groups of cores alternately in socket 0 and socket 1. For testing purposes, the cluster nodes IBM dx360 M3 consisting of two 6-core processors Intel E5645, 2.4GHz with 48 GB RAM were available. The nodes were connected by the InfiniBand interconnection network with the bandwidth of 40 Gbit/s per link and direction. In order to avoid interference of the tested simulation with other tasks being solved on the cluster, we reserved the whole computational nodes for the tested simulation exclusively.

IOP Conf. Series: Journal of Physics: Conf. Series **936** (2017) 012054 doi:10.1088/1742-6596/936/1/012054

Similarly to [2, 3], we decomposed the computational domain in the *x*-direction into m = 18, 24, 30, 36 cube computational meshes (denoted by 18M, 24M, 30M, 36M) with the 2 cm mesh resolution (Table 1 and Figure 3). As the chosen mesh resolution and the simulated fire scenario were the same as in [2, 3], the mesh sensitivity study performed in [3] for both the sequential and parallel simulations indicates that the chosen mesh resolution can be considered as fine enough.

Table 1. Series of simulations: N_P is the number of MPI processes, N_M is the number of computational meshes, N_C is the total mesh cells, N_{CpM} is the number of cells per mesh, N_{CC} is the number of required computational cores and N_{CN} is the number of required computational nodes.

	N_P	N_M	N_C	N_{CpM}	N_{CC}	N_{CN}
1M	-	1	2916000	2916000	1	1
6M	6	6	2916000	486000	6	1
12M	12	12	2916000	243000	12	1
18M	18	18	2916000	162000	18	2
24M	24	24	2916000	121500	24	2
30M	30	30	2916000	97200	30	3
36M	36	36	2916000	81000	36	3

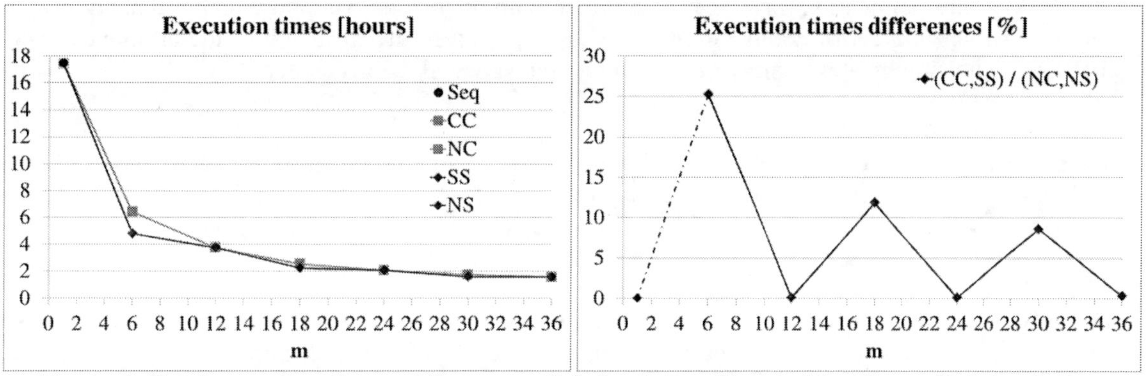

Figure 2. Total execution times of the sequential simulation (denoted by ●), parallel simulations carried out by CC and NC (denoted by ■) and parallel simulations carried out by SS and NS (denoted by ♦), where the values relating to those carried out by CC and NC are coloured grey (left), and their differences in % (right).

Table 2. Simulations efficiency: total execution times of the sequential and parallel simulations (in hours), where the values for the simulations carried out by CC and SS are coloured grey (columns 3-4) and for those carried out by NC and NS are coloured black (columns 5-9).

	1M	6M	12M	**18M**	**24M**	**30M**	**36M**
-	17.45						
CC / **NC**	-	6.393	3.744	2.551	2.073	1.740	1.576
SS / **NS**	-	4.778	3.748	2.249	2.075	1.591	1.581

Table 3. Simulations efficiency: differences between the total execution times from Table 2 in hours (row 2) and in % (row 3).

	6M	12M	**18M**	**24M**	**30M**	**36M**
CC,SS / **NC,NS**	1.615	0.004	0.302	0.002	0.149	0.005
CC,SS / **NC,NS**	25.262	0.107	11.838	0.096	8.563	0.317

In Table 2 and Figure 3, the total execution times (in hours) of the considered simulations 1M, 6M, 12M, 18M, 24M, 30M and 36M are shown. The simulations requiring a single computational node are coloured grey. In Table 3, the differences between the total execution times of the simulations listed in Table 2 are shown. One can make a similar observation about the allocation strategies NC and NS as it was described in [2, 3] about the CC and SS strategies. The total execution times of the simulations, where m equals to a multiple of the total number of cluster node cores (a multiple of 12), are very similar (see the execution times of the simulations 12M, 24M and 36M and their differences in Figure 3). In the case, where m does not equal to a multiple of the total cluster node cores number, the execution times of the simulations carried out by the NS strategy are smaller than of the ones carried out by the NS strategy (see the execution times of the simulations 18M and 30M and their differences).

5. Conclusion

In this paper, the efficiency of corridor fire simulation performed on the HPC cluster in Bratislava (Slovakia) is studied using the parallel MPI model of FDS and greater numbers of computational meshes. The strategies NC (`--map-by node, --bind-to core`) and NS (`--map-by node, --bind-to socket`) for allocation of the cluster computational resources were tested. The simulation results indicate that the NS strategy is more efficient than the NC strategy in cases where the number of computational meshes is not equal to a multiple of the total number of cluster node cores. As a fire simulation in large complex environments is computationally demanding task requiring greater number of computational meshes [5, 6, 7], further research of the efficiency of MPI parallelization using greater number of computational cores will be necessary.

Acknowledgment

This paper was partially supported by the Slovak Science Foundation (project VEGA 2/0165/17) and the Slovak Research and Development Agency (project APVV-15-0340).

6. References

[1] McGrattan K, Hostikka S, McDermott R, Floyd J, Weinschenk C and Overholt K 2015 Fire Dynamics Simulator (Version 6): User's Guide, *NIST Special Publication 1019-6*, NIST, Washington

[2] Valasek L and Glasa J 2016 A note on parallel efficiency of fire simulation on cluster *J. Phys.: Con. Ser.* 738 012129

[3] Valasek L and Glasa J 2016 Note on fire simulation efficiency realized on computer cluster In Proceedings of the EUROSIM Congress on Modelling and Simulation, Danvers, IEEE Computer Society, 671-676

[4] Documentation of OpenMPI, version 1.10.0, August 25, 2015, https://www.open-mpi.org/doc/v1.10/

[5] McGrattan K, Hostikka S, McDermott R, Floyd J, Weinschenk C and Overholt K 2017 Fire Dynamics Simulator Technical Reference Guide Volume 3: Validation, *NIST Special Publication 1018-3*, NIST, Washington.

[6] Valasek L and Glasa J 2017 On realization of cinema hall fire simulation using Fire Dynamics Simulator *Computing and Informatics* 35 4 in press

[7] Weisenpacher P, Glasa J, Halada L, Valasek L and Sipkova V 2014 Parallel computer simulation of fire in road tunnel and people evacuation *Computing and Informatics* 33 6 1237-1268

IOP Conf. Series: Journal of Physics: Conf. Series **936** (2017) 012055 doi:10.1088/1742-6596/936/1/012055

Simulation of the weakly interacting Bose gas relaxation for cases of various interaction types

P F Kartsev[1], **I O Kuznetsov**[2]
National research nuclear university MEPhI, 115409, Moscow, Kashirskoe shosse, 31
E-mail: PFKartsev@mephi.ru[1], ili-isk@mail.ru[2]

Abstract.
In this work, we investigate the role of interactions in the process of thermalization of a weakly interacting Bose gas. The system of kinetic equations based on the 'Fermi's golden rule' is solved numerically using special transformation for calculation efficiency. We study the distribution function for particles in various conditions, including interaction with phonon subsystem, i.e. energy exchange with thermal bath. The possibility to achieve the state of Bose-Einstein condensation with specific values of parameters, is also discussed.

1. Introduction

The phenomenon of Bose Einstein condensation (BEC) in various experimental systems attracts constant attention. Some of the new experiments are based on the study of the evolution of the Bose gas [1, 2] and Bose Einstein condensate [3]. Both experimental [4] and theoretical [5] works on the evolution of the Bose gas in various formulations of the problem are carried out.

It is a particular problem to determine the optimal parameters of the experiment with Bose gas, including that of achieving the state of BEC[6]. Among the processes determining the system evolution we distinguish the phenomenon of thermalization, i.e. relaxation to thermal equilibrium after perturbation. Thermalization is important for reaching BEC in many promising Bose systems, such as the cold gas of alkali atoms in an optical trap [7], exciton polaritons in a semiconductor resonator [8, 9], and probably for positronium gas in the recently proposed gamma laser scheme[10, 11].

The evolution of the many-particle system is governed by the interactions. The two-particle interaction affects the rate of thermalization in the system (as well as the spectrum of excitations of the system [12]), while the interaction with the phonon subsystem is important for energy exchange with the environment. To describe the behaviour of a complex interacting system, it is necessary to build a reliable theoretical model which takes into account all the important parameters of the problem and the interactions present in the system.

In this paper, we study numerically the effects of these two types of interaction on the relaxation characteristics in the weakly-interacting Bose gas. Successful simulation of many-particle kinetics would help to visualize important processes, and to plan further study.

2. Model

First, let us write the Hamiltonian of the weakly-interacting Bose gas of e.g. excitons $(\hat{a}, \hat{a}^\dagger)$ coupled with phonon subsystem $(\hat{b}, \hat{b}^\dagger)$:

$$\hat{H} = \sum_k \varepsilon_k \hat{a}_k^\dagger \hat{a}_k + \sum_q \tilde{\varepsilon}_q \hat{b}_q^\dagger \hat{b}_q + \sum_{k_1 k_2 k_3 k_4} U_{k_1 k_2 k_3 k_4} \hat{a}_{k_1}^\dagger \hat{a}_{k_2}^\dagger \hat{a}_{k_3} \hat{a}_{k_4} + \sum_{kq} M_{kq} \hat{b}_q^\dagger \hat{a}_k^\dagger \hat{a}_{k-q} + H.c. \quad (1)$$

Here ε_k and $\tilde{\varepsilon}_q$ are the dispersion laws of excitons and phonons, $U_{k_1 k_2 k_3 k_4}$ is the two-particle interaction and M_{kq} describes the particle transition with creation or anihilation of phonon. For the sake of simplicity, we restrict ourselves to the case of contact interaction $U_{k_1 k_2 k_3 k_4} \equiv U_0 \delta_{k_1+k_2, k_3+k_4}$ and constant matrix element $M_{kq} \equiv M_0$.

The phonon subsystem is considered as external bath with fixed temperature $T^{(\text{bath})}$ and classical field $\langle b_q \rangle \simeq \sqrt{n^{(\text{phon})}(\tilde{\varepsilon}_q)}$, $\left\langle b_q^\dagger \right\rangle \simeq \sqrt{n^{(\text{phon})}(\tilde{\varepsilon}_q)+1}$, with $n^{(\text{phon})}(\tilde{\varepsilon}) = \frac{1}{e^{\tilde{\varepsilon}/k_B T^{(\text{bath})}}-1}$. Then we study the kinetics of the main subsystem (e.g. excitons) only.

The evolution of bosonic system at high enough temperature can be described by the Boltzmann equation[12, 13]. Kinetic equation for the specific occupation number n_k is generated by the 'Fermi's golden rule' giving the rate of changes

$$\frac{1}{\tau} = \frac{2\pi}{\hbar} \sum_f |< i|\hat{V}|f >|^2 \, \delta(E_i - E_f), \quad (2)$$

where $|i\rangle$ and $|f\rangle$ are initial and final states of the quantum system connected by the perturbation operator $\hat{V}$ changing the occupation number n_k, with the conservation of the full system energy $E_i = E_f$. Later on, we assume dimensionless time and choose unities so that $\frac{2\pi}{\hbar} \equiv 1$.

In thermal equilibrium, occupation numbers are given by Bose-Einstein distribution function. If the stable state is disturbed, the occupations tend to relax and return to the equilibrium[14]. For small deviations from the equilibrium, the rate of relaxation should be proportional to $\langle V \rangle^2$.

The numerical simulation of kinetic equations can be done using any appropriate method. We simulate the evolution of the occupation numbers described by the system of kinetic equations using the especially developed numerical algorithm[14]. The main difficulty in the simulation is caused by the amount of calculations increasing drastically with the size of system and can be partially overcome using the auxiliary analytic transformation[15].

Let us consider the terms generated by interactions in (1).

Two-particle interaction. To calculate matrix elements, we need to neglect correlations between occupation numbers: $\langle n_p n_q \rangle = \langle n_p \rangle \langle n_q \rangle$, assuming the temperature be high enough (far from condensation temperature T_c). Then the equation (2) gives

$$\begin{aligned} \frac{dn_1^{(\text{e-e})}}{dt} &= \frac{2\pi}{\hbar} U_0^2 \sum_{234} \{(n_1+1)(n_2+1+\delta_{12}) n_3 (n_4 - \delta_{34}) - \\ &- n_1 (n_2 - \delta_{12})(n_3+1)(n_4+1+\delta_{34})\} \, \delta(\varepsilon_1 + \varepsilon_2 - \varepsilon_3 - \varepsilon_4). \end{aligned} \quad (3)$$

Here we denoted $k_1, \ldots k_4$ as $1, \ldots 4$. The Kronecker symbols δ_{12}, δ_{34} help to account the case of coinciding momenta of Bose particles.

Considering the nonlinear nature of two-particle interaction, the relaxation rate can be estimated as $\frac{1}{\tau} \sim U_0^2 N^2$, with N the count of bose particles.

Interaction with phonon bath. To take into account the finite width of phonon levels Γ, we introduce the Gaussian form factor $F(\Delta E) \sim \exp(-(\Delta E/\Gamma)^2)$:

$$\frac{dn_k^{(\text{e-ph})}}{dt} = P^{(+)} - P^{(-)},$$

$$P^{(+)} = M_0^2 \sum_q [(n_k+1)n_{k-q} \left[(n_q^{(\text{phon})}+1)F(\varepsilon_k - \varepsilon_{k-q} + \tilde{\varepsilon}_q) - n_q^{(\text{phon})}F(\varepsilon_k - \varepsilon_{k-q} - \tilde{\varepsilon}_q)\right], \tag{4}$$

$$P^{(-)} = M_0^2 \sum_q [n_k(n_{k-q}+1) \left[n_q^{(\text{phon})}F(\varepsilon_k - \varepsilon_{k-q} + \tilde{\varepsilon}_q) - (n_q^{(\text{phon})}+1)F(\varepsilon_k - \varepsilon_{k-q} - \tilde{\varepsilon}_q)\right].$$

Notice the two terms in both branches of particle addition and removal. The relaxation rate due to phonon field, can be estimated as $\frac{1}{\tau} \sim M_0^2$ (independent of N).

3. Results of numerical simulation

We investigate the effects of interparticle interaction and interaction with thermal bath separately. In both cases we expect that the excitation would decay and the state approaches thermal equilibrium. In the case of exponential decay, we try to extract its time constant τ.

As a starting (excited) state, we use the following particle distribution:

$$\begin{cases} n_k^{(\text{start})} = N^{(\text{inside})}, & |k-k_0| < R, \\ n_k^{(\text{start})} = N^{(\text{outside})}, & |k-k_0| \geq R, \end{cases} \tag{5}$$

where R is the size of initial "droplet" in momentum space, centered in the point k_0.

The simulations were carried for the system of size $L^3 = 16^3$ with dimensionless time step $\Delta t = 10^{-11}$. The problem parameters were taken $c = 1.0$, $\Gamma = 3.0$, $R^2 = 6.0$, $N^{(\text{outside})} = N^{(\text{inside})}/100$. The units were chosen so that dispersion laws were $\varepsilon_k = n_x^2 + n_y^2 + n_z^2$, $\tilde{\varepsilon}_q = c\sqrt{n_x^2+n_y^2+n_z^2}$ with n_α integer ($-L/2 \leq n_\alpha < L/2$). The number of time steps needed for relaxation, did not exceed the value of $3 \cdot 10^5$. The size $L = 16$ is large enough to see the needed behaviour.

Two-particle interaction. First, we put $U_0 = 1$, $M_0 = 0$. In Fig.1(a) we see how particle distribution changes in time. To confirm the Bose-Einstein form of the final distribution, we introduce the characteristic function $L(E) \equiv \ln\left(1 + \frac{1}{f(E)}\right)$, which equals to $(E-\mu)/k_BT$ in the case of Bose-Einstein function. In Fig.1(b), we see the tendency of $L(E)$ to become linear, confirming the fact that the system approaches the thermal distribution.

In Fig. 2(a), we show the temperature calculated by fitting the characteristic function with linear function in the range $E = 40..120$. Note that the final value of k_BT is reached very quickly. In Fig. 2(b), we show the mean squared deviation of $L(E)$ from the linear function decreasing to values as low as 10^{-11}, for N=50 000. The characteristic time of relaxation τ depending on N, is plotted in Fig. 3. The power law is almost N^{-2} as we expected.

So, for the thermalization due to interparticle interaction, we can see the following stages: (i) fast thermalization of particles in the middle range of energies; (ii) the temperature approaches the final value determined by full energy of the system; (iii) the range of energies where particles follow the thermal law, expands to all energies.

Interaction with phonon bath. Second, we put $U_0 = 0$, $M_0 = 10^3$. In Fig. 4 we show the graphs of $f(E)$ and $L(E)$ calculated for $N = 10^5$. The distribution approaches thermal function again, though the stages are slightlydifferent: (i) fast thermalization of particles in the limited range of energies; (ii) the temperature decreases to the final value determined by the bath $T \to T^{(bath)}$,

IOP Conf. Series: Journal of Physics: Conf. Series **936** (2017) 012055 doi:10.1088/1742-6596/936/1/012055

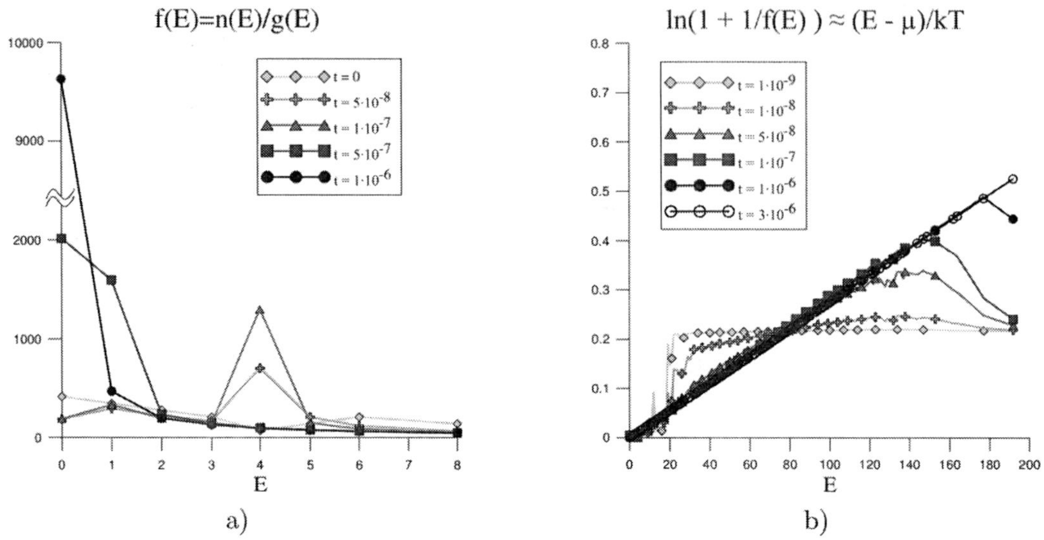

Figure 1. (a) Typical evolution of the particle distribution function $f(E) = n(E)/g(E)$ from the initial nonequilibrium state (5), calculated for $N = 5 \cdot 10^4$, $U_0 = 1$, in the case of two-particle interaction. Initially flat, the distribution transforms to emphasized peak at nonzero energy, which then shifts to $E = 0$. Time $t = 1 \cdot 10^{-6}$ appears enough to reach the Bose-Einstein function. (b) Evolution of the characteristic $L(E)$ calculated for data in Fig. 1(a). Notice the limited range of linear behaviour and gradual change of the slope $\sim 1/k_B T$. The range of linear behaviour starts from $E \simeq 40..120$ and slowly extends to all energy range. At the (dimensionless) time $t = 3 \cdot 10^{-6}$, the graph becomes very close to a linear law.

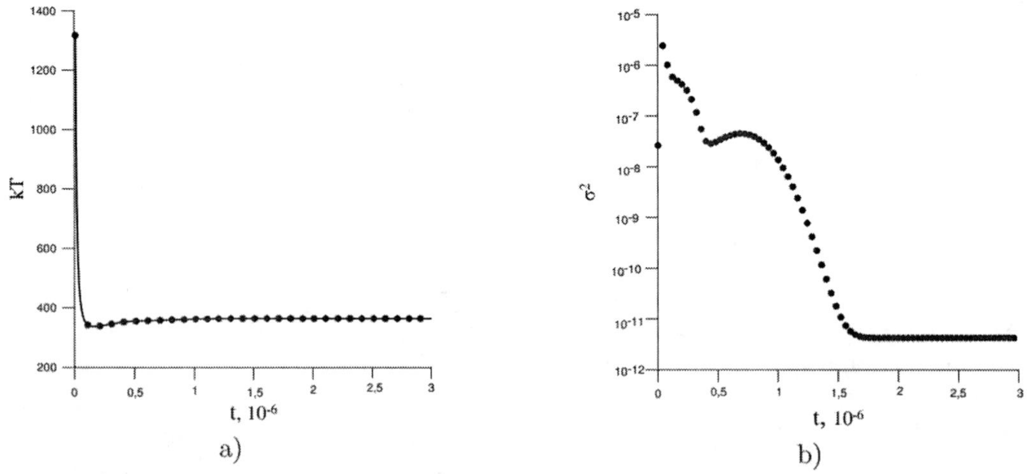

Figure 2. (a) Time dependence of the temperature calculated for data in Fig. 1(a). (b) Mean squared deviation of $L(E)$ from the linear law, calculated in the range $E = 40..120$, depending on time t. Minimal value is limited by $\sim 10^{-11}$ due to precision of numerical algorithm.

IOP Conf. Series: Journal of Physics: Conf. Series **936** (2017) 012055 doi:10.1088/1742-6596/936/1/012055

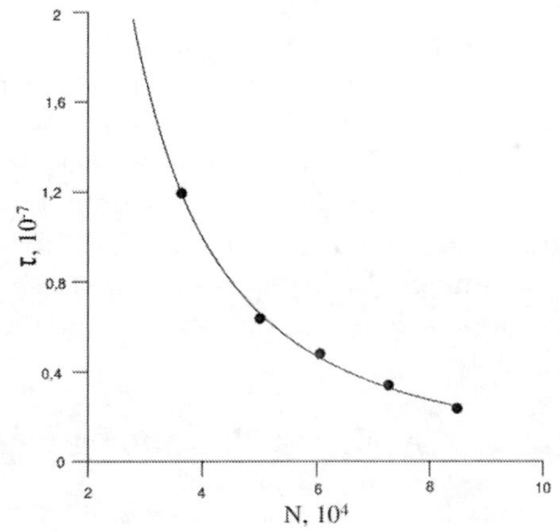

Figure 3. Characteristic time τ to establish equilibrium depending on the total number of particles N. Dots are the experimental data, the line shows the power-law fit. The power index -1.86 is close to expected -2. Relative error of τ is close to $\sim 5 \cdot 10^{-2}$.

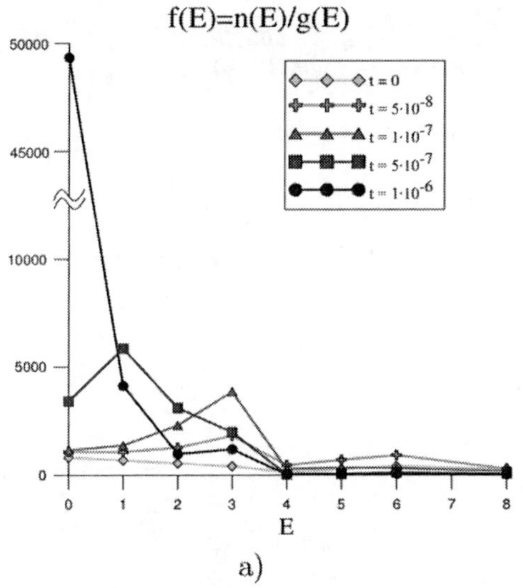

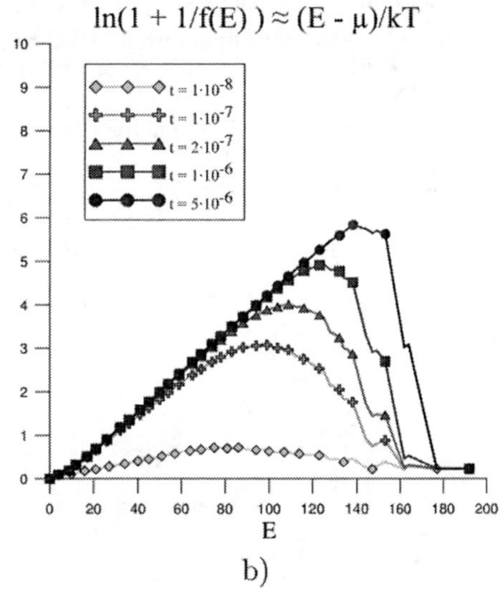

Figure 4. (a) Evolution of the particle distribution function in the case of interaction with phonon subsystem only, calculated for $N = 10^5$. (b) Evolution of the characteristic function $L(E)$ calculated for $N = 4.8 \cdot 10^4$.

simultaneously with expanding the range of energies. In our calculation, the time of calculation $t_{(max)} = 10^{-5}$ (10^6 time steps) was not enough to reach the completely linear law. Though the details can be different for various parameters of the phonon spectra, the characteristic time is determined by the exciton-phonon matrix element M_0 and does not depend on the number of particles N.

4. Conclusion

We studied the features of weakly-interacting Bose gas thermalization due to interparticle interaction and interaction with phonon subsystem. While the cooling of the gas can be realized

IOP Conf. Series: Journal of Physics: Conf. Series **936** (2017) 012055 doi:10.1088/1742-6596/936/1/012055

only with energy exchange with bath, internal redistribution and relaxation is more efficient with two-particle interaction. Both mechanisms should be taken into account when planning the experiments on cooling and condensation in novel Bose systems. The thermalization occurs in several stages, first in restricted range of energies, then the area of equilibrium is extended to wider range of energies.

The numerical algorithm [15] allows to simulate the kinetics of the finite-size interacting Bose systems, in particular the process of thermalization. In study of more complex systems for which the selection of optimal parameters for establishing equilibrium can be quite difficult, numerical simulation of this process is a promising direction of research.

The work is supported by Russian Foundation for Basic Research, grant No. 15-02-02764.

References

[1] Banuls M C, Cirac J I and Hasting M B 2011 *Phys. Rev. Lett.* **106** 050405
[2] Lin Ch-J and Motrunich Oi I 2017 *Phys. Rev.* A **95** 023621
[3] Berman O L, Gumbs G and Kezerashvili R Ya 2017 *Phys. Rev.* B **96** 014505
[4] Das A, Bhattacharya P, Banerjee A and Jankowski M 2012 *Phys. Rev.* B **85** 195321
[5] Wertz E, Amo A, Solnyshkov D D, Ferrier L, Liew T C H, Sanvitto D, Senellart P, Sagnes I, Lemaitre A, Kavokin A V, Malpuech G and Bloch J 2012 *Phys. Rev. Lett.* **109** 216404
[6] Straatsma C J E, Colussi V E, Davis M J, Lobser D S, Holland M J, Anderson D Z, Lewandowski H J and Cornell E A 2016 *Phys. Rev.* A **94** 043640
[7] Sturm M R, Schlosser M, Walser R and Birkl G 2017 *Phys. Rev.* A **95** 063625
[8] Voronova N S, Lozovik Y E 2012 *Phys. Rev.* B **86** 195305
[9] Cho Ch-H, Nukala P, Lambert K, Piccione B and Agarwal R 2011 *PNAS* **108** 10050
[10] Avetissian H K, Avetissian A K and Mkrtchian G F 2014 *Phys. Rev. Lett.* **113** 023904
[11] Wang Yi-Hs, Anderson B M and Clark Ch W 2014 *Phys. Rev.* A **89** 043624
[12] Griffin A, Nikuni T and Zaremba E 2009 *Bose-condensed gases at finite temperatures* (Cambridge: Cambridge University Press)
[13] Davis M J, Ballagh R J and Burnett K 2001 *J. Phys. B: Atomic, Molecular and Optical Physics* **34** 4487
[14] Kartsev P F, Kuznetsov I O 2016 J. Phys.: Conf. Ser. **737** 012033
[15] Kartsev P F 2017 *Proc. of the 5th Int. Workshop on OpenCL* (New York: USA/ACM) Article No. 28

IOP Conf. Series: Journal of Physics: Conf. Series **936** (2017) 012056 doi:10.1088/1742-6596/936/1/012056

To the problem of decomposition of the initial boundary value problems in mechanics

Mikhail U. Nikabadze

Department of Mathematics and Mechanics, Lomonosov Moscow State University, Leninskie Gory, 1, Moscow 119991 RUSSIA; and Bauman Moscow State Technical Uneversity

E-mail: `munikabadze@yandex.ru`

Abstract. We consider the questions on the decomposition of the initial boundary value problems of elasticity theory and thin bodies for some anisotropic media. In particular, the initial boundary problems of the micropolar (and classical) theory of elasticity are presented, i.e. to introduce tensor-block matrix operators (tensor-operators). In the case of an isotropic micropolar elastic medium (isotropic and transversally isotropic classical media) tensor-block matrix operators (tensor-operators) of cofactors corresponding to the tensor-block matrix operators (tensor-operators) of of the initial boundary value problems are obtained, which are splitting the initial boundary value problems. From three-dimensional decomposed initial boundary value problems the corresponding decomposed initial boundary value problems for the theory of thin bodies are obtained.

1. Equations of motion relative to the displacement and rotation vectors for an elastic material without a center of symmetry

The constitutive relations for the linear elastic nonhomogenous anisotropic material without a symmetry center for small displacements and rotations and also isothermal processes have the form

$$\underset{\sim}{\mathbf{P}} = \underset{\approx}{\mathbf{A}} \overset{2}{\otimes} \underset{\sim}{\gamma} + \underset{\approx}{\mathbf{B}} \overset{2}{\otimes} \underset{\sim}{\varkappa}, \quad \underset{\sim}{\mu} = \underset{\approx}{\mathbf{C}} \overset{2}{\otimes} \underset{\sim}{\gamma} + \underset{\approx}{\mathbf{D}} \overset{2}{\otimes} \underset{\sim}{\varkappa} \quad (\underset{\sim}{\gamma} = \nabla \mathbf{u} - \underset{\sim}{\mathbf{C}} \cdot \varphi, \quad \underset{\sim}{\varkappa} = \nabla \varphi), \tag{1}$$

where $\underset{\sim}{\mathbf{P}}$ and $\underset{\sim}{\mu}$ are tensors of stresses and couple stresses, respectively, $\underset{\sim}{\gamma}$ is the strain tensor, $\underset{\sim}{\varkappa}$ is the bending-torsion tensor, $\mathbf{u}$ is the displacement vector, φ is the rotation vector, $\underset{\approx}{\mathbf{A}}$, $\underset{\approx}{\mathbf{C}} = \underset{\approx}{\mathbf{B}}^{T}$ and $\underset{\approx}{\mathbf{D}}$ are the material tensors of the 4th rank, $\underset{\sim}{\mathbf{C}}$ is the discriminant 3rd rank tensor, the superscript T in the upper right corner of the quantities denotes transposition, $\overset{2}{\otimes}$ is the inner 2-product [1–10].

Substituting (1) in the equations of motion for small displacements and rotations

$$\nabla \cdot \underset{\sim}{\mathbf{P}} + \rho \mathbf{F} = \rho \partial_t^2 \mathbf{u}, \quad \nabla \cdot \underset{\sim}{\mu} + \underset{\sim}{\mathbf{C}} \overset{2}{\otimes} \underset{\sim}{\mathbf{P}} + \rho \mathbf{m} = \underset{\sim}{\mathbf{J}} \partial_t^2 \varphi$$

and introducing the tensor-block matrix operator of the equations of motion and the vector-columns of displacement and rotation vectors and vectors of volume forces and volume moments

$$\underset{\sim}{\mathbb{M}} = \begin{pmatrix} \underset{\approx}{\mathbf{A}} & \underset{\approx}{\mathbf{B}} \\ \underset{\approx}{\mathbf{C}} & \underset{\approx}{\mathbf{D}} \end{pmatrix}, \quad \mathbb{U} = \begin{pmatrix} \mathbf{u} \\ \varphi \end{pmatrix}, \quad \mathbb{X} = \begin{pmatrix} \rho \mathbf{F} \\ \rho \mathbf{m} \end{pmatrix}, \tag{2}$$

IOP Conf. Series: Journal of Physics: Conf. Series **936** (2017) 012056 doi:10.1088/1742-6596/936/1/012056

we obtain the equations of motion in displacement and rotation vectors

$$\underset{\sim}{\mathbb{M}} \cdot \mathbb{U} + \mathbb{X} = 0, \tag{3}$$

where the differential tensor-operators $\underset{\sim}{\mathbf{A}}$, $\underset{\sim}{\mathbf{B}}$, $\underset{\sim}{\mathbf{C}}$ and $\underset{\sim}{\mathbf{D}}$ have the following expressions:

$$\begin{gathered}
\underset{\sim}{\mathbf{A}} = \underset{\sim}{\mathbf{A}}' - \underset{\sim}{\mathbf{E}}\rho\partial_t^2, \quad \underset{\sim}{\mathbf{A}}' = \mathbf{r}_j\mathbf{r}_l(A^{ijkl}\nabla_i + \nabla_i A^{ijkl})\nabla_k, \\
\underset{\sim}{\mathbf{B}} = \mathbf{r}_j\mathbf{r}_l[(B^{ijkl}\nabla_i + \nabla_i B^{ijkl} - C^{l\,\cdot\,\cdot}_{\cdot\, mn}A^{mnkj})\nabla_k - C^{l\,\cdot\,\cdot}_{\cdot\, mn}\nabla_i A^{mnij}], \\
\underset{\sim}{\mathbf{C}} = \mathbf{r}_j\mathbf{r}_l(B^{klij}\nabla_i + \nabla_i B^{klij} + C^{j\,\cdot\,\cdot}_{\cdot\, mn}A^{mnkl})\nabla_k, \quad \underset{\sim}{\mathbf{D}} = \underset{\sim}{\mathbf{D}}' - \underset{\sim}{\mathbf{J}}\partial_t^2, \\
\underset{\sim}{\mathbf{D}}' = \mathbf{r}_j\mathbf{r}_l\{[D^{ijkl}\nabla_i + \nabla_i D^{ijkl} + (g_s^j g_t^l - g_s^l g_t^j)C^{s\,\cdot\,\cdot}_{\cdot\, mn}B^{mnkt}]\nabla_k - C^{l\,\cdot\,\cdot}_{\cdot\, pq}(A^{pqmn}C^{\cdot\,\cdot\, j}_{mn\,\cdot} + \nabla_i B^{pqij})\}.
\end{gathered} \tag{4}$$

Here and elsewhere $\underset{\sim}{\mathbf{E}}$ is the second-rank identity tensor, t is the time, ∂_t is the operator of the partial derivative with respect to time.

2. On static boundary conditions in the linear three-dimensional micropolar theory of elasticity. Tensor-operator of stress and couple stress

Taking into account (1), the static boundary conditions can be represented as follows

$$\underset{\approx}{\mathbf{T}}^{(1)} \cdot \mathbf{u} + \underset{\approx}{\mathbf{T}}^{(2)} \cdot \boldsymbol{\varphi} = \mathbf{P}, \quad \underset{\approx}{\mathbf{T}}^{(3)} \cdot \mathbf{u} + \underset{\approx}{\mathbf{T}}^{(4)} \cdot \boldsymbol{\varphi} = \boldsymbol{\mu}, \tag{5}$$

where $\mathbf{P}$ and $\boldsymbol{\mu}$ are the stress and couple stress vectors, respectively, given on the surface of the body and we introduce the following differential tensor operators

$$\begin{gathered}
\underset{\approx}{\mathbf{T}}^{(1)} = \mathbf{r}_j\mathbf{r}_l n_i A^{ijkl}\nabla_k, \quad \underset{\approx}{\mathbf{T}}^{(2)} = \mathbf{r}_j\mathbf{r}_l n_i B^{ijkl}\nabla_k - \mathbf{n} \cdot \underset{\approx}{\mathbf{A}} \overset{2}{\otimes} \underset{\approx}{\mathbf{C}}, \\
\underset{\approx}{\mathbf{T}}^{(3)} = \mathbf{r}_j\mathbf{r}_l n_i C^{ijkl}\nabla_k, \quad \underset{\approx}{\mathbf{T}}^{(4)} = \mathbf{r}_j\mathbf{r}_l n_i D^{ijkl}\nabla_k - \mathbf{n} \cdot \underset{\approx}{\mathbf{C}} \overset{2}{\otimes} \underset{\approx}{\mathbf{C}}.
\end{gathered}$$

Introducing the tensor-block matrix operator of stress and couple stress and also the vector-column of stress and couple stress vectors

$$\underset{\approx}{\mathbb{T}} = \begin{pmatrix} \underset{\approx}{\mathbf{T}}^{(1)} & \underset{\approx}{\mathbf{T}}^{(2)} \\ \underset{\approx}{\mathbf{T}}^{(3)} & \underset{\approx}{\mathbf{T}}^{(4)} \end{pmatrix}, \quad \mathbb{Q} = \begin{pmatrix} \mathbf{P} \\ \boldsymbol{\mu} \end{pmatrix},$$

the static boundary conditions (5) taking out of the second relation (2) we can write by the form

$$\underset{\approx}{\mathbb{T}} \cdot \mathbb{U}\big|_S = \mathbb{Q}. \tag{6}$$

We note that the kinematic boundary conditions can be written as

$$\mathbb{U}\big|_S = \mathbb{H} \quad \left(\mathbb{H} = \begin{pmatrix} \mathbf{f} \\ \boldsymbol{\psi} \end{pmatrix}\right), \tag{7}$$

mixed boundary conditions can be written as

$$\underset{\approx}{\mathbb{T}} \cdot \mathbb{U}\big|_{S_1} = \mathbb{Q}; \quad \mathbb{U}\big|_{S_2} = \mathbb{H},$$

and the initial conditions have the form

$$\mathbb{U}|_{t=t_0} = \mathbb{U}_0 \quad \mathbb{V}|_{t=t_0} = \mathbb{V}_0, \tag{8}$$

where

$$\mathbb{U}_0 = \begin{pmatrix} \mathbf{u}_0 \\ \boldsymbol{\varphi}_0 \end{pmatrix}, \quad \mathbb{V}_0 = \frac{d\mathbb{U}}{dt}\bigg|_{t=t_0} = \begin{pmatrix} \mathbf{v}_0 \\ \boldsymbol{\omega}_0 \end{pmatrix}, \quad \mathbf{v}_0 = \frac{d\mathbf{u}}{dt}\bigg|_{t=t_0}, \quad \boldsymbol{\omega}_0 = \frac{d\boldsymbol{\varphi}}{dt}\bigg|_{t=t_0}.$$

Here $\mathbf{f}$ and $\boldsymbol{\psi}$ are the displacement and rotation vectors given on the body surface; $\mathbf{u}_0$ and $\boldsymbol{\varphi}_0$ are the displacement and rotation vectors given for $t = t_0$; $\mathbf{v}_0$ and $\boldsymbol{\omega}_0$ are the velocity and angular velocity vectors given for $t = t_0$; S is the body surface, $S_1 \cup S_2 = S$, $S_1 \cap S_2 = S$.

3. Formulation of initial-boundary value problems

We introduce the definitions.

Definition 1. If the vectors of displacements and rotations (kinematic boundary conditions) are given on the boundary of the body S, then such conditions are called the first boundary conditions, and the problem of micropolar mechanics of deformable solids (MDS) using these conditions, as well as the initial conditions are called the first initial-boundary value problem.

In this case, the first initial-boundary value problem includes: the equations of motion (3), the kinematic boundary conditions (7) and the initial conditions (8).

Definition 2. If static boundary conditions (stress and couple stress vectors) are given on the body's boundary S, then such boundary conditions are called the second boundary conditions, and the problem of micropolar MDS, using them and the initial conditions, is called the second initial-boundary value problem.

In this case, the second initial-boundary value problem includes: the equations of motion (3), the stati c boundary conditions (6) and the initial conditions (8).

Definition 3. If static boundary conditions are given on one part of the body boundary S_1, and static boundary conditions are given on the rest of its part S_2, then such boundary conditions are called mixed boundary conditions, and the problem of micropolar MDS using them and initial conditions are called mixed initial-boundary value problem.

In this case, the mixed (third) initial-boundary value problem includes: the equations of motion (3), the kinematic boundary conditions (7) on one part of the body boundary and the static boundary conditions (6) on the other part of the body boundary and the initial conditions (8).

We note that, excluding the characteristics of the micropolar theory from the above definitions, we obtain the corresponding definitions for classical MDS.

We note also that the kinematic boundary conditions and the initial conditions do not need to be decompose, since they are set in a decomposed form. Hence, to decompose the first initial-boundary value problem, it is sufficient to decompose only the equations of motion, since, as already mentioned in the previous sentence, the kinematic boundary conditions and the initial conditions are decomposed. In this connection, the decomposition of the static boundary conditions represent of great interest. If the equations of motion (3) and the static boundary conditions (6) can be decompose under some conditions, then under the same conditions all the initial-boundary value problems formulated above can be decompose. Hence, it is necessary to establish the conditions under which the equations of motion (3) and the static boundary conditions (6) are decomposed.

4. Decomposition the equation of motion in case of a homogeneous isotropic micropolar medium

In this case, when many authors (see, for example, [11, 12]) considered, $\underset{\approx}{\mathbf{B}} = 0$ and differential operator tensors $\underset{\sim}{\mathbf{A}}$, $\underset{\sim}{\mathbf{B}}$, $\underset{\sim}{\mathbf{C}}$ and $\underset{\sim}{\mathbf{D}}$ (see (4)) have form

$$
\begin{aligned}
&\underset{\sim}{\mathbf{A}} = \underset{\sim}{\mathbf{E}} Q_2 + d\nabla\nabla, \quad \underset{\sim}{\mathbf{B}} = \underset{\sim}{\mathbf{C}} = -2\alpha \underset{\approx}{\mathbf{C}} \cdot \nabla, \quad \underset{\sim}{\mathbf{D}} = \underset{\sim}{\mathbf{E}} Q_4 + m\nabla\nabla, \\
&Q_2 = b\Delta - \rho\partial_t^2, \quad Q_4 = g\Delta - l - J\partial_t^2, \quad Q_1 = Q_2 + d\Delta, \quad Q_3 = Q_4 + m\Delta, \quad \underset{\sim}{\mathbf{J}} = J\underset{\sim}{\mathbf{E}}, \\
&d = \lambda + \mu - \alpha, \quad l = 4\alpha, \quad b = \mu + \alpha, \quad g = \delta + \beta, \quad m = \gamma + \delta - \beta,
\end{aligned}
$$

where Q_1, Q_2, Q_3 Q_4 are the wave operators, and the elasticity tensors have expressions

$$
\underset{\approx}{\mathbf{A}} = a_1 \underset{\approx}{\mathbf{C}}_{(1)} + a_2 \underset{\approx}{\mathbf{C}}_{(2)} + a_3 \underset{\approx}{\mathbf{C}}_{(3)}, \quad \underset{\approx}{\mathbf{D}} = d_1 \underset{\approx}{\mathbf{C}}_{(1)} + d_2 \underset{\approx}{\mathbf{C}}_{(2)} + d_3 \underset{\approx}{\mathbf{C}}_{(3)}. \tag{9}
$$

Here $\underset{\approx}{\mathbf{C}}_{(1)}$, $\underset{\approx}{\mathbf{C}}_{(2)}$ and $\underset{\approx}{\mathbf{C}}_{(3)}$ — are basic isotropic tensors of the 4th rank, and the notations for material constants are $a_1 = \lambda$, $a_2 = \mu$, $a_3 = \alpha$, $b_1 = \gamma$, $b_2 = \delta$ $b_3 = \beta$.

By

$$\underset{\sim}{\mathbb{M}}_* = \begin{pmatrix} \hat{\underset{\sim}{\mathbf{A}}} & \hat{\underset{\sim}{\mathbf{B}}}^{(1)} \\ \hat{\underset{\sim}{\mathbf{B}}}^{(2)} & \hat{\underset{\sim}{\mathbf{C}}} \end{pmatrix}$$

we denote the tensor-block matrix operator of cofactors for the tensor-block matrix operator $\underset{\sim}{\mathbb{M}}$ of the equation (3), and after simple calculations, we obtain

$$\begin{aligned}
&\hat{\underset{\sim}{\mathbf{A}}} = Q_3(Q_2Q_4 + 4\alpha^2\Delta)[\underset{\sim}{\mathbf{E}}Q_1Q_4 - (dQ_4 - 4\alpha^2)\nabla\nabla] \ (\hat{\underset{\sim}{\mathbf{A}}}^T = \hat{\underset{\sim}{\mathbf{A}}}),\\
&\hat{\underset{\sim}{\mathbf{B}}} = \hat{\underset{\sim}{\mathbf{B}}}^{(1)} = \hat{\underset{\sim}{\mathbf{B}}}^{(2)} = -2\alpha Q_1Q_3(Q_2Q_4 + 4\alpha^2\Delta)\underset{\approx}{\mathbf{C}} \cdot \nabla) \ (\hat{\underset{\sim}{\mathbf{B}}}^T = -\hat{\underset{\sim}{\mathbf{B}}}),\\
&\hat{\underset{\sim}{\mathbf{C}}} = Q_1(Q_2Q_4 + 4\alpha^2\Delta)[\underset{\sim}{\mathbf{E}}Q_2Q_3 - (mQ_2 - 4\alpha^2)\nabla\nabla] \ (\hat{\underset{\sim}{\mathbf{C}}}^T = \hat{\underset{\sim}{\mathbf{C}}}).
\end{aligned}$$

Let us introduce tensor-block matrix operators

$$\underset{\sim}{\mathbb{N}}^{(1)} = \begin{pmatrix} \underset{\sim}{\mathbf{R}} & \underset{\sim}{\mathbf{S}}^{(2)} \\ \underset{\sim}{\mathbf{S}}^{(1)} & \underset{\sim}{\mathbf{T}} \end{pmatrix}, \quad \underset{\sim}{\mathbb{N}}^{(2)} = \begin{pmatrix} \underset{\sim}{\mathbf{R}} & \underset{\sim}{\mathbf{S}}^{(1)} \\ \underset{\sim}{\mathbf{S}}^{(2)} & \underset{\sim}{\mathbf{T}} \end{pmatrix},$$

$$\begin{aligned}
&\underset{\sim}{\mathbf{R}} = \underset{\sim}{\mathbf{E}}Q_1Q_4 - (dQ_4 - 4\alpha^2)\nabla\nabla, \quad \underset{\sim}{\mathbf{S}}^{(1)} = Q_3\underset{\sim}{\mathbf{B}}, \quad \underset{\sim}{\mathbf{S}}^{(2)} = Q_1\underset{\sim}{\mathbf{B}},\\
&\underset{\sim}{\mathbf{B}} = -2\alpha\underset{\approx}{\mathbf{C}}\cdot\nabla, \quad \underset{\sim}{\mathbf{T}} = \underset{\sim}{\mathbf{E}}Q_2Q_3 - (mQ_2 - 4\alpha^2)\nabla\nabla.
\end{aligned}$$

Then if we will look for a solution of the equation (3) in the form (similar Galerkin)

$$\mathbb{U} = \underset{\sim}{\mathbb{N}}^{(1)T} \cdot \mathbb{V} \quad \left(\mathbb{U} = \begin{pmatrix} \mathbf{u} \\ \boldsymbol{\varphi} \end{pmatrix}, \quad \mathbb{V} = \begin{pmatrix} \mathbf{v} \\ \boldsymbol{\psi} \end{pmatrix}\right),$$

then we obtain the following decomposed equations:

$$Q_1(Q_2Q_4 + 4\alpha^2\Delta)\mathbf{v} + \rho\mathbf{F} = 0, \quad Q_3(Q_2Q_4 + 4\alpha^2\Delta)\boldsymbol{\psi} + \rho\mathbf{m} = 0.$$

If we apply the operator $\underset{\sim}{\mathbb{N}}^{(2)T}$ to the equation (3) on the left, then we have decomposed equations

$$\begin{aligned}
&Q_1[(Q_2Q_4 + 4\alpha^2\Delta)\mathbf{u} + 2\alpha(\underset{\approx}{\mathbf{C}}\cdot\nabla)\cdot(\rho\mathbf{m})] + [\underset{\sim}{\mathbf{E}}Q_1Q_4 - (dQ_4 - 4\alpha^2)\nabla\nabla]\cdot(\rho\mathbf{F}) = 0,\\
&Q_3[(Q_2Q_4 + 4\alpha^2\Delta)\boldsymbol{\varphi} + 2\alpha(\underset{\approx}{\mathbf{C}}\cdot\nabla)\cdot(\rho\mathbf{F})] + [\underset{\sim}{\mathbf{E}}Q_2Q_3 - (mQ_2 - 4\alpha^2)\nabla\nabla]\cdot(\rho\mathbf{m}) = 0.
\end{aligned} \tag{10}$$

For $\alpha = 0$ (the case of a reduced medium), the classical equation follows from the first equation (10), and the second equation has a similar form.

5. Decomposition of static boundary conditions

In case of an isotropic micropolar material without a center of symmetry, according to (9) and $\underset{\approx}{\mathbf{B}} = b_1\underset{\approx}{\mathbf{C}}_{(1)} + b_2\underset{\approx}{\mathbf{C}}_{(2)} + b_3\underset{\approx}{\mathbf{C}}_{(3)}$ we have

$$\begin{aligned}
&\underset{\sim}{\mathbf{T}}^{(1)} = a_2\underset{\sim}{\mathbf{E}}\mathbf{n}\cdot\nabla + a_1\mathbf{n}\nabla + a_3(\mathbf{n}\nabla)^T, \quad \underset{\sim}{\mathbf{T}}^{(2)} = b_2\underset{\sim}{\mathbf{E}}\mathbf{n}\cdot\nabla + b_1\mathbf{n}\nabla + b_3(\mathbf{n}\nabla)^T - (a_2 - a_3)\mathbf{n}\cdot\underset{\approx}{\mathbf{C}},\\
&\underset{\sim}{\mathbf{T}}^{(3)} = b_2\underset{\sim}{\mathbf{E}}\mathbf{n}\cdot\nabla + b_1\mathbf{n}\nabla + b_3(\mathbf{n}\nabla)^T, \quad \underset{\sim}{\mathbf{T}}^{(4)} = d_2\underset{\sim}{\mathbf{E}}\mathbf{n}\cdot\nabla + d_1\mathbf{n}\nabla + d_3(\mathbf{n}\nabla)^T - (b_2 - b_3)\mathbf{n}\cdot\underset{\approx}{\mathbf{C}}.
\end{aligned}$$

We note that some authors (see, for example, [11,12]) consider $\underset{\approx}{\mathbf{B}}$ is an asymmetric tensor, so in the case of an isotropic medium it is zero, as was done above. However, some authors (see, for example, [13]) prove that $\underset{\approx}{\mathbf{B}}$ is a symmetric tensor, therefore in the case of an isotropic medium it is not equal to zero and this tensor, like any isotropic tensor of the fourth rank, is generally determined by three parameters, as is customary in this case.

Next, it is easy to see, that

$$\underset{\sim}{\mathbf{T}}^{(2)}=\underset{\sim}{\mathbf{T}}^{(3)}-(a_2-a_3)\mathbf{n}\cdot\underset{\approx}{\mathbf{C}},\;\; \underset{\sim}{\mathbf{T}}^{(4)}=\underset{\sim}{\mathbf{T}}^{\prime(4)}-(b_2-b_3)\mathbf{n}\cdot\underset{\approx}{\mathbf{C}},\;\; \underset{\sim}{\mathbf{T}}^{\prime(4)}=d_2\underset{\sim}{\mathbf{E}}\mathbf{n}\cdot\nabla+d_1\mathbf{n}\nabla+d_3(\mathbf{n}\nabla)^T.$$

Assuming that the body has a piecewise-plain boundary and denoting by $\underset{\sim}{\mathbf{T}}_*^{(1)}$ and $|\underset{\sim}{\mathbf{T}}^{(1)}|$, $\underset{\sim}{\mathbf{T}}_*^{(3)}$ and $|\underset{\sim}{\mathbf{T}}^{(3)}|$, $\underset{\sim}{\mathbf{T}}_*^{\prime(4)}$ and $|\underset{\sim}{\mathbf{T}}^{\prime(4)}|$ differential tensor-operators of cofactors and determinants for the tensors-operators $\underset{\sim}{\mathbf{T}}^{(1)}$, $\underset{\sim}{\mathbf{T}}^{(3)}$ and $\underset{\sim}{\mathbf{T}}^{\prime(4)}$, after simple calculations we obtain

$$\begin{aligned}
\underset{\sim}{\mathbf{T}}_*^{(1)} &= [(a_1+a_2)(a_2+a_3)\underset{\sim}{\mathbf{E}}\mathbf{n}\cdot\nabla - a_3(a_1+a_2)\mathbf{n}\nabla - \\
&\qquad -a_1(a_2+a_3)(\mathbf{n}\nabla)^T]\mathbf{n}\cdot\nabla + a_1a_3[\nabla\nabla+(\mathbf{nn}-\underset{\sim}{\mathbf{E}})\Delta],\\
|\underset{\sim}{\mathbf{T}}^{(1)}| &= a_2[(a_1+a_2)(a_2+a_3)\mathbf{nnn}\overset{3}{\otimes}\nabla\nabla\nabla - a_1a_3\Delta\mathbf{n}\cdot\nabla] = \\
&= a_2[(a_1+a_2)(a_2+a_3)\mathbf{nn}\overset{2}{\otimes}\nabla\nabla - a_1a_3\Delta]\mathbf{n}\cdot\nabla,\\
\underset{\sim}{\mathbf{T}}_*^{(3)} &= [(b_1+b_2)(b_2+b_3)\underset{\sim}{\mathbf{E}}\mathbf{n}\cdot\nabla - b_3(b_1+b_2)\mathbf{n}\nabla - \\
&\qquad -b_1(b_2+b_3)(\mathbf{n}\nabla)^T]\mathbf{n}\cdot\nabla + b_1b_3[\nabla\nabla+(\mathbf{nn}-\underset{\sim}{\mathbf{E}})\Delta],\\
|\underset{\sim}{\mathbf{T}}^{(3)}| &= b_2[(b_1+b_2)(b_2+b_3)\mathbf{nn}\overset{2}{\otimes}\nabla\nabla - b_1b_3\Delta]\mathbf{n}\cdot\nabla,\\
\underset{\sim}{\mathbf{T}}_*^{\prime(4)} &= [(d_1+d_2)(d_2+d_3)\underset{\sim}{\mathbf{E}}\mathbf{n}\cdot\nabla - d_3(d_1+d_2)\mathbf{n}\nabla - \\
&\qquad -d_1(d_2+d_3)(\mathbf{n}\nabla)^T]\mathbf{n}\cdot\nabla + d_1d_3[\nabla\nabla+(\mathbf{nn}-\underset{\sim}{\mathbf{E}})\Delta],\\
|\underset{\sim}{\mathbf{T}}^{\prime(4)}| &= d_2[(d_1+d_2)(d_2+d_3)\mathbf{nn}\overset{2}{\otimes}\nabla\nabla - d_1d_3\Delta]\mathbf{n}\cdot\nabla.
\end{aligned}$$

Note that we want to obtain boundary conditions separately for $\mathbf{u}$ and $\boldsymbol{\varphi}$. Let us consider the case when $b_2=b_3$, $a_2=a_3$. Then $\underset{\sim}{\mathbf{T}}^{(2)}=\underset{\sim}{\mathbf{T}}^{(3)}$, $\underset{\sim}{\mathbf{T}}^{(4)}=\underset{\sim}{\mathbf{T}}^{\prime(4)}$ and the boundary conditions (6) can be written in form

$$\underset{\sim}{\mathbf{T}}^{(1)}\cdot\mathbf{u}+\underset{\sim}{\mathbf{T}}^{(3)}\cdot\boldsymbol{\varphi}=\mathbf{P},\quad \underset{\sim}{\mathbf{T}}^{(3)}\cdot\mathbf{u}+\underset{\sim}{\mathbf{T}}^{\prime(4)}\cdot\boldsymbol{\varphi}=\boldsymbol{\mu}.$$

In this case it is easy to obtain the boundary conditions separately for $\mathbf{u}$ and $\boldsymbol{\varphi}$

$$\Big(|\underset{\sim}{\mathbf{T}}^{\prime(4)}|\underset{\sim}{\mathbf{T}}_*^{(3)T}\cdot\underset{\sim}{\mathbf{T}}^{(1)} - |\underset{\sim}{\mathbf{T}}^{(3)}|\underset{\sim}{\mathbf{T}}_*^{\prime(4)T}\cdot\underset{\sim}{\mathbf{T}}^{(3)}\Big)\cdot\mathbf{u} = |\underset{\sim}{\mathbf{T}}^{\prime(4)}|\underset{\sim}{\mathbf{T}}_*^{(3)T}\cdot\mathbf{P} - |\underset{\sim}{\mathbf{T}}^{(3)}|\underset{\sim}{\mathbf{T}}_*^{\prime(4)T}\cdot\boldsymbol{\mu},$$

$$\Big(|\underset{\sim}{\mathbf{T}}^{(3)}|\underset{\sim}{\mathbf{T}}_*^{(1)T}\cdot\underset{\sim}{\mathbf{T}}^{(3)} - |\underset{\sim}{\mathbf{T}}^{(1)}|\underset{\sim}{\mathbf{T}}_*^{(3)T}\cdot\underset{\sim}{\mathbf{T}}^{\prime(4)}\Big)\cdot\boldsymbol{\varphi} = |\underset{\sim}{\mathbf{T}}^{(3)}|\underset{\sim}{\mathbf{T}}_*^{(1)T}\cdot\mathbf{P} - |\underset{\sim}{\mathbf{T}}^{(1)}|\underset{\sim}{\mathbf{T}}_*^{(3)T}\cdot\boldsymbol{\mu}.$$

6. Characteristic equation for a tensor-block matrix operator of second rank

The characteristic equation for the tensor-block matrix $\underset{\sim}{\mathbb{M}}$ has the form

$$\eta^6 - I_1(\underset{\sim}{\mathbb{M}})\eta^5 + I_2(\underset{\sim}{\mathbb{M}})\eta^4 - I_3(\underset{\sim}{\mathbb{M}})\eta^3 + I_4(\underset{\sim}{\mathbb{M}})\eta^2 - I_5(\underset{\sim}{\mathbb{M}})\eta + I_6(\underset{\sim}{\mathbb{M}}) = 0,$$

$$S_k = I_k(\mathbb{M}) = \frac{1}{k!}\begin{vmatrix} s_1 & 1 & \cdots & 0 & 0\\ s_2 & s_1 & \cdots & 0 & 0\\ \cdots & \cdots & \cdots & \cdots & \cdots\\ s_{k-1} & s_{k-2} & \cdots & s_1 & k\\ s_k & s_{k-1} & \cdots & s_2 & s_1 \end{vmatrix},\quad k=\overline{1,6}, \tag{11}$$

$$S_k = I_k(\mathbb{M}),\quad s_k = I_1(\mathbb{M}^k),\quad k=\overline{1,6},\quad \mathbb{M}^k = \overbrace{\mathbb{M}\cdot\mathbb{M}\cdot\ldots\cdot\mathbb{M}}^{k},$$

where $I_k(\mathbb{M})$, $k=\overline{1,6}$, we denote invariants of the tensor-block matrix operator $\mathbb{M}$.

It is easy to see, that the relations inverse to (11) are represented in the form

$$s_k = I_1(\mathbb{M}^k) = \begin{vmatrix} S_1 & 1 & 0 & \cdots & 0 \\ 2S_2 & S_1 & 1 & \cdots & 0 \\ \cdots & \cdots & \cdots & \cdots & \cdots \\ kS_k & S_{k-1} & S_{k-2} & \cdots & S_1 \end{vmatrix}, \quad k=\overline{1,6}.$$

Replacing here $\underset{\sim}{\mathbb{M}}$ by the tensor-block matrix operators of the equations of motion in displacements and rotations under various anisotropic media, we obtain the characteristic equations for them. If we find the roots (eigenvectors) of the obtained characteristic equations for the tensor-block matrix operators indicated above, then their determinants we can write as a product of eigenoperators (simple operators). Thus, the equations are split.

7. Decomposition of the equation of the classical theory of elasticity for a transversely-isotropic body

In this case the equations and boundary conditions are represented in the form

$$\underset{\sim}{\mathbf{L}} \cdot \mathbf{u} + \rho \mathbf{F} = 0, \quad \underset{\sim}{\mathbf{T}} \cdot \mathbf{u} = \mathbf{P} \quad (\underset{\sim}{\mathbf{L}} = \mathbf{r}_j \mathbf{r}_l A^{ijkl} \nabla_i \nabla_k, \ \ \underset{\sim}{\mathbf{T}} = \mathbf{r}_j \mathbf{r}_l n_i A^{ijkl} \nabla_k),$$

where the elastic tensor $\underset{\approx}{\mathbf{A}}$, the tensor-operator $\underset{\sim}{\mathbf{L}}$, the tensor-operator of cofactors $\underset{\sim}{\mathbf{L}}_*$ and the determinant $|\underset{\sim}{\mathbf{L}}|$ for $\underset{\sim}{\mathbf{L}}$ have the form

$$\begin{aligned}
&\underset{\approx}{\mathbf{A}} = A_1(\underset{\sim}{\mathbf{e}}_1\underset{\sim}{\mathbf{e}}_1 + \underset{\sim}{\mathbf{e}}_2\underset{\sim}{\mathbf{e}}_2 + \underset{\sim}{\mathbf{e}}_6\underset{\sim}{\mathbf{e}}_6) + A_2(\underset{\sim}{\mathbf{e}}_1\underset{\sim}{\mathbf{e}}_2 + \underset{\sim}{\mathbf{e}}_2\underset{\sim}{\mathbf{e}}_1 - \underset{\sim}{\mathbf{e}}_6\underset{\sim}{\mathbf{e}}_6) + \\
&\quad + A_3(\underset{\sim}{\mathbf{e}}_1\underset{\sim}{\mathbf{e}}_3 + \underset{\sim}{\mathbf{e}}_3\underset{\sim}{\mathbf{e}}_1 + \underset{\sim}{\mathbf{e}}_2\underset{\sim}{\mathbf{e}}_3 + \underset{\sim}{\mathbf{e}}_3\underset{\sim}{\mathbf{e}}_2) + A_4\underset{\sim}{\mathbf{e}}_3\underset{\sim}{\mathbf{e}}_3 + 2A_5(\underset{\sim}{\mathbf{e}}_4\underset{\sim}{\mathbf{e}}_4 + \underset{\sim}{\mathbf{e}}_5\underset{\sim}{\mathbf{e}}_5), \\
&\underset{\sim}{\mathbf{L}} = \mathbf{e}_1\mathbf{e}_1(1/2)[(A_1 - A_2)\Delta + (A_1 + A_2)\partial_1^2 + 2A_5\partial_3^2] + \\
&\quad + (\mathbf{e}_1\mathbf{e}_2 + \mathbf{e}_2\mathbf{e}_1))(1/2)(A_1 + A_2)\partial_1\partial_2 + (\mathbf{e}_1\mathbf{e}_3 + \mathbf{e}_3\mathbf{e}_1))(A_3 + A_5)\partial_1\partial_3 + \\
&\quad + \mathbf{e}_2\mathbf{e}_2(1/2)[(A_1 - A_2)\Delta + (A_1 + A_2)\partial_2^2 + 2A_5\partial_3^2] + \\
&\quad + (\mathbf{e}_2\mathbf{e}_3 + \mathbf{e}_3\mathbf{e}_2)(A_3 + A_5)\partial_2\partial_3 + \mathbf{e}_3\mathbf{e}_3(A_5\Delta + A_4\partial_3^2), \quad \Delta = \partial_1^2 + \partial_2^2, \\
&\underset{\sim}{\mathbf{L}}_* = L_{*11}\mathbf{e}_1\mathbf{e}_1 + L_{*22}\mathbf{e}_2\mathbf{e}_2 + L_{*33}\mathbf{e}_3\mathbf{e}_3 + L_{*12}(\mathbf{e}_1\mathbf{e}_2 + \mathbf{e}_2\mathbf{e}_1) + \\
&+ L_{*23}(\mathbf{e}_2\mathbf{e}_3 + \mathbf{e}_3\mathbf{e}_2) + L_{*31}(\mathbf{e}_3\mathbf{e}_1 + \mathbf{e}_1\mathbf{e}_3), \quad \partial_i = \partial/\partial x_i, \quad i = 1,2,3, \\
&2L_{*11} = 2A_1A_5\Delta^2 + (A_1 + A_2)A_5\Delta\partial_1^2 + \{2[A_1A_4 - A_3(A_3 + 2A_5)]\Delta - \\
&\quad - [(A_1 + A_2)A_4 - 2(A_3 + A_5)^2]\partial_1^2\}\partial_3^2 + 2A_4A_5\partial_3^4, \\
&2L_{*12} = \{-(A_1 + A_2)A_5\Delta + [2(A_3 + A_5)^2 - (A_1 + A_2)A_4]\partial_3^2\}\partial_1\partial_2, \\
&2L_{*13} = -[(A_1 - A_2)(A_3 + A_5)\Delta + 2(A_3 + A_5)A_5\partial_3^2]\partial_1\partial_3, \\
&2L_{*22} = 2A_1A_5\Delta^2 - (A_1 + A_2)A_5\Delta\partial_2^2 + \{2[A_1A_4 - A_3(A_3 + 2A_5)]\Delta - \\
&\quad - [(A_1 + A_2)A_4 - 2(A_3 + A_5)^2]\partial_2^2\}\partial_3^2 + 2A_4A_5\partial_3^4, \\
&2L_{*23} = -[(A_1 - A_2)(A_3 + A_5)\Delta + 2(A_3 + A_5)A_5\partial_3^2]\partial_2\partial_3, \\
&2L_{*33} = [(A_1 - A_2)A_1\Delta^2 + (3A_1 - A_2)A_5]\Delta\partial_3^2 + 2A_5^2\partial_3^4, \\
&2|\underset{\sim}{\mathbf{L}}| = (A_1 - A_2)A_1A_5\Delta^3 + \{(A_1 - A_2)[A_1A_4 + A_3(A_3 - 2A_5)] + 2A_1A_5\}\Delta^2\partial_3^2 + \\
&\quad + [(3A_1 - A_2)A_4A_5 - 2A_3A_5(A_3 + 2A_5)]\Delta\partial_3^4 + 2A_4A_5^2\partial_3^6,
\end{aligned}$$

$$\begin{aligned}
&\mathbf{e}_i \cdot \mathbf{e}_j = \delta_{ij}, \quad i,j = 1,2,3, \quad \underset{\sim}{\mathbf{e}}_m \overset{2}{\otimes} \underset{\sim}{\mathbf{e}}_n = \delta_{mn}, \quad m,n = 1,2,\ldots,6, \\
&\underset{\sim}{\mathbf{e}}_1 = \mathbf{e}_1\mathbf{e}_1, \quad \underset{\sim}{\mathbf{e}}_2 = \mathbf{e}_2\mathbf{e}_2, \quad \underset{\sim}{\mathbf{e}}_3 = \mathbf{e}_3\mathbf{e}_3, \quad \underset{\sim}{\mathbf{e}}_4 = (1/\sqrt{2})(\mathbf{e}_1\mathbf{e}_2 + \mathbf{e}_2\mathbf{e}_1), \\
&\underset{\sim}{\mathbf{e}}_5 = (1/\sqrt{2})(\mathbf{e}_2\mathbf{e}_3 + \mathbf{e}_3\mathbf{e}_2), \quad \mathbf{e}_6 = (1/\sqrt{2})(\mathbf{e}_3\mathbf{e}_1 + \mathbf{e}_1\mathbf{e}_3).
\end{aligned}$$

Here δ_{ij} is the Kronecker's delta-symbol, $\mathbf{e}_i$, $i, j = 1, 2, 3$ is the orthonormal basis.
Let us also note that

$$2I_1(\underset{\sim}{\mathbf{L}}) = (3A_1 - A_2 + 2A_5)\Delta + (A_4 + 2A_5)\partial_3^2,$$
$$2I_2(\underset{\sim}{\mathbf{L}}) = [(3A_1 - A_2)A_5 + A_1(A_1 - A_2)]\Delta^2 + [(3A_1 - A_2)A_4 - 2A_3(A_3 + 2A_5) + (2A_5 + 3A_1 - A_2)A_5]\Delta\partial_3^2 + 2(2A_4 + A_5)A_5\partial_3^4, \quad I_3(\underset{\sim}{\mathbf{L}}) = |\underset{\sim}{\mathbf{L}}|.$$

It is easy to see that the determinant of the tensor-operator $\underset{\sim}{\mathbf{L}}$ can be written in the form

$$|\underset{\sim}{\mathbf{L}}| = A\Delta^3 + B\Delta^2\partial_3^2 + C\Delta\partial_3^4 + D\partial_3^6 = k(\Delta + k_1\partial_3^2)(\Delta + k_2\partial_3^2)(\Delta + k_3\partial_3^2);$$
$$k = A, \ k_1 + k_2 + k_3 = B/A, \ k_1k_2 + k_1k_3 + k_2k_3 = C/A, \ k_1k_2k_3 = D/A,$$
$$A = (1/2)(A_1 - A_2)A_1A_5, \ B = (1/2)\{(A_1 - A_2)[A_1A_4 + A_3(A_3 - 2A_5)] + 2A_1A_5\},$$
$$C = (1/2)[(3A_1 - A_2)A_4A_5 - 2A_3A_5(A_3 + 2A_5)], \ D = A_4A_5^2.$$

The split equations will have the form

$$|\underset{\sim}{\mathbf{L}}|\mathbf{u} + \underset{\sim}{\mathbf{L}}_* \cdot (\rho\mathbf{F}) = 0 \quad \text{or} \quad \big(\text{if we will look for } \mathbf{u} \text{ in form } \mathbf{u} = \underset{\sim}{\mathbf{L}}_* \cdot \mathbf{v}, \text{then } |\underset{\sim}{\mathbf{L}}|\mathbf{v} + \rho\mathbf{F} = 0\big). \tag{12}$$

Applying, for example, to the split equations in brackets from (12), the operator of moments of the k-th order [3,8], the equations for prismatic bodies in moments with respect to any system of orthogonal polynomials have the form

$$A\Delta^3\overset{(k)}{\mathbf{v}} + B\Delta^2\overset{(k)}{\mathbf{v}}{}'' + C\Delta\overset{(k)}{\mathbf{v}}{}^{IV} + D\overset{(k)}{\mathbf{v}}{}^{VI} + \rho\overset{(k)}{\mathbf{F}} = 0, \ k = \overline{0, \infty};$$

where in the application of the system of Legendre polynomials, the expressions for $\overset{(k)}{\mathbf{v}}{}''$, $\overset{(k)}{\mathbf{v}}{}^{IV}$ and $\overset{(k)}{\mathbf{v}}{}^{VI}$ are defined by the following relation:

$$\overset{(n)(2m)}{\mathbf{v}}(x') = (2n+1)\sum_{k=1}^{\infty} C_{k+2m-2}^{2m-1} \prod_{s=1}^{2m-1} (2n+2k+2s-1) \overset{(n+2k+2m-2)}{\mathbf{v}}, \ n \in \mathbb{N}_0, \ m \in \mathbb{N}. \tag{13}$$

8. Decomposition of static boundary conditions of the classical theory of elasticity in the case of a transversely isotropic body

In this case, the boundary conditions can be written in the form $\underset{\sim}{\mathbf{T}} \cdot \mathbf{u} = \mathbf{P}$ ($\underset{\sim}{\mathbf{T}} = \mathbf{r}_j\mathbf{r}_l n_i A^{ijkl}\nabla_k$), where the tensor operator of stress operator $\underset{\sim}{\mathbf{T}}$, the tensor-operator of the cofacors $\underset{\sim}{\mathbf{T}}_*$ for $\underset{\sim}{\mathbf{T}}$ and the determinant $|\underset{\sim}{\mathbf{T}}|$ have the form

$$\underset{\sim}{\mathbf{T}} = \mathbf{e}_1\mathbf{e}_1[A_1n_1\partial_1 + 1/2(A_1 - A_2)n_2\partial_2 + A_5n_3\partial_3] +$$
$$+\mathbf{e}_1\mathbf{e}_2[1/2(A_1 - A_2)n_2\partial_1 + A_2n_1\partial_2] + \mathbf{e}_1\mathbf{e}_3(A_3n_1\partial_3 + A_5n_3\partial_1) +$$
$$+\mathbf{e}_2\mathbf{e}_1[1/2(A_1 - A_2)n_1\partial_2 + A_2n_2\partial_1] + \mathbf{e}_2\mathbf{e}_2[A_1n_2\partial_2 + 1/2(A_1 - A_2)n_1\partial_1 +$$
$$+A_5n_3\partial_3] + \mathbf{e}_2\mathbf{e}_3(A_3n_2\partial_3 + A_5n_3\partial_2) + \mathbf{e}_3\mathbf{e}_1(A_3n_3\partial_1 + A_5n_1\partial_3) +$$
$$+\mathbf{e}_3\mathbf{e}_2(A_3n_3\partial_2 + A_5n_2\partial_3) + \mathbf{e}_3\mathbf{e}_3[A_4n_3\partial_3 + A_5(n_1\partial_1 + n_2\partial_2)], \ \underset{\sim}{\mathbf{T}}^T \neq \underset{\sim}{\mathbf{T}};$$
$$|\underset{\sim}{\mathbf{T}}| = AA_5\{[-A_2 + (A_2 - A_3)n_3^2]\Delta + (A_1 + A_2)\mathbf{n}^0\mathbf{n}^0 \overset{2}{\otimes} \nabla^0\nabla^0\}\mathbf{n}^0 \cdot \nabla^0 +$$
$$+\{A_2(A_5^2 - AA_4) + AA_3^2 - [(A_2 + A_3)A_5^2 + A(A_3^2 - A_2A_4)]n_3^2\}n_3\Delta\partial_3 +$$
$$+[2A(A_5^2 - A_3^2) + (A_1 + A_2)AA_4]n_3\mathbf{n}^0\mathbf{n}^0 \overset{2}{\otimes} \nabla^0\nabla^0\partial_3 +$$
$$+A_5[-AA_3 + (AA_3 + 3AA_4 + A_2A_4 - A_3^2)n_3^2]\mathbf{n}^0 \cdot \nabla^0\partial_3^2 - A_5^2[A_3 - (A_3 + A_4)n_3^2]n_3\partial_3^3,$$
$$T_{*11} = A_5(An_1\partial_1 + A_1n_2\partial_2)\mathbf{n}^0 \cdot \nabla^0 - A_3A_5n_3^2\partial_2^2 + [(AA_4 + A_5^2)n_1\partial_1 + (A_1A_4 - A_3^2)n_2\partial_2]n_3\partial_3 +$$
$$+(A_4n_3^2 - A_3n_2^2)A_5\partial_3^2,$$
$$T_{*12} = -A_5(A_2n_1\partial_2 + An_2\partial_1)\mathbf{n}^0 \cdot \nabla^0 + [(A_3^2 - A_2A_4)n_1\partial_2 + (A_5^2 - AA_4)n_2\partial_1]n_3\partial_3 + A_3A_5(n_3^2\partial_1\partial_2 + n_1n_2\partial_3^2),$$
$$T_{*13} = (A_2A_5n_3\partial_2 - AA_3n_2\partial_3)\underset{\sim}{\mathbf{C}} \overset{2}{\otimes} \mathbf{n}^0\nabla^0 - A(A_5n_3\partial_1 + A_3n_1\partial_3)\mathbf{n}^0 \cdot \nabla^0 - A_5^2n_3^2\partial_1\partial_3 - A_3A_5n_1n_3\partial_3^2,$$

$$\begin{aligned}
&T_{*21}=-A_5(A_2n_2\partial_1+An_1\partial_2)\mathbf{n}^0\cdot\nabla^0+[(A_3^2-A_2A_4)n_2\partial_1+(A_5^2-AA_4)n_1\partial_2]n_3\partial_3+A_3A_5(n_3^2\partial_1\partial_2+n_1n_2\partial_3^2),\\
&T_{*22}=A_5(An_2\partial_2+A_1n_1\partial_1)\mathbf{n}^0\cdot\nabla^0-A_3A_5n_3^2\partial_1^2+[(AA_4+A_5^2)n_2\partial_2+(A_1A_4-A_3^2)n_1\partial_1]n_3\partial_3+\\
&+(A_4n_3^2-A_3n_1^2)A_5\partial_3^2,\\
&T_{*23}=-(A_2A_5n_3\partial_1-AA_3n_1\partial_3)\underset{\sim}{\mathbf{C}}\overset{2}{\otimes}\mathbf{n}^0\nabla^0-A(A_5n_3\partial_2+A_3n_2\partial_3)\mathbf{n}^0\cdot\nabla^0-A_5^2n_3^2\partial_2\partial_3-A_3A_5n_2n_3\partial_3^2,\\
&T_{*31}=(AA_3n_3\partial_2-A_2A_5n_2\partial_3)\underset{\sim}{\mathbf{C}}\overset{2}{\otimes}\mathbf{n}^0\nabla^0-A(A_3n_3\partial_1+A_5n_1\partial_3)\mathbf{n}^0\cdot\nabla^0-A_3A_5n_3^2\partial_1\partial_3-A_5^2n_1n_3\partial_3^2,\\
&T_{*32}=-(AA_3n_3\partial_1-A_2A_5n_1\partial_3)\underset{\sim}{\mathbf{C}}\overset{2}{\otimes}\mathbf{n}^0\nabla^0-A(A_3n_3\partial_2+A_5n_2\partial_3)\mathbf{n}^0\cdot\nabla^0-A_3A_5n_3^2\partial_2\partial_3-A_5^2n_2n_3\partial_3^2,\\
&T_{*33}=-AA_2(1-n_3^2)\Delta+A(A_1+A_2)\mathbf{n}^0\mathbf{n}^0\overset{2}{\otimes}\nabla^0\nabla^0+(A_1A_5+AA_3)n_3\mathbf{n}^0\cdot\nabla^0\partial_3+A_5^2n_3^2\partial_3^2,\\
&\Delta=\partial_1^2+\partial_2^2,\ \mathbf{n}^0=n_I\mathbf{e}_I,\ \nabla^0=\mathbf{e}_I\partial_I,\ \mathbf{n}^0\mathbf{n}^0\overset{2}{\otimes}\nabla^0\nabla^0=n_In_J\partial_I\partial_J,\ A=1/2(A_1-A_2),\ \underset{\sim}{\mathbf{C}}=C_{IJ}\mathbf{e}_I\mathbf{e}_J.
\end{aligned}$$

The split boundary conditions have the form: $|\underset{\sim}{\mathbf{T}}|\mathbf{u}=\underset{\sim}{\mathbf{T}}_*^T\cdot\mathbf{P}$.

We note that the dynamic equations (the equilibrium equations) in displacements for any homogeneous anisotropic body we have the form

$$\underset{\sim}{\mathbf{M}}\cdot\mathbf{u}+\rho\mathbf{F}=0\ \ (\underset{\sim}{\mathbf{L}}\cdot\mathbf{u}+\rho\mathbf{F}=0),\qquad \underset{\sim}{\mathbf{M}}=\underset{\sim}{\mathbf{L}}-\underset{\sim}{\mathbf{E}}\rho\partial_t^2,\ \underset{\sim}{\mathbf{L}}=A_{piqj}\mathbf{e}_i\mathbf{e}_j\partial_p\partial_q. \tag{14}$$

It is easy to see, that

$$\begin{aligned}
&I_3(\underset{\sim}{\mathbf{M}})=|\underset{\sim}{\mathbf{M}}|=I_3(\underset{\sim}{\mathbf{L}})-I_2(\underset{\sim}{\mathbf{L}})\rho\partial_t^2+I_1(\underset{\sim}{\mathbf{L}})\rho^2\partial_t^4-\rho^3\partial_t^6,\\
&\underset{\sim}{\mathbf{M}}_*=\underset{\sim}{\mathbf{M}}^2-I_1(\underset{\sim}{\mathbf{M}})\underset{\sim}{\mathbf{M}}+I_2(\underset{\sim}{\mathbf{M}})\underset{\sim}{\mathbf{E}}=\underset{\sim}{\mathbf{L}}_*+[\underset{\sim}{\mathbf{L}}-I_1(\underset{\sim}{\mathbf{L}})\mathbf{E}]\rho\partial_t^2+\underset{\sim}{\mathbf{E}}\rho^2\partial_t^4=\\
&=\underset{\sim}{\mathbf{L}}^2-\underset{\sim}{\mathbf{L}}[I_1(\underset{\sim}{\mathbf{L}})-\rho\partial_t^2]+\underset{\sim}{\mathbf{E}}[I_2(\underset{\sim}{\mathbf{L}})-I_1(\underset{\sim}{\mathbf{L}})\rho\partial_t^2+\rho^2\partial_t^4],\ \underset{\sim}{\mathbf{L}}_*=\underset{\sim}{\mathbf{L}}^2-I_1(\underset{\sim}{\mathbf{L}})\underset{\sim}{\mathbf{L}}+I_2(\underset{\sim}{\mathbf{L}})\underset{\sim}{\mathbf{E}}.
\end{aligned}$$

So, the split equations have the form

$$|\underset{\sim}{\mathbf{M}}|\mathbf{u}+\underset{\sim}{\mathbf{M}}_*\cdot(\rho\mathbf{F})=0,\ \ \big(|\underset{\sim}{\mathbf{L}}|\mathbf{u}+\underset{\sim}{\mathbf{L}}_*\cdot(\rho\mathbf{F})=0\big).$$

Further, we represent $|\underset{\sim}{\mathbf{M}}|$ as a product of simple operators

$$\begin{aligned}
&|\underset{\sim}{\mathbf{M}}|=I_3(\underset{\sim}{\mathbf{L}})-I_2(\underset{\sim}{\mathbf{L}})\rho\partial_t^2+I_1(\underset{\sim}{\mathbf{L}})\rho^2\partial_t^4-\rho^3\partial_t^6=(a-\rho\partial_t^2)(b-\rho\partial_t^2)(c-\rho\partial_t^2)=\\
&=abc-(ab+ac+bc)\rho\partial_t^2+(a+b+c)\rho^4\partial_t^4-\rho^3\partial_t^6;\\
&I_1(\underset{\sim}{\mathbf{L}})=a+b+c,\ \ I_2(\underset{\sim}{\mathbf{L}})=ab+ac+bc,\ \ I_3(\underset{\sim}{\mathbf{L}})=abc.
\end{aligned}$$

It is clear that a, b and c are eigenoperators for $\underset{\sim}{\mathbf{L}}$, and $a-\rho\partial_t^2$, $b-\rho\partial_t^2$ and $c-\rho\partial_t^2$ are the eigenoperators (values) for $\underset{\sim}{\mathbf{M}}$.

Now consider the case of an isotropic medium. In this case, the tensor-operators and the differential operators are represented in the form

$$\begin{aligned}
&\underset{\sim}{\mathbf{M}}=\underset{\sim}{\mathbf{E}}Q_2+(\lambda+\mu)\nabla\nabla,\ \underset{\sim}{\mathbf{M}}_*=Q_2\underset{\sim}{\mathbf{N}},\ \underset{\sim}{\mathbf{N}}=\underset{\sim}{\mathbf{E}}Q_1-(\lambda+\mu)\nabla\nabla;\ ,|\underset{\sim}{\mathbf{M}}|=Q_1Q_2^2,\ Q_2=\mu\Delta-\rho\partial_t^2,\\
&Q_1=(\lambda+2\mu)\Delta-\rho\partial_t^2,\ \underset{\sim}{\mathbf{M}}\cdot\underset{\sim}{\mathbf{M}}_*=\underset{\sim}{\mathbf{M}}_*\cdot\underset{\sim}{\mathbf{M}}=\underset{\sim}{\mathbf{E}}|\underset{\sim}{\mathbf{M}}|,\ \underset{\sim}{\mathbf{M}}\cdot\underset{\sim}{\mathbf{N}}=\underset{\sim}{\mathbf{N}}\cdot\underset{\sim}{\mathbf{M}}=\underset{\sim}{\mathbf{E}}Q_1Q_2.
\end{aligned}$$

Applying the operator $\underset{\sim}{\mathbf{M}}_*$ and $\underset{\sim}{\mathbf{N}}$ to the equation (the first equality (14)) on the left, we obtain the following split equations:

$$Q_2[Q_1Q_2\mathbf{u}+\underset{\sim}{\mathbf{N}}\cdot(\rho\mathbf{F})]=0,\ \ Q_1Q_2\mathbf{u}+\underset{\sim}{\mathbf{N}}\cdot(\rho\mathbf{F})=0.$$

It can be seen that the second equation has two orders of magnitude less than the first. This, of course, gives him an advantage over the first.

We consider the eigenvalue problem for $\underset{\sim}{\mathbf{M}}$, which consists of the following: find all the vectors $\mathbf{a}$ that satisfy the equation $\underset{\sim}{\mathbf{M}}\cdot\mathbf{a}=\lambda\mathbf{a}$, where λ is a scalar operator.

The characteristic equation and its solutions have the form

$$\big(\det(\underset{\sim}{\mathbf{M}} - \lambda\underset{\sim}{\mathbf{E}}) = 0\big) \Leftrightarrow \big(\lambda^3 - I_1(\underset{\sim}{\mathbf{M}})\lambda^2 + I_2(\underset{\sim}{\mathbf{M}})\lambda - I_3(\underset{\sim}{\mathbf{M}}) = 0\big);$$
$$I_1(\underset{\sim}{\mathbf{M}}){=}Q_1{+}2Q_2,\ \ I_2(\underset{\sim}{\mathbf{M}}){=}Q_2(2Q_1{+}Q_2),\ \ I_3(\underset{\sim}{\mathbf{M}}){=}Q_1Q_2^2,\ \ \lambda_1 = Q_1,\ \ \lambda_2 = \lambda_3 = Q_2.$$

It is clear that the wave operators are the eigenoperators for the tensor-operator $\underset{\sim}{\mathbf{M}}$. In this case, the second wave operator is the double root of the characteristic equation. To determine the eigenvectors, we obtain equations

$$(\underset{\sim}{\mathbf{E}}\Delta - \nabla\nabla)\cdot\mathbf{a}_1 = 0,\ \ \nabla\nabla\cdot\mathbf{a}_2 = 0,$$

whose solutions are the following eigenvectors

$$\mathbf{a}_1 = \nabla\varphi,\ \ \mathbf{a}_2 = \nabla\times\boldsymbol{\psi} + 1/3A\mathbf{r}.$$

9. Quasistatic problem of the micropolar theory of elasticity in displacements and rotations

Let us consider an isotropic material with a center of symmetry. Then in case of quasistatics from (10) we have equations

$$\begin{aligned}&[(\lambda+2\mu)(\mu+\alpha)(\delta+\beta)\Delta^3 - 4\alpha\mu(\lambda+2\mu)\Delta^2]\mathbf{u} + \mathbf{S}^* = 0,\\ &\big\{(\gamma+2\delta)(\mu+\alpha)(\delta+\beta)\Delta^3 - 4\alpha[\mu(\gamma+2\delta)+(\mu+\alpha)(\delta+\beta)]\Delta^2 + 16\alpha^2\mu\Delta\big\}\boldsymbol{\varphi} + \mathbf{H}^* = 0,\end{aligned} \tag{15}$$

$$\begin{aligned}&\mathbf{S}^* = 2\alpha Q_1^*(\underset{\approx}{\mathbf{C}}\cdot\nabla)\cdot(\rho\mathbf{m}) + [\underset{\sim}{\mathbf{E}}Q_1^*Q_4^* - (dQ_4^* - 4\alpha^2)\nabla\nabla]\cdot(\rho\mathbf{F}),\\ &\mathbf{H}^* = 2\alpha Q_3^*(\underset{\approx}{\mathbf{C}}\cdot\nabla)\cdot(\rho\mathbf{F}) + [\underset{\sim}{\mathbf{E}}Q_2^*Q_3^* - (mQ_2^* - 4\alpha^2)\nabla\nabla]\cdot(\rho\mathbf{m}),\\ &Q_1^* = (b+d)\Delta,\quad Q_2^* = b\Delta,\quad Q_3^* = (g+m)\Delta - l,\quad Q_4^* = g\Delta - l,\\ &d = \lambda+\mu-\alpha,\quad l = 4\alpha,\quad b = \mu+\alpha,\quad m = \gamma+\delta-\beta,\quad g = \delta+\beta.\end{aligned} \tag{16}$$

For $\alpha = 0$, that is, in the case of a reduced medium from (15) and (16) we obtain

$$\begin{aligned}&\Delta^2\mathbf{u} + \mathbf{G} = 0,\quad \mathbf{G} = \frac{1}{\mu(\lambda+2\mu)}[\underset{\sim}{\mathbf{E}}(\lambda+2\mu)\Delta - (\lambda+\mu)\nabla\nabla]\cdot(\rho\mathbf{F});\\ &\Delta^2\boldsymbol{\varphi} + \mathbf{H} = 0,\quad \mathbf{H} = \frac{1}{(\delta+\beta)(\gamma+2\delta)}[\underset{\sim}{\mathbf{E}}(\gamma+2\delta)\Delta - (\gamma+\delta-\beta)\nabla\nabla]\cdot(\rho\mathbf{m}).\end{aligned} \tag{17}$$

Note that the first of the equations (17) is a classical equation, and the second equation has a similar form.

10. The quasistatic problem of the micropolar theory of prismatic bodies of constant thickness in displacements and rotations and in moments of displacement and rotation vectors

Let us consider a prismatic body of constant thickness $2h$, and let us take the middle plane as the base plane. Then in this case $g_M^{\hat{P}} = \delta_M^P$, $g_{\hat{P}}^3 = 0$, $g^{33} = h^{-2}$ and nabla-operator and Laplacian $\bar{\Delta}$ and $\hat{\Delta}^2$ and $\hat{\Delta}^3$ will be presented in the form

$$\begin{aligned}&\hat{\nabla}\mathbb{F} = (\mathbf{r}^P\partial_P + \mathbf{r}^3\partial_3)\mathbb{F} = (\mathbf{r}^P\partial_P + h^{-1}\mathbf{n}\partial_3)\mathbb{F},\quad -1 \le x^3 \le 1\\ &\hat{\Delta}\mathbb{F} = \hat{\nabla}^2\mathbb{F} = (g^{PQ}\nabla_P\nabla_Q + g^{33}\partial_3^2)\mathbb{F} = (\bar{\Delta} + h^{-2}\partial_3^2)\mathbb{F},\quad \bar{\Delta} = g^{PQ}\nabla_P\nabla_Q,\\ &\hat{\Delta}^2 = \bar{\Delta}^2 + 2h^{-2}\bar{\Delta}\partial_3^2 + h^{-4}\partial_3^4,\quad \hat{\Delta}^3 = \bar{\Delta}^3 + 3h^{-2}\bar{\Delta}^2\partial_3^2 + 3h^{-4}\bar{\Delta}\partial_3^4 + h^{-6}\partial_3^6.\end{aligned} \tag{18}$$

By the corresponding formulas (18), the equations (15) for the theory of prismatic bodies of costant thickness in displacements and rotations have the form

$$\begin{aligned}&[\bar{\Delta}^3 + A\bar{\Delta}^2 + h^{-2}(3\bar{\Delta} + 2A)\bar{\Delta}\partial_3^2 + h^{-4}(3\bar{\Delta} + A)\partial_3^4 + h^{-6}\partial_3^6]\hat{\mathbf{u}} + \hat{\mathbf{S}}^{**} = 0,\\ &[\bar{\Delta}^3 + (B\bar{\Delta}+A)\bar{\Delta} + h^{-2}[(3\bar{\Delta}+2B)\bar{\Delta}+C]\partial_3^2 + h^{-4}(3\bar{\Delta}+B)\partial_3^4 + h^{-6}\partial_3^6]\hat{\boldsymbol{\varphi}} + \hat{\mathbf{H}}^{**} = 0;\end{aligned} \tag{19}$$

$$\hat{\mathbf{S}}^{**} = \frac{\hat{\mathbf{S}}^{*}}{(\lambda + 2\mu)(\mu + \alpha)(\delta + \beta)}, \quad \hat{\mathbf{H}}^{**} = \frac{\hat{\mathbf{H}}^{*}}{(\gamma + 2\delta)(\mu + \alpha)(\delta + \beta)}, \quad A = -\frac{4\alpha\mu}{(\mu + \alpha)(\delta + \beta)},$$

$$B = -\frac{4\alpha[\mu(\gamma + 2\delta) + (\mu + \alpha)(\delta + \beta)]}{(\gamma + 2\delta)(\mu + \alpha)(\delta + \beta)}, \quad C = \frac{16\alpha^2\mu}{(\gamma + 2\delta)(\mu + \alpha)(\delta + \beta)}.$$

Applying the operator of moments of the k-th order of some system of orthogonal polynomials (Legendre, Chebyshev) to the equations (19), we obtain the following equations for the micropolar theory of prismatic bodies of constant thickness in the moments of the displacement and rotation vectors:

$$\begin{aligned}&[\bar{\Delta}^3 + (B\bar{\Delta}+A)\bar{\Delta}]\overset{(k)}{\boldsymbol{\varphi}} + h^{-2}[(3\bar{\Delta}+2B)\bar{\Delta}+C]\overset{(k)}{\boldsymbol{\varphi}}{}'' + h^{-4}(3\bar{\Delta}+B)\overset{(k)}{\boldsymbol{\varphi}}{}^{IV} + h^{-6}\overset{(k)}{\boldsymbol{\varphi}}{}^{VI} + \overset{(k)}{\mathbf{H}}{}^{**} = 0,\\ &[\bar{\Delta}^3 + A\bar{\Delta}^2]\overset{(k)}{\mathbf{u}} + h^{-2}(3\bar{\Delta}+2A)\bar{\Delta}\overset{(k)}{\mathbf{u}}{}'' + h^{-4}(3\bar{\Delta}+A)\overset{(k)}{\mathbf{u}}{}^{IV} + h^{-6}\overset{(k)}{\mathbf{u}}{}^{VI} + \overset{(k)}{\mathbf{S}}{}^{**} = 0, \ k \in \mathbb{N}_0.\end{aligned} \tag{20}$$

Having the equations (20), by the formula (13), in which the vector $\mathbf{v}$ is replaced by $\mathbf{u}$, it is easy to obtain systems of equations of any approximation at moments with respect to the system of Legendre polynomials. The equations of the 5th (in the classical case) and the 8th (in the micropolar case) approximations in the moments are obtained in the articles [8,9] for an isotropic material.

We note that using the canonical expressions for the tensor and the tensor-block matrix of any even rank [3, 9, 10], the initial-boundary value problems can be represented in canonical form and then split. In addition, I want to note that, for the theory of thin bodies, the split equations at equilibrium are equations of elliptic type of high order [9]. Using the method of I.N. Vekua [15], for them it is possible to write out analytical solutions.

The work was supported by Shota Rustaveli National Science Foundation, project no. DI-2016-41 and the RFBR, project no. 15-01-00848-a.

References

[1] Vekua I N 1978, *Fundamentals of tensor analysis and covariant theory.* Nauka, Moscow (in Russia)

[2] Nikabadze M U 2013, *Several Questions of Tensor Calculus with Applications to Mechanics.* Manuscript in VINITI Ross. Akad. Nauk, Issue 231–V2013, 242 p. (in Russian).

[3] Nikabadze M U 2014, "Construction of Eigentensor Columns in the Linear Micropolar Theory of Elasticity", *Moscow Univ. Mech. Bull.* **69**:1, 1–9. DOI: 10.3103/S0027133014010014

[4] Nikabadze M U 2014, *The Method of Orthogonal Polynomials in the Classical and Micropolar Mechanics of Elastic Thin Bodies.* I Manuscript in VINITI Ross. Akad. Nauk, Issue 135–V2014, 278 pp. (in Russian).

[5] Nikabadze M U 2014, *The Method of Orthogonal Polynomials in the Classical and Micropolar Mechanics of Elastic Thin Bodies.* I Manuscript in VINITI Ross. Akad. Nauk, Issue 136–V2014, 218 pp. (in Russian).

[6] Nikabadze M U 2014, *Development of the Method of Orthogonal Polynomials in the Classical and Micropolar Mechanics of Elastic Thin Bodies.* Moscow University Press, 515 pp. (in Russian). http://istina.msu.ru/media/publications/book/707/ea1/6738800/Monographiya.pdf

[7] Nikabadze M U 2014, *The Method of Orthogonal Polynomials in the Classical and Micropolar Mechanics of Elastic Thin Bodies.* Diss. Dr. Phys. Math. Sci. Moscow, 384 pp. (in Russian). http://istina.msu.ru/media/dissertations/dissertation/659/37c/6738997/dissertatsiya.pdf

[8] Nikabadze M U 2015, *On Several Issues of Tensor Calculus with Applications to Mechanics.* Contemporary Mathematics. Fundamental Directions. **55**, 3–194, (in Russian). http://mi.mathnet.ru/rus/cmfd/v55/p3, http://istina.msu.ru/media/publications/book/e25/00c/10117043/M.U.Nikabadze.pdf

[9] Nikabadze Mikhail U 2016 *Eigenvalue Problems of a Tensor and a Tensor-Block Matrix (TMB) of Any Even Rank with Some Applications in Mechanics*, 279–317, in: Generalized Continua as Models for Classical and Advanced Materials, Advanced Structured Materials, Vol. 42, Eds.: H. Altenbach and S. Forest, Springer. DOI 10.1007/978 – 3 – 319 – 31721 – 2_14

[10] Nikabadze M U 2017, *Topics on tensor calculus with applications to mechanics.* J. Math. Sci. **225**:1. 194 p.

[11] Eringen A C 1999, *Microcontinuum Field Theories. 1. Foundation and solids.* Springer-Verlag. 341 pp.

[12] Nowacki V 1975, *Theory of Elasticity.* oscow, Mir. 872 pp.

[13] Baskakov V A, Bestuzheva N P and Konchakova N A 2001, *Linear dynamic theory of thermoelastic medium with microstructure.* Voronezh. Voronezh University Press. 162 pp.

[14] Nikabadze Mikhail U and Ulukhanyan Armine R 2016, *Analytical Solutions in the Theory of Thin Bodies*, 319–361, in: Generalized Continua as Models for Classical and Advanced Materials, Advanced Structured Materials, Vol. 42, Eds.: H. Altenbach and S. Forest, Springer. DOI 10.1007/978 – 3 – 319 – 31721 – 2_15

[15] Vekua I N 1948, *New methods for solving elliptic equations.* OGIZ, Moscow, 296 pp. (in Russian).

Quasi-coherent structures and flows in turbulent transport

M Vlad[1], **F Spineanu**
National Institute of Laser, Plasma and Radiation Physics,
Atomistilor 409, Magurele 077125, Bucharest, Romania

E-mail: madalina.vlad@inflpr.ro

Abstract. The nonlinear effects that appear in test particle (tracer) transport in two-dimensional incompressible velocity fields are discussed. Test particle trajectories show both random and quasi-coherent aspects. The coherent motion is associated with trapping or eddying in the structure of the stochastic field. It generates quasi-coherent trajectory structures. The random motion leads to diffusive transport while the structures determine a micro-confinement process. The strength of each of these aspects depends on the parameters of the turbulence. The transport coefficients and the average size of the quasi-coherent structures are determined as function of the characteristics of the turbulence. The transport process is completely different in the presence of structures in the sense that the dependence on the parameters is different. The quasi-coherence of the motion can also be represented by the generation of flows. We show that tracers flows yield in a stochastic potential with space-dependent amplitude.

1. Introduction

Tracer transport in stochastic velocity fields or the stochastic advection process is of fundamental importance in plasma physics, astrophysics, fluid mechanics and atmospheric and oceanic sciences. The trajectories are obtained from equations that are nonlinear due to the space-dependence of the velocity, and they can have rather complex structures in important special cases. One example is the turbulence in incompressible media that is dominantly two dimensional, which has a self-organizing character [1], [2]. It consists of the generation of quasi-coherent structure (vortices). This property appears at the basic level of particle (tracer) trajectories. They are random sequences of trapping or eddying events and long jumps. The trapping process [3]-[5] strongly modifies the diffusion coefficients and leads to non-Gaussian distribution of displacements. The direct numerical simulations largely dominate the analytical advance in this case.

The trapping events appear when the trajectory arrives around the maxima or the minima of the stream function (potential). The process is quasi-coherent because it affects all the particles situated in these regions in the same way. These trajectories form intermittent quasi-coherent structures similar to fluid vortices and represent eddying regions. The trapping events alternate with large displacements that correspond to large size contour lines of the potential. These large jumps are random. The transport is essentially determined by them, while the trapping events have negligible contributions.

The transport coefficients and the characteristics of the quasi-coherent structures are determined as function of the statistical parameters of the turbulence. The transport process is strongly modified in the presence of trajectory structures. We show that quasi-coherent flows can also be generated due to trapping. Our instrument, the decorrelation trajectory method [6], [7], is essentially analytical.

[1] M. Vlad

2. Trajectory trapping or eddying in stochastic potential fields

The origin of trapping is the Hamiltonian structure of the equation of motion

$$\frac{d\mathbf{x}(t)}{dt} = \mathbf{v}(\mathbf{x},t), \qquad v_i(\mathbf{x},t) = \varepsilon_{ij}\frac{\partial\phi(\mathbf{x})}{\partial x_j}, \tag{1}$$

where $\mathbf{x}(t)$ is the trajectory in rectangular coordinates $\mathbf{x} = (x_1, x_2)$, $\varepsilon_{12} = 1, \varepsilon_{21} = -1, \varepsilon_{ii} = 0$, and $\phi(\mathbf{x},t)$ is the stream function or the potential that defines the two-dimensional velocity field $\mathbf{v}(\mathbf{x},t)$. The two components of the trajectory are couplet Hamiltonian variables.

The trajectories are periodic and they wind on the contour lines of the potential when the latter is time independent. This defines the state of permanent trapping. The Lagrangian potential is invariant in these conditions and the transport is subdiffusive because the particles are tied on the fixed contour lines of the potential.

If the motion is weakly perturbed (by time variation of the potential or by other components of the motion that can by introduced in Eq. (1)), the trapping is temporary and alternates with large jumps. The statistical importance of the trapping events depends on the strength of the perturbation that is represented by dimensionless factors, which are usually ratios of characteristic times. Two characteristic times are defined in (1): the correlation time of the potential τ_c that gives the scale of the time variation of $\varphi(\mathbf{x},t)$ and the time of flight of the particles $\tau_{fl}=\lambda_c/V$, where λ_c is the correlation length of $\varphi(\mathbf{x},t)$ and V is the amplitude of the velocity. These characteristic times define the Kubo number $K=\tau_c/\tau_{fl}$, which is a measure of trapping in time dependent potentials. Namely, the permanent trapping corresponds to $K\rightarrow\infty$, temporary trapping exists for $K>1$, and the statistical relevance of trapping is a decreasing function of K. Any other perturbation of the basic Hamiltonian motion defines a characteristic time τ_d for the decorrelation of the trajectory from the potential, and a dimensionless parameter similar to Kubo number, the decorrelation number $K_d=\tau_d/\tau_{fl}$.

The trapping events appear when the trajectory arrives around the maxima and the minima of the potential. The process is quasi-coherent because it affects all the particles situated in these regions in the same way. The high degree of coherence of the trapped trajectories is evidenced in a study of the statistical properties of the distance δx between neighbor trajectories [7]. The time evolution of $\langle\delta x^2(t)\rangle$ is very slow, which shows that neighbor particles have a coherent motion for a long time compared to τ_{fl}. They are characterized by a strong clump effect. The increase of $\langle\delta x^2(t)\rangle$ is slower than the Richardson law. These trajectories form intermittent quasi-coherent structures similar to fluid vortices and represent eddying regions. Their average size and life-time depend on the characteristic of the turbulence.

The particles that evolve on contour lines corresponding to small absolute values of the potential have much larger displacements. The large displacements are caused by the large size of these contour lines and also by the decorrelation, which allows transitions between neighbor potential cells. These large jumps are random. The transport is essentially determined by them, while the trapping events have negligible contributions.

Trapping was associated to a process of micro-confinement [8]. At a given moment, the micro-confinement affects a fraction of the particle, and, in a large time interval, every particle is subjected to micro-confinement events during a fraction of this time. The micro-confinement determines the existence of transport reservoirs [8] in the sense that diffusion can be strongly enhanced when these trapped particles are released by the increase of the perturbation strength.

Typical trajectories obtained from Eq. (1) for a potential with K=1 and 10 are shown in Figure 1. One can see that the trapping events and the degree of coherence of the motion have significant increase at K=10 compared to the example for K=1.

The aim of the paper is to analyze the statistical characteristics of these complex trajectories and to identify the importance of the random and quasi-coherent aspects for given parameters of the potential. We show that the coherence of the trajectories can manifest also as quasi-coherent flows.

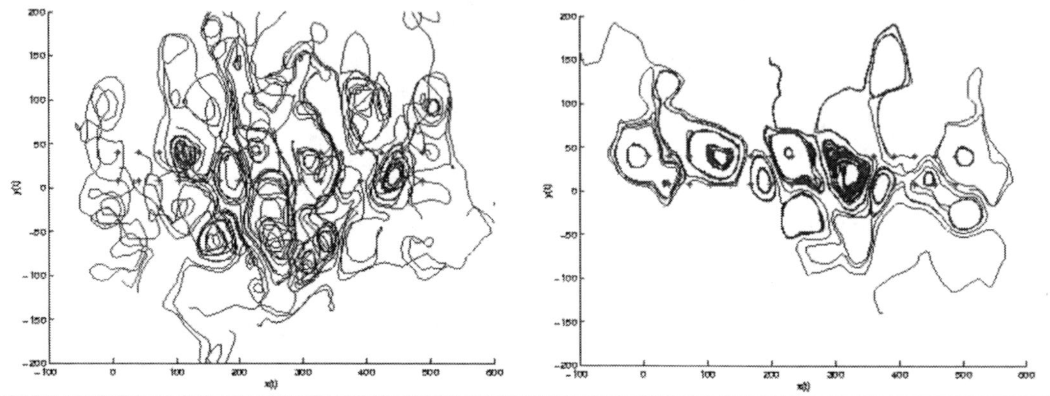

Figure 1. Typical trajectories obtained from numerical solutions of Eq. (1) for K=1 (left panel) and K=10 (right panel).

3. Tracer statistics in stochastic potential fields

The Eulerian and the Lagragian correlations are the main concepts in test particle transport.

The two-point Eulerian correlation (EC) of the potential is

$$E(\mathbf{x},t) = \langle \phi(\mathbf{x}_1,t_1)\,\phi(\mathbf{x}_2,t_2)\rangle$$

where $\mathbf{x} = \mathbf{x}_1 - \mathbf{x}_2$, $t = t_1 - t_2$ and $\langle\ \rangle$ is the statistical average over the realizations of the stochastic field or the space average over $\mathbf{x}_1$. The main statistical parameters of the velocity field are evidenced by this function. The correlation length λ_c and time τ_c correspond to the decay to zero of the function $E(\mathbf{x},t)$ and the amplitude of the stochastic potential is $\Phi^2 = E(\mathbf{0},0)$. The ECs for the components of the velocity are derivatives of $E(\mathbf{x},t)$. The EC describes the space structure and the time variation of the stochastic velocity field.

The Lagrangian velocity correlation (LVC) is a time dependent function that describes statistical properties of particle motion in the stochastic velocity field. In most cases the LVC decays to zero since the Lagrangian velocity becomes statistically independent on the initial velocity (it decorrelates). The characteristic time τ_L of the LVC decay is the measure of the memory of the stochastic Lagrangian velocity. The Lagrangian time τ_L is the decorrelation time of the potential τ_c in the case of the stochastic process described by Eq. (1).

Besides this important effect of memory, the LVC determines the main quantities related to test particle transport. As shown by Taylor [9], the mean square displacement (MSD) $\langle x_i^2(t)\rangle$ and its derivative, the running diffusion coefficient $D_i(t)$, are integrals of the LVC. The time dependent diffusion coefficients provide the "microscopic" characteristics of the transport process.

The diffusion at the transport space-time scales, which are much larger than λ_c and τ_c, is described by the asymptotic values D_i. The decorrelation of the Lagrangian velocity after the memory time τ_c leads to the possibility of decomposing the statistics of the displacements at large time $t \gg \tau_c$ in a sequence of independent small-scale processes of time τ_c. The probability of the small scale displacements (the micro-probability), $P^m(\mathbf{x},t)$, determines the elementary step Δ of the transport process as the MSD. The statistics of the displacements at large times $t \gg \tau_c$ is Gaussian, with the exception of the processes that have $\Delta \to \infty$ or $\tau_c \to \infty$ (see [10]).

The diffusion coefficients and the characteristics of the stochastic transport are determined by the micro-probability, which contains all statistical information on the trajectories at small time $t \lesssim \tau_c$. The use of the ergodic theory is not necessary.

IOP Conf. Series: Journal of Physics: Conf. Series **936** (2017) 012057 doi:10.1088/1742-6596/936/1/012057

Most of the theoretical methods are essentially based on Corrsin hypotheses [11], which assume that the micro-probability $P^m(\mathbf{x},t)$ is Gaussian and that the displacements are statistically independent on the velocity field. They lead to diffusive transport in the limit of static velocity fields, and thus they cannot apply to the special case of zero divergence two-dimensional velocity fields. It can be shown that a diffusive transport is obtained even when the second assumption is eliminated, which suggest that the micro-probability is not Gaussian in the presence of trapping. The first theoretical approach that finds subdiffusive transport in this special case [12], [13] does not use the Gaussian assumption, but it is based on the percolation theory. It only determines the scaling of the asymptotic diffusion coefficients. The detailed statistical information that is contained in the LVC was first obtained by the decorrelation trajectory method (DTM) [6]. This semi-analytical method was developed and validated by the nested subensemble approach (NSA) [7]. Both methods are in agreement with the statistical consequences of the invariance of the potential.

The main idea of this approach [6-8] is to determine the Lagrangian averages not on the whole set of trajectories but to group together trajectories that are similar, to average on them and then to perform averages of these averages. Similar trajectories are obtained by imposing supplementary initial conditions besides the necessary one. Particularly important initial conditions are provided by the conserved quantities (the potential in this case). The value of the initial potential determines the average path of the trajectories. The supplementary initial conditions define a set of subensembles of the realizations of the potential. Conditional averages lead to space-time dependent average potential and velocity in each subensemble and to simple trajectories that represent average particle motion: the decorrelation trajectories (DTs). The DTs are the main ingredient of DTM and NSA. They are smooth, simple trajectories determined from the EC of the stochastic fields. They are much different from particle trajectories as they saturate after the decorrelation time. The DTs represent the average evolution of the particles through the correlated zone of the potential. They describe the decorrelation process and provide characteristics of the transport.

The statistics of the trajectories is represented by weighted averages along the DTs. The weighting factors are the probabilities of the subensembles that correspond to the DTs. The LVC, $D_i(t)$, the probability of the small scale displacements $P^m(\mathbf{x},t)$ and other Lagrangian correlations are evaluated using the DTs. The detailed effective calculations are presented in [6-8]. Applications of the DTM for more complicated transport processes can be found in [14]-[19] for plasma turbulence and in [20]-[21] for space plasmas.

4. Quasi-coherent structures and transport

The parameters of the quasi-coherent structures are determined using the DTM as time dependent functions [14]. The fraction of trapped trajectories $n_{tr}(t)$ counts the number of trajectories that are closed at time t in a frozen potential. The average sizes of the structures $S_i(t)$ are determined by the maxima of the displacements on the closed DTs.

The results presented in the figures are obtained for spectra with two maxima as in [15], which correspond to positive and negative cells with similar weight that yield zero space-integral of the EC. The time t in Figure 2 and 3 is normalized with the time of flight.

Figure 2a shows that the sizes of the structures and the time dependent diffusion coefficients for a time-inependent potential. The structures appear at times of the order of the time of flight (t=~1) and they continuously increase. The fraction of trapped trajectories grows asymptotically towards one, and the sizes $S_x(t)$, $S_y(t)$ are increasing functions. Figure 2b demonstrates the time-correlation between the decrease of the fraction of free trajectories $n_{fr}(t)=1-n_{tr}(t)$ and the decay of the diffusion coefficients at t>1. The decrease of $D_i(t)$ at large times is produced by the quasi-coherent structures that trap a part of the trajectories. However, the diffusion coefficients are not proportional with $n_{fr}(t)$, which shows that the process is more complicated and depend on the size of the structures.

Figure 3 illustrates the effects of an average velocity $V_2\mathbf{e}_2$ superposed on a time independent stochastic potential. One can see in Figure 3a that the characteristics of the structures saturate. The fraction of trapped trajectories is limited, which shows that the average velocity destroys a part of the structures. The released trajectories contribute to the increase of the diffusion (Figure 3b). It is interesting to underline that the transport along the average velocity strongly increases and it is superdiffusive of ballistic type $\langle x_2^{\ 2} \rangle \approx t^2$, while the perpendicular transport remains subdiffusive. The shape of the quasi-coherent structures is modified. They are elongated in the in the direction of the average velocity.

The existence of an average velocity in the equation of motion determines modifications of the DTs in the nonlinear regime [8], [15]. A new type of DTs appears. They are opened trajectories along the direction of the average velocity. Their number increases as V_2 increases and, when V_2 is much large than the stochastic velocity, all DTs are of this type. Thus, the average velocity eliminates progressively (as Vd increases) the quasi-coherent structures.

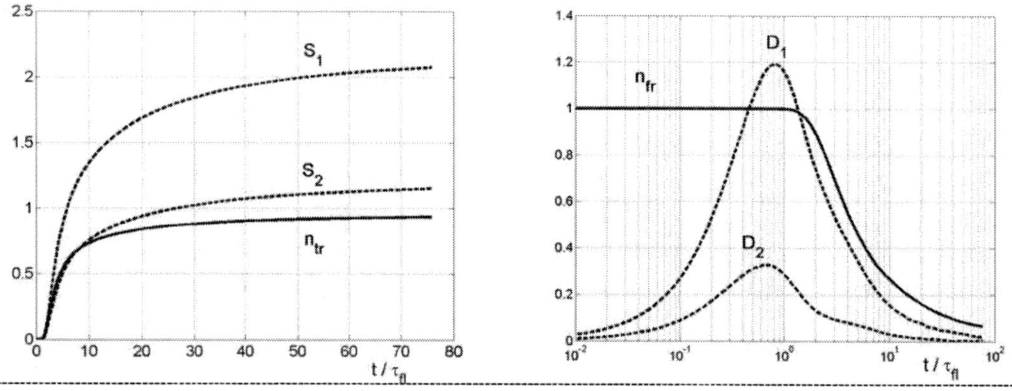

Figure 2. The time dependence of the parameters of the structure and of the diffusion coefficients obtained for an anisotropic potential. Left panel: the fraction of trapped trajectories $n_{tr}(t)$ and the sizes of the structures $S_x(t)$, $S_y(t)$. Right panel: $D_x(t)$ and $D_y(t)$ compared to the ratio of free trajectories $n_{fr}(t)$.

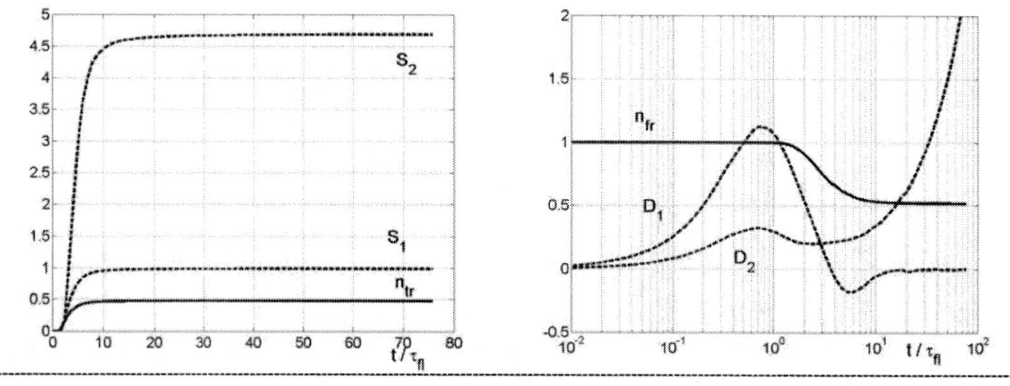

Figure 3. The effects of an average velocity $V_2 = 0.3\ \Phi / \lambda_y$ added to the stochastic potential of Figure 2. Left panel: the fraction of trapped trajectories $n_{tr}(t)$ and the sizes of the structures $S_x(t)$, $S_y(t)$. Right panel: $D_x(t)$, $D_y(t)$ and the ratio of free trajectories $n_{fr}(t)$.

5. Quasi-coherent flows

A stochastic velocity field generates random steps of the tracers that yield diffusive transport. They have zero average and are determined as the square root of the mean square displacement during the decorrelation time (obtained with the short-time probability $P^m(\mathbf{x},t)$). In special conditions, orientated steps of the tracers can appear even if the stochastic velocity field has zero average. They lead to quasi-coherent flows (also named pinches or direct transport). One mechanism for the generation of pinches in turbulent plasmas is related to the space variation of the confining magnetic field. It is a ratchet-type process that yields pinch velocities along the gradient of the magnetic field [16]. This flow is influenced by the presence of trajectory structures, which can reverse the direction of the flow from anti-parallel to parallel to the gradient.

A new mechanism of generation of quasi-coherent flows is presented here. The drive is the gradient of the amplitude of the stochastic potential. The potential is not a homogeneous stochastic field, but it can be described by

$$\phi(\mathbf{x},t) = \Phi(x_1)\bar{\phi}(\mathbf{x},t),$$

where $\bar{\varphi}$ is a homogeneous stochastic field with the amplitude one and $\Phi(x_1)$ is a space-dependent function $\Phi(x_1) = \Phi^0 \exp\left(-\frac{x_1}{L}\right) \cong \Phi^0\left(1-\frac{x_1}{L}\right)$. The characteristic length L of the gradient of $\Phi(x_1)$ is large compared to the correlation length of the potential. The components of the velocity in the first order in 1/L are

$$v_1(\mathbf{x},t) = -\Phi(x_1)\frac{\partial}{\partial x_2}\bar{\phi}(\mathbf{x},t), \quad v_2(\mathbf{x},t) = \Phi(x_1)\frac{\partial}{\partial x_1}\bar{\phi}(\mathbf{x},t) - \frac{\Phi^0}{L}\bar{\phi}(\mathbf{x},t).$$

A supplementary term appears in the second component of the velocity at finite L. It determines a quasi-coherent flow along the gradient.

We have determined the diffusion coefficient and the velocity of the quasi-coherent flow using the DTM. The results show that the diffusion is weakly influenced by the gradient of $\Phi(x_1)$. The main effect consists of the average displacement X that yields in this stochastic process, which determines the average velocity V_f. The flow is in the direction of the gradient of the amplitude $\Phi(x_1)$. As seen in Figure 4, X and V_f depend on the Kubo number of the stochastic potential. They are increasing functions of K for K<<1 and decreasing functions at large K. The maximum of the velocity of the flow appears around K=1, in the range of the transition from quasilinear to nonlinear transport. This shows that the quasi-coherent structures that appear at K>1 determine the decrease of the velocity. They actually produce an average velocity of opposite direction, which reduces the pinch.

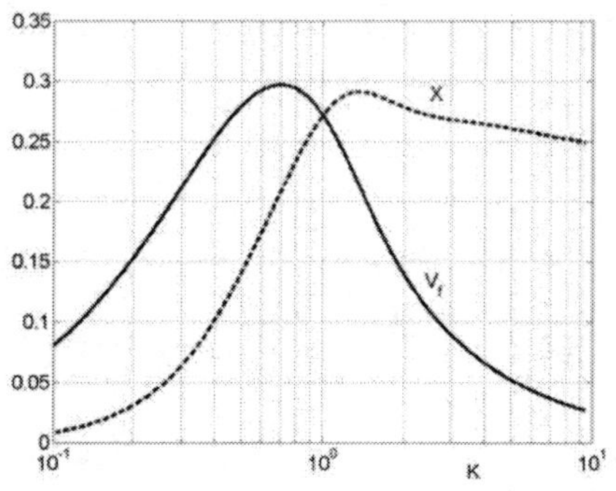

Figure 4. The average displacement X and the velocity V_f of the quasi-coherent flow as functions of the Kubo number K. The stochastic potential is as in Figure 2 and the gradient length L=5.

6. Conclusions

The nonlinear effects that appear in test particle transport in two-dimensional incompressible turbulence are analyzed using the decorrelation trajectory method. Tracer trajectories show both random and quasi-coherent aspects. They consist of random sequences of trapping or eddying and large jumps. The strength of each of these aspects depends on the parameters of the turbulence.

The trapping generates quasi-coherent trajectory structures that have a micro-confinement effect represented by the decrease of the time dependent diffusion coefficients. A clear correlation between the decrease of the fraction of free trajectories and of the time dependent diffusion coefficients is evidenced.

An average velocity superposed on the stochastic potential has strong effects on the quasi-coherent structures. A part of them is destroyed and the released trajectories strongly increase the transport in the direction of the average velocity, which becomes superdiffusive. The remaining structures are have modified shapes elongated in the direction of the velocity.

The quasi-coherence of the trajectories appears in special conditions as flows of the tracers. We discuss here the effects of a large-scale gradient of the amplitude Φ of the stochastic potential. The tracers acquire an average velocity along the gradient of Φ. This shows that in the space dependent stochastic potential, the tracers have the tendency to accumulate on the maxima of the amplitude.

7. References

[1] Provenzale A 1999 *Annu. Rev. Fluid Mech.* **31** 55
[2] Kraichnan R H and Montgomery D 1980 Rep. Progr. Phys. 43 547
[3] Balescu R 2005 Aspects of Anomalous Transport in Plasmas, IoP, Bristol and Philadelphia
[4] Kraichnan R H 1970 Phys. Fluids 19 22
[5] McComb W D 1990 The Physics of Fluid Turbulence, Clarendon, Oxford
[6] Vlad M, Spineanu F, Misguich J H and Balescu R 1998 *Phys. Rev. E* **58** 7359
[7] Vlad M and Spineanu F 2004 *Phys. Rev. E* **70** 056304
[8] Vlad M and Spineanu F 2017 New J. Phys. **19** 025014
[9] Taylor G I 1921 *Proc. London Math. Soc.* **20** 196
[10] Bouchaud J-P and Georges A 1990 *Phys. Reports* **195** 127
[11] Corrsin S 1959, in Atmospheric Diffusion and Air Pollution, ed. F. Frenkel & P. Sheppard (Advances in Geophysics, Vol. 6, Academic, New York), 161
[12] Gruzinov A V, Isichenko M B and Kalda Y L 1990 *Sov. Phys.-JETP* **70** 263
[13] Isichenko M B 1991 *Plasma Phys. Control. Fusion* **33** 809
[14] Vlad M and Spineanu F 2015 *Chaos Solitons Fractals* **81** 463
[15] Vlad M and Spineanu F 2013 *Phys. Plasmas* **20** 122304
[16] Vlad M, Spineanu F and Benkadda S 2006 *Phys. Rev. Lett.* **96** 085001
[17] Vlad M and Spineanu F 2016 *Nuclear Fusion* **56** 092003
[18] Croitoru A M, Palade D I, Vlad M and Spineanu F 2017 *Nuclear Fusion* **56** 036019
[19] Vlad M and Spineanu F 2015 *Phys. Plasmas* **22** 112305
[20] Vlad M and Spineanu F 2014 *Astrophys. J.* **791** 56
[21] Vlad M, Spineanu F and Croitoru A 2015 *Astrophys. J.* **815** 11

A cryptographic hash function based on chaotic network automata

Jeaneth Machicao[1] **and Odemir M Bruno**[1]

[1] São Carlos Institute of Physics, University of São Paulo. Av. Trabalhador São Carlense, 400, São Carlos, SP - Brazil

E-mail: bruno@usp.br

Abstract.
Chaos theory has been used to develop several cryptographic methods relying on the pseudo-random properties extracted from simple nonlinear systems such as cellular automata (CA). Cryptographic hash functions (CHF) are commonly used to check data integrity. CHF "compress" arbitrary long messages (input) into much smaller representations called hash values or message digest (output), designed to prevent the ability to reverse the hash values into the original message. This paper proposes a chaos-based CHF inspired on an encryption method based on chaotic CA rule B1357-S2468. Here, we propose an hybrid model that combines CA and networks, called network automata (CNA), whose chaotic spatio-temporal outputs are used to compute a hash value. Following the Merkle and Damgård model of construction, a portion of the message is entered as the initial condition of the network automata, so that the rest parts of messages are iteratively entered to perturb the system. The chaotic network automata shuffles the message using flexible control parameters, so that the generated hash value is highly sensitive to the message. As demonstrated in our experiments, the proposed model has excellent pseudo-randomness and sensitivity properties with acceptable performance when compared to conventional hash functions.

1. Introduction

Cellular automata has been widely applied in many domains of science including physics, biological systems modelings, image processing, and Cryptography. Some CAs present chaotic properties (such as ergodicity, high sensitivity to initial conditions, deterministic dynamics, structural complexity and long periodicity) that can meet desired confusion/diffusion properties and random-like behavior for cryptographic purposes. This close relation between chaos and cryptography have encouraged many researchers to develop chaos-based cryptographic primitives such as hash functions [1, 2, 3]. A CHF is a one-way function that maps an arbitrary finite length message to a fixed-length output, leading a fundamental role in digital signature, document compression, document authentication, etc [4].

In previous work [5], the chaotic properties of the life-like CA with rule B1357/S2468 have shown to be a good source of pseudo-randomness, being appropriate for the construction of cipher algorithms. These CAs were originally built into a 2D lattice, whose dynamics is governed by a rule and its local neighborhood that defines the states of its cells at each time step. However, in order to prevent weaknesses and cryptanalytic attacks of CHFs (by predicting the chaotic orbits), further enhancement can be introduce by considering a non-spatial neighborhood, i.e. to use a network structure as the tessellation of the CA, called as a network automata [6].

In this paper, we propose a cryptographic hash function scheme based on the chaotic properties of network automata. Thus, the hashing processing relies on the Merkle-Damgård [7] scheme: (i) by mixing a block of message together with the outputs from the evolution of the CNA and (ii) to feedback the previous outputs as entrance for the hashing processing with the next block of messages, in a iterative manner. Thus, through these iterations, the final hash value is obtained by extracting 512 bits from the ultimate state of the network automaton.

2. Chaotic network automata

The classical two-dimensional CA gave the inspiration for the network automata model (see Fig. 1). Instead of a lattice, a network becomes the tessellation of the automaton, where each vertex v_i represent a cell c_i that is connected to its k_i neighbors, and a local rule specifies how the NA evolves in time t. According to the nomenclature of the Life-Like rules family, for example rule B1357/S2468 (Eq. 1) implies that a cell c_i "born" when it is surrounding by $1 \vee 3 \vee 5 \vee 7$ **alive** neighbors. Contrarily, a cell "survives" when there exists $2 \vee 4 \vee 6 \vee 8$, otherwise dies.

$$s(c_i, t+1) = \begin{cases} 1, & \text{if } \sum_{i=1}^{k_i} s(c_i, t) = 1 \vee 3 \vee 5 \vee 7 \quad \Rightarrow \text{born (B)} \\ 1, & \text{if } \sum_{i=1}^{k_i} s(c_i, t) = 2 \vee 4 \vee 6 \vee 8 \quad \Rightarrow \text{survive (S)} \\ 0, & \text{otherwise} \end{cases}, \tag{1}$$

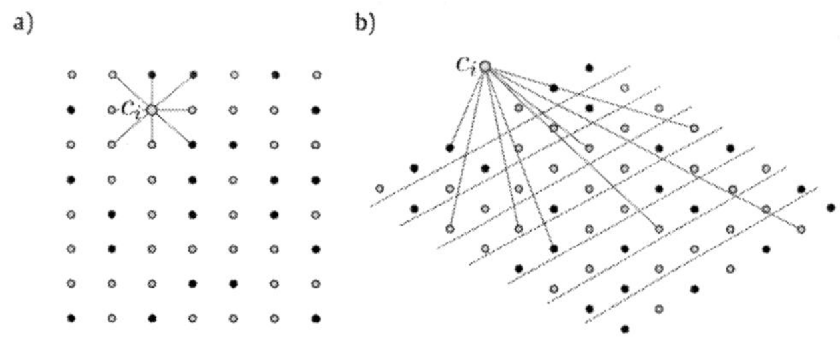

Figure 1. Differences between a) the two-dimensional CA, represented by a lattice and b) the network automata, whose tessellation is represented by a network. The black and white colors correspond to the binary states of their cells c_i, which are connected to their k_i a) local neighbors or b) non-spatial neighborhood, respectively.

3. Proposed cryptographic hash function

We propose a cryptographic hash function based on the chaotic spatio-temporal outputs from network automata. In general, the design of CHF consists of two components [8]: (i) a **compression function** $H : \{0,1\}^m \rightarrow \{0,1\}^n$, that maps a fixed-length message to a fixed-length output and (ii) the **domain extender** algorithm, which is a generic construction that transforms the former function to a hash function taking a message of arbitrary length $\mathcal{M} \in \{0,1\}^*$ and returning an output of fixed-length n [8]. Thus, we follow the MerkleDamgård construction [7], which is one of the simplest and most used domain extender in literature.

The propose CHF is illustrated in Fig. 2. First, a message $\mathcal{M}$ is broken into m-bit blocks (with padding if necessary), such that $\mathcal{M} = \{\mathcal{M}_1, \mathcal{M}_2, \ldots, \mathcal{M}_b\}$. Here we considered $m = 512$, as many conventional algorithms do. The initial conditions of the NA at $t = 0$ is setup by considering the binary elements from the first block $\mathcal{M}_1$. Then, iteratively, every input block $\mathcal{M}_i$ is xored with the outputs states H_1 from the evolving CNA and the resulting output is used to perturb the system for the next iteration, until a final hash value $H(\mathcal{M})$ is composed by the n-bits corresponding to the ultimate states of the network automaton, with $n = 512$.

Regarding the configuration of the NA, we considered a network with 512 nodes, each of them is connected to 8 non-local neighbors, which are randomly chosen. This connectivity parameter

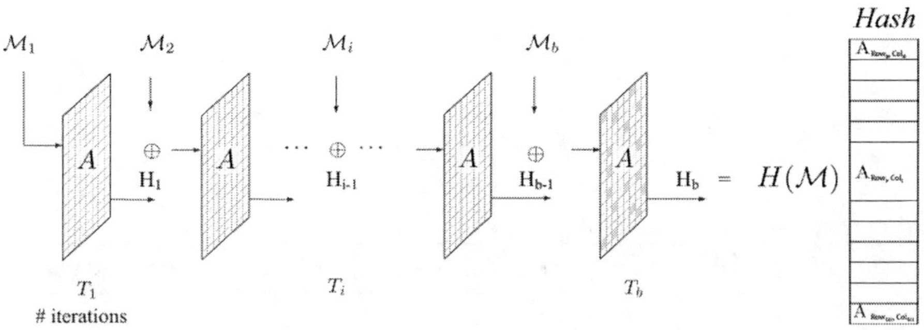

Figure 2. Proposed cryptographic hash function based on a chaotic network automaton.

brings a large number of possible combinations. For our experiments, we considered a random fixed neighborhood, however many other approaches could be considered, such as introducing a hash key to determine the configuration of this parameter. We use a NA with rule B1357/S2468, as this rule has been proven to be chaotic by means of the positive Lyapunov exponent explored in previous work [5], which is maintained since the proposed CNA also contains 8 neighbors.

4. Experimental results and discussion

To show that our proposed method ensures that even a small change on the message can causes a substantial difference on the hash value. We have compared the number of bits changed among the hash values of two close messages: $\mathcal{M}$ = "Schrodinger's cat" and $\mathcal{M}'$ = "schrodinger's cat"

$H(\mathcal{M})$ = 53e300f4a7c550b453d85b16cb2d118189932d5f7db343a46aa40cdf1aace3d7 084d52c1ce1148d7b94c60d5877a6c54285575306d828877c8ba5c5e3885c526 and

$H(\mathcal{M}')$ = 1801cb79c18a9800adbcb4ce5641f42cc830c003611424e6792a7a225975409b4 a907769086a25bbd281b1d8698803cdc6ab2d56319f06bf166b422c650c19ae2, which differs in 257 bits.

In general, when a minimal disturbance on the original message is introduced, an ideal CHF should produce a hash value with 50% of changing probability in each bit (avalanche effect). A well-known security test [1, 2, 3] consists to perform statistical analysis on a set of ensembles denoted by $\mathcal{E} = \{R^e \mid e = 1, \ldots, n\}$, where R represents a replica of the original message $\mathcal{M}$ except that the e-th bit is changed, i.e. $R^e = \overline{\mathcal{M}_e}$ (Boolean complement).

Similar to the former experiment, we systematically calculated the distance between two hash values $\mathcal{E}_e = d(H(\mathcal{M}), H(R^i))$. These tests were run for several lengths, using different number of blocks. The following measures were calculated: the mean number of changed bits $\bar{B}$, the minimum and maximum number of changed bits ($B_{\min}$ and $B_{\max}$), standard deviation ΔB and the mean changed probability: $P = \frac{\bar{B}*100\%}{n}$. The results for this security test are summarized in Table 1. Here, we can observe very similar results when comparing our proposal with conventional hash algorithms such as Whirlpool algorithm [9]. This result reinforces the potential of the proposed algorithm and its good level of security.

The performance of the algorithm was also measured in terms of the average computing time (in seconds) depending on the size of the input message. For this experiment, we applied the proposed CHF using 10 random messages with different numbers of message blocks $b = \{1\times10^5, \ldots, 10\times10^5\}$. Fig. 3 presents a time performance comparison between the proposed CHF and the Whirlpool algorithm [9].

5. Conclusions

A cryptographic hash function based on a network automata was proposed, which takes advantages from rule B1357/S2468 which has a chaotic behavior. The proposed CHF possesses

Table 1. Statistical analysis comparison of the proposed CHF and Whirlpool algorithm.

	# blocks	$\bar{B}$	ΔB	$B_{\min}$	$B_{\max}$	P
Proposed CHF	$b = 2$	255.6	3.8	198	323	49.9%
	$b = 4$	255.8	3.7	189	315	50.0%
	$b = 6$	255.9	3.6	197	313	50.0%
	$b = 8$	255.2	3.7	190	319	49.9%
	$b = 10$	255.2	3.7	199	318	49.8%
Whirlpool [9]	$b = 10$	255.9	2.9	216	298	50.0%

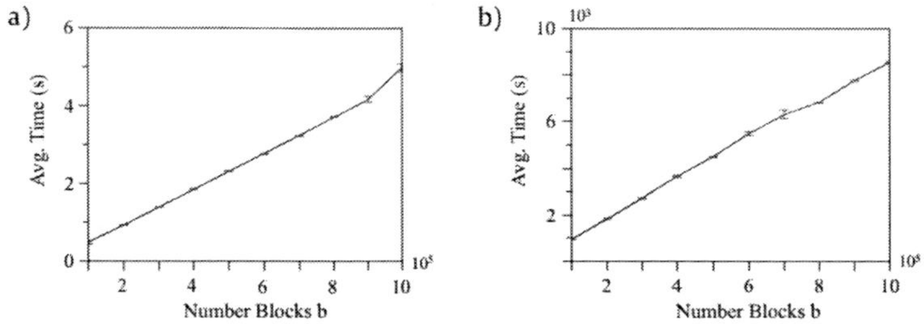

Figure 3. Time performance comparison between the proposed CHF and the Whirlpool algorithm [9] regarding to the number of blocks messages.

good statistical properties, high flexibility (network connectivity and CNA initial conditions), and good collision resistance, since it was shown that given any change in the original message, even minimal, it will eventually generate a significant change in the final hash result. Although, the algorithm has a lower performance when compared to a classical algorithm, it does not underestimate its equivalent security, flexibility and easy construction.

Acknowledgements

J.M. acknowledges the Brazilian agency CAPES (PROEX). O.M.B. acknowledges support from CNPq (Grant #307797/2014-7 and Grant #484312/2013-8) and FAPESP (grant #14/08026-1).

References

[1] Kanso A and Ghebleh M 2013 *Communications in Nonlinear Science and Numerical Simulation* **18** 109–123
[2] Yang H and et al 2009 *Chaos, Solitons & Fractals* **41** 2566–2574
[3] Xiao D, Liao X and Deng S 2005 *Chaos, Solitons & Fractals* **24** 65–71
[4] Bakhtiari S, Safavi-naini R and Pieprzyk J 1995 Cryptographic hash functions: A survey Technical report 95-09 Department of Computer Science, University of Wollongong New South Wales, Australia
[5] Machicao J, Marco A G and Bruno O M 2012 *Expert Systems with Applications* **39** 12626–12635
[6] Miranda G, Machicao J and Bruno O M 2016 *Scientific Reports* **6**
[7] Coron J S and et al 2005 Merkle-damgård revisited: How to construct a hash function *Advances in Cryptology: CRYPTO 05 Proceedings, Springer-Verlag* pp 430–448
[8] Mironov I 2005 Hash functions: Theory, attacks, and applications Tech. Rep. MSR-TR-2005-187 Microsoft Research Moffett Field, CA, USA
[9] Barreto P and Rijmen V 2000 *First open NESSIE Workshop* URL `http://www.larc.usp.br/ pbarreto/WhirlpoolPage.html`

Modeling the assessment of the economic factors impact on the development of social entrepreneurship

T Absalyamov[1,2,3], R Kundakchyan[2], L Zulfakarova[2] and Z Zapparova[2]
[2] Kazan Federal University, Kazan, 420008, 18 Kremlyovskaya str., Russia

[3]E-mail: abstimur@yandex.ru

Abstract
The article deals with the research of modern trends in the development of social entrepreneurship in Russia. The results of the research allow the authors to identify a system of factors that affect the development of entrepreneurship in the modern Russian economy. Moreover, the authors argue the regional specificity of the development of social entrepreneurship. The paper considers specific features and formulates the main limitations of the development of entrepreneurship and the competitive environment in the social sphere. The authors suggest an econometric model for assessing the influence of economic factors on the development of socially-oriented entrepreneurship and present an algorithm for calculating its components. The results of the econometric analysis identify the main factors of the change in the performance indicators of entrepreneurial activity and determine the degree of their impact on social entrepreneurship. The results and conclusions can serve as an estimation of the socioeconomic consequences of the sustainability disruption of the entrepreneurial potential realization in the social sphere.
Keywords: econometric model, economic factors, entrepreneurial potential, social entrepreneurship

1. Introduction

In recent times the problems and prospects of entrepreneurship development and the innovative activities in Russia have been receiving increasing attention. Today, a number of programs for the development and support of entrepreneurship have been developed and are being implemented in Russia. However, unfortunately, they do not fully take into account the specifics of the development of Russian business and the consequences of changing the external environment for Russian business [1].

Russian researchers are trying to find an explanation for the specifics of the current stage of development of entrepreneurship both in the Russian economy and in its regions. Of particular importance is the study of the influence of the national cultural and socio-economic characteristics of the region on the development of private entrepreneurship, the specifics of the formation of the economic culture of Russian entrepreneurship and its impact on entrepreneurial activity [2].

More and more studies are devoted to a comparative assessment and analysis of the differences between Russian and international entrepreneurship, based on different methods [3]. Regional differentiation of entrepreneurial activity with the use of key criteria indicators for assessing

[1] To whom any correspondence should be addressed.

IOP Conf. Series: Journal of Physics: Conf. Series **936** (2017) 012059 doi:10.1088/1742-6596/936/1/012059

entrepreneurial activity in the Volga regions was carried out in the works of M.A. Sarancha and E.A. Rubleva [4]. Particular attention was paid to the HSE study of innovative activity of the regions [5].

Almost all of the studies made it possible to classify Tatarstan as a region that occupies a leading position in the development of innovative and entrepreneurial activity among Russian regions. In this connection, the goal of the work was to identify and study the key components of the national model for the development of innovation activity and entrepreneurship in the Republic of Tatarstan, to identify the specifics of the region's participation in the political, economic and cultural life of the country in order to develop recommendations for other Russian regions and transfer experience to other countries.

2. Materials and methods

To conduct the present research we used a methodological approach, based on the allocation and use of the dialectical method, which determines the system of principles and methods of research; General scientific methods (systemic, structural-functional, synergetic, observation, deduction, etc.), which allow to adapt philosophical methods to fundamental theoretical and methodological positions of special sciences; Private-scientific methods (methods of social and human sciences) as a set of methods, research techniques and procedures that are applied taking into account the content of the interdisciplinary subject of research (global and expert interviews, the method of included observation, interviewing, etc.).

To write this article, we relied on the analysis of existing research in the field of entrepreneurship in Russia and abroad, international reports Global Entrepreneurship Monitor, Global Competitiveness Report, Global Competitiveness Index, Report on Innovative Activity of Regions (HSE) and current situation in the Russian economy. We have used the statistical data of recent years, as well as data from surveys of enterprise managers.

3. Results

As shown by the analysis of statistical data and the results of numerous surveys of entrepreneurial and innovative activity in the regions, the Republic of Tatarstan firmly stands in top-3 among the regions of the Russian Federation. Moreover, the study of innovation activity in the regions, conducted in 2016 by the Higher School of Economics (HSE), allowed Tatarstan to take the first place among the regions (fig. 1).

Figure 1. Russian Regional Innovation Index, source: HSE.

The study distributes the regions on the basis of cluster analysis in four groups based on the value of the integral indicator. In 2014, Tatarstan managed to overtake Moscow and strengthen its leading position, moving from second place to first. The integral indicator "Russian Regional Innovation Index" is calculated on the basis of four subindexes - "Social and Economic Conditions of Innovative Activity" (ISEC), "Scientific and Technical Potential" (ISTP), "Innovation Policy Quality" (IIPQ) and "Innovative Activity" (IIA).

The study of the state of innovation activity on the part of subindexes showed that the republic is one of the three leaders in three main subindexes (table 1).

Table 1. Tatarstan subindexes ranks, source: HSE.

	ISEC rank	ISTP rank	IIPQ rank	IIA rank
Tatarstan	3	17	2	1

High values of the first three indices show the balance of various aspects of the innovative development of the Republic of Tatarstan, which is not common for other regions.

The maximum value of the subindex "Quality of innovation policy" of the Republic of Tatarstan (0.7895) belongs traditionally to the full range of innovative policy instruments used to support entrepreneurial and innovative activity in the region. The republic has established the State Educational Institution "Center for the Implementation of Programs for Support and Development of Small and Medium-Sized Enterprises of the Republic of Tatarstan", the NGO "Entrepreneurship Support Fund of the Republic of Tatarstan", the "Guarantee Fund of the Republic of Tatarstan", the Center for Support of Entrepreneurship, the Coordination Center for Support of Export-oriented enterprises, the Center for Innovations in the Social Sphere, the Business Incubation Center, as well as industrial parks, industrial sites, regional engineering centers, NGO "Kama center of cluster development of small and medium-sized businesses", a special economic zone of the industrial-production type "Alabuga", Special economic zone Innopolis, etc.

Due to the creation of an effective infrastructure for supporting entrepreneurship and implementing a well-thought-out innovation policy, the Republic of Tatarstan became the leader of the National rating of the investment climate in the regions of the Russian Federation for the third consecutive time. Results of the rating were announced in the framework of the St. Petersburg International Economic Forum [6].
The effective innovative policy and the creation of institutions for supporting entrepreneurship allowed the republic to take the second position and on the other subindex - "Innovation activity".

The conducted calculations show a tangible gap between the regions of the Russian Federation in terms of the level of development of innovation activity. If the Republic of Tatarstan has a given indicator of 0.5747, Kalmykia - the most passive subject of the Russian Federation in this field - 0.0417 (minimum non-zero value). As can be seen, the gap is 14 times.

The republic also has a high rank of the subindex "Socio-economic conditions of innovation activity" (ISEU), which is 18% higher than the national average. Tatarstan has very high (exceeding the average Russian) values for six of the eight rating indicators. And only the level of employment in the knowledge-intensive sectors of the service sector and the availability of Internet in households the republic does not achieve average estimates for the country.

The conducted researches showed that a number of factors influenced the innovative and entrepreneurial activity of the republic, which formed a phenomenon called the Tatarstan model of entrepreneurship and innovation activity development. These factors are:

1. Active creation of institutions to support the development of entrepreneurship and innovative activity, including various Centers, Entrepreneurship and Innovation Support Funds, Special economic zones and Advanced Development Territories.

2. Resource potential of the region, the republic's strength in energy resources and agricultural lands, which contributed to the active development of entrepreneurship in the areas of mining and agriculture [7]. These areas, and primarily oil production and processing, have traditionally contributed to the country's achievement of high indicators of socio-economic development.

3. The industrial potential of the region, the presence of heavy industry, including the defense industry enterprises, which today contributes to the active introduction of innovative technologies, the development of entrepreneurship in the high-tech sector.

4. Scientific potential and staff. The region is distinguished, first of all, by the high level of provision of science and technology with financial and human resources, as well as the achievement of the most significant scientific and technical results relative to other regions [8]. Availability of educational institutions of the Federal level in the territory of the republic - Kazan Federal University, KNRTU (Kazan national research technical university), KAI (Kazan aviation Institute), KSUAE (Kazan state university of architecture and engineering) etc.

5. Cultural factors, the presence of objects included in the list of cultural heritage of UNESCO. Implementation of mega cultural and sports projects, which contributed to the active development of entrepreneurship in the tourism sector, in the field of tourism infrastructure [9]. National and religious features of the region, the presence of two main denominations - Islam and Orthodoxy, as well as the national mentality.

The factors listed above made it possible to present the Tatarstan model of entrepreneurship development in the following form:

$$\text{Entrepreneurship in Tatarstan} = f(\text{ISEI}, \text{RPR}, \text{IDP}, \text{SPHR}, \text{NCF})$$

Where

ISEI - Institutes for support of entrepreneurship development and innovation activities;
RPR - Resource potential of the region;
IDP - Industrial development potential;
SPHR - Scientific potential and human resources;
NCF - National and cultural factors.

4. Conclusion

IOP Conf. Series: Journal of Physics: Conf. Series **936** (2017) 012059 doi:10.1088/1742-6596/936/1/012059

To sum up, the study of the specifics of the development of entrepreneurship and innovation activities in the regions of the Russian Federation made it possible to attribute the Republic of Tatarstan to the leaders among its subjects. In the presented work, the factors that form the basis of the Tatarstan model of entrepreneurship development were identified and make it the most effective policy in the field of entrepreneurship development and support of innovative activity.

The presented model can serve as the objects of further research with a view to transferring best practices to other regions.

References

[1] Global Entrepreneurship Monitor – GEM Report 2016/2017 http://www.gemconsortium.org/report [accessed on 4.04.2017]

[2] Troshikhin V.V. *Ekonomicheskaja kul'tura rossiyskogo predprinimatel'stva* Vestnik Belgorodskogo universiteta kooperacii, jekonomiki i prava. 2017 №2.:64-77.

[3] Safiullin, M. R., Elshin, L. A., & Prygunova, M. I. *Ustoychivost' i perspektivy razvitija delovoj aktivnosti malogo i srednego predprinimatel'stva v regional'noy ekonomike rossiyskoy federacii.* Kazanskiy ekonomicheskiy vestnik, 2015. (5), 46-53..

[4] Sarancha, M. A., & Rubleva, E. A. *Ocenka masshtabov i urovnja razvitija malogo predprinimatel'stva v Privolzhskom federal'nom okruge.* Vestnik Udmurtskogo universiteta. 2014

[5] Rejting innovacionnogo razvitija sub#ektov Rossijskoj Federacii. Vypusk 4 / pod red. L. M. Gohberga; Nac. issled. un-t «Vysshaja shkola ekonomiki». – M.: NIU VShJe, 2016. – 248 s. – 450 ekz. – ISBN 978-5-7598-1508-2(v obl.).

[6] National rating of investment climate conditions in Russian Fedearation subjects 2017 https://asi.ru/investclimate/rating/ (accesed on May 25th 2017)

[7] Zulfakarova L. F., Kundakchyan R. M., Astafieva L. K. ACTUAL PROBLEMS OF INCREASED EFFICIENCY OF USE OF THE REGION'S RESOURCE POTENTIAL //Journal of Economics and Economic Education Research. – 2016. – T. 17. – C. 74.

[8] Sakhapov, R. L., and S. G. Absalyamova. "The use of telecommunication technologies in education network." *Remote Engineering and Virtual Instrumentation (REV), 2015 12th International Conference on*. IEEE, 2015.

[9] Absalyamov, T. 2015. The influence of cultural and sport mega-events on sustainable development of the city. Procedia-Social and Behavioral Sciences,188, 197-201.

IOP Conf. Series: Journal of Physics: Conf. Series **936** (2017) 012060 doi:10.1088/1742-6596/936/1/012060

Enhancement of the space of solutions of the kinetic-conformal Hořava theory

Alvaro Restuccia[1], **Jorge Bellorín**[2], **Adrián Sotomayor**[3]

[1] Departamento de Física, Universidad de Antofagasta, Antofagasta, Chile
[1] Departamento de Física, Universidad Simón Bolívar, Caracas, Venezuela
[3] Departamento de Matemáticas, Universidad de Antofagasta, Antofagasta, Chile

E-mail: `arestu@usb.ve`, `jorgebellorin@usb.ve`, `adrian.sotomayor@uantof.cl`

Abstract. We show that in the Horava theory at the kinetic-conformal point there are more solutions of the field equations in the Hamiltonian formalism than in the original Lagrangian formalism. To hold this we add all the constraints, primary and secondary, to the Hamiltonian. There are certain configurations for the Lagrange multipliers associated to the secondary constraints that lead to solutions that cannot be found in the original Lagrangian formulation. We show specific examples in vacuum and with a source. The solution with the source has homogeneous and isotropic spatial hypersurfaces. Here we relax one condition on one Lagrange multiplier, showing the existence of more special solutions on the Hamiltonian side.

1. Introduction

The kinetic-conformal formulation [1] of the (nonprojectable) Hořava theory [2, 3] is defined in $3+1$ dimensions by setting the value $\lambda = 1/3$, where λ is the coupling constant arising in the kinetic term of the Lagrangian. The particularity of this value is that, when it is set, two additional second-class constraints arise, implying that the extra scalar mode is eliminated. Therefore, the theory shares the same number of degrees of freedom with General Relativity, but with different dynamics and symmetries. The name kinetic-conformal is inspired by the fact that at the $\lambda = 1/3$ value the kinetic term of the Lagrangian gets an anisotropic conformal symmetry [2], although the whole theory is not conformally invariant.

Previous studies on the Hamiltonian formulation of the theory can be found in Refs. [1, 4]. Within the features of the Hamiltonian formulation, in this paper we point out that in the kinetic-conformal theory there are more classical solutions in the Hamiltonian formulation than in the original Lagrangian formulation. In the Hamiltonian formulation we add all the constraints, primary and secondary, to the Hamiltonian with Lagrange multipliers. An special configuration for a Lagrange multiplier allows the enhancement of the space of solutions. We give concrete examples by studying vacuum (i. e., purely gravitational) configurations and configurations with matter sources. In the first case we find examples with static solutions. In the second case we incorporate a homogeneous and isotropic perfect fluid. We present a solution that has homogeneous and isotropic spatial slices with a scale factor for the spatial metric that is time dependent. These solutions are given by fixing another Lagrange multiplier different to the one that leads to the special Hamiltonian solutions. Here we relax this, and also discuss a more general setting for the Lagrange multipliers within the special Hamiltonian solutions. In this case we show the existence of more special Hamiltonian solutions.

IOP Conf. Series: Journal of Physics: Conf. Series **936** (2017) 012060 doi:10.1088/1742-6596/936/1/012060

The enhancement of the space of solutions on the Hamiltonian side may have incidence on the problem of cosmological backgrounds in the kinetic-conformal theory. It is known that a fully homogeneous and isotropic spacetime coupled to an homogeneous and isotropic perfect fluid admits as its only solution an empty fluid, i. e., zero density and pressure (for the flat case) [3, 5, 6] (see also [7, 8] for an analogous result in the Einstein-aether theory). However, this result is based on the Lagrangian field equations. The result we present here leaves open the possibility that more homogeneous and isotropic configurations cand be found in the Hamiltonian side.

2. Hamiltonian of the kinetic-conformal Hořava theory

The Hořava theory is written in terms of the ADM variables N, N_i and g_{ij}. We consider the nonprojectable version where the lapse function N is a function of the time and the space. We denote by π^{ij} the canonically conjugate momentum of g_{ij} and by P_N the one of N, which is zero. We consider only the terms that are of second order in spatial derivatives (the theory is of second order in time derivatives).

The complete Hamiltonian of the kinetic-conformal theory, with all the constraints added, is

$$H = \int d^3x \left[\sqrt{g} N \left(\frac{1}{g} \pi^{ij}\pi_{ij} - \beta R - \alpha a_i a^i \right) + N_i \mathcal{H}^i + \sigma P_N + \mu \pi + A_1 \mathcal{C}_1 + A_2 \mathcal{C}_2 \right] \quad (1)$$

where R is the spatial scalar curvature and $a_i = \partial_i \ln N$. β and α are coupling constants. σ, μ, A_1 and A_2 are the Lagrange multipliers of the constraints

$$P_N = 0 \,, \quad \pi = 0 \,, \quad (2)$$

$$\frac{1}{\sqrt{g}} \mathcal{C}_1 \equiv \frac{2}{g} \pi^{ij}\pi_{ij} - \gamma_1 \frac{\nabla^2 N}{N} = 0 \,, \quad (3)$$

$$\frac{1}{\sqrt{g}} \mathcal{C}_2 \equiv \gamma_2 \frac{\nabla^2 N}{N} - \alpha a_i a^i - \beta R = 0 \,, \quad (4)$$

where

$$\gamma_1 \equiv 2\beta - \alpha \,, \quad \gamma_2 \equiv \beta + \frac{3\alpha}{2} \,. \quad (5)$$

$\mathcal{H}^i$ is the momentum constraint,

$$\mathcal{H}^j \equiv -2\nabla_i \pi^{ij} + P_N \partial^j N = 0 \,. \quad (6)$$

By applying Dirac's procedure for the time preservation of the constraints, we obtain that the preservation of the four second-class constraints (2 - 4) leads to the following four equations for the Lagrange multipliers,

$$0 = \nabla^2 B + 2\alpha a^k \nabla_k \left(\frac{A_2}{N} \right) + \frac{\gamma_1}{2} \frac{\nabla^2 N}{N^2} (2A_1 - A_2) \,, \quad (7)$$

$$0 = 4\beta \frac{\nabla^2 A_2}{N} - a_k \nabla^k B - \frac{\nabla^2 N}{N^2} \left(3\gamma_1 A_1 + \gamma_2 A_2 \right) \,, \quad (8)$$

$$\begin{aligned} 0 = {} & \gamma_1 \frac{\sqrt{g}}{N} \left(\nabla^2 \sigma - \frac{\nabla^2 N}{N} \sigma \right) + \left(\frac{3}{\sqrt{g}} \pi^{ij}\pi_{ij} + \frac{\gamma_1}{2} \sqrt{g} a_k a^k \right) \mu \\ & - \left\{ \mathcal{C}_1, \left\langle (N + 2A_1) \frac{\pi^{ij}\pi_{ij}}{\sqrt{g}} - \sqrt{g}(N + A_2) \left(\beta R + \alpha a_k a^k \right) + \sqrt{g} B \nabla^2 N \right\rangle \right\} \,, \end{aligned} \quad (9)$$

$$\begin{aligned} 0 = {} & \gamma_2 \frac{\sqrt{g}}{N} \left(\nabla^2 \sigma - \frac{\nabla^2 N}{N} \sigma \right) + 2\alpha \frac{\sqrt{g}}{N} a^k \left(\partial_k \sigma + a_k \sigma \right) + 2\beta \sqrt{g} \nabla^2 \mu \\ & - \frac{\sqrt{g}}{2} \left(\beta R + (\alpha - \gamma_2) a_k a^k \right) \mu + \left\{ \mathcal{C}_2, \left\langle (N + 2A_1) \frac{\pi^{ij}\pi_{ij}}{\sqrt{g}} \right\rangle \right\} \,, \end{aligned} \quad (10)$$

where

$$B \equiv -\frac{1}{N}\left(\gamma_1 A_1 - \gamma_2 A_2\right) . \tag{11}$$

We now present the evolution equations. Taking variations with respect to N reproduces exactly the Eq. (7). Variations with respect to π^{ij}, g_{ij} and P_N yield, respectively, the equations

$$\dot{g}_{ij} = \frac{2}{\sqrt{g}}(N+2A_1)\pi_{ij} + \mu g_{ij} , \tag{12}$$

$$\begin{aligned} \frac{\dot{\pi}^{ij}}{\sqrt{g}} &= -\frac{1}{g}(N+2A_1)(2\pi^{ik}\pi_k{}^j - \frac{1}{2}g^{ij}\pi^{kl}\pi_{kl}) - \mu\frac{\pi^{ij}}{\sqrt{g}} - \nabla^{(i}N\nabla^{j)}B + \frac{1}{2}g^{ij}\nabla_k N\nabla^k B \\ &\quad + \left[\beta(\nabla^{ij} - g^{ij}\nabla^2) - \beta(R^{ij} - \frac{1}{2}g^{ij}R) - \alpha(a^i a^j - g^{ij}a_k a^k)\right](N+A_2) , \end{aligned} \tag{13}$$

$$\dot{N} = \sigma . \tag{14}$$

In these equations we have fixed the gauge condition $N_i = 0$.

3. Existence of special Hamiltonian solutions

The key for the special solutions is the vanishing of the coefficient of π^{ij} in the evolution Eq. (12). This leads to the lacking of the invertibility of the Legendre transformation. Therefore, these special configurations are characterized by the relation $A_1 = -\frac{1}{2}N$. Let us analyze the existence of them. We first require that also A_2 is proportional to N, $A_2 = k_2 N$ (further details on this assumption can be found in [9]). With these settings for $A_{1,2}$ the variable B becomes constant, and we obtain that the two Eqs. (7) and (8) are completely solved if $k_2 = -1$, i. e., $A_{1,2}$ are given in terms of N by

$$N + 2A_1 = N + A_2 = 0 . \tag{15}$$

The Eqs. (9 - 10) result

$$0 = \frac{\gamma_1}{N}\left(\nabla^2 - \frac{\nabla^2 N}{N}\right)\sigma + \left(\frac{3}{g}\pi^{ij}\pi_{ij} + \frac{\gamma_1}{2}a_k a^k\right)\mu , \tag{16}$$

$$0 = \frac{\gamma_2}{N}\left(\nabla^2 - \frac{\nabla^2 N}{N}\right)\sigma + \frac{2\alpha}{N}a^k\left(\partial_k + a_k\right)\sigma + 2\beta\nabla^2\mu - \frac{1}{2}\left(\beta R + (\alpha - \gamma_2)a_k a^k\right)\mu . \tag{17}$$

These two equations form a system of homogeneous partial differential equations for σ and μ. A particular solution, without requiring any further condition, is $\sigma = \mu = 0$. We shall use this solution on the next section. On more general grounds, the system (16 - 17) becomes elliptic if the matrix of coefficients of the terms of second order in derivatives is positive definite. This sets two conditions on the space of coupling constants, namely $\beta > 0$, $\gamma_1 = 2\beta - \alpha > 0$. Therefore, if we assume these condition holds, there exist solutions of the Eqs. (16 - 17), each one corresponding to a given boundary condition. The equations for the evolution of the canonical variables evaluated on these configurations take the simple form

$$\dot{g}_{ij} = \mu g_{ij} , \quad \dot{\pi}^{ij} = -\mu\pi^{ij} , \quad \dot{N} = \sigma . \tag{18}$$

Therefore, we have that a class of interesting special solutions exists. They represent a flow of the three-dimensional metric on a given topology. These solutions are characterized by $2A_1 = A_2 = -N$. The remaining Lagrange multipliers σ and μ are determined by the equations (16 - 17), and the initial data on the canonical variables must solve the constraints (2 - 4) and (6). The evolution equations are (18). Notice that the evolution flows of g_{ij} and π^{ij} are coupled each other due to the factor μ. This class of solutions cannot be obtained from the original Lagrangian field equations since there is no direct relation between $\dot{g}_{ij}$ and π^{ij}.

We may study the existence of more general solutions on the Hamiltonian side by relaxing the condition on the Lagrange multiplier A_2. In this case we focus on the constraints and the equations that determine the stationary points of the canonical action; these are the Eqs. (2 - 4), (6), (7) and (12 - 14) [Equations (8 - 10) are needed only for the time preservation of the second-class constraints]. By substituting the condition $A_1 = -\frac{1}{2}N$ that leads to the noninvertibility of the Legendre transformation, equation (7) takes the form

$$\gamma_2 \nabla^2 \left(\frac{A_2}{N}\right) + 2\alpha \frac{\nabla^k N}{N} \nabla_k \left(\frac{A_2}{N}\right) - \frac{\gamma_1}{2}\frac{\nabla^2 N}{N}\left(\frac{A_2}{N}\right) = \frac{\gamma_1}{2}\frac{\nabla^2 N}{N} \tag{19}$$

If the condition $\gamma_2 > 0$ is satisfied, this is an elliptic equation for the variable A_2/N. The numerical bounds on the coupling constants can be extracted from the discussion in Ref. [10]. Observational constraints relative to the solar-system tests demands that α is strongly close to the expression $\alpha = 2(\beta - 1)$. Hence we expect that the value of γ_2 (5) is close to $\gamma_2 = 4\beta - 3$. Cherenkov radiation imposes the bound $\beta \geq 1$. With this information we can take γ_2 to be positive safely on physical grounds. Therefore, Eq. (19) is an inhomogeneous elliptic partial differential equation for A_2. It can be solved by giving boundary conditions. As can be easily seen from (19), one of these solutions is the case $A_2 = -N$ we discussed above, but there are more solutions, and each of them leads to a special Hamiltonian configuration since they cannot be found in the Lagrangian formalism. These configurations must satisfy the constraints (2 - 4) and (6), as well as the evolution equations (12 - 14), which take the form

$$\dot{g}_{ij} = \mu g_{ij}\,, \tag{20}$$

$$\begin{aligned}\frac{\dot{\pi}^{ij}}{\sqrt{g}} =& -\mu\frac{\pi^{ij}}{\sqrt{g}} - \gamma_2 \nabla^{(i} N \nabla^{j)}\left(\frac{A_2}{N}\right) + \frac{\gamma_2}{2} g^{ij}\nabla_k N \nabla^k \left(\frac{A_2}{N}\right) \\ &+ \left[\beta(\nabla^{ij} - g^{ij}\nabla^2) - \beta(R^{ij} - \frac{1}{2}g^{ij}R) - \alpha(a^i a^j - g^{ij}a_k a^k)\right](N + A_2)\,,\end{aligned} \tag{21}$$

$$\dot{N} = \sigma\,, \tag{22}$$

where $B = \gamma_2 \dfrac{A_2}{N} + \dfrac{\gamma_1}{2}$. The conditions on the couplings constants α and β that are necessary to have elliptic equations for μ and σ, which we discussed above, are also satisfied once the observational bounds are imposed.

4. Example: static solutions

We start with the equations for the Lagrange multipliers σ and μ, Eqs. (16 - 17). Since these equations are homogeneous, an obvious solution is $\sigma = \mu = 0$. With this the right-hand sides of Eqs. (18) vanish, which implies that the configurations have static canonical fields, $\dot{g}_{ij} = \dot{\pi}^{ij} = \dot{N} = 0$.

It is convenient to use the constraint $\mathcal{C}_2$ to solve N. To this end the following change of variables is useful,

$$N = W^\gamma\,, \quad \gamma \equiv \frac{2\gamma_2}{2\beta + \alpha}\,. \tag{23}$$

With this change the constraint $\mathcal{C}_2$ given in (4) becomes

$$\nabla^2 W = \chi_1 R W\,, \quad \chi_1 \equiv \frac{\beta(2\beta + \alpha)}{2\gamma_2^2}\,, \tag{24}$$

which we regard as an equation for W. We now use (24) in the constraint $\mathcal{C}_1$ given in (3), obtaining

$$\frac{1}{g}\pi^{ij}\pi_{ij} = 2\chi_2 \frac{\partial_i W \partial^i W}{W^2} + 2\chi_3 R \,, \tag{25}$$

where

$$\chi_2 \equiv \frac{\alpha\gamma_1\gamma_2}{(2\beta+\alpha)^2} \,, \quad \chi_3 \equiv \frac{\beta(2\beta+\alpha)}{4\gamma_2} \,. \tag{26}$$

To further advance, we assume that we are in a region on which the static spatial metric g_{ij} is a flat metric. Then the Eq. (24) takes the form

$$\nabla^2 W = 0 \,, \tag{27}$$

where now ∇^2 is the flat Laplacian. Therefore, any harmonic function W on the given region satisfying some boundary condition solves the constraint (27). There remain the constraints on π^{ij}, which we sumarize here,

$$\begin{aligned} \pi &= 0 \,, & (28)\\ \nabla_i \pi^{ij} &= 0 \,, & (29)\\ \frac{1}{g}\pi^{ij}\pi_{ij} &= 2\chi_2 \frac{\partial_i W \partial^i W}{W^2} \,. & (30) \end{aligned}$$

We may give a class of static solutions of the system (27 - 30) with the flat metric in the following way. We implement Cartesian coordinates on the spatial slices. The solutions start with the assumption that one is given with a harmonic function W that can be expressed in separate variables,

$$W = X^1(x^1)X^2(x^2)X^3(x^3) \,. \tag{31}$$

For the momentum field we assume that all its diagonal components vanish, $\pi^{11} = \pi^{22} = \pi^{33} = 0$, such that the constraint (28) is automatically solved. The momentum constraint (29) is completely solved if the off-diagonal components have the following dependence on the coordinates:

$$\pi^{12}(x^3) \,, \quad \pi^{13}(x^2) \,, \quad \pi^{23}(x^1) \,. \tag{32}$$

Finally, the constraint (30) is solved if these components of the momentum are given in terms of the components of the harmonic function in the form

$$\pi^{12} = \frac{\sqrt{\chi_2}}{X^3}\frac{dX^3}{dx^3} \,, \quad \pi^{13} = \frac{\sqrt{\chi_2}}{X^2}\frac{dX^2}{dx^2} \,, \quad \pi^{23} = \frac{\sqrt{\chi_2}}{X^1}\frac{dX^1}{dx^1} \,. \tag{33}$$

5. Coupling to sources: homogeneity and isotropy

Paralellizing the standard approach of GR, we consider a perfect fluid that has a relativistic energy-momentum tensor,

$$T_{\mu\nu} = \rho u_\mu u_\nu + P(g_{\mu\nu} + u_\mu u_\nu) \,. \tag{34}$$

We also consider that the perfect fluid is at rest in the chosen reference frame, $u^\mu = (N^{-1}, 0, 0, 0)$, such that the constraint $u_\mu u^\mu = -1$ is satisfied.

We summarize the resulting system of field equations,

$$\begin{aligned} \pi &= 0 \,, & (35)\\ \nabla_i \pi^{ij} &= 0 \,, & (36)\\ \frac{1}{\sqrt{g}}\mathcal{C}_1 &\equiv \frac{2}{g}\pi^{ij}\pi_{ij} - \gamma_1 \frac{\nabla^2 N}{N} + \rho + 3P = 0 \,, & (37) \end{aligned}$$

$$\frac{1}{\sqrt{g}}\mathcal{C}_2 \equiv \gamma_2 \frac{\nabla^2 N}{N} - \alpha a_i a^i - \beta R + \frac{3}{2}(\rho - P) = 0\,, \tag{38}$$

$$\dot{g}_{ij} = \frac{2}{\sqrt{g}}(N + 2A_1)\pi_{ij} + \mu g_{ij}\,, \tag{39}$$

$$\begin{aligned}\frac{\dot{\pi}^{ij}}{\sqrt{g}} =& -\frac{1}{g}(N+2A_1)(2\pi^{ik}\pi_k{}^j - \frac{1}{2}g^{ij}\pi^{kl}\pi_{kl}) - \mu\frac{\pi^{ij}}{\sqrt{g}} - \nabla^{(i}N\nabla^{j)}B + \frac{1}{2}g^{ij}\nabla_k N\nabla^k B \\ &+ \left[\beta(\nabla^{ij} - g^{ij}\nabla^2) - \beta(R^{ij} - \frac{1}{2}g^{ij}R) - \alpha(a^i a^j - g^{ij}a_k a^k)\right](N + A_2) \\ &- \frac{1}{2}g^{ij}\left[(A_1 + \frac{3}{2}A_2)\rho - (2N - 3A_1 + \frac{3}{2}A_2)P\right]\,,\end{aligned} \tag{40}$$

$$0 = \nabla^2 B + 2\alpha a^k \nabla_k\left(\frac{A_2}{N}\right) + \frac{\gamma_1}{2}\frac{\nabla^2 N}{N^2}(2A_1 - A_2)\,. \tag{41}$$

We fix again the Lagrange multipliers A_1 and A_2 according to $2A_1 = A_2 = -N$, such that Eq. (41) is completely solved. We consider that the source variables ρ and P depend only on the time coordinate. We introduce the ansatz on which the spatial metric is homogeneous and isotropic. For the sake of simplicity we only consider the flat case. In Cartesian coordinates, this ansatz is impleted by the spatial metric

$$ds^2_{(3)} = a^2(t)\left((dx^1)^2 + (dx^2)^2 + (dx^3)^2\right)\,. \tag{42}$$

Notice that we restrict the conditions of homogeneity and isotropy to the spatial metric, but not to the lapse function. We can do so because in this theory the diffeomorphisms that preserve the foliation do not mix the lapse function with the spatial metric, and observationally these conditions are restricted to the spatial metric. We also consider that the momentum π^{ij} depends only on the time coordinate. With these settings the momentum constraint (36) is automatically solved.

Equation (39) reduces to

$$2a\dot{a}\delta_{ij} = \mu a^2\delta_{ij}\,, \tag{43}$$

and its solution is

$$\mu = \frac{2\dot{a}}{a}\,. \tag{44}$$

Equation (40) becomes

$$\dot{\pi}^{ij} + \frac{2\dot{a}}{a}\pi^{ij} = \delta^{ij}aN(\rho + P)\,. \tag{45}$$

Since constraint (35) demands that π^{ij} is traceless, from (45) we extract the equation of state $P = -\rho$. In turn, this implies that the right hand side of Eq. (45) is equal to zero. This equation can be directly integrated, its solution is

$$\pi^{ij} = \frac{m^{ij}}{a^2}\,, \tag{46}$$

where m^{ij} is a symmetric traceless constant matrix, such that the constraint (35) is automatically solved.

There remain the constraints $\mathcal{C}_1$ and $\mathcal{C}_2$, given in (37) and (38), to be solved. Under the conditions we are considering, these constraints take, respectively, the form

$$\frac{2m^2}{a^4} - (2\beta - \alpha)\frac{\partial^2 N}{N} = 2a^2\rho\,, \tag{47}$$

$$(\beta + \frac{3\alpha}{2})\frac{\partial^2 N}{N} - \alpha\frac{\partial_i N\partial_i N}{N^2} = -3a^2\rho\,, \tag{48}$$

where $m^2 \equiv m^{ij}m^{ij}$ and ∂^2 is the flat Laplacian in Cartesian coordinates, $\partial^2 = \partial_i\partial_i$. We introduce the following anstaz for N,

$$N = \exp\left(f_0(t) + f_1(t)x^1 + f_2(t)x^2 + f_3(t)x^3\right) , \tag{49}$$

where $f_{0,1,2,3}(t)$ are arbitrary functions of time. Then the Eq. (48) reduces to

$$f_i f_i = -\frac{6a^2\rho}{2\beta + \alpha} , \tag{50}$$

where $f_i f_i = f_1^2 + f_2^2 + f_3^2$. Consistency of this equation requires that the coupling constants satisfy the bound $2\beta + \alpha < 0$ (assuming $\rho > 0$). By using Eq. (50) in Eq. (47), we obtain that (47) becomes an algebraic equation for a and ρ, whose solution is

$$\rho = ka^{-6} , \quad k \equiv \frac{|2\beta + \alpha| m^2}{4(\beta - \alpha)} . \tag{51}$$

(If we assume $\beta > 0$ and $2\beta + \alpha < 0$, then $\beta - \alpha > 0$).

Conclusions

We have shown how in the kinetic-conformal formulation of the nonprojectable Hořava theory the space of solutions of the Hamiltonian formulation is bigger than the corresponding space in the original Lagrangian formulation. A key role is played by one of the Lagrange multipliers associated to the secondary constraints, namely A_1, since there is a configuration of this multiplier for which the Legendre transformation cannot be inverted.

We have presented a large class of evolving solutions governed by the flow of a three-dimensional metric, new static vacuum solutions and a new solution with a source. The solution with the source, which specifically is a homogeneous and isotropic perfect fluid with a relativistic energy-momentum tensor, has the interesting feature of possessing a homogeneous and isotropic spatial metric with nontrivial (time dependent) scale factor. This solution signals how more homogeneous and isotropic configurations arise when, first, the solutions are studied in the Hamiltonian formalism and, second, the condition of homogeneity and isotropy is restricted to the spatial metric, whereas the lapse function is allowed to have a more general dependence on the space. All these examples are characterized by a specific setting of one of the Lagrange multipliers, namely A_2. This is a condition additional to the setting that leads to the noninvertibility of the Legendre transformation, which is a condition on A_1. We have extended the discussion by relaxing the restriction on A_2, and we have shown that more special Hamiltonian solutions exists in this more general case.

Acknowledgments

A. R. and A. S. are partially supported by Project Fondecyt 1161192, Chile.

References

[1] J. Bellorín, A. Restuccia and A. Sotomayor, *A consistent Hořava gravity without extra modes and equivalent to general relativity at the linearized level*, Phys. Rev. D **87** (2013) 084020 [arXiv:1302.1357 [hep-th]].
[2] P. Hořava, *Quantum Gravity at a Lifshitz Point*, Phys. Rev. D **79** (2009) 084008 [arXiv:0901.3775 [hep-th]].
[3] D. Blas, O. Pujolàs and S. Sibiryakov, *Consistent Extension Of Hořava Gravity*, Phys. Rev. Lett. **104** (2010) 181302 [arXiv:0909.3525 [hep-th]].
[4] J. Bellorín and A. Restuccia, *Quantization of the Hořava theory at the kinetic-conformal point*, Phys. Rev. D **94** (2016) 064041 [arXiv:1606.02606 [hep-th]].

[5] T. Jacobson, *Undoing the twist: The Hořava limit of Einstein-aether theory*, Phys. Rev. D **89** (2014) 081501 [arXiv:1310.5115 [gr-qc]].
[6] B. Audren, D. Blas, M. M. Ivanov, J. Lesgourgues and S. Sibiryakov, *Cosmological constraints on deviations from Lorentz invariance in gravity and dark matter*, JCAP **1503** (2015) 016 [arXiv:1410.6514 [astro-ph.CO]].
[7] D. Mattingly and T. Jacobson, *Relativistic gravity with a dynamical preferred frame*, Proceedings of the Second Meeting on CPT and Lorentz Symmetry, Ed. V. Alan Kostelecky (Singapore, World Scientific, 2002) [arXiv:gr-qc/0112012 [gr-qc]].
[8] S. M. Carroll and E. A. Lim, *Lorentz-violating vector fields slow the universe down*, Phys. Rev. D **70** (2004) 123525 [arXiv:hep-th/0407149 [hep-th]].
[9] J. Bellorín and A. Restuccia, *On the space of solutions of the Hořava theory at the kinetic-conformal point*, (2017) [arXiv:1705.10161 [hep-th]].
[10] J. Bellorín and A. Restuccia, *Einstein's quadrupole formula from the kinetic-conformal Hořava theory*, (2017) to be published in IJMPD [arXiv:1612.04414 [gr-qc]].

Numerical Investigations of the Thermal, Pressure and Size Effects on 2D Spin Crossover Nanoparticles

C. Harlé[1], S.E. Allal[1]., D. Sohier[2], T. Dufaud[2,3], R. Caballero[4], F. de Zela[4], P. R. Dahoo[1], K. Boukheddaden[5], J. Linares [4,5]
[1]LATMOS, Université de Versailles-Saint-Quentin-en-Yvelines, CNRS-UPMC-UVSQ (UMR 8190), 11 boulevard d'Alembert, 78280 Guyancourt, France
[2] LI-PaRAD, Université de Versailles Saint-Quentin-en-Yvelines, 45 Av. des Etats-Unis, EA 7432, 78035 Versailles Cedex, France
[3] Maison de la Simulation, Université de Versailles Saint-Quentin-en-Yvelines, CNRS - USR 3441, Bâtiment 565 - Digiteo CEA Saclay, 91191 Gif-sur-Yvette cedex, France
[4] Departamento de Ciencias, Sección Física, , Pontificia Universidad Católica del Perú, Apartado 1761 - Lima, Peru
[5] GEMAC, Université de Versailles Saint-Quentin-en-Yvelines, 45 Av. des Etats-Unis, CNRS-UMR 8635, 78035 Versailles Cedex, France

Abstract: In the framework of the Ising-like model, the thermal and pressure effects on the spin crossover systems are evaluated through two-states fictitious spin operators σ with eigenvalues $\sigma = -1$ and $\sigma = +1$ respectively associated with the low-spin (LS) and high-spin (HS) states of each spin-crossover (SCO) molecule. Based on each configurational state, the macroscopic SCO system, is described by the following variables: $m=\Sigma \sigma_i$, $s=\Sigma \sigma_i \sigma_j$ and $c=\Sigma \sigma_k$ standing respectively for the total magnetization, the short-range correlations and surface magnetization. To solve this problem, we first determine the density of macrostates d[m][s][c], giving the number of microscopic configurations with the same m, s and c values. In this contribution, two different ways have been performed to calculate this important quantity: (i) the entropic sampling method, based on Monte Carlo simulations and (ii) a new algorithm based on specific dynamic programming. These two methods were tested on the 2D SCO nanoparticles for which, we calculated the average magnetization $<\sigma>$ taking into account for short-, long-range interactions as well as for the interaction between surface molecules with their surrounding matrix. We monitored the effect of the pressure, temperature and size on the properties of the SCO nanoparticles.

1. Introduction

Spin transition compounds are part of inorganic coordination complexes of transition metal ions with $3d^4$- $3d^7$ electron configurations, namely Fe(II), Fe(III), Co(II), Mn(II), and Ni(II), located in an octahedral ligand field, the strength of which induces the competition between spin states[1-9]. These compounds are usually known as bistable materials, are able to switch between the diamagnetic low-spin state (LS), stable at low-temperature, and the paramagnetic high spin-state (HS), stable state at high-temperature, due to its larger entropy. Due to the local volume change accompanying each spin transition of the molecules, long- and short-range range elastic interactions[10-13] take place between the SCO units causing cooperative hysteretic first-order thermal transitions, driven by the large entropy (originating from spin and vibrations) of the HS state compared to that of the LS. In addition to the temperature[1-2], the spin state can also be switched by the application of pressure[14-17], by a magnetic field[18], or by an irradiation in the visible light[19]. In this work, we will focus exclusively on thermal and pressure effects. From the experimental side, the switching between the LS and HS states results in large changes in magnetic, optical, and dielectric properties of SCO compounds, which make them potential candidates as promising nano-materials for innovative applications such as ultra-sensitive sensors, in data storage or as display devices. From the theoretical side, the modelling of SCO phenomena helps to get new insights on the physical properties of SCO nanoparticles, particularly those displaying phase transitions, whose behaviour as function of their size and their interaction parameters play an important role in the control of the switching features of these complex materials, with the ultimate goal to help to design novel ones with specific ones with tailored properties.

Furthermore, the modelling of phase transitions makes possible the understanding of the intimate processes of cooperativity in SCO solids, and how these interactions between molecules are influenced by the various constraints applied to the material, like light, pressure or temperature. On the other hand, it is now widely admitted that long range interactions between SCO molecules are of elastic origin and that depending on their strength compared to the ligand field energy, the high spin fraction curve can display different shapes, whose features can change gradually, abruptly, or in two-step with or without hysteresis. In SCO nanoparticles, these trends are intrinsically related to the cooperativity between the SCO units but also to the size of the nanoparticles as well as to other external constraints influencing the transition, as it will be demonstrated in the present work.

The manuscript is organized as follows: section (2) devoted to the presentation of the model Hamiltonian, then to the methods of resolution and finally to the results of the numerical simulations conducted on the 2D lattices. The discussion of the behaviour of the high-spin fraction obtained using several stimuli will be realized in this section. In section (3) we conclude and outline some possible extensions of the present work.

2. Results and discussion

The different simulations we have carried out in the present investigations were based on the Ising-like model[20-22], in which we analysed the behaviour of the average net magnetization $<\sigma>$ or HS fraction, $N_{HS} = \frac{1+<\sigma>}{2}$, that is to say the probability of obtaining of HS molecules in the material under study. We have considered the following lattice configurations of size: 4x4 and 12x12 in 2D, and 1x12 and 1x500 for the 1D case.

To deeply explore the thermal and pressure properties of SCO nanoparticles, we simulated N_{HS} as a function of temperature and pressure thanks to various parameters such as: (i) the ligand-field Δ which represents the energy gap between the LS and HS states of isolated spin-crossover units, (ii) the respective short- and long-range interaction parameters, J and G, (iii) the interaction between the molecules at the surface and their environments, L, introduced by Linares *et al* [23-24], (iv) ln (g) with g = gHS / gLS corresponding to the ratio of degeneracies of the 2 spin states.

The specific Ising-like designed to mimic the SCO nanoparticles under various constraints writes under the following form [15-16]:

$$H = \frac{\Delta + \Delta V.P \; - k_B T \ln g}{2} \sum_{i=1}^{N} \sigma_i - G \sum_{i=1}^{N} \sigma_i < \sigma > - J \sum_{<i,j>} \sigma_i \sigma_j - L \sum_{i=1}^{M} \sigma_i \qquad (1)$$

Hamiltonian (1) has the advantage to make possible to simulate the material's behaviour in an environment that is characterized by interactions that occur not only in the core of the material, but also at the edge.

The first term in (1) results from one site contributions and concerns all lattice atoms, the second term accounts for long-range interactions and it is written here under the form a mean-field in which $< \sigma >$

is the average magnetization per site. The third term expresses short-range interactions between the SCO species, while the last term only acts on surface molecules due to their specific environment.
In addition, Hamiltonian (1) includes the contribution of an external isotropic pressure, whose effect is to renormalize the ligand field energy, the effective expression of which becomes, $\Delta + \Delta V.P$, where P is the applied external pressure and $\Delta V = V_{HS} - V_{LS}$ is the volume change of the material between the HS and the LS states. Here, N is the total number of molecules in the lattice and M the number of molecules at the surface ("edge").

The following macroscopic parameters will be useful in the description of the thermodynamic properties:

$$m = \sum_{i=1}^{N} \sigma_i \qquad s = \sum_{<i,j>} \sigma_i \sigma_j \qquad c = \sum_{k=1}^{M} \sigma'_k \qquad (2)$$

Where σ'_k represents the molecules on the edge. There, m and c are the total and surface magnetizations, s is the sum of two spins correlations in the lattice.

Then the Hamiltonian can be cast as:

$$H = \left(\frac{\Delta + \Delta V.P - k_B T \ln g}{2} - G < \sigma > \right) m - Js - Lc \qquad (3)$$

where, $\frac{m}{N} = < \sigma >$. Eq. (3) clearly indicates that the knowledge of the density of states $d(m, s, c)$, that represents the number configuration with the same m, s and c values, allows the determination of all thermodynamic properties of this system.

The average value of operator σ is given by:

$$< \sigma > = \frac{\sum_{i=1}^{NL} \frac{m_i}{N} d(m_i s_i c_i) \exp\left(-\frac{1}{k_B T} (-hm_i - Js_i - Lc_i) \right)}{\sum_{i=1}^{NL} d(m_i s_i c_i) \exp\left(-\frac{1}{k_B T} (-hm_i - Js_i - Lc_i) \right)} \qquad (4)$$

where NL is the number of possible configurations $\{m, s, c\}$, and h is expressed as :

$$h = -\left(\frac{\Delta + \Delta V.P - k_B T \ln g}{2} - G < \sigma > \right) \qquad (5)$$

The density of macrostates, $d(m, s, c)$, and the different configurations $\{m, s, c\}$, have been calculated using Monte Carlo Entropic Sampling[25] and also by a new dynamic programming algorithm.

Now we introduce the assumption that the transition temperature of the system is the result of a null total effective ligand-field. So, the transition temperature, T_{eq}, for a square lattice $N_x \times N_x$ is the solution of the following equation:

$$\frac{\Delta + \Delta V.P - k_B T_{eq} \ln g}{2} \times (N_x - 2)^2 + \frac{\Delta + \Delta V.P - 2L - k_B T_{eq} \ln g}{2} \times 4(N_x - 1) = 0 \qquad (6)$$

where, $(N_x - 2)^2$ and $4(N_x - 1)$ are respectively the number of the core and the edges sites.
Solving this linear equation gives the analytical expression of transition temperature as function of the lattice size and the bulk, T_{eq}^{bulk}, and surface, T_{eq}^{surf}, transition temperatures:

$$T_{eq} = \frac{N_c}{N} T_{eq}^{bulk} + \frac{N_s}{N} T_{eq}^{surf} \qquad (7)$$

where the analytical expressions of the transition temperatures associated to the bulk and the surface, are:

$$T_{eq}^{bulk} = \frac{\Delta + \Delta V.P}{k_B lng} \; and \; T_{eq}^{surf} = \frac{\Delta + \Delta V.P - 2L}{k_B lng} \qquad (8)$$

The present investigations are mainly focussed on the role of the interaction between the matrix and the molecules at the surface, i.e. L (which is one of the most relevant parameters is the behaviour of SCO nanoparticles) on the thermal properties of SCO nanoparticles. At this end, we simulate the thermal behaviour of SCO system in isobaric conditions as well as its pressure dependence in isothermal conditions, for L= 0 and L≠ 0.

The 2D case:

We first start with the situation L≠0. The results of the dynamic algorithm programming leading to the exact density of states, from which we derived the thermodynamic properties of the studied lattices, are presented in Fig 1, which summarizes the temperature dependence of the HS fraction for a fixed pressure value. For the parameter values, Δ/k_B = 1300 K, G/k_B = 172.2 K, J/k_B = 13.5 K, L/k_B = 120 K, ln(g) = 6.01, we clearly see that the square lattice 12x12 does not show any thermal hysteresis. In contrast, when the lattice's size is reduced to a 4x4, a thermal hysteresis appears. To understand this trend, one has to analyse carefully the role of surface effects. In fact, by decreasing the lattice's size, the surface effects become dominant, the transition temperature approaches that of the surface, $(\Delta-2L)/k_B\ln(g)$ ~ 176 K. To see whether, the first order transition occurs or not at the surface, it is important to evaluate the order-disorder transition, T_{OD}, of the system under short- and long-range interactions J and G, for Δ=0, L=0 and g=1. In the case J=0, the order-disorder transition is equal to that of the mean-field model $T_{OD}=G/k_B$, while in the case G=0 and J≠0, the order-disorder is given by the exact Onsager's expression, Tc=2.269J=30,63K. When the short- and long-range interactions are nonzero, the transition order-disorder transition can be approximately estimated to Tc = $(2.269J+G)/k_B$ ~ 203 K, a value in quite good agreement with that found by our numercial simulations on the same system. Compared to the previous value of Tc (203 K), we have T_{eq}^{surf} (~ 176 K)< Tc, while T_{eq}^{bulk} (=216 K) > Tc. The latter conditions lead to a gradual transition, while the former implies the existence of a thermal hysteresis. So decreasing the size, enhances the surface effects for which the conditions of appearance of first-order transition are fulfilled.

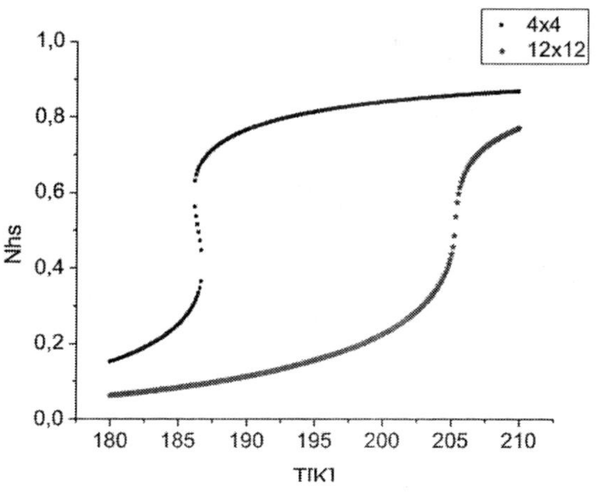

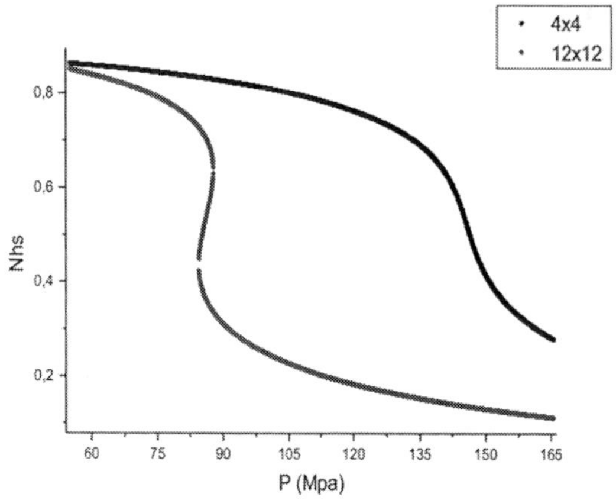

(1) (2)

Figure 1. Thermal evolution of the high spin fraction for the two lattice sizes: 4x4 (black square) and 12x12 (red star). The parameters value are: $\Delta/k_B = 1300$ K, $G/k_B = 172.2$ K, $J/k_B = 13.5$ K, $L/k_B = 120$ K, ln(g) = 6.01.
Figure 2. Evolution of the high spin fraction versus the pressure for 2 sizes: 4x4 (black square) and 12x12 (red circle); the parameters value are: Δ/kB = 1978.6 K, G/kB = 172.2 K, J/kB = 48 K, L/kB = 120 K, T = 300 K, ln(g) = 6.906, $\Delta V = 100$ Å^3.

Now, we inspect the effect of a hydrostatic pressure on the thermal properties of the SCO material. According to Eq. (8), we can easily see that pressure appears a conjugate parameter of temperature. Under pressure, the transition temperature of the SCO lattice increases as a result of the increase of the ligand field, which stabilizes the LS state, which has a smaller volume, compared to that of the HS state. Thus, increasing pressure is somehow equivalent to reduce temperature, which leads to switch the SCO system between the HS and the LS states.

Thus, once again, as seen in Fig. 2, the thermal behaviour and the pressure behaviour are reversed concerning the hysteresis since the system switches from the HS to the LS state by increasing pressure. Although not discussed, this isothermal transition crucially depends on temperature, whose variation affects the width of the pressure hysteresis and even the character (first-order or gradual) of the transition.

The case L=0:

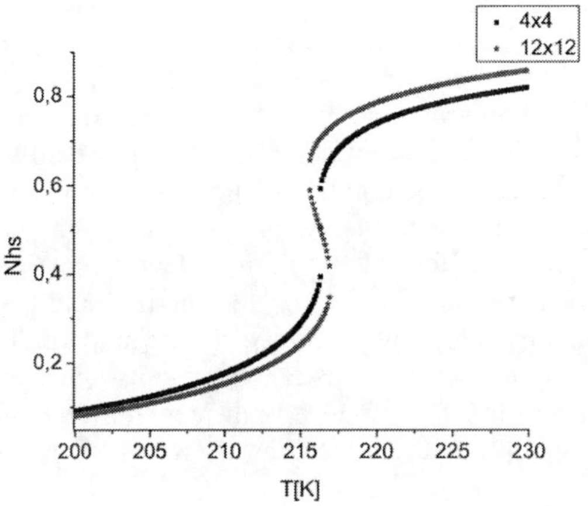

(3)

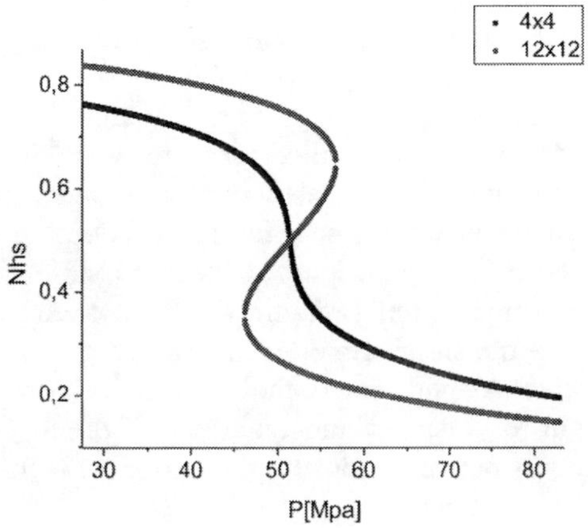

(4)

Figure 3. Thermal evolution of the high spin for two lattice sizes: 4x4 (black square) and 12x12 (red star) at zero applied pressure. The parameters value are: Δ/kB = 1300 K, G/kB = 172.2 K, J/kB = 16.8 K, L/kB = 0 K, ln(g) = 6.01.

Figure 4. Pressure-dependence of the HS fraction in isothermal conditions (T=300 K) for two lattice sizes: 4x4 (black square) and 12x12 (red circle). The parameters value are: Δ/kB = 1978.6 K, G/kB = 172.2 K, J/kB = 48, L/kB = 0 K and ln(g) = 6.906.

Unlike the case L ≠ 0, we note that the behaviour is almost identical: the hysteresis witdh increases when the size is increased. Moreover, it can be noticed for the thermal behaviour that the equilibrium temperature which corresponds to N_{hs}= ½ is the same whatever the size of the material. Indeed, when L=0 is set, the lattice does not experience any surface effects and consequently, according to Eqs. (7) and (8), the transition temperature, T_{eq}, remains independent on the lattice size. In contrast, the order-disorder transition T_{OD} (obtained for Δ = 0 and g=1) increases with lattice size, which clearly explains the reason for which the thermal hysteresis becomes wider (see Fig. 3) as the lattice size increases.
Applying an isotropic pressure on the system, in isothermal conditions at T=300K, results in the pressure hysteresis of Fig. 4. Interestingly, the latter is much wider than that of Fig. 2 with a pressure hysteresis width equal to ΔP ~ 13 MPa against 7 MPa for Fig.2 and a pressure transition $P_{1/2}^{bulk}$ = $(k_B T \ln(g)-\Delta)/\Delta V$ ~ 52 MPa, which is smaller than that of Fig. 2 which is affected by the surface effects.

3. Conclusion

In this paper, we analysed the thermal and pressure behaviour of 1D and 2D spin-crossover compounds with particular attention paid to the effect of the interactions between molecules at the edges of the nanoparticles and their local environment, which were taken into account in the coupling parameter, L. When L = 0, the thermal and pressure behaviour are identical and the increase in size is characterized by the appearance of a hysteresis. However, for L different from zero, the behaviours due to the applied pressure and the temperature are reversed with respect to the modification of the size of the material. This result, which is directly linked to the thermal behaviour, as the only property that is reversed when L is modified, is explained by a quantitative relationship between the equilibrium temperature and the order-disorder temperature of the same Hamiltonian with a degeneracy ratio g=1 and without ligand field, i.e. Δ=0 and L=0. Differently, we observed that the effect of pressure does not depend significantly on L, except for the very small sizes. Indeed, the effect of pressure combines both surface and bulk effects and so its impact on the surface depends on the ration surface/volume. Moreover, since the ligand field at the surface of the lattice is weaker than that of the bulk, due to matrix's contribution, the effect of pressure will be clearly different on the surface and the volume. Furthermore, the elastic properties of the material are different in the surface and the volume[12-13] as well as the phonon spectra, which may constitute an interesting target for future investigations. In addition, one may underline the fact the interaction parameters (J and G) themselves may depend on the applied pressure, since the latter are elastic in nature and then should be influenced by the change of the elastic constants under high pressure, for example. This interesting point will be also investigated in a near future, by adopting simple expansions of the interaction parameters with the applied pressure. Despite of these possible extensions, the effect of the various factors, such as the internal interaction strength, the local environment or the system's size, on the thermal and pressure induced transition, reported in the present work, are consistent with several experimental findings in this area. As underlined, the current work opens the way to many interesting extensions which will be helpful for a deeper understanding of the SCO properties under various stimuli, which are mandatory prerequisites for the development of new technological applications based on the switchable properties of these materials.

Acknowledgments
CHAIR Materials Simulation and Engineering, UVSQ, Université Paris Saclay is gratefully acknowledged. The present work has also been supported by the French "Ministère de la Recherche", Université de Versailles St. Quentin en Yvelines, CNRS and ANR BISTA-MAT: ANR-12-BS07-0030-01, whose support is highly appreciated.

References

[1] Hauser A., Coord. Chem. Rev., 111, 275-290, (1991)
[2] Gütlich P, Gaspar AB and Garcia Y, Beilstein J. Org. Chem. 9 342, (2013)
[3] Varret F., Bleuzen A., Boukheddaden K., Bousseksou A., Codjovi E., Enachescu C., Goujon A., Linares J., N. Menendez N. and M. Verdaguer M., Pure Appl. Chem., 74, 2159-2168, (2002)
[4] Sy M. , Garrot D., Slimani A., Paez-Espejo M., Varret F., Boukheddaden K., Ang. Chem, 128, 1787-1791, (2016)
[5] Enachescu C., Linares J. , Varret F. , J. Phys.:Condens. Matter, 13,2481-2495, (2001)
[6] Jureschi C., Linares J., Boulmaali A., Dahoo P.R., Rotaru A. and Garcia Y., Sensors,16, 187, (2016)
[7] Seredyuk M, Gaspar AB, Ksenofontov V, Reiman S, Galyametdinov Y, Haase W, Rentschler E, Gütlich P, Chem. Mater. 18 2513, (2006)
[8] Mishra V., Mukherjee R., Linares J., Balde C., Desplanches C., Létard J.F., Collet E., Toupet L., Castro M., Varret F. , Inorg Chemistry, 47, 7577-7587, (2008)
[9] Kamel Boukheddaden, Isidor Shteto, Benoit Hôo, and François Varret, Phys. Rev. B 62, 14796, (2000)
[10] Ahmed Slimani, Kamel Boukheddaden, François Varret, Hassane Oubouchou, Masamichi Nishino, and Seiji Miyashita, Phys. Rev. B 87, 014111, (2013)
[11] Miguel Paez-Espejo, Mouhamadou Sy, and Kamel Boukheddaden, J. Am. Chem. Soc., 138, 3202–3210, (2016)
[12] Kamel Boukheddaden, Phys. Rev. B 88, 134105, (2013)
[13] Masamichi Nishino, Kamel Boukheddaden, and Seiji Miyashita, Phys. Rev. B 79, 012409, (2009)

[14] Klokishner S., Linares J., Varret F., Chemical. Physics, 255, 317-323, (2000)
[15] Morscheidt W, Jeftic J, Codjovi E, Linares J, Bousseksou A, Constant-Machado H, Varret F, Meas. Sci.Techn., 9, 1311-1315, (1998)
[16] Rotaru A, Varret F, Gîndulescu A, Linares J, Stancu A, Létard JF, Forestier T, Etrillard C, Eur. Phys. J. B 84, 439-449, (2011)
[17] Jeftic J, Menéndez N, Wack A, Codjovi E, Linarès J, Goujon A, Hamel G, Klotz S, Syfosse G, Varret F , Meas. Sci. Techn., 10, 1059-1064, (1999)
[18] Bonhommeau, S.; Molnar, G.; Galet, A.; Zwick, A.; Real, J.A.; McGarvey, J.J.; Bousseksou, A. . Angew. Chem. Int. Edit., 44, 4069–4073, (2005)
[19] Bousseksou, A.; Boukheddaden, K.; Goiran, M.; Consejo, C.; Boillot, M.L.; Tuchagues, J.P. . Phys. Rev. B, 65, (2002)
[20] Wajnflasz J, Pick R, J. Phys.Colloques, 32, 91-92, (1971)
[21] Bousseksou A, Nasser J, Linares J, Boukheddaden K, Varret F, J. Physique I 2 1381, (1992)
[22] J. Linares, J. Nasser, A. Bousseksou,K. Boukheddaden, , F. Varret, J. Mag. Mag. Mat. 140-144, 1503-1504, (1995)
[23] Linares J, Jureschi C, Boukheddaden K, Magnetochemistry, 2, 24, (2016)
[24] Linares J, Jureschi C, Boulmaali A, Boukheddaden K, Physica B, 486, 164–168, (2016)
[25] Shteto I., Linares J., Varret F., Physical Review E 56, 5128-5137, (1997)

IOP Conf. Series: Journal of Physics: Conf. Series **936** (2017) 012062 doi:10.1088/1742-6596/936/1/012062

On damping of an edge dislocation vibrations in a dissipative crystal: limiting cases

V V Dezhin

Voronezh State Technical University, Moskovsky prospect 14, Voronezh 394026, Russian Federation

E-mail: viktor.dezhin@mail.ru

Abstract. The expression for the generalized susceptibility of the dislocation obtained earlier was used. The electronic drag mechanism of dislocations is considered. The study of small dislocation oscillations was limited. The contribution of the attenuation of low-frequency bending dislocation vibrations to the overall coefficient of dynamic dislocation drag in the long-wave and short-wave limits is calculated. The damping of short-wave bending dislocation vibrations caused by an external action of an arbitrary frequency has been investigated. The contribution of long-wave bending dislocation vibrations damping in total drag coefficient at an arbitrary frequency is found.

1. Introduction

It is known that the influence of dissipative processes will lead to the drag of a moving dislocation. One of the first theoretical works on the drag of a moving dislocation was carried out by Eshelby [1]. He considered in it the damping of a rectilinear dislocation oscillations due to thermoelastic energy dissipation and radiation of elastic waves. Subsequently, a large number of specific mechanisms calculations for the dislocations drag were carried out. Reviews of these mechanisms in [2, 3] are given. Along with this, any dissipative processes in the most general form can be taken into account in the equations of motion by including friction forces in the right-hand side of the Lagrange equation. Such an approach, based on dissipative processes accounting through the dispersion of elastic modules, was developed in [4]. In [6], on the basis of [5] results generalization, the equations for the small oscillations of a crystal with a dislocation taking into account the dissipative processes were obtained, and the generalized susceptibility of a dislocation in a dissipative crystal is found. Also in the work [6], the cases $\omega = 0$ and $k_z = 0$ are considered, where ω is frequency, k_z is wave vector component along dislocation line. In [7], using the results of [5, 6], a dislocation amplitude-independent internal friction was found. In recent papers [8-11], the studies of the dislocation oscillations damping were continued. In this paper, on the basis of the expression for the dislocation generalized susceptibility in a dissipative crystal [6], various limiting cases of edge dislocation vibrations damping are investigated.

2. Damping of low-frequency bending dislocation vibrations

Damped bending vibrations of an edge dislocation lying along the axis Oz were investigated. The case of small oscillations of a dislocation near the equilibrium position was limited. In the low-frequency limit, from [6] for the imaginary part of the edge dislocation inverse generalized susceptibility, taking into account only the terms corresponding to the dislocation bending vibrations, we obtain

$$\operatorname{Im} g_e^{-1}(k_z,\omega) = -\frac{\rho b_e^2}{2\pi}\omega\int_{|k_z l|}^{k_m l} dx\left[4\xi\frac{(k_z l)^2}{x^3}\gamma_t - 2\xi^2\frac{(k_z l)^2}{x^3}\gamma_l + (1-2\xi)\frac{(k_z l)^4}{x^5}\gamma_t + \xi^2\frac{(k_z l)^4}{x^5}\gamma_l\right]. \quad (1)$$

Here ρ is density of the crystal, b_e is edge component of the Burgers vector, k_m is maximum wave number, l is electron mean free path, $x = kl$ is dimensionless variable, $\xi = c_t^2/c_l^2$, c_t and c_l are the transverse and longitudinal sound velocities in a nondissipative crystal, $\gamma_t(x)$ and $\gamma_l(x)$ are the damping coefficients of transverse and longitudinal sound waves in a dissipative medium. As a concrete damping mechanism for oscillations, electronic drag was considered, for which the damping coefficients of the transverse and longitudinal sound are expressed in the form [12]:

$$\gamma_t(x) = \gamma^0\left(\frac{2x^3}{3[(1+x^2)\operatorname{arctg}x - x]} - 1\right) \quad \text{and} \quad \gamma_l(x) = \gamma^0\left(\frac{x^2\operatorname{arctg}x}{3(x - \operatorname{arctg}x)} - 1\right), \quad (2)$$

where γ^0 is a constant depending on the material. For the purpose of analytical integration in formula (1), the damping coefficients (2) were approximated as follows:

$$\gamma_t(x) \approx \frac{4}{3\pi}\gamma^0\frac{x^2}{x + 20/3\pi} \quad \text{and} \quad \gamma_l(x) \approx \frac{\pi}{6}\gamma^0\frac{x^2}{x + 5\pi/8}. \quad (3)$$

Substituting expressions (3) into formula (1) and performing the integration, we obtain

$$\operatorname{Im} g_e^{-1}(k_z,\omega) = -\frac{\rho b_e^2}{2\pi}\gamma^0\omega\left\{\frac{4}{5}\xi(k_z l)^2\ln\frac{|k_z l| + 20/3\pi}{|k_z l|} - \frac{8}{15}\xi^2(k_z l)^2\ln\frac{|k_z l| + 5\pi/8}{|k_z l|}\right.$$

$$+\frac{4}{3\pi}(1-2\xi)(k_z l)^4\left[\left(\frac{3\pi}{20}\right)^3\ln\frac{|k_z l| + 20/3\pi}{|k_z l|} - \left(\frac{3\pi}{20}\right)^2\frac{1}{|k_z l|} + \frac{3\pi}{40}\frac{1}{(k_z l)^2}\right]$$

$$\left.+\frac{\pi}{6}\xi^2(k_z l)^4\left[\left(\frac{8}{5\pi}\right)^3\ln\frac{|k_z l| + 5\pi/8}{|k_z l|} - \left(\frac{8}{5\pi}\right)^2\frac{1}{|k_z l|} + \frac{4}{5\pi}\frac{1}{(k_z l)^2}\right]\right\}. \quad (4)$$

We now consider two limiting cases: long-wave and short-wave, using expression (4). In the long-wave limit ($|k_z l| << 1$) we obtain

$$\operatorname{Im} g_e^{-1}(k_z,\omega) = -\frac{\rho b_e^2}{10\pi}\gamma^0\omega\left[\left(4\xi - \frac{8}{3}\xi^2\right)(k_z l)^2\ln\frac{1}{|k_z l|} + \left(\frac{1}{2} - \xi + \frac{2}{3}\xi^2\right)(k_z l)^2\right]. \quad (5)$$

In the short-wave limit ($1 << |k_z l| << k_m l$) we obtain

$$\operatorname{Im} g_e^{-1}(k_z,\omega) = -\frac{\rho b_e^2}{18\pi}\gamma^0\omega\left[\frac{4}{\pi} + \frac{40}{\pi}\xi - \frac{5\pi}{2}\xi^2\right]|k_z l|. \quad (6)$$

It is known that the imaginary part of the generalized susceptibility is proportional to the average energy dissipated per unit time in the system [13]. Consequently, the imaginary part of the inverse generalized susceptibility of a dislocation, found in the present section, determines an additional dissipation of energy due to bending dislocation vibrations.

3. Damping of short-wave bending dislocation vibrations

In the short-wave case ($1 << |k_z l| << k_m l$), the sound waves damping coefficients (2) can be approximately written in the form $\gamma_t(x) \approx (4/3\pi)\gamma^0 x$ and $\gamma_l(x) \approx (\pi/6)\gamma^0 x$. The expression for the imaginary part of the edge dislocation inverse generalized susceptibility in a dissipative crystal [6], taking into account the major terms with respect to k terms, we write:

$$\operatorname{Im} g_e^{-1}(k_z, \omega) = -\frac{\mu b_e^2}{2\pi}\omega \int_{|k_z|}^{k_m} kdk \frac{\omega^2 \gamma_l^2}{(c_l^2 k^2 - \omega^2)^2 + 4\omega^2 \gamma_l^2}\left(\frac{2c_t^2}{c_l^2 \gamma_l} - \frac{4c_t^2 \gamma_l}{c_l^4 k^2} - \frac{2c_t^2 k_z^2}{\omega^2 \gamma_l}\right.$$

$$\left. - \frac{4\gamma_t^2}{c_t^2 k^2 \gamma_l} - \frac{2\gamma_t}{\gamma_l^2} + \frac{4k_z^2 c_l^2 \gamma_t}{\omega^2 \gamma_l^2} + \frac{8\gamma_t}{c_l^2 k^2}\right). \tag{7}$$

Here taking into account only the terms corresponding to the dislocation bending vibrations, μ is shear modulus of the crystal. Using the dimensionless variable $x = kl$ in (7) and neglecting the terms with $\gamma^0 l / c_{t,l} << 1$, we obtain

$$\operatorname{Im} g_e^{-1}(k_z, \omega) = -\rho b_e^2 \omega \gamma^0 \xi \left[\left(\frac{8}{3\pi^2} - \frac{1}{6}\xi\right)(k_z l)^2 - \left(\frac{4}{3\pi^2} - \frac{1}{6}\xi\right)\frac{\omega^2}{\omega_l^2}\right] \int_{|k_z l|}^{k_m l} \frac{x^2 dx}{\left(x^2 - \frac{\omega^2}{\omega_l^2}\right)^2 + \left(\frac{\pi}{3}\frac{\omega \gamma^0}{\omega_l^2} x\right)^2},$$

where $\omega_l = c_l / l$. After integrating, we obtain, taking into account the main summands

$$\operatorname{Im} g_e^{-1}(k_z, \omega) = -\frac{3}{\pi}\xi \rho b_e^2 \omega_l^2 \left[\left(\frac{8}{3\pi^2} - \frac{1}{6}\xi\right)(k_z l)^2 - \left(\frac{4}{3\pi^2} - \frac{1}{6}\xi\right)\frac{\omega^2}{\omega_l^2}\right] \arctan \frac{\pi \omega \gamma^0}{3\omega_l^2 |k_z l|}. \tag{8}$$

At low frequency values, the expression (8) is written as follows

$$\operatorname{Im} g_e^{-1}(k_z, \omega) = -\left(\frac{8}{3\pi^2} - \frac{1}{6}\xi\right)\xi \rho b_e^2 \gamma^0 \omega |k_z l| . \tag{9}$$

The result (9) practically coincides with the result (6) obtained in section 2.

4. Damping of long-wave bending dislocation vibrations

In the long-wave case ($|k_z l| << 1$), we assume that $\omega/\omega_t << |k_z l|$, where $\omega_t = c_t / l \sim 10^{10}\, s^{-1}$. Then the expression for the imaginary part of the edge dislocation inverse generalized susceptibility in a dissipative crystal [6] takes the form

$$\operatorname{Im} g_e^{-1}(k_z, \omega) = -\frac{\mu b_e^2}{2\pi}\omega \int_{|k_z|}^{k_m} kdk \left[\left(\frac{1}{\omega_t^4 (kl)^4} + \frac{2\omega^2}{\omega_t^6 (kl)^6}\right)\left(\frac{c_t^2 k_z^4 \gamma_t}{k^2} + 4\frac{k_z^4 \omega^2 \gamma_t^3}{c_t^2 k^6} - 2\frac{k_z^4 \omega^2 \gamma_t}{k^4}\right)\right.$$

$$+ \left(\frac{1}{\omega_l^4 (kl)^4} + \frac{2\omega^2}{\omega_l^6 (kl)^6}\right)\left(2\frac{c_t^2 \omega^2 \gamma_l}{c_l^2} - 4\frac{c_t^2 \omega^2 \gamma_l^3}{c_l^4 k^2} - 2c_t^2 k_z^2 \gamma_l + \frac{k_z^4 c_t^2 \gamma_l}{k^2} + 8\frac{k_z^2 \omega^2 \gamma_t^2 \gamma_l}{c_t^2 k^4} - 4\frac{\omega^2 \gamma_t^2 \gamma_l}{c_l^2 k^2}$$

$$-4\frac{k_z^4\omega^2\gamma_t^2\gamma_l}{c_t^2k^6}-2\omega^2\gamma_t+4k_z^2c_l^2\gamma_t-4\frac{k_z^2\omega^2\gamma_t}{k^2}+8\frac{\omega^2\gamma_t\gamma_l^2}{c_l^2k^2}-2\frac{k_z^4c_l^2\gamma_t}{k^2}+2\frac{k_z^4\omega^2\gamma_t}{k^4}\Bigg)\Bigg]. \quad (10)$$

Substituting expressions (3) into formula (10) and integrating under the condition $|k_z l| << 1$, we obtain, taking into account the major terms:

$$\operatorname{Im} g_e^{-1}(k_z,\omega) = -\frac{\rho b_e^2}{10\pi}\gamma^0\omega\left[\left(4\xi-\frac{8}{3}\xi^2\right)(k_z l)^2\ln\frac{1}{|k_z l|}+\left(\frac{1}{2}-\xi+\frac{2}{3}\xi^2\right)(k_z l)^2\right.$$

$$\left.-\left(2\xi-\frac{8}{3}\xi^2\right)\frac{\omega^2}{\omega_l^2}\ln\frac{1}{|k_z l|}+\left(\frac{3}{2}\xi-2\xi^2\right)\frac{\omega^2}{\omega_l^2}-\left(\frac{\pi}{2}\xi-\frac{64}{9\pi}\xi^2\right)\frac{\omega^2}{\omega_l^2}|k_z l|\right]. \quad (11)$$

The result (11) in the low-frequency limit $\omega \to 0$ coincides with the result (5) obtained in section 2.

5. Conclusion

The results obtained in this paper are most applicable to metals in the low temperature range. Numerical estimates show that the contribution of edge dislocation bending vibrations drag reaches 10% in the total coefficient of an edge dislocation drag. The investigations carried out in this work can be used to study the influence of external effects on the crystal plasticity.

References

[1] Eshelby J D 1949 *Proc. Roy. Soc. London A.* **197** 396
[2] Kaganov M I, Kravchenko V Ya and Natsik V D 1974 *Sov. Phys. Usp.* **16** 878
[3] Alshits V I and Indenbom V L 1986 *Dislocations in Solids* vol 7, ed F R N Nabarro (Amsterdam: Elsevier Science Publishers) p 43
[4] Kosevich A M and Natsik V D 1966 *Sov. Phys. Solid State* **8** 993
[5] Bataronov I L, Dezhin V V and Roshchupkin A M 1993 *Bulletin of the Russian Academy of Sciences: Physics* **57** 1947
[6] Roshchupkin A M, Bataronov I L and Dezhin V V 1995 *Bulletin of the Russian Academy of Sciences: Physics* **59** 1648
[7] Bataronov I L and Dezhin V V 2006 *Vestnik Voronezhskogo gosudarstvennogo tekhnicheskogo universiteta* **2** (8) 15 (In Russian)
[8] Dezhin V V 2016 *Vestnik Tambovskogo Universiteta. Ser.: Yestestv. i Tekhnich. Nauki* **21** 956 (In Russian)
[9] Dezhin V V 2016 *Vestnik Tambovskogo Universiteta. Ser.: Yestestv. i Tekhnich. Nauki* **21** 959 (In Russian)
[10] Dezhin V V 2016 *Aktual'nyye problemy prikladnoy matematiki, informatiki i mekhaniki*: Proc. Int. Conf. (Voronezh: Voronezh State University) p 345 (In Russian)
[11] Dezhin V V 2017 *Informatika: problemy, metodologiya, tekhnologii*: Mater. XVII Int. Conf. vol 2 (Voronezh: Voronezh State University) p 154 (In Russian)
[12] Kittel C 1963 *Quantum Theory of Solids* (New York: Wiley)
[13] Landau L D and Lifshitz E M 1980 *Statistical Physics* Part 1 (*Course of Theoretical Physics* vol 5) (Oxford: Butterworth-Heinemann)

IOP Conf. Series: Journal of Physics: Conf. Series **936** (2017) 012063 doi:10.1088/1742-6596/936/1/012063

Data re-arranging techniques leading to proper variable selections in high energy physics

Václav Kůs, Petr Bouř

Department of Mathematics, Faculty of Nuclear Sciences and Physical Engineering, Czech Technical University in Prague, Trojanova 13, 12000 Praha 2, Czech Republic

E-mail: vaclav.kus@fjfi.cvut.cz, petr.bour@fjfi.cvut.cz

Abstract. We introduce a new data based approach to homogeneity testing and variable selection carried out in high energy physics experiments, where one of the basic tasks is to test the homogeneity of weighted samples, mainly the Monte Carlo simulations (weighted) and real data measurements (unweighted). This technique is called 'data re-arranging' and it enables variable selection performed by means of the classical statistical homogeneity tests such as Kolmogorov-Smirnov, Anderson-Darling, or Pearson's chi-square divergence test. P-values of our variants of homogeneity tests are investigated and the empirical verification through 46 dimensional high energy particle physics data sets is accomplished under newly proposed (equiprobable) quantile binning. Particularly, the procedure of homogeneity testing is applied to re-arranged Monte Carlo samples and real DATA sets measured at the particle accelerator Tevatron in Fermilab at DØ experiment originating from top-antitop quark pair production in two decay channels (electron, muon) with 2, 3, or 4+ jets detected. Finally, the variable selections in the electron and muon channels induced by the re-arranging procedure for homogeneity testing are provided for Tevatron top-antitop quark data sets.

1. Particle physics tasks and data structure

In this paper, we deal with the top-antitop quark pair production after annihilated collisions of accelerated proton and antiproton beams at Tevatron synchrotron within DØ experiment in Fermilab [1] at 2TeV centered energy. The data set (DATA) measured is provided with the Monte Carlo simulation set (MC). The top-antitop quark decay process, together with the levels of contamination of the relevant signal S with respect to different background decays B, are shown in Figure 1. The dimensions of MC and DATA are both limited to 46 selected physical variables of various ranges and supports listed in Table 1.

Table 1. Short names of variables in DØ top quark decay channel.

1	Apla	9	M0nl	17	Centr	25	LepEta	33	DRJetJet2	41	Jteta3
2	Spher	10	M1nl	18	DRminejet	26	Jetm	34	DRJetJet3	42	Jteta4
3	HTL	11	MT0nl	19	DiJetDrmin	27	Metper	35	Jtpt1	43	JtMVA1
4	JetMt	12	Met	20	Ht	28	Metpar	36	Jtpt2	44	JtMVA2
5	HT3	13	Mtt	21	Ht20	29	DphiJetJet1	37	Jtpt3	45	JtMVA3
6	MEvent	14	Mva_max	22	Ktminp	30	DphiJetJet2	38	Jtpt4	46	JtMVA4
7	MT1NL	15	Wmt	23	Lepdphimet	31	DphiJetJet3	39	Jteta1		
8	M01mall	16	Wpt	24	LeppT	32	DRJetJet1	40	Jteta2		

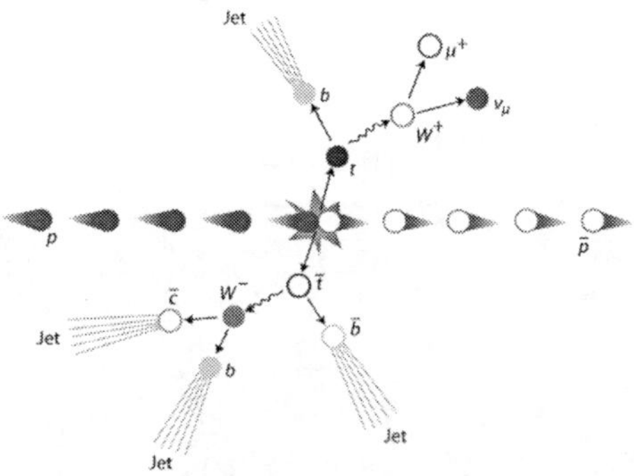

#jets	Electron channel	Muon channel
2	2011176	2001522
3	719484	668072
4+	295932	277214
#jets	S/B ratio (%)	S/B ratio (%)
2	1.13%	0.88%
3	12.24%	10.98%
4+	39.09%	39.34%

Figure 1. Number of generated samples in Top-antitop quark pair production from $p\bar{p}$ annihilation at Tevatron and signal (S) to background (B) ratio in MC simulations.

Prior to any supervised physical task based on MC samples [2], it is vital to guarantee the homogeneity of MC and DATA. However, in most cases, the MC samples are re-weighted according to the actual detection ability of the accelerator. In TMVA ROOT package, the homogeneity testing procedures are simply based on weighted histograms found for equidistant binning in each single dimension (variable). We propose the homogeneity testing approach based on equiprobable quantile binning and re-arranged MC sample in Sections 2 and 3, which is applicable to arbitrary weighted observations. Then we apply the tests to the DØ top quark production at Tevatron. This data set of 46 variables represents a bunch of wide-ranging and significantly complex array of observations, which are sometimes positive, negative, bounded, discrete or continuous, etc. When this homogeneity is not possible to accept, the variable selection is consequently induced, i.e., we are forced to considerable dimensionality reduction in all six top quark decay channels, as it is treated in Section 4.

2. Tests of homogeneity for weighted samples

Let X_1 denote the random variable representing the selected MC physical variable distributed by F and X_2 be the corresponding DATA associated variable distributed by G. Then the homogeneity testing means the statistical test of hypothesis $H_0 : F = G$ at a given significance level $\alpha \in (0, 1)$. Let $\boldsymbol{X}_1 = (X_1^{(1)}, \ldots, X_{n_1}^{(1)})$ denote repeated real valued random variables identically and independently distributed (i.i.d.) from a cumulative distribution function (cdf) F and let $(w_1^{(1)}, \ldots, w_{n_1}^{(1)})$ be their corresponding nonnegative weights. Then we define *weighted empirical distribution function* (wedf) of $\boldsymbol{X}_1$ by

$$F_{n_1}^{W_1}(x) = \frac{1}{W_1} \sum_{i=1}^{n_1} w_i^{(1)} I_{(-\infty,x]}(X_i^{(1)}), \qquad \text{where } W_1 = \sum_{i=1}^{n_1} w_i^{(1)}, \tag{1}$$

and $I_A(\cdot)$ denotes the characteristic function (indicator) of the set A. Assume further that $\boldsymbol{X}_2 = (X_1^{(2)}, \ldots, X_{n_2}^{(2)})$ represents repeated i.i.d. variables under cdf G with the corresponding nonnegative weights $(w_1^{(2)}, \ldots, w_{n_2}^{(2)})$. Then the same definition (1) gives us the wedf $G_{n_2}^{W_2}$ with $W_2 = \sum_{j=1}^{n_2} w_j^{(2)}$. In our HEP representation, $\boldsymbol{X}_2$ corresponds to DATA and so all the weights $w_j^{(2)} = 1$, $W_2 = n_2$, and $G_{n_2}^{W_2} = G_{n_2}$ is the usual edf.

In this paper, we use classical two sample Kolmogorov-Smirnov (K-S) test [3] and two sample Anderson-Darling (A-D) test [4, 5], both applied to wedf's. Further, the χ^2 divergence test of homogeneity [6] is applied to weighted histograms, which does not employ the concept of wedf's. First of all, it is necessary to set up the number of bins k and the subsequent binning $t_0 < \ldots < t_k$. The most frequently used binning is the equidistant one made within a chosen interval $[a, b]$

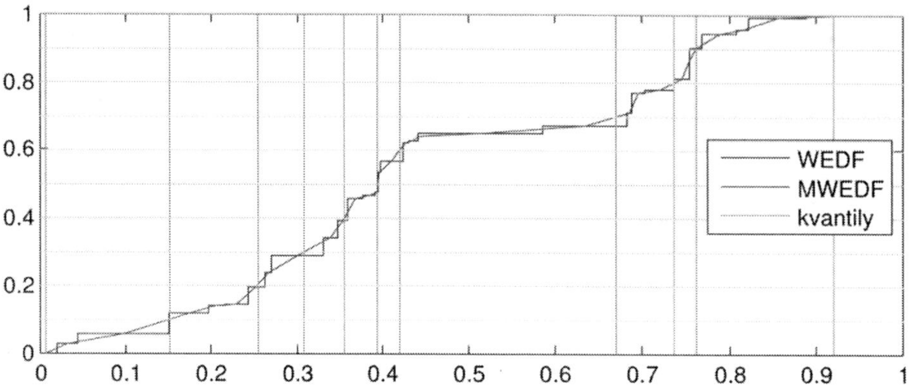

Figure 2. Quantile binning technique for weighted pooled sample $\{\boldsymbol{X}_1, \boldsymbol{X}_2\}$ through mwedf.

spread out over the data sample. However, this approach can be generally insufficient due to poor diversification of data into the equidistant bins, thus the local distribution of data can be lost. Therefore, for a given fixed k, we propose using of the so called equiprobable *quantile binning*, $t_0 < \ldots < t_k$, developed in [7], which assigns approximately the same portion of (weighted) observations to every cell $(t_{j-1}, t_j]$. This method of binning initially consists in finding out the wedf of pooled sample $\{\boldsymbol{X}_1, \boldsymbol{X}_2\}$ with $W = W_1 + W_2$ weighted observations, then creating a modified continuous version of wedf (say mwedf) by linear interpolation of all neighbouring middle points of the constant segments of wedf. Consequently, equi-balanced (j/k)-th quantiles t_j of mwedf are evaluated for all $j = 1, \ldots, k-1$, as illustrated in Figure 2. Thus the quantile binning guarantees the uniformly distributed information contained in weighted data sample since approximately the same sum of weights is concentrated in every cell.

After we have defined the equiprobable pattern of quantile binning $t_0 < \ldots < t_k$ for a fixed k, we make a judicious choice of k. Because of the large number of observations in DATA and large sums of weights in MC samples, the χ^2 divergence test could potentially loose its power with excessively increasing number of bins k, as was shown in [8]. Thus, we decided to choose the following wise choice of the number of bins $k = \lceil 1 + \log_2 W \rceil$ proposed and tested in [9].

Table 2. Sample sizes and sums of weights W_1, W_2 in MC and DATA measurements in top quark decay channel at Tevatron, $W = W_1 + W_2$, and number of bins $k = \lceil 1 + \log_2 W \rceil$.

Top quark decay		MC samples $\boldsymbol{X}_1$		DATA samples $\boldsymbol{X}_2$		Weights	#Bins
	#jets	n_1	W_1	n_2	W_2	W	k
	2	2011176	59118.87	59121	59121	118239.87	17
Electron	3	719484	11904.55	11905	11905	23809.55	15
	4+	295932	3006.98	3007	3007	6013.98	13
	2	2001522	44736.64	44736	44736	89472.64	17
Muon	3	668072	9098.06	9098	9098	18196.06	15
	4+	277214	2325.02	2325	2325	4650.02	13

Now, we apply the three homogeneity tests K-S, A-D, and χ^2 to the set of all 46 variables from top quark pair decay production at Tevatron introduced in Section 1. The factual sample sizes with corresponding sums of weights W and number of bins k are given in Table 2. Figure 3 provides us with the comparison of weighted χ^2 divergence test of homogeneity with K-S and A-D tests based on wedf's. Notice that divergence tests produce generally slightly higher p-values compared with K-S and A-D tests.

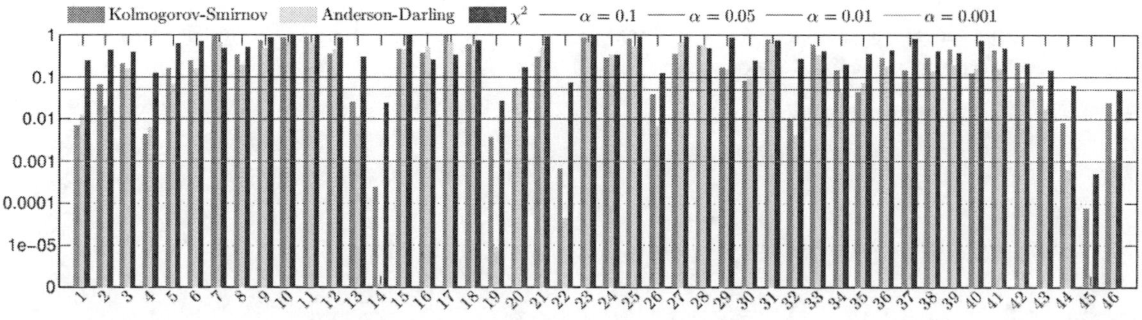

Figure 3. P-values of weighted homogeneity tests K-S, A-D, and χ^2, of MC and DATA distributions for all $m = 46$ variables in Electron 4+ Jets channel. (log-scale)

3. Re-arranging technique for DØ top quark data

The regular asymptotic properties derived for classical unweighted K-S, A-D, and χ^2 tests motivate us to plug into the testing an unweighted data set instead of weighted MC samples. Therefore, we propose a certain transformation of the weighted MC data file into an aggregated unweighted $MC_\dagger$ data array. We make two requirements for such a re-arrangement. First, we desire preserving or even exploiting the information contained in MC weighting, since the weights linked to certain observations refer the distribution layout in close neighbourhood of these observations. Secondly, we require the sum of weights in MC to correspond to the number of observations in the unweighted re-arranged $MC_\dagger$.

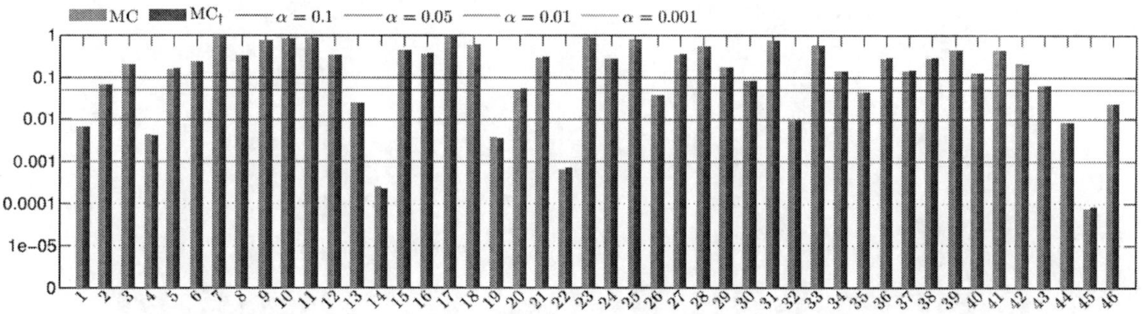

Figure 4. P-values of K-S tests for all $m = 46$ variables in MC and $MC_\dagger$ data sets of Electron 4+ Jets channel. (log-scale)

For simplicity, let us denote by $\boldsymbol{X} = \big(X_{(1)}, \dots, X_{(n)}\big)$ the ordered MC sample with the corresponding ordered weights $(w_{(1)}, \dots, w_{(n)})$ and let $W = \sum_{i=1}^{n} w_i$. Let $\widetilde{n} = \lfloor W \rfloor + 1$ denote the desired minimal number of observations in the new transformed file $MC_\dagger$, where $\lfloor W \rfloor$ denotes the biggest integer smaller than W. Afterwards, we construct the special weighted averages $\boldsymbol{Y}$ from $\boldsymbol{X}$, presuming that $0 \leq w_{(i)} \leq 1$ for all $i \in 1, \dots, n$. To define the first weighted average $Y_{(1)}$, we use the smallest possible number of observations $\big(X_{(1)}, \dots, X_{(k_1)}\big)$ such that $1 \leq \sum_{i=1}^{k_1} w_{(i)} < 2$. Thereby, for all $l < k_1$ it holds that $\sum_{i=1}^{l} w_{(i)} < 1$. The residual portion of weight $w_{(k_1)}$ of the observation $X_{(k_1)}$ equal to $\sum_{i=1}^{k_1} w_{(i)} - 1$ is denoted as r_{k_1}. Thereafter the first $MC_\dagger$ observation $Y_{(1)}$ can be defined as the following weighted average

$$Y_{(1)} = \frac{\sum_{i=1}^{k_1} w_{(i)} X_{(i)} - r_{k_1} X_{(k_1)}}{\sum_{i=1}^{k_1} w_{(i)} - r_{k_1}} = \sum_{i=1}^{k_1 - 1} w_{(i)} X_{(i)} + (w_{(k_1)} - r_{k_1}) X_{(k_1)}, \tag{2}$$

where we simply substituted for r_{k_1} in (2). The residual portion r_{k_1} of $X_{(k_1)}$ will be included into the next weighted average $Y_{(2)}$. If this re-arranging procedure is applied recursively to the successive observations in $\boldsymbol{X}$, we obtain for all $j = 1, \ldots, \widetilde{n} - 1$,

$$r_{k_j} = \sum_{i=k_{j-1}+1}^{k_j} w_{(i)} - r_{k_j-1} - 1 \qquad (r_{k_0} \triangleq 0), \tag{3}$$

$$Y_{(j)} = r_{k_{j-1}} X_{(k_{j-1})} + \sum_{i=k_{j-1}+1}^{k_j-1} w_{(i)} X_{(i)} + (w_{(k_j)} - r_{k_j}) X_{(k_j)}, \tag{4}$$

$$Y_{(\widetilde{n})} = r_{k_{\widetilde{n}}} X_{(n)}, \qquad \text{where } r_{k_{\widetilde{n}}} = \begin{cases} 0, & \text{if } w_{(n)} - r_{k_{\widetilde{n}}-1} < 1/2; \\ w_{(n)} - r_{k_{\widetilde{n}}-1}, & \text{otherwise.} \end{cases} \tag{5}$$

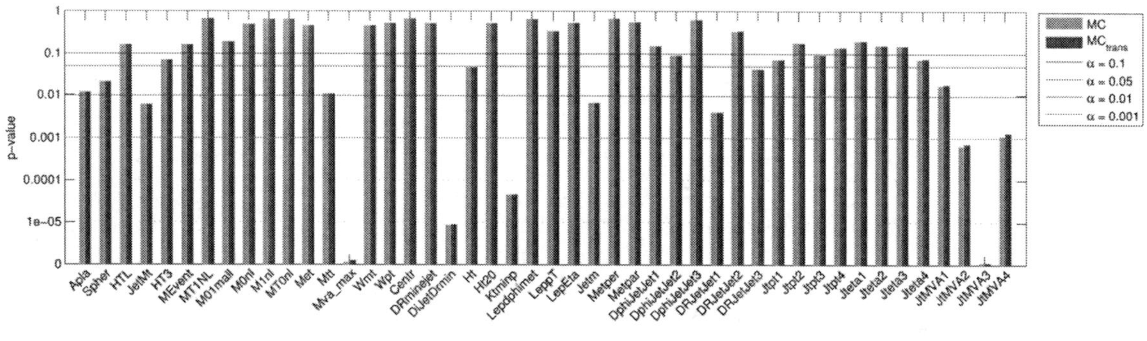

Figure 5. P-values of A-D tests for all $m = 46$ variables in MC and MC$_\dagger$ data sets of Electron 4+ Jets channel. (log-scale)

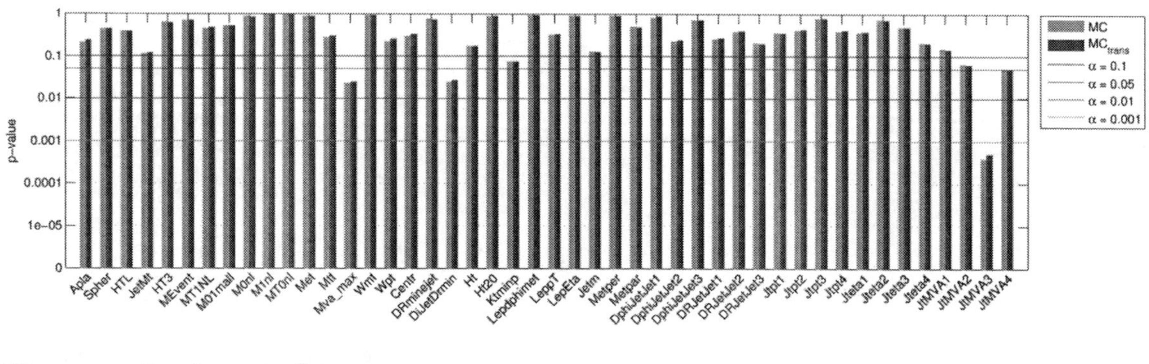

Figure 6. P-values of χ^2 homogeneity tests for all $m = 46$ variables in MC and MC$_\dagger$ data sets of Electron 4+ Jets channel. (log-scale)

Thus we have transformed the original weighted MC array $\boldsymbol{X} = (X_{(1)}, \ldots, X_{(n)})$ into the new unweighted MC$_\dagger$ vector $\boldsymbol{Y} = (Y_{(1)}, \ldots, Y_{(\widetilde{n})})$ by re-distributing the original weights from MC sample to all the unit weights of vector $\boldsymbol{Y}$. Therefore, we are authorized to apply the standard unweighted theoretical asymptotic properties of K-S, A-D, and χ^2. Indeed, the resulting unweighted p-values (red bars) in Figures 4, 5, 6 from all standard K-S, A-D, χ^2 tests, performed over MC$_\dagger$, remarkably matches with p-values (green bars) obtained by the weighted tests performed over original MC sample. This coincidence holds true even for small magnitudes of p-values.

4. Induced variable selection for DØ top quark data

Finally, we applied all the three homogeneity tests to carry out the variable selection of the complete 46 re-arranged MC and pure DATA samples measured in DØ top quark pair decay channel at Tevatron accelerator. In Table 3 we present the list of variables for which the hypotheses H_0 of homogeneity were rejected at the significance level $\alpha = 0.01$ for at least two out of three applied tests of homogeneity. The results differ for our six decay channels. Notice that 16 different variables were rejected in Muon 3 Jets channel, whilst only 5 different variables were rejected in Muon 4+ Jets decay channel. It means that we are forced to reduce considerably the dimension of Muon 3 Jets data sets to only 30 variables accepted by the homogeneity test procedure. Similar variable selection to only 31 reduced dimensions is induced by the homogeneity rejections in Muon 2 Jets channel. Moreover, the variables Jtpt1-3, Ht, Ht20, HT3, HTL were rejected only in Muon 2 Jets and Muon 3 Jets channels while they work quite well in all Electron channels. Also, Jteta3 and Centr variables are not properly generated only in the case of 3 jets channels (Electron or Muon). On the contrary, the variable DRminejet should be used in a consequent HEP analysis only in 4+ Jets channels (Electron or Muon) and the worst result was achieved for Mva_max variable, which is not allowed to enter any HEP analysis since it was rejected in all top quark channels considered in Table 3.

Table 3. Variables rejected ($\star$) by at least two of three considered K-S, A-D, χ^2 homogeneity tests of MC versus DATA at significance level $\alpha = 0.01$

	Jets	3	4	5	8	10	14	16	17	18	19	20	21	22	23	24	25	27	32	33	35	36	37	41	43	44	45	46
	4+						⋆																		⋆	⋆	⋆	⋆
Muon	3	⋆	⋆	⋆			⋆		⋆	⋆		⋆	⋆		⋆				⋆		⋆	⋆	⋆	⋆	⋆		⋆	
	2	⋆	⋆		⋆		⋆			⋆		⋆	⋆	⋆	⋆		⋆	⋆			⋆	⋆			⋆	⋆		
	4+		⋆			⋆	⋆	⋆			⋆			⋆												⋆	⋆	
Elec	3						⋆		⋆	⋆	⋆									⋆				⋆		⋆	⋆	
	2						⋆			⋆						⋆	⋆								⋆	⋆		

Acknowledgments

This work was supported by the grants LG15047 (MYES), LM2015068 (MYES), SGS15/214/OHK4/3T/14 (CTU), and GA16-09848S (GACR).

References

[1] Kvita J 2009 *Measurement of Differential Cross-Sections in the ttbar to Lepton+Jets Channel at sqrt(s) = 1.96 TeV with the D0 Experiment at Fermilab* (Prague: Charles University)
[2] Bouř P, Kůs V and Franc J 2016 Statistical classification techniques in high energy physics (SDDT algorithm) *J. Phys.: Conf. Series* **738** 012034
[3] Smirnov N J 1944 Approximate laws of distribution of random variables from empirical data *Usp. Mat. Nauk* **10** 179–206
[4] Pettitt A N 1976 A two-sample Anderson-Darling rank statistic *Biometrika* **1** 161–168
[5] Engmann S and Cousineau D 2011 Comparing distributions: the two-sample Anderson-Darling test as an alternative to the Kolmogorov-Smirnoff test *Journal of Applied Quantitative Methods* **6** 1–17
[6] Pardo L 2006 *Statistical Inference Based on Divergence Measures* (Chapman & Hall/CRC)
[7] Bouř P 2016 *Statistical nonparametric and divergence methods for data processing in HEP* (Disertation thesis, FNSPE CTU in Prague (in progress))
[8] Mann H B and Wald A 1942 On the Choice of the Number of Class Intervals in the Application of the Chi Square Test *Annals of Mathematical Statistics* **13** 306–317
[9] Kececioglu D B 1993 *Reliability and Life Testing Handbook, 1st ed* (Prentice Hall)

IOP Conf. Series: Journal of Physics: Conf. Series **936** (2017) 012064 doi:10.1088/1742-6596/936/1/012064

The Use of Modeling Approach for Teaching Exponential Functions

L. F. Nunes[1], D. B. Prates[2], J. M. da Silva[3],
UFVJM-PROFMAT[1], UFVJM-ICET[2], UFVJM-ICET[3] - Teófilo Otoni - Minas Gerais - Brazil

E-mail: linferreira86@gmail.com[1], derekbomfim@hotmail.com[2] and jaqueline.silva@ufvjm.edu.br[3]

Abstract. This work presents a discussion related to the teaching and learning of mathematical contents related to the study of exponential functions in a freshman students group enrolled in the first semester of the Science and Technology Bachelor's (STB of the Federal University of Jequitinhonha and Mucuri Valleys (UFVJM). As a contextualization tool strongly mentioned in the literature, the modelling approach was used as an educational teaching tool to produce contextualization in the teaching-learning process of exponential functions to these students. In this sense, were used some simple models elaborated with the GeoGebra software and, to have a qualitative evaluation of the investigation and the results, was used Didactic Engineering as a methodology research. As a consequence of this detailed research, some interesting details about the teaching and learning process were observed, discussed and described.

1. Introduction

Mathematics gradually developed with the intellectual and practical needs of the society. The growth of this science had, as one of its precursors, the thought of Euclides, which aimed at the logical and axiomatic construction of Geometry, in addition to the useful inventions of Archimedes. With the emergence of new scientific challenges, new mathematical theories have been built to aid in problem-solving theories. Subsequently, these new theories were used to study and represent similar or entirely different problems. As an example, the logarithms that were commonly used only as an aid for procedural calculations, and today, are used to model several biological and physical phenomena, such as the study of the alcohol elimination from the organism, [1, 2, 3].

In [4], Burak and Klüber use psychology to discuss and emphasize the importance of student interest and their interaction along the teaching-learning process. This process, when using modeling approach as a didactical tool, is carried out in a shared knowledge way with students, unlike in the usual way, where the process is held only by the teacher. The modeling approach provides a greater interaction with the students, especially because it works with their interest, which generates an advance in the learning process.

There are several models in the literature that describe the process of alcohol elimination from the body. These models, according to their complexity, can address several specific parameters of the phenomenon, such as the concentration of alcohol in the stomach after ingestion, the rate of absorption of alcohol into the bloodstream, and the rate at which the liver metabolizes the alcohol [5]. For a better understanding of the ingested alcohol elimination model, there is a rather interesting discussion on the subject in [6] and [7]. Also, [8] discusses the metrological control of alcohol meters, instruments for measuring the amount of alcohol in the blood.

To study these models it is important to know and understand the basic logarithmic and exponential functions concepts. In Brazil, the educational secondary institutions are the means by which the learning of these functions by the students traditionally occurs. In this way, this work started with some questions about how the teaching-learning process of these functions happens in the public shools. To analyze the teaching of these functions and to propose didactical interventions that can provide improvements in this process, we used Didactic Engineering (DE) as a research methodology. DE is more detailed in [9] and [10].

In this work, we made an in-depth discussion derived from [11], analysing teaching of exponential and logarithmic functions process. We also discussed the way these functions are addressed in Mathematics textbooks. Thus, we used modeling approach as a tool in the teaching-learning process, through the application of a model of alcohol elimination, which provides a relevant proposal for the freshman students of STB. The reader can find more details about STB in [12], [13] and [14].

An important factor to be highlighted is the fact that only 2% to 10% of the absorbed alcohol is eliminated, the body oxidizes the excess [7]. We adapted the initial hypotheses of the model, according to the proposal, to simplify the model and the students' understanding.

2. The Modelling Approach and the Research Methodology

The research methodology used in this work will be presented in four stages. We will not describe the main characteristics of this methodology, but they are more detailed in [10].

In the first stage, which is called **previous analysis**, we investigated the exponential functions fundamental concepts. We analyzed these functions in the textbooks provided by the Brazilian education system. Then, we checked the deficiencies in the contextualization of this content. Finally, we detected that these gaps negatively impacted students' understanding of the subject. More details about the procedures of this stage can be found detailed in [11].

The second stage is called the ***a priori* analysis**. At this stage, we choose the model 2 as a didactical approach with a computational graph produced with GeoGebra software (see Figure 1). We described the didactical approach as well as the actions we were expecting from the students on each stage of the research. In section 3 we will detail the two main goals that the teacher was trying to achieve.

In this second stage, we analyze the system (3) and observe that it represents the model (2). For the didactical modeling approach used in this research, we considered that the rate of elimination of alcohol ingested at a given time t is proportional to the amount of alcohol ingested at the same time. Considering C the alcohol concentration, measured in g/l (Grams per liter of alcohol) and $\frac{dC}{dt}$ the rate of elimination of ingested alcohol measured in $g/l/min$ (Grams per liter per minute), there is α such as:

$$\frac{dC}{dt} = \alpha \cdot C \qquad (1)$$

Since the amount of alcohol decreases along the time, there is has a problem of decay in the phenomenon of the elimination of the alcohol. Then, considering $\alpha = -k$:

$$\frac{dC}{dt} = -k \cdot C \qquad (2)$$

As the elimination of alcohol occurs exclusively by the lungs, we considered the rate of elimination of alcohol as $k = 0,0075\ g/l/min$ [7]. Considering the concentration as $C = C(t)$ and the initial concentration in t_0 as $C(t_0) = C_0$, we have a Cauchy problem:

$$\begin{cases} \dfrac{dC}{dt} = -kC \\ \\ C(t_0) = C_0 \end{cases} \tag{3}$$

The solution of the system (3) is expressed as an exponential function:

$$C(t) = C_0 e^{-0{,}0075t} \tag{4}$$

Although the domain of the exponential functions is the real numbers set ($\mathbb{R}$), for the application of the modeling approach, we considered only the real positives numbers set ($\mathbb{R}^+$). GeoGebra was used to represent graphically the model (3), as appears in Figure 1.

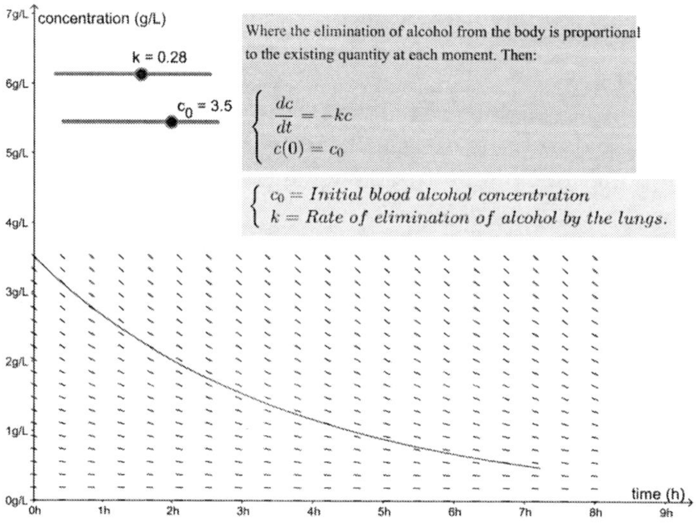

Figure 1. Graphical representation of the model (3) produced by the teacher using GeoGebra.

In the third stage, called **experimentation**, the teacher executed his planning in the classroom and some diagnostic questionnaries were introduced to the students to evaluate their understanding about the subject ando also about the didactical approach. Figure 2 shows the students producing their models. The reader can find all the questionnaires detailed on [11].

Figure 2. Students working in the experimentation stage.

In the last stage, called ***a posteriori* analysis and validation**, the obtained results were compared with the goals that were related to the second stage, the *a priori* analysis stage. Along this stage, we perfomed a stratification of the results obtained with the application of the didactic sequences. The collected information was separated and organizated in groups. It is important that this analysis reaches the reality of the students' production [15].

3. Discussions about the Research Methodology

A careful analisy was made with the data collected from the students answers after the application of some diagnostics questionnaires. Using direct observation and direct observation, two hypotheses were analyzed:

Hypothesis A: The use of computers, specially GeoGebra software as a didactic resource, encourages the possibilities of visualization, calculation, and interpretation.

Hypothesis B: The study of exponential and logarithmic functions, through growth and decay models, improves the understanding of these contents.

Regarding the construction of the slope field (Figure 3) related to the equation (2) in the analysis *a priori* stage, it was expected that the students would connect the visualization of the slope field to the possible solutions of the system (3). We also expected that they could see and understand how the concentration decreasing occurs.

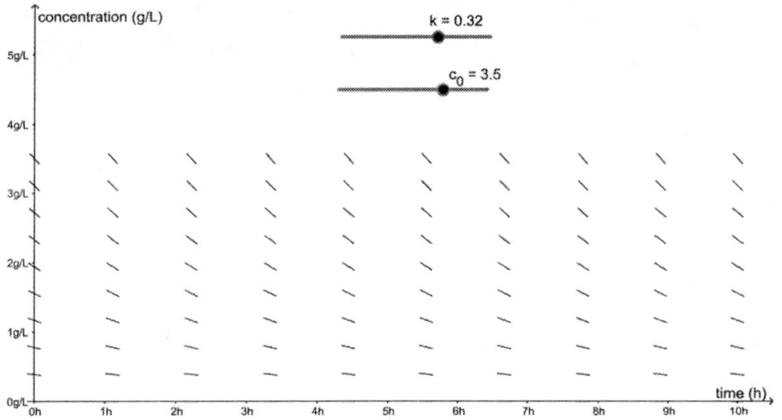

Figure 3. Slope field produced by a student using GeoGebra.

In the model presentation and discussion process, the students were able to visualize the behavior of the concentration of the alcohol C as a function that depends on the time, observing the slope field and already imagining which family of functions would be the solutions of the system. Particularly, one of the students described the phenomena behavior as a decreasing exponential function. So, with this report, it is possible to consider that the hypotheses **A** and **B** were validated.

It was expected from the students in the *a priori* analysis, that after observing the models graphical solutions, they realized that as higher the value of C, the higher the alcohol concentration will be, and the faster this concentration will decay.

It was detected in the *a posteriori* analysis that the students perceived the relationship between the value of the alcohol concentration and the decay speed. It can be seen in Figure 4, a graphical construction produced by a student. In this construction, we can observe that he was varying the value of k. In this way the hypotheses **A** and **B** were validated.

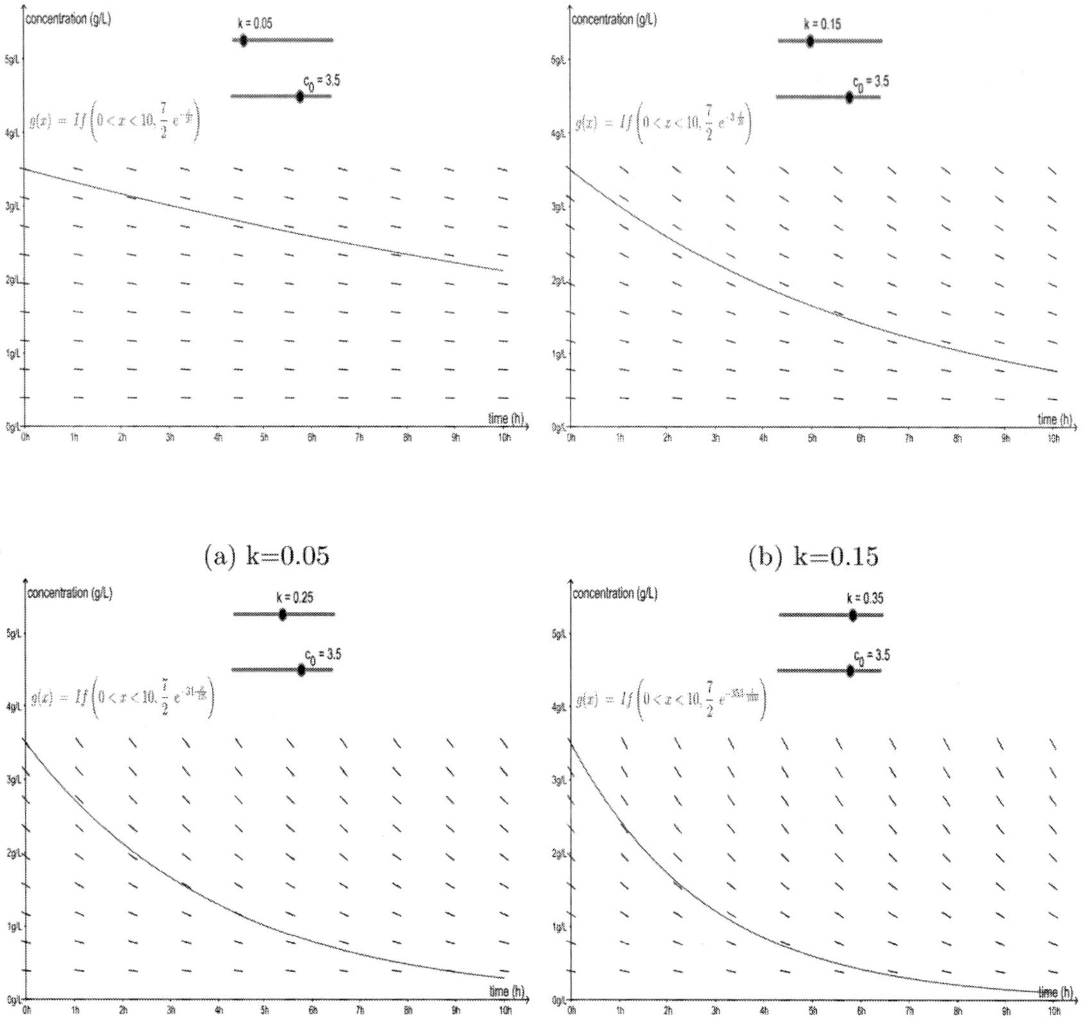

(a) k=0.05

(b) k=0.15

(c) k=0.25

(d) k=0.35

Figure 4. Slope fields fixing $C_0 = 3.5$ and the variation of k

Finally, in the *a priori* analysis, it was expected that the students were able to relate the decay of the body's alcohol concentration to the exponential function with base e. It was also expected that the student would realize the advantage of writing the final solution as a function of the exponential type in the form $f(x) = be^{\alpha x}$, because this expression displays not only the initial value $b = f(0)$, but also the coefficient α, which is closely connected to the growth/decay rate of the function.

In the *a posteriori* analysis was detected that the students realized that the natural-based exponential function appeared as a solution to represent the behavior of the alcohol elimination from the organism. The construction produced by the student is presented in Figure 5. In this way, we consider that the hypothesis **B** was validated.

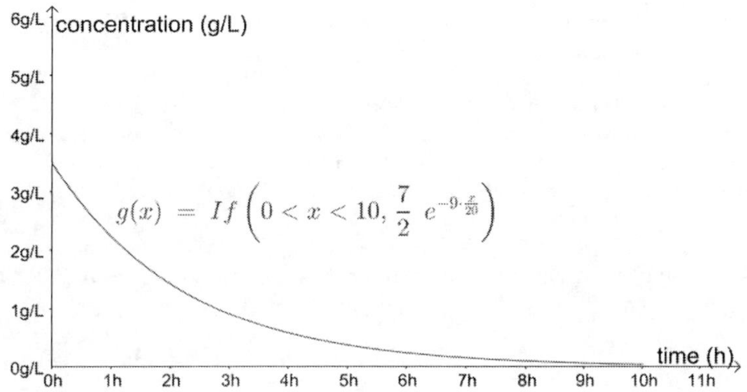

Figure 5. Exponential function solution of the model produced by one of the students

4. Final considerations

As is possible is possible to observe in [11] research, this work opened some possibilities of proposals for didactic interventions that can actually adds to the classroom some experiences that can contribute to the understanding of the mathematical specificities related to these functions by the students.

The modeling approach has proven to be an efficient and effective tool to teach exponential functions, especially when it awakens students' interest in working with models that describe natural phenomena. Besides, it enables students to understand the behavior of functions dynamically; this is possible through the sliding commands that the GeoGebra presents, which makes it possible to comprehend the functions when their parameters vary.

This research can also serve as a suggestion of classroom applications, keeping the specificities of each teaching environment, as well as to start new investigations using modeling approach, GeoGebra software and DE. It is very important to emphasize the relevance of a proper and detailed planning of the activities as well as the organization of the information by the teacher to this research progress.

Acknowledgments

Thanks to Capes for the fellowship to one of the authors that belong to the PROFMAT post graduating program at UFVJM. Also thanks to FAPEMIG for the financial support.

References

[1] Lima E L 2013 *Logaritmos* (Rio de Janeiro: SBM)

[2] Beltro M E P and Igliori S B C 2010 Modelagem matemtica e aplicaes: uma abordagem para o ensino de funes *Educao Matemtica Pesquisa* **12**

[3] Tatsch K J S and Bisognin V 2007 Modelagem matemtica no ensino mdio: alimentao, obesidade e desnutrio *Vidya* **24**

[4] Burak D and Klber T E 2008 Concepes de modelagem matemtica: Contribuies tericas. *Educao Matemtica Pesquisa* **10** 17–34

[5] Ludwin C 2011 Blood alcohol content *Undergraduate Journal of Mathematical Modeling: One+ Two* **3** 1

[6] Sanches A D B and Rodrigues R N 2015 *Proceeding Series of the Brazilian Society of Computational and Applied Mathematics* vol 3 (So Carlos: SBMAC)

[7] Bassanezi R C 2014 *Ensino-aprendizagem com modelagem matemtica: uma nova estratgia.* 4th ed (So Paulo: Contexto)
[8] Ferreira M C, Cruz A and Gião R A 2006 *Proceedings of the 2a Conferência Nacional SPMET* vol 17
[9] Artigue M 1996 *Didctica das Matemticas* ed BRUN J (Lisboa: Instituto Piaget) pp 193–217
[10] Artigue M 1989 Ingnierie didactique *Recherches en Didactique des Mathmatiques* **9** 281–308
[11] Nunes L F 2017 *Modelos de crescimento e decaimento aplicados ao ensino de funções exponenciais e logarítmicas* Master's thesis Universidade Federal dos Vales do Jequitinhonha e Mucuri. PROFMAT
[12] Silva J M, Jardim D F, Carius A C and Silva A C 2017 *Conference Proceedings. New Perspectives in Science Education* (Florence: Libreriauniversitaria.it) p 119
[13] Jardim D F, Silva J M, Pereira M M, Soares E A, Nepomucena T V and Pinheiro T 2015 Estudando limites com o GeoGebra *Revista Vozes dos Vales* **4**
[14] Silva J M, Jardim D F and Carius A C 2016 O ensino e a aprendizagem de conceitos de clculo usando modelos matemticos e ferramentas tecnolgicas. *Revista de Ensino de Engenharia* **35** 70–80
[15] Pais L C 2015 *Didtica da Matemtica: uma anlise da influncia francesa* (Belo Horizonte: Autntica)

IOP Conf. Series: Journal of Physics: Conf. Series **936** (2017) 012065 doi:10.1088/1742-6596/936/1/012065

Hydrodynamic instabilities at decay flow of nematic liquid crystals through the plane capillary of a variable gap

D Shmeliova, S Pasechnik, A Girgidov and G Maksimochkin
Moscow Technological University (MIREA), Moscow, Russia

e-mail: pasechnik@mirea.ru

Abstract. In this paper we consider the possibility to study hydrodynamic instabilities (HDI) in a specific type of a decay flow existing at a rise of a nematic liquid crystal (NLC) in the flat capillary of a variable gap. In this case the capillary can be replaced by a number of parallel channels of different gaps with different instant values of a pressure gradient, which induces a flow. The last feature makes such flow to be different from the previously studied decay flow [1] with a constant value of a pressure gradient, applied to the channels of different thickness.

1. INTRODUCTION

Nematic liquid crystals (NLC) can be considered as anisotropic fluids with the same mechanism of translational molecular motions as in isotropic liquids, which is responsible for high fluidity and zero values of static shear elastic modules. The main difference from isotropic liquids arises due to long range orientational order described in terms of the director n, which corresponds to the average directions of long molecular axes, and the scalar order parameter S referred to the mean declination of molecules from the overall direction n. While order parameter S is useful for a description of phase transition NLC – isotropic liquid, the unite vector n plays the role of an additional hydrodynamic parameter, which has to be incorporated into hydrodynamic equations of nematic liquid crystals. The connection between translation motions, and a director field **n**(r, t) can be considered as fundamental physical property of liquid crystals. It is responsible for a number of specific phenomena existing in liquid crystal media. In particular, it results in non-Newtonian behavior and a number of specific hydrodynamic instabilities (HDI), arising in shear flows of nematic liquid crystals [2]. The realization of the given instability drastically depends on the initial hydrodynamic state, characterized by a space distribution of a velocity and orientation. For example, in the case of a simple shear flow of NLC layer of thickness h, confined between two plates, the initial homogeneous orientation, normal to the flow plane, becomes unstable at some critical value sc of the shear rate $s = v_o /h$ (v_o – the velocity of a relative motion of the plates). The similar instability arises also in a Poiusellie flow at a critical value G_c of the pressure gradient $G=dP/dx$, which induces the flow of NLC confined between two motionless plates. Such instabilities was investigated in details both experimentally and theoretically [2]. At the same time, the initial orientation of NLC normal to the plates can be realized only in the absence of flows. So, the specific orientation instability (the escape of a director from the flow plane), arises in a Poiusellie flow at the initially non homogeneous orientation fields of the velocity and orientation [2,3]. It is important, that for both cases mentioned above, the critical values of a shear rate and a pressure gradient essentially depend on the thickness of LC layer. It provided effective usage of a wedge- like LC cell for visualization and study of the mentioned above escape of a director from the flow plane. In this case, the cell was considered as a number of capillaries of different thickness with the same instant value of a pressure gradient, slowly decreasing with time (decay flow).

The goal of this paper is to consider the alternative type of a decay flow, produced by a rise of NLC in vertically oriented wedge-like cell (see figure 1). The theoretical estimates for different initial geometries make possible to predict the dynamic of a capillary rise and time variations of local orientation of NLC layer, which can be used for explanation of the new experimental results, obtained at study of a capillary rise of NLC. .

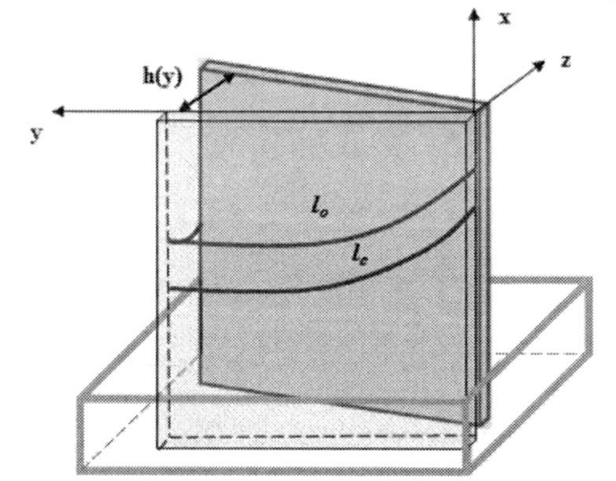

Figure 1. Experimental geometry.

2. Decay flow and hydrodynamic instabilities in a wedge-shaped capillary

At filling of a wedge-shaped cell such capillary with a liquid crystal one can observe a specific type of the Poiseuille flow with a permanently decreasing velocity. It allows us to make some analogy with the previously studied decaying Poiseuille flow in a wedge-shaped cell. At the same time, there are significant differences between these two cases. In particular, in the previously investigated case, a decay character of the flow resulted due to a decrease with time of the hydrostatic pressure drop applied to a wedge-shaped capillary of constant length previously filled with a liquid crystal. This geometry is equivalent to a set of variable-gap capillaries connected in a parallel scheme with the same pressure gradient applied to different capillaries. In our case the flow arises as a result of the motion of the LC on a previously dry surface, while the capillary pressure causing the flow depended on the local gap of the capillary. It led, in particular, to the dependence of the instant height of the meniscus on the coordinate y. In this connection, a corresponding modification of the previously considered theoretical models of linear and nonlinear phenomena in the decay Poiseuille flow of liquid crystals is necessary.

In particular, the question of the possibility of using this type of flow to study the hydrodynamic instabilities of the orientational structure is of interest. These instabilities manifest themselves as a sharp change in the orientation structure at a certain threshold value of the pressure gradient $(\Delta P/\Delta x)$ c or the average value of the velocity gradient (shear rate) sc. In particular, the instability of the simplest type with the initial orientation, normal to the flow plane, is realized when the dimensionless Ericksen number

$$Er = \frac{\Delta P/\Delta x}{\eta_3}\sqrt{\frac{\alpha_2\alpha_3}{K_{11}K_{22}}}\cdot\left(\frac{h}{2}\right)^3 = \overline{s}^{av}\sqrt{\frac{\alpha_2\alpha_3}{K_{11}K_{22}}}\left(\frac{h}{2}\right)^2. \tag{1}$$

increases up to the critical value $Er_c = 17.3$. In expression (1) α_2 and α_3 – the Leslie's coefficients, K_{11} и K_{22} – the Frank's modules of orientation elasticity, η_3 – the shear viscosity coefficient (Miesovich viscosity), corresponding to the initial orientation, s^{av} – the average value of the velocity gradient.

From eq.(1) it follows a strong dependence (h^{-2}) of the threshold value s^{av} c on the local thickness h. It is worthwhile to notice, that the same kind of functional dependence is valid for the mentioned above hydrodynamic instability connected with the escape of the director from the flow plane, which is realized at the initially normal orientation respectively to LC layer. This fact allows us to consider, at a qualitative level, the appearance of both types of instabilities for the wedge-shaped capillary in a similar way.

Taking into account equations (1), the generalized equation that determines the critical value s^{av}_c can be written in the form:

$$s_c^{av}\left(h/2\right)^2 B = M \tag{2}$$

where M is the critical value of the Eriksen number corresponding to the given geometry, B is the combination of the material parameters of the liquid crystal.

In particular, for the simplest geometry, the expressions for B and M are easily obtained by comparing formulas (1) and (2):

$$B = \left[(\alpha_2\alpha_3)/(K_{11}K_{22})\right]^{1/2}, M = Er_c = 17{,}5 \tag{3}$$

The instant value of s can be calculated from the average linear velocity (v^{av}) of motion of the meniscus in the capillary.

To make such calculations we will consider a liquid crystal as a conventionally Newtonian liquid (Stewart) with the effective shear viscosity, independent on the velocity gradient. Such approximation holds for nematic liquid crystals with orientation, stabilized by strong magnetic (electric) fields or surfaces [2]. So, some results concerning to a rise of isotropic Newtonian liquids in the vertical capillaries of different cross sections and sizes [4] can be applied for a capillary flow of liquid crystals.

In particular, the well-known Poiseullie formula, describing the volumetric rate of a flow $Q = dV/dt$ induced by a pressure difference ΔP in a capillary of a constant gap with a length l is valid. For a plane capillary with a large value of the aspect ratio R= A/h (A and h – the width and the gap of a capillary) the Poiseullie equation is written as:

$$\frac{dV}{dt} = \frac{h^3 A}{12\eta}\sum\frac{\Delta P}{l} \tag{4}$$

where $\sum\Delta P$ – the resulting pressure difference, which includes a hydrostatic pressure and a capillary pressure:

$$P_g = \rho g l \tag{5}$$

$$P_\sigma = \frac{2\sigma\cos\theta}{h} \tag{6}$$

In these equations a density ρ and a shear viscosity η are the parameters describing bulk properties of a liquid, whereas a surface tension coefficient σ and a contact angle θ are the parameters which characterize the interaction of a liquid with an air and a solid surfaces.

Taking into account the trivial expression

$$dV = hAdl \tag{7}$$

one can get from(4-6) the next differential equation, describing a motion of a contact line:

$$\frac{dl}{dt} = \frac{\sigma h \cos\theta}{6\eta l} - \frac{h^2 \rho g}{12\eta} \tag{8}$$

This equation can be considered as a simplified (and modified) version of well-known Washburn equation [4] proposed to describe the contact line motion in a capillary of a circular cross section.

Taking into account equations (1), (4) and (7), it is simple to obtain the connection between the averaged value of the velocity gradient and the velocity of the contact line motion (*dl/dt*):

$$s^{av} = 3v^{av} / (h / 2) = 6h \;\; 1(dl / dt) \tag{9}$$

where $v^{av} = Q_i / A_i h_i$, index *i* corresponds to the given elementary capillary of width A_i and thickness h_i. Inserting value (*dl/dt*), defined by (8) into (9) results in the next expression:

$$[(\sigma cos\Theta)/(\eta l_c) - (h\rho g)/\mp 2\eta)] \cdot (h/2) \cdot 2B \quad M \tag{10}$$

Where l_c determines the critical value of the local height of the meniscus rise corresponding to the occurrence of the instability, η is the effective viscosity value corresponding to the given geometry. Using (10), we obtain the following equation for l_c as a function of the local thickness of the LC layer:

$$lc(h) = l_0(h)\{1 + (8M \;\; \eta)/(\rho g B h^3)\}^{-1} \tag{11}$$

where the dependence of the maximum lift height on the gap $l_0(h)$ is determined by the well known expression:

$$l_0 = \frac{2\sigma\cos\theta}{\rho g h} \tag{12},$$

obtained from the equilibrium condition:

$$P\sigma = Pg \tag{13}$$

The usage of the linear dependence of the local thickness on the coordinate *y*

$$h = h_0 \left[1 + \left(h/A_0\right) \left(y/A\right)\right] \tag{14}$$

($\Delta h = h_{max} - h_0$, A is the cell size with respect to the y-coordinate), makes possible to calculate the maximum height l_0 of the meniscus' rise and the critical height of the meniscus l_c as a function of the y coordinate. It is interesting to define the time t_c needed for the meniscus to reach a critical height l_c for different values of the local thickness h, defined by a coordinate y. The corresponding equation, derived from expression (9), (11), reads as:

$$t_c = -\tau(h) ln\left\{\left[1 - l_c(h)/l_0(h)\right]\right\} \tag{15}$$

where the characteristic decay time of the flow $\tau\ (h)$, defined as:

$$\tau = \frac{24\sigma\eta\cos\theta}{h^3(\rho g)^2} = \frac{12\eta l_0}{h^2\rho g} \tag{16}$$

enters into the solution of equation (8):

$$y = \frac{l_0 - l}{l_0}\exp\left(\frac{l}{l_0}\right) = \exp\left(\frac{t}{\tau}\right) \tag{17},$$

which was obtained taking the time independent value of contact angle θ_e . Numerical calculations were performed for the instability of the simplest type mentioned above. In this case, equation (11) can be written as:

$$l_c = l_0(h_0)\left\{\left[1 + (\Delta h/h_0)(y/L)\right]2\right\} / \left\{\left[1 \quad (\Delta h/h_0)(y/A)\right]3 + \left[(8Er_c\ \eta_3)/(h_0^3 B\rho g)\right]\right\} \tag{18}$$

In calculations, we assumed a gap ranging from $h_0 = 60$ µm to $h_{max} = 100$ µm ($\Delta h = 40$ µm) in a cell of width A and used the material parameters 5CB [5]: $\alpha_2 = -0.0812$ Pa·s, $\alpha_3 = -0.0036$ Pa·s, $K_{11} = 6.2\cdot10^{-12}$ N and $K_{22} = 3.9\cdot10^{-12}$ N, $\eta_3 = 0.0326$ Pa·s, $\eta_2 = 0.0204$ Pa·s, $\sigma = 0.0326$ N/m. The value of $cos\theta = 0.820$ was determined from the results of experimental measurements of the value $l_0 = 0.0532$ m in a planar cell with a constant gap of h=100 µm and a director perpendicular to the flow direction. The detailed of these experiments will be published elsewhere.

Table 1. The characteristic times τ and t_c for two values of the effective shear viscosity η

y/l	0	0.25	0.5	0.75	1
h (µm)	60	70	80	90	100
l_0 (cm)	8.86	7.60	6.64	5.90	5.32
lc (cm)	5.51	5.50	5.29	5.02	4.70
τ (s) , $\eta=\eta_2= 0.0204$ (Pa·s)	600	378	253	177	127
τ (s), $\eta=\eta_3= 0.0326$ (Pa·s)	961	605	405	284	207
t_c (s), $\eta=\eta_2= 0.0204$ (Pa·s)	210	212	200	188	162
t_c, (s), $\eta=\eta_3= 0.0326$ (Pa·s)	337	338	319	301	261
τ_n (s)	3.6	4.9	6.4	8.1	10

The characteristic times τ and t_c, presented in Table 1 and were calculated for two values of the effective shear viscosity η. It reflects a rather complicated scenario of the director variations under a decay flow. On the initial stage of a capillary rise the values of a pressure gradient are essentially higher than the critical values. It has to induce fast reorientation of LC from the initial state, characterized by a Miesovich viscosity η_3, towards the flow direction with corresponding viscosity $\eta \approx \eta_2$. After some time t_c the flow decays to the critical value of a pressure gradient and backward relaxation to the initial state takes place. So, on practice, the effective viscosity can be varied in the range $\eta_2 < \eta < \eta_3$ on different stages of a decay flow.

In this table we also present the calculated values of the characteristic time τ_n, describing backward reorientation of LC director in the absence of external fields, defined as [6]:

$$\tau_n = \left(\gamma_1 h^2\right) / \left(\pi^2 K_{22}\right) \tag{19}$$

where $\gamma_1 = \alpha_3 - \alpha_2$ – the rotational viscosity. The values of τ_n are essentially lower than corresponding values of a decay time τ. It means that a director follows the variations of the velocity film without phase shift after decreasing of the pressure gradient to the critical value, defined by equation (1). So, possible optical study of a director reorientation provides registration of the flow induced instability and experimental checking of theoretical estimates made above. The corresponding experiments are in a progress now.

The calculation of the instability threshold associated with the escape of a director from the flow plane for the initial homeotropic orientation is a much more complicated problem, the solution of which was performed for methoxybenzylidene-p-n-butylaniline (MBBA) by numerical methods both in the absence and in the presence of destabilizing electric fields [2]. It should be noted that in this case, the torque acting on the director from the flow is maximal and the shear flow can cause both primary and secondary instabilities at relatively small values of the pressure gradient (of the order of 30 Pa/cm at a thickness of the LC layer of 100 μm). The values of the pressure gradient in our experiments at the initial stage of capillary filling exceed by about an order of magnitude the given value. In this case, the flow can be considered sufficiently strong for the orientation of the entire sample at a small angle θ_{fl} , $tg^2\theta_{fl} = \alpha_3 / \alpha_2$.

For the values of the material parameters 5CB given above, we obtain $\theta_{fl} = 11{,}8^0$. This, in turn, means that the effective value of the shear viscosity coefficient is close of the minimum Mesovich viscosity ($\eta \approx \eta_2$). Table 2 shows the results of calculating the parameters, which determine the dynamics of the meniscus' motion, for $\eta = \eta_2$ and $\eta = \eta_3$. In this case, the value of $cos\theta = 0{,}468$, obtained from the experimentally determined value of l_0, corresponds to the contact angle $\theta = 62.1^0$, which is significantly larger than those for the case of a planar orientation.

It is worthwhile to mention, that instabilities of higher orders are characterized by a very slow backward relaxation (up to some hours) which prevent the optical study of the primary instability [2] The photos (in crossed polarizers) of the initially homeotropic wedge-like cell on the stage of a backward relaxation after filling with 5 CB confirm this conclusion. The situation can be partly improved by additional usage of stabilizing electric fields, which results in suppression of hydrodynamic instabilities and decreasing of the relaxation's time of a director (see Table 2) defined as [2]:

$$\tau_n = \left(\gamma_1 h^2\right) / \left(\varepsilon_0 \, \Delta\varepsilon \, \mathrm{U}^2 - \pi^2 K_{33}\right) \tag{20}$$

where $\Delta\varepsilon \approx 11$ – dielectric permittivity anisotropy, U – applied voltage, $K_{33} = 8.2 \cdot 10^{-12}$ N – Frank's module. The experiments of such kind are in a progress now.

Table 2. The characteristic times τ and t_c for two values of the effective shear viscosity η

y/l	0	0.25	0.5	0.75	1
h (μm)	60	70	80	90	100
l_0 (cm)	5.95	5.10	4.46	3.96	3.57
lc (cm)	3.70	3.69	3.55	3.37	3.15
τ (s) , η=η_2= 0.0204 (Pa·s)	403	254	170	119	87
τ (s), η=η_3= 0.0326 (Pa·s)	645	406	272	191	139
t_c (s), η=η_2= 0.0204 (Pa·s)	141	142	134	126	109
t_c, (s), η=η_3= 0.0326 (Pa·s)	226	227	214	202	175
τ_n (s), U=0 (V)	3.6	4.9	6.4	8.1	10
τ_n (s), U=5 (V)	0.14	0.20	0.26	0.32	0.4

3. CONCLUSION

In this paper the capillary decay flow of a nematic liquid crystal in a wedge-shape capillary was considered and applied to calculate the critical values of the height of a capillary rise and of the time of the rise, which corresponds to the hydrodynamic instabilities of two types. The explicit calculations for the simplest hydrodynamic instability were performed taking into account the characteristic time of a director motion and the decay time. It is shown, that the long living instabilities of higher order can prevent the observation of the primary instability, connected with an escape of a director from the flow plane. In this case the observations can be provided via additional usage of a stabilizing electric field.

Acknowledgements: This work was supported by Ministry of Education and Science of Russian Federation - grant 3.9585.2017/8.9.

References

[1] Tsvetkov V A., Tsvetkov O V., Balandin V A. 1999 *Molecular Crystals and Liquid Crystals* 329 (1) 305-312
[2] Pasechnik S V; Chigrinov V G; Shmeliova D V 2009 *Liquid crystals: viscous and elastic properties in theory and applications* (John Wiley & Sons)
[3] Guyon E and Pieranski P,1975 *J. de physique* N3,T3, c1-203
[4] Heshmati M and Pir M, 2014 *Langmuir* 2014, 30, 14151−14162
[5] Stewart I W 2004 *The Static and Dynamic Continuum Theory of Liquid Crystals* (London: Tayor & Francis)
[6] P.G. de Gennes 1974 *The physics of liquid crystals* (Oxford: Clarendon Press)

Model of personal consumption under conditions of modern economy

D K Rakhmatullina[1,3,6], E R Akhmetshina[1,4] and O A Ignatjeva [2,5]

[1] Department of Economic Methodology and History, Kazan Federal University, Russian Federation, Republic of Tatarstan, 420008, Kazan, Kremlyovskaya str., 18
[2] Department of Economic theory, Kazan Federal University, Russian Federation, Republic of Tatarstan, 420008, Kazan, Kremlyovskaya str., 18
[3] E-mail: na_nulya_@list.ru
[4] E-mail: kiyamova-ksfei09@yandex.ru
[5] E-mail: dip789@mail.ru
[6] Author to whom any correspondence should be addressed.

Abstract. In the conditions of the modern economy, in connection with the development of production, the expansion of the market for goods and services, its differentiation, active use of marketing tools in the sphere of sales, changes occur in the system of values and consumer needs. Motives that drive the consumer are transformed, stimulating it to activity. The article presents a model of personal consumption that takes into account modern trends in consumer behavior. The consumer, making a choice, seeks to maximize the overall utility from consumption, physiological and socio-psychological satisfaction, in accordance with his expectations, preferences and conditions of consumption. The system of his preferences is formed under the influence of factors of a different nature. It is also shown that the structure of consumer spending allows us to characterize and predict its further behavior in the market. Based on the proposed model and analysis of current trends in consumer behavior, conclusions and recommendations have been made that can be used by legislative and executive government bodies, business organizations, research centres and other structures to form a methodological and analytical tool for preparing a forecast model of consumption.

1. Introduction

Consumption in the modern economy is characterized by an expressive individuality of every consumer. It is demonstrated by their goods and services choice [1, 2] which is being formed in accordance with the factors and conditions of consumption and their influences and forcing the model of behavior. At the same time the modern economy provides a wide range of opportunities for consumers to defend their individuality, erasing the absolute limit of resources.

Russian scientists have made the great contribution to the development of the theory of consumption within the neoclassical approach framework and other directions of economic scientific research. In their works consumer demand, the factors influencing on it and the process of consumption are analyzed by the economic-sociological approach (V. V. Radaev [3], Y. D. Krasovsky [4], M. V. Ryzhkova [5], L. N. Safiullin [2], N. Z. Safiullin [2], etc.), by the economic and socio-psychological approach (S. P. Bogachev [6], Yu. a. Zimerman [7], etc.), by the new institutional Economics (G. N. Ismagilova [1], A. E. Shastitko [8], M. Yakushev [9], etc.) methodologies. In their own way they characterize the basic formation criterions and factors of the consumption process based on the changes of the conditions of the modernizing economic system.

Despite the creation of many research works about various theoretical aspects of consumption, many questions have not found a complete solution. For example: the development of the system of needs based on transformation of the consumption process; the features of the formation of the consumer budget in emerging economy of welfare, etc.

2. Theory

Everyone can make an exclusive assessment of any product relevance based on the kind of utility that it brings to him. Depending on the nature of utility from the consumption, it is possible to identify three types of goods:

- the consumption of which brings only the physiological utility for a consumer;
- the consumption of which brings the social and psychological utility for a consumer;
- the consumption of which brings satisfies the physiological and socio-psychological needs of a consumer.

Moreover, the same goods might be differently categorized, that depends on the individual needs and satisfaction of different consumers and their attitude to the goods.

We have found three types of demand: functional, non-functional and extended functional.

Functional consumer demand – demand for goods, based on only the physiological needs of the individual. In this case the goods are solely purchased for the intended purpose.

Non-functional consumer demand – demand for goods (often luxury good) with the purpose of getting psychological satisfaction and minimum satisfying physiological needs. To this type it is possible to identify the ostentatious consumption, the speculative consumption and any irrational consumption (just want), etc.

Advanced functional consumer demand – the demand for mixed goods, which can bring both types of satisfaction: physiological and psychological. In this case the goods are more expensive, fashionable. Often the subjects of advanced functional demand are imports. [10] Moreover, this type of demand associated with the growing needs and requirements of today's consumers and the expanding grocery market that promotes some goods to become very important for life support despite the fact that they were previously unknown and non-used.

Assume that we can represent some function *f* of dependence between the total utility level of the physiological and socio-psychological consumer satisfaction that is obtained from using the good or the set of the goods.

$$U_{total} = f(U_{phys}; U_{socio-psych}) \tag{1}$$

In equation (1) U_{total} *is* the total level of utility from the consumption of goods (in percent, maximum =100% (or fractional =1), is determined in accordance with the sensations and feelings of consumer), U_{phys} is physiological utility (% or share of complete satisfaction of the physiological needs) and $U_{socio\text{-}psych}$ is socio-psychological utility (% or share of complete satisfaction of the socio-psychological needs).

Our research work has shown that the level of satisfaction of consumption and the level of the utility simply determined by consumers and is estimated at 10, 20, 30, 40, 50, 60, 70, 80, 90 or 100% in accordance with the existing needs and the possible for the individuals' minds maximum pleasure and the full satisfaction that they are expecting for and to which they are aspiring.

We suggest introducing the coefficients of the importance of utility showing shares of each types of utility in the whole consumption set so that the overall level of satisfaction is the highest:

- i_{ph} is the ratio of the importance of physiological satisfaction;
- $i_{s\text{-}p}$ is the coefficient of the importance of socio-psychological satisfaction.

The total satisfaction (100%=1) includes the satisfaction of physiological needs accounting its importance and the satisfaction of socio-psychological needs accounting its importance. Accordingly, we can conclude that

$$i_{ph} + i_{s-p} = 1. \tag{2}$$

There are eleven possible types of consumers, the coefficients of importance of utilities of which are respectively equal to:

1) i_{ph} =1, i_{s-p} =0
2) i_{ph} =0,9, i_{s-p} =0,1
3) i_{ph} =0,8, i_{s-p} =0,2
4) i_{ph} =0,7, i_{s-p} =0,3
5) i_{ph} =0,6, i_{s-p} =0,4
6) i_{ph} =0,5, i_{s-p} =0,5
7) i_{ph} =0,4, i_{s-p} =0,6
8) i_{ph} =0,3, i_{s-p} =0,7
9) i_{ph} =0,2, i_{s-p} =0,8
10) i_{ph} =0,1, v =0,9
11) i_{ph} =0, i_{s-p} =1.

The total level of utility that is brought by some set of goods relating to a particular type in accordance with the targets of the consumption (based on the list of goods of the consumer basket) and perception of utility by the consumer in accordance with his belonging to one of eleven types will be calculated by the following equation:

$$U_{total} = (U_{phys})^{i_p} \cdot (U_{socio-psych})^{i_{s-p}} \tag{3}$$

In equation (3) U_{total} is the total level of utility from the consumption of goods, U_{phys} – the physiological utility and $U_{socio\text{-}psych}$ – socio-psychological utility, i_p is the coefficient of the importance of physiological satisfaction, i_{s-p} is the coefficient of the importance of socio-psychological satisfaction.

The consumption will be recognizing as rational if consumer spending his own and borrowed funds feels most satisfied and happy. Then the total utility level of consumption, which involves socio-psychological and physiological components, approaches 100%.

The income of the consumer is used in two ways: the expenses for the consumption and the savings.

$$Y = C + S \tag{4}$$

In equation (4) Y is the total income of the consumer, C – the consumer expenses and S – the savings.

It is necessary to consider how the income (Y) is being spent by today's consumers taking into account their features. The consumer expenses (C) include the acquisition costs of goods and services to satisfy the physiological needs and functional consumption (C_f), for the acquisition of the mixed utility goods and services– advanced functional consumption (C_{af}) (both of physiological and socio-psychological utilities); for the purchase of socio-psychological orientation goods and services – non-functional consumption (C_{nf}). This situation can be represented by the equation:

$$C = C_f + C_{af} + C_{nf} \quad (5)$$

The other part of the income goes to savings; they constitute monies of the delayed demand. Depending on the purpose of savings and the level of satisfaction of the consumer at the time of making savings we can distinguish the following types of savings:

- the funds in excess of the required budget, when the consumer is satisfied at one hundred percent and has no more desire and need to spend more. In this case, as a rule, the money is kept aimlessly and has no value at the moment for the consumer. He may give it everyone, help family, friends and loved ones, do charity or may be do nothing. These funds provide the consumer the state of confidence, stability, sense of well-being and reliability. It is not necessary to be this consumer a man of high wealth, because maybe he needs the absolute minimum budget goods for a complete total satisfaction. Or the purpose of this kind of savings may be the future satisfaction of the different nature needs of other people. For example, parents save money for their children's future, their education, for purchase an apartment for them, etc. Accordingly, the type of utility that carry these savings is socio – psychological and they satisfy the consumer's non-functional demand (S_{nf});
- the funds derived from the budget, which is essential for full satisfaction. In this case, the consumer is not absolutely satisfied, because he cannot purchase any good or service that he needs at the moment. The cause of these savings may be the consumer's uncertainty that he won't be able to reach the enough satisfaction in the future, for example, unstable income, possible illness, changing of the residence, etc. Respectively, the socio-psychological satisfaction of this consumer at this stage is not high and with the aim of it increasing the funds are deposited. The total utility is higher when the allocation of income is with the generated savings than the utility from fully spent available income. It is possible to identify three types of savings: the savings of functional demand (S_f) – the consumer saves money for his future expenditure for the purchase of goods of the physiological purposes (both primary and secondary); savings of advanced functional demand (S_{af}) is for the purchase of the mixed goods; savings of non-functional demand (S_{nf}) is for the purchase of goods of the socio-psychological orientation;
- "the savings for the savings". Some consumers, regardless of their satisfaction level of their real needs, are unable to achieve any level of total utility from the consumption set, not saving a part of their income. This is about consumers like i_{ph} =0,1, $i_{\text{s-p}}$ =0,9 or i_{ph} =0, $i_{\text{s-p}}$ =1 for whom the socio-psychological type is the most important utility that they get from making savings. The utility from the really acquired goods will be reduced under the influence of socio-psychological dissatisfaction in the absence of savings. This kind of accumulation also can be considered the savings of non-functional demand (S_{nf}) as they bring only the socio-psychological utility. So we can get the following equations:

$$S = S_f + S_{af} + S_{nf} \quad (6)$$

$$Y = C + S = C_f + C_{af} + C_{nf} + S_f + S_{af} + S_{nf} \quad (7)$$

Savings can be defined as the good and the consumer spends his money to purchase it. Ultimately, all savings will be ever turned into real products or services, and at the time of their formation, they already are an independent good, providing the consumer the socio-psychological satisfaction. The structure of expenditure of the consumer's income can be summarized into three components: functional expending costs, advanced functional and non-functional expenses:

$$Y = C + S = (C_f + S_f) + (C_{af} + S_{af}) + (C_{nf} + S_{nf}), \quad (8)$$

$$Y = C_f + C_{af} + C_{nf}. \quad (9)$$

Knowing the structure of the consumer expenditure by the nature of purchases, his level of satisfaction from the set of consumed goods, you can determine what type of consumer he is, what will be his future desires and, therefore, what purposes will determine his next choice.

3. Conclusion

The transformation of the existing theory of the consumption allows to take into account new important determinants of individual consumer choice and satisfaction in the modern economic system, which facilitates a theoretical justification of the actual consumption.

The proposed methodology of the consumption research allows analyzing the quality of life of the population. After exploring of the people desires, their needs and the level of satisfaction, you can recognize the features of the society and the stage of its development, compare and characterize the consumption process and its outcome in different countries, allowing to identify the advantages and disadvantages that are presented in a particular society, predict the future direction of the development of the social relations.

4. References

[1] Ismagilova G N, Bagautdinova N G and Safiullin L N 2015 Demand in the conditions of heterogeneity of goods and consumers *J. Asian Social Science.* **11, 11** 271-6

[2] Safiullin L N, Oduntsova J L and Safiullin N Z 2015 The theory of demand in the conditions of heterogeneity of goods and consumers *J. Procedia Economics and Finance.* **24** 288-95

[3] Radajev V V 2005 The sociology of consumption: basic approaches *J. Socis.* **1** 5 – 18.

[4] Krasovsky J D 2004 The phenomenology of the economic behavior *J. Socis.* **1** 28 – 33.

[5] Ryjkova M V 2003 *The social-economic factors of consumer behavior* (Tomsk: Tomsk University) p103

[6] Bogachev S P 2003 The value as the fundament of the needs systematization *The Moscow University Herald. Economics ser.* **1** 22 –3.

[7] Zimerman Y A 2007 Conspicuous consumption in conditions of reformed Russia *Economic theory in 21 century. National economy and socium* **5(12)** *605-8*

[8] Shastitko A E 2006 *Models of human in economic theory* (Moscow: INFRA-M) p141

[9] Yakushev M M 2006 Institutional determinants of consumer demand in the modern economy *The bulletin of Saratov Socio-Economic Institute* **14(3)** 159-63

[10] Larionova N and Varlamova J 2016 Consumption of import evidence from industrialized countries *J. Academy of Marketing Studies* **20 SI 1** 44-8

Semantic similarity and analysis of the word frequency dynamics

V V Bochkarev, Yu S Maslennikova, A A Svetovidov

Kazan Federal University, Kremlevskaya str.18, Kazan 420018, Russia

E-mail: svetovidov1994@gmail.com

Abstract. In this study a similarity in changes of frequencies dynamics for semantically related words was analyzed using word statistics extracted from more than 4.5 million books written over a period of 205 years. The approach is based on the correlation analysis of 1-grams frequency dynamics. We analyzed the frequencies correlation of synonym pairs, their corresponding antonymous groups and random words pairs. Also, we compared several metrics to find the most effective for assessing the degree of similarity in the dynamics of use of different words. Comparing differences between logarithmic rank variations in pairs of synonyms and random word pairs, significant differences are found, though they are smaller than it could be expected.

1. Introduction

Evolution of written language is significantly influenced by the wide range of cultural and historical factors. Social and political events also influence on stylistics and semantic context of written speech. Furthermore, choice of specific word forms in the text is determined by grammar requirements. With the lapse of time, these external factors bring certain contribution into lexicon usage statistics, as some word forms become more prevalent for use, while the others become less preferable. Therefore, it is expected that related in meaning concepts and word forms will have similar word usage frequency dynamics. However, obsolete concepts and archaisms may have opposite usage frequency trends to concepts, which replace the first ones in use. Today an increasing number of works are dedicated to statistical analysis of semantically related word frequencies. For example, the approach called "relative cosine similarity" is proposed in [1]. Distributed word vectors were analyzed, the cosine distance was used to estimate the synonymic similarity for different part-of-speech separately. This approach gave a significant improvement the accuracy of synonyms recognition.

The similar study is proposed in [2], where a synonym acquisition trajectory is analyzed, chosen by non-native English speakers learning English. It is found that with the growth of language proficiency the number of correctly selected synonyms increases, but even advanced learners face difficulties with synonyms in context-called-for unique constructions. The study [3] develops the functional systems theory (politics, religion etc.), analyzing also the frequency graphs for words «money», «power» and «love». It was noticed that frequency changes trends for concepts «love» and «money» are rather close to each other in English. However, the authors analyzed the words without consideration of their synonyms or words from the same semantic groups. Therefore, our research could supplement achieved results of the above work. Resent results [4] shows that semantically related words show strong phase coherence. It is noted that found patterns in the statistics of language may be a consequence of changes in the cultural framework that influences the thematic focus of writers.

Authors of this study analyzed only a core vocabulary of 5,630 common English nouns using only correlation metrics. The purpose of this work is to figure out, whether there is a correlation between frequency usage changes of semantically related words based on pre-verified lists, and also to determine the most informative measure frequency series proximity.

Our study is based on the 2012 version of the Google Ngram database [5]. We focus on 1-grams data, which consist of single word frequency counts for every year independently. The 2012 version of the database is annotated with parts-of-speech tags [6], which allow the extraction of particular word classes. Additionally, WordNet database was used [7]. WordNet is one of the largest lexical databases, which provides access to more than 140 000 words from 4 parts of speech (nouns, adjectives, verbs, adverbs) and the information about their semantic relationships. For our study, we considered groups of common nouns and adjectives.

2. Methods

For English language, our starting database was the 1-gram counts from 1800 year because, although the available data start at 1520, the data for the first 300 years is rather sparse. Using the parts-of-speech tag included in the database [6], we extracted the 1-gram information for the noun and adjective classes. This procedure finally left a core vocabulary of 493 nouns and 584 adjectives. Figure 1 demonstrates examples of normalized time series of the group of synonyms and antonyms.

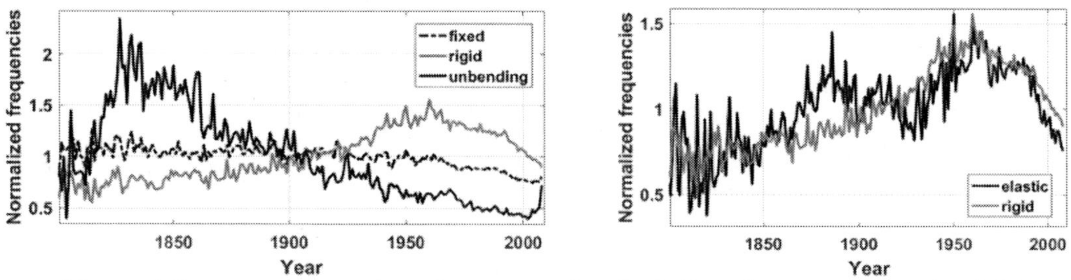

Figure 1 Frequency dynamics of synonyms (on the left) and antonyms (on the right)

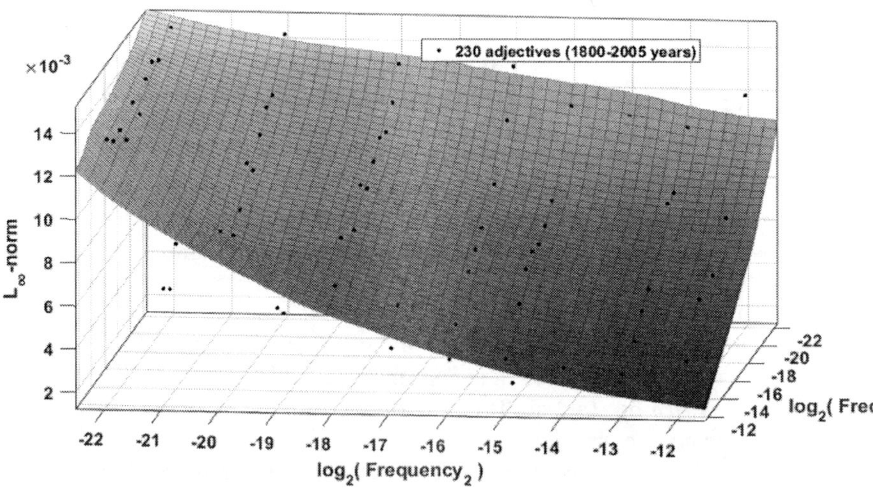

Figure 2 L_∞-metric between random pairs of adjectives with different words frequencies

To acquire information about semantic similarity degree through comparison of word-use data, correlation analysis was executed. Following list of pairwise distances was used: Euclidean distance (L_2-metric), City block metric (L_1-metric), Chebychev distance (L_∞-metric), cosine distance, Pearson's correlation and Spearman's rank correlation. Correlation metrics and cosine distance only reflect changes in forms of researched series, however, distances for the first three metrics also depend on absolute values of frequencies. We have been interested in the analysis of words frequency dynamics, that's why time series of frequencies was normalized by the mean for the considered period (1800-2005). Also, we compared logarithmic rank variations, i.e. logarithms of numbers from a word frequencies list, as in [4].

To estimate the significance of acquired distances between semantically related words, estimation of distances between random words (reference pairs) were also executed. Reference pairs from similar parts of speech were chosen, because several papers demonstrate steady difference in usage frequency trends, for example, between nouns and adjectives, which should be taken into consideration [6]. While choosing pairs of random words the average usage frequencies were considered. It is well known that time series of rare words frequencies are extremely noisy. As it is shown in [8], relative value of words frequency variations has a power-law dependence on frequency: $\sigma_f/f \sim 1/f^{0.316}$. This fact leads to increasing of acquired distance values between rare words frequencies. Figure 2 shows, how the average Chebychev distance varies according to changes of logarithms of random adjective pairs usage frequencies. Therefore, reference pairs with words frequencies that are similar to analyzed synonym and antonym groups were selected for our research.

3. Results

At the first step, similarities between the normalized frequencies dynamics of synonyms and reference pairs were estimated. However, the analysis of the distribution for distances between groups of synonyms shows that it is almost same as the distribution of distances between random pairs of words. Differences that are more significant were obtained by comparing the logarithmic rank variations as in [4]. For each pair of synonyms (the total number of pairs was 1457 for adjectives and 1484 - for nouns), we selected 40 pairs of randomly selected words of the same part of speech with sufficiently close frequencies. Then, the logarithmic ratio of distances in reference pairs and pairs of synonyms was calculated. In Figure 3 distributions of the logarithmic ratio using L_1-metrics are shown for adjectives and nouns separately. The positive values of the logarithmic ratio show that distances between synonyms is less than distances in reference pairs. The observed asymmetry of both distributions (with a significant prevalence of positive values) indicates that semantically related words are close to each other with comparison of reference random pairs.

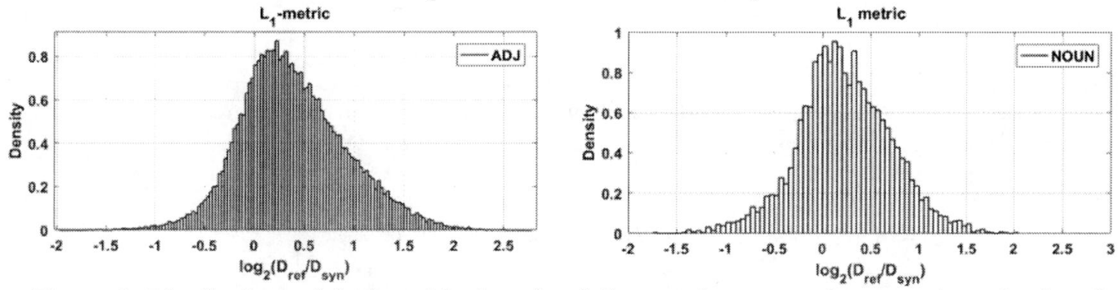

Figure 3. Distributions of the logarithmic ratio of distances between reference pairs and pairs of synonyms for adjectives (on the left) and nouns (on the right)

To estimate the significance of the found effect, the null hypothesis about the homogeneity of samples in reference pairs and pairs of synonyms was tested. According to the sign test, p-value was $6.84 \cdot 10^{-191}$ for adjectives and $3.48 \cdot 10^{-203}$ for nouns. Different metrics were compared based on analysis of the accuracy for semantically related words recognition using different pairwise distance metrics. Thus, it can be concluded that the found effect is significant.

To compare different metrics, we estimated the accuracy of semantically related words recognition. To do this, we compared distances in synonyms pairs D_{syn} with distances in reference pairs D_{ref} and considered the number of cases, when $D_{syn}<D_{ref}$. If the analyzed pair consists of semantically unrelated words, then the number of cases will be any value from the interval from 0 to N, where N is the number of reference pairs. If analyzed pair consists of semantically related words, then this number

will be larger with high probability, as follows from Figure 3. In our study, the threshold value for make the decision was chosen so that errors of the first and second kind would be equal. The results for adjectives and nouns are shown on the Figure 4. We can see that the L_1-metric is the most accurate and shows minimal errors: 0.29 - for adjectives and 0.32 - for nouns.

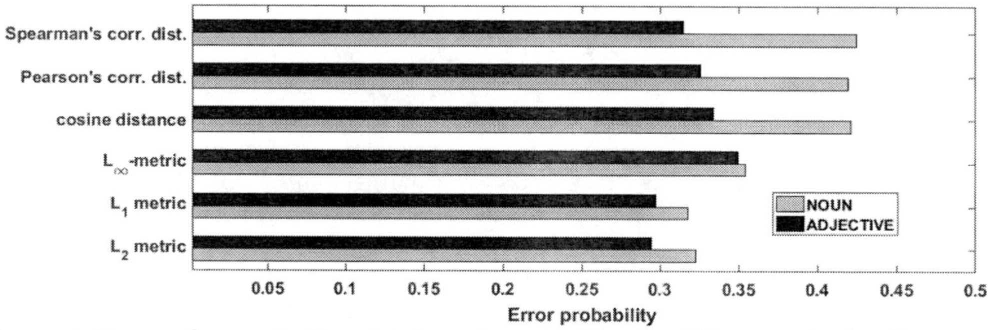

Figure 4. Errors of semantically related words recognition for different pairwise distance metrics

4. Conclusion

Comparing distances between the series of normalized frequencies of semantically related words and similarly obtained distances for random word pairs, no significant differences are observed. On the contrary, comparing differences between increment series of ranks in pairs of synonyms and random word pairs, significant differences are found, though they are smaller than it could be expected. Apparently, a sort of competition between words and word substitution in groups of synonyms play an important role in dynamics of language use. It should also be taken into account that there are words which on average don't have significant change in frequency of use over the last two centuries (for example, the word "unbending" in Figure1). When assessing the degree of similarity in the dynamics of use of different words, the most effective metrics are L_1 and L_2.

This work was supported by the Russian Foundation for Basic Research, Grant no. 15-06-07402. The research of the first author was supported by the Russian Government Program of Competitive Growth of Kazan Federal University.

References

[1] Leeuwenberg A, Vela M, Dehdari J, Genabith J A 2016 Minimally Supervised Approach for Synonym Extraction with Word Embeddings *The Prague Bulletin of Mathematical Linguistics* 111-142

[2] Liu D, Zhong S 2016 L2 vs. L1 use of synonymy: An empirical study of synonym use/acquisition *Applied Linguistics*. **37 Issue 2** 239-261

[3] Roth, Steffen et al. 2016 The Fashionable Functions Reloaded: An Updated Google Ngram View of Trends in Functional Differentiation (1800-2000) *In: Mesquita, A. (Ed.) Research Paradigms and Contemporary Perspectives on Human-Technology Interaction. Hershey: IGI-Global*

[4] Montemurro M A, Zanette D H 2016 Coherent oscillations in word-use data from 1700 to 2008 *Palgrave communications*

[5] Michel J, Shen Y, Aiden A et al. 2011 Quantitative Analysis of Culture Using Millions of Digitized Books *Science* **331 (6014)** 176-182

[6] Lin Y, Michel J-B, Aiden EL, Orwant J, Brockman W and Petrov S 2012 *Syntactic annotations for the google books ngram corpus, in 'Proceedings of the ACL*

[7] Fellbaum C. 1998 WordNet: An Electronic Lexical Database *Cambridge, MA: MIT Press*

[8] Bochkarev V, Solovyev V, Wichmann S 2014 Universals versus historical contingencies in lexical evolution *J. R. Soc. Interface* **11: 20140841**

IOP Conf. Series: Journal of Physics: Conf. Series **936** (2017) 012068 doi:10.1088/1742-6596/936/1/012068

Modelling Plane Geometry: the connection between Geometrical Visualization and Algebraic Demonstration

L. R. Pereira[1], D. F. Jardim[2] and J. M. da Silva[3]
UFVJM-PROFMAT[1], UFVJM[2], UFVJM-ICET[3] - Teófilo Otoni-Minas Gerais - Brazil
E-mail: llucasrp@gmail.com[1], deborah.farago.@ufvjm.edu.br[2] and jaqueline.silva@ufvjm.edu.br[3]

Abstract. The teaching and learning of Mathematics contents have been challenging along the history of the education, both for the teacher, in his dedicated task of teaching, as for the student, in his arduous and constant task of learning. One of the topics that are most discussed in these contents is the difference between the concepts of proof and demonstration. This work presents an interesting discussion about such concepts considering the use of the mathematical modeling approach for teaching, applied to some examples developed in the classroom with a group of students enrolled in the discipline of Geometry of the Mathematics curse of UFVJM.

1. Introduction
Mathematics science presents several factors that make it hardly understandable by the students when is taught traditionally. This traditional teaching focused on the teacher as a transmitter of knowledge leads the students to reach their goals by they own effort. In this way, the mathematical subjects, the teaching methodologies and the interpersonal relationship between teacher and students do not have any relation with the student's academic life, without any contextualization.

Is very known in the literature that the use of mathematical modeling approach to teach Plane Geometry helps the students in their geometrical visualization process. However, we must emphasize the contributions of this visualization process to the students without distancing them to the algebraic thinking. Although some dynamic geometry software can not perform demonstrations, the didactical technique of modeling the behavior of the geometrical construction lead the students to search for the proof of a given theorem.

In this context, based on the concepts of Didactic Engineering (DE) as a research methodology, this work presents the teaching of area concept in Plane Geometry using mathematical modeling approach and GeoGebra, seeking to emphasize the importance of understanding the distinction between geometric visualization and algebraic demonstration. We divided the activities of this research into three stages. A diagnostic questionnaire was initially applied in the first stage, followed by the intervention and the student submission to some dynamical activities with the software in the second stage, and finally, was applied a diagnostic questionnaire for a better analysis of the data in the third and last stage.

2. Modelling Approach and Research Methodology

2.1. Investigating the Problem

According to [1], the definition of area is a mathematical perception that makes possible to verify and measure the occupation of a particular surface. The teaching of this concept, in plane surfaces, has shown to be of a great social relevance and thus, contributed to the formation of the individual. However, [2] and [3] affirm that this topic is worked on schools demanding from the student's some competencies that involve calculations.

The teaching of mathematics related to the memorization of mathematical concepts does not value the construction process of knowledge, such as observation, investigation, experimentation, and formalization of ideas. According to [4], the conceptual difficulties expressed by the students present several factors related to their learning process. Among them, the errors made in the geometric and numerical conceptions are evidenced, as was also observed in this research, through diagnostic questionnaires applied in the classroom. Further details about these questionnaires can be found in [8].

This research, which was part of the first stage of DE, showed that the main mistakes made by the students participating in this research were related to the association of graphic representation and their properties. The Plane Geometry class, offered in the first year of the Mathematics course in UFVJM, is one of the first disciplines that introduces the abstract thinking, focused on theorems demonstrations, definitions and propositions. In this way, students were expected to have difficulties in writing a text using appropriate formal language.

Due to these difficulties, it is necessary to discuss new pedagogical practices aimed to improving students' learning, allowing them to establish cognitive correspondences between the formulas and the graphical and geometrical representations of plane figures. In this sense, the computational resource through modeling approach can represent the bridge between the formal treatment of the concepts associated with the figures and their graphic representations.

Researchers in Mathematics Education, such as Bassanezi and D 'Ambrsio, discuss the improvement and renewal of teaching mathematics in the schools as well as the way in which it can be developed, seeking justifications for the insertion of modeling approach in the students' curriculum [5]. Among the mathematical subjects that can be inserted in this proposal, Geometry stands out as an exponent, since it presents several options of contextualization. In this way, the use of technological resources for teaching Mathematics stands out as an excellent contribution to the teaching-learning process, particularly in the topic area of plane figures, which is the subject discussed below.

2.2. Didactical Activities with Modeling Approach

The didactical activities were elaborated and planned with detail, based on the principles of modeling approach, with the help of GeoGebra software. In these models, the visual resources obtained from the constructions with GeoGebra are used with profusion since there are several ways to calculate the area of a plane figure. In this case, the geometric verification is necessary so the teacher can reach the main objective, improving the learning of concepts and the formalization of students' ideas.

The proposed model is composed of theoretical and practical contents discussed by the students. The theoretical contents portray the definitions and propositions related to each chosen figure, besides proposing ideas for the demonstration of propositions. Only one practical exercise was included since most of the time was devoted to the demonstrations of the polygons areas. Also, the models were designed and elaborated to lead the student from the experimentation stage to the concepts formalization.

The visual resources from the GeoGebra constructions have been extensively explored, since there are several ways of calculating the area of a plane figure, that is, the geometric proof is necessary to achieve the primary objective of improving the learning of concepts and formalization of the students' ideas. The following two activities present the use of modeling approach in problems that deals with areas of plane figures subject.

Activity 1

We made a distinction between congruent surfaces and equivalent surfaces, concepts that the students had difficulties to visualize geometrically. Two congruent triangles were constructed, and the students verified, through the command "Area [< *Polygon* >]", that the triangles had the same area. As an example of equivalent surfaces, we constructed two quadrilaterals composed of two triangles T_1 and T_2, interconnected by distinct segments, as can be observed in Figure 1.

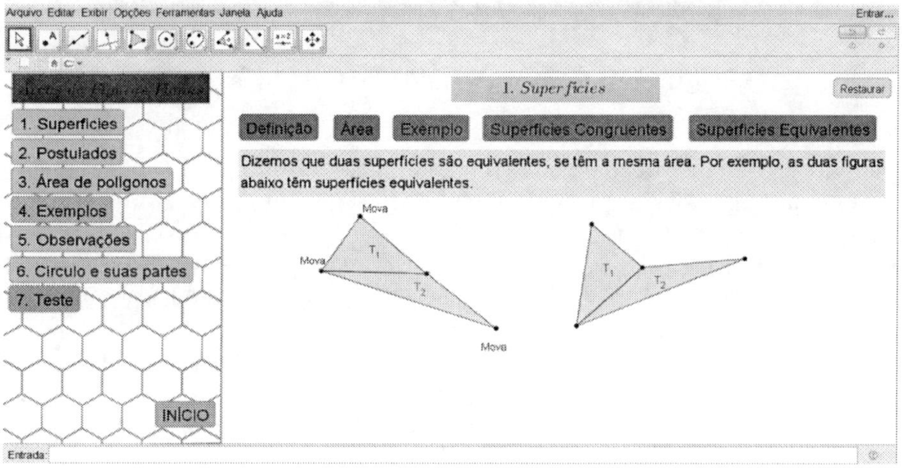

Figure 1. Models of two equivalent surfaces

During this activity, it was suggested that the students change the arrangement of the points indicated in the Figure. With this suggestion, they found that, although the quadrilaterals are distinct, they are also equivalent surfaces. In addition, they reported that with the aid of the modeling approach there was a better understanding when the main differences between the concepts of congruence and equivalence were approached practically and dynamically.

Activity 2 :

In this activity was presented the formula to calculate the rectangle area and an idea of how to obtain the respective equation. Students were asked if they would be able to obtain the formula for the area using a formal demonstration method, initially without the aid of the software. Later, the modeling, shown in Figure 3, was used to integrate the use of the software and its tools with the demonstration of geometric propositions.

On the proposed questioning, one of the students considered that it was enough to subdivide the rectangle into unit squares and soon after conjecture the formula using the postulates. The demonstration, presented in Figure 2, was done on the board by the student himself so that the other students could share the idea, exchange information and to interact. After completing the explanation, the students were asked if the demonstration was correct and some realized that there was an error.

The students pointed out that the mistake was to consider that m and n are both natural and for that they used as an example a rectangle whose sides measure was $3.65cm$. Therefore, in the form that was performed, where only the case where the segments are commensurable

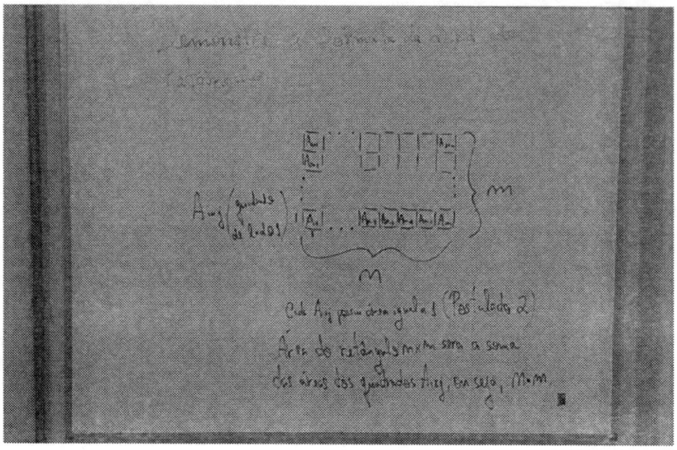

Figure 2. Demonstration produced by a student

was considered, the demonstration would be incomplete. To finish the demonstration suggested by the student, we must consider the case where the segments are incommensurate.

Another way of proving the formula of the rectangle area was presented, using the previous postulates and with the aid of mathematical modeling approach through some auxiliary constructions, as shown in Figure 3. Using the GeoGebra tool, which gives dynamism to the model. In this case, the students considered that the demonstration was conducted more simply.

The students' observation was expected since the activity was based on a research that reinforces the need to correlate the manipulation of dynamic geometry software with the demonstration of geometric propositions in the teaching and learning process [6].

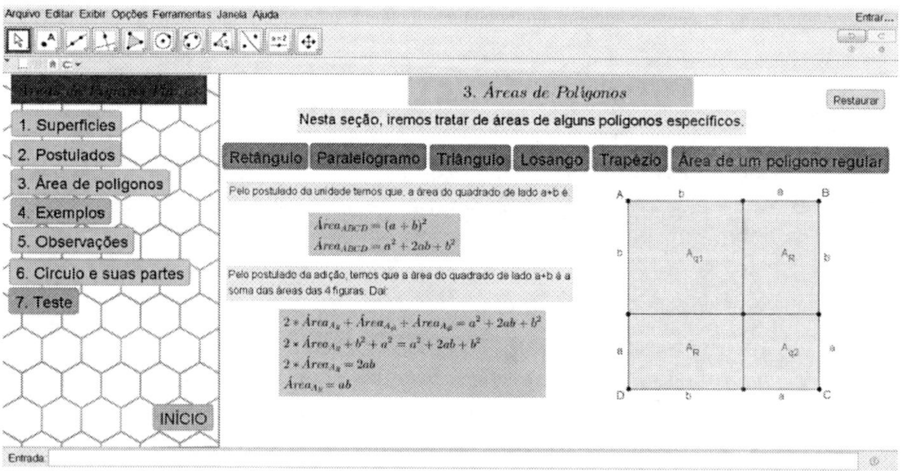

Figure 3. Support Model - Rectangle Area

Activity 3

In this activity was presented an alternative to the demonstration of the circle area, through modeling approach and the concept of infinity. In the model elaborated for this didactic activity, the student should first click on the "Animate" button and verify that as the number of sides of the polygon inscribed in the circle increases, the polygon area tends to circle area, as can be seen in Figure 4.

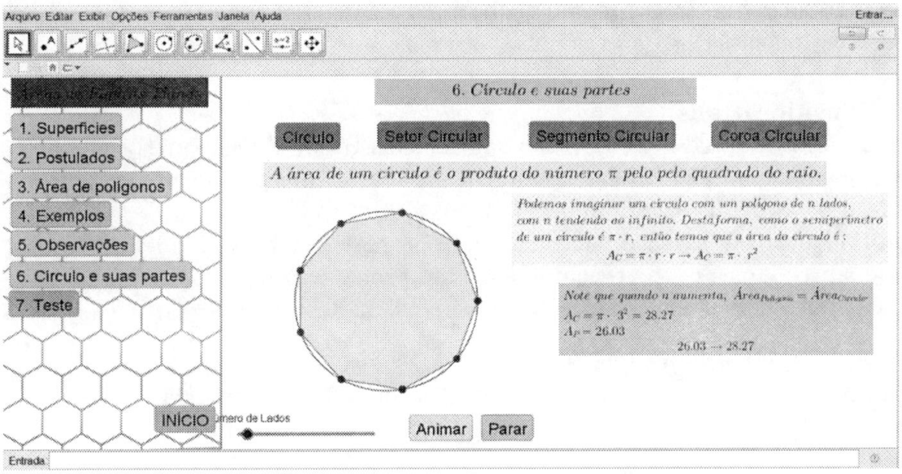

Figure 4. Support Model - Rectangle Area

It should be emphasized that while demonstrating the formulas, students were instructed to manipulate the constructions carried out in the software to verify the veracity of the statements and to propose new ideas for future demonstrations.

Subsequently, were taken some construction steps with the students:

- Create a slider. In this case: name "n", initial value 3, final value 30, and increment 1.
- In the input field, enter the command: sequence "(cos(360°/ n*i),sin(360°/n*i)), i, 1, n)".
- Animate the slider "n".

We emphasized that in demonstrating the formulas of the remaining figures, students were instructed to manipulate the constructions carried out in the software to verify the veracity of the affirmations and to propose new ideas for the future demonstrations.

For example, during the demonstration of the area formula of a circular segment, a student questioned whether the fact that the central angle was acute, obtuse or straight could interfere in the final equation. These questions arise through the manipulation of the constructed figures, which would hardly be done in the classroom without the use of computational resources.

3. Results Discussion

Along Activity 1, by changing the arrangement of the points indicated in the models built in GeoGebra, the students were enthusiastic to verify that, although the quadrilaterals are different, they are also similar surfaces. In addition, they reported that with the aid of mathematical modeling approach there was a better understanding when the main differences between the concepts of congruence and equivalence were approached in a practical and dynamic way.

In general, it was found that the objective of the proposed activities was successfully achieved because, when requested, the students responded promptly and safely. The improvement of the formal mathematical writing is a factor of extreme importance since in the classroom the students presented great difficulty in developing a demonstration.

One of the main benefits gained by the use of mathematical modeling approach and the application of didactic sequences in the teaching of Geometry are precision and visualization, exploration and discovery and proof of theorems. Although dynamic geometry software can not perform demonstrations, the technique of experimenting with assumptions leads the student to search for the proof of a theorem.

It was verified that the dynamics that GeoGebra brought to the classes of Plane Geometry, promoted the communication of the students among themselves, and between the teacher and the students. Collective work proved to be beneficial, allowing students to share information about construction processes and demonstrations.

4. Final Considerations

From the results generated in observations in the classroom, it was found that the main obstacles faced by students regarding the teaching of areas are due to the precipitous introduction of formulas lead to apparent reasoning.

In addition, it was verified that the use of mathematical modeling approach allied to computational resources in the teaching-learning process of Plane Geometry, in the investigated group , contributed significantly in several aspects, such as:

- Greater interaction among students;
- The development of skills and competencies related to Geometric Design;
- The development of demonstration techniques;
- Significant improvement as to the essay questions, appropriate to the formal mathematical writing.
- The benefit of the process of graphical and geometric visualization, without there being a distancing of the algebraic thought;

In addition to what has been presented, it is worth mentioning that the experiences acquired in this research, confers the training of researchers and readers albeit in a restricted way, contributing to future discussions about the use of Mathematical Modeling in the teaching of Plane Geometry.

Acknowledgments

Thanks to FAPEMIG for the financial support.

References

[1] Baltar, P M. 1996. Enseignement et apprentissage de la notion d'aire de surfaces planes: une étude de l'acquisition des relations entre les longueurs et les aires au collège. Universidade Joseph Fourier.

[2] Chiumo, A *et all.* 1998. O conceito de áreas de figuras planas: capacitação para professores do ensino fundamental. VI Encontro Nacional de educação Matemática. 697–698.

[3] Baldini, L A F and Póla, M C R. 2004. Construção do conceito de área e perímetro: uma seqüência didática com o auxílio do software de Geometria dinâmica. VIII ENEM–Encontro Nacional de Educação Matemática, Recife/PE.

[4] Belleimain, P M B and Bittar, M. 2002. O ensino da geometria e a teoria do campos conceituais. 25ª reunião anual da ANPEd.

[5] Bassanezi, R. C. Modelagem matemática - uma disciplina emergente nos programas de formação de professores. In: XXII Congresso Nacional de Matemática Aplicada e Computacional. Biomatemática IX. Santos: IMECC, 1999. v. 9, p. 922.

[6] Gravina, M A. Geometria Dinâmica: uma nova abordagem para o aprendizado da Geometria. Anais do VII Congresso Brasileiro de Informática na Educação. Belo Horizonte. 1996

[7] Pereira, L R, Silva J M, Jardim D F. 2017. Practices for Geometry Teaching Using Geogebra. In Conference Proceedings. New Perspectives in Science Education, page 237. libreriauniversitaria. it Edizioni.

[8] Pereira, L R. Práticas de Ensino em Geometria Plana. 2017. 171 f. Masther's thesis. Universidade Federal dos Vales do Jequitinhonha e Mucuri. PROFMAT.

IOP Conf. Series: Journal of Physics: Conf. Series **936** (2017) 012069 doi:10.1088/1742-6596/936/1/012069

A study about teaching quadratic functions using mathematical models and free software

T. V. Nepomucena, A. C. da Silva, D. F. Jardim, J. M. da Silva

Institute for Science, Engineering and Technology. Federal University of Jequitinhonha and Mucuri Valleys. Campus Mucuri. Rua do Cruzeiro, 01. Jardim São Paulo. Teófilo Otoni. Minas Gerais. Brazil

E-mail: `thamaranepo@gmail.com`

Abstract. In the face of the reality of teaching Mathematics in Basic Education in Brazil, specially relating teach functions focusing their relevance to the student's academic development in Basic and Superior Education, this work proposes the use of educational software to help the teaching of functions in Basic Education since the computers and software show as an outstanding option to help the teaching and learning processes. On the other hand, the study also proposes the use of Didactic Transposition as a methodology investigation and research. Along with this survey, some teaching interventions were applied to detect the main difficulties in the teaching process of functions in the Basic Education, analyzing the results obtained along the interventions in a qualitative form. Considering the discussion of the results at the end of the didactic interventions, it was verified that the results obtained were satisfactory.

1. Introduction

The mathematical teaching in basic education in most of the Brazilian public schools do not reach their goal. Several researchers have shown that students finish their high school studies and start their studies in the university without any basic knowledge about functions. The difficulties that secondary students experience in applying mathematics to real-life or context-based tasks are a long-standing problem in educational research and that limitation to use mathematics, as Paulos [1] observed, limits an individuals career aspirations, social well-being, and financial security. In this sense, [2] shows in Freudenthal that understands mathematization as the local structuring of mathematical and nonmathematical fields employing mathematical tools for which the direction from reality to mathematics is highly relevant.

The lack of interest can be one of the main factors that cause a small achievement whether they are students children, young people or adults and this can be strongly related to the teaching strategy.

In general, the traditional teaching method does not stimulate the mathematical learning, especially because it deals with a new students generation that is used to working with technology in their everyday life. In this sense, computational software and mathematical models have become essential tools to help the students learning and also can facilitate not only the teaching of quadratic functions but also other mathematical basic contents.

GeoGebra, for example, as a dynamic mathematical software very known in mathematical teaching, is a great teaching tool for quadratic functions, especially because it contains interactive features, such as the slider, which enables dynamic and instantaneous parameter changes, as well as the visualization of the graphical features, clarifying and refining students' mental models of the situation providing them work with the maximum and minimum points concepts of the quadratic function more easily.

The purpose of this work is to understand and discuss the student's difficulties related to the quadratic functions. As a secondary purpose, it analyzes the student's comprehension of the fundamental concepts and to evaluate the use of mathematical and computational tools with modeling approach as a teaching tool capable of providing an understanding of some problems that are presented in secondary school textbooks. The work was developed with a group of students enrolled in the second year of a public school in the state of Minas Gerais and a mathematics teacher, the second author of this paper.

After analyzing the students difficulties, based on the didactic transposition process, the models were used to stimulate learning in some intervention activities with the objective of complementing the teaching. The discussion is focused in two activities related to physics, one that investigates the launch of an object from the ground (Activity 1) and another that examines the peace of metal in a dilatation phenomena (Activity 2).

2. The use of modeling approach to teaching quadratic functions

2.1. Modelling Approach associated with Didactic Transposition

Some mathematical concepts are commonly considered quite abstract and the use of mathematical and computational models for teaching brings to the students some mathematical problems applied to their reality. Kaiser and Sriraman pointed in [3] the importance to use mathematics as science and humanistic ideals of education with focus on the ability of learners to create relations between mathematics and reality

Traditional teaching, especially in Brazil, where classrooms are quite numerous and with a few pedagogical resources, no longer attracts the attention of young students, especially because in these days they are used to deal with technological resources. On the other hand, many teachers who complete their undergraduate courses are still poorly trained and often fail to transform the mathematical contents presented in the textbooks into knowledge to be understood by the students. Thus, the choice of an adequate didactic approach for teaching mathematics can help in the process of mediating the knowledge between teacher and students.

An interesting research proposal about a better adequacy for teaching, in the sense, so the teacher language can be understood by the students, is the process called Didactic Transposition (DT), proposed by Michel Verret in 1975 and thoroughly researched by Yves Chevallard in 1991. The computational resource, through the modeling of situations-problems described by mathematics, according to the principles of DT, can give a more didactic dressing to the content explored by the teacher in the classroom.

The computational modeling of some activities taken from textbooks of the basic education allows us to simulate situations that go beyond what the exercise proposes, exploring different concepts and transforming "taught knowledge", which is limited to the bibliographic content used by the teacher, in "used knowledge" whose object is the student [4] and[5]. For this purpose, GeoGebra software has proved to be an excellent tool, easy to use by students, very intuitive and allows the development of dynamic interactive models [8], [9] and [10].

2.2. Investigating the teaching of functions in secondary school

A careful research about the difficulties of a small group of students of the second year of the middle level at the Geraldo de Souza Norte school pointed out a deficiency in their comprehension about functions. Some students were able to associate the functions with situations of their everyday life, but they were not able to manipulate these functions conveniently, to associate the values of maximum and minimum with the particularities of each case, to understand the implications of the differences between the concavities, among other things.

Quadratic functions are approached in different ways in the Brazilian school's textbooks, but usually, the books have an exhaustive amount of activities only for fixation of the contents. Usually, these activities seek to bring real situations to be discussed in the classroom, but not always the teacher can transform the information into knowledge so the students can finally comprehend it. Therefore, working with these activities in a more attractive way, with the use of mathematical and computational modeling approach, should contemplate DT.

There are interesting applications of quadratic functions in physics, which could be worked with the help of the modeling process and the visual resource of GeoGebra software, allowing to explore the interdisciplinary character between mathematics and physics. For example, the phenomena of launching an object under the gravitational effect, use two quadratic functions. One of these two quadratic functions deals with the behavior of the object trajectory in the plane and the other quadratic function refers to the vertical position of the object with respect to the time.

In mathematics textbooks it is always possible to find such questions, as it can be found in [11], [12], [13] and [14]. But whatever exercise is taken from the textbook, to be modeled through software, it will be necessary to evaluate how the process of DT can be the bridge that leads to the students learning.

3. Activities with the Modelling Approach

3.1. Activity proposed to the students

The first stage of this research discusses the concepts of the quadratic function of an object launched from the ground, according to the following proposal:

Activity 1 *Consider a ball thrown vertically upwards from the ground. The ball position at each instant follows the function* $h(t) = 40t - 5t^2$*, where* h *is expressed in meters and* t *in seconds. Determine: (a) The height at which the ball lies 1 second after the throw; (b) The instant at which it is 75m from the ground; (c) The maximum height reached by the ball and (d) The instant when the ball returns to the ground.*

The students would have to answer the questions proposed in the exercise without the use of modeling process. The items (a) and (b) were easily answered by the students because it was necessary only a simple substitution of values and the determination of the other parameter. However, the students presented difficulties to solve especially the item (c), which needed the use of a concept of the maximum point of a quadratic function for the resolution. The item (d) was solved by some students, but most of them did not understand that they needed to determinate the roots of the function to solve the problem.

With the use of modeling process as a teaching tool, it is possible to deal with the issues that were misunderstood in the problem above and to explore it a little more, as shown in Figure 1 below.

Figure 1 shows that the computational model simulates the problem. Through the "Play" button, it shows the ball going up and down while the graph of the function $h(t)$ is constructed. In this sense, it is possible to explore the concept of maximum point of the function and also show that it is associated with the maximum height that the ball reaches. Likewise, the roots of the function can be observed in the graph and also associated with the moment when the ball lies on the ground just before climbing and at the exact moment it returns to the ground.

Then, with the use of the modeling approach, it is possible to work on the domain of the quadratic function and the constraints from the real situations, showing the student the importance of understanding how to use Mathematics as a tool for solving real problems. For example, if you wish to analyze only the ball going up, or just descending, the visual feature of

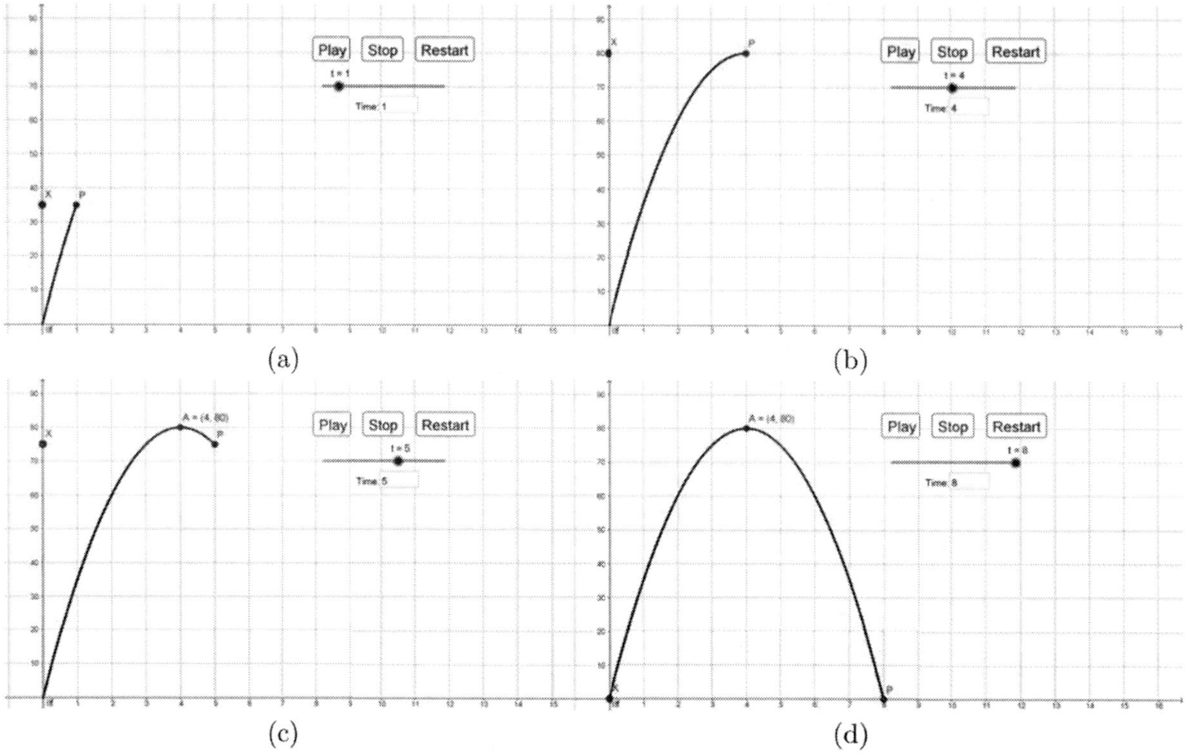

Figure 1. The picture represents the simulation for the vertical upward launch of a ball. It was built with the ball movement. In (a) the ball in the upward movement; (b) the ball reaches the maximum high; (c) the ball is going down and (d) it returns to the ground. The pictures show and explore the mathematical concepts discussed in the classroom activity.

the model shows that the domain of the function is restricted to the ball behavior in the real world, even though mathematically the domain of the function is the real numbers set.

3.2. Activity produced by the students

The following model, presented in Figure 2, was constructed by one of the students along the classroom activities, using the following proposed activity:

Activity 2 *The dilation and contraction behavior of metal, measured in a laboratory with a test specimen. It was exposed to a sudden temperature variation along 10 minutes, is in accordance with the function* $T(t) = t^2 - 12t + 32$*, where* T *is the temperature measured in Celsius and is* t *measured in minutes. What is the lowest temperature recorded in this experiment and at what time this temperature occur?*

The idea of putting students in the condition of pro activity, building their model, allowed us to verify how this activity could stimulate their thinking and increase their interest in study mathematics, especially when is related to the content of the functions.

As expected, with the computational resource, the students presented no difficulty in determining that the lowest temperature was -4 Celsius and that this occurred when the time was equal to 6 min. Thus, it was possible to approach the concept of maximum and minimum of a quadratic function and the students realized that the minimum value that the function could

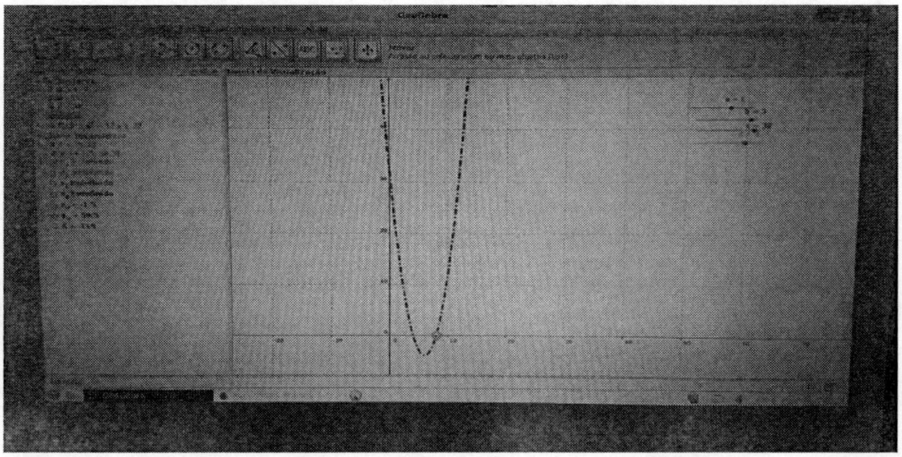

Figure 2. The model constructed by the students. It represents the behavior of the temperature in the metal under dilation and contraction phenomena observed along the time.

assume was the value corresponding to the image of the parabola that describes the function behavior.

Besides, the visual resource allowed the students to determine the time interval in which the temperature presented negative values. With the free parameter "t" for the time representation, it was possible to determine the exact values of this interval, that is, 4 and 8 minutes after the start of the experiment. It was also possible to analyze that if the temperature becomes negative only after four minutes, the only possibility is to have a parabola with concavity upwards, having the function a minimum and not a maximum point.

Then, the use of sliders, whose free parameters in the modeling are "a", "b" and "c", allow us to evaluate, for example, if the problem would accept a negative value for the coefficient of the quadratic term. Then the parabola would have concavity down, and what that would represent. It is also possible to open a discussion with the students to evaluate the changes that would occur in the proposed activity, dynamically changing these parameters through sliders and verifying their implications.

The discussion for this case would go a long way towards showing that even though it is possible mathematically, physical explanation does not always allow for any situation, often without even a real interpretation, requiring the student to reflect on each subject, as in the case of obtaining a negative value for the time.

4. Results and Discussions

The difficulty presented by the students about the understanding of quadratic functions motivated the study, and the results are presented in this work. The investigation about the students' knowledge related to the basic concepts of quadratic functions, as the minimum or maximum points, the function roots and the implications of these concepts analyzed in real problems, showed that these fixation activities found in the books and handled by the teacher in the classroom were not enough efficient to promote the students comprehension of these concepts.

On the other hand, the search of an updated teaching methodology, using the technological resource, led to different activities with the GeoGebra software producing interactive models based on the exercises of the books. That was a good result because the teacher could finally work with the quadratic functions using real-world applications.

IOP Conf. Series: Journal of Physics: Conf. Series **936** (2017) 012069 doi:10.1088/1742-6596/936/1/012069

The activity performed for the students, described in section 3.1, generated some expectations regarding the use of modeling approach with the teaching of functions, both by the teacher and by the students. The visual resource using technology as an educational tool is fundamental when you want to explore something that using only pencil and paper is not possible. The model of the ball that was going up and down and the graph of the function produced simultaneously, provides a dynamical effect with a positive impact on the students, growing their curiosity and rousing their attention.

The teacher, using the modeling approach as a teaching tool, stimulates the students about the learning of mathematics, promotes interdisciplinary application and makes the process of teaching, in the DT sense, closer to the student's reality.

In general, with the modeling activities described in this work, it was observed that the students understood the implication of the quadratic function signal. Indeed, since they managed to identify, even intuitively, the interval at which the function, particularly in the exercise described in Section 3.2, assumes negative or positive values, what means, the range which the temperature was present.

Therefore, the modeling process allowed students to quickly identify the minimum or maximum points of a quadratic function and their respective values, as well as to better understand concepts such as the function roots, the signs and also the concavities of the quadratic function, even if this does not have used technical terms. So, as Kaiser says in [3], the central character of the realistic or applied perspective formulated by [6], can be stated that modeling is understood as an activity to solve authentic problems and not as a development of mathematical theory.

5. Final Considerations

The search for different methods to approach the basic contents of mathematics for secondary education has led the scientific community to propose new didactic and teaching practices. In particular, DT is understood as a process that acts to provide the reflection about the best teaching practice in the classroom (taught knowledge) that leads to understanding by the student (used knowledge)[4]. In this case, modeling, as long as associated with well-developed pedagogical activities, plays the role of restructuring the "hard" context of the textbooks into contents that can be learned by the student, helping him to achieve a necessary level of knowledge to use mathematics effectively in real-life situations. This modeling association with pedagogical activities is a pedagogical and psychological goal mostly discussed in [7] although the author had identified and described the nuances differently.

The use of technology in the classroom, no matter if is only as a demonstrative tool, where only the teacher manipulates the model and explores it, or if is an activity in which the student constructs his model, is fundamental in the present times. where the students learn quickly to deal with the computer. In this case, the modeling gains more dynamism with the resources of the GeoGebra software, being possible to work concepts that are hardly fully exploited only by the traditional method, with the use of pencil and paper. But it is important to say that, as [16] pointed, the modeling activity is this changed by the presence of the software.

Therefore, the modeling process, particularly if is used to teach quadratic functions and problems involving this content, complements an outstanding teaching pedagogical practice in this modern world in which technology is an indisputable reality.

Acknowledgments

Thanks to Capes for the fellowship to one of the authors that belong to the PROFMAT post graduating program at UFVJM. Thanks to FAPEMIG for the financial support.

References

[1] Paulos J A (2000). Innumeracy: mathematical illiteracy and its consequences. London: Penguin.

[2] Freudenthal H (1973). Mathematics as an Educational Task. Dordrecht: Reidel.

[3] Kaiser G and Sriraman B. A global survey of international perspectives on modelling in mathematics education. ZDM - The International Journal on Mathematics Education. 2006. 38(3), 302310.

[4] Chevallard Y. On Didactic Transposition Theory: Some Introductory Notes. Communication à lÍnternational Symposium on Selected Domains of Research and Development in Mathematics Education (Bratislava, 3-7 aot 1988). Paru dans les Proceedings de ce symposium (Bratislava, 1989), pp. 51-62.

[5] Chevallard Y. La Transposition Didactique: Du Savoir Savant au Savoir Ensigné. Grenoble, La pensnée Sauvage, 1991.

[6] Haines C P and Crouch R. (2005). Getting to grips with real world contexts: Developing research in mathematical modelling. (To appear in) Proceedings of the 4th European Congress of Mathematics Education, in St. Feliu de Guixols, Spain, Feb 16-22, 2005.

[7] Blum W. (1996). Anwendungsbezüge im Mathematikunterricht Trends und Perspektiven. In G. Kadunz, H. Kautschitsch, G. Ossimitz and E. Schneider (Ed.), Trends und Perspektiven (pp. 15-38). Wien: HölderPichler-Tempsky.

[8] Silva J M, Jardim D F, Carius A C (2016). O ensino e a aprendizagem de conceitos de Cálculo usando modelos matemáticos e ferramentas tecnológicas. Revista de Ensino de Engenharia, 35(02):7080.

[9] Silva J M, Jardim D F, Carius A C, Silva A C. (2017). Teaching derivatives concepts with computational techniques. In Conference Proceedings. New Perspectives in Science Education, page 237. libreriauniversitaria. it Edizioni.

[10] Jardim D F, Silva J M, Pereira M M, Junior E A S, Nepomucena T V, Pinheiro, T R. (2015). Estudando Limites com o GeoGebra. Vozes dos Vales, 08(08):119.

[11] Dante L R. (2014). Matemática Contexto e aplicações, volume 01. Ática, São Paulo, 2a edition.

[12] Iezzi G, Dolce O, Degenszajn D, Périgo R, Almeida N. (2013). Matemática Ciência e Aplicações, volume 01. Saraiva, São Paulo, 7 edition.

[13] Souza J. (2013). Coleção: Novo Olhar-Matemática., volume 1. FTD, São Paulo.

[14] Diniz M and Stocco K. (2004). Matemática (Ensino M´dio), volume 1. Saraiva, São Paulo, 4a ed. reformulada edition.

[15] Araujo J L, Campos I S. (2015). Negotiating the Use of Mathematics in a Mathematical Modeling Project. In: Stillman, G. and Blum, W. and Bibemgut, M. S. and (Eds). Mathematical Modelling in Education Research and Practice. New York: Springer. Cap.23, p. 283-292.

[16] Stillman G A, Blum W., Biembengut M S. (2015). Cultural, Social, Cognitive and Research Influences on Mathematical Modelling Education. In: Stillman, G. and Blum, W. and Bibemgut, M. S. and (Eds). Mathematical Modelling in Education Research and Practice. New York: Springer. Cap.1, p. 1-32.

IOP Conf. Series: Journal of Physics: Conf. Series **936** (2017) 012070 doi:10.1088/1742-6596/936/1/012070

Modeling the dynamics of multipartite quantum systems created departing from two-level systems using general local and non-local interactions

Francisco Delgado

Departamento de Física y Matemáticas, Escuela de Ingeniería y Ciencias, Tecnológico de Monterrey, Campus Estado de México, Atizapán, Estado de México, CP. 52926, México.

E-mail: `fdelgado@itesm.mx`

Abstract. Quantum information is an emergent area merging physics, mathematics, computer science and engineering. To reach its technological goals, it is requiring adequate approaches to understand how to combine physical restrictions, computational approaches and technological requirements to get functional universal quantum information processing. This work presents the modeling and the analysis of certain general type of Hamiltonian representing several physical systems used in quantum information and establishing a dynamics reduction in a natural grammar for bipartite processing based on entangled states.

1. Introduction

Quantum computation [1, 2, 3] and quantum cryptography [4, 5] are based on disruptive quantum phenomena: superposition and entanglement. Binary nature of computer science is connected with quantum computation using two-level quantum systems through the qubit, a unipartite minimum element of information adopting two simultaneous states of existence. Multipartite systems obtained from the combination and interaction of qubits exhibit a notable dynamics complexity. There, Heisenberg-Ising model written in the Bell states basis reproduces quantum dynamics for magnetic systems exhibiting $SU(2)$ block decomposition for two-qubits [6, 7]. Hilbert space $\mathcal{H}^2$ becomes a direct sum of two subspaces, each one spanned by a pair of Bell states and $U \in U(1) \times SU(2)^2$, reducing the dynamics to well-known procedures in control theory [8, 9]. This paper demonstrates that such decomposition is achievable for larger systems. The second section states the Hamiltonian and the basis to reach the decomposition. The third section discusses the architecture and the structure of U. Last section includes the conclusions.

2. Hamiltonian for $2d$−partite two level systems and basis to reach $SU(2)$ reduction

Heisenberg-Ising model is exploited in developments as D-Wave and IBM-Q quantum computers because it reproduces the exchange information on spins. With $SU(2)$ reduction, a set of universal gates has been achieved with a notable degree of fidelity [10, 11, 12]. As first approach:

$$H = \sum_{\{i_k\}} h_{\{i_k\}} \bigotimes_{k=1}^{n} \sigma_{i_k} = \sum_{\mathcal{I}=0}^{4^n-1} h_{\mathcal{I}_4^n} \bigotimes_{k=1}^{n} \sigma_{\mathcal{I}_{4,k}^n} \tag{1}$$

for n qubits, with $\{i_k\} = \{i_1, i_2, ..., i_n\}$ and $i_k \in \{0,1,2,3\}$. There, $h_{\{i_k\}}$ is a set of time-dependent real functions. $\{i_k\}$ could be represented as the number $\mathcal{I} \in \{0,1,...,4^n-1\}$ in base-4 with n digits, $\mathcal{I}_4^n$. Then, $\mathcal{I}_{4,k}^n = i_k$ for $k = 1,2,...,n$. σ_{i_k} are the Pauli matrices in the computational basis $|0\rangle, |1\rangle \in \mathcal{H}$ for each qubit k. Hamiltonian obeys the Schrödinger equation for the evolution operator U. This Hamiltonian contains some unphysical terms, but we will restrict H to pair spin coupling like interactions H_{nl_i} together with local operations H_{l_i} generated by fields on each qubit, for an even number of them $n = 2d$. The Bell basis fits the evolution in the $U(1) \times SU(2)^2$ decomposition of $SU(4)$ for bipartite systems [6]. Here, the most obvious candidates are the generalized Bell states (GBS) for $n = 2d$ [13], a tensor product of Bell states settling an orthogonal basis of non-maximal entangled states for $2d$ qubits. Such elements are:

$$\left|\Psi_{\mathcal{I}_4^d}\right\rangle = \frac{1}{\sqrt{2^d}} \sum_{\mathcal{E},\mathcal{D}=0}^{2^d-1} (\tilde{\sigma}_{i_1} \otimes ... \otimes \tilde{\sigma}_{i_d})_{\mathcal{E}_2^d,\mathcal{D}_2^d} \left|\mathcal{E}_2^d\right\rangle \otimes \left|\mathcal{D}_2^d\right\rangle \tag{2}$$

At this point, $\tilde{\sigma}_i$ is a unitary factor of the traditional Pauli matrices (in fact, $\tilde{\sigma}_{i\neq 2} = \sigma_i, \tilde{\sigma}_2 = i\sigma_2$) [13]. $\mathcal{E}_2^d, \mathcal{D}_2^d$ are numbers in base-2 with d digits ($\mathcal{E}, \mathcal{D} \in \{0,1,...,2^d-1\}$) representing $\{\epsilon_1,...,\epsilon_d\}, \{\delta_1,...,\delta_d\}$ respectively, where $\epsilon_s, \delta_s \in \{0,1\}$ (digits appear reversed explicitly in $\mathcal{E}_2^d$ or $\mathcal{I}_4^d$ expressions). Noting $\left\langle \Psi_{\mathcal{I}_4^d} | \sigma_{j_1} \otimes ... \otimes \sigma_{j_{2d}} | \Psi_{\mathcal{K}_4^d} \right\rangle = \frac{1}{2^d} \prod_{s=1}^d \mathrm{Tr}(\tilde{\sigma}_{i_s}^* \sigma_{j_{d+s}} \tilde{\sigma}_{k_s}^T \sigma_{j_s}^T)$, then:

$$\left\langle \Psi_{\mathcal{I}_4^d} | H | \Psi_{\mathcal{K}_4^d} \right\rangle = \frac{1}{2^d} \sum_{\mathcal{J}=0}^{4^{2d}-1} h_{\mathcal{J}_4^{2d}} \prod_{s=1}^{d} \mathrm{Tr}(\tilde{\sigma}_{i_s}^* \sigma_{j_{d+s}} \tilde{\sigma}_{k_s}^T \sigma_{j_s}^T) \tag{3}$$

with $\mathcal{J} \in \{0,1,...,4^{2d}-1\}$. $\tilde{\sigma}_{i_s}^* \sigma_{j_{d+s}} \tilde{\sigma}_{k_s}^T \sigma_{j_s}^T$ has some properties inherited from Pauli matrices because $\sigma_1, \sigma_2, \sigma_3$ are traceless and $\sigma_i^T = \pm\sigma_i$ (negative sign for $i = 2$). $\mathrm{Tr}(\tilde{\sigma}_{i_s}^* \sigma_{j_{d+s}} \tilde{\sigma}_{k_s}^T \sigma_{j_s}^T)$ is non-zero only if i_s, j_{d+s}, k_s, j_s are: a) completely different among them, or b) equal by pairs. For non-local interactions, if there are non-local interactions between pairs s and $s+d$, for $s = 1,2,...,d$, then $\sigma_{j_s} = \sigma_{j_{d+s}}$, implying $\tilde{\sigma}_{i_s} = \tilde{\sigma}_{k_s}$ to get $\left\langle \Psi_{\mathcal{I}_4^d} | H | \Psi_{\mathcal{K}_4^d} \right\rangle \neq 0$. Then these terms are diagonal. These pairs will be called *correspondent* pairs in the following. For local interactions, only one matrix between $\sigma_{j_s}, \sigma_{j_{d+s}}$ will be different from σ_0, namely σ_i, then the remaining factors will be σ_j, σ_k, where i, j, k are a permutation from $1,2,3$. It implies they will give $\left\langle \Psi_{\mathcal{I}_4^d} | H_{\mathrm{l}_i} | \Psi_{\mathcal{K}_4^d} \right\rangle \neq 0$ only if $\mathcal{I}_4^d$ and $\mathcal{K}_4^d$ in $\left|\Psi_{\mathcal{I}_4^d}\right\rangle, \left|\Psi_{\mathcal{K}_4^d}\right\rangle$ differ in only one digit, following the exchange rules: $0 \leftrightarrow i$ or $j \leftrightarrow k$. This aspect, together with the diagonal entries of non-local interactions generate the 2×2 blocks in H (and U as a consequence).This is only possible if direction i is unique for the local interactions, otherwise the block structure becomes destroyed. Some remarks about the notation are convenient: a) $\tilde{\sigma}_i = \alpha_i \sigma_i$, $\alpha_i \in \{1, i\}$; b) $\sigma_i^T = \beta_i \sigma_i$, $\beta_i \in \{-1,1\}$; c) $\sigma_i \sigma_j = \gamma_{i,j} \sigma_j \sigma_i$, $\gamma_{i,j} \in \{-1,1\}$. Then, $2c_{j_s,j_{d+s}}^{i_s,k_s} \equiv \mathrm{Tr}(\tilde{\sigma}_{i_s} \sigma_{j_{d+s}} \tilde{\sigma}_{k_s}^T \sigma_{j_s}^T) = \alpha_{i_s} \alpha_{k_s} \beta_{j_s} \beta_{k_s} \gamma_{k_s j_s} \gamma_{k_s i_s} \mathrm{Tr}(\sigma_{i_s} \sigma_{k_s} \sigma_{j_{d+s}} \sigma_{j_s}) \in \{0, \pm 2, \pm 2i\}$. Formulas for $\alpha_i, \beta_i, \gamma_{i,j}, c_{j_s,j_{d+s}}^{i_s,k_s}$ can be obtained from the Pauli matrices properties.

3. Hamiltonian conditions to get $SU(2)$ reduction and related dynamics

For larger systems, analytical formulas for $\left\langle \Psi_{\mathcal{I}_4^d} | H_{\mathrm{nl}_i} | \Psi_{\mathcal{K}_4^d} \right\rangle$ and $\left\langle \Psi_{\mathcal{I}_4^d} | H_{\mathrm{l}_i} | \Psi_{\mathcal{K}_4^d} \right\rangle$ are useful for computer simulation optimality. We get them for the $SU(2)$ reduction case, with:

$$H = \bigoplus_{i=1}^{2^{2d-1}} \mathbf{S}_{Hi} \quad \Rightarrow \quad U = \bigoplus_{i=1}^{2^{2d-1}} \mathbf{S}_{Ui} \tag{4}$$

IOP Conf. Series: Journal of Physics: Conf. Series **936** (2017) 012070 doi:10.1088/1742-6596/936/1/012070

with each $\mathbf{S}_{Hi}$ a 2×2 matrix. The structure is preserved under matrix products, then inherited to U due to the time ordered integral: $\mathbf{S}_{Ui} = \tau\{e^{-\frac{i}{\hbar}\int_0^t \mathbf{S}_{Hi} dt'}\} \equiv e^{-i\Delta_i^+ t}\tau\{e^{-\frac{i}{\hbar}\int_0^t \mathbf{S}_{H_i^0} dt'}\}$, where $\mathbf{S}_{H_i^0} = \mathbf{S}_{Hi} - \Delta_i^+ \mathbf{I}_i$ is the part of $\mathbf{S}_{Hi}$ free of $\mathbf{I}_i$. $\mathbf{S}_{Ui} \in U(1) \times SU(2)$ as it will be seen. To get the $SU(2)$ decomposition and a closer approach to the Heisenberg-Ising model in [6], we state:

$$H^{(j,k')} = H_D + H_{ND}^{(j,k')} \quad \text{with}: \quad H_{ND}^{(j,k')} = \sum_{t'=0}^{1} h_{(j4^{k'}+dt'-1)_4^{2d}} \bigotimes_{s=1}^{2d} \sigma_{(j4^{k'}+dt'-1)_{4,s}^{2d}} \tag{5}$$

$$H_D \equiv \sum_{i'=1}^{3}\sum_{k=1}^{d} h_{(i'(4^{k-1}+4^{k+d-1}))_4^{2d}} \bigotimes_{s=1}^{2d} \sigma_{(i'(4^{k-1}+4^{k+d-1}))_{4,s}^{2d}}$$

generating the $SU(2)$ blocks with the diagonal entries coming from non-local interactions between correspondent parts and the non-diagonal terms from local interactions (only applied on the pair $[k', k'+d]$). Figure 1a shows the architecture being considered. Introducing $\delta_{\mathcal{IK}}^{\mathrm{S}} \equiv \prod_{s \notin S} \delta_{i_s k_s}$, with the set S of scripts excluded in the product and defining the exchange factor in the diagonal-off entries (Einstein summation convention understood):

$$\mathcal{F}_{i,k'}^{j,k} = \delta_{i_{k'}0}\delta_{k_{k'}i}c_{j,k}^{0,i} + \delta_{i_{k'}i}\delta_{k_{k'}0}c_{j,k}^{i,0} + \epsilon_{ii'i''}(\delta_{i_{k'}i'}\delta_{k_{k'}i''}c_{j,k}^{i',i''} + \delta_{i_{k'}i''}\delta_{k_{k'}i'}c_{j,k}^{i'',i'}) \tag{6}$$

by using the properties $c_{0,0}^{i_s,i_s} = 1$ and $c_{i,i}^{i_{k''},i_{k''}} = (-1)^{\delta_{i,2}+(1-\delta_{i,i_{k''}})(1-\delta_{0,i_{k''}})}$, then:

$$\begin{aligned}\left\langle \Psi_{\mathcal{I}_4^d} | H^{(j,k')} | \Psi_{\mathcal{K}_4^d} \right\rangle = {} & \delta_{\mathcal{IK}} \sum_{i'=1}^{3}\sum_{k''=1}^{d} c_{i',i'}^{i_{k''},i_{k''}} h_{(i'(4^{k''-1}+4^{k''+d-1}))_4^{2d}} + \\ & \sum_{t'=0}^{1} \delta_{\mathcal{IK}}^{\{k'\}} \mathcal{F}_{j,k'}^{j\delta_{0,t'},j\delta_{1,t'}} h_{(j4^{k'}+dt'-1)_4^{2d}} \equiv H_{D\mathcal{IK}} + H_{ND}^{(j,k')}{}_{\mathcal{IK}}\end{aligned} \tag{7}$$

$H_{D\mathcal{IK}}, H_{ND}^{(j,k')}{}_{\mathcal{IK}}$ denotes the diagonal and non-diagonal terms of the whole interaction. This formula shows that the pair of diagonal entries for each $SU(2) = \{h_{ij}\}, i,j = 1,2$ block are different: it is formed switching only one index $i_{k''}$ in the rows labels, then for $i' \neq j$ the terms in $H_{D\mathcal{IK}}$ have a sign change, implying $h_{11} \neq h_{22}$. We should note that the exchange of entanglement will be allowed only between correspondent pairs through the driven local interactions.

Hilbert space $\mathcal{H}^{2d}$ becomes the direct sum of 2^{2d-1} subspaces generated by pairs of GBS basis $\{\left|\Psi_{\mathcal{I}_4^d}\right\rangle, \left|\Psi_{\mathcal{K}_4^d}\right\rangle\}$ (those having only one index k' exchanged with the rules $0 \leftrightarrow i$, $j \leftrightarrow k$). As a consequence, only two different blocks are independent through the complete dynamics. There, the dynamics mixes the probabilities leaving unchanged those among subspaces. If $|\psi_0\rangle = \sum_{s=1}^{2^{2d-1}} |\psi_{0s}\rangle$ is the initial state with amplitudes $|\psi_{0s}\rangle = \psi_{0s,\mathcal{I}_4^d}\left|\alpha_{\mathcal{I}_4^d}\right\rangle + \psi_{0s,\mathcal{K}_4^d}\left|\alpha_{\mathcal{K}_4^d}\right\rangle$, then the evolved component for each $s = 1, 2, ..., 2^{2d-1}$ fulfills $\| |\psi_{ts}\rangle \| = \|\mathbf{S}_{Us}|\psi_{0s}\rangle\| = \| |\psi_{0s}\rangle \|$. Figure 1b depicts the dynamics evolving separately in Bloch spheres for each pair. To extend the exchange of entanglement between different pairs, a rearrangement among parts conforming the correspondent pairs should be generated, remixing the original pair probabilities. Exchange dynamics operates on the information states, not directly on physical parts of the system. Blocks in the rows $\mathcal{I}$ and $\mathcal{I}'$ have the form (for time independent Hamiltonian):

$$\mathbf{S}_{H\mathcal{I},\mathcal{I}'} = \frac{h_{11}+h_{22}}{2}\mathbf{I}_{\mathcal{I},\mathcal{I}'} + \mathbf{n}\cdot\mathbf{s}_{\mathcal{I},\mathcal{I}'} \tag{8}$$

$$\mathbf{S}_{U\mathcal{I},\mathcal{I}'} = e^{i\mathbf{S}_{H\mathcal{I},\mathcal{I}'}\frac{t}{\hbar}} = e^{i\frac{h_{11}+h_{22}}{2\hbar}t}(\cos\omega t + i\sin\omega t\mathbf{n}\cdot\mathbf{s}_{\mathcal{I},\mathcal{I}'}) \in U(1)\times SU(2) \tag{9}$$

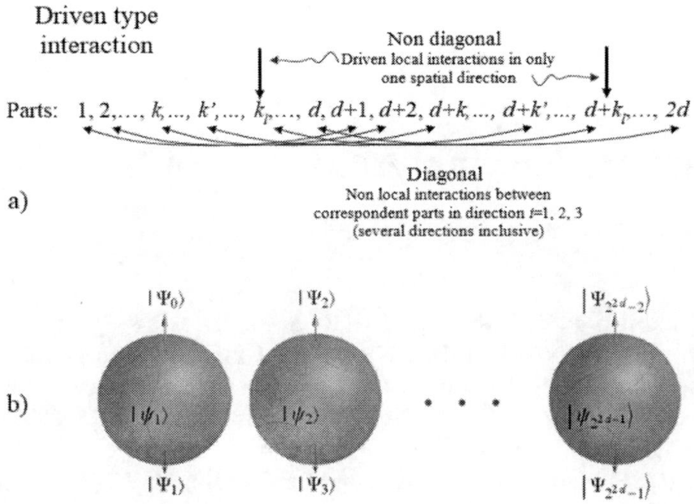

Figure 1. a) Interactions to achieve the $SU(2)$ decomposition, and b) probability exchange in temporary pairs evolving in separate Bloch spheres.

Here, $\{\mathbf{I}_{\mathcal{I},\mathcal{I}'}, \mathbf{X}_{\mathcal{I},\mathcal{I}'}, \mathbf{Y}_{\mathcal{I},\mathcal{I}'}, \mathbf{Z}_{\mathcal{I},\mathcal{I}'}\}$ is a Pauli basis for the $SU(2)$ block, $\mathbf{s}_{\mathcal{I},\mathcal{I}'} = (\mathbf{X}_{\mathcal{I},\mathcal{I}'}, \mathbf{Y}_{\mathcal{I},\mathcal{I}'}, \mathbf{Z}_{\mathcal{I},\mathcal{I}'})$, $\hbar\omega\mathbf{n} = (\mathrm{Re}(h_{12}), -\mathrm{Im}(h_{12}), \frac{h_{11}-h_{22}}{2})$ and $\mathbf{n}$ is a unitary vector defining ω. $\mathcal{F}_{j,k'}^{j_d,j_{d+s}}$ is imaginary only if j_d or j_{d+s} are 2, then only one n_1 or n_2 is non-zero. It completes the dynamics description.

4. Conclusions

The generic Hamiltonian modeled in this work comprises characteristic elements of the systems used for quantum information processing. Here, only a special kind of systems exhibiting the $SU(2)$ reduction have been modeled through of Hamiltonian (5), but still other possible interactions including this property should be considered as extension [14]. $SU(2)$ reduction lets understand the quantum processing as a simple series of operations manipulating directly the quantum information underlying the system. GBS basis works as a universal grammar for the group of analyzed systems having adaptable control elements as the driven local interactions presented here. Additional research to develop this grammar should be still developed.

Acknowledgments

The support from Tecnológico de Monterrey to develop this research is acknowledged.

References

[1] Feynman R 1982 *Int. J. Theor. Phys.* **21**, 467.
[2] Deutsch D 1985 *Proc. R. Soc. London* Ser. A **400** 97
[3] Steane A 1996 *Phys. Rev. Lett* **77** 793
[4] Bennett C and Brassard G 1984 *Proc. IEEE Intl. Conf. on Comp.* **175** 475
[5] Ekert A 1991 *Phys. Rev. Lett.* **67** 661
[6] Delgado F 2015 *Int. J. of Quant. Inf.* **13** (7) 1550055
[7] Delgado F 2016 *J. of Phys. Conf. Ser.* **648** 012024
[8] D'Alessandro D and Dahleh M 2001 *IEEE Transactions on Automatic Control* **46** (6) 866
[9] Boscain U and Mason P 2006 *J. Math. Phys.* **47** 062101
[10] Delgado F 2017 *Quant. Inf. and Comp.* **17** (9) 0721
[11] Delgado F 2017 *J. of Phys. Conf. Ser.* **839** 012014
[12] Bermúdez D and Delgado F 2017 *J. of Phys. Conf. Ser.* **839** 012016
[13] Sych D and Leuchs G 2009 *New Journal of Physics* **11** 013006
[14] Delgado F 2017 *Quantum Information and Measurement (QIM) 2017*, OSA Technical Digest (online) (Optical Society of America) paper QT6A.23

IOP Conf. Series: Journal of Physics: Conf. Series **936** (2017) 012071 doi:10.1088/1742-6596/936/1/012071

Investigating CH_4 Thermal Activation in Clathrate Nano-cages

A. Lakhlifi[1], P. R. Dahoo[2,3*], C. Meis[3,4] and J. D. Gale[3,5]

[1] Institut UTINAM, UMR CNRS 6213, Université Bourgogne Franche-Comté, Observatoire de Besançon, 25010 Besançon Cedex, France.

[2] LATMOS /IPSL, UVSQ Université Paris-Saclay, UPMC Univ. Paris 06, CNRS, F-78280, Guyancourt, France.

[3] CHAIR Materials Simulation and Engineering, UVSQ, Université Paris Saclay, 78035 Versailles Cedex, France.

[4] INSTN- CEA, Université Paris Saclay – 91191 Gif-sur-Yvette, France.

[5] Nanochemistry Research Institute/Curtin Institute for Computation, Department of Chemistry, Curtin University, PO Box U1987, Perth, WA 6845, Australia.

E-mail: pierre-richard.dahoo@uvsq.fr

Abstract. The energy levels of methane molecule trapped, at low temperature, in small (*s*) and large (*l*) nano-cages of cubic sI clathrates are calculated in the Born-Oppenheimer approximation using the Extended Lakhlifi-Dahoo model based on pairwise atom-atom effective interaction potentials. In the *s* cage, the center of mass of CH_4 exhibits a slightly asymmetrical 3D oscillation motion with small amplitude around the cage center. Two methods were used to calculate the frequencies of such a motion: a 3D harmonic treatment and a 1D Discrete Variable Representation (DVR) treatment in the *X*, *Y* and *Z* directions. They give approximately the same values of, respectively, 133 cm^{-1}, 108 cm^{-1} and 120 cm^{-1}. In the *l* cage, the oscillations are anharmonic and characterized by large amplitude motions with frequencies of 63 cm^{-1}, 52 cm^{-1} and 47 cm^{-1}. In the *s* and *l* nano-cages, the molecule exhibits strongly perturbed rotational motion. The rotational level schemes are quite different from that of the molecular free rotational motion, and for each nano-cage, the obtained levels are described as combination of the free rotation levels.

1. Introduction

A great number of research activities in the field of astrophysics are devoted to clathrates, crystalline solids formed by a compact assembly of nano-cages. Clathrates exist as three different kinds of Archimedes polyhedra characterized by two types of cubic structures termed sI and sII with a cage-like structure of water molecules surrounding atoms or small molecules and a hexagonal structure-H (sH) [1], which forms in the presence of large guest molecules. The unit cell of sI consists of 46 water molecules with two types of cages; 2 small 5^{12} cages (*s*), and 6 large $5^{12}6^{2}$ cages (*l*) [4]. In the sII structure unit cell there are 136 water molecules which form 16 small 5^{12} and 8 large $5^{12}6^{4}$ cages. In this work, we investigate the effect on the thermal properties of clathrates when they trap methane (CH_4) molecules. This effect is studied through the relaxation channels through which an IR excited CH_4 redistributes its energy. These channels are connected to the structure of the energy levels related to its low frequency motions and which are determined by solving the Schrodinger's eigen-value equation based on the

Hamiltonian of the encaged molecule. In the following, the Extended Lakhlifi-Dahoo model [2][3] based on pairwise atom-atom effective interaction potentials is applied to CH_4 in sI and sII types clathrate and the results discussed by comparison to previous results obtained on CO_2.

2. The Site inclusion model and interaction potential energy

To simulate the spectra of the CH_4 molecule trapped in clathrate nano-cages, the theoretical model is built as an extension of the site inclusion one applied to study C_2, CO_2 and N_2O molecules [4-6] trapped in rare gas matrices. Electrostatic interactions between atoms of CH_4 and H_2O, are modelled by the usual 12-6 Lennard-Jones (LJ) interatomic potentials at short range and mid-range distances and by effective charges placed on the atoms of the molecules at long range distances. The validity of the Born-Oppenheimer approximation is assumed to separate the vibrational degrees of freedom (high frequency motions) from the orientational and translational degrees of freedom (low frequency motions) of the system formed by the molecule and the nano-cage. Then the total potential energy of the trapped molecule inside the nano-cage is written as:

$$V = \sum_{j} V_{Mj}\left(\mathbf{r}_{0j}\right) + \sum_{jj', j\neq j'} V_{jj'}\left(\mathbf{r}_{jj'}\right) \tag{1}$$

where the term V_{Mj} is the molecule-cage interaction and $\mathbf{r}_{0j}$ is the distance vector between the centre of mass (c.m.) of CH_4 and the jth atom of the cage (s or l). The second term $V_{jj'}$ represents the interaction potential between two atoms of the cage: *i*) a jth atom, of one H_2O molecule, located at position $\mathbf{r}_j$ and *ii*) a j'^{th} atom, of another H_2O molecule of the cage, located at position $\mathbf{r}_{j'}$ with respect to an absolute frame (O, **X**, **Y**, **Z**) tied to the cage.

The interaction potential energy V_{Mj} between the CH_4 molecule and an atom of the cage is the sum of 12-6 LJ potentials representing the repulsion-dispersion contributions and the electrostatic interaction from effective charges q_i (i=1,..., 5 for CH_4) and q_j on the atoms of the H_2O molecules of the s or l cage, such that :

$$V_{Mj}\left(\mathbf{r}_{0j}\right) = \sum_{i=1}^{5}\left[4\varepsilon_{ij}\left\{\left(\frac{\sigma_{ij}}{r_{ij}}\right)^{12} - \left(\frac{\sigma_{ij}}{r_{ij}}\right)^{6}\right\} + \frac{1}{4\pi\varepsilon_0}\frac{q_i q_j e^2}{r_{ij}}\right] \tag{2}$$

where ε_{ij} and σ_{ij} are the mixed LJ potential parameters from the Lorentz-Berthelot combination rules $\varepsilon_{ij} = \sqrt{\varepsilon_{ii}\varepsilon_{jj}}$ and $\sigma_{ij} = \dfrac{\sigma_{ii} + \sigma_{jj}}{2}$. The distance vector $\mathbf{r}_{ij}$ between i and j atoms can be expressed in the molecular frame (G, **x**, **y**, **z**) or the crystal frame (O, **X**, **Y**, **Z**) as given in reference [2]. The orientational degrees of freedom, $\mathbf{\Omega} = (\varphi, \theta, \chi)$, of the molecule and the position vector of any atom, j, are defined in the absolute crystal frame (O, **X**, **Y**, **Z**).

In the rigid crystal approximation, the potential energy hypersurface of the molecule in its cage, n (n = s or l), can then be written as :

$$V_M^{(n)}\left(\xi^e, \mathbf{u}_0, \mathbf{\Omega}, \{Q\}\right) = V_M^{(n)e}\left(\xi^e\right) + V_M^{(n)}\left(\xi^e, \mathbf{u}_0\right) + V_M^{(n)}\left(\xi^e, \{Q\}\right) + V_M^{(n)}\left(\xi^e, \mathbf{\Omega}\right) + \Delta V_M^{(n)}\left(\xi^e, \mathbf{u}_0, \mathbf{\Omega}, \{Q\}\right) \tag{3}$$

where $V_M^{(n)e}(\xi^e)$ is the minimum of the potential energy corresponding to the equilibrium configuration; $V_M^{(n)}(\xi^e,\mathbf{u}_0)$ is the c.m. translation dependence in terms of the dynamical coordinate vector; $V_M^{(n)}(\xi^e,\{Q\})$ is the vibrational dependence in terms of normal coordinates $\{Q\}$; $V_M^{(n)}(\xi^e,\mathbf{\Omega})$ is the orientational dependence in terms of Euler angles $\mathbf{\Omega}=(\varphi,\theta,\chi)$; $\Delta V_M^{(n)}(\xi^e,\mathbf{u}_0,\mathbf{\Omega},\{Q\})$ characterizes the translation-orientation-vibration dynamic coupling term, which can induce the relaxation of vibrational modes onto the translational and orientational modes; and ξ^e represents the equilibrium configuration displacement vector of the c.m. of H_2O molecules of the cage and of the trapped CH_4, determined on the basis of the Green's functions of the perfect crystal.
From the potential energies calculated for the low frequency motions, the Hamiltonian can be solved for the eigen-values of the allowed translation and orientation in each cage (*s* or *l*);

$$H_M^{(n)} = H_{vib}^{(n)} + T_{rot}^{(n)} + T_{trans}^{(n)} + V_M^{(n)}(\xi^e,\mathbf{u}_0,\mathbf{\Omega},\{Q\}) \tag{4}$$

where the sum of the first three terms, $H_{vib}^{(n)} + T_{rot}^{(n)} + T_{trans}^{(n)}$, corresponds to the free CH_4 molecular Hamiltonian.

3. Results and discussion
Calculations were performed for the CH_4 molecule trapped in clathrate matrices: sI containing 4x4x4 unit cells (up to 2,944 water molecules) and sII with 3x3x3 unit cells (up to 3,672 water molecules). In the *s* cage of sI, the equilibrium configuration corresponds to a minimum energy value of -266 meV (-2,145 cm^{-1}), the molecular c.m. being located nearly at the cage center and oscillating with small amplitude. In the *l* cage, the minimum energy value is of -263 meV (-2,123 cm^{-1}), and the molecule is displaced from the cage center by about 0.3 Å in a direction parallel to the two hexagonal faces and exhibiting a translational motion of large amplitude. In the sII structure, the minimum energy values are of -238 meV and of -229 meV, respectively, for the *s* and *l* cages. These results show that trapping of CH_4 in sI is more favourable than in sII. As a result, we focus our quantum mechanical (molecular internal vibrational and orientational modes) computations on the CH_4-clathrate sI system, only.
Using the Born-Oppenheimer approximation, the c.m. of the rigid molecule is fixed at its equilibrium position in the cages, the orientational part $V_M^{(n)}(\varphi,\theta)$ of the interaction potential energy minimized with respect to the χ proper rotation (spinning) angle, is analysed. Figure 1 gives the 2D contour plot of the energy calculated for CH_4 trapped in the small (a) and large (b) cages.

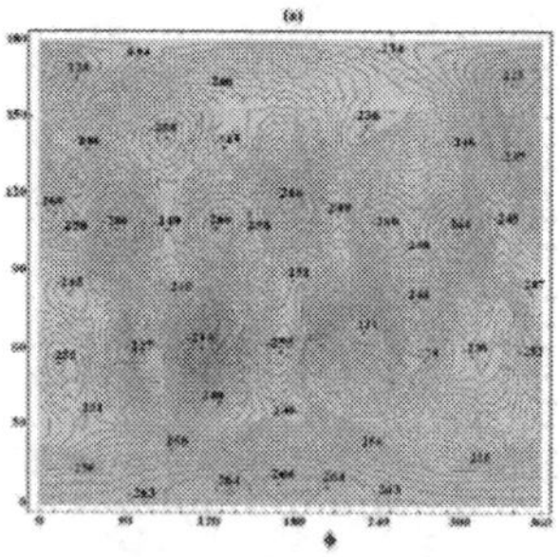

Figure 1 a). Potential energy of CH_4 as a function of orientation in cage *s*.

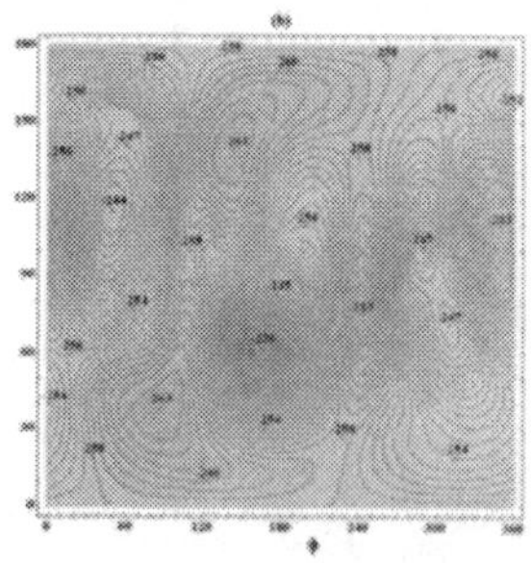

Figure 1 b). Potential energy of CH_4 as a function of orientation in cage *l*.

IOP Conf. Series: Journal of Physics: Conf. Series **936** (2017) 012071 doi:10.1088/1742-6596/936/1/012071

The Figures show that the molecule exhibits a rather perturbed rotational motion such that the calculated rotational eigen-vectors are linear combinations of eigen states linked to the free rotation of the molecule. It is then necessary to define new quantum numbers to label the resulting eigen-states as given by:

$$\left|j\gamma\right\rangle^{(n)} = \sum_{JMK} \Lambda_{JMK}^{j\gamma} \left|JMK\right\rangle \tag{5}$$

where j is an integer number ($j = 0, 1, 2, \ldots$) and γ takes values: *i)* $\gamma = 0$, for simply degenerate states, for the values $K = 0, \pm 3, \pm 6, \ldots$ of the spinning number and *ii)* $\gamma = \pm$, doubly degenerate states, for the other K values. This is due to the symmetry C_{3v} of the CH_4 molecule. Finally, when the transition elements of the IR rovibrational active modes ν_3 and ν_4 of CH_4 are calculated, the IR spectra are comprised of an intense Q line associated with vibrational transition surrounded by many R branch (blue shift) and P branch (red shift).

The level schemes in the small (Figure 2a) and in the large (Figure 2b) cages are quite different from that of the CH_4 free rotation motion as shown in Figure 2c.

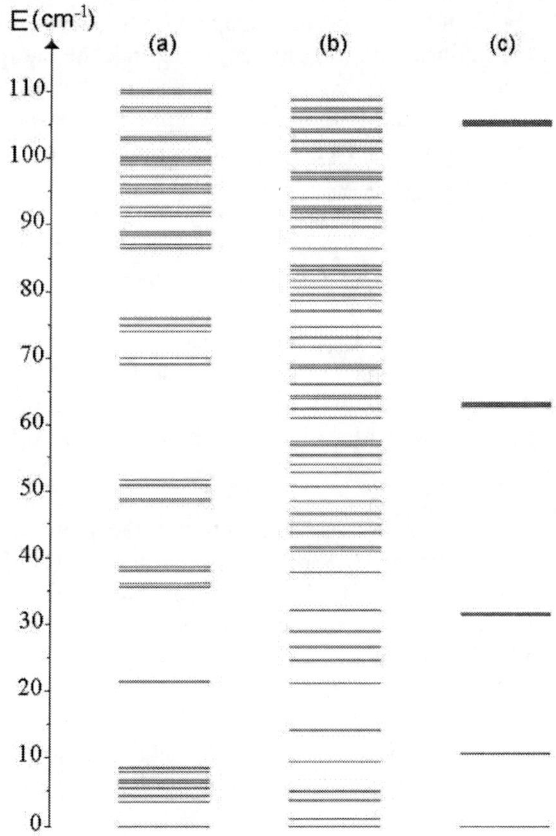

Figure 2. Calculated rotational level schemes of CH_4 in the small (a) and large (b) cages of sI structure. Level scheme (c) corresponds to that of the free molecule. Green levels are doubly degenerate and red ones are simply degenerate.

In the small cage, the orientational levels are roughly grouped in packets of levels in the range of 2 to 4 cm^{-1}, separated by energy gaps of 12 to 18 cm^{-1} in an interval of 110 cm^{-1}. In the large cage, the maximum energy gap between the levels is about 7 cm^{-1}, and in the same interval of 110 cm^{-1}, the levels appear as almost continuous compared to the small cage case. In comparison, calculations performed on CO_2 [2] led to fewer levels within the same range, with 3 levels separated by a first energy gap of 55 cm^{-1} and the next one of about 40 cm^{-1} in the small cage and only 8 levels separated by 30 cm^{-1} for the first two levels and less than 20 cm^{-1} for the other levels. To analyze these structures in terms of thermal properties of the clathrate trapping CH_4 or CO_2, it is necessary to link the redistribution of energy from the excited states of these structures to the solid environment.

Thermal effects are due to the excitation of the phonons that constitutes the quantized thermal bath which absorbs the energy dissipated in the condensed phase matter during a non-radiative transition of an excited molecule. The vibrational or orientational energy of the molecule is thus transferred to this reservoir in the form of kinetic energy to the solid environment thus increasing the temperature of the surroundings. The number of phonons involved in this process is proportional to the energy difference between the levels involved in the transition. This phenomenon has been modelled from two different approaches: one that is based on the short range repulsive forces between the molecule and its environment and assumed to be responsible for the transfer to phonons [7-13] possibly assisted by the rotational motion [14] and one based on the binary collision model [15] initially developed to interpret vibration-translation (V-T) transfers in liquids [16-17] and gases [18].

In the case of CO_2, observations from energy transfer experimental studies carried on low temperature CO_2-rare gas solid systems by laser induced fluorescence and double resonance techniques have shown that radiative relaxation was the main relaxation channel by which the energy from excited vibration-orientational levels was redistributed. Calculations performed on such systems led to energy gaps between the orientational levels corresponding to more than 5 phonons of the different rare gas matrices. Non-radiative relaxation were therefore less probable than radiative relaxation from upper excited levels. Based on these results, it was postulated that since the calculated orientational levels of CO_2 in clathrates had structures similar to those determined in rare gas matrices, then radiative relaxation were the main pathways for redistribution of the energy of excited levels of CO_2 and that the thermal properties of clathrates containing trapped CO_2 were not significantly changed [2].

In the case of CH_4, the situation is quite different because the energy level structures of CH_4 and CO_2 are very different in clathrates. Then, the energy gaps determined for CH_4 should involve 3 or less phonons, which is in favor of non-radiative relaxation in contrast to the CO_2 case which involves more than 5 phonons. The thermal properties of CH_4-clathrate systems should therefore be significantly modified and the nano-cages may free the CH_4 molecules if the temperature rise is significant.

4. Conclusions

In this work, the low frequency energy states of CH_4 trapped in the *s* and *l* nano-cages of cubic sI type clathrate are numerically calculated from a theoretical model and analysed by comparison to former studies on CO_2 trapped in clathrates. The results show that although the molecule is rather isolated in its cage, non-radiative relaxation processes are expected to be more important than radiative ones, in contrast to results obtained for CO_2-clathrate systems.

Acknowledgments

The authors would like to thank CHAIR Materials Simulation and Engineering, UVSQ, Université Paris Saclay for funding. Calculations were performed on computers from UTINAM Institute.

References

[1] E. D. Sloan, 2004, *Am. Mineral.* **89**, 1155
[2] A. Lakhlifi, P. R. Dahoo and E. Chassefière, 2017, *J.Q.S.R.T.* **187**, 124-134
[3] P. R. Dahoo, R. Puig, A. Lakhlifi, C. Meis and J. D. Gale, 2016, *J.P.C.S.* **738**, 012072
[4] Lakhlifi A., Dahoo P. R., Vala M., Szczepanski J. and Ekern S., 1997 *Chem. Phys.* **222**, 241.

[5] Lakhlifi A., Chabbi H., Dahoo P. R. and Teffo J. L., 2000 *Eur. Phys. J. D.* **12**, 435.
[6] Dahoo P. R., Lakhlifi A., Chabbi H. and Coanga J. M., 2006 *J. of Mol. Struc.* **786,** 157.
[7] Nitzan A., Jortner J., « Line shape of a molecular resonance », *Mol. Phys.*, vol. 24, p. 109-131, 1972.
[8] Nitzan A., Mukamel S., Jortner J., « Some features of vibrational relaxation of a diatomic molecule in a dense medium », *J. Chem. Phys.*, vol. 60, p. 3929, 1974.
[9] Nitzan A., Silbey R.J., « Relaxation in simple quantum systems », *J. Chem. Phys.*, vol. 60, 1974, p. 4070, 1974.
[10] Mukamel S., Jortner J., « Lifetimes for resonance fluorescence and near resonance Raman scattering », *J. Chem. Phys.*, vol. 62, p. 3609-3615, 1975.
[11] Berkowitz M., Gerber R.B., « Vibrational relaxation of molecules in solids: The role of rotational and of translational local modes », *Chem. Phys. Lett.*, vol. 49, p. 260-264, 1977.
[12] Gerber R.B., Berkowitz M., « Role of Rotational and Translational Local Modes in Vibrational Relaxation in Solids: A Study of NH and ND in Solid Ar », *Phys. Rev. Lett.*, vol. 39, p. 1000, 1977.
[13] Berkowitz M., Gerber R.B., « Rotational mode participation in long-distance vibrational energy transfer in solids », *Chem. Phys. Lett.*, vol. 56, p. 105-108, 1978.
[14] Legay F., Chemical and Biochemical Applications of Lasers, vol. 2, C.B. Moore, New York, 1977.
[15] Sun H.Y., Rice S.A., « Conjecture on the Rate of Vibrational Relaxation of a Diatomic Molecule in a Monatomic Lattice », *J. Chem. Phys.*, vol. 42, p. 3826, 1968.
[16] Zwanzig R.W., « Theory of Vibrational Relaxation in Liquids », *J. Chem. Phys.*, vol. 34, 1961, p. 1931.
[17] Litovitz T.A., « Theory of Ultrasonic Thermal Relaxation Times in Liquids », *J. Chem. Phys.*, vol. 26, p. 469, 1967.
[18] Schwartz R.N., Slawsky Z.I., Herzfeld K.F., « Calculation of Vibrational Relaxation Times in Gases », *J. Chem. Phys.*, vol. 20, p. 1591, 1952.

IOP Conf. Series: Journal of Physics: Conf. Series **936** (2017) 012072 doi:10.1088/1742-6596/936/1/012072

ASYMPTOTIC SYMMETRIES IN 3-DIM GENERAL RELATIVITY: THE B(2,1) CASE

Evangelos Melas

University of Athens, Department of Economics, Unit of Mathematics and Informatics, Sofokleous 1, 10559 Athens, Greece

E-mail: emelas@econ.uoa.gr

Abstract. The ordinary Bondi–Metzner–Sachs (BMS) group B is the common asymptotic symmetry group of all asymptotically flat Lorentzian space–times. As such, B is the best candidate for the universal symmetry group of General Relativity (G.R.). Strongly continuous unitary irreducible representations (IRs) of $B(2,1)$, the analogue of B in three space–time dimensions, are analysed in the Hilbert topology. It is proved that all IRs of $B(2,1)$ are induced from IRs of compact 'little groups', which are the closed subgroups of $SO(2)$. It is proved that *all* IRs of $B(2,1)$ are obtained by Wigner–Mackey's inducing construction notwithstanding the fact that $B(2,1)$ is not locally compact in the employed Hilbert topology.

1. Introduction

The Bondi–Metzner–Sachs (BMS) group B is the common asymptotic group of all curved real Lorentzian space–times which are asymptotically flat in future null directions [1, 2], and is the best candidate for the universal symmetry group of G.R..

In 1939 Wigner laid the foundations of special relativistic quantum mechanics [3] and relativistic quantum field theory by constructing the Hilbert space strongly continuous unitary irreducible representations (IRs) of the (universal cover) of the Poincare group P.

The universal property of B for G.R. makes it reasonable to attempt to lay a similarly firm foundation for quantum gravity by following through the analogue of Wigner's programme with B replacing P. Some years ago McCarthy constructed explicitly [4, 5, 6, 7, 8, 9, 10, 11] the IRs of B for exactly this purpose. This work was based on G.W.Mackey's pioneering work on group representations [3, 12, 13, 14, 15]; in particular McCarthy's work extended G.W.Mackey's work to the relevant infinite–dimensional case.

It is difficult to overemphasize the importance of Piard's results [16, 17] who soon afterwards proved that *all* the IRs of B, when this is equipped with the Hibert topology, are derivable by the inducing construction. This proves the exhaustivity of McCarthy's list of representations and renders his results even more important.

Here, we follow this programme for 3–dim G.R. and construct in the Hilbert topology the IRs of $B(2,1)$, the analogue of B in three space–time dimensions. It is proved that all IRs of $B(2,1)$ are induced from IRs of *compact* 'little groups'. It follows that some IRS of $B(2,1)$ are controlled by IRs of the finite symmetry groups of regular polygons in ordinary euclidean 2–space. It is proved that *all* IRS of $B(2,1)$ are induced by the IRs of its little groups notwithstanding the fact that $B(2,1)$ is *not* locally compact in the employed Hilbert topology. The paper closes with the explicit construction of the IRs of $B(2,1)$.

2. The group $B(2,1)$

We turn now to the study of to $B(2,1)$, the analogue of B in three space–time dimensions.

2.1. The group $B^{2,1}(N^+)$

Recall that the $2+1$ Minkowski space is the vector space R^3 of row vectors with 3 real components, with the inner product defined as follows. Let $x, y \in R^3$ have components x^μ and y^μ respectively, where $\mu = 0, 1, 2$. Define the inner product $x.y$ between x and y by

$$x.y = x^0 y^0 - x^1 y^1 - x^2 y^2. \tag{1}$$

Then the $2+1$ Minkowski space, sometimes written $R^{2,1}$, is just R^3 with this inner product. The "2,1" refers to the one plus and two minus signs in the inner product. Let $SO(2,1)$ be the (connected component of the identity element of the) group of linear transformations preserving the inner product. Matrices $\Lambda \in SO(2,1)$ are taken as acting by matrix multiplication from the right, $x \longmapsto x\Lambda$, on row vectors $x \in R^{2,1}$.

The future null cone $N^+ \subset R^{2,1}$ is just the set of nonzero vectors with zero length and $x^0 > 0$:

$$N^+ = \left\{ x \in R^{2,1} | x.x = 0, x^0 > 0 \right\}. \tag{2}$$

Let R^*_+ denote the multiplicative group of all *positive* real numbers. Obviously, if $x \in N^+$, then $tx \in N^+$ for any $t \in R^*_+$. Let $F_1(N^+)$ denote the vector space (under pointwise addition) of all functions $f : N^+ \to R$ satisfying the homogeneity condition

$$f(tx) = tf(x) \tag{3}$$

for all $x \in N^+$ and $t \in R^*_+$. Define a representation T of $SO(2,1)$ on $F_1(N^+)$ by setting, for each $x \in N^+$ and $\Lambda \in SO(2,1)$,

$$(T(\Lambda)f)(x) = f(x\Lambda). \tag{4}$$

Now let $B^{2,1}(N^+)$ be the semi–direct product

$$B^{2,1}(N^+) = F_1(N^+) \rtimes_T SO(2,1). \tag{5}$$

That is to say, $B^{2,1}(N^+)$ is, as a set, just the product $F_1(N^+) \times SO(2,1)$, and the group multiplication law for pairs is

$$(f_1, \Lambda_1)(f_2, \Lambda_2) = (f_1 + T(\Lambda_1)f_2, \Lambda_1\Lambda_2). \tag{6}$$

2.2. The double cover $B^{2,1}(N^+)_c$

Let $\mathrm{SL}(2,R)$ be the group of all real 2×2 matrices with determinant one. $\mathrm{SL}(2,R)$ is sometimes denoted by G below. Let $\mathrm{M}_s(2,R)$ be the set of all 2×2 symmetric real matrices. We define a right action of G on $\mathrm{M}_s(2,R)$ by $\mathrm{M}_s(2,R) \times G \to \mathrm{M}_s(2,R)$ with

$$(m, g) \mapsto g^\top m g, \tag{7}$$

where the superscript $\top$ means transpose. Clearly any element $\mu \in \mathrm{M}_s(2,R)$ can be parameterized as follows:

$$\mu = \begin{bmatrix} x^o - x^1 & x^2 \\ x^2 & x^o + x^1 \end{bmatrix}$$

where $x^o, x^1, x^2 \in R$. We now consider the map $b: R^3 \to \mathrm{M}_s(2, R)$ defined by

$$b(x) = \begin{bmatrix} x^o - x^1 & x^2 \\ x^2 & x^o + x^1 \end{bmatrix}, \tag{8}$$

where the x^μ are the components of $x \in R^3$. This map is a linear bijection, so the right action of G on $\mathrm{M}_s(2, R)$ induces a linear right action of G on R^3. Since

$$\det(b(x)) = x \cdot x \tag{9}$$

and the G action preserves determinants (indeed $\det g = 1$) in $\mathrm{M}_s(2, R)$, G acts as transformations from $SO(2, 1)$. In fact, this construction gives an homomorphism

$$\gamma : G \to SO(2, 1) \tag{10}$$

which is onto, and has kernel $Z_2 = \{Id, -Id\}$ in G, Id denoting the identity element of G. Thus γ identifies G as the double cover of $SO(2, 1)$

$$G = SO(2, 1)_c. \tag{11}$$

Therefore, the double cover of the group $B^{2,1}(N^+)$, given in (5), has the form

$$B^{2,1}(N^+)_c = F_1(N^+) \rtimes_T \mathrm{SL}(2, R). \tag{12}$$

Strictly speaking, "T" should read "$T\gamma$", but the notation is simpler as above.

2.3. The group $B(2, 1)$

So far, the supertranslation space $F_1(N^+)$ has been defined as a space of truly *arbitrary* homogeneous functions of degree one. This has been merely for clarity; for physical applications, it is necessary to give this space additional structure. For reasons discussed in detail in McCarthy [18], we now give a new realization of $B^{2,1}(N^+)_c$ where the supertranslation space is restricted to be the separable Hilbert space $L^2(P_1(R), \lambda, R)$ of real−valued functions defined on $P_1(R) \simeq S^1$; functions square integrable with respect to the standard normalized (Lesbegue) measure λ on $P_1(R) \simeq S^1$; $P_1(R) \equiv S^1/Z_2$ is the one−dimensional real projective space (the circle quotient the antipodal map).

In particular in [19] the following Theorem is proved:

Theorem 1 *The group $B^{2,1}(N^+)_c$ can be realised as*

$$B(2, 1) = L^2(P_1(R), \lambda, R) \textcircled{s}_T G \tag{13}$$

with semi−direct product specified by

$$(T(g)\alpha)(x) = \kappa_g(x)\alpha(xg), \tag{14}$$

where $G = \mathrm{SL}(2,R)$, $g \in G$, $\alpha \in L^2(P_1(R), \lambda, R)$. Moreover, if

$$g = \begin{bmatrix} a & b \\ c & d \end{bmatrix}, \tag{15}$$

then the components x_1, x_2 of $\mathbf{x} \in R^2 - \{0\}$ transform linearly under g, so that the ratio $x = x_1/x_2$, $x_2 \neq 0$, transforms fraction linearly under g. Writing xg for the transformed ratio,

$$xg = \frac{(\mathbf{x}g)_1}{(\mathbf{x}g)_2} = \frac{x_1 a + x_2 c}{x_1 b + x_2 d} = \frac{xa + c}{xb + d}. \tag{16}$$

The ratio $x = x_1/x_2$, $x_2 \neq 0$, is a local inhomogeneous coordinate of $P_1(R)$. We denote our final realization of our group by $B(2,1)$ to distinguish it from the previous realizations $B^{2,1}(N^+)$ and $B^{2,1}(N^+)_c$. In analogy to B, it is natural to choose a measure λ on $P_1(R)$ which is invariant under the maximal compact subgroup $SO(2)$ of G; we choose λ to be the standard normalized Lesbegue measure $\mathrm{d}\lambda = \frac{\mathrm{d}\theta}{2\pi}$.

The factor $\kappa_g(x)$ on the right hand side of (14) is defined by

$$\kappa_g(x) = \frac{(xb+d)^2 + (xa+c)^2}{1+x^2}. \tag{17}$$

It is well known [20] that the topological dual of a Hilbert space can be identified with the Hilbert space itself, so that we have $L^{2'}(P_1(R), \lambda, R) \simeq L^2(P_1(R), \lambda, R)$. In fact, given a continuous linear functional $\phi \in L^{2'}(P_1(R), \lambda, R)$, we can write, for $\alpha \in L^2(P_1(R), \lambda, R)$

$$(\phi, \alpha) = < \phi, \alpha >, \tag{18}$$

where the function $\phi \in L^2(P_1(R), \lambda, R)$ on the right is uniquely determined by (and denoted by the same symbol as) the linear functional $\phi \in L^{2'}(P_1(R), \lambda, R)$ on the left. The representation theory of $B(2,1)$ is governed by the dual action T' of G on the topological dual $L^{2'}(P_1(R), \lambda, R)$ of $L^2(P_1(R), \lambda, R)$. The dual action T' is defined by:

$$< T'(g)\phi, \alpha > = < \phi, T(g^{-1})\alpha > \cdot \tag{19}$$

A short calculation gives

$$(T'(g)\phi)(x) = \kappa_g^{-2}(x)\phi(xg). \tag{20}$$

Now, this action T' of G on $L^{2'}(P_1(R), \lambda, R)$, given explicitly above is, like the action T of G on $L^2(P_1(R), \lambda, R)$, continuous. The 'little group' L_ϕ of any $\phi \in L^{2'}(P_1(R), \lambda, R)$ is the stabilizer

$$L_\phi = \{g \in G \mid T'(g)\phi = \phi\}. \tag{21}$$

By continuity, $L_\phi \subset G$ is a closed subgroup.

3. Representation theory

Let A and $\mathcal{G}$ be topological groups, and let T be a given homomorphism from $\mathcal{G}$ into the group of automorphisms $\mathrm{Aut}(A)$ of A. Suppose A is abelian and $\mathcal{H} = A \rtimes_T \mathcal{G}$ is the semi−direct product of A and $\mathcal{G}$, specified by the continuous action $T : \mathcal{G} \longrightarrow \mathrm{Aut}(A)$. In the product topology of $A \times \mathcal{G}$, $\mathcal{H}$ then becomes a topological group. It is assumed that it becomes a separable locally compact topological group.
In order to give the operators of the induced representations explicitly it is necessary ([3], [12], [13], [14], [15] and references therein) to give the following information:

(i) An irreducible unitary representation U of L_{ϕ_o} on a Hilbert space D for each L_{ϕ_o}.

(ii) A $\mathcal{G}$ −quasi−invariant measure μ on each orbit $\mathcal{G}\phi \approx \mathcal{G}/L_{\phi_o}$, where L_{ϕ_o} denotes the little group of the base point $\phi_o \in A'$ of the orbit $\mathcal{G}\phi_o$; A' is the topological dual of A.

Let D_μ be the space of functions $\psi : \mathcal{G} \to D$ which satisfy the conditions

$$(a) \qquad \psi(gl) = U(l^{-1})\psi(g) \qquad (g \in \mathcal{G},\ l \in L_\phi)$$

$$(b) \quad \int_{\mathcal{G}\phi_o} < \psi(q), \psi(q) > \mathrm{d}\mu(q) < \infty,$$

where the scalar product under the integral sign is that of D. Note, that the constraint (a) implies that $<\psi(gl),\psi(gl)>=<\psi(g),\psi(g)>$, and therefore the inner product $<\psi(g),\psi(g)>$, $g\in\mathcal{G}$, is constant along every element q of the coset space $\mathcal{G}/L_{\phi_o}\approx\mathcal{G}\phi_o$. This allows to assign a meaning to $<\psi(q),\psi(q)>$, where $q=gL_{\phi_o}$, by defining $<\psi(q),\psi(q)>:=<\psi(g),\psi(g)>$. Thus the integrand in (b) becomes meaningful due to the condition (a). A pre–Hilbert space structure can now be given to D_μ by defining the scalar product

$$<\psi_1,\psi_2>=\int_{\mathcal{G}\phi_o}<\psi_1(q),\psi_2(q)>\mathrm{d}\mu(q), \tag{22}$$

where $\psi_1,\psi_2\in D_\mu$. It is convenient to complete the space D_μ with respect to the norm defined by the scalar product (22). In the resulting Hilbert space, functions are identified whenever they differ, at most, on a set of μ–measure zero. Thus our Hilbert space is

$$D_\mu=L^2(\mathcal{G}\phi_o,\mu,D). \tag{23}$$

Define now an action of $\mathcal{H}=A\rtimes_T\mathcal{G}$ on D_μ by

$$(g_o\psi)(q) = \sqrt{\frac{\mathrm{d}\mu_{g_o}}{\mathrm{d}\mu}(q)}\psi(g_o^{-1}q), \tag{24}$$

$$\alpha\psi(q) = e^{i<g_o\phi_o,\alpha>}\psi(q) \tag{25}$$

where, $g_o\in\mathcal{G}$, $q\in\mathcal{G}\phi_o$, and $\alpha\in A$. Eqs. (24) and (25) define the IRs of $B(2,1)$ induced for *each* $\phi_o\in A'$ and *each* irreducible representation U of L_{ϕ_o}. The 'Jacobian' $\frac{\mathrm{d}\mu_{g_o}}{\mathrm{d}\mu}$ of the group transformation is known as the *Radon–Nikodym* derivative of μ_{g_o} with respect to μ and ensures that the resulting IRs of $B(2,1)$ are unitary.
The central results of induced representation theory ([3], [12], [13], [14], [15] and references therein) are the following:

(i) Given the topological restrictions on $\mathcal{H}=A\rtimes_T\mathcal{G}$ (separability and local compactness), any representation of $\mathcal{H}$, constructed by the method above, is irreducible if the representation U of L_{ϕ_o} on D is irreducible. Thus an irreducible representation of $\mathcal{H}$ is obtained for *each* $\phi_o\in A'$ and *each* irreducible representation U of L_{ϕ_o}.

(ii) If $\mathcal{H}=A\rtimes_T\mathcal{G}$ is a regular semi–direct product (i.e., A' contains a Borel subset which meets each orbit in A' under $\mathcal{H}$ in just one point) then *all* of its irreducible representations can be obtained in this way.

4. Obstructions and resolutions

Two remarks are in order regarding the representations of $B(2,1)$ obtained by the above construction:

(i) As it is explained in [19] the subgroup $L^2(P_1(R),\lambda,R)$ of $B(2,1)=L^2(P_1(R),\lambda,R)\rtimes_T G$ is topologised as a (pre) Hilbert space by using a natural measure on $P_1(R)$ and by introducing a scalar product into $L^2(P_1(R),\lambda,R)$. If R^4 is endowed with the natural metric topology then the group $G=SL(2,R)$, considered as a subset of R^4, inherits the induced topology on G. In the product topology of $L^2(P_1(R),\lambda,R)\times G$, $B(2,1)$ is a non–locally compact group (the proof follows, without substantial change, Cantoni's proof [21], see also [4]). (In fact the subgroup $L^2(P_1(R),\lambda,R)$, and therefore the group $B(2,1)$ can be employed with many different topologies. The Hilbert type topology employed here appears to describe quantum mechanical systems in asymptotically flat space–times [10]). Since in the Hilbert type topology $B(2,1)=L^2(P_1(R),\lambda,R)\rtimes_T G$ is not locally compact the theorems dealing with the irreducibility of the representations obtained by the above construction no longer apply (see e.g. [13]). However, it can be proved that the induced representations obtained above *are* irreducible. The proof follows very closely the one given in [7] for the case of the original BMS group B.

(ii) Here it is assumed that $B(2,1)$ is equipped with the Hilbert topology. It is of outmost significance that it can be proved [19] that in this topology $B(2,1)$ is a regular semi−direct−product. The proof follows the corresponding proof [16, 17] for the group B. Regularity amounts to the fact [12] that $L^{2'}(P_1(R), \lambda, R)$ can have no equivalent classes of quasi−invariant measures μ such that the action of $\mathcal{G}$ is strictly ergodic with respect to μ. When such measures μ do exist it can be proved [12] that an irreducible representation of the group, with the semi−direct−product structure at hand, may be associated with each that is not equivalent to any of the IRs constructed by the Wigner−Mackey's inducing method. In a different topology it is not known if $B(2,1)$ is a regular or irregular semi−direct−product. Irregularity of $B(2,1)$ in a topology different from the Hilbert topology would imply that there are IRs of $B(2,1)$ that are not equivalent to any of the IRs obtained above by the inducing construction. Strictly ergodic actions are notoriously hard to deal with even in the locally compact case. Indeed, for locally compact non−regular semi−direct products, there is no known example for which all inequivalent irreducibles arising from strictly ergodic actions have been found. For the other 41 groups defined in [18] regularity has only been proved for B [16, 17] when B is equipped with the Hilbert topology. Similar remarks apply to all of them regarding IRs arising from strictly ergodic actions in a given topology.

5. Results

In [19] it is proved that when $B(2,1)$ is employed with the Hilbert topology *all* little groups of $B(2,1)$ are compact. In particular the following Theorem is proved

Theorem 2 *The little groups for $B(2,1)$ are precisely the closed subgroups of $K = SO(2)$ which contain the element $-I$, I being the identity element of G. These are (A) K itself, and (B) the cyclic groups of even order.*

Moreover in [19] it is shown that the Wigner−Mackey's inducing construction is exhaustive despite the fact that $B(2,1)$ is not locally compact in the employed Hilbert topology. This result is rather important because other group theoretical approaches to quantum gravity which invoke Wigner−Mackey's inducing construction (see e.g. [22]) are typically plagued by the non−exhaustiveness of the inducing construction which results precisely from the fact that the group in question is not locally compact in the prescribed topology. Exhaustiveness is not just a mathematical nicety: If the inducing construction is not exhaustive one cannot simply know if the most interesting information or part of it is coded in the irreducibles which cannot be found by the Wigner−Mackey's inducing procedure. These results, i.e. compactness of the little groups and exhaustiveness of the inducing construction, not only are significant for the group theoretical approach to quantum gravity advocated here, but also they have repercussions [19] for other approaches to quantum gravity.

References

[1] Bondi H, Van Der Berg M G J and Metzner A W K 1962 *Proc. R. Soc. Lond. A* **269** 21
[2] Sachs R K 1962 *Proc. R. Soc. Lond. A* **270** 103
[3] Wigner E 1939 *Annals of Mathematics* **40** 149
[4] McCarthy P J 1972 *Proc. R. Soc. Lond. A.* **330** 517
[5] McCarthy P J 1972 *J. Math. Phys.* **13** 1837

[6] McCarthy P J 1973 *Proc. R. Soc. Lond. A* **333** 317
[7] McCarthy P J and Crampin M 1973 *Proc. R. Soc. Lond. A.* **335** 301
[8] McCarthy P J 1975 *Proc. R. Soc. Lond. A* **343** 489
[9] McCarthy P J 1978 *Proc. R. Soc. Lond. A* **358** 141
[10] McCarthy P J and Crampin M 1974 *Phys. Rev. Lett.* **33** 547
[11] McCarthy P J 1972 *Phys. Rev. Lett.* **29** 817
[12] Mackey G W 1968 *Induced representations of groups and quantum mechanics* Benjamin
[13] Mackey G W 1955 *The theory of group representations* The University of Chicago Press
[14] Simms D J 1968 *Lie groups and quantum mechanics* (Bonn notes) Berlin: Springer
[15] Isham C J 1984 Topological and quantum aspects of quantum theory in *Relativity groups and topology* DeWitt B S and Stora R eds
[16] Piard A 1977 *Rep. Math. Phys.* **11** 259
[17] Piard A 1977 *Rep. Math. Phys.* **11** 279
[18] McCarthy P J 1992 *Soc. Lond. A.* **338** 271
[19] Melas E The Bondi–Metzner–Sachs group $B(2,1)$ in the Hilbert topology (unpublished)
[20] Rudin W 1991 *Functional Analysis* McGraw-Hill
[21] Cantoni V 1967 On the representations of the Bondi–Metzner–Sachs group Ph.D. Thesis, University of London
[22] Isham C J and Kakas A C 1984 *Class. Quantum Grav.* **1** 633

Design of Oscillator Networks for Generating Signal with Prescribed Statistical Property

Tatsuo Yanagita
Department of Engineering Science, Osaka Electro-Communication University, Neyagawa 572-8530, Japan
E-mail: yanagita@osakac.ac.jp

Abstract. We design oscillator networks to generate a signal with a prescribed power spectrum. We consider networks of identical sin-wave oscillators and the Kuramoto order parameter as an output signal of the system. We use the Kullback-Leibler entropy as a measure of the distance between the power spectrum of the output signal and those of desired one. By optimizing the connection network through the Markov chain Monte Carlo method with replica exchange, we found that even oscillator network with a small number of elements can be generate a variety of time signals. The output signals include periodic and quasi-periodic signals with prescribed periods, and aperiodic signals with prescribed power spectrums.

1. Introduction

Generations of time signal with prescribed statistical properties is important for engineering and living organisms, e.g., controlling machines and robots, and waking and sleeping rhythms of animals and so on. The control signals for the motions of animals and machines is not only periodic rhythms , it sometimes need becomes aperiodic or stochastic. Complex autonomous rhythms are ubiquitous in living organisms, and signal generators are crucial roles to control organisms and machines. In many engineering applications, signals with prescribed character are usually generated by special electrical circuit, and we have to manufacture a specialized circuit for each purpose. If a simple circuit can generated a diversity of signals with prescribed properties by changing parameters, it is versatile importance for many engineering fields.

Recently, the oscillator network show varieties of dynamics, from synchronization to chaos [1, 2, 3, 4]. Even in the system with a small number of oscillator elements, it exhibits a diversity of dynamics, e.g., stationary, periodic, quasi-periodic and chaotic, depending on the coupling strength and the connection structures [5, 6]. Here, we treat oscillator network as a signal generator, and consider an inverse problem that is how to construct oscillator networks which produce output signals with a given characteristics. By designing such oscillator networks, we show that the simple dynamical system can be used as versatile signal generators with prescribed statistical properties.

Application of evolutionary learning is one possible approach to constructing networks with prescribed dynamical properties. On this issue, several methods have been explored, where network structure has been modified in response to selection pressure via learning algorithms in such a way that the system evolved towards a specified goal [7, 8, 9, 10]. In previous studies, we have applied the Markov chain Monte Carlo (MCMC) method with replica exchange to design synchronization-optimized networks [9, 10]. In these studies, we constructed large ensembles of optimal networks for synchronization, which was evaluated by the Kuramoto order parameter, and analyzed their common statistical properties. For a network of heterogeneous phase oscillators, we found the transition from the linear to bipartite-like networks by increasing the number of links [9]. For a network of identical phase oscillators under uncommon noise, the most optimal network is a star-like structure when the number of links is small, while an interlaced structure is preferable when the number of links is large.

The most of previous works treat the synchronization property of oscillator networks. In this study, we design oscillator networks, which generate time signal with a desired statistical property. As a statistical property, we consider the periodgram and power spectrum for an output signal of the oscillator network. The paper is organized as follows. In Sec. 2, we introduce oscillator network which is a model of oscillators occupying the nodes of a asymmetrically coupled network, and define the Kuramoto order parameter as the output signal of the system. The Kullback-Leibler entropy introduces as a measure between the periodgram (or power spectrum) of the output signal and those of desired one. The sampling method, the Markov chain Monte Carlo, is also introduced in this section. Then, construction of the learned networks and the generated time signals and their periodgram are performed in Sec. 3 while, finally, the results are discussed in Sec. 4.

2. Model and Method

We consider the following network of identical oscillators as an autonomous signal generator,

$$\dot{\phi}_i = \omega_0 + \frac{\epsilon}{N}\sum_j A_{i,j}\sin(\phi_j - \phi_i + 2\pi\alpha), \tag{1}$$

where ϵ is the coupling strength and $A_{i,j}$ specify connectivity between oscillators, and $\alpha \in [0,1)$ is a phase shift parameter. Because the rotation frequencies of all the oscillators are the same, we can always go into the rotational frame $\phi_i \mapsto \phi_i - \omega_0 t$ and, thus, eliminate the term with ω_0. Hence, without any loss of generality, we can set $\omega_0 = 0$ in Eq. (1). Is is known that when the $\alpha > \alpha_c = 0.25$, the oscillator networks shows chaotic dynamics [5, 6]. The connections between oscillators are asymmetrical and weighted, and $A_{i,j}$ takes any of $2L+1$ number of values, i.e., $A_{i,j} \in \{-1, (-L+1)/L, \cdots, -2/L, -1/L, 0, 1/L, 2/L \cdots, (L-1)/L, 1\}$.

Collective motion of oscillator networks is usually described by the Kuramoto order parameter,

$$R(t) = \frac{1}{N}\left|\sum_{i=1}^{N}\exp(-i\phi_i)\right|. \tag{2}$$

We observe the real part of the Kuramoto order parameter at each time interval τ, and consider the time series,

$$\{x_k\} = \{x_k \| \Re(R(T_0 + n\tau)) \ (k = 0, \cdots, K-1)\}, \tag{3}$$

as the output signal of the system, where the initial transient time T_0 is discarded. For the time sequence $\{x_i\}$, we require a prescribed statistical property, and determine the network $A_{i,j}$, the phase shift α, and the sampling interval τ. Thus the output signal is determined by the combination of system parameters, $\mathbf{s} = (\{A_{i,j}\}, \alpha, \tau)$, which we will call the "state" of the model. As a desired property for the signal, we consider the normalized periodgram $p(\omega_m)$ of the time sequence $\{x_n\}$,

$$p(\omega_m) = \frac{|f(\omega_m)|^2}{\sum_{k=0}^{K-1}|f(\omega_k)|^2}, \tag{4}$$

where

$$f(\omega_m) = \sum_{k=0}^{K-1} x_k \exp(i\omega_m k), \ (k = 0, 1, \cdots, K-1) \tag{5}$$

is the Fourier transformation of the time sequence, where $\omega_m = \frac{2\pi m}{K}$. For a chaotic signal, we also consider the power spectrum as an average of the periodgram,

$$S(\omega_m) = \langle p(\omega_m) \rangle), \tag{6}$$

where $\langle \cdots \rangle$ represents time average over J realizations of time sequences starting from a given initial condition. When $J = 1$, $S(\omega_m)$ equals to the periodgram $p(\omega_m)$. We represent the target of the power spectrum as $\tilde{S}(\omega_m)$, and we assume the distance between these spectrums is defined by the Kullback-Leibler entropy,

$$D_{KL}(\tilde{S}|S) = \sum_{k=0}^{K-1} \tilde{S}(\omega_k) \log \frac{\tilde{S}(\omega_k)}{S(\omega_k)}. \tag{7}$$

Using the the Kullback-Leibler entropy, we define an "energy" or "cost function" of the time series generated by the oscillator network with state $\mathbf{s}$ as

$$E(\mathbf{s}) = \log \frac{D_{KL}(\tilde{S}|S)}{|S|}, \tag{8}$$

where $|S| = \sum_m S(\omega_m)$ is the total power of the out signal. The total power $|S|$ in the energy function prevents that the designed signal becomes too weak . In other words, it works as a penalty for the system.

Since our goal is to construct oscillator network having smaller energy, it is convenient to borrowed from the idea of statistical mechanics in which lower energy states are studied . We sample the state $\mathbf{s}$ (the parameters for the oscillator network and the sampling interval) from the canonical ensemble, i.e.,

$$q(\mathbf{s}) = \frac{\exp[-\beta E(\mathbf{s})]}{\sum_{\mathbf{s}} \exp[-\beta E(\mathbf{s})]}. \tag{9}$$

The application of the canonical ensemble to a network is also considered in [11, 12], in which E is called the graph Hamiltonian. The sampling is carried out by the Markov chain Monte Carlo with replica exchange method, which has been previously applied to constructing the oscillator networks and several dynamical systems [13, 14, 15, 16, 17, 18, 19, 20, 9, 10].

3. Numerical Results

In this section, we designed oscillator networks with prescribed statistical properties. In order to estimate the energy of a network oscillator, we integrate the differential equation (1) numerically, and generate time series and calculate it's periodgram. Using the periodgram, the estimation of energy $E(\mathbf{s})$ with a state s is obtained. By using the $E(\mathbf{s})$, we sample the network $A_{i,j}$ from the Gibbs distribution $\exp(-\beta E)$ with the inverse temperature β by the Markov exchange Monte Carlo with replica exchange method. 24 replicas are used, and each replica has the following inverse temperatures $\beta_k = \kappa^{k-1} - 1$ $(k = 1, \cdots, 24)$.

The several types of target periodgrams or power spectrums are considered. We consider the target periodgrams with several characteristic frequencies. We also use uniform power spectrum as target spectrum in which the designed oscillator network exhibits chaotic. We have mainly carried out the numerics with the following parameters, $N = 5, L = 8, T_0 = 500$, and $K = 5000$. The details of the target spectrums and numerical results are shown as follows.

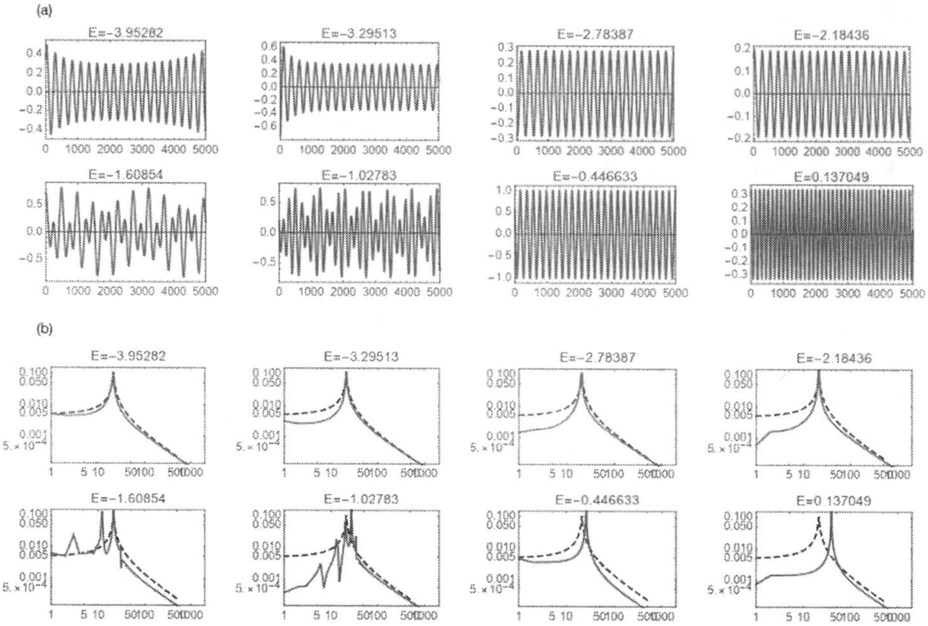

Figure 1. Typical time series and the corresponding periodgrams generated by the oscillator networks design by MCMC with prescribed periodgram are shown. The value at the top of each figure is the energy estimated through the Kullbuck-Leibler entropy. The time series produced by the learned oscillator network are shown in (a), and the corresponding periodgrams are shown in (b). The red broken lines are the prescribed or target periodgram. The parameters of the target periodgram in the equation (10) are $M = 1, \alpha = 1.0, \tilde{\omega_1} = 20$. The number of elements of the oscillator network is $N = 3$ and the number of wighted values of the link is $L = 8$.

3.1. Design of periodic and quasi-periodic signal generators

For periodic and quasi-periodic signal generators having prescribed frequencies, we use a sum of the Lorenzian type weight function as target periodgrams as follows,

$$\tilde{S}(\omega_m) = \sum_{i=1}^{M} \frac{a_i}{|\omega_m - \tilde{\omega}_i|^\alpha + \delta}, \tag{10}$$

where M is the number of characteristic frequencies and a_i represents the strength of the target signal with frequency $\tilde{\omega}_i$, and δ and α determine the the singularity of the peak of the target periodgram at $\tilde{\omega}_i$. The typical parameters for the oscillator network are $N = 5, L = 8$, and $\epsilon = 1.5$. In Figure 1, we show typical time series and their periodgrams of the designed oscillator networks for a given specified target periodgram. The value at the top of each figure is "energy" estimated through the Kullback-Leibler entropy between the obtained and prescribed periodgram. The parameters of the target periodgram in the equation (10) are $M = 1, a_1 = 1.0, \tilde{\omega_1} = 20, \alpha = 1.0$, and $\delta = 1.0$. In this figure, a typical time series and their corresponding periodgrams for a different energy levels are shown. We observe that the periodgrams with lower energies level are fitted well with the prescribed periodgram, which is shown in the red broken lines in Figure 1(b). The top left periodgram obtain by the "learned" oscillator network with $N = 3$ having smallest energy (E=-3.95282) is good agreement with the target periodgram. In particular, the value of the characteristic frequency are matched well with the target one.

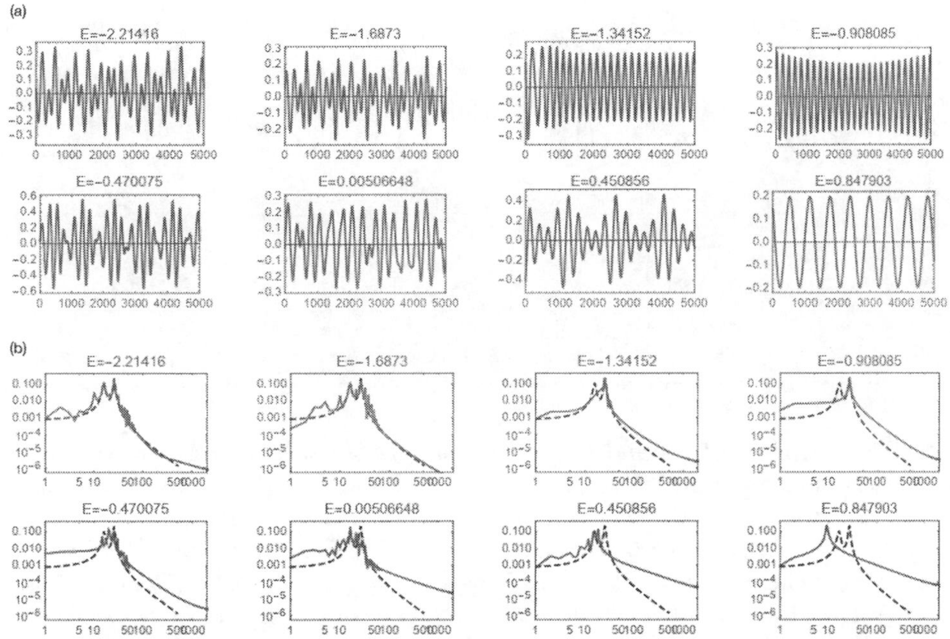

Figure 2. The same graphs as shown in figure 1 for the target periodgram with two characteristic frequencies. The parameters for the target periodgram are $M = 2, \tilde{\omega_1} = 10, \tilde{\omega_2} = 20, \tilde{\omega_3} = 30, a_1 = 1.0, a_2 = 0.5, a_3 = 0.33$. Other parameters are the same as in Fig. 1

The results for a target periodgram having two characteristic frequencies with $M = 2, \tilde{\omega_1} = 10, \tilde{\omega_2} = 20$ and the corresponding strength $a_1 = 0.5, a_2 = 0.5$, and $\alpha = 2.0, \delta = 1.0$, are shown in Figure 2. The periodgram generated by the "learned" oscillator network having the smallest energy, i.e., $E = -2.21416$ also is good agreement with the target periodgram. In this case, we use the oscillator network with 5 elements.

We also consider more complex target periodgram having three and four characteristic frequency. For the target periodgram with three characteristic frequencies, we use the following parameters, $M = 3, \tilde{\omega_1} = 10, \tilde{\omega_2} = 20, \tilde{\omega_3} = 30$, and the corresponding Fourier coefficients are $a_1 = 1.0, a_2 = 0.5, a_3 = 0.33$, respectively, and other parameters are $\alpha = 2.0, \delta = 1.0$. For the target periodgram with four characteristic frequencies, we use the following parameters, $M = 4, \tilde{\omega_1} = 5, \tilde{\omega_2} = 10, \tilde{\omega_3} = 20, \tilde{\omega_4} = 40$, and the corresponding Fourier coefficients are $a_1 = 0.125, a_2 = 0.25, a_3 = 0.50, a_4 = 1.0$, respectively, and other parameters are $\alpha = 1.0, \delta = 1.0$. The times series and the corresponding periodgrams are shown in Figure 3. For target periodgram with three characteristic frequency, the periodgram generated by the "optimal" oscillator network is fitted well with the target one as shown in Figure 3(a). It is, however, that in the case with four characteristic frequency, the periodgram obtain by the "optimal" oscillator network is not good agreement with the target periodgram. In this case, time series generated by the optimal oscillator network does not include the low frequency component, i.e., $\tilde{\omega_1} = 5$, as shown in the lower panel in Figure 3(b). It is because that the number of elements of the oscillator network is too small to describe the complex time series. The time series and the corresponding periodgram generated by the optimal oscillator network with $N = 12$ is also shown in Figure 3(c). In this Figure, we notice that small modulation with long period appears in the time series and the peak at the lower characteristic frequency also emerges in the periodgram. However, the energy, $E = -0.157259$ in the case of $N = 12$, is larger

IOP Conf. Series: Journal of Physics: Conf. Series **936** (2017) 012073 doi:10.1088/1742-6596/936/1/012073

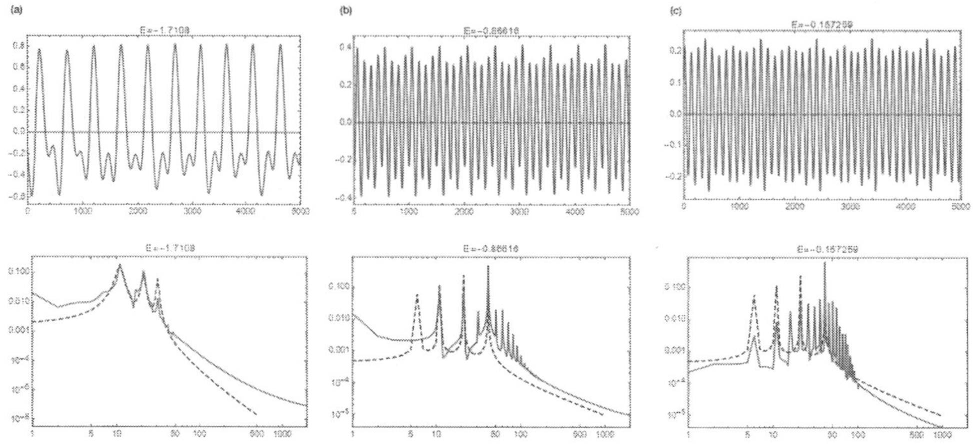

Figure 3. Typical time series and the corresponding periodgrams generated by the optimal oscillator networks for the quasi-periodic target periodgrams are shown. (a) The parameters for the target periodgrams with three characteristic frequencies are $M = 3, \tilde{\omega_1} = 10, \tilde{\omega_2} = 20, \tilde{\omega_3} = 30$, $a_1 = 1.0, a_2 = 0.5, a_3 = 0.33, \alpha = 2.0, \delta = 1.0$. (b) For the target periodgram with four characteristic frequencies, we use $M = 4, \tilde{\omega_1} = 5, \tilde{\omega_2} = 10, \tilde{\omega_3} = 20, \tilde{\omega_4} = 40, a_1 = 0.125, a_2 = 0.25, a_3 = 0.5, a_4 = 1.0, \alpha = 1.0, \delta = 1.0$. (c) In the case of the oscillator network with $N = 12$. Other parameters are the same as in (b).

than the energy $E = -0.86616$ in the case of $N = 5$, and thus the learned oscillator network with smaller number of elements perform better than that with larger number of elements. This is because that numerical cost increases with the number of elements in the oscillator network since both the time integration for the oscillator network increases and the Monte Carlo sampling time increases. Thus there is a trade off between the learning capability of oscillator network which increases with N and computational cost.

3.2. Designing oscillator network as white noise generator

In this subsection, we consider white noise as a typical broad-ranged target power-spectrum. The modulus of the Fourier coefficient for white noise does not depend on frequencies. Thus the power of any frequency should equivalently be included in the signal, and thus, we set the target weight of power spectrum equals one for all frequency, i.e.,

$$\tilde{S}(\omega_m) = 1/|S|. \tag{11}$$

Typical time series and the power spectrum generated by the learned oscillator networks for different energy levels are shown in Figure 4. It is clearly seen that periodic times series varies to chaotic one with decease of energy level, corresponding to how oscillator network learns. For oscillator network having smallest energy, i.e., the optimal oscillator networks exhibit chaotically and the broad-ranged power spectrum is obtained. When the number of oscillators increases, the power spectrum of the optimal oscillator network becomes broader, which means that power of frequencies involved in the time series equivalently be included. It is, however, that there is a trade off between the system size and the numerical cost for optimization as noticed previously.

4. Conclusions

We have shown that the oscillator network with a small number of elements can generate variety of time signal, i.e., periodic, quasi-periodic, and aperiodic when the connection network

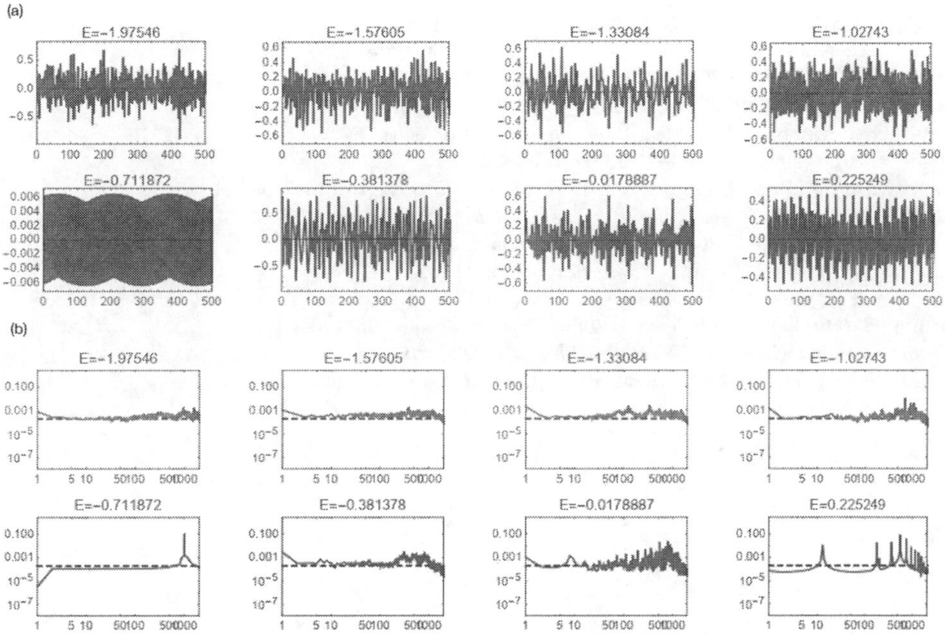

Figure 4. Typical time series and corresponding power spectrums are shown for different energy levels. The time series (a) and the corresponding power spectrums (b). The parameters of the oscillator networks are $N = 10, L = 8$, and $\epsilon = 2.0$. Other parameters are the same as in Fig. 3.

is designed so as to fit with a given target periodgram or a power spectrum through the Markov chain Monte Carlo method with replica exchange. Although the elements considered here are identical and sinusoidal-oscillator, the designed oscillator networks with only five elements can generate quasi-periodic time series with three characteristic frequencies. Furthermore, the designed oscillator-network also can produce aperiodic time-series whose power of any frequency is almost equivalently.

It should be noted that the simple and small oscillator networks can generate given characteristic frequencies only by changing the connection network with preserving the dynamics of oscillator elements. In other words, when the realtime leaning applied, we can obtain the desired time signals with same architecture by changing the system parameters. Such a "learning capability" of oscillator networks can be capable of signals generator for many engineering applications

Acknowledgments
This study has been partially supported by JSPS KAKENHI Grant No. 15K05221 and the Volkswagen Foundation (Germany).

References
[1] Nakagawa N and Kuramoto Y 1993 *Progress of Theoretical Physics* **89** 313–323
[2] Topaj D and Pikovsky A 2002 *Physica D: Nonlinear Phenomena* **170** 118 – 130 ISSN 0167-2789
[3] Maistrenko Y, Popovych O and Tass P 2005 Desynchronization and chaos in the kuramoto model *Dynamics of Coupled Map Lattices and of Related Spatially Extended Systems* (*Lecture Notes in Physics* vol 671) (Springer Berlin Heidelberg) pp 285–306 ISBN 978-3-540-24289-5
[4] Popovych O V, Maistrenko Y L and Tass P A 2005 *Phys. Rev. E* **71**(6) 065201
[5] Maistrenko Y, Popovych O, Burylko O and Tass P A 2004 *Phys. Rev. Lett.* **93**(8) 084102

[6] Ashwin P, Burylko O and Maistrenko Y 2008 *Physica D: Nonlinear Phenomena* **237** 454 – 466 ISSN 0167-2789
[7] Ipsen M and Mikhailov A S 2002 *Phys. Rev. E* **66** 046109
[8] Moyano L G, Abramson G and Zanette D H 2001 *Eur. Phys. J. B* **22** 223–228
[9] Yanagita T and Mikhailov A S 2010 *Phys. Rev. E* **81** 056204
[10] Yanagita T and Mikhailov A S 2012 *Phys. Rev. E* **85** 056206
[11] Park J and Newman M E J 2004 *Phys. Rev. E* **70**(6) 066117
[12] Garrido A 2011 *Symmetry* **3** 1–15 ISSN 2073-8994
[13] Cho A E, Doll J D and Freeman D L 1994 *Chem. Phys. Lett.* **229** 218–224
[14] Bolhuis P G, Dellago C and Chandler D 1998 *Faraday Discuss.* **110** 421–436
[15] J H Vlugt T and Smit B 2001 *PhysChemComm* **4** 11–17
[16] Kawasaki M and Sasa S I 2005 *Phys. Rev. E* **72** 037202
[17] Sasa S I and Hayashi K 2006 *Europhys. Lett.* **74** 156–162
[18] Giardiná C, Kurchan J and Peliti L 2006 *Phys. Rev. Lett.* **96** 120603
[19] Tailleur J and Kurchan J 2007 *Nature Physics* **3** 203–207
[20] Yanagita T and Iba Y 2009 *J. Stat. Mech.* P02043

Methods of verification of compliance with an asymptotically power law

I A Belashova, V V Bochkarev, V A Tyurin

Kazan Federal University, Kremlevskaya str.18, Kazan 420018, Russia

E-mail: vbochkarev@mail.ru

Keywords: power-law distribution, robust criteria, runs test, Zipf's law

Abstract This paper describes methods of the power law verification using different empirical data. We do not analyse the value of deviations from the model but try to found out whether these deviations are regular or random. The suggested approach is based on the idea of finding local power approximation of the considered series for each range of ranks, after which one or another trend criterion is applied to the obtained series of local exponents. Application of the runs test is also discussed. The suggested methods were tested using 10 sets of empirical data, which are available for free. It was shown that compliance with the power law is satisfactory only in one case.

1. Introduction

It was found out that the power law manifests itself in many practical cases related to different fields of science and technology [1]. However, the conclusion about distribution of empirical data by the power law was not strictly justified for all cases in terms of statistics. A large number of data sets from previous works was analysed in [2] and it was shown that the conclusion about compliance with the power law was statistically justified only in 11 cases out of 24. In [2], the verification was performed assuming that the measured values precisely correspond to power distribution, at list in some relatively large range of values. The paper describes the case when the distribution strictly complies with the power law:

$$1-F(x)=A\,x^{-\gamma} \tag{1}$$

Here $F(x)=P(\xi \leq x)$ is a cumulative distribution function. However, only asymptotic approximation to the power law can be expected in many practical cases. As an example, let us consider the distribution of word frequency. Zipf's law established in 1949 was one of the first examples of the power law in empirical data. According to this law, frequencies of words f_r are characterised by power dependence on their rank r (the numbers in a list of words sorted by frequency reduction):

$$f_r \sim r^{-\gamma} \tag{2}$$

As the sample cumulative distribution function is equal to $1-r/n$, where n is the sample size, Zipf's law is equivalent to the assumption that the word frequencies are distributed according to the power law (1), while the exponents in formulas (1) and (2) are related by an obvious relation $\gamma=\alpha^{-1}$.

Probability models, such as monkey model and Markov models are traditionally used to explain Zipf's law. The monkey model assumes that a text is a random sequence of symbols (including a white-space) and the probability of selecting the subsequent symbol doesn't depend on the previous symbol. Dependence of the subsequent symbol on the previous one is taken into account in Markov models. In [3, 4], it is proved that for a typical case both models provide distribution which satisfies the following inequation:

$$A_1 x^{-\gamma} \le 1 - F(x) \le A_2 x^{-\gamma} \qquad (3)$$

It is clear that if we apply a statistical criterion based on model (1) to a set of empirical data with a distribution law which satisfies equations (3) (but not an exact relation (1)), the correct underlying hypothesis will certainly be rejected if a sample is relatively large.

This paper describes methods which allow us to perform correct verification for cases when the power law is satisfied only asymptotically, in other words when the distribution satisfies inequations (3). Testing is performed using 10 sets of empirical data, which were considered in [2] and are available for free.

2. Methods

Assume that the data are described by the following dependence:

$$f(r) = A(r)\, r^{-\gamma(r)} \qquad (4)$$

Functions *A(r)* and *γ(r)*, which the mentioned equation contains, are considered to be slowly varying. Observing the behaviour of these functions, it can be concluded whether there is convergence to a power law or not. When the distribution asymptotically tends to the power distribution, *A(r)* and *γ(r)* should tend to constant values with increasing of the argument. If a trend is detected in these sequences, this will mean that there is a systematic deviation from the power law.

The proposed algorithm includes the following sequence of steps:

- The variation range *r* is divided into a sufficient number of intervals. Then, data are separately fitted at each interval by a power curve, estimating the amplitude and the exponent.
- This or that trend criterion is applied to the obtained series A_i and γ_i.

We thus assume that intervals can be divided in such a way that the deviations from the average values γ_i (and A_i) will be uncorrelated series. It is clear that to do this, the intervals should not be less than the characteristic scale of variations *A(r)*, *γ(r)*. The choice of dividing into intervals is left to a researcher in each case. Of course, there are some cases when this method doesn't give the desired result. In particular, these are all cases when *A(r)* and *γ(r)* monotonically approach the limit value (from above or below). Nevertheless, the proposed approach is completely correct for a wide range of cases when *A(r)* and *γ(r)* are around the limiting value.

If the number of sample values is small, it can be difficult to perform fitting for some intervals. In this case, the following method can be used. The value $-\partial \log f / \partial \log r$ gives the value of the exponent for the exact power-law dependence. Calculating the derivative using the method of finite differences we obtain

$$\gamma_i = -\frac{\log f_{i+1}/f_i}{\log r_{i+1}/r_i} \qquad (5)$$

To verify the presence of a trend in the series A_i and γ_i, it is preferable to use nonparametric criteria, for example, the runs test. Run are sequences of points for which the error has the same sign. If the deviations of the empirical data from the approximating curve are random and weakly correlated, the probability of occurrence of long runs is small. Thus, the presence of long runs indicates systematic deviations of empirical data from the model. The Ramachandran-Ranganatan criterion [5] is used in this paper. According to this criterion, the decision is made depending on the value of the sum of the squares of the lengths of the runs:

$$R = \sum_i l_i^2 \qquad (6)$$

Here l_i is the length of the i-th run. P-values, corresponding to different values of R, are found by statistical modelling.

3. Data analysis

The following data sets are considered [2]:

1) The frequency of occurrence of unique words in the novel Moby Dick by Herman Melville.
2) The severity of terrorist attacks worldwide from February 1968 to June 2006, measured as the number of deaths directly resulting.
3) The number of species per genus of "recent" mammals.
4) The numbers of customers affected in electrical blackouts in the United States between 1984 and 2002.
5) The human populations of US cities in the 2000 US Census.
6) The sizes in acres of wildfires occurring on US federal land between 1986 and 1996.
7) Peak gamma-ray intensity of solar flares between 1980 and 1989.
8) The intensities of earthquakes occurring in California between 1910 and 1992, measured as the maximum amplitude of motion during the quake.
9) The frequencies of occurrence of US family names in the 1990 US Census.
10) The number of links to web sites found in a 1997 web crawl of about 200 million web pages, represented as a simple histogram.

All these data sets and brief information about these cases can be found on the page of one of the authors of the work [2] (http://tuvalu.santafe.edu/~aaronc/powerlaws//data.htm). Due to the limited size, we present the results of the tests only for the series of local indices γ_i. The range of rank variations was divided into uniform intervals in a logarithmic scale. The value $\log r_{i+1}/r_i$ was chosen as 0.25 for examples 2-10, the value 0.45 was chosen for the first example. The calculation results for the series of local exponents are given in Table 1. The sample size, the number of intervals for which the values γ_i were found, the value of the statistics R and the corresponding p-value are given for each example.

Table 1. The results of the Ramachandran-Ranganatana test for the series γ_i.

	The number of samples	The number of intervals	R	p-value
1	18855	20	182	$9.64\cdot10^{-4}$
2	9101	32	266	$5.43\cdot10^{-4}$
3	5731	33	311	$1.81\cdot10^{-4}$
4	211	21	91	0.0661
5	19447	39	303	$5.11\cdot10^{-4}$
6	203785	46	974	$< 10^{-7}$
7	12773	37	423	$1.96\cdot10^{-4}$
8	19302	39	305	$4.75\cdot10^{-4}$
9	2753	30	248	$6.82\cdot10^{-4}$
10	14480	36	182	0.017

As can be seen from the table, the hypothesis about the power law is confirmed only in one case (example 4). It should be noted that we have the smallest sample for this example. In another case (example 10), the p-value is equal to 0.017 and the solution will depend on the accepted level of significance.

Let us consider Example 7 in more detail (the intensity of gamma rays during solar flares). Figure 1.a shows (in a log-log scale) the dependence of empirical values sorted in a list in a descending order on the rank. The shape of this graph suggests that the power law describes the empirical data well, at least in a relatively wide range of ranks (approximately from 20 to 2000). But Figure 1.b, which shows change of local exponents γ_i, doesn't confirm this information. It should be noted that no horizontal sections can be seen in the graph (otherwise, the presence of horizontal sections would indicate that the power-law is realised in the selected range), on the contrary, significant systematic variations of exponents are observed. Selecting any range, we obtain 2-3 long runs and a large value of statistics (6), which results in the rejection of the hypothesis about the power-law. The above-mentioned interval 20-2000 which is characterised by systematic growth of $\gamma(r)$ is the most interesting. This means that in the given case it is necessary to choose a model distribution that decreases more rapidly than the power distribution.

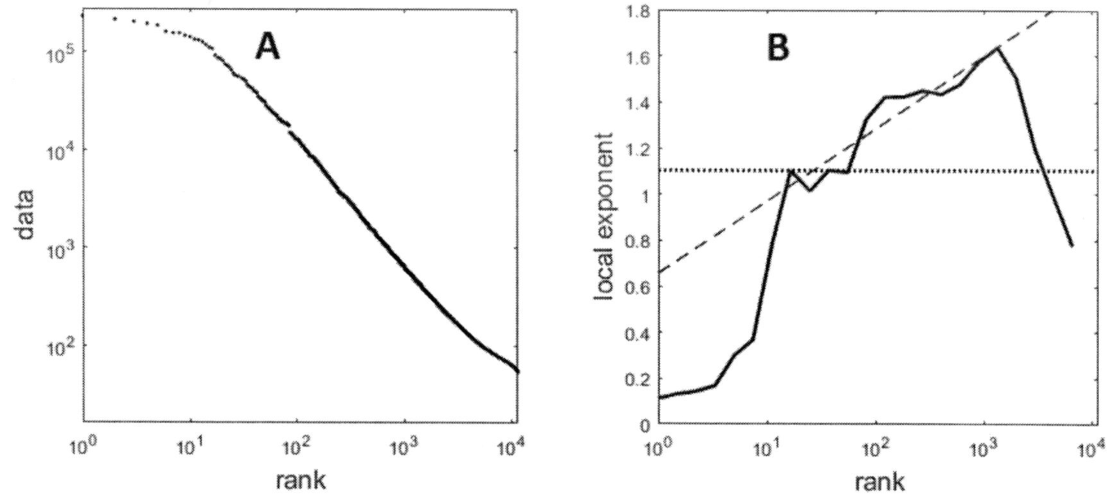

Fig. 1 a) Distribution of peak gamma-ray intensity of solar flares between 1980 and 1989 as a rank function. b) Local exponents depending on the rank. The dotted line shows the median level, the dashed line shows a trend schematically

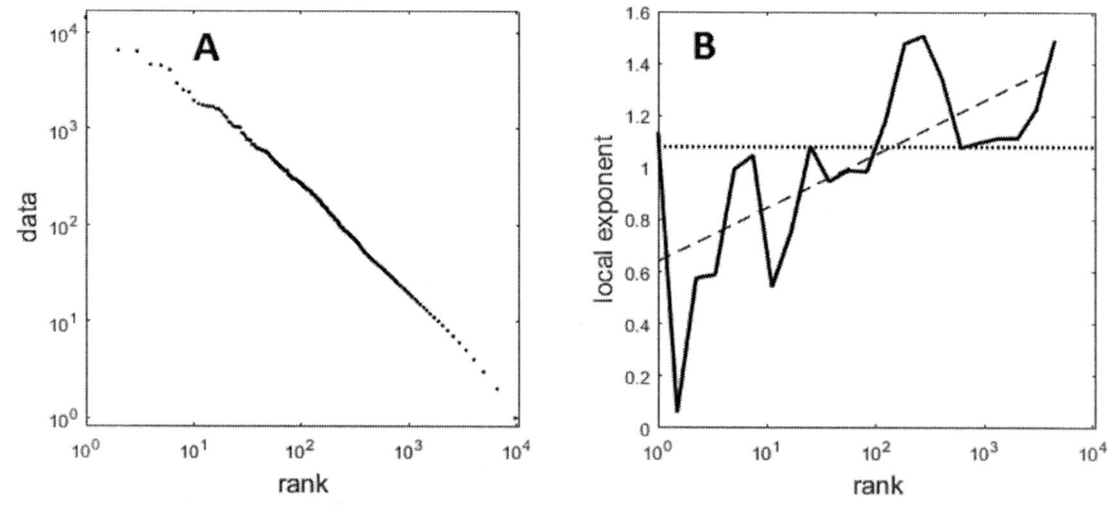

Fig. 2 a) Distribution of the frequency of occurrence of unique words in the novel "Moby Dick" by Herman Melville. b) Local exponents depending on the rank. The dotted line shows the median level, the dashed line shows a trend schematically

Let us consider example 1. Figure 2.a shows dependence of the frequency of words in the novel "Moby Dick" on their rank, Figure 2.b shows the corresponding dependence for local exponents. Variations of local exponents are more significant in this case than in the previous one. However, a systematic increase of exponents is observed with increase of a rank. However, there are no indicators that the exponent $\gamma(r)$ can tend to some limit with a further increase in rank. Thus, the assumption about the power -law (even in the expanded treatment (3)) should be rejected.

A similar behaviour is observed for other examples considered (except for example 4 and, possibly, example 10): it is a systematic increase of the exponent $\gamma(r)$ with increase of the rank.

4. Conclusion

In this paper, we proposed and tested a criterion that allows one to verify the correspondence of empirical data to an asymptotically power-law distribution which satisfies inequalities of the form (3). It was shown that compliance with the power law can be regarded satisfactory only in one of the ten cases considered. For three cases (examples 1, 2, 5), our conclusions differ from the conclusions of [2]. In these cases, deviations from the distribution (1) may seem small, but they are of a regular nature, which results in rejection of the hypothesis of a power law. It should also be noted that the analysis of the series of local exponents $\gamma(r)$, both quantitative and visual, allows us to make more valid conclusions about the correspondence of empirical data to the power law.

This work was supported by the Russian Foundation for Basic Research, Grant no. 15-06-07402. The research of the second author was supported by the Russian Government Program of Competitive Growth of Kazan Federal University.

References

[1] M. E. J. Newman Power laws, Pareto distributions and Zipf's law. Contemporary Physics 46, 323-351 (2005)

[2] A. Clauset, C. R. Shalizi, M. E. J. Newman Power-law distributions in empirical data. SIAM Review 51, 661-703 (2009)

[3] Bochkarev, V.V., Lerner, E.Y. The Zipf law for random texts with unequal probabilities of occurrence of letters and the Pascal pyramid. Russian Mathematics (2012) 56: 25. doi:10.3103/S1066369X12120031.

[4] Bochkarev, Vladimir V. and Lerner, Eduard Yu.. (2014), "Strong power and subexponential laws for an ordered list of trajectories of a Markov chain", Electronic Journal of Linear Algebra, Volume 27. pp. 534-556. doi: 10.13001/1081-3810.1917

[5] David M. Himmelblau Process Analysis by Statistical Methods. John Wiley & Sons Inc, 1970.

IOP Conf. Series: Journal of Physics: Conf. Series **936** (2017) 012075 doi:10.1088/1742-6596/936/1/012075

The continuous-time quantum walk with transition moments of the Gaussian distribution as a mathematical model of rotational excitation in molecules by an optical pulse

L Matsuoka[1] **and H Lavička**[2]

[1] Graduate School of Engineering, Hiroshima University, Higashi-Hiroshima, Hiroshima 739-8527, Japan

[2] Institute for Theoretical Physics, KU Leuven, Leuven 3001, Belgium

E-mail: `leo-matsuoka@hiroshima-u.ac.jp`

Abstract. The time evolution of the rotational distribution of diatomic molecules induced by a train of broadband optical pulses can be modeled as a class of continuous-time quantum walk (CTQW) on a half line. As the transition moments asymptotically decrease to zero like Gaussian distribution, the rotational distribution cannot reach infinity far from the initial state and exhibits a localization-like behavior where the upper limit of the distribution is not trivial. We investigated a time evolution of the density distribution with the transition moments of the Gaussian distribution peaked at the boundary of the half line for the initial state of J=0. Even though the time evolution exhibited oscillatory motion inside a certain region repeatedly, we observed a clear time-averaged distribution that is well fitted to the stretched exponential distribution with parameters close to the Gaussian distribution by numerical simulation. By varying the coefficients of the Gaussian distribution of the transition moments, the coefficients of the stretched exponential distribution can be determined by regression.

1. Introduction

Quantum mechanics of a rotator has been of interest for mathematics and physics for many decades, thereby constructing a foundation of atomic or molecular physics. The quantum kicked rotor (QKR) is one of the standard models, which describe the development of angular momentum by periodical kicking of delta pulses[1]. Recently, the dynamics of the QKR was experimentally implemented on cold atoms[2, 3, 4] and diatomic molecules[5, 6]. The characteristics of the QKR was studied on real molecules, such as the so-called quantum resonance[7] and the localization similar to Anderson localization or Bloch oscillation[8, 9]. Currently, the series of rotational states in diatomic molecules is considered as a one-dimensional network of quantum states connected by optical transition, which provides a testing field of fundamental quantum transport dynamics. The characterization of localization in a broad sense is also important for practical applications such as isotope separation by molecular selective excitation by using a coherent light source[7, 10].

The time evolution of the rotational distribution in the QKR can be described as a class of continuous-time quantum walk (CTQW)[11]. The CTQW is a well-studied mathematical

IOP Conf. Series: Journal of Physics: Conf. Series **936** (2017) 012075 doi:10.1088/1742-6596/936/1/012075

model[12, 13, 14]. To develop simple mathematical models of the localization in the QKR from the viewpoint of the CTQW, we classify the localization into four types and attempt to characterize each type of localization independently[15]. We derived an empirical unified parameter to characterize the range of localization for the localizations, namely interval mismatch (Type 1) and centrifugal distortion (Type 2), caused by rotational coherence[16]. The remaining types are the lack of spectral amplitude (Type 3) and the augmentation of spectral amplitude (Type 4), which are caused by the relative distribution of transition moments. The CTQW with the transition moments of the Gaussian distribution is a practical model to investigate Type 3 independently, because the distribution of the spectral amplitude of optical pulses are frequently approximated by Gaussian functions. However, it is not discussed explicitly, because the independent implementation of such a process is difficult in real molecules.

In this paper, we discuss the asymptotic behavior of a density distribution on the CTQW on a half line with the transition moments of the Gaussian distribution. The model emulates the time evolution of the rotational distribution with a broadband optical pulse where other effects, such as multiphoton processes or interferences relevant to the rotational coherence are neglected. To develop a foundation of mathematical physics of the localization of the rotational distribution in molecules, we performed the numerical analysis of the simplest case in symmetry.

2. The model and method

The modeling of molecular rotational excitation by optical pulses was shown in our previous studies[16, 17]. By neglecting the effect of pulse–interval mismatch and centrifugal distortions, the time-dependent Schrödinger equation system for molecular rotational distribution in optical pulses can be expressed as follows:

$$-i\frac{d}{dt}C_J(t) = \sqrt{\omega_{J-1}}C_{J-1}(t) + \sqrt{\omega_J}C_{J+1}(t), \tag{1}$$

where $C_J(t)$ is the complex amplitude of the rotational states J at time t, $\sqrt{\omega_{J-1}}$ is the transition moment from J to $J+1$. The probability density distribution of state J is given by $|C_J(t)|^2$. The transition moments of the Gaussian distribution peaked at the boundary of a half line can be expressed as follows:

$$\sqrt{\omega_J} = \begin{cases} 0 & (J < 0) \\ A_\mathrm{g} \exp\left[-\left(\frac{\sqrt{\log 2}}{H_\mathrm{g}}J\right)^2\right] & (J \geq 0), \end{cases} \tag{2}$$

where H_g is the half width half maximum (HWHM) of the Gaussian distribution. The parameter A_g is fixed at 1.0. This parameter makes no influence on the time evolution except the speed of the evolution. The initial state is fixed as follows:

$$C_J(0) = \delta_{J,J_0}. \tag{3}$$

In this study, we considered only $J_0 = 0$.

The time evolution was evaluated by solving Eqs.(1–3) numerically using the fourth-order Runge–Kutta method. The parameter H_g was varied from 5.0 to 50.0 for regression. Some examples of the distribution of the transition moments are shown in Figure 1. The number of J states considered in the calculation was $10H_\mathrm{g}$, which was sufficient to ignore the influence of the upper boundary of the system. The evolution was simulated till $t_\mathrm{end} = 500H_\mathrm{g}$ by a step of 0.025. The numerical error estimated from the sum of probability distribution was in the order of 10^{-5}. The time-averaged distribution was obtained by using the following approximation:

$$\rho(J) = \lim_{t\to\infty}\frac{\int_0^t |C_J(t)|^2 dt}{t} \simeq \frac{\sum_{t=0}^{t_\mathrm{end}} |C_J(t)|^2}{t_\mathrm{end}}, \tag{4}$$

where only integer t was included in the summation.

IOP Conf. Series: Journal of Physics: Conf. Series **936** (2017) 012075 doi:10.1088/1742-6596/936/1/012075

3. Results and Discussion

We executed the computer simulations of the model described by Eq.(1). The examples of the transition moments that were used in simulations are shown in Figure 1. The result of time evolution of probability distribution obtained by the simulation with H_g=20 is shown in Figure 2. Note that this figure shows only the first one-tenth of the simulation. Around t=0, the population appears as if moving toward the high J states at a nearly linear propagation speed. However, a massive reflection behavior is observed around J=20. One of the primary parts of the population approaches the region around J=60 while producing many branching reflections. The entire time evolution shows interference-like patterns caused by many reflection waves. The upper limit of the propagation observed in the linear scale is about J=60, which is a position similar to one where the relative transition moment approaches zero. The characteristics observed here were common regardless of input H_g except for the width of the distribution.

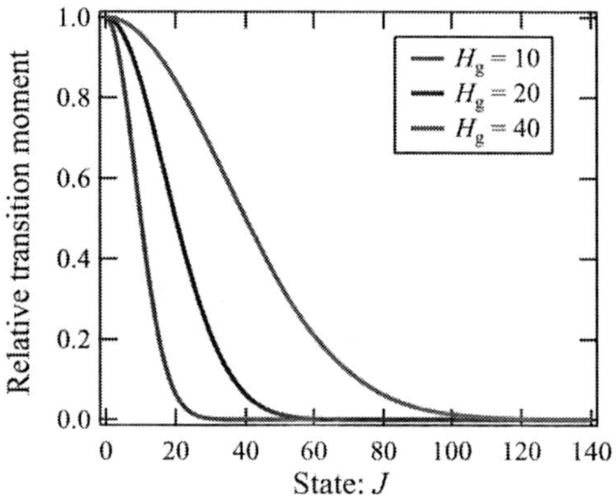

Figure 1. Examples of Gaussian distribution of relative transition moments obtained with H_g=10, 20, and 40.

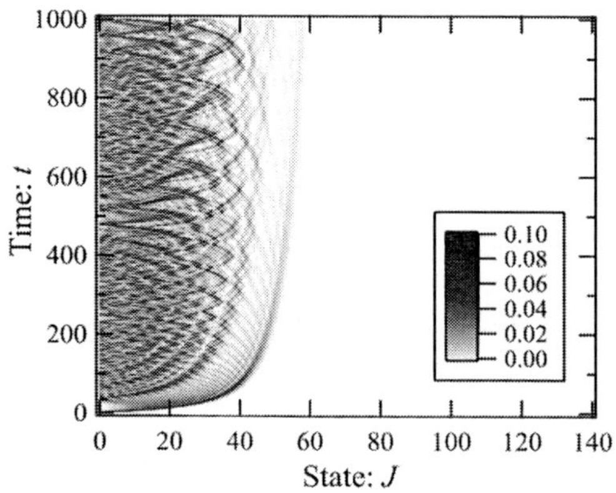

Figure 2. Example of time evolution of $|C_J(t)|^2$ simulated with H_g = 20.0 and $J_0 = 0$.

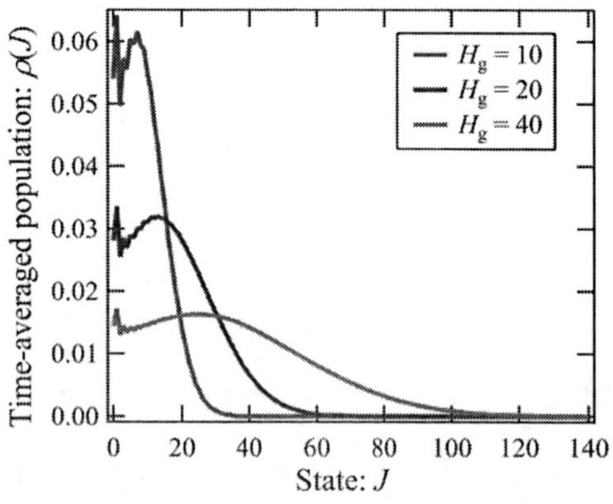

Figure 3. Examples of time-averaged distribution $\rho(J)$ simulated with H_g=10, 20, and 40.

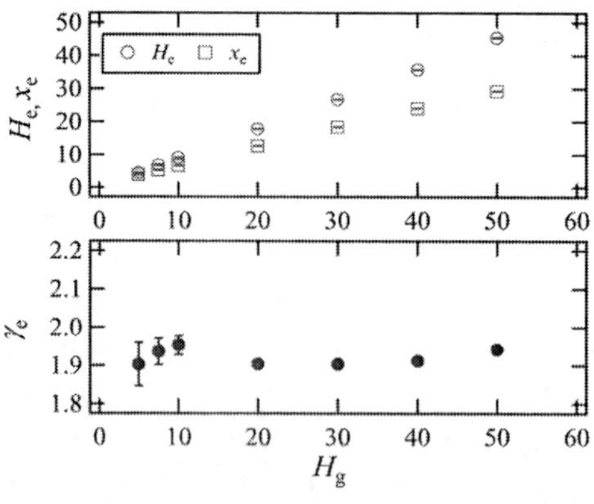

Figure 4. Coefficients of time-averaged distributions simulated with various H_g fitted by stretched exponential function.

Figure 3 shows the examples of time-averaged distribution $\rho(J)$ obtained by simulation. Even though the time evolutions were oscillatory, smooth structures were observed. The distributions comprise two peaks. One is the sharp peak at J=1, and the other is the broad peak with properties similar to the Gaussian distribution around the maximum. A small interference pattern was observed between the two peaks, which is also a common characteristic.

We focus on the shape of the broad peak, because it determines the upper limit of the propagation. We fitted the broad peak by using the stretched exponential function given by

$$f(J) = A_\mathrm{e} \exp\left[- \left| \frac{(\log 2)^{\frac{1}{\gamma_\mathrm{e}}}}{H_\mathrm{e}} (J - x_\mathrm{e}) \right|^{\gamma_\mathrm{e}} \right], \tag{5}$$

where H_e provides the HWHM for any γ_e. $f(J)$ provides the Gaussian distribution for γ_e=2.0. A_e is not very important, because it is nearly the same as the normalization factor.

Figure 4 shows the coefficients of the time-averaged distribution fitted by Eq. (5). The fitting errors were not more than 0.1, which are not displayed except γ_e for small H_g. A linear scaling rule for H_g was observed for H_e and x_e. We arrived at the following relation by regression of the obtained data:

$$H_\mathrm{e} = A_1 H_\mathrm{g} + B_1 \quad (A_1 = 0.904 \pm 0.006, B_1 = -0.132 \pm 0.177), \tag{6}$$

$$x_\mathrm{e} = A_2 H_\mathrm{g} + B_2 \quad (A_2 = 0.572 \pm 0.005, B_2 = 1.020 \pm 0.152). \tag{7}$$

By considering the coefficients and the error obtained, we propose a hypothesis where the coefficients can be expressed in a simple analytical form:

$$H_\mathrm{e} \simeq \frac{2}{\pi \log 2} H_\mathrm{g}, \quad x_\mathrm{e} \simeq \frac{1}{\sqrt{\pi}} H_\mathrm{g} + 1. \tag{8}$$

Almost no dependence was observed on H_g for γ_e. The average value was γ_e=1.922±0.008, which was evidently less than γ_e=2.0 of the Gaussian distribution. We observed that the stretched exponential function is essential to describe the distribution analytically. However, the Gaussian distribution still seems to be a good approximation function for expressing the distributions, because it is in good agreement with the obtained distribution.

4. Conclusion

We performed numerical analysis of a simple mathematical model of the CTQW with the transition moments of the Gaussian distribution, which describes one of the characteristics of the molecular rotational excitation by an optical pulse. We observed smooth time-averaged distributions that can be fitted by the stretched exponential function. The order of the stretched exponential γ_e was close to the Gaussian distribution. However, the evident difference from the Gaussian was evaluated numerically. The coefficients obtained from regression were analytically expressed. However, those should be validated by mathematical analysis using approximate methods, because the model is not solvable. The upper limit of the population propagation in linear scale was not considerably different from the intuitive estimation from the distribution of the relative transition moment. However, the shape of the time-averaged distribution had not been known as far as authors' knowledge, although this model expresses fundamental characteristics of population transfer by optical transitions.

The peak position of the Gaussian distribution and the initial state should be varied to obtain the general tendency of the asymptotic behavior of this model. In fact, the case investigated in this study was a special case from the viewpoint of the symmetry. Moreover, the interaction between other localizations should be investigated for application in real molecules. In particular, the interaction between the localizations induced by the centrifugal distortion is of interest, because it exhibits nearly the same range of localization as the time-averaged distribution obtained in this study.

Acknowledgments
We thank Etsuo Segawa and Osamu Ogurisu for their useful discussion. This work was supported by JSPS KAKENHI Grant Number JP16K14538 and JP17H04978.

References

[1] Casati G and Chirikov B 2006 *Quantum Chaos: Between Order and Disorder* (Cambridge: Cambridge University Press)
[2] Moore F L, Robinson J C, Bharucha C F, Sundaram B and Raizen M G, 1995 *Phys. Rev. Lett.* **75** 4598
[3] Sadgrove M and Wimberger S 2011 *Adv. At. Mol. Opt. Phys.* **60** 315
[4] Behinaein G, Ramareddy V, Ahmadi P and Summy G S 2006 *Phys. Rev. Lett.* **97** 244101
[5] Cryan J P, Bucksbaum P H and Coffee R N 2009 *Phys. Rev.* A **80** 063412
[6] Floß J and Averbukh I Sh 2012 *Phys. Rev.* A **86** 021401(R)
[7] Zhdanovich S, Bloomquist C, Floß J, Averbukh I Sh, Hepburn J W and Milner V 2012 *Phys. Rev. Lett.* **109** 043003
[8] Floß J, Kamalov A, Averbukh I Sh and Bucksbaum P H 2015 *Phys. Rev. Lett.* **115** 203002
[9] Floß J and Averbukh I Sh 2016 *J. Phys. Chem.* A **120** 3206
[10] Ichihara A, Matsuoka L, Segawa E and Yokoyama K 2015 *Phys. Rev.* A **91** 043404
[11] Matsuoka L, Kasajima T, Hashimoto M and Yokoyama K 2011 *J. Korean Phys. Soc.* **59** 2897
[12] Farhi E and Gutmann S 1998 *Phys. Rev.* A **58** 915
[13] Konno N 2005 *Phys. Rev.* E **72** 026113
[14] Venegas-Andraca S E 2012 *Quantum Inf. Process.* **11** 1015
[15] Matsuoka L and Segawa E 2017 *Interdiscipl. Inform. Sci.* **23** 51
[16] Matsuoka L 2015 *Phys. Rev.* A **91** 043420
[17] Matsuoka L, Segawa E, Yuki K, Konno N and Obata N 2017 *Phys. Lett.* A **381** 1773

Object recognition of real targets using modelled SAR images

D.A. Zherdev[1]
Samara University, 34, Moskovskoye shosse, Samara, 443086, Russia Federation

Abstract. In this work the problem of recognition is studied using SAR images. The algorithm of recognition is based on the computation of conjugation indices with vectors of class. The support subspaces for each class are constructed by exception of the most and the less correlated vectors in a class. In the study we examine the ability of a significant feature vector size reduce that leads to recognition time decrease. The images of targets form the feature vectors that are transformed using pre-trained convolutional neural network (CNN).

1. Introduction

This research is a continuation of ideas and methods of work [1], in which the recognition method based on computation of conjugation indices was considered. The conjugation index is a function that uses a feature vector and matrix as the input values. The class matrix is formed using vectors of a class. In the output of this function there is a rate of similarity between inputs. In particular, we examine the support subspace formation algorithm. The support subspace presents the matrix that is a best in a case of recognition using the conjugation index. For this purpose, we used the algorithm of exhaustive search which requires many computations.

The goal of this work is computational complexity reduce with the similar or higher recognition quality. It is obtained by a procedure of vector exception from classes and with previous feature extraction of 64x64 pixel images. We train the convolutional neural network on this dataset of images and in result of convolutions we receive the vector of features with 120 components. Therefore, the computational complexity on the training stage is raised, but on the test stage the algorithm executes faster besides without drop of recognition quality.

2. Modelling technology

The SAR images are more useful in the object recognition problem as a form of radar data in a case of a more detailed object representation. We developed the algorithm that based on the similar SAR modelling principles that described in [2-4]. The SAR image is formed in the process of phase history recording in the result of airborne radar movement on a specified trajectory with sequentially propageted and received radar impulses. Therefore, we improved the resolution along a track of synthesis, which is larger than a real antenna length with a less resolution. The SAR image is obtained using the received signals which stored in a phase and amplitude form.

We can write the general expression for computation of a SAR image value in a pixel with the slant and cross range x_u, y_u based on the presented structure of the general model of the radar observation:

[1] E-mail: denis.zherdev.91@mail.ru.

$$I(x_u, y_u) = \iint\limits_{-\infty}^{+\infty} v(x, y)\chi(x, y, x_u, y_u)\, dx\, dy\,.$$

The slant and cross range are coordinates of an obtained SAR image. Figure 1 shows the examples of modelled and real image. The model was formed in the similar condition of the real experiment, therefor the images can be matched.

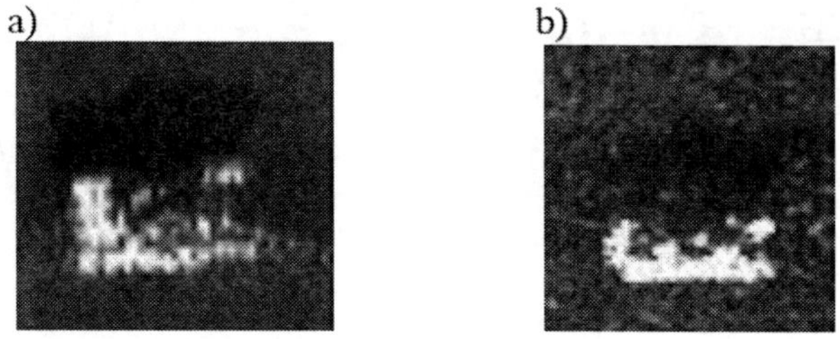

Figure 1. SAR images of (**a**) modelled and (**b**) real target.

The program of a modelling has such parts as: map editor, 3D viewer with plug-in editor, relief generator, and common interface. There are two main stages to model: forming a relief that with some parameters and construction of a 3D object model which then is placed on this relief with specified properties. The 3D object is stored in a memory as a set of facets. The material properties are belonged to facets that reflect the radar impulses with different aspect angles.
The cost of real recognition experiments can be reduced by the using discussed modelling technology at the training stage. Moreover the results of recognition are improved by the fact that we can model specified object in the different conditions such as: ground properties, object reflection properties and etc.

3. Recognition and dimension reducing method

In this work we use the same method and algorithm that described in work [1]. There is a briefly present of the main part of method. The feature vector of size $N\times 1$ presents as:

$$\mathbf{x}_j = [x_1, x_2, \ldots, x_i, \ldots, x_N]$$

For a class can be constructed a matrix that consists of all vectors from this class:

$$\mathbf{X}_k = [\mathbf{x}_1(k), \mathbf{x}_2(k), \ldots, \mathbf{x}_j(k), \ldots, \mathbf{x}_M(k)],\ k = \overline{1, K}$$

Then we can form a matrix $\mathbf{Q}_k$ of size $N \times N$:

$$\mathbf{Q}_k = \mathbf{X}_k \left[\mathbf{X}_k^T \mathbf{X}_k\right]^{-1} \mathbf{X}_k^T, \quad k = \overline{1, K},$$

for whole possible K classes.
Follow this notation we can make a decision and calculate the conjugation index:

$$R_k(\mathbf{x}_j) = \frac{\mathbf{x}_j^T \mathbf{Q}_k \mathbf{x}_j}{\mathbf{x}_j^T \mathbf{x}_j}, \quad k = \overline{1, K}.$$

Then we searching the best subspaces that combined from class vectors and have higher recognition quality on the training set. Unfortunately, if the size or number of vectors are high then the algorithm will execute slowly. Thereby, in Section 4 we will present the technique of feature extraction and in this section we will discuss how to reduce the size of class matrix that constructed using training vectors.

We calculate correlations between vectors within a class in a goal to achieve the best separability of classes. The vectors with high and low correlation are excluded from the training set. The mean of this operation is that the new set without the most distant and typical vectors of class will be more effective in the recognition process. With the next step we train obtained set using algorithm of the support subspace construction.

4. CNN and feature vector size reducing

In the present time the CNN method is a fast grown method in the object recognition area. It finds many applications in the recognition of information in images [5,6]. One advantage of the CNN is in aggregating of most informative features in training set and representing it in more compact form. We are following the work [7] and explore the CNN advantage for achieving better recognition quality. Moreover, in the results of sequential convolutions the feature vector of less size is obtained. In a sense, it is similar to the fractal compression technique that authors used in [8] for reducing vector dimension and getting the less computational complexity of the support subspace method.
We implement CNN training to MSTAR database images. Figure 2 shows the general structure of the applied CNN. All images 128x128 were previously processed as described in [1] and finally have size of 64x64 pixel images. Our CNN implementation consists of two convolution layers, one max-pooling layer and one fully-connected layer.

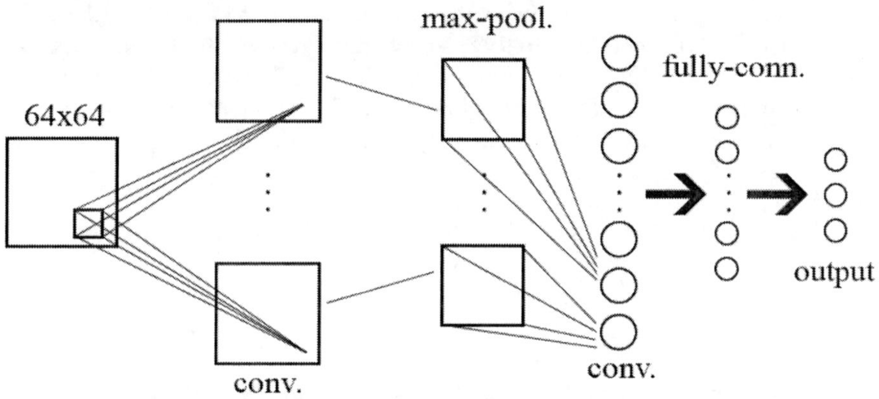

Figure 2. The CNN structure for MSTAR dataset.

The structure of layers is organazed as follows. First layer is consists of twenty features with a kernel size of 13x13. The output of the layer contains the images of the size 52x52 pixels. Next layer is the downsampling with the factor of four. In the result on the third layer input twenty downsampled images of the size 13x13. The third layer proceed the convolution with 120 kernels size of 13x13 and in the output we obtain the vector 120x1. On the next stage of our work we use the fully-connected layer of 40 neurons. In the result on final output layer the class of current vector are estimated.

5. Results and discussion

In our experiments we used the MSTAR dataset with three objects: BMP2, BTR70, T72. The training and testing sets consists of 1618 and 1365 SAR images obtained under 17° and 15° of depression angle accordingly. The training stage include three main parts.
Firstly, we train the CNN that structure described in Section 4 using backpropagation algorithm. The weights and biases of network at initialisation generated randomly. The training dropout factor was 0.85. After training process we obtained the result 98.02% of correct recognitions on testing set. We saved the resulting weights and biases that necessary to reducing the size of feature vector. Then the new dataset of training and testing vectors with sizes of 120 components were formed using the convolutions of three first CNN layers described in Section 4.

Secondly, following the work [1] we clustered all classes using the conjugated index. The best result of recognition was obtained by clustering the BMP2 on 32, the BTR70 on 8 and T72 on 8 subclasses. We clustered the BMP2 class on the higher number because that leads to better separation of class on the recognition stage.
Finally, we trained our method of support subspaces on the datasets of new vectors. Moreover, we used the exclusion of high and low correlated vectors within a class that described in Section 3. Table 1 presents the confusion matrix of the recognition experiment. We reached the result of 98.15% that is higher than result was obtained by CNN. In addition experiment we obtained the result of 76.50% using modelled images on the training stage.

Table 1. Confusion matrix of recognition using support subspaces method.

		Estimated class			
		BMP2	BTR70	T72	
Actual class	BMP2	572	5	5	98.28%
	BTR70	5	191	0	97.45%
	T72	10	0	577	98.30%

In this work the recognition program made less errors in BMP2 class than that shown in [1]. It means that the clustering process of BMP2 on 32 subclasses have shown some influence. The obtained result of MSTAR database recognition using our method is a much close to result of the CNN in contrast of the work [7] where authors is kept close to 99% correct recognition rate. It can be explained by the fact that they used SVM with specified kernel function that has nonlinear transform of the feature space. In further works we would study the effect of nonlinear transformation of our subspaces. Moreover, it is a big interest in extremely extension of the training set with the modelled images with training using CNN.

Acknowledgments
The publication is realised with support of the Ministry of Education and Science of Russian Federation and RFBR project №17-29-03112.

References

[1] Fursov V., Zherdev D., Kazanskiy N. 2016 *International Journal of Advanced Robotic Systems.* **13**(5)
[2] Franceschetti G., Iodice A., Perna S., Riccio D. 2006 *IEEE Transactions on Geoscience and Remote Sensing* **44**(10) pp. 2851-2860
[3] Franceschetti G., Guida R., Iodice A., Riccio D., Ruello G. 2004 *IEEE Transactions on Geoscience and Remote Sensing* **42** (11) pp. 2385-2396
[4] Xie H., An D., Huang X., Zhou Z. 2016 *IEEE Transactions on Geoscience and Remote Sensing* **54**(6) pp. 3360-3377
[5] Spitsyn V.G., Bolotova Yu.A., Phan N.H., Bui T.T.T. 2016 *Computer optics* **40**(2) pp. 249-257 (in Russian)
[6] Vizilter Y.V., Gorbatsevich V.S., Vorotnikov A.V., Kostromov N.A. 2017 *Computer optics* **41**(2) pp. 254-265 (in Russian)
[7] Wagner S.A. 2016 *IEEE Transactions on Aerospace and Electronic Systems* **52**(6) pp. 2861-2872
[8] Minaev Y., Fursov V. 2016 *Proceedings of the second International conference Information Technology and Nanotechnology* pp. 379-385

Evaluation of the implementation of entrepreneurial potential (on the example of the Republic of Tatarstan)

G T Guzelbaeva[1,3,6], D K Rakhmatullina[2,4] and E R Akhmetshina[2,5]

[1] Department of Finance Management, Kazan Federal University, Russian Federation, Republic of Tatarstan, 420008, Kazan, Kremlyovskaya str., 18
[2] Department of Economic Methodology and History, Kazan Federal University, Russian Federation, Republic of Tatarstan, 420008, Kazan, Kremlyovskaya str., 18
[3] E-mail: guzelbaeva@rambler.ru
[4] E-mail: na_nulya_@list.ru
[5] E-mail: kiyamova-ksfei09@yandex.ru
[6] Author to whom any correspondence should be addressed.

Abstract. Increase in the number of small businesses and the lack of effectiveness of existing support at local and regional level have led to curb the development of entrepreneurial activity. The article presents the methodological tools for assessing the effectiveness, which allows to identify existing municipalities in the advantages and disadvantages for the implementation of entrepreneurial potential, including the effectiveness of state and municipal support measures, as well as calculate the degree of progress in the implementation of entrepreneurial potential. In order to evaluate the implementation of the business potential of the Republic of Tatarstan and the impact of their public support at the regional level was calculated indexes of business activity in the municipal districts (MD) of the republic. It should be noted that a major breakthrough in share of small and medium-sized businesses is important to the whole ecosystem. The method of evaluation of the implementation of entrepreneurial potential which is presented in this paper can be used by every competent organization to analyze and form the effective programs of the economic and finance development.

1. Introduction

Promoting entrepreneurship in Russia has been a core objective of the public policy ever since 9 years ago President Medvedev voiced the objective that by 2020 60-70% of the active population of the country should be engaged in entrepreneurial activity. Today this figure is significantly smaller, so an increasing need in the qualitative development of the functioning of business, and quantitative growth of new business entities.

The research goals of the paper are to find the method for evaluation of the implementation of entrepreneurial potential of regions and on the example of the republic of Tatarstan try it in the analyses practice.

2. Theory/calculation

In Russian Federation (RF) entrepreneurial activity in the presence of positive dynamics is characterized by a low level and pace. Thus, if the average index of overall business activity is 4.8 - 5.2%, then by region its value ranges from 1.2% to 10%. The most difficult is the state of the business sector in the regions where a small number of "potential" or "nascent" entrepreneurs.

State developed countries consider SMEs as a basis of economic policy. In this case, big business is not opposed to small, they are only mutually complement each other. The development of entrepreneurship has important benefits, both economically and socially. In the Report of European Commission 2012 states that Entrepreneurship is not only a driving force for job creation, competitiveness and growth; it also contributes to personal fulfillment and the achievement of social objectives. [1] A major role of entrepreneurship in the social and economic development of the country and the factors that determine its weakness in comparison with large enterprises, determine the need for state regulation of its development. Bulk of the work on development of business relates to the regional level.

Republic of Tatarstan, according to the magazine Forbes, heads the list of the top 30 regions in Russia for business. Tatarstan is one of the most investment-attractive regions in Russia and among the top ten in the number of small and medium-sized businesses. There are various elements of business support infrastructure - technology parks, business incubators, industrial parks and playgrounds, cluster development centers and others. And what is important - not only in cities but also in areas of the Republic. In this case, the occupancy rate of technology parks and business incubators today is almost 100% - entrepreneurs actively utilize the opportunities available to them the business support infrastructure in the Republic.

However, the statistics and the results of sociological studies show that the entrepreneurial potential of the economy of the Tatarstan is far from being fully utilized. Today, small and medium business in Tatarstan has more than 140 thousand business entities, which employs about 500 thousand people. Meanwhile, studies show that the business would like to have a significantly greater number of citizens of the republic, and more than a quarter of them (28.6%) - on a permanent professional basis. However, more than half of the respondents (51.4%) said that, as part of their profession from doing business there is.

In developed countries such as USA, Japan, Canada, Germany, the share of small enterprises in the economy is 97 - 99 percent, and only 1 - 2 per cent is attributed to large holdings. Share in GDP is on average 50 percent. In addition MSEs attracts large share of foreign investments. Direct foreign investments are an important factor to intensify territorial development, because in addition to increasing the GRP, they contribute to the growth of the regional local investments. [2]

In Tatarstan, the index of the share of small and medium-sized businesses in the GRP is only 27%. GRP RT is formed mainly from the large business of the real sector of the economy (mining and industry). In the small and medium business as the bulk of the republic does not take the manufacturing sector (in 2011 the share of turnover of small and medium-sized enterprises (including micro-enterprises) all activities in the turnover of organizations of all activities was 23.3%. Greatest contribution to the turnover made small and medium-sized enterprises operating in the wholesale and retail trade (46.6%), construction (19.6%), in the area of operations with real estate, renting and business services (11.5%) and only 10.6% in the manufacturing sector. In this regard, for the past few years, all the main directions and measures of state support of entrepreneurship focused on the development, namely, the production sector.

Table 1. Levers and tools for action and regulation of business activity in the region.

Methods of direct action	Methods of indirect actions.
Regional order	Price limit
Preferential taxation	Customs tariffs
Concessional lending	Lobbying for the interests of the region
Warranty administration	Federal investments and leasing
Grants and interest-free loans	The state order
Compensation	Fiscal policy
Accelerated depreciation	
Examination and Licensing	
Preferential regional legislation	
Antitrust policy	
Lleasing policy	

Training and consulting services
Competitive bidding
Restructuring of trading and payments
The bankruptcy procedure

The main methods of influence on entrepreneurial activity at the regional level are shown in table1.

Problems in the regions of Russia in general and in the Republic of Tatarstan in particular are (ranked in descending Relevance - first problem that can be solved at the municipal - the regional level, then the problems that can be solved only at the federal or inter-regional level, or it is impossible at this stage development):

- The unavailability or high cost of resources, the cost of construction, equipment, connectivity
- Problems of access to affordable financing (unfavorable)
- Lack of qualified staff
- Lack of information and legal support
- Lack of an integrated approach to the start-up and consistent system of state support for the stages of development of micro - small - medium
- Lack of a differentiated approach to SME companies of all sizes (selective support measures): micro-business (grants, training, business incubation); small (subsidies, innovation, equipment, resources); medium (here already can only support innovation, the manufacturing sector) and different fields of activity (industrial, rural business innovation)
- Administrative barriers and difficulties, long-term registration of all necessary permits and approvals
- High levels of corruption and inequitable distribution of resources and demands, hence - unfair competition
- Inequality before the law of all market participants, changing and contradictory legislation
- Ttax policy does not stimulate research and development business, production and deep processing, does not stimulate the growth of official wage fund
- Monopolization of industries (in particular, trade, energy), and unlimited high growth rates for energy resources (fuel, electricity and obtaining additional capacity)
- Unstable economic situation, does not have to start investment projects

Activities in the Republic of Tatarstan measures to support SMEs for 2011-2013 covered areas such as preferential loans SMEs providing guarantees to small and medium-sized businesses Guarantee Fund RT program "Leasing-Grand" (subsidizing the cost of the lease agreement), the provision of financial assistance in establishing their own business, grants for the development of farms, partial compensation tuition., legal and information support of business.

To improve the human resource capacity of the business in Tatarstan regular events aimed at training of entrepreneurs and their employees. They include both direct training and various types of seminars, meetings, meetings of the "round table" gallery with the invitation as representatives of the state bodies of the republican and federal level, as well as prominent members of the business community, business coaches. There is a program "Virtual University Entrepreneurs" - educational site on entrepreneurship, bringing together as operators of all universities and consulting firms engaged in this specificity. Human capital density measures the share of highly-qualified individuals in the regional labor force. A high human capital density may be seen as an indicator that a region is rich in knowledge. There is extensive empirical evidence for a positive relationship between regional human capital and entrepreneurial activities. [3] So the interaction of the business community with the universities is one of the key conditions for the realization of entrepreneurial potential. Moreover, as formal entrepreneurship education was found to have a motivating rather than a qualifying impact it should be complemented by internships and on-the-job training (The Interplay between Entrepreneurship Education and Regional Knowledge Potential in Forming Entrepreneurial Intentions)

In order to evaluate the implementation of the business potential of the Republic of Tatarstan and the impact of their public support at the regional level was calculated indexes of business activity in

IOP Conf. Series: Journal of Physics: Conf. Series **936** (2017) 012077 doi:10.1088/1742-6596/936/1/012077

the municipal districts (MD) of the republic. Business activity is estimated using relative indicators such as the number of SMEs in the 10 000 residents output by 1 company, and others., To achieve comparability of data and evaluate the business activity of all MP, regardless of their size. The present work is based on using only the official statistics, excluding the effect of business impact of unresolved elements of work (underground economy).

Table 2 shows metrics for calculating the business activity of municipalities of the Republic of Tatarstan.

Table 2. The indicators included in the composite index of business activity.

Parameters for calculating the index of business activity	The weight of factor.
Number of economically active SMEs per 10,000 population MR RT (total growth of over 5 years)	0.3
Number of jobs per enterprise (total growth of over 5 years)	0.2
Revenue from sale of goods (works, services) for 1 company (the total increase over 5 years)	0.2
The volume of investments by an average of 1 company (total growth of over 5 years)	0.3

An index is calculated by the formula (1).

$$BA = \sum [X_i * P_i] \tag{1}$$

where BA - the total weighted assessment of business activity of a particular municipality (index);
X (i) - the total numerical score of the i-th factor for the municipality;
P (i) - i-th weight factor.

The results vary from 0 to 1, where 1 is the best practice. Results of calculation are grouped by six economic zones (table 3), which allowed to identify the most developed areas of the country.

Table 3. Business Activity Index.

Economic zones	Index of business activity	Economic zones	Index of business activity
Kazan agglomeration		Predvolzhskaya economic zone	
Vysokogorsky MD	0.31	Kamsko-Ustynsky MD	0.31
Kazan city	0.34	Buinsky MD	0.35
Verkhneuslonsky MD	0.38	Tetyushsky MD	0.4
Atninsky MD	0.42	Drojanovsky MD	0.41
Pestrechinsky MD	0.45	Apastovsky MD	0.48
Zelenodolsky MD	0.48	Kaybitsky MD	0.23
Laishevsky MD	0.5		
Predkamsky economic zone		Zakamskaya economic zone	
Arsky MD	0.35	Aksubaevsky MD	0.36
Rybno-Slobodskoy MD	0.36	Chistopoksky MD	0.38
Kukmorsky MD	0.38	Nurlatsky MD	0.38
Tyulachinsky MD	0.44	Novosheshminsky MD	0.39
Baltasinsky MD	0.47	Alkeevsky MD	0.41
Mamadyshsky MD	0.52	Alekseevsky MD	0.43
Sabinsky MD	0.56	Spassky MD	0.5
Naberezhnochelninskaya agglomeration		Yugovostochnaya economic zone	
Agrizsky MD	0.26	Sarmanovsky MD	0.09
Mendeleevsky MD	0.27	Bavlinsky MD	0.34
Menzelinsky MD	0.34	Yutazinsky MD	0.34

Zainsky MD	0.34	Aznakaevsky MD	0.38
Naberezhny Chekny city	0.38	Leninogorsky MD	0.4
Nizhnekamsky MD	0.43	CheremshanskyMD	0.44
Muslumovsky MD	0.44	BugulminskyMD	0.48
Elabyzhsky MD	0.46	Almeyevsky MD	0.49
Tukaevsky MD	0.48		
Aktanyshsky MD	0.50		

In the whole country during the analyzed period is marked increase in the number of economically active small businesses, with a minimal increase in production of goods and services per facility. Quantitative growth has not led to an increase in the share of "manufacturing industries." A marked reduction in the number of medium-sized enterprises, accompanied by an increase in turnover, including one company at the same time revealed a decrease overall investment activity of medium-sized enterprises questioned the increase in business activity in the following period. The totality of the established indicators of municipalities of the Republic of Tatarstan conditionally divided into 3 groups with high (BA index > 0.45), medium (0.38< BA index < 0.44), and low activity (BA index < 0.38). The group with high business activity attributed 13 municipal districts, 17 - to a group with an average of business activity, and 15 - the group with low business activity.

The role of public support evident in the municipal districts of the Republic of Tatarstan, but there are areas that showed minimal activity with a significant amount of state support. In the process of monitoring manifested question of lack of information in the public space of the results of independent studies on the effectiveness of public support for SMEs. A slowdown in the development of entrepreneurship and the quality has been adversely affected insufficient focus management (at all levels) for technological innovations to modernize and blurring policy priorities in support of SMEs.

According to this study, small and medium business is actively developed in areas traditionally being subsidized. In terms of state support for SMEs is the leader Kazan - more than 80% of all funds allocated for this activity in the country. Next come Naberezhnye Chelny, Tukaevsky and Sabinsry MD. In addition, attention is drawn to the following trend: the largest volumes of state prepared areas with high and medium business activity. Least of all was the Kamsko-Ustynsky MD - 400 thousand. Rubles for these five years.

Summarizing, we can say that the problem areas of support of SMEs in Tatarstan are irrational, and lack of support structure (preponderance towards infrastructure), lack of municipal funding programs absence of some of the institutions of SMEs lack of a number of effective measures to support. The business climate in almost all municipalities have significant potential to improve and opportunities to use other benefits for enterprise development. According Ulesov, Safiullin et al. [4], due to lack of funding, many structures provide services in the free market prices are not available for small businesses. There is a deformation of infrastructures to support small businesses as a result of the lack of financial and resource assistance and improper personnel policy.

Incomplete implementation of entrepreneurial potential in Republic, as well as the existing problems in the field of SME support identified the need for change in the approach to the solution of these problems. To date, in the Republic of Tatarstan adopted the "Roadmap" development of small and medium-sized businesses in the 2014-2016 " [5], in the preparation of which take into account the experience of developed countries and regions of the Russian Federation. The main goal of this project was to identify and create a balanced system of interaction between business community and state enforcement authorities in the assessment of the formation, development and protection of the environment conducive to the sustainable development of small and medium-sized enterprises in the Republic of Tatarstan.

3. Conclusion

A major breakthrough in share of small and medium-sized businesses is important to the whole ecosystem. Speaking of small and medium business, we mean financial support in the form of grants and subsidies, as well as infrastructure and consulting. In addition, the development of a network of

IOP Conf. Series: Journal of Physics: Conf. Series **936** (2017) 012077 doi:10.1088/1742-6596/936/1/012077

industrial sites at the municipal level, the work of the Guarantee Fund, the Center for Export Support, Development Corporation RT contribute to the development of entrepreneurship.

4. References

[1] European Commission 2012 *Entrepreneurship in the EU and beyond: Report, Flash Eurobarometer 354 June - August 2012.*

[2] Hejazi W and Pauly P 2003 The impact of FDI on Economic Growth under Foreign Trade Regimes: A Case Study of Pakistan *J. of International Business Studies*. 282-9.

[3] Bates T 1990 Entrepreneur human capital inputs and small business longevity *Review of Economics & Statistics* **72(4)** 551-9.

[4] Ulesov D V, Shigabieva A M, Maratkanova E M and Shaidullin R N 2013 Information Infrastructure of Small Business Development *J. World Applied Sciences Journal* **27(13)** 193-6.

[5] Chamber of Commerce and Industry of Republic of Tatarstan 2013 *"Roadmap" development of small and medium-sized businesses in Republic of Tatarstan in the 2014-2016.*

[6] Sascha G, Walter and Dirk Dohse 2009 The Interplay between Entrepreneurship Education and Regional Knowledge Potential in Forming Entrepreneurial Intentions *Kiel Working Paper September 2009* **1549**

IOP Conf. Series: Journal of Physics: Conf. Series **936** (2017) 012078 doi:10.1088/1742-6596/936/1/012078

Hadronization via gravitational confinement

C.G. Vayenas[1,2,*] and D. Grigoriou[1]
[1]LCEP, 1 Caratheodory St., University of Patras, Patras GR 26504, Greece
[2]Division of Natural Sciences, Academy of Athens, 28 Panepistimiou Ave., GR-10679 Athens, Greece

E-mail: cgvayenas@upatras.gr

Abstract. We analyze baryogenesis using a Bohr-type rotating neutrino model in which the strong force is modeled as the gravitational force between ultrarelativistic neutrinos. The kinetic energy plus rest energy of these neutrinos constitutes the rest energy of the gravitationally confined composite neutrino structure. In this work we derive a simple expression for the gravitational mass of the rotating neutrinos in terms of their rest mass and their total energy. It is found, both analytically and graphically, that when the gravitational mass reaches the Planck mass, then the relativistic mass of the neutrinos equals the effective mass of u and d quarks. The model contains no adjustable parameters and leads to semiquantitative (within 1%) agreement with the masses of baryons. It also shows that hadronization is easier to achieve with light particles such as neutrinos rather than with heavier particles. The model can also be extended by considering ultrarelavivistic rotating positron or electron-neutrino pairs, triplets or quartets whose mass is found to match within 2% those of $W^{\pm}$, Z^o and H^o bosons.

1. Introduction

Since the pioneering 1905 special relativistic paper of Einstein [1] it is well known that the inertial mass, m_i, of any body moving on a straight line is larger than its rest mass, m_o, and the two are related via

$$m_i = \gamma^3 m_o, \tag{1}$$

where γ is the Lorentz factor given by

$$\gamma = (1 - \mathrm{v}^2/c^2)^{-1/2}, \tag{2}$$

where v is the speed of the body. Equation (1) is obtained [2] by combining the definition of force, F, i.e.

$$F = \frac{dp}{dt} = \frac{d(\gamma m_o \mathrm{v})}{dt} = \gamma^3 m_o \frac{d\mathrm{v}}{dt}, \tag{3}$$

where p is the momentum, with Newton's 2nd law of motion, i.e.

$$F = m_i \frac{d\mathrm{v}}{dt}. \tag{4}$$

According to the equivalence principle [3], the inertial mass, m_i, equals the gravitational mass, m_g, and thus, from (1), (3) and (4), it follows

$$m_g = \gamma^3 m_o, \tag{5}$$

which is valid for any linear motion.

Using the approach of instantaneous reference frames [2] in which the arbitrary motion of a particle, $\mathbf{d}\underline{\mathbf{x}}$, at any point of the trajectory, is attributed to the action of a force, $\mathbf{d}\underline{\mathbf{F}}$, parallel to $\mathbf{d}\underline{\mathbf{x}}$, it is easy to show that equation (5) remains valid for arbitrary particle motion. Consequently, as shown in references [4] and [5] the inertial, and thus gravitational, mass of a body equals its so called longitudinal mass [1] which is invariant along the direction of motion, and thus is an inertial mass, while the so called transverse mass [1] is not invariant and thus is not an inertial mass [2]. Invariant means that the same force value is perceived by the laboratory observer and by an observer on the moving particle [2]. These are simple and well established concepts of special relativity [2], yet the importance of the resulting equation (5) has been only recently recognized. Starting from Newton's universal gravitational law

$$F = \frac{Gm_{g,1}m_{g,2}}{d^2}, \tag{6}$$

where d is the distance, and using equation (5), one obtains

$$F = \frac{Gm_{1,o}m_{2,0}\gamma_1^3\gamma_2^3}{d^2}. \tag{7}$$

It is interesting that more than 110 years after Einstein's pioneering 1905 paper [1], we still use Newton's universal gravitational law always with $\gamma_1 = \gamma_2 = 1$. However in doing so we neglect the important increase in gravitational mass and gravitational attraction with particle velocity as expressed by equation (5). In a book [4] and in several publications [5, 6, 7, 8, 9, 10, 11, 12] we have presented Bohr-type rotating lepton models for hadrons [4], and bosons [6, 7, 11], based on equation (7) in conjunction with the de Broglie wavelength equation, which is the basis of quantum mechanics. These models contain no adjustable parameters and allow for the computation of the masses of composite particles formed by the rotating leptons. An example is given in Table 1 for the case of three rotating neutrinos forming a composite rotational state with the mass of the proton. The electric charge of the latter, as well as the fractional charges of the quarks, induced by charge-induced dipole interactions [6, 7], can be explained if a positron is present at the center of the rotating neutrinos ring [6, 7]. The literature of neutrino electromagnetic interactions has been reviewed recently [13].

It is worth noting that the Bohr-type rotating lepton models for hadrons presented in [4, 5, 6, 7, 8, 9, 10, 11, 12] and further analyzed here, are fully consistent with the bagel shape of protons computed by Miller [14] via model proton wave functions constructed with Poincaré invariance for quark spin antiparallel to the proton spin and quark momenta of 1 GeV/c [14]. This value is in close agreement with the momenta (0.939 GeV/c) of the neutrinos in the rotating neutrino proton model computed in [4, 5], as shown in Table 1 using the value of 0.0437 eV/c^2 for the rest mass of neutrinos [4, 5, 15, 16, 17, 18] which is in good agreement with the recent results of the Kamiokande experiments [15, 17].

In the present work we first derive a simple equation and corresponding graphical method for determining the masses of gravitationally confined hadron structures and then we discuss the key advantageous property of neutrinos which makes them suitable for hadronization, i.e. for the generation of mass.

2. Results

2.1. Gravitational mass and its dependence on particle velocity and total energy

According to the weak equivalence principle of Einstein and Eötvös [3], which has been confirmed by the most modern torsion balance measurements to at least 1 part in 10^{12} [3, 4], it is

$$m_g = m_i, \tag{8}$$

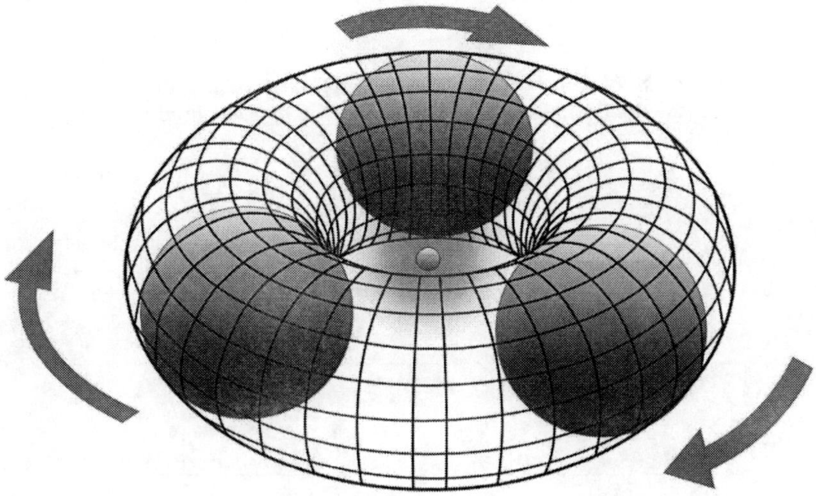

Figure 1. Schematic comparison and synthesis of the bagel shape of protons computed in [14] via model wave functions, constructed with Poincaré invariance, and of the three-rotating neutrino model of protons [4, 5, 6]. Particle size represents gravitational mass. The central particle is a positron of negligible speed, thus negligible gravitational mass [6].

i.e. gravitational mass equals inertial mass and therefore, as already stated, it follows from equation (1) that

$$m_g = \gamma^3 m_o, \tag{9}$$

where m_o is the rest mass.

Also, from Einstein's special relativity [1, 2], the relativistic mass, m_r, and the total (i.e. rest plus kinetic) energy, E, of a particle are related to the rest mass m_o via

$$m_r = \gamma m_o \quad ; \quad E = m_r c^2 = \gamma m_o c^2. \tag{10}$$

Upon eliminating γ between (9) and (10) one obtains

$$m_g = (1/m_o^2) m_r^3 \quad ; \quad m_g = (1/m_o^2 c^6) E^3. \tag{11}$$

These equations, and particularly the latter one, are found to be quite useful since they enable one to express directly the gravitational mass of a particle in terms of its total energy.

2.2. Forces between fast neutrinos

The potential importance of gravitational interactions between neutrinos and gravitational fields had been discussed already in 1957 by Dieter and Wheeler [19], long before it was established that neutrinos have mass [17].

Figure 2 is obtained from equations (11) and shows, in the left hand side axis, the dependence of the gravitational mass, $m_{g,\nu}$, of a neutrino with rest mass $m_{o,\nu}(= 0.0437\ \mathrm{eV/c^2})$ on its relativistic mass, $m_{r,\nu}$, and on its total energy $E(= m_{r,\nu}c^2)$. The slope of 3 on the $\log m_{g,\nu}$ vs $\log m_{r,\nu}$ and $\log m_{g,\nu}$ vs E plots of Figure 2 is the result of the 3d power in equations (11).

The right hand of axis of Figure 2 shows the dependence on $m_{r,\nu}$ and E of the ratio of the gravitational force, F_G, between two neutrinos with rest mass $m_{o,\nu}$ each, computed from equation

Table 1. Comparison of the Bohr model for the H atom and of the rotating neutrino model for baryons [4, 5]

Bohr model for the H atom	**Bohr-Einstein model for baryons [4,5]**
Electron as *particle* $m_e v^2 / r = e^2 / \varepsilon r^2$ ↑ Newton's 2nd law ↑ Coulomb law	Neutrino as *particle* $\gamma_\nu m_{o,\nu} \frac{v^2}{r} = \frac{G m_{o,\nu}^2 \gamma^6}{\sqrt{3}\, r^2}$ ↑ Relativistic equation of motion for circular motion ↑ Newton's gravitational law accounting for special relativity ($m_i = \gamma^3 m_o$) and for equivalence principle ($m_g = m_i$)
Electron as *wave* $\frac{\hbar}{m_e v} = r$ De Broglie (for n=1)	Neutrino as *wave* $\frac{\hbar}{\gamma_\nu m_{o,\nu} v} \approx \frac{\hbar}{\gamma_\nu m_{o,\nu} c} = r$ De Broglie (for n=1)
<u>Results</u> $v = \alpha c \;;\; \gamma \approx 1$ $r = \frac{\hbar}{\alpha m_e c} = 0.5292 \cdot 10^{-10}\text{m}$ $V_e = -27.2$ eV $T = 13.6$ eV $\mathcal{H} = T + V_e = -13.6$ eV $\Delta(RE) \equiv \Delta m c^2 = T = (1/2)\alpha^2 m_e c^2$ $= 13.6$ eV	<u>Results</u> (with $m_{o,\nu}$=0.043723 eV/c^2) $v \approx c \;;\; \gamma = 3^{1/12}(m_{Pl} / m_{o,\nu})^{1/3} = 7.163 \cdot 10^9$ $r = \frac{\hbar}{\gamma_\nu m_{o,\nu} c} = 0.630 \cdot 10^{-15}\text{m}$ $V_G = -(5/3) m_n c^2 = -1565.9$ MeV $T = m_n c^2 = 939.565$ MeV $p = m_n c = 939.565$ MeV/c $\mathcal{H} = T + V_G = -(2/3) m_n c^2 = -626.4$ MeV $\Delta(RE) = \Delta m c^2 = T = m_n c^2 = 3^{13/12}(m_{Pl} m_o^2)^{1/3}$ $= 939.565$ MeV Per particle: $T_p = 313.2$ MeV $\mathcal{H}_p = -208.8$ MeV

(7), and of the Coulombic electrostatic force, F_e, between a positron, e^+, and an electron, e^-, at the same distance. This force is computed from Coulomb's law, i.e. from

$$F_e = \frac{e^2}{\epsilon r^2}, \tag{12}$$

with $\varepsilon = 4\pi\varepsilon_o = 1.11 \times 10^{-10}$ C^2/Nm, where ε_o is the permittivity of vacuum.

Surprisingly, as shown in Figure 2, this ratio, which is computed directly from equations (7) and (12), i.e.

$$\frac{F_G}{F_e} = \frac{\epsilon G m_{o,\nu}^2}{e^2} \gamma_\nu^6 = \frac{\epsilon G m_{o,\nu}^2}{e^2} \left[\frac{E}{m_{o,\nu} c^2}\right]^6 = \frac{\epsilon G E^6}{e^2 m_{o,\nu}^4 c^{12}}, \tag{13}$$

exceeds unity for relativistic neutrino masses above 130 MeV/c^2 or momenta above 0.13 MeV/c., i.e. for total neutrino energies above 130 MeV. Consequently, for neutrino energies above 130 MeV, Newtonian gravity with gravitational masses is stronger than Coulombic attraction or repulsion. As shown in Figure 2, for a neutrino energy of 313 MeV (corresponding closely to the effective mass of u and d quarks [20, 21]), the F_G/F_e ratio takes the value $3^{1/2}/\alpha \approx 237$.

Clearly, for such neutrino energies, their gravitational attraction is as strong as the Strong Force itself. As shown in the left axis of Figure 2, at the same neutrino energy of 313 MeV the

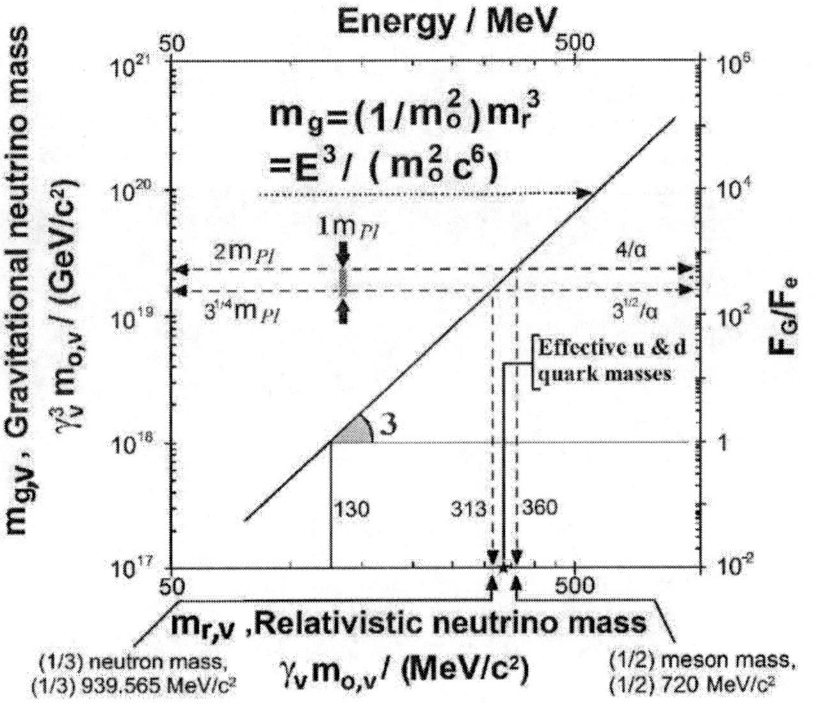

Figure 2. Plot of equations (11) for electron neutrinos ($m_{o,\nu}$ = 0.0437 eV [4, 5, 15, 16]) showing that baryons and mesons form from neutrinos when the gravitational mass, $m_{g,\nu}$, of the latter reaches the Planck mass. (more precisely when it reaches $2m_{Pl}$ for mesons, $3^{1/4}m_{Pl}$ for baryons). At these points the F_G/F_e ratio reaches $4/\alpha(\approx 548)$ and $3^{1/2}/\alpha(\approx 237)$ respectively; $\alpha = e^2/\epsilon c\hbar \approx 1/137.035$.

neutrino gravitational mass, $\gamma^3 m_o$, reaches the value of $3^{1/4}$ m_{Pl}, i.e. it exceeds the Planck mass ($= \hbar c/G = 1.221 \times 10^{19}$ GeV/c^2) by 31.6%.

This has been shown to be exactly the value of the gravitational mass of a neutrino rotating with two other partners around their gravitational mass center in circular motion [4, 5].

Indeed, upon combining the equation of motion

$$\gamma_\nu m_{o,\nu} c^2/r = \frac{G m_{o,\nu}^2 \gamma_\nu^6}{\sqrt{3} r^2} \tag{14}$$

with the de Broglie equation

$$r = \frac{\hbar}{\gamma_\nu m_{o,\nu} c} \tag{15}$$

one obtains

$$3\gamma_\nu m_{o,\nu} = 3^{13/12}(m_{Pl} m_{o,\nu}^2)^{1/3} = 939\ MeV/c^2, \tag{16}$$

which is the proton mass and

$$\gamma_\nu^3 m_{o,\nu} = 3^{1/4} m_{Pl} \tag{17}$$

IOP Conf. Series: Journal of Physics: Conf. Series **936** (2017) 012078 doi:10.1088/1742-6596/936/1/012078

which as presented in Figure 2 is the gravitational mass corresponding to the effective u and d quark mass of 313 MeV. This shows clearly that the effective mass of quarks is that of relativistic neutrinos.

It is worth noting in equation (13), that for fixed particle energy E, the ratio F_G/F_e increases dramatically with decreasing neutrino rest mass $m_{o,\nu}$. This is a direct consequence of equations (11) which show that, for fixed E, the gravitational mass of a particle increases significantly with decreasing rest mass $m_{o,\nu}$. This in turn shows that gravitational confinement of neutrinos can take place with moderate (e.g. less than 500 MeV) energies (Fig. 2) because the rest mass of neutrinos is so small ($\sim 0.0437\ eV/c^2$) [4, 5, 15]. Indeed the condition for gravitational confinement, i.e.

$$m_g \geq m_{Pl}, \tag{18}$$

implies in view of equations (11) that

$$E_\nu \geq (m_{Pl} m_{o,\nu}^2)^{1/3} c^2 = 286\ GeV, \tag{19}$$

or equivalently

$$m_{o,\nu} \leq \frac{E^{3/2}}{m_{Pl}^{1/2} c^3}, \tag{20}$$

which for $E_\nu = 0.313\ GeV$ gives

$$m_{o,\nu} \leq 0.05\ eV/c^2. \tag{21}$$

This limit practically coincides with the largest mass of electron neutrinos [4, 5, 15, 16].

Equation (19) written for electrons, which have a rest mass $1.17{\cdot}10^7$ times larger than that of neutrinos, leads to an E_ν value of 14.3 TeV. This shows why lighter particles such as neutrinos, are ideal for building composite particles such as hadrons.

3. Discussion and Conclusions

This paper presents a simple expression for the gravitational mass of particles in terms of their rest mass and their relativistic mass or their total energy.This expression, which is based entirely on the special relativistic definitions of rest, relativistic, inertial and gravitational mass, is then used in conjunction with the laws of Newton and Coulomb to show that the gravitational attraction between relativistic neutrinos with energies above 130 MeV is stronger than the electrostatic attraction between a positron and an electron at the same distance. This is because inertial and gravitational mass is accounted to increase with γ^3, as first derived by Einstein already in 1905 [1], whereas relativistic mass and total energy increase with γ [1].

The same expression (equation (5)) shows that the energy (313 MeV) for which the neutrino gravitational mass reaches the Planck mass, corresponds to the effective mass of quarks, i.e. to one third the mass of protons or neutrons. This, in conjunction with the fact that the Newtonian gravitational attraction between two Planck masses equals $\hbar c/d^2$, shows that the Strong Force, which confines protons and neutrons, is the relativistic gravitational force. Thus, this same equation allows for the graphical determination of the mass of gravitationally confined rotating lepton composite particles as shown in Figure 2.

One interesting aspect of the rotating lepton model which is shown explicitly by equations (11) and by the derived equations (19) to (21) is that for any given particle energy, E, hadronization occurs easier with particles which have smaller rest mass. This explains why neutrinos are ideal as hadronization components.

The present results confirm that the use of the gravitational mass, $\gamma^3 m_o$, in Newton's gravitational law leads to excellent agreement with experimental results in the microcosmos of hadrons. They show that gravity generates mass via gravitaional confinement and concomitant trapping of neutrinos in ultrarelativisic orbits. The kinetic energy of these trapped neutrinos is the rest energy of the composite rotational hadronic structures. The gravitational field generated by these ultrarelativistic neutrinos has the basic property of the Higgs field, i.e. it generates mass. The kinetics of this mass generation have been discussed in [9].

It is thus important to establish the consistency of the relativistic mass Newtonian equation (7) with the theory of general relativity (GR) since the latter is the most well known and generally accepted gravitational theory [22, 23]. Indeed the effective force computed from the effective potential of the Schwarzschild geodesics of neutrinos rotating around Planckian masses matches that computed via the present special relativistic approach [5, 8], which provides further support for the rotating neutrino model. In the GR treatment [5, 8] gravitational collapse of the composite particle to a singularity is prevented by the Heisenberg uncertainty principle. This is consistent with the hypothesis that protons and neutrons can be also viewed as microscopic black holes [8, 25, 26, 27] whose gravitational collapse to a singularity is prevented by the Heisenberg uncertainty principle, expressed via the de Broglie wavelength equation, leading to the generalized uncertainty principle and generalized event horizon concepts. Furthermore, equation (7) has been shown recently to give the same results with GR regarding the perihelion precession of Mercury [24], which was one of the key initial tests of GR. Consequently, equation (7) appears to be in good agreement with the GR treatment both in the microcosmos of quarks and in the macrocosmos of celestial motion.

Acknowledgments
We thank Professor A. Fokas for numerous invaluable discussions.

References

[1] Einstein, A.: Zür Elektrodynamik bewegter Körper. Ann. der Physik., Bd. XVII, S. **17**, 891-921 (1905); English translation On the Electrodynamics of Moving Bodies (*http : //fourmilab.ch/etexts/einstein/specrel/www/*) by G.B. Jeffery and W. Perrett (1923).
[2] French AP (1968) Special relativity. W.W. Norton and Co., New York.
[3] Roll PG, Krotkov R, Dicke RG (1964) The equivalence of inertial and passive gravitational mass. Annals of Physics 26(3):442-517
[4] Vayenas CG & Souentie S: (2012), Gravity, special relativity and the strong force: A Bohr-Einstein-de-Broglie model for the formation of hadrons. Springer, ISBN 978-1-4614-3935-6.
[5] Vayenas CG, Souentie S & Fokas A (2013) A Bohr-type model with gravity as the attractive force. arXiv:1306.5979v1 [physics.gen-ph]; (2014) Physica A 405:360-379.
[6] Vayenas CG, Fokas AS, Grigoriou D (2016) On the structure, masses and thermodynamics of the $W^{\pm}$ bosons. Physica A 450:37-48.
[7] Fokas AS, Vayenas CG (2016) On the structure, mass and thermodynamics of the Z^o bosons. Physica A 464:231-240.
[8] Vayenas CG and Grigoriou D (2015) Microscopic black hole stabilization via the uncertainty principle. J. Phys.: Conf. Ser. 574, 012059(1-8).
[9] Vayenas CG, Fokas AS, Grigoriou D (2017) Catalysis and autocatalysis of chemical synthesis and of hadronization, Appl. Catal. B, 203:582-590.
[10] Vayenas CG, Fokas AS, Grigoriou D (2016) Computation of masses and binding energies of some hadrons and bosons according to the rotating lepton model and the relativistic Newton equation. J. Phys.: Conf. Ser. 738, 012080.
[11] Fokas AS, Vayenas CG, Grigoriou D (2017) On the mass and thermodynamics of the Higgs boson, submitted.
[12] Vayenas CG, Fokas AS, Grigoriou D (2016) Relations between the baryon quantum numbers of the Standard Model and of the rotating neutrino model. arXiv:1606.09570.
[13] Giunti C, Studenikin A (2015) Neutrino electromagnetic interactions: a window to new physics. Rev. Mod. Phys. 87:531; arXiv:1403.6344v3 [hep-ph].
[14] Miller GA (2003) Shapes of the proton, Physical Review C: 68:022201(R).
[15] Mohapatra RN et al (2007) Theory of neutrinos: a white paper. Rep Prog Phys 70:1757-1867
[16] Fritzsch H (2015) Neutrino masses and flavor mixing. Modern Physics Letters A, 30(16):1530012(1-6).
[17] Takaaki Kajita(2009) Neutrino oscillations: Discovery, current status, future directions. Int. J. Mod. Phys. A, 24:3437-3446.

[18] Winter W (2015) Neutrino mass hierarchy: Theory and phenomenology. AIP Conf. Proc. 1666, 120001; http://dx.doi.org/10.1063/1.4915577.
[19] Dieter RB and Wheeler JA (1957) Interaction of Neutrinos and Gravitational Fields. Rev. Mod. Phys. 29(3):465-479.
[20] Griffiths D (2008) Introduction to Elementary Particles. 2nd ed. Wiley-VCH Verlag GmbH & Co. KgaA, Weinheim.
[21] Tully CG (2011) Elementary Particle Physics in a Nutshell, ISBN13: 978-0-691-13116-0, Princeton University Press.
[22] Will CM (1993) Theory and experiment in gravitational physics. Cambridge University Press, Cambridge .
[23] Aretakis S (2015) Horizon instability of extremal black holes. Advances in Theoretical and Mathematical Physics 19:507-530.
[24] Fokas AS, Vayenas CG and Grigoriou D (2016) Analytical computation of the Mercury perihelion precession via the relativistic gravitational law and comparison with general relativity. arXiv:1509.03326v2 [gr-qc].
[25] Carr B, Modesto L, Prémont-Schwarz I (2011) Generalized Uncertainty Principle and Self-dual Black Holes. arXiv:1107.0708v1 [gr-qc].
[26] Carr BJ (2014) The Black Hole Uncertainty Principle Correspondence. arXiv:1402.1427v1 [gr-qc].
[27] Das S and Vagenas EC (2009) Phenomenological Implications of the Generalized Uncertainty Principle. Can. J. Phys. 87:233-240.

IOP Conf. Series: Journal of Physics: Conf. Series **936** (2017) 012079 doi:10.1088/1742-6596/936/1/012079

Dynamics of nanoparticle concentration in nanofluids under laser light field

A I Livashvili[1], V V Krishtop[1], P V Vinogradova[1], G V Kostina[1] and T N Bryukhanova[2]

[1]Far Eastern State Transport University, 47, Seryshev st., Khabarovsk, Russia,

[2]Pacific National University, 136, Tikhookeanskaya st., Khabarovsk, Russia.

E-mail: krishtop@list.ru

Abstract. The dynamics of the concentration of nanofluids placed in a light field with a Gaussian intensity profile is studied theoretically. The investigation is based on the analytical and numerical solutions of the system of linearized heat conduction and convection-diffusion equations. The convection-diffusion equation contains terms that correspond both to the Soret effect and to the transfer of nanoparticles, caused by the action of a light field on them (electrostriction). The dependence of the coefficient of thermal conductivity of the medium on the concentration is taken into account. It is shown that under these conditions single travelling waves appear in the medium, the velocity of which depends not only on the thermal physical parameters of the medium, but also on the polarization of the particles. Conditions under which the stratification of the medium is possible are found.

1. Introduction

Colloidal suspensions or, as they are now called, nanofluids, are widely used in various areas of modern technology. For example, magnetic fluids are used to polish optical components, and suspensions of silica particles in liquid crystals significantly improve the characteristics of optical storage devices [1-2]. We also note their use in chemical processes (catalysis), in the creation of new drugs, lubricants and so on [3-7]. With the growth of the productivity of electronic devices and the development of high-energy technologies, it has become necessary to create efficient cooling systems and control large heat flows [8]. Promising developments have emerged related to molecular computers based on switchable bistable molecules or their aggregates. One way to intensify heat transfer is to increase the thermal conductivity of a liquid by adding solid particles with high thermal conductivity [9-12].

Particularly interesting for the creation of such suspensions are nanoparticles. Recent studies have shown that liquid-phase media, in which nanoparticles from wide-gap semiconductors or dielectrics are used as a dispersive component, are very effective for the realization of a number of nonlinear optical effects [13]. In these media, unlike homogeneous media, the nonlinear optical response arises from the change in the refractive index induced by the light wave and the absorption coefficient due to the phenomena of thermal diffusion and electrostriction of particles [14-18]. At the same time, in our opinion, physical mechanisms connected in particular with the processes of heat and mass transfer in such media require additional investigation. It should be added that a number of mathematical

IOP Conf. Series: Journal of Physics: Conf. Series **936** (2017) 012079 doi:10.1088/1742-6596/936/1/012079

problems associated with analytical methods for solving the corresponding equations are also far from being resolved.

2. Theoretical model

The paper considers a liquid-phase medium with nanoparticles irradiated by a light beam with a Gaussian intensity profile. As a result of the action of the light field in the medium, gradients of temperature and concentration arise, which cause the processes of heat and mass transfer. These phenomena are described by a system of balance equations for temperature and particles, taking into account the concentration convection [19], written in one-dimensional form (without using the Boussinesq approximation):

$$C_p \rho \frac{\partial T}{\partial t} = \frac{\partial}{\partial x}(\lambda(C)\nabla_x T) + \alpha_0 \tilde{I}_0 \exp(-\frac{x^2}{x_0^2}), \tag{1}$$

$$\frac{\partial C}{\partial t} = D\nabla_x^2 C - v\nabla_x C + D_T \nabla_x \left(C(1-C)\nabla_x T\right) - \gamma \frac{\partial}{\partial x}(C\frac{\partial \tilde{I}}{\partial x}). \tag{2}$$

In the heat equation, the term responsible for the Dufour effect is omitted, in view of its smallness. The processes of sedimentation are also neglected. Here, the following designations are used: T is the temperature of the medium, $C = C(r,t) = m_0/m$ is the mass concentration of the particles (m_0 is the mass of the particles, m is the mass of the whole medium), C_p, ρ are the thermophysical constants of the fluid, λ (C) is the coefficient of thermal conductivity, I_0 is the light intensity, the absorption coefficient of the medium, and the diffusion and thermal diffusion coefficients, respectively, and $v = \eta/h$ is the rate of concentration convection, which we assume to be constant. In this case, η is the kinematic viscosity, and h is the characteristic distance (in our case, we take $h \approx 10x_0$, $\gamma = 4\pi\beta D/(cn_{ef}k_B T)$, β is the polarizability of the particles, k_B is Boltzmann's constant, c is the speed of light in vacuum, n_{ef} is the effective refractive index of the medium. The last term expresses the contribution from the particle flux and is related to the effect of the gradient force from the side of the light field (electrostriction of particles).

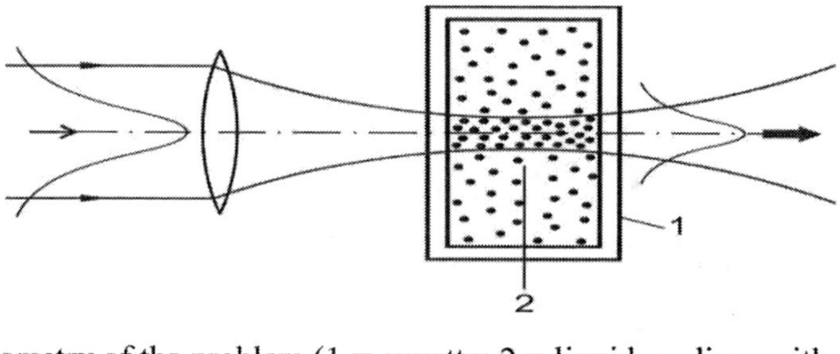

Figure 1. Geometry of the problem (1 = cuvette; 2 = liquid medium with nanoparticle).

Let us take into account the fact that the temperature establishment processes go faster than the diffusion processes. This makes it possible to study the latter against a background of a stationary temperature. We will consider the case of small concentrations: C << 1.

Further, the concentration dependence of the thermal conductivity can be represented in the form:

$$\lambda(C) = \lambda_0 + \beta C = \lambda_0(1 + pC),$$

where $p = \frac{\beta}{\lambda_0} > 1$. This kind of dependence was theoretically found in [20] and was experimentally confirmed in publications [21, 22]. In the equations of heat conduction and diffusion, divergent terms are taken in the form:

$$\frac{\partial}{\partial x}\left(\lambda(C)\frac{\partial T}{\partial x}\right)\approx\lambda(C)\frac{\partial^2 T}{\partial x^2},\quad \gamma\frac{\partial}{\partial x}\left(C\frac{\partial I}{\partial x}\right)\approx\gamma C\frac{\partial^2 \tilde{I}}{\partial x^2},\tag{4}$$

The validity of this approximation for the first formula is based on the use of equality (3), which leads to a reasonable inequality, $pC<1$, and for the second, it is justified by direct estimates of the quantities. Next, we linearize equation (2), step by step. First, we assume:

$$(\lambda(C))^{-1}=(\lambda_0+\beta C)^{-1}=(\lambda_0)^{-1}(1-pC).(pC<<1)\tag{5}$$

Assuming that C << 1, we represent the concentration in the form:

$$C(x,t)=C_0+\Delta C(x,t)=C_0(1+C'(x,t))\qquad C'(x,t)=\frac{\Delta C}{C_0},\tag{6}$$

where $\Delta C(x,t)$ is the disturbed part of the concentration, and $|\Delta C|<<|C_0|$ approximations (3), (4) and its linearization with respect to the function $u(x,t)=1+C'(x,t)$, we obtain the problem. After substituting expressions (5) and (6) into equation (2), taking into account the:

$$\frac{\partial u}{\partial t}=D\frac{\partial^2 u}{\partial x^2}-v\frac{\partial u}{\partial x}-\frac{D_T\alpha_0\tilde{I}_0}{\lambda_0}(1-pC_0)\exp(-\frac{x^2}{x_0{}^2})u+\frac{4\gamma\tilde{I}_0}{x_0{}^2}(1-\frac{x^2}{x_0{}^2})\exp(-\frac{x^2}{x_0{}^2})u\tag{7}$$

$$u(x,0)=\exp(-\frac{x^2}{x_0{}^2}),\quad \frac{\partial u(0,t)}{\partial t}=0,\qquad 0\le x<\infty.\tag{8}$$

We rewrite the problem (7)–(8) in dimensionless variables:

$$\frac{\partial u}{\partial\tau}=b\frac{\partial^2 u}{\partial\rho^2}-Peb\frac{\partial u}{\partial\rho}+(q(1-\rho^2)-1)\exp\left(-\rho^2\right)u,\tag{9}$$

$$u(\rho,0)=\exp(-\rho^2)\ ,\ \frac{\partial u(o,\tau)}{\partial\rho}=0,\ 0\le\rho<\infty,\quad 0\le\tau<\infty,\tag{10}$$

where $\tau=\frac{S_T\alpha_0\tilde{I}_0 D}{\lambda_0}(1-pC_0)t,\ \rho=\frac{x}{x_0},\ Pe=\frac{vx_0}{D}$ – Peclet number, $b=\frac{\lambda_0}{S_T\alpha_0\tilde{I}_0(1-pC_0)x_0{}^2}$, $q=\frac{4\gamma\tilde{I}_0}{D}b$,

It is not possible to find the exact analytic solution of equation (7) with initial-boundary conditions (8). The Peclet number estimate for typical values of the quantities in our case gives $Pe\ge 10^3$.

This value indicates the predominance of convection processes over molecular diffusion. In view of this fact, we rewrite problem (7)–(8) in a truncated form:

$$\frac{\partial u}{\partial\tau}=-Peb\frac{\partial u}{\partial\rho}+(q(1-\rho^2)-1)\exp\left(-\rho^2\right)u,\tag{11}$$

Its exact solution can be written in the form:

$$u(\rho,\tau)=A\exp[-(\rho-Peb\tau)^2]\exp\{\frac{1}{4Peb}[\sqrt{\pi}(q-2)erf(\rho)+2q\rho\exp(-\rho^2)]\},\tag{12}$$

where $erf(\rho)$ is the error function (Gauss). The constant A can be determined from the condition of conservation of the number of particles in the system.

IOP Conf. Series: Journal of Physics: Conf. Series **936** (2017) 012079 doi:10.1088/1742-6596/936/1/012079

Figure 2 shows the graphs of the function *u (ρ)* at increasing times τ.

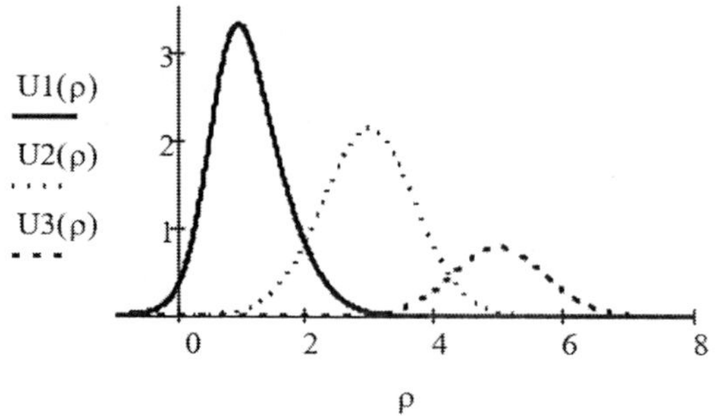

Figure 2. The graphs of the solution u (ρ, τ) presented for different instants of time (as they increase).

Thus, accounting for the concentration convective flow, even in the one-dimensional approximation, leads to amplitude-modulated travelling waves.

Let us now consider the state of mechanical equilibrium of a nanofluid, that is, when its macroscopic flow is absent and the equilibrium distributions of temperature and concentration are described by the equations:

$$\nabla^2 T + \frac{\alpha_0 \tilde{I}_0}{\lambda_0}\exp(-\frac{x^2}{x_0^2}) = 0, \qquad T(0) = T_0, \tag{13}$$

$$\frac{dC}{dx} + (S_T\frac{dT}{dx} - \frac{\gamma}{D}\frac{dI}{dx}) = 0. \tag{14}$$

The solution of this system is the expression:

$$C(x) = B\exp[-S_T T_0 + n\rho erf(\rho) + (n+m)\exp(-\rho^2), \tag{15}$$

where $n = \frac{\alpha_0 \tilde{I}_0 S_T}{2\lambda_0} x_0^2$, $m = \frac{2\tilde{I}_0 \gamma}{D}$.

The constant B in formula (15) can be determined from the condition of constancy of the average concentration.

3. Discussion

Note that the convective processes in a colloidal suspension of nanoparticles taking into account sedimentation (in the Benard scheme) were studied in [23-25]. The mode of running concentration waves was detected. In our approach, the effect of radiation due to the action of gradient forces on the side of the light field, being an additional source of convection, leads to a completely different spatial distribution of concentration. We note that the amplitudes of the modulated waves in solution (12) are damped and have a spatially complex structure. In particular, an analysis of the exponent shows that with positive thermal diffusion (b > 0) and a change in the sign of the polarization of the particles β, its structure significantly changes (figure 3). These observations indicate the possibility of controlling the optical properties of nanofluids, since the nonlinear part of the refractive index of such a medium depends on the concentration.

IOP Conf. Series: Journal of Physics: Conf. Series **936** (2017) 012079 doi:10.1088/1742-6596/936/1/012079

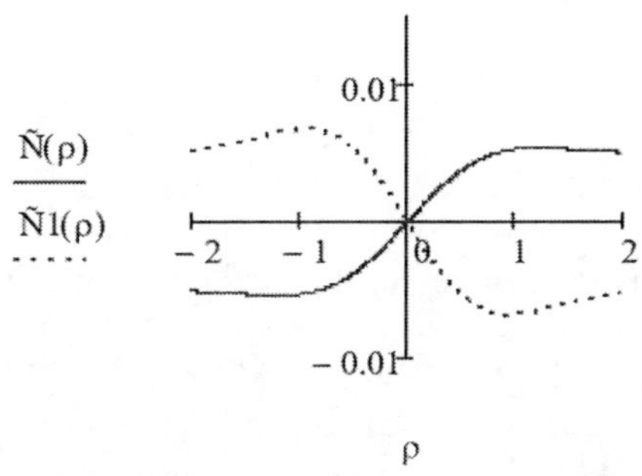

Figure 3. Dependence of the amplitude of the solution (12) on the sign of the polarization.

Solution (15) shows that in our case the convection of a nanofluid has two processes: thermal diffusion and electrostriction, which can be either unidirectional or multidirectional. It is easy to see from (15) that if the contribution from electrostriction is small (m << n), then in the conditions of mechanical equilibrium, the resulting stratification (stratification) is associated with thermal diffusion.

We understand that a number of questions require more detailed study, for example, analysis of convection stability, numerical modelling of these processes and their study in a 2D measurement. We will devote our attention to these problems in future research.

This work was supported by Ministry of Education and Science of Khabarovsk krai (Russia).

References

[1] Bardakhanov S P, Novopashin S A and Serebrjakova M A 2012 Study conductivity nanofluids nanoparticle alumina *Nanosystems: Phys. Chem. Math.* **3** 27–33

[2] Kreuzer M, Tschudi T, de Jeu W H and Eidenschink R 1993 New liquid crystal display with bistability and selective erasure using scattering in filled nematics *Appl. Phys. Lett.* **62** 17121–30

[3] Apter B, Guilatt O and Efron U 2011 Ring-type plasmon resonance in metallic nanoshells *Appl. Opt.* **50** 5457–64

[4] Krasnikov I, Popov A, Seteikin A and Myllylä R 2011 Influence of titanium dioxide nanoparticles on skin surface temperature at sunlight irradiation *Biomed. Opt. Express* **2** 3278–83

[5] Lee G J, Attri P, Choi E H, Kwon Y-W, Krasnikov I and Seteikin A 2014 Optical and structural properties of nanobiomaterials *J. Nanosci. Nanotechnol.* **14** 221–49

[6] Goreva O V, Illarionov A I and Gorev D V (2016). Frequency conversion of broadband laser radiation distorted by spherical aberration in uniaxial crystals. *Bulletin of the Russian Academy of Sciences: Physics* **80** 28-31

[7] Krishtop V, Doronin I and Okishev K 2012 Improvement of photon correlation spectroscopy method for measuring nanoparticle size by using attenuated total reflectance *Opt. Express* **20** 25693–99

[8] Bulte W M and Modo M 2008 Introduction: The emergence of nanoparticles as imaging platform in biomedicine *Nanoparticles in Biomedical Imaging* (Heidelberg: Springer), 1-5.

[9] Jun Q, Tao, F, Qiuqiang Z and Sailing H 2010 Using some nanoparticles as contrast agents for optical bioimaging *IEEE Journal of Selected Topics in Quantum Electronics* **16** 672

[10] Allison R R, Mota H C, Bagnato V S and Sibata C H 2008 Bio-nanotechnology and photodynamic therapy—state of the art review *Photodiagnosis and Photodynamic Therapy* **5** 19–28

[11] Putnam S A, Cahill D G and Braun P V 2006 Thermal conductivity of nanoparticle suspensions *Journal of Applied Physics.* **99** 084308

[12] Rambidi, N. (2006) TALKING POINTS-Nanotechnology and Molecular Computing Devices. *Science in Russia*, (6), 38.

[13] Kul'chin Y N, Shcherbakov A V, Dzyuba V P, Voznesenskiy S S and Mikaelyan G T 2008

Nonlinear-optical properties of heterogeneous liquid nanophase composites based on high-energy-gap Al_2O_3 nanoparticles *Quant. Electron.* **38** 154–8
[14] Livashvili A I, Krishtop V V and Yakunina M I 2013 Electrostrictive self-action mechanism of radiation in nanofluids *Adv Condens Matter Phys* 591087
[15] Livashvili A I, Krishtop V V, Bryukhanova T N and Kostina G V 2015 Concentration dynamics of nanoparticles under a periodic light field *Phys. Procedia* **73** 156–8
[16] Livashvili A I, Krishtop V V, Karpets Y M., Bryuhanova T N and Kireeva N M 2016 Laser beam-induced bistability of concentration in nanofluids *Journal of Physics: Conference Series* **737** 012011
[17] Livashvili A I, Krishtop V V, Karpets Y M and Kireeva N M 2016 Formation of bistable medium state in nanofluids under the influence of laser radiation *Asia-Pacific Conference on Fundamental Problems of Opto-and Microelectronics* (International Society for Optics and Photonics) pp. 1017605–1017605
[18] Livashvili A I, Krishtop V V, Karpets Y M and Kireyeva N M 2017 The formation process of a bistable state in nanofluids. *IOP Conference Series: Materials Science and Engineering* **168** 012077
[19] De Groot S R and Mazur P 1964 *Non-equilibrium thermodynamics* (Moscow: Mir)
[20] Rudyak V Y, Belkin A A and Tomilina E A 2010 On the thermal conductivity of nanofluids *Tech. Phys. Lett.* **36** 660–2
[21] Putnam S A, Cahill D G and Braun P V 2006 Thermal conductivity of nanoparticle suspensions *J. Appl. Phys.* **99** 084308
[22] Hong T, Yang H and Choi C J 2005 Study of the enhanced thermal conductivity of Fe nanofluids *J. Appl. Phys.* **97**, 064311-1–064311-4
[23] Shliomis M I and Smorodin B L 2005 Onset of convection in colloids stratified by gravity *Phys. Rev.E* **71** 036312
[24] Cherepanov I N and Smorodin B L 2013 Convection of a stratified colloidal suspension *JETP* **144** 1090–1106
[25] Smorodin B L and Cherepanov I N 2015 Convection in a colloidal suspension in a closed horizontal cell *JETP* **147** 363–71

IOP Conf. Series: Journal of Physics: Conf. Series **936** (2017) 012080 doi:10.1088/1742-6596/936/1/012080

The Use of Logistic Model in RUL Assessment

R Gumiński [1], **S Radkowski**

Institute of Vehicles, Warsaw University of Technology
Narbutta 84, 02-524 Warsaw, Poland

E-mail: rgumin@simr.pw.edu.pl

Abstract. The paper takes on the issue of assessment of remaining useful life (RUL). The goal of the paper was to develop a method, which would enable use of diagnostic information in the task of reducing the uncertainty related to technical risk. Prediction of the remaining useful life (RUL) of the system is a very important task for maintenance strategy. In the literature RUL of an engineering system is defined as the first future time instant in which thresholds of conditions (safety, operational quality, maintenance cost, etc) are violated. Knowledge of RUL offers the possibility of planning the testing and repair activities. Building models of damage development is important in this task. In the presented work, logistic function will be used to model fatigue crack development. It should be remembered that modeling of every phase of damage development is very difficult, yet modeling of every phase of damage separately, especially including on-line diagnostic information is more effective. Particular attention was paid to the possibility of forecasting the occurrence of damage due to fatigue while relying on the analysis of the structure of a vibroacoustic signal.

1. Introduction

Defining (estimating) the remaining useful life is a key task in the maintenance system of an object or a group of objects. Determination of RUL is connected with several issues: limited time of use as a result of degradation processes, often of random character, taking place. Knowledge of the damage phase, its progressing development, allows for estimation of the useful life. Information about the damage can be generated by means of the appropriately prepared model of damage development. In real life, it proves to be the case that prediction on the basis of prepared models renders results with a wide margin of uncertainty. It is caused by varying conditions of object's operation, which correlates with initiation time and then, with the pace of the damage development. Therefore, in order to diminish the range of uncertainty of the conducted analyses, parameter updating of the assumed model or of the form of the model itself is necessary. Another issue worth emphasising is relating the information on the damage development to the remaining useful life. This task requires determining the limit value of the damage size or the related probability of damage and the scale of the damage, i.e. the level of acceptable risk.

The alternative to the damage development model and the need to its updating is the use of diagnostic information; the advantage of this approach being the information about the actual object and not the model object. Such an approach requires developing diagnostic parameters but eliminates uncertainty connected with the model.

[1] To whom any correspondence should be addressed.

A different approach uses statistical models [1, 2] describing the object's behaviour. Such an approach requires an access to statistical data, which is often a substantial restraint, especially if a small group of objects is taken into consideration. Additionally, inference with the help of statistical methods renders good results in the case of the maintenance system of a large group of objects. However, in the case of maintenance of individual units or even a single object, it can lead to great uncertainty. Inference about population on the basis of not particularly large sample size, and inversely, expressing opinion about a single element on the basis of statistical data can lead to major errors. Therefore, in such cases an individualised approach is necessary, i.e. prediction of the reliability indicators on the basis of the general model of the "personalised" damage development for the specific group of objects or even a specific technical object, for example working for one end-user, under defined conditions, technical culture, etc. Such an approach requires determining the model parameters for the group of objects or a specific object but as a consequence offers a more precise prediction. Particularly good results can be obtained when prognostics is enhanced with diagnostic data conveying information about a potential damage. In this context, several issues should be raised: possibility of detection of early phases of damage development/initiation, monitoring of the development, determination of the acceptable level of damage.

A review of the methods for determination of RUL should be conducted in the context of uncertainty [3] and particularly of its sources and methods of their elimination. This is particularly vital because of the fact that, as often proved by the maintenance practice, inaccuracies of RUL estimates range from a few tens to several hundred percent which leads to the disastrous aftermath of the damage or an overly early putting an object out of operation. Both cases create the cost increase, nonetheless of different character – random occurrence of high costs (in form of the failure effects) or the continuous increase in the object's maintenance costs (in form of too frequently performed maintenance routines and preventive repairs). In the case of the occurrence of the estimate uncertainty of reliability characteristics, and RUL as a consequence, taking maintenance-related decision can be seen as an optimization task between the occurrence of the increased maintenance costs and the increased risk level.

It should be kept in mind that detection of damage initiation does not necessarily mean putting the object out of operation, which depends on the adopted/required level of acceptable risk. For a better demonstration of the subject of the influence reliability characteristics have on the maintenance decisions, the curve of the exemplary reliability function with the assumed uncertainty range has been shown. The uncertainty range of the characteristics translates into the uncertainty range of the estimated time (with the assumed probability of damage occurrence). It should be emphasised that the width of the uncertainty range of the maintenance time depends on the steepness of the reliability function.

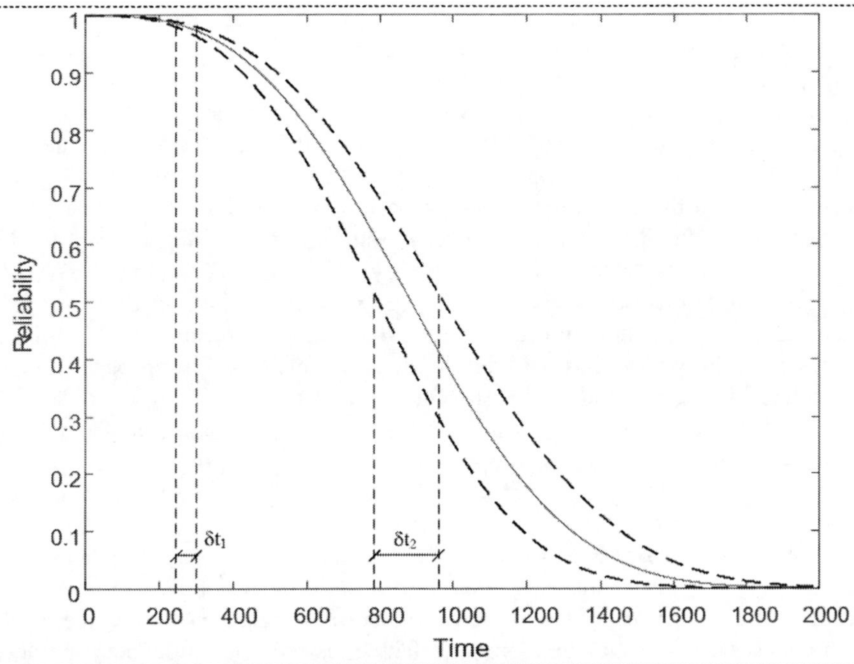

Figure 1. Reliability function with the uncertainty range and time uncertainty ranges marked

Another aspect that needs to be discussed is multimodal damaging, i.e. every possible type of damage and occurring of successive phases of its development is related to a specific curve of the reliability function. A fact which complicates the inference under the conditions of multimodal damaging is the possibility of initiation of subsequent types of damage at different times and their development with varying intensity, which as a result can lead to a situation, in which for example the damage initiated as the first one can develop with little intensity, however, after a longer period of use, another or a different type of damage can be initiated, which will develop with greater intensity and in spite of its occurring relatively late can be decisive, i.e. it can reach the threshold value earlier. The abovementioned issues regarding the process of determining the RUL result in various methods of its estimation.

2. Review of RUL estimation methods

Generally, the RUL prediction methods should be divided into: based on the data and based on models [4].

Physical model of an object usually describes the dynamic behaviour and kinematic relations. Having been built, it requires selection of parameters and validation of the whole model, which is connected with conducting the appropriate experiments [5]. Moreover, as mentioned before, such a model can require updating so as to diminish the discrepancy in the results obtained from the model and on the basis of the actual object's behaviour. In order to reduce the uncertainty of the simulation results it is worthwhile to model additional phenomena taking place on a real object, which often leads to building multidomain models. Modelling and verification of such models is both time- and cost consuming but leads to obtaining results burdened with lesser uncertainty. Additionally, such an approach causes better understanding of phenomena and mutual relations taking place in a real object, which is valued highly in dynamic states of work.

Methods based on knowledge, most frequently use historical data recorded in data bases. Customarily, using the data focuses on describing relations between observation of incidents (undesirable) and the data recorded in the data base [6]. Relations built on the basis of historical data can be used to create prognostics of the future damage occurrences with the simultaneous taking into

account the current measurements or historical and current measurements (for example using the trend analysis). Description of the abovementioned relation by means of function relationships can lead to designing of models for changes in diagnostic parameters as a function of time, numerous cycles, etc.
Considering the fact that data can be burdened with high risk of error and indeterminacy, a substantial dynamics of phenomena occurs so as a consequence, also of the data; methods based on knowledge often use expert systems (based on artificial intelligence). It is caused also by the fact that building mathematical model is expensive and often impossible to carry out at the required level of detailedness. Referring to the definition, systems based on artificial intelligence have the feature of self-teaching that causes flexible adjusting to changing conditions. A good proposition in inference with the foundation of a knowledge base is employing the theory of fuzzy sets [7], offering the possibility of taking decisions on the basis of the uncertain input data. In order to make the building process of the fuzzy logic system more efficient, they are often supported with neural networks (neuro-fuzzy systems). In explorations of knowledge and decision-taking process regarding operation, the Bayesian probability is also used, especially Bayes networks.
In recent times, methods bordering on physical models and models based on knowledge, the so-called **hybrid methods** have been gaining popularity. Such an approach aims at eliminating the disadvantages of one method (e.g. expensive building of a very detailed physical model), by means of providing support with expert-system based methods.
The simplest case of RUL determination takes place when there is a one-dimension correlation of the diagnostic parameter with the remaining useful life, such a case however, occurs rarely, i.e. the correlation is often multi-dimensional, and the character of the relationship changes with occurrence of the subsequent phases of damage development. The most recent tendencies in modelling damage development take into consideration a multi-phase development of the damage [8], i.e. because of the changing character of the phenomena taking place at the stage of initiation and damage development, each of the stages is described by a separate model, which leads to a greater precision of results obtained from the simulation. An important issue in application of such approach is the decision about changing the model being used.
In the following section of this work, the analysis will be shown of the results of an experiment consisting in fatigue breaking of the gear tooth. An attempt was made to describe changes in the parameter of a vibration signal using the logistic function determined by the relationship (1):

$$PD = \frac{a}{1+be^{-ct}} \tag{1}$$

Estimation of parameters of the logistic function was conducted by means of linearisation of the relationship (1) and by juxtaposing it with the equation of the straight line. Line coefficients were determined on the basis of measuring data using the method of the least squares.

3. The test stand

The experiment was conducted at the FZG back to back test-bed. The test-bed consists of two toothed gears operating in a revolving power setup and it enables examination of both toothed wheels as well as gear lubricants. The diagram showing the test-bed is presented in Figure. 2.
The shaft connecting the pinions is divided, which enables rotating one of its sections versus the other and thus introducing relevant meshing forces. Strain gauges are affixed to the shaft and they enable measuring the torque. Wheels with straight teeth are installed in the examined gear, while wheels with helical teeth are installed in the closing gear. Thanks to such a set up it was the examined toothed gear that was subject to defect-development during the experiment.
Toothed wheels made of 20H2N4A carburized steel, hardened to 60 HRC hardness were used for the research (Figure 3). They were subjected to accelerated fatigue test. It should be emphasised that the examined toothed wheel did not have any defects that were previously made, such as tooth cuts. During the experiment, the vibration accelerations of the tested gear housing were registered, measured on the cover, directly over the pinion shaft bearing, in the direction X – radial – horizontal, and Z – radial – vertical. A detailed description of the test stand was given in [9].

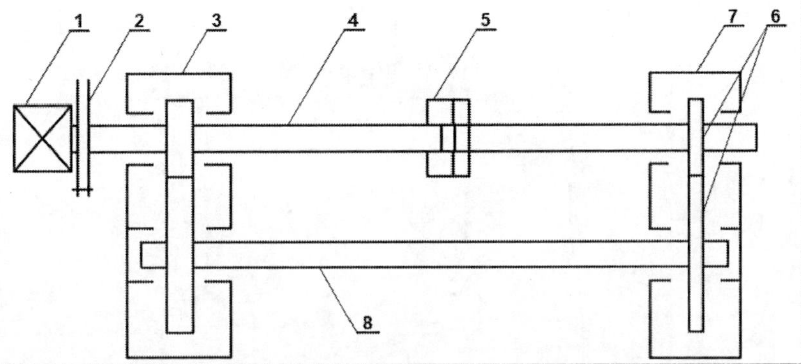

Figure 2. Test-bed diagram: 1 - motor, 2 - clutch, 3 - closing gear, 4 - coupling shaft, 5 - prestressed clutch, 6 - examined toothed wheels, 7 - examined gear, 8 - shaft.

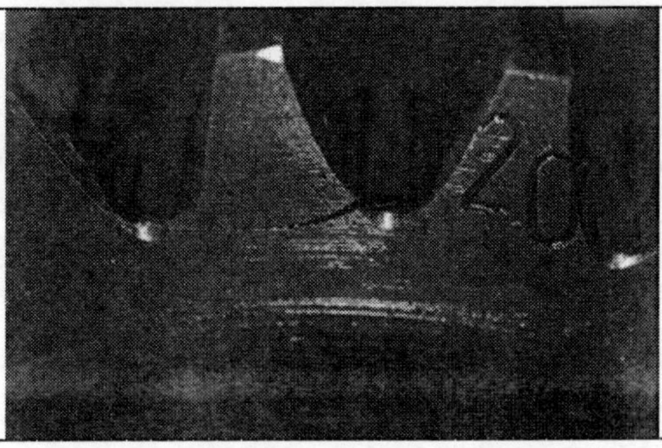

Figure 3. Pinion with fatigue crack.

Using the measurements of vibration accelerations conducted in such a way, the Fourier transform was designated (spectrum of vibration accelerations), on the basis of which changes in vibration energy were determined in the course of the experiment, i.e. starting from installation of the new toothed wheel up until the fatigue tooth breaking. Figure 4 shows the curve of the vibration energy and the logistic function describing energy.

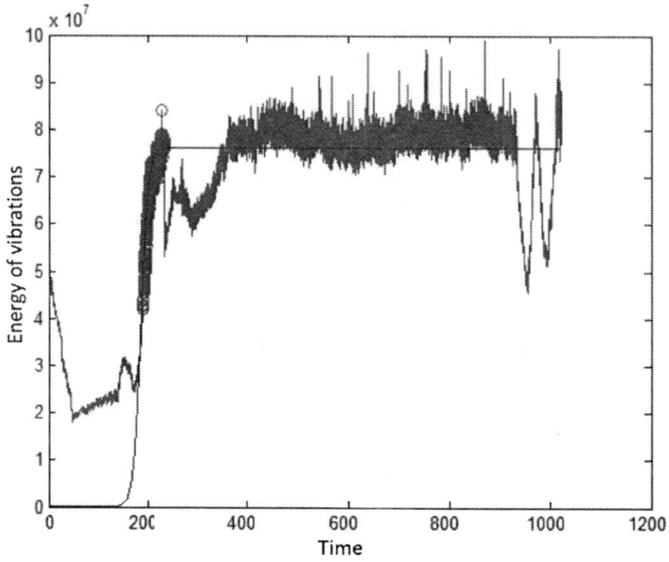

Figure 4. Energy of vibrations in the course of experiment.

Unfortunately, an attempt to describe changes in energy by means of one model throughout the whole experiment time, i.e. during occurrence of the successive phases of the fatigue breaking, leads to big errors. Because of the fact that the successive phases of damage are related to the occurrence of phenomena of different character, an attempt was made to describe the curve of vibration energy, dividing the operation time into two sub-periods (Figure 5).

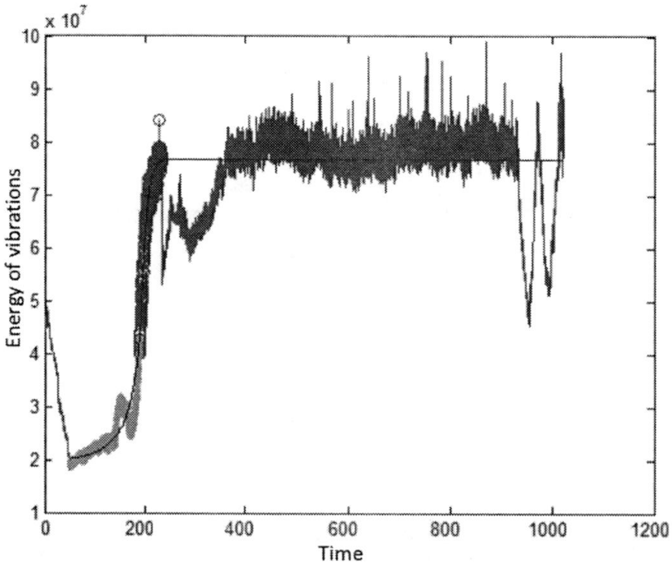

Figure 5. Energy of vibrations in the course of experiment (with division into two phases).

Parameter estimation of the logistic curve in two phases causes a substantial decrease in the model error in relation to the values of energy obtained from measurement. This relation is of the repetitive character towards other examined gearwheels. Table 1 shows estimated parameters of the logistic curve for the exemplary gearwheels.

Table 1. Logistic function parameters.

	Phase I		Phase II	
Wheel no.1	*a*	2,0E+07	*a*	7,6E+07
	b	-9,1E-03	*b*	4,6E+10
	c	-2,2E-03	*c*	1,3E-02
Wheel no.2	*a*	5,5E+07	*a*	2,8E+08
	b	-4,0E-02	*b*	5,4E+01
	c	-3,0E-03	*c*	4,1E-03
Wheel no.3	*a*	8,5E+07	*a*	2,8E+08
	b	-4,0E-02	*b*	7,8E+01
	c	-3,0E-03	*c*	5,6E-03

Comparing values of the obtained parameters, the following regularities can be observed: all three parameters in Phase II assume much greater values than in Phase I, parameters *b* and *c* additionally have the sign changed from negative to positive.

Analysing the obtained curves of energy and the logistic model curves, it should be noticed that the first part (slow increase) is relatively short and as a consequence the second part (rapid value increase) takes place early. What ensues is that monitoring energy allows for a relatively early detection of the changes taking place, which is an undisputed advantage. This causes a significant utility of a proposed parameter in the systems with the required high level of safety. The possibility of description of changes in vibration energy with a function relationship offers a possibility of prediction. On the other hand, long period of saturation of the parameter value and of the function relationship describing it, causes much smaller usefulness in the case of application in managing the system operation, where the acceptable level of safety is lower and the maintenance and repair activities undertaken too early are avoided so as to reduce the operation costs. In such a situation, in order to avoid prediction errors, other diagnostic methods should be referred to.

4. Conclusions

The article presents a suggestion of using the energy of vibrations in a toothed gear as a diagnostic parameter correlated with the fatigue crack propagation at the base of a tooth in the toothed gear wheel.

An attempt was made to describe changes in vibration energy as a function of time using the logistic curve. The conducted analyses indicated the need of multi-phase modelling of the occurring changes, which results from the physics of the fatigue breaking phenomenon. The problem of decision-taking regarding the change within the model, i.e. moving from Phase I to Phase II remains unresolved. However, the obtained results are promising and are of great utility especially in the operation of systems with demanding requirements defined by the acceptable level of safety.

5. References

[1] Banjevic D, 2009, Remaining useful life in theory and practice, Metrika March 2009, Volume 69, Issue 2–3, pp 337–349,

[2] Si X, Wang W, Hu Ch, Zhou D, 2011, Remaining useful life estimation – A review on the statistical data driven approaches, European Journal of Operational Research, Volume 213, Issue 1, pp. 1-14,

[3] Sankararaman S and Goebel K, 2013 Why is the Remaining Useful Life Prediction Uncertain?, Annual Conference of the Prognostics and Health Management Society 2013,

[4] Mosallam A, Medjaher K and Nourredine Zerhouni, 2013, Bayesian Approach for Remaining Useful Life Prediction, Chemical Engineering Transactions 33, pp. 139-144, DOI: 10.3303/CET1333024,

[5] Chmielewski A, Gumiński R, Mączak J (2017): Selected Properties of the Dynamic Model of the Piston-Crankshaft Assembly in Stirling Engine Combined with the Thermodynamic Submodel, International Journal of Structural Stability and Dynamics, vol. 17, nr 9, 2017, ss. 1740009-1 - 1740009-24, DOI:http://dx.doi.org/10.1142/S0219455417400090,

[6] Mączak J, 2013, Local meshing plane analysis as a source of information about the gear quality, Mechanical Systems and Signal Processing, Volume 38, Issue 1, pp 154-164,

[7] Prognostic modelling options for remaining useful life estimation by industry. Sikorska J Z, Hodkiewicz M, Ma L, 2011, Mechanical Systems and Signal Processing, Volume 25, Issue 5, Pages 1803-1836,

[8] Ciavarella M, Monno F, 2006, On the possible generalizations of the Kitagawa–Takahashi diagram and of the El Haddad equation to finite life, International Journal of Fatigue, Volume 28, Issue 12, pp 1826-1837,

[9] Jasiński M., Radkowski S. Use of Bispectral-Based Fault Detection Method in the Vibroacoustic Diagnosis of the Gearbox, Proceedings of the 4th World Congress on Engineering Asset Management, Athens, Greece, 2009.

IOP Conf. Series: Journal of Physics: Conf. Series **936** (2017) 012081 doi:10.1088/1742-6596/936/1/012081

Interplay of frustrations, interaction length, and dilution on magnetic transitions in vector models

Petr D. Andriushchenko[1,2], Alexei A. Peretyatko[1], Vitalii Yu. Kapitan[1,2], Alexander V. Perzhu[1], Konstantin V. Nefedev[1,2]

[1] Department of Computer Systems, School of Natural Sciences, Far Eastern Federal University, Vladivostok, 690950, 8 Sukhanova St., Russian Federation

[2] Institute of Applied Mathematics, Far Eastern Branch, Russian Academy of Science, Vladivostok, 690041, 7 Radio St., Russian Federation

E-mail: `andriushchenko.pd@dvfu.ru`

Abstract. The research of one-dimensional Ising model with the different number of neighbors and the different sign of the exchange integral were performed. The clustered order parameter is proposed for the description of the magnetic transition in spin systems regardless of the sign of the exchange integral and the number of neighbors. The method of calculation of critical field to change the magnetic states was presented. For diluted 1D antiferromagnetic model two plateaus in magnetization behavior are observed, and explained. Influence of length of interaction plays critical role for phase transition of dipoles system. Dipoles on hexagonal lattice show the phase transition only in model of long range interaction, and there isn't critical behavior for short range. The approach to calculation of spin system in Heisenberg model is proposed.

1. Introduction

The influence of computer modeling in the physics of condensed matter is increasing, which allows the development of new methods for the construction of complex nanostructures and which promotes the discovery of new phenomena and materials [1, 2, 3, 4, 5, 6]. The study of the collective behavior of magnetic systems will allow to predict and manage the properties of new materials, and will also make a significant contribution to the development of fundamental knowledge about the nanoworld and ongoing processes.

Today may seem from the side, that the theory of ordering reached its peak and now is reaping the fruits of their approaches on new structures, elaborating the critical indexes and the phase transition temperature. But theorists and experimenters publishing the evidence of the opposite opinion that current approaches were ill-suited to describe the order in frustrated systems or systems with quenched disorder [7, 8, 9, 10, 11]. But for better understanding the processes occurring in such complex systems, it is needed to show such processes on the simpler models. That is why finding a common approach to ordering need to start from simple models and move on to more complex ones.

Here we presented our resent results of the research well-known models of magnetism. And in our opinion, these old Ising and dipoles models of magnetics allow throwing the light on frustrations phenomena [12], description of phase transitions [13, 14], the influence of interaction length on the existence of critical phenomena [1], plateau appearance in antiferromagnetic models [5] and other interesting phenomena, which attract the attention of researchers in last time.

In this work there were used all units of temperature as dimensionless, i.e. $kT/J \to T$, and magnetic fields units were $h/J \to h$.

2. Ordering in 1D Ising model

2.1. 1D Ising model with two nearest neighbors

Ising model with two nearest neighbors is probably oldest model of magnetism. But it can be used to answer on the questions about order parameter and plateau in magnetization behavior of the diluted spin model [15].

Root mean square magnetization $< M^2 >$ or its module is usually used to describe transitions in magnets models. This parameter can be used to study phase transitions in ferromagnets. However, as soon as we change the sign of the exchange integral with "+" to "-", the order parameter becomes 0 for any T and $N \to \infty$. Therefore, the problem of determining the universal order parameter, especially for systems with a negative or alternating exchange interaction is an actual task [16, 17].

Therefore, we have proposed an approach to the determination of the order parameter to describe the transition from a disordered phase to an ordered (or partially ordered) [13]. Order parameter can be a ratio of the maximum size of the cluster formed by spins in the ground state (lowest energy) to the total number of particles (Fig.1).

The simplest magnetic model is the one-dimensional Ising model [18]. Classical 1D Ising model with Hamiltonian (1) was investigated, where n - total number of spins, J_{ij} - exchange integral and z - the number of nearest neighbors, which interact with spin. The model with $z = 2, 3, 4$ considered in this paper.

$$\mathcal{H} = -\sum_{i=1}^{n}\sum_{j=1}^{z} J_{ij}S_iS_j - h\sum_{i=1}^{n} S_i \tag{1}$$

The 1D Ising model with $n = 1000$ spins with two nearest neighbors and periodic boundary conditions (PBC) (Fig. 2) was investigated.

First of all, the ferromagnetic model with a positive exchange integral was calculated. From now on the simulation was performed with using Monte Carlo (MC), the number of MC steps for every $0.02T$ was 10^{10}. As expected, the temperature dependence of the clustered order parameter (COP) γ_1 is almost identical to the behavior of magnetization. That is, in the case

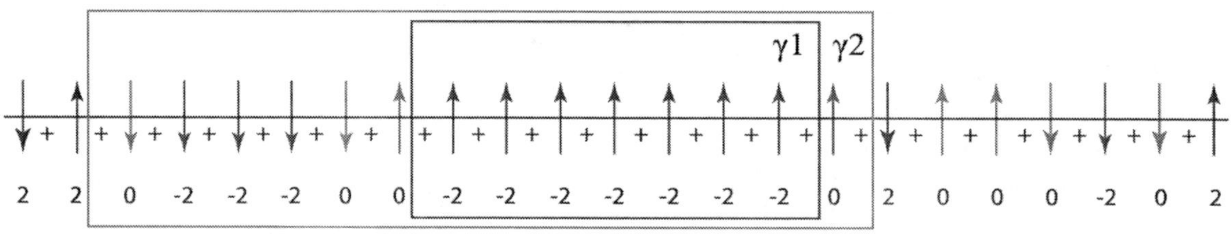

Figure 1. An example of calculating the clustered order parameter on ferromagnet ($J > 0$) 1D Ising spin chain with two nearest neighbors. Each spin in the chain has a different color (blue, green and red) depending on its interaction energy respectively $(-2, 0, 2)$. Red line highlights the biggest cluster of spins in the lowest energy state (-2, that is blue). The ratio of the spins in ground state in the maximal cluster to the total number of spins in a system is the order parameter γ_1. In this sample number of the spins in ground state in the maximal cluster is 7, $n = 23$, $\gamma_1 = 7/23 \approx 0.3$

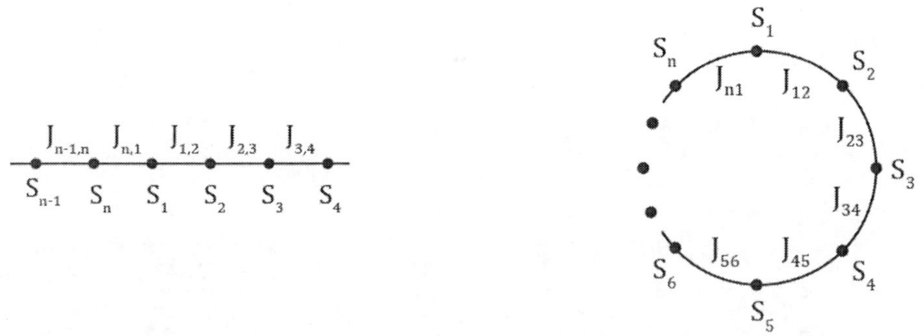

Figure 2. 1D Ising model with two close neighbors and PBC.

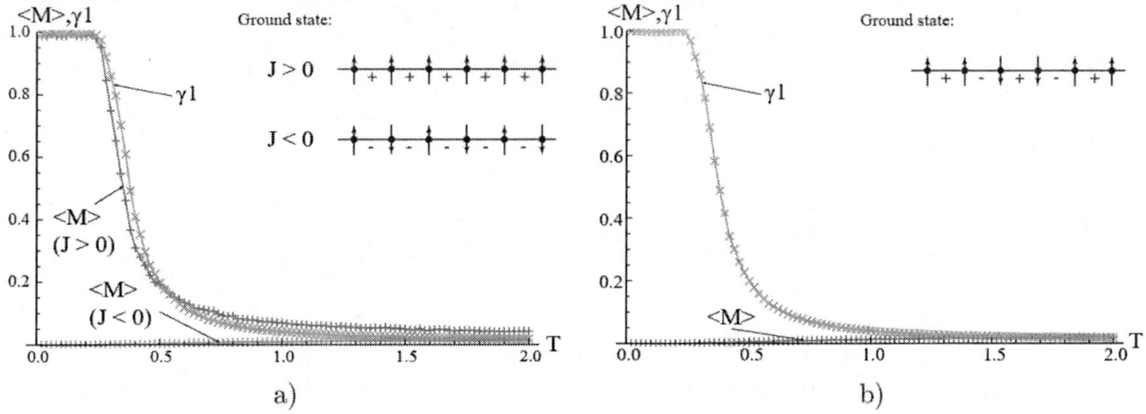

Figure 3. The temperature behavior of COP γ_1 and magnetization of 1D Ising model (1000 spins) with two nearest neighbors and PBC. a) Blue denotes the mean magnetization of the ferromagnetic model with $J > 0$, green - magnetization of the antiferromagnetic model with $J < 0$, orange COP (the same for these models). b) Blue denotes the magnetization with alternating exchange integral, orange - the same COP for such model.

of ferromagnetic ($J > 0$) the COP γ_1 describes a transition from disordered phase to ordered like magnetization.

If you change the sign of the exchange integral J to the negative, the magnetization becomes zero at any temperature, as the ground state of the system will be antiferromagnetic (Fig. 3a). Despite this, COP γ_1 continues to describe the transition to the ordered phase (Fig.3a).

If we consider a system with alternating exchange integral (each spin has one positive link and negative second one), it becomes clear that such system is not much different from the antiferromagnet, the magnetization of such system is also equal to 0 for any T, but the COP describes the ordering process (Fig.3b).

2.2. Microstates in antiferromagnetic 1D Ising model with dilution

Consider a chain of five magnetic moments. The field is applied perpendicular to the axis of the array. The Hamiltonian of the system:

$$\mathcal{H} = S_1S_2 + S_2S_3 + S_3S_4 + S_5S_1 - h\sum_{j=1}^{5} S_j, \tag{2}$$

where $S_j = \pm 1$.

IOP Conf. Series: Journal of Physics: Conf. Series **936** (2017) 012081 doi:10.1088/1742-6596/936/1/012081

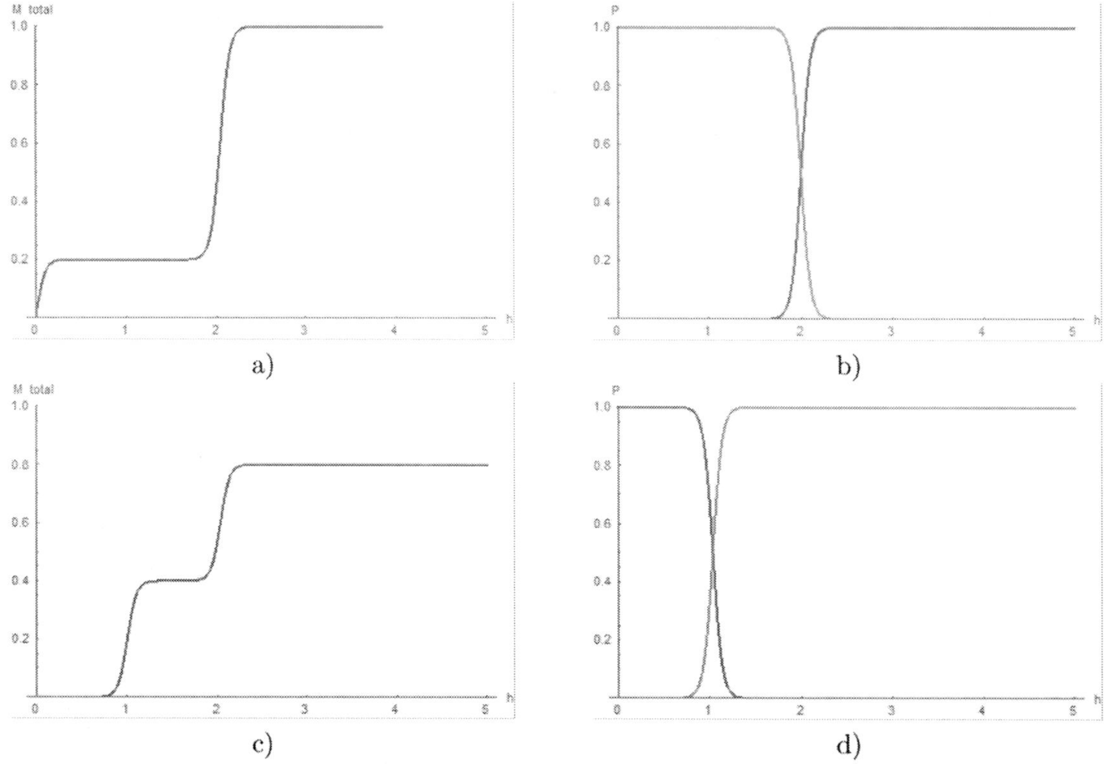

Figure 4. a) Magnetization of a chain of five magnetic moments at $T = 0.1$. b) The probability of configurations in the chain of five magnetic moments at $T = 0.1$. Blue line is probability of 3 up ([3/0]), yellow is probability of two up one down ([2/1]). c) Magnetization of diluted chain (concentration of dilution x = 0.2) at $T = 0.1$. d) The probability of configurations in the chain with one vacancy at $T = 0.1$. Blue line is probability of one up and one down ([1/1]), yellow is probability of two up ([2/0]).

The magnetization of the chain of five magnetic moments at a temperature $T = 0.1$ is shown in Fig. 4a. There are two plateaus in magnetization curve. Consider the reason for the appearance of the plateau of magnetization. The reversal of one spin will change the energy of interaction with the two neighboring spins. Let us consider the probability of the appearance of different configurations of tree spins in a system of five spins.

As can be seen in Figure 4b, the probability of a configuration of two up one down ([2/1]) drops sharply at a field value of $h_c = 2$, while the probability of the configuration of 3 up ([3/0]) sharply increases at the same value of the field $h_c = 2$. There were no other configurations. Thus, the jump in the magnetization curve is caused by a change in the probabilities of microstates.

Lets remove one spin from the array (concentration of dilution $x = 0.2$), then the Hamiltonian of the system:

$$\mathcal{H} = S_1 S_2 + S_5 S_1 - h \sum_{j=1}^{5} S_j + h S_3. \tag{3}$$

The magnetization curve for the "diluted" system is shown in Fig. 4c.

Consider the cause of additional magnetization plateau in the case of dilution concentration $x = 0.2$. At such a dilution concentration, a revolution of one spin will change the energy of interaction with the two neighboring spins or only with one, in the case of proximity to the vacancy. The probability of a configuration with three spins has not changed and is shown in Fig. 4a, but at the same time a configuration appeared in which there are 2 spins and one vacancy. The probability of configurations with one vacancy is shown in Fig. 4d.

Table 1. The energies of microstates.

Configuration	number of spins	Energy
3/0	3	-2+3h
2/1	3	2+h
2/0	2	-1-2h
1/1	2	1
1/0	1	h
0/1	1	-h

Using Table 1, we can calculate the values of h_c at which the transition occurs from one configuration to another. For example, in the transition from 2/1 to 3/0:

$$-2 + 3h = 2 + h \tag{4}$$

$$-4 = -2h \tag{5}$$

$$h_c = 2 \tag{6}$$

Similarly, the remaining critical transition fields can be calculated.

In this way the jumps on the graph of the magnetization of the diluted sample coincide with the jumps in the probability curves of the configurations. The appearance of an additional plateau on the magnetization schedule is caused by an increase in the number of possible microstates.

2.3. 1D Ising model with three nearest neighbors

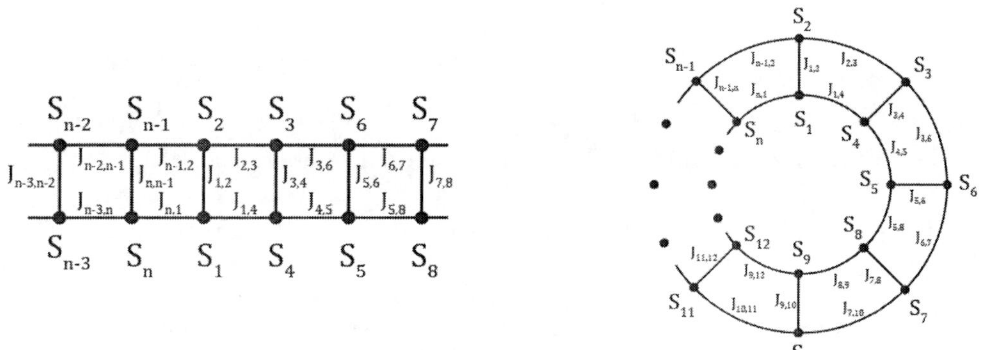

Figure 5. 1D Ising model with three close neighbors and PBC.

Also, the 1D Ising model with $n = 1000$ spins with three nearest neighbors and PBC (Fig. 5) was investigated. As in the model with two neighbors, the ferromagnetic model (all exchange integrals are positive $J > 0$) shows the coincidence of the temperature behavior of the magnetization and COP γ_1 (Fig. 6a). At the same time in the model with $J < 0$ the average magnetization is equal to 0, but COP γ_1 showed the transition to order. As well as in model with alternating exchange integral (all vertical $J_v < 0$, and horizontal $J_h > 0$, or vice versa) COP γ_1 showed transition to order while magnetization is equal to 0 (Fig. 6b).

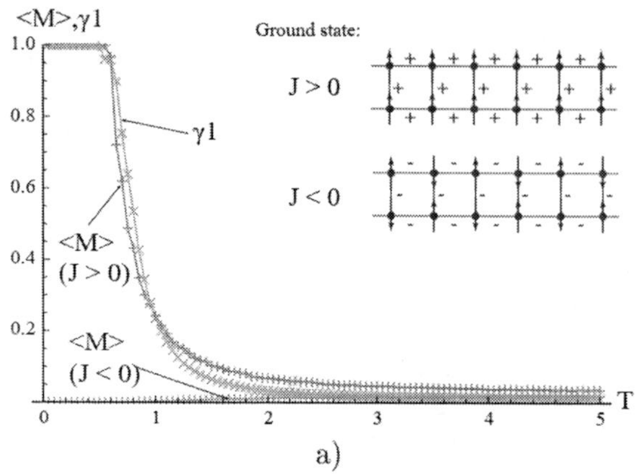

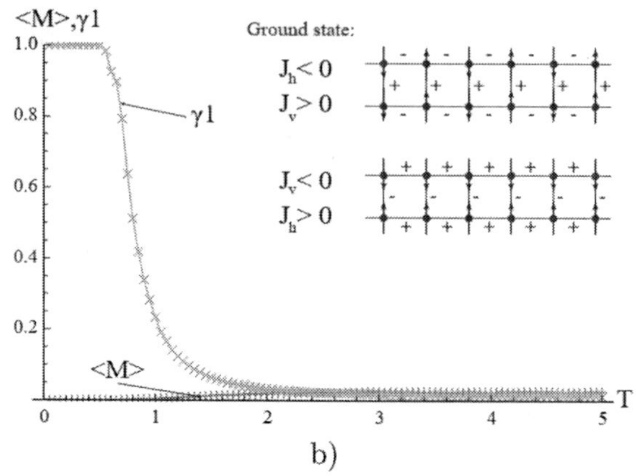

Figure 6. The temperature behavior of COP γ_1 and magnetization of 1D Ising model (1000 spins) with three nearest neighbors and PBC. a) Blue denotes the magnetization of the ferromagnetic model with $J > 0$, green - magnetization of the antiferromagnetic model with $J < 0$, orange - COP (the same for both models). b) Blue denotes the magnetization with alternating exchange integral (each horizontal bonds of the spins are positive $J_h > 0$, and the vertical are negative $J_v < 0$ or vice versa), orange - the same COP γ_1 for such model.

2.4. 1D Ising model with four nearest neighbors

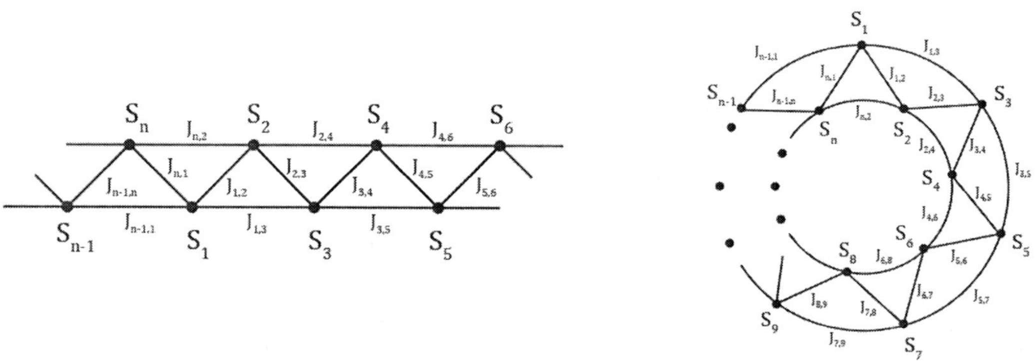

Figure 7. 1D Ising model with four close neighbors and PBC.

The 1D Ising model with $n = 1000$ spins, with four nearest neighbors and PBC (Fig. 7) was investigated. In case of ferromagnet interaction ($J > 0$) COP γ_1 coincides with the magnetization (Fig. 8a). In model with alternating exchange integral $J_h > 0$ and $J_v < 0$ magnetization equal to 0, but COP γ_1 shows the ordering (the same for both model).

But when we change the exchange integral to negative $J < 0$ or to alternating with negative horizontal $J_h < 0$ and positive vertical $J_v > 0$ the frustration appears (Fig. 9).

The COP γ_1 does not working for frustrated systems because spins can not minimize its energy, which results to the impossibility of ground state cluster formation. However, under the proposed approach a more general clustered order parameter γ_2 for frustrated systems has been proposed.COP γ_2 represents a ratio of the maximum size of the cluster formed by spins not only in the ground state but also with energy closest to the ground state to the total number of particles (highlighted by the green line in Fig. 1).

Figure 8b shows that γ_2 describes ordering process in frustrating systems. The behavior of γ_2

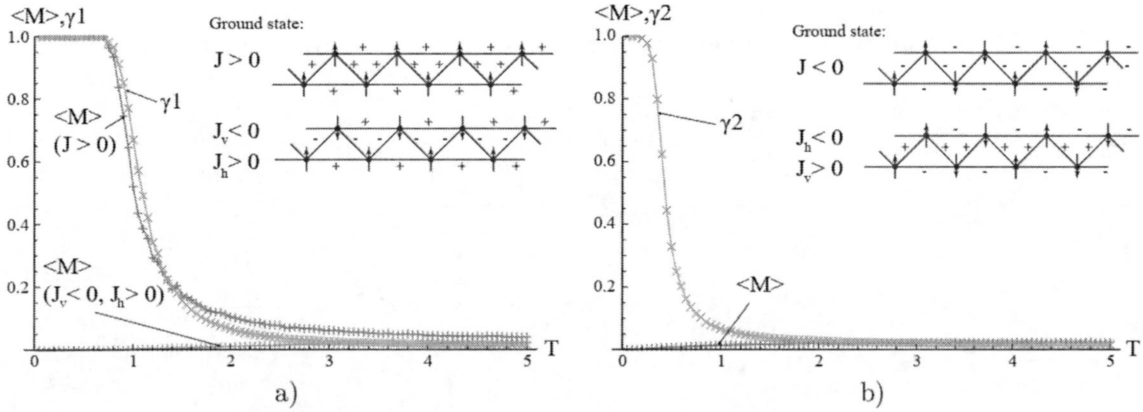

Figure 8. The temperature dependencies of COP γ_1, γ_2 and magnetization of 1D Ising model (1000 spins) with four nearest neighbors and PBC. a) Blue denotes the magnetization of the model with $J > 0$, green - the magnetization of the model with $J_h > 0$ and $J_v < 0$, orange - COP γ_1 same for these two models. b) Orange denotes COP γ_2 of frustrated systems with negative exchange integral $J < 0$ and with alternating exchange integral with $J_h < 0$ and $J_v > 0$. Magnetization of these model is denoted as blue

Figure 9. Frustrations in 1D Ising model with four nearest neighbors a) in the antiferromagnet model with $J < 0$ b) in the model with alternating exchange integral $J_h < 0$ and $J_v > 0$.

is the same for two frustrated systems mentioned previously. Also, the clustered order parameter γ_2 can describe the ordering in other frustrated systems.

A more detailed description of the approach can be found in the paper [13].

3. Honeycomb Spin Ice Heat Capacity

As was shown above, frustrations introduce new interesting effects and unusual behavior of the system. The most interesting frustrated system is artificial spin ice (ASI) of various geometries and one of the most discussed - a model of hexagonal artificial spin ice. Depending on the Hamiltonian, there are different approaches to modeling such systems. Some calculate the model including only the interactions of the nearest neighbors with Hamiltonian (1) [19]. Others taking into account only the dipole-dipole interaction with (or without) cut-off radius [20, 21]. And some taking into account both the interaction of the nearest neighbors and the dipole-dipole interaction, trying to find the correct coefficient of their proportionality [22, 23, 24].

$$\mathcal{H} = Da^3 \sum_{i<j} \frac{\vec{m_i}\vec{m_j}}{r^{3/2}} - 3\frac{(\vec{m_i}\vec{r_{ij}})(\vec{m_j}\vec{r_{ij}})}{r^{5/2}} \tag{7}$$

We decided to consider a model with a pure dipole interaction (7). Effect of long- and short-range interactions on the thermodynamics of dipolar spin ice with different geometries was considered here [1].

Two series of samples which are consisting of various numbers of particles formed in hexagonal spin ice arrays, square and hexagonal ("from the center") forms with the different radius of interaction were simulated. Models were considered with long-range (LR) and short-range (SR) dipole-dipole interaction. At LR interaction, every spin interacted with everyone. At SR interaction, every spin interacted with the 4 nearest neighbors (less on the borders). The simulation was performed by Wang-Landau method [25, 26, 27]. Thermodynamics of the samples was calculated using 24 WL-steps. The heat capacity temperature behavior of the hexagonal spin ice (hexagon sample) with the number of particles (N) = 30, 72, 132, 210, 306 with LR and SR interactions shown in fig.10.

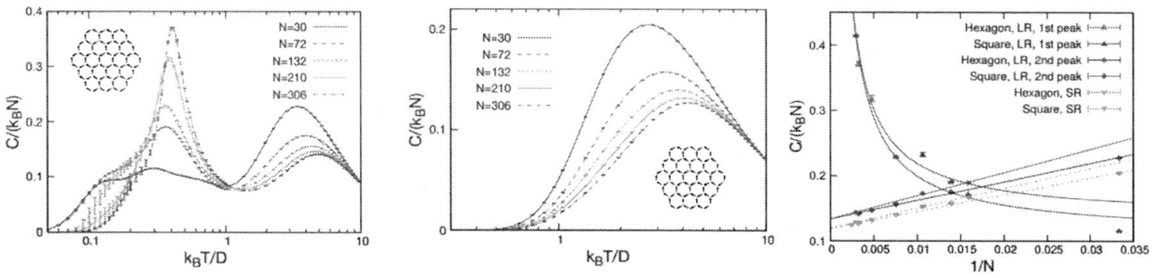

Figure 10. The heat capacity temperature behavior of the hexagonal spin ice (hexagon sample) with LR (a) and SR (b) interactions. (c) Comparing of the heat capacity peaks depending on the number of particles in the system for honeycomb spin ice with square and hexagonal shapes. At $N \to \infty$ with LR interaction the temperature behavior of the heat capacity reveals the singularity, the second peaks converge to $0.133D/k_B$, and with SR peaks converge to $0.119D/k_B$.

In LR (Fig. 10a) interaction models the temperature dependence of the heat capacity shows anomalous character, there are two temperature peaks, ie, critical behavior of the system of dipoles in the region of T_c significantly changes. The first peak with increasing number of particles increases and the second peak reduces. As can be seen from Fig. 10b, the heat capacity peak of models with SR interaction with increasing number of particles is reduces. The temperature behavior of the specific heats square shape samples with LR and SR interactions repeats the behavior of the specific heats hexagonal shape samples, described above. The comparing of the heat capacity peaks depending on the number of particles in the system for honeycomb spin ice with square and hexagonal shapes shown in Fig. 10c. The solid lines in this figure show the approximation. From Fig. 10c follows that at $N \to \infty$ the influence of the boundaries on the critical phenomena in T_c can be neglected. There is the convergence of specific heat peaks for square and hexagonal shapes. At SR interaction, peaks values converge to $0.119D/k_B$. At LR interaction heat capacity reveals the singularity at low temperatures and the height of the second peak with increasing system size convergences to the $0.133D/k_B$. In models with SR interaction, the phase transition is absent, while models with LR interactions show phase transition. The question of how much neighbors should be taken into consideration to not lose the basic thermodynamic properties of the system is necessary to investigate further.

4. 3D Heisenberg model

A further development of this study may be the transition to the use of the classical Heisenberg model. The Hamiltonian of the spin system in frame of Heisenberg model was set as follows:

$$\mathcal{H} = -J \sum_{<i,j>} \vec{S}_i \vec{S}_j - A_z \sum_i \vec{S}_i - h \sum_i \vec{S}_i \tag{8}$$

where $\vec{S_i}$ is the atomic spin at the i-th lattice site, J - ferromagnetic short-range exchange interaction, $|A|$ - constant of magnetic anisotropy, h external magnetic field. The sum is multiplied by 1/2 in order to avoid double summation, but some write it without the 1/2 factor; it's alright as long as one remembers that the summation is over i and j so that $i < j$. A spin $\vec{S_i} = \{S_i^x, S_i^y, S_i^z\}$ is introduced as a three-dimensional unit vector in accordance with formulas:

$$\begin{cases} x = r\sin(\phi)\cos(\theta) \\ y = r\sin(\phi)\sin(\theta) \\ z = r\cos(\phi) \end{cases} \tag{9}$$

One may note that in the case of ferromagnets the above Heisenberg Hamiltonian (8) predicts the parallel alignment of atomic spins, but doesnt specify a preferential direction of alignment, see the first two terms in Eq. 8. Thus, in such case the Heisenberg Hamiltonian is called the isotropic Heisenberg Hamiltonian. However, in real crystals the isotropy is broken by other magnetic effects that were neglected in the original Hamiltonian, like the dipolar interactions and spin-orbit coupling. Also, an external magnetic field can be applied so that the isotropy is broken by introducing a certain direction (the direction of the field). Thus, the whole Eq. 8 describes the anisotropic Heisenberg Hamiltonian.

The developed software for simulation of systems in frame of Heisenberg model is based on the new, promising programming language Rust. Rust is a new experimental programming language developed by Mozilla. Rust supports functional, parallel, procedural and object-oriented programming, that is, almost the entire range of paradigms actually used in applied programming. From the above, it could be concluded that this language combines all the properties needed for numerical simulation. Also this language supports MPI for high performance computing.

The results obtained by the created software was verified in comparison with known results for the Heisenberg model [28], e.g. convergence of critical temperature $T_c = 1.44(2)$ has been reached.

5. Summary

In this paper, the investigation of some Ising spins and dipoles models was performed.

We show that the COP is more universal than magnetization. In general, these models show similar results. Magnetization shows the transition only at positive J, but COP describes the transition, regardless the exchange integral sign and the number of neighbors.

The open question is leave for us to develop the approach of the cluster order parameter on ASI systems with dipolar Hamiltonian (7). In such systems there are a lot of neighbors for each dipole, that can present some troubles for approach. But knowing the order parameter is necessary for calculation of critical indexes of phase transition.

The observed anomalous behavior of the heat capacity in the arrays must be investigated in detail to find out the reason behind the variation of critical behavior near the critical phase transition temperature. Here two directions of research are possible with the purpose of theoretical description of the second-order phase transition.

The first can obviously be due to the fact that the classical concept of a second-order phase transition assumes a mutual transformation of the two phases, which is usually explained in terms of percolation theory, i.e. the appearance of the "percolation cluster". The presence of one or several peaks in T_c region can be due to the existence of several transitions between possible coexisting phases, the creation and sequential transformation of a percolation cluster under the conditions of long-range dipole interaction. The question of the concept of "percolation cluster" in systems with alternating long-range interaction also remains open.

The second direction of research can be connected with the definition of critical behavior, which is no longer described in the framework of universal and simple power laws, i.e. the hypothesis of the universality of the critical behavior of second order phase transitions requires verification.

This work was supported RFBR according to the research project No 16 − 32 − 00202 and the Ministry of Education and Science of the Russian Federation in the frame of the state task #3.7383.2017/8.9, the scholarship of the President of the Russian Federation for young scientists and postgraduate students performing advanced research and development in priority areas of modernization of the Russian Economics for 2015-2017 years (SP-118.2015.5, SP-1675.2015.5, order #184, 10/03/2015).

References

[1] Shevchenko Y, Makarov A and Nefedev K 2017 *Physics Letters A* **381** 428–434
[2] Soldatov K, Nefedev K, Kapitan V Y and Andriushchenko P 2016 *Journal of Physics: Conference Series* vol 741 (IOP Publishing) p 012199
[3] Belokon V, Kapitan V and Dyachenko O 2016 *Journal of Magnetism and Magnetic Materials* **401** 651–655
[4] Belokon V, Dyachenko O and Kapitan V Y 2015 *Izvestiya, Physics of the Solid Earth* **51** 622–629
[5] Peretyatko A, Nefedev K and Okabe Y 2017 *Physical Review B* **95** 144410
[6] Prudnikov P V, Prudnikov V V, Menshikova M A and Piskunova N I 2015 *Journal of Magnetism and Magnetic Materials* **387** 77–82
[7] Morgan J P, Stein A, Langridge S and Marrows C H 2011 *Nature Physics* **7** 75–79
[8] Kapaklis V, Arnalds U B, Farhan A, Chopdekar R V, Balan A, Scholl A, Heyderman L J and Hjörvarsson B 2014 *Nature nanotechnology* **9** 514–519
[9] Daunheimer S A, Petrova O, Tchernyshyov O and Cumings J 2011 *Physical review letters* **107** 167201
[10] Morgan J P, Akerman J, Stein A, Phatak C, Evans R, Langridge S and Marrows C H 2013 *Physical Review B* **87** 024405
[11] Ahlberg M, Andersson G and Hjörvarsson B 2011 *Physical Review B* **83** 224404
[12] Shevchenko Y, Nefedev K and Okabe Y 2017 *Physical Review E* **95** 052132
[13] Andriushchenko P D and Nefedev K V 2013 *Advanced Materials Research* vol 718 (Trans Tech Publ) pp 166–171
[14] Soldatov K, Nefedev K, Komura Y and Okabe Y 2017 *Physics Letters A* **381** 707–712
[15] Ivanov V and Goroshko O 2016 *Journal of Superconductivity and Novel Magnetism* **29** 2579–2585
[16] Antal T, Droz M and Rácz Z 2004 *Journal of Physics A: Mathematical and General* **37** 1465
[17] Gambardella P, Dallmeyer A, Maiti K, Malagoli M, Eberhardt W, Kern K and Carbone C 2002 *Nature* **416** 301–304
[18] Fan Y 2011 *European Journal of Physics* **32** 1643
[19] Otsuka H, Takatsu H, Goto K and Kadowaki H 2014 *Physical Review B* **90** 144428
[20] Perrin Y, Canals B and Rougemaille N 2016 *Nature* **540** 410
[21] Mól L, Pereira A and Moura-Melo W 2012 *Physical Review B* **85** 184410
[22] Rougemaille N, Montaigne F, Canals B, Duluard A, Lacour D, Hehn M, Belkhou R, Fruchart O, El Moussaoui S, Bendounan A *et al.* 2011 *Physical Review Letters* **106** 057209
[23] Chioar I, Rougemaille N and Canals B 2016 *Physical Review B* **93** 214410
[24] Canals B, Chioar I A, Nguyen V D, Hehn M, Lacour D, Montaigne F, Locatelli A, Menteş T O, Burgos B S and Rougemaille N 2016 *Nature communications* **7**
[25] Wang F and Landau D 2001 *Physical review letters* **86** 2050
[26] Middleton T F and Wales D J 2003 *The Journal of chemical physics* **118** 4583–4593
[27] Landau D, Tsai S H and Exler M 2004 *American Journal of Physics* **72** 1294–1302
[28] Chen K, Ferrenberg A M and Landau D 1993 *Physical Review B* **48** 3249

IOP Conf. Series: Journal of Physics: Conf. Series **936** (2017) 012082 doi:10.1088/1742-6596/936/1/012082

An analytic relation between the fractional parameter in the Mittag–Leffler function and the chemical potential in the Bose–Einstein distribution through the analysis of the NASA COBE monopole data

Minoru Biyajima[1], Takuya Mizoguchi[2] and Naomichi Suzuki[3]

[1] Department of Physics, Shinshu University, Matsumoto 390-8621, Japan
[2] National Institute of Technology, Toba College, Toba 517-8501, Japan
[3] Matsumoto University, Matsumoto, 390-1295, Japan

E-mail: suzuki@matsu.ac.jp

Abstract. To extend the Bose-Einstein (BE) distribution to fractional order, we turn our attention to the differential equation, $df/dx = -f - f^2$. It is satisfied with the stationary solution, $f(x) = 1/(e^{x+\mu} - 1)$, of the Kompaneets equation, where μ is the constant chemical potential. Setting $R = 1/f$, we obtain a linear differential equation for R. Then, the Caputo fractional derivative of order p ($p > 0$) is introduced in place of the derivative of x, and fractional BE distribution is obtained, where function e^x is replaced by the Mittag–Leffler (ML) function $E_p(x^p)$. Using the integral representation of the ML function, we obtain a new formula. Based on the analysis of the NASA COBE monopole data, an identity $p \simeq e^{-\mu}$ is found.

1. Introduction

The COBE FIRAS experiments have shown that the cosmic microwave background (CMB) radiation spectrum is well described by the Planck distribution with temperature, $T = 2725.0 \pm 1$ mK [1, 2]. Furthermore, a slight distortion from the Planck distribution in the photon number distribution, $f(x)$, is observed. It is expressed by

$$f(x) = 1/(e^{x+\mu} - 1), \quad x = h\nu/(kT), \tag{1}$$

where μ is the dimensionless constant chemical potential and ν is the frequency of photon. The measured value is $\mu = (-1 \pm 4) \times 10^{-5}$ or $|\mu| < 9 \times 10^{-5}$ with 95% confidence [1, 2].

Equation (1) is known as a stationary solution of the Kompaneets equation[3]:

$$\frac{\partial f}{\partial t} = \frac{kT_e}{m_e c^2} \frac{n_e \sigma_e}{c} {x_e}^{-2} \frac{\partial}{\partial x_e} {x_e}^4 \Big(\partial f/\partial x_e + f + f^2\Big), \tag{2}$$

where n_e is the electron density, σ_e is the Thomson scattering cross-section and $x_e = h\nu/(kT_e)$. Equation (2) describes the photon distribution, which obeys the Planck distribution at the initial stage, and is affected by the elastic e-γ scatterings in the expanding universe.

In the theories of stochastic processes, in order to take a sort of memory effect into account, fractional calculus is introduced [4, 5, 6, 7]. Based on the Caputo derivative [4], Ertik et al. proposed a generalized BE distribution [8], $f(x) = 1/(E_p(x) - 1)$, where $E_p(x)$ denotes the Mittag-Leffler (ML) function defined by

$$E_p(x) = \sum_{n=0}^{\infty} x^n/\Gamma(np+1), \quad p > 0. \tag{3}$$

To extend the BE distribution to fractional order, we turn our attention to the equation,

$$df(x)/dx = -af(x) - bf(x)^2, \tag{4}$$

where a and b are constant. If $a = b = 1$, Eq.(4) reduces to the equation which is adopted by Planck [9, 10] to derive the blackbody radiation law, and is satisfied with the stationary solution (1) of the Kompaneets equation (2). Putting $f = 1/R$, we obtain the linear differential equation,

$$dR/dx = aR + b. \tag{5}$$

In Appendix A, the Caputo fractional derivative is introduced into Eq. (5) in place of the derivative x, and a fractional BE and other distributions are obtained.

In [11], we have applied the Riemann–Liouville fractional derivative to obtain a fractional BE distribution $f(x) = 1/(E_p(x^p) - 1)$, and we have investigated the NASA COBE monopole data using BE and fractional BE distributions. The photon spectrum given from Eq.(1) is written as

$$U^{\mathrm{BE}}(x, \mu) = C_B/(e^{x+\mu} - 1), \tag{6}$$

where $x = h\nu/(kT)$ and $C_B = 2h\nu^3/c^2$. On the other hand, the photon spectrum in the Universe, based on the fractional calculus, is given by

$$U(x, p) = C_B/(E_p(x^p) - 1). \tag{7}$$

From the analysis of NASA COBE monopole data [2] , the following values of parameters are estimated [11]: from Eq.(6), $T = 2.72501 \pm 0.00002$ K and $\mu = (-1.1 \pm 3.2) \times 10^{-5}$, and from Eq.(7), $T = 2.72501 \pm 0.00003$ K and $p - 1 = (1.1 \pm 3.5) \times 10^{-5}$. Then, we estimated a relation between μ and p as $\mu \approx 1 - p$.

In the present study, the COBE monopole data [2] is analyzed by the use of an integral representation of the ML function [5, 12]:

$$E_p(x^p) = e^x/p + \delta(x, p), \tag{8}$$

$$\delta(x, p) = -\frac{\sin(p\pi)}{\pi} \int_0^{+\infty} \frac{y^{p-1} e^{-xy}}{y^{2p} - 2y^p \cos(p\pi) + 1} dy. \tag{9}$$

Function $\delta(p, x)$ for $0 < p < 2$ and $0 \le x$ satisfies the relation, $|\delta(p, x)| \le |\delta(p, 0)| = |p - 1|/p$.

2. Analysis of the COBE monople data by Eqs.(8) and (9)

By the use of the integral representation of the ML function, Eq. (8), Eq. (7) is written as

$$U(x, p) = C_B/(e^x/p - 1 + \delta(x, p)), \tag{10}$$

with two parameters, T and p. At first we analyze the COBE monopole data using Eq.(10). The results are shown in Table 1 and in Fig. 1. As is seen from Table 1, conditions that $|p - 1| << 1$ and $|\delta(p, x)| \le |p - 1|/p << 1$ are satisfied. Then, we can expand Eq. (10) as

$$U(x, p) = C_B/(e^x/p - 1) - C_B\, \delta(x, p)/(e^x/p - 1)^2. \tag{11}$$

Table 1. Analysis of the NASA COBE monopole data by Eq.(10).

T (K)	$(p-1)$	χ^2/N F
$2.72501 \pm 3 \times 10^{-5}$	$(1.1 \pm 3.5) \times 10^{-5}$	45.0/41

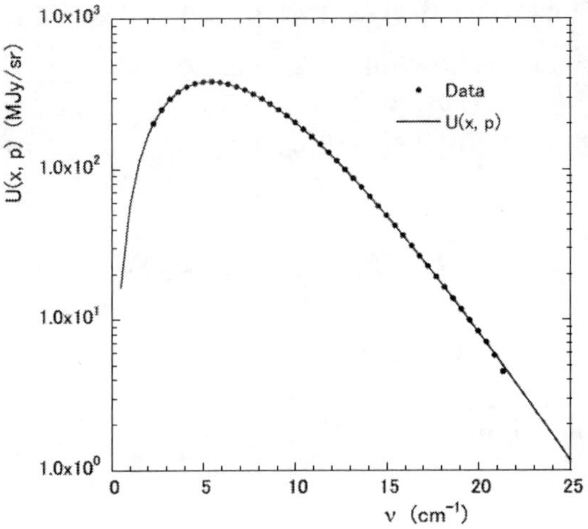

Figure 1. Analysis of the COBE monopole data by Eq. (10). $x = 0.528\nu$.

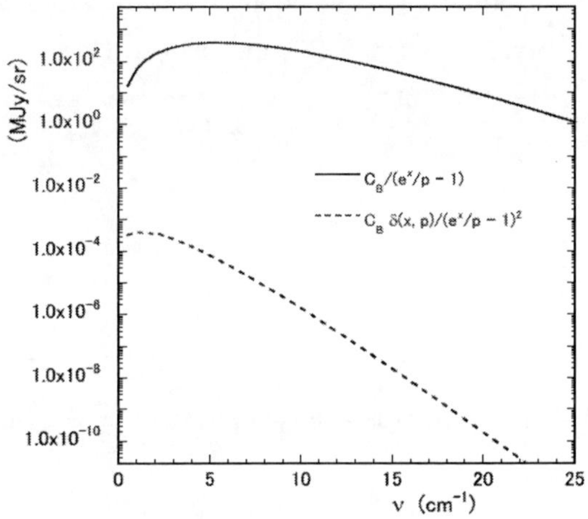

Figure 2. Contribution of first and second terms on the right hand side of Eq. (11).

From the Analysis with Eq.(11), we obtain the same results with those in Table 1. Contribution from the first and second terms on the right hand side of Eq. (11) with parameter values in Table 1 are shown in Fig. 2.

As the ratio of the second term to the first term on the right hand side of Eq. (11) becomes $\delta(x,p)/(e^x/p-1) < 2 \times 10^{-6}$ over the range of the COBE monopole data, $1.20 \leq x \leq 11.26$, we can approximate Eq. (11) as,

$$U(x,p) \simeq C_B/(e^x/p-1) = C_B/(e^{x-\ln p}-1). \quad (12)$$

Comparing Eq.(1) and Eq.(12), we obtain an analytic relation, $\mu = -\ln p$.

3. Concluding remarks

1) If the Caputo fractional derivative is introduced into Eq. (4), contrary to the case of Riemann-Liouville fractional derivative [11], we have fractional Bose-Einstein, Fermi-Dirac and Maxwell-Boltzmann distributions, where function e^x is replaced by the ML function, $E_p(x^p)$.

2) Under the condition that $|p-1| << 1$, we can show that the analytic relation, $\mu = -\ln p$, is satisfied. In other words, the fractional parameter p, where a kind of memory effect of the expanding universe would be included, has a role of inverse fugacity to the dimensionless chemical potential μ.

3) Extension of statistical distributions has already been investigated from the non-extensive statistical approach [10], where parameter q is included. We would like to study how fractional parameter p is related to q and other approaches.

Appendix A. Application of the Caputo fractional derivative to Eq. (5)

The Caputo fractional derivative [4, 5] of function $f(x)$ for $m = 1, 2, \ldots$ is defined as

$$ {}_0^C D_x^p f(x) = \frac{1}{\Gamma(m-p)} \int_0^x (x-\tau)^{m-p-1} f^{(m)}(\tau) d\tau, \quad m-1 < p < m, \tag{A.1}$$

where $f^{(m)}(\tau) = d^m f(\tau)/d\tau^m$. We consider the following equation,

$$ {}_0^C D_x^p R(x) = aR(x) + b. \tag{A.2}$$

The Laplace transform of function $R(x)$ is defined as, $\tilde{R}(s) = \mathcal{L}[R(x); s] = \int_0^\infty \mathrm{e}^{-sx} R(x) dx$. Applying the Laplace transform to Eq. (A.2), we obtain the following equation,

$$ \tilde{R}(s) = b/\{s(s^p - a)\} + \sum_{k=0}^{m-1} R^{(m-k-1)}(0) s^{k-\nu}/(s^p - a). \tag{A.3}$$

Using the formula [5],

$$ \mathcal{L}[x^{\beta-1} E_{\alpha,\beta}(\gamma x^\alpha); s] = s^{\alpha-\beta}/(s^\alpha - \gamma), \quad Re(s) > |a|^{1/\alpha}, \tag{A.4}$$

$$ E_{\alpha,\beta}(z) = \sum_{k=0}^{\infty} z^k/\Gamma(\alpha k + \beta), \quad \alpha > 0, \ \beta > 0, \tag{A.5}$$

where $E_{\alpha,\beta}(z)$ is the two parameter ML function, we have

$$ R(x) = b\{E_p(ax^p) - 1\} + R(0)E_p(ax^p) + \sum_{k=1}^{m-1} R^{(m-k)}(0) x^{m-k-1} E_{p,m-k+1}(ax^p). \tag{A.6}$$

Solutions $R(x)$ according to the values of a, b, and the initial conditions are shown in Table A1.

Table A1. Solutions of Eq.(A.2).

(a, b)	Initial conditions	$R(x)$	$f(x) = 1/R(x)$
$(1, 1)$	$R(0) = \cdots = R^{(m-1)}(0) = 0$	$E_p(x^p) - 1$	$1/(E_p(x^p) - 1)$
$(1, -1)$	$R(0) = 2,\ R^{(1)}(0) = \cdots = R^{(m-1)}(0) = 0$	$E_p(x^p) + 1$	$1/(E_p(x^p) + 1)$
$(1, 0)$	$R(0) = 1,\ R^{(1)}(0) = \cdots = R^{(m-1)}(0) = 0$	$E_p(x^p)$	$1/E_p(x^p)$

References

[1] Mather J C et al. 1994 *Astrophys. J.* **420** 439; Fixsen D J and Mather J C 2002 *Astrophys. J.* **581** 817
[2] COBE/FIRAS CMB monopole spectrum, May 2005, http://lambda.gsfc.nasa.gov/product/cobe/firas_monopole_get.cfm
[3] Kompaneets A S 1957 *Sov. Phy. JETP* **4** 730; Weymann R 1965 *Phys. Fluids* **8** 2112
[4] Caputo M and Mainardi F 1971 *Riv. Nuov. Cim.* (Ser. II), **1** 61
[5] Podlubny I 1999 *Fractional Differential Equations* (San Diego: Academic Press)
[6] Metzler R and Klafter J 2000 *Phys. Rep.* **339** 1
[7] West B J, Bologna M and Grigolini P 2003 *Physics of Fractal Operators* (Springer-Verlag)
[8] Ertik H et al 2009 *Physica* **A 388** 4573
[9] M. Planck M 1900 *Verh. Deutsch. Phys. Ges.* **2** 202 and 237.
[10] Tsallis C 2009 *Introduction to Nonextensive Statistical Mechanics* (Springer Science+Business media LLC).
[11] Biyajima M, Mizoguchi T and Suzuki N 2015 *Physica* A **440** 129.
[12] Gorenflo R , Loutchko J and Luchko Y 2002 *Frac. Calc. Appl. Anal.* **5** 491

IOP Conf. Series: Journal of Physics: Conf. Series **936** (2017) 012083 doi:10.1088/1742-6596/936/1/012083

Generalized Bell states map physical systems' quantum evolution into a grammar for quantum information processing

Francisco Delgado
Departamento de Física y Matemáticas, Escuela de Ingeniería y Ciencias, Tecnológico de Monterrey, Campus Estado de México, Atizapán, Estado de México, CP. 52926, México.
E-mail: fdelgado@itesm.mx

Abstract. Quantum information processing should be generated through control of quantum evolution for physical systems being used as resources, such as superconducting circuits, spin-spin couplings in ions and artificial anyons in electronic gases. They have a quantum dynamics which should be translated into more natural languages for quantum information processing. On this terrain, this language should let to establish manipulation operations on the associated quantum information states as classical information processing does. This work shows how a kind of processing operations can be settled and implemented for quantum states design and quantum processing for systems fulfilling a $SU(2)$ reduction in their dynamics.

1. Introduction
Synergy between quantum mechanics and computer science is generating disruptive developments. In the gate array version of quantum processing, appropriate logical gates are essential. They adopt forms inspired by classical computation, despite their construction is based on the behavior of quantum systems and normally is based on universal sets of quantum gates with which any processing task can be expressed [1, 2]. Commonly, some universal gates still should be constructed as iterative steps of controlled physical evolutions. Methods as Cosine-Sine decomposition [3], multiplexor gates [4], Gray codes basis [5] or $q-$deformed algebras [6] are used as approaches. In the $SU(2)$ reduction approach [7, 8] for two qubits, which has been extended to multipartite systems [9], unitary factorization and natural gates for two-qubits has been obtained, letting to define basic operations closer to classical computation. This paper exploits the $SU(2)$ decomposition [10] to address the evolution into basic processing tasks for state design. The second section presents the technical details of $SU(2)$ reduction to give a generic processing gate. The third section presents key operations for quantum processing. The fourth section presents examples of quantum state design. Last section set the conclusions.

2. $SU(2)$ decomposition technical details
$SU(2)$ decomposition was developed for a generic Hamiltonian combining $2d$ two-level interacting subsystems resembling the Heisenberg-Ising interaction and including local driven operations [11]. It is reached with an appropriate basis to reduce the Hamiltonian into a 2×2-block matrix and as a consequence, the evolution matrix in a block matrix with blocks in $SU(2)$:

$$H = \left(\begin{array}{c|c|c|c} \mathbf{S}_{H1} & 0 & \dots & 0 \\ \hline 0 & \mathbf{S}_{H2} & \dots & 0 \\ \hline \vdots & \vdots & \ddots & \vdots \\ \hline 0 & 0 & \dots & \mathbf{S}_{H2^{2d-1}} \end{array}\right) \Rightarrow U = \left(\begin{array}{c|c|c|c} \mathbf{S}_{U1} & 0 & \dots & 0 \\ \hline 0 & \mathbf{S}_{U2} & \dots & 0 \\ \hline \vdots & \vdots & \ddots & \vdots \\ \hline 0 & 0 & \dots & \mathbf{S}_{U2^{2d-1}} \end{array}\right) \quad (1)$$

the inheritance for U is due to the group properties of block matrices via time ordered integral $\mathbf{S}_{Ui} = \tau\{e^{-\frac{i}{\hbar}\int_0^t \mathbf{S}_{Hi} dt'}\}$. Basis elements should be rearranged to show the blocks as in (1). For:

$$H^{(j,k')} = H_D + H_{ND}^{(j,k')} \quad \text{with}: \quad H_{ND}^{(j,k')} = \sum_{t'=0}^{1} h_{(j4^{k'+dt'-1})_4^{2d}} \bigotimes_{s=1}^{2d} \sigma_{(j4^{k'+dt'-1})_{4,s}^{2d}} \quad (2)$$

$$H_D \equiv \sum_{i'=1}^{3}\sum_{k=1}^{d} h_{(i'(4^{k-1}+4^{k+d-1}))_4^{2d}} \bigotimes_{s=1}^{2d} \sigma_{(i'(4^{k-1}+4^{k+d-1}))_{4,s}^{2d}}$$

as Hamiltonian ($h_{\{i_k\}}$ are time-dependent real functions), including a) non-local interactions H_D (spin-spin couplings like) between all paired parts $[k, k+d]$ (called *correspondent*) in three directions $i' = 1, 2, 3$, giving the diagonal entries; and b) $H_{ND}^{(j,k')}$, a couple of local interactions on the correspondent pair $[k', k'+d]$ in direction j, giving the diagonal-off entries [9]. For:

$$\left|\Psi_{\mathcal{I}_4^d}\right\rangle = \frac{1}{\sqrt{2^d}} \sum_{\mathcal{E},\mathcal{D}=0}^{2^d-1} (\tilde{\sigma}_{i_1} \otimes \dots \otimes \tilde{\sigma}_{i_d})_{\mathcal{E}_2^d,\mathcal{D}_2^d} \left|\mathcal{E}_2^d\right\rangle \otimes \left|\mathcal{D}_2^d\right\rangle \quad (3)$$

the generalized Bell states (GBS) basis with $\tilde{\sigma}_{i\neq 2} = \sigma_i, \tilde{\sigma}_2 = i\sigma_2$ [12], the decomposition is achieved. $\mathcal{E}_2^d, \mathcal{D}_2^d$ are base-2 numbers with d digits ($\mathcal{E}, \mathcal{D} \in \{0, 1, ..., 2^d - 1\}$) representing $\{\epsilon_1, ..., \epsilon_d\}, \{\delta_1, ..., \delta_d\}$, with $\epsilon_s, \delta_s \in \{0, 1\}$ (digits appear reversed in $\mathcal{E}_2^d$ or $\mathcal{I}_4^d$), giving:

$$\left\langle \Psi_{\mathcal{I}_4^d} | H^{(j,k')} | \Psi_{\mathcal{K}_4^d} \right\rangle = \delta_{\mathcal{I}\mathcal{K}} \sum_{i'=1}^{3}\sum_{k''=1}^{d} c_{i',i'}^{i_{k''},i_{k''}} h_{(i'(4^{k''-1}+4^{k''+d-1}))_4^{2d}} +$$

$$\sum_{t'=0}^{1} \delta_{\mathcal{I}\mathcal{K}}^{\{k'\}} \mathcal{F}_{j,k'}^{j\delta_{0,t'},j\delta_{1,t'}} h_{(j4^{k'+dt'-1})_4^{2d}} \equiv H_{D\mathcal{I}\mathcal{K}} + H_{ND}^{(j,k')}{}_{\mathcal{I}\mathcal{K}} \quad (4)$$

with the coefficients $c_{i',i'}^{i_{k''},i_{k''}}$, $\mathcal{F}_{j,k'}^{j\delta_{0,t'},j\delta_{1,t'}}$ reported in [11]. The most important aspect is the establishment of blocks. Due to the coefficient properties, they are located in the rows $\mathcal{I} = \{i_1 i_2 ... i_n\}$ and $\mathcal{I}' = \{i'_1 i'_2 ... i'_n\}$ differing in only one subscript $i'_k \neq i_k$ with the correspondence rule: $0 \leftrightarrow j$, $i \leftrightarrow k$, being j the direction of local driven interactions in (2) and i, j, k a permutation of $1, 2, 3$. This is the relation rule for the GBS basis elements in the blocks. Hilbert space $\mathcal{H}^{2d}$ becomes the direct sum of 2^{2d-1} subspaces generated by these GBS basis elements. There, the dynamics mixes the probabilities only inside each subspace. Because states in the basis are entangled (non-maximal), each block $\mathbf{S}_{Ui}$ in (1) includes the controlled gate operations. For the time-independent case, in terms of $\mathbf{S}_{H\mathcal{I},\mathcal{I}'} = \{h_{ij}\}$ in (4):

$$\mathbf{S}_{U\mathcal{I},\mathcal{I}'} = e^{i\frac{h_{11}+h_{22}}{2\hbar}t} \begin{pmatrix} \cos\omega t + i\frac{h_{11}-h_{22}}{2\hbar\omega}\sin\omega t & i\frac{h_{12}}{\hbar\omega}\sin\omega t \\ i\frac{h_{12}^*}{\hbar\omega}\sin\omega t & \cos\omega t - i\frac{h_{11}-h_{22}}{2\hbar\omega}\sin\omega t \end{pmatrix} \quad (5)$$

with: $\hbar\omega = \sqrt{|h_{12}|^2 + \frac{1}{4}|h_{11} - h_{22}|^2}$. Clearly $\mathbf{S}_{U\mathcal{I},\mathcal{I}'} \in U(1) \times SU(2)$. $\mathcal{F}_{j,k'}^{j_d,j_{d+s}}$ in (4) is imaginary only if j_d or j_{d+s} are 2, then only one n_1 or n_2 is non-zero.

IOP Conf. Series: Journal of Physics: Conf. Series **936** (2017) 012083 doi:10.1088/1742-6596/936/1/012083

3. Stating notable manipulation operations in quantum information

Note $\mathbf{S}_{U\mathcal{I},\mathcal{I}'}$ byself is a general $SU(2)$ operation in the form of a mixing matrix. Other notable operations are easily achievable. The first one with $\frac{h_{11}+h_{22}}{2\hbar\omega} = (\frac{\alpha}{m} - 1)\pi, \omega t = m\pi; n, m \in \mathbf{Z}$, the quasi-identity gate $\mathbf{S}_{U\mathcal{I},\mathcal{I}'} = e^{i\alpha\pi}\mathbf{I}_{\mathcal{I},\mathcal{I}'} \equiv I^{\alpha}_{\mathcal{I},\mathcal{I}'}$. The second one is a family of operations achieved with $\omega t = \frac{2n+1}{2}\pi$, $\frac{h_{11}-h_{22}}{2\hbar\omega} = \epsilon$, $\frac{h_{12}}{\hbar\omega} = i^c\delta$, $\frac{h_{11}+h_{22}}{2\hbar\omega} = 2(m - \frac{1}{2}); n, m \in \mathbf{Z}, c \in \{0,1\}, \delta \in \mathbf{R}$, where $\epsilon^2 + \delta^2 = 1$. c depends on the direction of local interaction. We get for $\mathbf{S}_{U\mathcal{I},\mathcal{I}'}$ on the GBS basis:

$$\mathbf{H}^c_m(\delta,\epsilon)_{\mathcal{I},\mathcal{I}'} = (-1)^m \begin{pmatrix} \epsilon & i^c\delta \\ (-i)^c\delta & -\epsilon \end{pmatrix} \tag{6}$$

It is widely versatile: a) if $s_\epsilon \epsilon = \delta = \frac{1}{\sqrt{2}}$ ($s_\epsilon = \text{sign}(\epsilon)$, $\pm$ in the notation), a Hadamard-like gate $H^{m,c,\text{sign}(\epsilon)}_{\mathcal{I},\mathcal{I}'}$ is obtained; b) if $\delta = 1$, the exchange-like gate $E^{m,c}_{\mathcal{I},\mathcal{I}'}$ is obtained [7], a limit case for the independent-time case when $h_{12} \gg h_{11} - h_{22}$ (5). Despite there are only two independent blocks in the evolution matrix, they have $3d+3$ parameters (including time and Hamiltonian terms involved). By rearranging the parts of correspondent pairs and controlling non-local and local strengths as well as their directions, global operations could be made on the entire quantum information space of the system.

4. Quantum states design: maximal entangled states

Quantum design and manipulation become reduced to operations transforming indexes and changing probability amplitudes in the GBS basis elements. Depending on the path of the state to manipulate, this task could be still hard. Despite one index exchange on the states are limited to reach general states on $2d-$qubits, by rearranging the pairing of qubits, we can extend the entanglement manipulation. To illustrate the procedure for four qubits ($d = 2$), we show the construction of $|GHZ\rangle^4$ and $|W\rangle^4$, which could be expressed in the GBS basis as:

$$|GHZ\rangle^4 = \frac{1}{\sqrt{2}}(|\Psi_{0,0}\rangle + |\Psi_{3,3}\rangle) = \frac{1}{\sqrt{2}} \sum_{\mathcal{I}\in\{0,15\}} |\Psi_\mathcal{I}\rangle \tag{7}$$

$$|W\rangle^4 = \frac{1}{2}(|\Psi_{1,0}\rangle + |\Psi_{0,1}\rangle + |\Psi_{3,1}\rangle + |\Psi_{1,3}\rangle) = \frac{1}{2} \sum_{\mathcal{I}\in\{1,4,7,13\}} |\Psi_\mathcal{I}\rangle \tag{8}$$

In terms of the previous operations and departing from $|0000\rangle$, the steps to achieve $|GHZ\rangle^4$ are:

$$\begin{aligned}
|0000\rangle \quad & \underset{H^{0,0,+}_{0,3} \oplus H^{0,0,+}_{12,15}}{\xrightarrow{H^{(3,1)}}} \quad \frac{1}{\sqrt{2}}|\Psi_0\rangle_1 \otimes (|\Psi_0\rangle_2 + |\Psi_3\rangle_2) \\
& \underset{H^{0,0,+}_{0,12}}{\xrightarrow{H^{(3,2)}}} \quad |\Psi_0\rangle_1 \otimes |\Psi_0\rangle_2 = |\Psi_0\rangle = \frac{1}{2}\sum_{i=0}^{3} |\Psi_i\rangle_{1'} \otimes |\Psi_i\rangle_{2'} \\
& \underset{H^{0,0,+}_{0,3} \oplus H^{0,0,+}_{12,15} \oplus H^{0,0,+}_{5,6} \oplus H^{0,0,+}_{9,10}}{\xrightarrow{H^{(3,1')}}} \quad \frac{1}{\sqrt{2}}(|\Psi_3\rangle_1 \otimes |\Psi_0\rangle_2 + |\Psi_1\rangle_1 \otimes |\Psi_1\rangle_2) \\
& \underset{E^{0,0}_{0,3} \oplus I^0_{5,6}}{\xrightarrow{H^{(3,1)}}} \quad \frac{1}{\sqrt{2}}(|\Psi_0\rangle_1 \otimes |\Psi_0\rangle_2 + |\Psi_1\rangle_1 \otimes |\Psi_1\rangle_2) = \frac{1}{\sqrt{2}}(|\Psi_0\rangle + |\Psi_5\rangle) \\
& \underset{E^{0,1}_{5,7} \oplus I^0_{0,2}}{\xrightarrow{H^{(2,1)}}} \quad \frac{1}{\sqrt{2}}(|\Psi_0\rangle_1 \otimes |\Psi_0\rangle_2 + i\,|\Psi_3\rangle_1 \otimes |\Psi_1\rangle_2) \\
& \underset{E^{1,1}_{7,15} \oplus I^0_{0,8}}{\xrightarrow{H^{(2,2)}}} \quad \frac{1}{\sqrt{2}}(|\Psi_0\rangle_1 \otimes |\Psi_0\rangle_2 + |\Psi_3\rangle_1 \otimes |\Psi_3\rangle_2) = |GHZ\rangle^4
\end{aligned} \tag{9}$$

where symbol above the arrow point to the Hamiltonian being used, while symbol below points to the concrete operations being used. Two notations for the kets are used by convenience: $|\Psi_k\rangle_j$ is the Bell state $|\Psi_k\rangle$ on the jth correspondent pair, while $|\Psi_{i_1,i_2,\ldots,i_d}\rangle = |\Psi_{\mathcal{I}}\rangle =$ is the $\mathcal{I} = 4^{d-1}i_d + \ldots + 4i_2 + i_1$ element in the GBS basis. Note that pairs 1 and 2 contains the parts $[1,2]$ and $[3,4]$ respectively, while $1'$ and $2'$ contains the parts $[1,4]$ and $[2,3]$ respectively. This temporary rearrangement lets enlarge the entanglement into the whole system. Finally, departing from $\frac{1}{\sqrt{2}}(|\Psi_0\rangle + |\Psi_5\rangle)$ (included as step in the previous process), we can achieve $|W\rangle^4$:

$$\begin{aligned}
\frac{1}{\sqrt{2}}(|\Psi_0\rangle + |\Psi_5\rangle) \quad & \underset{E_{0,4}^{0,0}\oplus E_{1,5}^{0,0}}{\overset{H^{(1,2)}}{\longrightarrow}} \quad \frac{1}{\sqrt{2}}(|\Psi_0\rangle_1 \otimes |\Psi_1\rangle_2 + |\Psi_1\rangle_1 \otimes |\Psi_0\rangle_2) \\
& \underset{H_{4,7}^{0,0,+}\oplus I_{1,2}^{2p}}{\overset{H^{(3,1)}}{\longrightarrow}} \quad \frac{1}{\sqrt{2}}(\frac{1}{\sqrt{2}}(|\Psi_0\rangle_1 + |\Psi_3\rangle_1) \otimes |\Psi_1\rangle_2 + |\Psi_1\rangle_1 \otimes |\Psi_0\rangle_2) \\
& \underset{I_{4,8}^{2q}\oplus I_{7,11}^{2r}\oplus H_{1,13}^{0,0,+}}{\overset{H^{(3,2)}}{\longrightarrow}} \quad \frac{1}{2}((|\Psi_0\rangle_1 + |\Psi_3\rangle_1) \otimes |\Psi_1\rangle_2 + |\Psi_1\rangle_1 \otimes (|\Psi_0\rangle_2 + |\Psi_3\rangle_2)) \\
& \qquad = \frac{1}{2}(|\Psi_4\rangle + |\Psi_7\rangle + |\Psi_1\rangle + |\Psi_{13}\rangle) = |W\rangle^4
\end{aligned} \tag{10}$$

with $p, q, r \in \mathbf{Z}$. Block independence has been applied justifying simultaneous operations, despite only two are possible (but repeated through different blocks). This couple of examples shows briefly the manipulation model of quantum information applied to state design.

5. Conclusions

While quantum system is the object where interactions work, quantum information is the associated element containing the sensible data of the history and the existence properties of it. The generic Hamiltonian in this work comprises representative systems in quantum information processing. $SU(2)$ reduction lets understand quantum processing as a simple series of operations manipulating directly the underlying quantum information while GBS basis works as a universal grammar there. Still, the general state design process is an open problem, in particular those containing the complexity of entanglement creation in all their possible levels between separability and maximal entanglement.

Acknowledgments

The support from Tecnológico de Monterrey to develop this research is acknowledged.

References

[1] Barenco A, Bennet C, Cleve R, Di Vincenzo D, Margolus N, Shor P, Sleator T, Smolin J and Weinfurter H 1995 *Phys. Rev. A* **52** 3457
[2] Boykin P, Mor T, Pulver M, Roychowdhry V and Vatan F 2000 *Inf. Proc. Lett.* **75** 101
[3] Möttönen M, Vartiainen J, Bergholm V, Salomaa M 2004 *Phys. Rev. Lett.* **93** 130502
[4] Schende V, Bullock S and Markov I 2006 Synthesis of Quantum Logic Circuits *IEEE Transaction on computer-aided design of integrated circuits and systems* **25** 6
[5] Vartianen J, Möttönen M and Salomma M 2004, *Phys. Rev. Lett.* **92** (17) 177902.
[6] Altintas A, Ozaydin F, Yesilyurt C, Bugu S and Arik M 2014 *Quant. Inf Proc.* **13** (13) 1035
[7] Delgado F 2017 *Quant. Inf. and Comp.* **17** (9) 0721
[8] Delgado F 2017 *J. of Phys. Conf. Ser.* **839** 012014
[9] Delgado F 2017 *Quantum Information and Measurement (QIM) 2017*, OSA Technical Digest (online) (Optical Society of America) paper QT6A.23
[10] Delgado F 2015 *Int. J. of Quant. Inf.* **13** (7) 1550055
[11] Delgado F 2017 *Modeling the dynamics of multipartite quantum systems created departing from two-level systems using general local and non-local interactions* To appear.
[12] Sych D and Leuchs G 2009 *New Journal of Physics* **11** 013006

Modelling of magnetic elastomer deformation dependence on magnetic field

D A Solodkiy and L L Afremov

Department of Theoretical and Nuclear Physics, The School of Natural Sciences, Far Eastern Federal University, 8, Sukhanova St., 690950, Vladivostok, Russian Federation

E-mail: dm_solodkiy@mail.ru, afremov.ll@dvfu.ru

Abstract. The model of magnetic deformable system was developed. The modelling of magnetic elastomer deformation and magnetization dependence on external magnetic field was calculated within the method of movable cellular automata (MCA). It was shown that ordering of magnetic particles substantially effects on elastomer saturation deformation. The maximum orderly elastomer saturation deformation is twice higher the saturation deformation of system with chaotic ordering of magnetic particles.

1. Introduction

Magnetic elastomers are materials that mechanic properties are depended on external magnetic field. They are non-magnetic environment (matrix) including magnetic micro- and nanoparticles.

Magnetic elastomers can be used in engineering field due to their specific properties. For example, they can be used as the framework for synthesis of managed elements that controlling adsorption of impacts. Thus magnetic elastomers are used as adaptive stiffness elements in vibration absorbers [1]. Besides magnetic elastomers are applied in biomedicine [2,3], microelectronics and electrical engineering [4].

This work is aimed to model external magnetic field influence on magnetic elastomer magnetization and deformation.

The modelling of composite elastomers physical properties was calculated within the method of movable cellular automata (MCA).

2. Magnetic elastomer model

Magnetic elastomer is N interacting with each other elements (spherical particles); the particles size is much lower system size. Each system element is movable cellular automata [5,6]. Automata is homogenous system, thus the center of automata mass coincides with its geometrical center. Besides particle-automata describes by set of the physical parameters (density, elastic and shear modules, saturation magnetization and coercive force) that characterize real environments. Some groups of particles can be under special conditions, for example, surface layer (group of particles) is rigidly fixed in initial position while another group is being effected by magnetic or external forces. This approach allows to describe electromagnetic and mechanic properties of composite materials quite simply.

- The state of the automata i is set by its coordinates $\boldsymbol{r}^i$ and velocity $\boldsymbol{v}^i$.

- The communication of pair adjacent automata is determined by overlap parameter $h^{ij} = r^{ij} - r_0^{ij}$, where $r^{ij} = |\boldsymbol{r}^i - \boldsymbol{r}^j|$ and $r_0^{ij} = (d^i + d^j)/2$ are current and equilibrium distances between automata i and j, respectively. Where d^i is the diameter of the i-th automata. If $h^{ij} \leq h_{\max}$, the pair of automata is interconnected, otherwise ($h^{ij} > h_{\max}$) automata are unconnected. Moreover, sign of h_{ij} determines sign of the interaction force between automata.
- It's supposed that N_m automata are magnetic and single domain particles (from total N automata).
- The dynamics of automata is describing by differential equations of translational and rotational movements:

$$m^i \frac{d^2 \boldsymbol{r}^i}{dt^2} = \boldsymbol{F}^i, \quad J^i \frac{d^2 \theta^i}{dt^2} = \boldsymbol{K}^i, \tag{1}$$

where m^i, $\boldsymbol{J}^i$, $\boldsymbol{r}^i$, θ^i are mass, moment of inertia, linear and angular coordinates of automata i, respectively, $\boldsymbol{F}^i$ is resulting force acting on automata i, $\boldsymbol{K}^i$ is moment of resulting force. The resulting force $\boldsymbol{F}^i$ is set by superposition of short-range mechanical forces acting from adjacent automata $\sum_j \boldsymbol{F}^{ij}$, long-range magnetic forces $\boldsymbol{F}^i_{magn}$ and external forces $\boldsymbol{F}^i_{ex}$: $\boldsymbol{F}^i = \sum_j \boldsymbol{F}^{ij} + \boldsymbol{F}^i_{magn} + \boldsymbol{F}^i_{ex}$. In its turn force $\boldsymbol{F}^{ij}$ is set by superposition of central force $\boldsymbol{P}^{ij}$ and viscous force $\boldsymbol{F}^{ij}_v$: $\boldsymbol{F}^{ij} = \boldsymbol{P}^{ij} + \boldsymbol{F}^{ij}_v$.

3. Algorithms of deformation and magnetization calculation

- Numerical integration of differential equations system (1) has been calculated by velocity Verlet algorithm:

$$\begin{aligned}
\boldsymbol{r}^i_{n+1} &= \boldsymbol{r}^i_n + \boldsymbol{v}^i_n \Delta t + \frac{\boldsymbol{F}^i_n}{2m^i} \Delta t^2, \\
\boldsymbol{v}^i_{n+1} &= \boldsymbol{v}^i_n + \frac{\boldsymbol{F}^i_n + \boldsymbol{F}^i_{n+1}}{2m^i} \Delta t, \\
\boldsymbol{\theta}^i_{n+1} &= \boldsymbol{\theta}^i_n + \boldsymbol{\omega}^i_n \Delta t + \frac{\boldsymbol{K}^i_n}{2J^i} \Delta t^2, \\
\boldsymbol{\omega}^i_{n+1} &= \boldsymbol{\omega}^i_n + \frac{\boldsymbol{K}^i_n + \boldsymbol{K}^i_{n+1}}{2J^i} \Delta t,
\end{aligned} \tag{2}$$

  where step size (time step) $\Delta t = d/5c_p$ is determined by automata size d and sound velocity c_p. With next timestep (iteration) automata change their states synchronically, depending on their and adjacent automata states on the previous timestep. It can be accompanied by changes automata relations, position and orientation in space under the influence of external and internal forces, and also change of automata magnetic moments, depending on the resulting magnetic field in the sample.
- According to spherical form of magnetic automata, their magnetic field are dipole ones. Thus magnetic automata i is influenced by external field $\boldsymbol{H}_0$ and, in addition, total field of other magnetic particles-automata:

$$\boldsymbol{H}(\boldsymbol{r}^i) = \boldsymbol{H}_0 + \sum_{j \neq i} \left(-\frac{\boldsymbol{\mu}^i}{(r^{ij})^3} + \frac{3(\boldsymbol{\mu}^i, \boldsymbol{r}^{ij})\boldsymbol{r}^{ij}}{(r^{ij})^3} \right), \tag{3}$$

  so acting on automata i magnetic force $\boldsymbol{F}^i_{\text{magn}} = \nabla\left(\boldsymbol{\mu}^i, \boldsymbol{H}(\boldsymbol{r}^i)\right)$:

$$\boldsymbol{F}^i_{\text{magn}} = \sum_{j \neq i, j=1}^{N_m} 3\left(-\frac{(\boldsymbol{\mu}^i, \boldsymbol{r}^{ij})\boldsymbol{\mu}^j + (\boldsymbol{\mu}^j, \boldsymbol{r}^{ij})\boldsymbol{\mu}^i + (\boldsymbol{\mu}^i, \boldsymbol{\mu}^j)\boldsymbol{r}^{ij}}{r^5_{ij}} + \frac{5(\boldsymbol{\mu}^i, \boldsymbol{r}^{ij})(\boldsymbol{\mu}^j, \boldsymbol{r}^{ij})\boldsymbol{r}^{ij}}{r^7_{ij}} \right). \tag{4}$$

- For calculation of magnetic automata interaction force (4) fields of interaction (3) are recalculated on every integration step. Fields of interaction effect on magnetic moments orientation. However calculation of magnetic moments distribution in field $\boldsymbol{H}(\boldsymbol{r}^i)$ is non-trivial independent problem, in this work it is not considered. Thus it is supposed that
  1) automata magnetic moments can be oriented only along or against certain axis 0y, i.e. automata magnetic moment: $\boldsymbol{\mu}^i = \{0, \mu_y^i, 0\}$;
  2) the magnetic particle, which chosen from the bunch of total magnetic particles, has maximum value of resulting magnetic field module;
  3) if magnetic particle field is more than critical one $(|\boldsymbol{H}(\boldsymbol{r}^i)|_{\max} > h_{cr})$ and magnetic moment of particle $\boldsymbol{\mu}^i$ is directed against the field, then $\boldsymbol{\mu}^i$ will be oriented along the field.

4. Modelling results

Modelling was based on system corresponding environment consisting on 2500 particles-automata (including single domain ferrum particles) with elastic module equals $2.04*10^9$ dyn/cm^2 and Poisson's ratio equals 0.44.

The magnetic elastomer magnetization and deformation dependences on external magnetic field are shown in figure 1 (a) and (b), respectively (inset figures show chaotic distribution of magnetic particles).

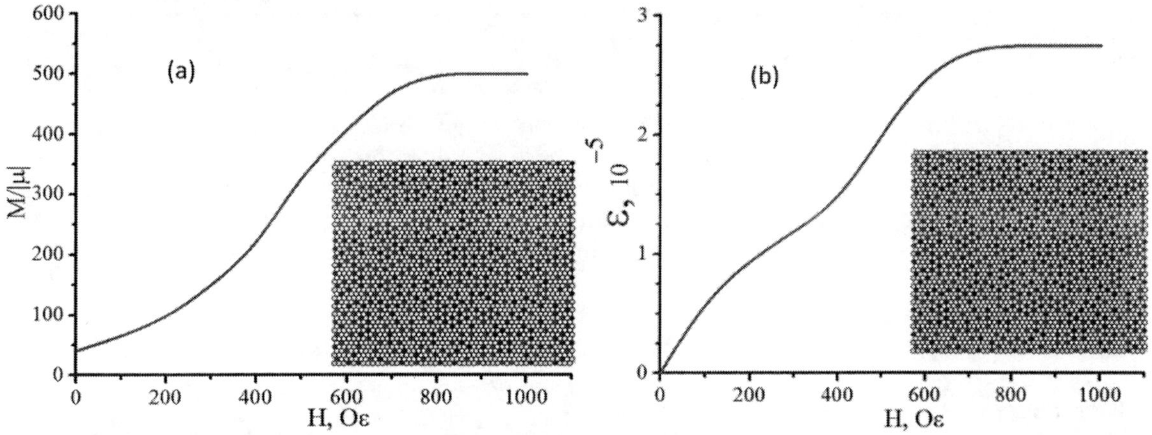

Fig 1. Magnetic elastomer reduced magnetization M/μ (a) and deformation ε (b) dependences on external magnetic field. The model is 2500 automata with 20% single domain ferrum particles. Inset figures show non-magnetic particles-automata (clear) and particles-automata with directed oppositely magnetic moments (colored).

Figure 2 shows saturation deformation of composite dependence on number of magnetic automata and magnetic particles distribution. It is seen that saturation deformation of composite with ordered distribution of particles is much more deformation in system with random distribution.

Thus distribution of magnetic particles in material has great effect on its elastic properties.

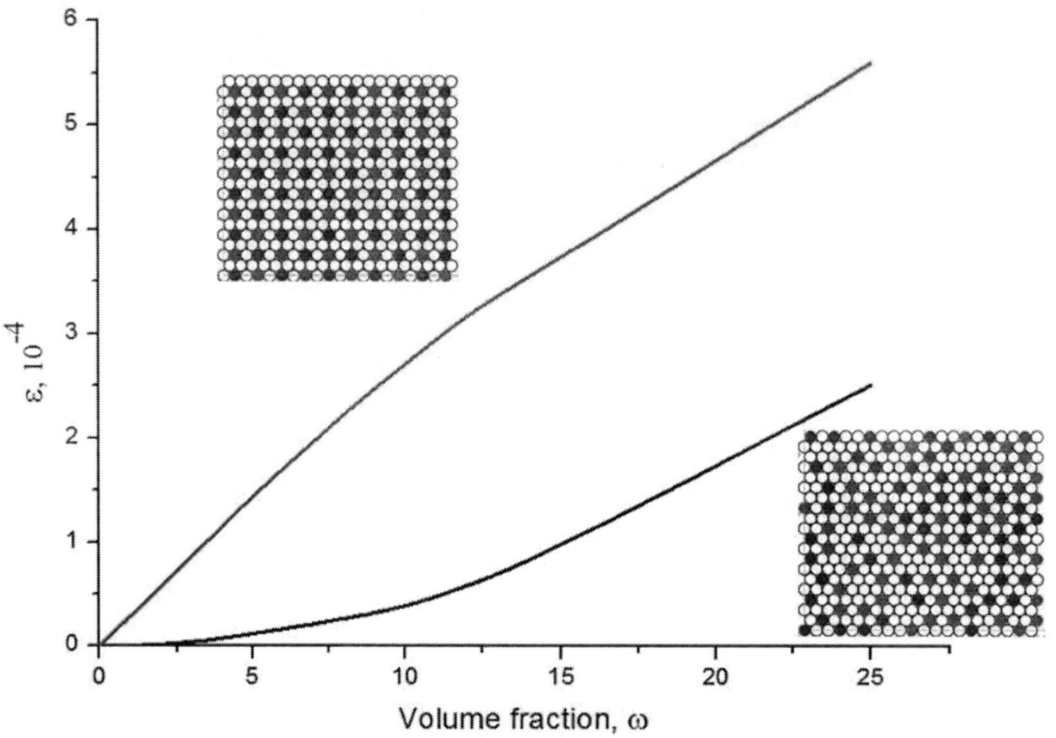

Fig. 2. Saturation deformation ε dependence on volume fraction magnetic automata ω and distribution of magnetic particles. Red curve corresponds to ordered distribution, black curve corresponds to random one. The upper inset figure shows ordered distribution of non-magnetic particles-automata (clear) and particles with oppositely directed magnetic moments (colored). The bottom inset figure shows chaotic distribution of particles-automata.

5. Acknowledgments

The work was supported by financial support in the framework of state task 3.7383.2017/8.9

References

[1] Deng H and Gong X 2007 *Journal of intelligent material systems and structures* **18** 1205-1210
[2] Falk M H and Issels R D 2001 *Int J Hyperthermia* **17** (1) 1–18
[3] Glöckl G, Hergt R, Zeisberger M, Dutz S, Nagel S and Weitschies W 2006 *J Phys Condens Matter* **18** (38) 2935–2949
[4] Zhou G and Wang Q 2005 *Smart Materials and Structures* **14** (4) 1154
[5] Psakhie S G, Ostermeyer G P, Dmitriev A I, Shilko E V, Smolin A Yu and Korostelev S Y 2000 Method of movable cellular automata as a new trend of discrete computational mechanics. I. Theoretical description *Physical Mesomechanics* **3** 5–12
[6] Psakhie S G, Horie Y, Ostermeyer G P, Korostelev S Yu and Smolin A Yu 2001 Movable cellular automata method for simulating materials with mesostructure *Theoretical and Applied Fracture Mechanics* **37** 311-334

Geometry from stomata networks at leaves of the *Ctenanthe oppenheimiana*

Humberto Antunes de Almeida Filho, Jeaneth Machicao and Odemir Martinez Bruno

Scientific Computing Group, Sao Carlos Institute of Physics, University of Sao Paulo, PO Box 369, 13560-970, Sao Carlos, SP, Brazil

E-mail: beto@ursa.ifsc.usp.br

Abstract. We measure the stomatic plasticity at *Ctenanthe oppenheimiana* leaves based on distance between stomata pairs in microscopic images. Theoretical graphs were built according with distance thresholds as connectivity parameter. The theoretical networks based at distances between stomata depends on how the connectivity can vary according with the threshold distances. Plants exposed to extreme light irradiation times from 24 to 4 hours per day, presented discernible relations between stomatic distances. The graphs related to average distances between stomata pairs revealed a powerful tool to predict changes of the geometric distribution of stomata at live *Ctenanthe openheimiana* plants.

1. Introduction

Stomata are cells that act as valves localized at plant leaves and stems performing the exchange of gas with the atmosphere [1]. They control the relation between the carbon gain and water loss to adjust photosynthetic needs with environmental and atmospheric conditions [2]. The plant epidermis is highly plastic during environmental adaptation, thus the number of stomata per leaf area can be highly variable and adapted according to the environmental conditions. [3]. Many works have have measured of the distribution of stomata from the epidermis as parameter to build theoretical graphs, based on distance intervals between pairs of stomates at the plant epidermis [4].

Therefore the distribution of stomata in the epidermis can be a very valuable quantitative parameter in the analysis of plasticity as it provides many details of distance that can change at plant adaptation to certain environmental conditions.

In this study, we attempt to develop computational metrics based on distance between stomata pairs located in the microscopic images from the extant *Ctenanthe oppenheimiana* plants using different conditions of acclimatization to light as environmental change, and developed theoretical graphs as a tool to further characterization of the stomatic networks based on distance between stomata pairs.

2. Methods

2.1. Optic microscopy and plant acclimation

We performed optic microscopy with *Ctenanthe oppenheimiana* leaves from plants artificially adapted at controlled growth chambers. The microscope used was Axio-Lab A1, Zeiss, adapted with cameras axiocam ERc5s, the magnification used was 100× . The plants used at microscopy were adapted with continuous 55% relative humidity, photon flux from $100 \mu mol/m^2 \times s$, temperature $25^o c$ and light photo periods from 24 and 4 hours per day during a period of 60 days.

2.2. Model to build graphs with stomata from microscopic images

For each microscopic image a graph with stomata over the leaf can be modeled. The segmented stomata (s_i) from the leaf image represents a node in the graph. The criterion for establishing the edges is based on the distance between pairs of stomata, where a threshold radius (T_r) determines whether two stomata get connected with each other. The threshold parameter controls the connectivity of the network. For instance, as T_r increases, more connections will be established between the centroids of stomata, and therefore, the density of the network will be higher [4]. Moreover, note that, through this threshold criterion, a great number of networks can be built by modifying the range condition, which allows to analyze several independent networks.

A stomatic network is defined by the adjacency matrix A, which receives values of zero or ones, whether the distance values between pair of stomata $d(s_i, s_j)$ are within the limited range T_r as follows:

$$A_{i,j} = 1, \text{ if } d(s_i, s_j) \leq T_r$$
$$0, \text{otherwise}$$

The construction of a stomata network from microscopic images is illustrated in Figure 1.

A)

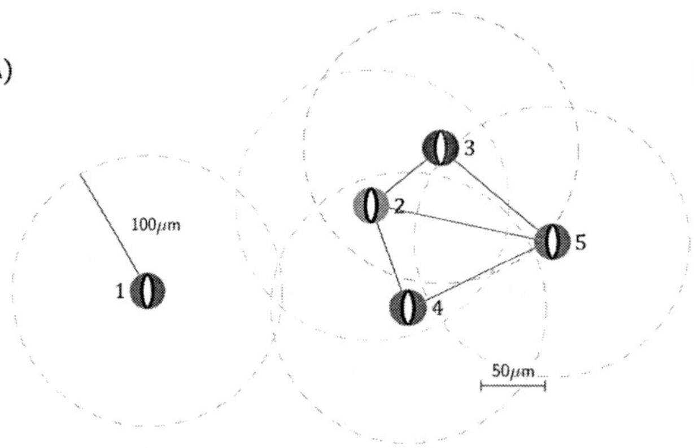

B)

	1	2	3	4	5
1	0	0	0	0	0
2	0	0	1	1	1
3	0	1	0	0	1
4	0	1	0	0	1
5	0	1	1	1	0

Figure 1. Model of a stomatal network. A) The stomata are delimited by a circumference with a given threshold radius of $T_r = 100\ \mu m$. If the distance between stomata pairs is less than or equal to $100 \mu m$, then an edge (filled line) is established between them. Observe that each stomata is at the center of a circumference depicted with same color, which defines the connection radius from one stomata with each other on the network. B) Corresponding adjacency matrix of the stomatic network depicted in A).

2.3. Network parameters

Several measures that characterize the statistics and topology associated with the stomatic network can be calculated as soon as an adjacency matrix of each image is established. The number of stomata per sample is represented by the number of nodes (N), and the number of edges (E) refers to all possible stomata present inside the threshold radius (T_r). The degree or connectivity k_i of a stomata s_i is defined as the number of neighbors that s_i posses, which is given by

$$k_i = \sum_{j=1}^{N} A_{ij}. \tag{1}$$

The connectivity is also usually measured in global terms as

$$\langle k \rangle = \frac{1}{N} \sum_{i=1}^{N} \sum_{j=1}^{N} A_{ij}. \tag{2}$$

that is the mean degree to all stomata of the network in each microscopic image.

3. Results

3.1. Images from extant leaves

Microscopic images collected from fresh leaves of the plant showed an interesting pattern of stomata completely visible on the abaxial epidermis (figure 2). It revealed a contrast between the stomata of the epidermis (green) and the pavement cells (purple), with greenish cellular walls between the pavement cells.

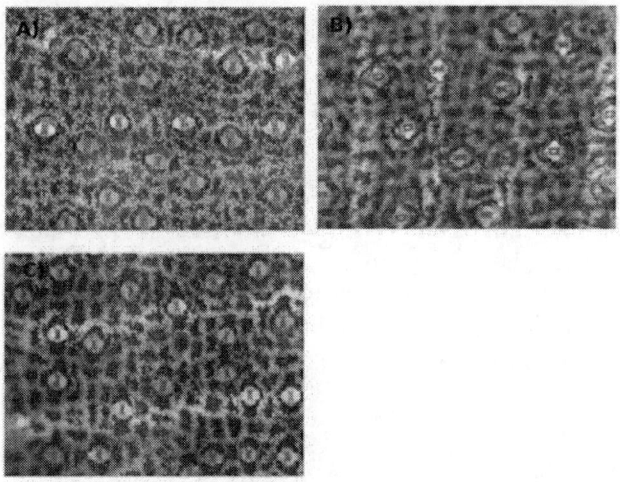

Figure 2. Microscopic images of the abaxial epidermis from plant submitted to photo periods of 24 and 4 hours per day with artificial illumination and solar photo periods. Magnification of 200X. A) Plant exposed to photo period of the 24 hours per day. B) Plant exposed to photo period of the 4 hours per day. C) Plant exposed to solar photo periods. Size scales in red

3.2. Stomatic networks

Distance thresholds (T_r) were used as a condition to construct the graphs, in agreement with what is described in the methods (figure 1). The network connectivity depend on the distance between the stomates in the images and the distances vary according to the stomatic densities. Based on it the network connectivity can be compared according with the thresholds adopted at network percolation.

The number of edges added in the networks of plants exposed to 24h of daily illumination is proportionally greater than in the plants exposed to 4 hours.

The figure below (figure 3) illustrates the relation between the average values of the mean degree $\langle\langle k\rangle\rangle$ in function of network percolation at several distance thresholds (T_r) in 18 plants exposed to 4 and 24 hours of daily illumination.

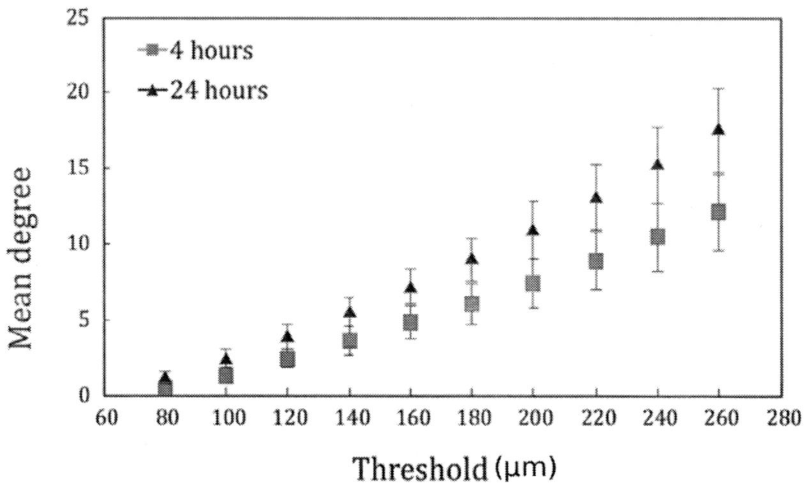

Figure 3. Dispersion of the the mean values of the mean degree $\langle\langle k\rangle\rangle$ in 10 distance thresholds in plants submitted to 4 (grayish square), and 24 hours (dark triangle) of daily illumination, the error bars shows the range of variation relative to the mean values of images of the 18 plants.

4. Conclusions

We showed here that the stomata distribution over the leaf surfaces (*Ctenanthe oppenheimiana*) varies according with the light acclimatization conditions. The quantification of changes at the stomata plasticity over the leaves can be performed by means of graphs based at distances between stomata. Therefore the precise measures of the graph properties at stomatic networks, specially at the *Ctenanthe oppenheimiana*, can be useful as a sensor to environmental changes that plants can be exposed in general.

5. Acknowledgements

Humberto Antunes de Almeida Filho acknowledges support from CNPQ (Conselho Nacional de Desenvolvimento Cientfico e Tecnolgico) (Grants No. 135137/2013-4).

6. References

[1] COWAN, I. R.; TROUGHTON, J. H. The relative role of stomata in transpiration and assimilation. Planta, v. 97, n. 4, p. 325-336, 1971.
[2] KRAMER, Paul J.; BOYER, John S. Water relations of plants and soils. Academic press, 1995.
[3] HETHERINGTON, Alistair M.; WOODWARD, F. Ian. The role of stomata in sensing and driving environmental change. Nature, v. 424, n. 6951, p. 901-908, 2003.
[4] FLORINDO, Joao Batista et al. Analysis of stomata distribution patterns for quantification of the foliar plasticity of tradescantia zebrina. In: Journal of Physics: Conference Series. IOP Publishing, 2015. p. 012113.

QKD optical scheme calibration system

A.Duplinskiy[1,2], V. Ustimchik[1], A. Kanapin[1,3], V. Kurochkin[1] and Y.Kurochkin[1]
[1]Russian Quantum Center (RQC), Business Center «Ural», 100, Novaya st., Skolkovo, Moscow reg., 143025, Russian Federation
[2]Moscow Institute of Physics and Technology, 9 Institutskiy per., Dolgoprudny, Moscow Region, 141700, Russian Federation
[3]Lomonosov Moscow State University, GSP-1, Leninskie Gory, Moscow, 119991, Russian Federation

E-mail: a.duplinsky@rqc.ru

Abstract. One-way quantum key distribution (QKD) optical schemes require active stabilization algorithms for continuous operation. To achieve this a mathematical model for a new optical scheme has been developed. The model uses Jones calculus to describe the polarization state evolution within the optical scheme. The developed algorithms have been successfully tested experimentally.

1. Introduction

QKD is currently the most practical part of the quantum information technology. It allows unconditional secrecy of sharing keys between two distant users – the transmitter (Alice) and the receiver (Bob), the security being guaranteed by fundamental laws of quantum physics. In QKD, Alice and Bob generate a secret key by transmitting information encoded in the states of single photons or weak coherent pulses.

One-way polarization encoding optical scheme that is used in the current work requires polarization state stabilization both within the devices of Alice and Bob and the quantum channel between them [1].

2. Modelling of the scheme

For correct operation all parts of the scheme should apply a correct transformation of the polarization state. To formalize the requirements we use Jones matrixes and vectors of polarization. Note that Jones calculus are the same for both classical and quantum states of polarization. The quantum states that are implemented in the current scheme fulfil the BB84 protocol requirements (Fig. 1) [1]:

$$\begin{aligned}
\psi_1 &= \frac{1}{\sqrt{2}}\left(|\leftrightarrow\rangle + e^{i\varphi_1}\,|\updownarrow\rangle\right) \\
\psi_2 &= \frac{1}{\sqrt{2}}\left(|\leftrightarrow\rangle + e^{i(\varphi_1+\pi)}\,|\updownarrow\rangle\right) \\
\chi_1 &= \frac{1}{\sqrt{2}}\left(|\leftrightarrow\rangle + e^{i(\varphi_1+\pi/2)}\,|\updownarrow\rangle\right) \\
\chi_2 &= \frac{1}{\sqrt{2}}\left(|\leftrightarrow\rangle + e^{i(\varphi_1+3\pi/2)}\,|\updownarrow\rangle\right)
\end{aligned} \tag{1}$$

Note that actually φ_1 does not have to be zero and in general case the states that fit our requirements are two pairs of orthogonal ellipses. This fact significantly simplifies the procedure of guiding the light into the modulator and is exploited within the experimental setup that we present.

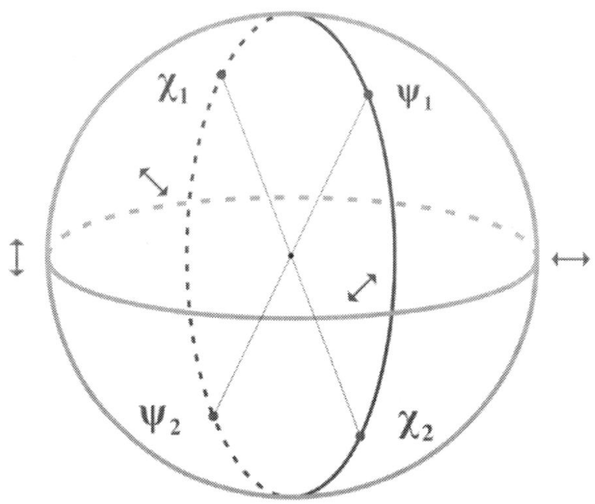

Fig. 1 States generated by Alice on the Poincare sphere.

3. Calibration algorithm

As we cannot directly measure the elements of the Jones matrices for every section, we establish a set of observed values to implement the calibration. We use the counts of the single photon detectors for calibration of all the piezo-driven polarization controllers within the scheme.

For every controller a parameter for minimization is chosen. We went through several iterations of the algorithm for minimization before we chose one that fitted our needs best.

The first version of the minimization process was through scanning the whole range of the parameter while changing voltages applied to the polarization controllers. The minimal value was stored in memory aswell as its position, and after all the values of voltage had been checked, the algorithm started to search for a value close enough to the found minimum. "Close enough" is an allowed percentage of deviance, chosen empirically to balance precision and search time. The problem with this version was that due to the existence of a random noise component in our parameter, and a non-zero search time during which the polarization could drift, the algorithm could often not find a value close enough to the minimum on its second passing (Fig. 2), which resulted in the repetition of the process, often with the same results.

For the same reasons our second version also didn't suit us – we tried returning to the position of the minimum in terms of voltage, ignoring its value. But again, after the search time, the minimum would often shift from its old position, enough that the desired precision required to reach a low error rate would not be met.

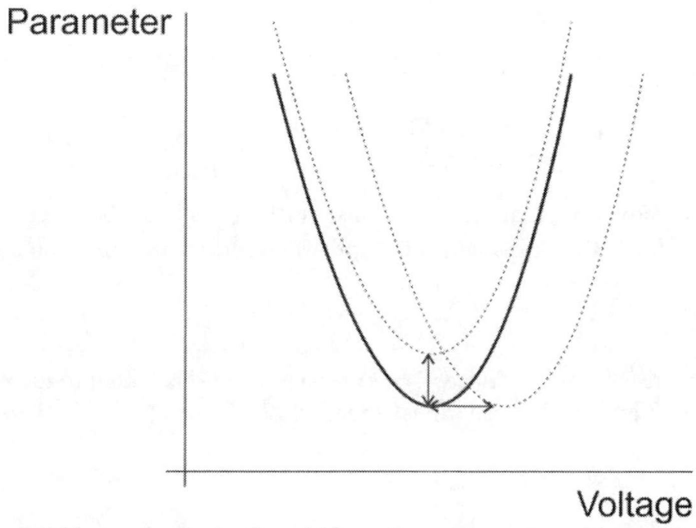

Fig. 2 Effects of polarization drift on the value and position of the minimum.

Finally, we chose the gradient descent approach to find the minimal value. It does not require a second passing over the voltage values, which means that the minimal parameter value is found in real time. But because of the same existing noise components and drift over time, certain problems arose, limiting the stop conditions that we could use to find the minimum effectively.

In gradient descent, the following conditions are used most often to decide when the algorithm should stop: 1) when the change in parameter becomes lower than a certain threshold; 2) when the direction of the search changes repeatedly more than a certain amount of times; 3) a fixed number of steps is applied. Since the change in parameter has a random component, the first and second approaches cannot be applied, since the stop condition might be reached either too early, or it will never trigger even close enough to the minimum. Through experiment we found that option 3 is the most optimal. The algorithm runs a fixed amount of steps on each voltage channel, repeating the process if necessary until the global stop condition (an acceptable QBER) is reached.

4. Requirements for calibration.

Let us briefly summarize the requirements for the SOPs described above:

1) At the input of Alice's phase modulator, the components along ordinary and extraordinary axes are equal
2) At the input of Bob's modulator, the components swap
3) Bob's measurements differentiate BB84 orthogonal states with extinction higher than 98%.

These requirements can be formulated mathematically by defining the Jones matrices for each of the scheme's sections.

The first section connects the laser source and Alice's phase modulator. We assume the incident light to be linearly polarized. To fulfil the criterion 1 the transformation of this section has to be:

$$\frac{\sqrt{2}}{2}\begin{Vmatrix} e^{i\varphi_1} & 1 \\ 1 & e^{-i\varphi_1} \end{Vmatrix} \tag{2}$$

The second section lies between Alice's and Bob's modulators and includes QC. According to the condition 2 the overall transformation of this part should be:

$$\begin{Vmatrix} 0 & 1 \\ e^{i\varphi_2} & 0 \end{Vmatrix} \tag{3}$$

Finally, the condition 3 is implemented by section between Bob's modulator and the PBS:

$$\frac{\sqrt{2}}{2}\begin{Vmatrix} e^{i\varphi_3} & 1 \\ 1 & e^{-i\varphi_3} \end{Vmatrix} \tag{4}$$

All three sections may introduce arbitrary relative phase shifts $(\varphi_1, \varphi_2, \varphi_3)$ between the polarization components, but in order for PBS to distinguish the states correctly their sum should be divisible by 2π:

$$(\varphi_1 + \varphi_2 + \varphi_3) \vdots 2\pi \tag{5}$$

This is required as production of the matrices (2), (3) and (4) should be equal to identity matrix. Note that φ_1 mentioned in sec. 2 is a result of the transformation (2) that is applied in the first section.

5. Conclusions

We have provided the necessary mathematical conditions and algorithm required for the operation of a single-pass polarization-encoding QKD scheme of our own design. The conditions are based on the Jones matrix formalism used to describe the polarization state of light. The algorithm of choice is the gradient descent method, although not all of its versions can be used in experimental conditions, due to the existence of a random noise component in the parameters for minimization and their drift over time.

The stabilization feedback system based on the calibration algorithm proved itself reliable both for laboratory and urban communication channels. The system has also been used as a link within the quantum network [2]. The calibration procedure was performed automatically and maintained QBER below 5% for hours, applying recalibration if needed. Average time that the system has spent in the data transfer mode is about 80%, while the other 20% has been required for recalibrations.

6. Acknowledgements

The support from Ministry of Education and Science of the Russian Federation in the framework of the Federal Program (Agreement 14.582.21.0009, ID RFMEFI58215X0009) is acknowledged.

7. References

[1] Duplinskiy A, Ustimchik V, Kanapin A and Kurochkin Y 2016 *Proc. SPIE* **10224** 102242W
[2] Kiktenko E, Pozhar N, Duplinskiy A, Kanapin A, Sokolov A, Vorobey S, Miller A, Ustimchik V, Anufriev M, Trushechkin A, Yunusov R, Kurochkin V, Kurochkin Y, Fedorov A, "Demonstration of a quantum key distribution network across urban fiber channels", *arXiv:1705.07154*

Modelling of pulsed electron beam induced graphite ablation: Sublimation versus melting

Muddassir Ali and Redhouane Henda

Bharti School of Engineering, Laurentian University, 935 Ramsey Lake Road, Sudbury, ON P3E 2C6, Canada

E-mail: mx1_ali@laurentian.ca

Abstract. Pulsed electron beam ablation (PEBA) has recently emerged as a very promising technique for the deposition of thin films with superior properties. Interaction of the pulsed electron beam with the target material is a complex process, which consists of heating, phase transition, and erosion of a small portion from the target surface. Ablation can be significantly affected by the nature of thermal phenomena taking place at the target surface, with subsequent bearing on the properties, stoichiometry and structure of deposited thin films. A two stage, one-dimensional heat conduction model is presented to describe two different thermal phenomena accounting for interaction of a graphite target with a polyenergetic electron beam. In the first instance, the thermal phenomena are comprised of heating, melting and vaporization of the target surface, while in the second instance the thermal phenomena are described in terms of heating and sublimation of the graphite surface. In this work, the electron beam delivers intense electron pulses of ~100 ns with energies up to 16 keV and an electric current of ~400 A to a graphite target. The temperature distribution, surface recession velocity, ablated mass per unit area, and ablation depth for the graphite target are numerically simulated by the finite element method for each case. Based on calculation findings and available experimental data, ablation appears to occur mainly in the regime of melting and vaporization from the surface.

1. Introduction

Over the past few years, short powerful electron pulses have gained significant attention owing to technologically important applications in various fields ranging from thin film deposition to electron beam lithography [1,2]. Recently, pulsed electron beam ablation (PEBA) has arisen as a relatively novel and potential alternative, due to many superior technical capabilities, to the well-established pulsed laser ablation (PLA) method. In PEBA, electron pulses, instead of pulses of photons, of high current and voltage are employed to ablate the target. Energetic electron pulses during PEBA are generated by a beam source known as channel-spark discharge [2-4]. The resulting electron beam is characterized by short pulse duration (~100 ns) and high power density (10^8 W/cm^2). The beam is directed at a target surface by low-energy electrons (~10-20 keV), penetrating into a target surface. This causes a quick vaporization, as a result of non-equilibrium heating, of a small portion of the target surface and ensuing plasma plume expansion toward a substrate. Accordingly, thin films with superior properties can be deposited at a considerably lower substrate temperature compared to other deposition techniques such as single stage co-evaporation or sputtering [5]. PEBA offers several attractive advantages over PLA, but one of the most distinguishing capabilities is its ability to ablate a wide range of materials, including optically transparent or reflective materials, owing to the electrons shorter absorption depth as compared to photons [4-6].

It is has been demonstrated experimentally that pulsed electron beam ablation can be an efficient technique for the synthesis of thin films, nanostructures and clusters [7-9]. Despite PEBA growing experimental developments, the mechanism of surface ablation during beam-target interaction is still not well understood. It has been well established that ablation induced during nanosecond PLA can typically take place through three types of thermal processes: (i) explosive boiling (phase explosion), (ii) normal boiling and (iii) normal vaporization [10]. Depending on pulse duration and laser intensity, any kind of thermal mechanism can take place during laser ablation. When laser irradiance is in the range of 10^9 W/cm^2, normal vaporization is one of the major dominating thermal processes in PLA [11]. Electron beam ablation of a target surface is a complex process that comprises absorption of pulsed electron beam energy, heating, phase transition and ablation of the target surface. Owing to the short spatial and temporal scales involved in beam-target interaction, ablation is not always easily amenable to experimental studies as in situ characterizations of the ablation process can be quite challenging. Accordingly, due to the inherent limitations and complications involved in the ablation process, mathematical modeling of the phenomena involved during ablation can prove to be an efficient tool to gain a better understanding of electron-beam induced ablation. In the present work, a brief discussion is given of the thermal model used to describe ablation during electron beam interaction with the graphite target. The cases of melting and sublimation of a graphite target are considered and the results are discussed in terms of temperature distribution, surface recession velocity, ablated mass per unit, and ablation depth.

2. Mathematical model

The theoretical description of electron beam interaction with a graphite target is based on the models described in detail elsewhere [12,13]. Target heating upon interaction with a nanosecond pulsed electron beam can be defined in the solid phase by the heat conduction equation in a one-dimension as,

$$\rho_s C_{p_s} \frac{\partial T}{\partial t} = \nabla \cdot (k_s \nabla T) + Q(x,t) \qquad (0 < t \le t_{SM}) \tag{1}$$

where ρ_s, C_{p_s}, and k_s are the solid phase density, specific heat capacity, and thermal conductivity of the target material, respectively. The term Q is the heat source, x is the distance from the target surface along the incident direction, and t_{SM} is the time required by the target surface to attain the sublimation or melting point.

In the case of fluid (vapor/liquid) phase, thermodynamics of the surface target can be defined by introducing the recession velocity in the one-dimensional heat conduction equation [14],

$$\rho_f C_{p_f} \left(\frac{\partial T}{\partial t} - u_s \frac{\partial T}{\partial x}\right) = \nabla \cdot (k_f \nabla T) + Q(x,t) \qquad (t_{SM} < t \le \tau) \tag{2}$$

where ρ_f, C_{p_f}, and k_f are the target material fluid phase density, specific heat capacity, and thermal conductivity, respectively. The terms u_s and τ are the surface recession velocity and duration of the electron beam pulse, respectively.

When vaporization becomes substantial at higher surface temperature, velocity due to surface recession during this stage can be calculated by Hertz-Knudsen equation [14],

$$u_s = (0.82)\frac{P_S}{\rho}\sqrt{\frac{m}{2\pi k_B T_S}} \tag{3}$$

where P_S is the saturated vapor pressure of the ablated particles, k_B is Boltzmann constant, T_S is the surface temperature, and m is the average mass of target particle. The saturated vapor pressure at the target surface can be determined according to the Clausius-Clapeyron equation [14],

$$P_S = P_0 exp\left[\frac{mL_v}{k_B}\left(\frac{1}{T_{SB}} - \frac{1}{T_S}\right)\right] \quad (4)$$

where P_0 and T_{SB} are the normal atmospheric pressure and sublimation/boiling temperature under P_0, respectively. The heat source, Q, term used in this study has been described somewhere else thoroughly in detail [12,13].

3. Results and discussion

Due to the rapid absorption of electron beam irradiation onto the target surface, the target temperature, which is initially at ambient condition (298 K), rises very quickly. The temperature profiles of the target surface calculated at an accelerating voltage of 16 kV and beam efficiency of 0.6 for the cases of melting and sublimation are illustrated for the entire pulse width in figure 1. For both cases, it can be observed that the surface temperature rises slightly quickly initially, reaches a maximum value of ~7700 K, then falls off towards the end of the pulse. The decline in surface temperature is significantly slower than the initial temperature rise. As it can be seen in figure 1, the surface temperature is still higher than the boiling/sublimation temperature at the end of the pulse. This can be most likely explained in light of the polyenergetic character of the electron beam [15,16]. Beam electrons appearing at the end of the pulse convey a power density of ~10^6 W/cm^2 owing to high beam current and short penetration depths in the target, which, in turn, sustains the ablation process. Furthermore, it can be observed from figure 1 that the maximum temperature attained by the surface is well below 0.9T_c (where T_c is the critical thermodynamic temperature) for graphite. Accordingly, calculated surface temperature profiles suggest no indication of phase explosion (explosive boiling) during PEBA. Normal vaporization is the most likely thermal process involved in the ablation of the target surface. Around the two cusps exhibited by the temperature profile, as shown in figure 1, the difference in surface temperature corresponding to melting and sublimation reaches a maximum value of ~110 K.

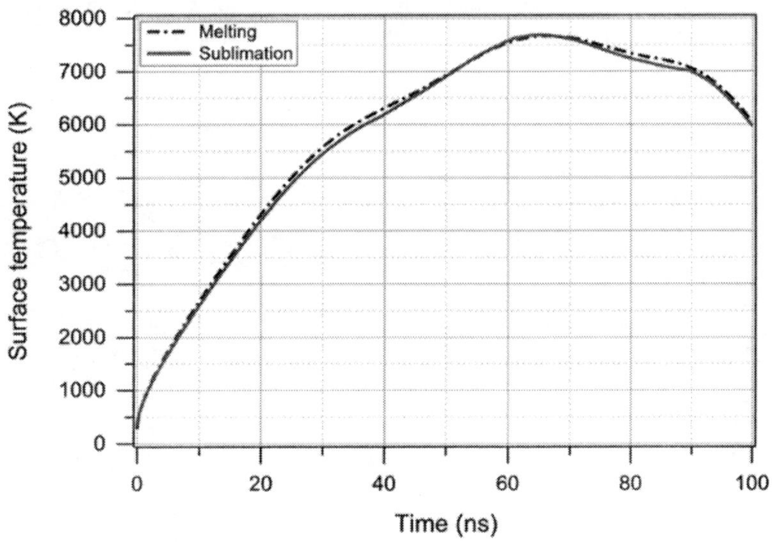

Figure 1. Target surface temperature corresponding to melting and sublimation (indicated in the inset) as a function pulse duration.

Surface recession velocity calculated for the different thermal phenomena for the pulse width of 100 ns is depicted in figure 2. As can be observed, for sublimation the calculated recession velocity from the target surface increases rapidly and reaches a maximum value of 95 m/s at about 66 ns, i.e., when the target surface temperature is at maximum (7700 K). For the case of melting, the recession velocity increases very gradually over the period of the pulse duration and reaches up to 23 m/s at about 66 ns.

Upon reaching the maximum value, and for both cases, it can be observed that the surface recession velocity decreases abruptly until the end of the pulse.

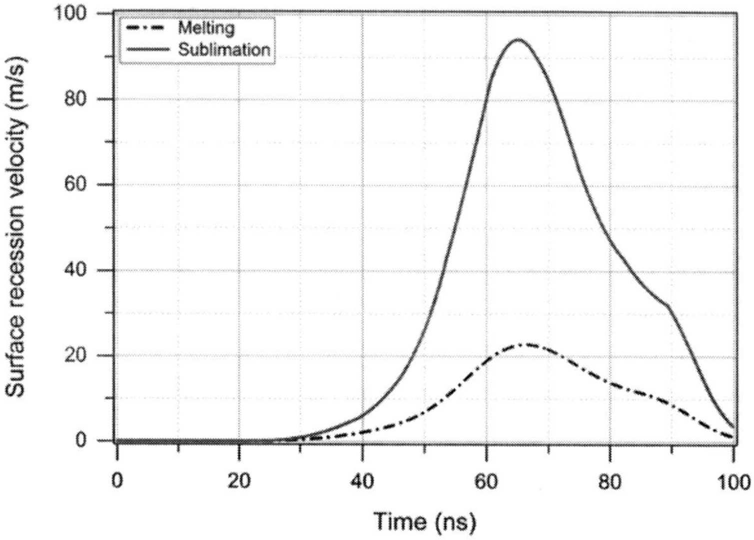

Figure 2. Surface recession velocity profiles corresponding to melting and sublimation as a function pulse duration.

Surface recession velocity can be integrated over the pulse duration to calculate ablated depth and ablation mass per unit area for both thermal cases. These are reported for the entire pulsed width in figures 3 and 4. As a result of the maximum surface recession velocity reaching maximum value for the case of sublimation, it can be expected that the magnitudes of ablation depth and ablated mass per unit area in this case, i.e., sublimation, will be higher than their counterparts in the case of melting, as can be appreciated in figures 3 and 4.

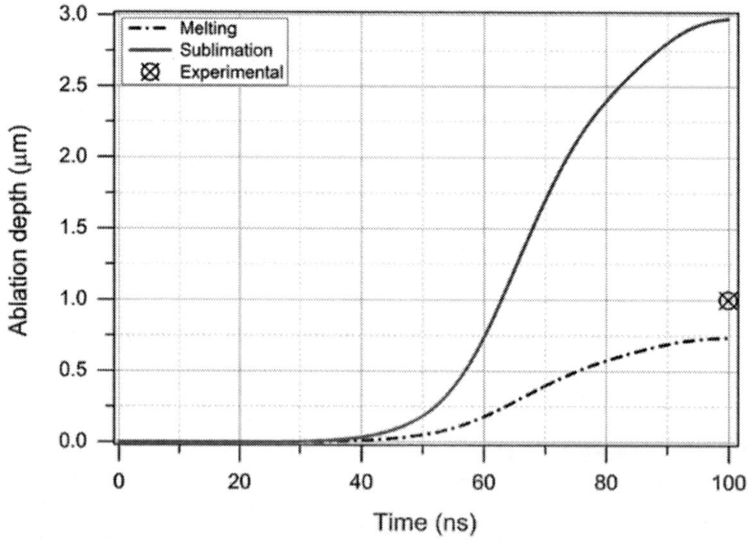

Figure 3. Ablation depth profile corresponding to melting and sublimation as a function pulse duration. The experimental ablation depth is reported for comparison.

4. Model assessment

To assess the validity of the model, appropriate parameters calculated from the model have been contrasted with available experimental results, under the prevailing conditions of this study,_in the literature [5,17-19]. The reported experimental value of the ablation depth has been estimated to be around 1.0 μm [5,17]. This is in good accordance with the ablation depth of 0.75 μm obtained from the model based on melting as illustrated in figure 3. Experimental observations have indicated that the ablated mass per unit area via PEBA is about 5-10 times greater than its counterpart via PLA for a similar energy fluence [5,18]. The reported experimental value of ablated mass per unit area via PLA has been estimated to be around 0.30 $\mu g/mm^2$ for a graphite target [18]. The calculated ablated mass per unit area from the model accounting for melting is about five times greater than the reported experimental value, as illustrated in figure 4. According to the results obtained from the present model calculations and from appropriate previous experimental studies [17,19], target ablation seems to take place primarily in the regime of melting and vaporization from the surface.

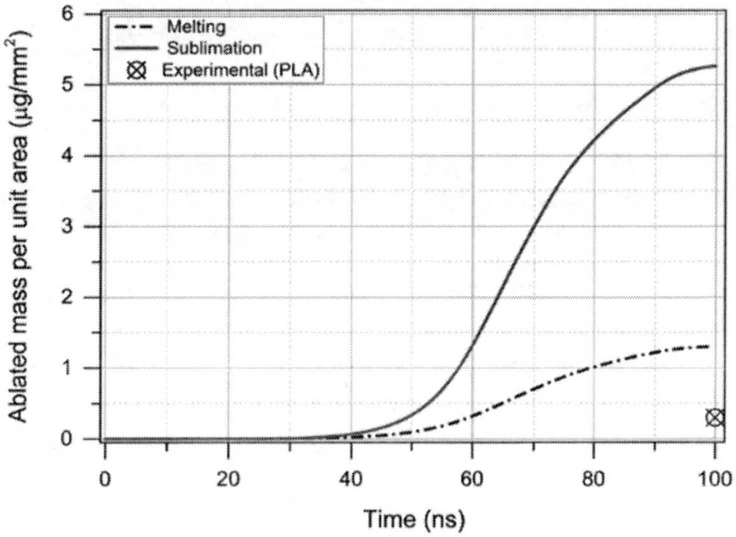

Figure 4. Ablated mass per unit area profile corresponding to melting and sublimation as a function pulse duration. The experimental ablation depth for PLA is reported for comparison.

5. Conclusions

A two stage, one-dimensional heat conduction model is presented to investigate thermal phenomena responsible for ablation in PEBA. A finite element analysis is applied to model equations to estimate the temporal dependence of the surface temperature for a graphite target. Surface recession velocity, ablation depth and ablated mass per unit area are also calculated. Calculation results are compared with available experimental results to assess the validity of the model. It is concluded that the target ablation via PEBA takes place via melting instead of sublimation.

Acknowledgments

The authors would like to acknowledge the financial support received from the Natural Sciences and Engineering Research Council (NSERC) of Canada.

References

[1] Lee P, Feng X, Zhang G X, Liu M H and Lee S 1995 *Plasma Sources Sci. Technol.* **6** 343
[2] Jiang Q B, Matacotta F C, Konijnenberg M C, Müller G and Schultheiss C 1994 *Thin Solid Films* **241** 100

[3] Dediu V I, Jiang Q B, Matacotta F C, Scardi P, Lazzarino M, Nieva G and Civale L 1995 *Supercond. Sci. Technol.* **8** 160
[4] Müller G, Konijnenberg M, Krafft G and Schultheiss C 1995 Thin film deposition by means of pulsed electron beam ablation *Scienc and Technology of Thin Films* ed F C Matacotta and G Ottaviani (Singapore: World Scientific) 89-119
[5] Harshavardhan K S and Strikovski M 2005 Pulsed electron–beam deposition of high temperature superconducting films for coated conductor applications *Second Generation HTS Conductors* ed A. Goyal, (Boston: Kluwer Academic Publishers) 109-133
[6] Gilgenbach R M, Kovaleski S D, Lash J S, Ang L K and Lau Y Y 1999 *IEEE Trans. Plasma Sci.* **27** 150
[7] Henda R, Wilson G, Gray-Munro J, Alshekhli O and McDonald A M 2012 *Thin Solid Films* **520** 1885
[8] Alshekhli O and Henda R 2014 *ECS J. Solid State Sci. Technol.* **3** M21
[9] Ali A, Henda R and Fagerberg R 2017 *Appl. Surf. Sci.* **422** 1082
[10] Kelly R and Miotello A 1996 *Appl. Surf. Sci.* **96-92** 205
[11] Lu Q, Mao S S, Mao X and Russo R E 2002 *Appl. Phys. Lett.* **80** 3072
[12] Ali M and Henda R 2015 *ECS J. Solid State Sci. Technol.* **4** P369
[13] Ali M and Henda R 2017 *Appl. Surf. Sci.* **396** 67
[14] Peterlongo A, Mintello A and Kelly R 1994 *Phys. Rev.* E: *Stat. Nonlinear Soft Matter Phys.* **50** 4716
[15] Kowalewicz R and Redel T 1995 *IEEE Trans. Plasma Sci.* **25** 270
[16] Tricot S, Semmar N, Lebbah L and Boulmer–Leborgne C 2010 *J. Phys. D: Appl. Phys.* **43** 065301
[17] Hobel M, Geerk J, Linker G and Schultheiss C 1990 *Appl. Phys. Lett.* **56** 973
[18] Bulgakova N M and Bulgakov A V 2001 *Appl. Phys. A: Mater. Sci. Process.* **73** 199
[19] Müller G and Schultheiss C 1994 *10th International Conference on High-Power Particle Beams (San Diego)* **2** 833

IOP Conf. Series: Journal of Physics: Conf. Series **936** (2017) 012088 doi:10.1088/1742-6596/936/1/012088

Approximate Bayesian Computation in the estimation of the parameters of the Forbush decrease model

A. Wawrzynczak[1,2], **P. Kopka**[2]

[1]Institute of Computer Sciences, Siedlce University, Poland,
[2]National Centre for Nuclear Research, Poland.

E-mail: `awawrzynczak@uph.edu.pl`, `piotr.kopka@ncbj.gov.pl`

Abstract. Realistic modeling of the complicated phenomena as Forbush decrease of the galactic cosmic ray intensity is a quite challenging task. One aspect is a numerical solution of the Fokker-Planck equation in five-dimensional space (three spatial variables, the time and particles energy). The second difficulty arises from a lack of detailed knowledge about the spatial and time profiles of the parameters responsible for the creation of the Forbush decrease. Among these parameters, the central role plays a diffusion coefficient. Assessment of the correctness of the proposed model can be done only by comparison of the model output with the experimental observations of the galactic cosmic ray intensity. We apply the Approximate Bayesian Computation (ABC) methodology to match the Forbush decrease model to experimental data. The ABC method is becoming increasing exploited for dynamic complex problems in which the likelihood function is costly to compute. The main idea of all ABC methods is to accept samples as an approximate posterior draw if its associated modeled data are close enough to the observed one. In this paper, we present application of the Sequential Monte Carlo Approximate Bayesian Computation algorithm scanning the space of the diffusion coefficient parameters. The proposed algorithm is adopted to create the model of the Forbush decrease observed by the neutron monitors at the Earth in March 2002. The model of the Forbush decrease is based on the stochastic approach to the solution of the Fokker-Planck equation.

1. Introduction and background information

The Forbush decrease (Fd) of galactic cosmic ray (GCR) intensity are identified as a short period decrease (5-12 days) in the stream of the GCR particles reaching the Earth [1]. The Fds occur as an outcome the substantial disturbances in the interplanetary space occurring due to the powerful coronal mass ejecta and solar flares on the Sun. It was shown [2, 3, 4, 5] that Fd of GCR intensity observed by super neutron monitors can provide beneficial information about the structure of the interplanetary magnetic field (IMF) turbulence. In connection with this, the theoretical modeling plays an important role to explain the information obtained from Fds observed by neutron monitors. In general modeling of the GCR transport in the heliosphere is a quite challenging task due to complicated interaction of four main processes: convection by the solar wind, diffusion on irregularities of interplanetary magnetic field (IMF), particle drifts in the non-uniform magnetic field and adiabatic cooling (e.g. [6]). The interplay of these processes represents the Parker transport equation (PTE) [7] being the second order parabolic type partial differential equation.

To compose the complete model of the Fd is complex problem relating with the complex structure of the disturbed vicinity of the interplanetary space responsible for the Fd of the GCR intensity. To take into account the simultaneous changes of the solar wind velocity, structure of the heliospheric neutral sheet (HNS), regular and turbulence components of the IMF in the restricted disturbed vicinity of the interplanetary space is, basically impossible.

In papers [2, 3, 4, 5] was shown that during the Fd is observed the increase of the exponent ν of the power spectral density (PSD) of the components of the IMF ($PSD \propto f^{-\nu}$, where f is the frequency). From the other side according to the quasi-linear theory [8, 9] the diffusion coefficient K is proportional to the rigidity R of the GCR particles as,$K \propto R^{\gamma}$, where $\gamma = 2-\nu$. Thus, during Fd we observe a decrease of the diffusion coefficient due to the increase if the IMF turbulence. However, direct insertion of the exponent ν calculated based on the experimental IMF data is not possible. One reason is that in the model we should consider the spatial distribution of the exponent ν while the IMF data cover only one point of space where the spacecraft recording the IMF is placed. The second is that we observe the increase of ν mainly for the IMF components perpendicular to the field line i.e B_y and B_z, while in the model we should include the overall ν. The third problem is that we cannot calculate the value of ν with arbitrary resolution. The reason is the requirement of employing relatively long data series to attain the values of the exponent ν in low-frequency range $f \in < 10^{-6}, 10^{-5} >$ responsible for the scattering of the GCR particles to which neutron monitor and ground meson telescopes respond. This usually was done (e.g. [2, 3]) based on the comparison of the single data series with length predetermined by the duration of the Fd; i.e. there were calculated the PSDs for three periods, before, during and after the Fd.

In this paper we employ the Approximate Bayesian Computation (ABC) method to compose the model of the Fd best fitted to the Fd recorded by the neutron monitors in the period of 18 March-2 April 2002. (Fig. 1). We assume that Fd occurs due to decrease of the diffusion coefficient K in the vicinity of space where the Fd is observed. The parameter γ responsible for the K decrease will be fitted by the ABC method.

2. Model of the Forbush decrease

We model the transport of GCR in the heliosphere by the stochastic approach. In this approach the individual particle motion is described as a Markov stochastic process, and the system evolves probabilistically. Consequently the Parker equation is brought into a form of the backward Fokker-Planck equation of the form:

$$\frac{\partial f}{\partial t} = A_1\frac{\partial^2 f}{\partial r^2} + A_2\frac{\partial^2 f}{\partial \theta^2} + A_3\frac{\partial^2 f}{\partial \varphi^2} + A_4\frac{\partial^2 f}{\partial r\partial \theta} + A_5\frac{\partial^2 f}{\partial r\partial \varphi} + A_6\frac{\partial^2 f}{\partial \theta\partial \varphi} + A_7\frac{\partial f}{\partial r} + A_8\frac{\partial f}{\partial \theta} + A_9\frac{\partial f}{\partial \varphi} + A_{10}\frac{\partial f}{\partial R} \quad (1)$$

where $f = f(\vec{r}, R, t)$ is an omnidirectional distribution function of three spatial coordinates $\vec{r} = r(r, \theta, \varphi)$, particles rigidity R and time t; The coefficients $A_1, ..., A_{10}$ are dependent on the solar wind velocity $\vec{U}$, the drift velocity components $\vec{v}_d$, and the 3D anisotropic diffusion tensor K_{ij} of GCR particles [11]. Based on the Ito formulation the corresponding set of the stochastic differential equations (SDE) is solved (e.g., [12]). In the case of the GCR transport the set of SDEs in three dimensional heliocentric spherical space have a form

$$\begin{aligned} dr &= A_7 \cdot dt + [B \cdot dW]_r \\ d\theta &= A_8 \cdot dt + [B \cdot dW]_\theta \\ d\varphi &= A_9 \cdot dt + [B \cdot dW]_\varphi \\ dR &= A_{10} \cdot dt, \end{aligned} \quad (2)$$

where $\vec{r}$ is the trajectory of individual pseudoparticle in the phase space and dW_i is the Wiener process. All coefficients and the details of the numerical solution of the set Eq. 2 are given in [13, 14, 15].

In the proposed model we assume that Fd takes place due to established corotating heliolongitudinal disturbances in the interplanetary space. In this region the IMF turbulence is increased resulting in a gradual diminishing of the diffusion at the Earth orbit, causing larger scattering of the GCR particles, and in effect, fewer GCR particles reach the Earth (e.g., [3, 5]). We simulate this process by the gradual decrease and then the increase of the diffusion coefficient at the Earth orbit with respect the heliolongitudes. The diffusion coefficient K_{II} of cosmic ray particles has a form:

$$\begin{aligned} &K_{II} = K_0 \cdot K(r) \cdot K(R,\gamma) \\ &K(r) = 1 + 0.5 \cdot (r/1AU) \\ &K(R,\gamma) = R^{\gamma} \\ &\gamma = \begin{cases} \alpha + \beta sin(\varphi - \zeta); for(r < 30AU) and (90^{\circ} \leq \varphi \leq 270^{\circ}), \\ \alpha; otherwise \end{cases} \end{aligned} \tag{3}$$

The parameters α, β and ζ values are selected by the ABC method so that the Fd model matches the observational amplitude of Fd in March 2002 (see Fig. 1).

3. Approximate Bayesian Computation algorithm setup

Let λ be a parameter vector, given the prior distribution $\pi(\lambda)$. The goal of Bayesian inference is to approximate the posterior distribution, $\pi(\lambda|d_{obs}) \propto \pi(d_{obs}|\lambda)\pi(\lambda)$, where $\pi(d_{obs}|\lambda)$ is the likelihood of λ given the observed data d_{obs}. The main idea of ABC methods is to accept λ as an approximate posterior draw if its associate (modeled) data d is close enough to the observed data d_{obs}. Accepted parameters are a sample from $\pi(\lambda|\rho(d, d_{obs}) < \epsilon)$ where the $\rho(d, d_{obs})$ is the chosen measure of discrepancy, and ϵ is a threshold defining 'closeness margin'. If ϵ is sufficiently small then the distribution $\pi(\lambda|\rho(d, d_{obs}) < \epsilon)$ will be a good approximation for the posterior distribution $\pi(\lambda|d_{obs})$. It is often difficult to define an adequate distance function $\rho(d, d_{obs})$ between the simulated and observed data. So, in many cases it is replaced with a distance defined by summary statistics, $\rho(S(d), S(d_{obs}))$. However, as here we consider the Fd amplitudes in specific time points, we can compare those data directly without a use of summary statistics.

In ABC methods, Sequential Monte Carlo (SMC) is used to automatically, sequentially 'clean' approximation of posterior distribution to be used to generate proposals for further steps. In ABC SMC method, the set of samples with weights, called particles sampled from the prior distribution $\pi(\lambda^1)$, are propagated through a sequence of intermediate posterior distributions $\pi(\lambda^t|\rho(d, d_{obs}) < \epsilon_t)$, $t = 1, ..., T$, until it represents a sample from the target distribution, $\pi(\lambda^T|\rho(d, d_{obs}) < \epsilon_T)$. These methods aim to generate draws from $p(\lambda^t|\rho(d, d_{obs}) < \epsilon_t)$, at each of a series of sequential steps t, where ϵ_t define a series of thresholds. In [16] the authors proposed strategies called ABC SMC with Adaptive Weights (ABC SMC AW). In the first step algorithm 1 initializes the threshold schedule $\epsilon_1 > \epsilon_2 >, ..., > \epsilon_T$. Then N samples are simulated based on the predefined a priori distribution $\pi(\lambda^1)$ and the corresponding acceptance condition $\rho(d, d_{obs}) < \epsilon_1$. Next, the initial uniform weights are calculated. Samples, denoted by a tilde are drawn from the previous generation with probabilities w_j^{t-1}. Using perturbation kernel $K_{\lambda,t}(\lambda_i^t|\tilde{\lambda}_i)$ new 'fresh' samples λ_i^t are obtained, with the veracity of the condition $\rho(d, d_{obs}) < \epsilon_t$. The weights are calculated according to the formula in stage (11); in stage (12) the weights are normalized and the time step is increased $t = t + 1$. This procedure is repeated until $t \leq T$.

In this paper the scanned parameters vector λ is

$$\lambda \equiv (\alpha, \beta, \zeta). \tag{4}$$

with the following priori distribution on particular parameters:

$$\pi(\lambda^1) \equiv \begin{cases} \alpha \sim U(0.7, 1.5), \\ \beta \sim U(0.1, 0.5) \\ \zeta \sim U(70, 100) \end{cases}$$

To estimate the fit of the Fd model to the observations as a distance measure we used normalized approximation error between all the data. This measure is called Fractional Bias (FB) is used to indicate

Algorithm 1 ABC SMC AW

1. Initialize threshold schedule $\epsilon_1 > \epsilon_2 >, ..., > \epsilon_T$
2. Set $t = 1$

for $i = 1$ to N **do**
 3. Simulate $\lambda_i^t \sim \pi(\lambda^t)$ and $d \sim \pi(d|\lambda_i^t)$
 4. Until $\rho(d, d_{obs}) < \epsilon_t$
 5. Set $w_i^t = \frac{1}{N}$

end for

for $t = 2$ to T **do**
 for $i = 1$ to N **do**
  8. Pick $\tilde{\lambda}_i$ from the set $\{\lambda_j^{t-1}\}_{1\leq j\leq N}$ with probabilities $\{v_j^{t-1}\}_{1\leq j\leq N}$
  9. Draw $\lambda_i^t \sim K_{\lambda,t}(\lambda_i^t|\tilde{\lambda}_i)$ and $d \sim \pi(d|\lambda_i^t)$
  10. Until $\rho(d, d_{obs}) < \epsilon_t$
  11. Compute new weights as $w_i^t \propto \frac{\pi(\lambda_i^t)}{\sum_j v_j^{t-1} K_{\lambda,t}(\lambda_i^t|\lambda_j^{(t-1)})}$
  12. Normalize weights w_i^t for $i = 1, ..., N$
 end for

end for

a bias towards underprediction or overprediction of data by the model. We have slightly modified the classical FB measure $\rho(d, d_{obs})$ as:

$$\rho(d, d_{obs}) = \frac{1}{days}\sum_{i=1}^{days}\frac{|d_i - \hat{d}_i|}{|d_i + \hat{d}_i|}, \tag{5}$$

under additional definition, that $\frac{|d_i - \hat{d}_i|}{|d_i + \hat{d}_i|} = 0$ when $d_i = 0$ and $\hat{d}_i = 0$.

The most commonly used adaptive scheme for threshold choice is based on the quantile of the empirical distribution of the distances between the simulated data and observations (e.g. [18]). This method determines ϵ_t in the t time-iteration by sorting the measure $\{\rho(d_{\theta i}^{t-1}, d_{obs})\}_{1<i\leq N}$ and setting ϵ_t such that predetermined percent α_t of the simulated data fulfill the condition $\rho(d, d_{obs}) < \epsilon_t$. The basic problem with this approach is that we are not able to estimate the value of the measure $\rho(d, d_{obs})$ based on the values obtained in iteration $t-1$. In this paper, we propose an efficient adaptive mechanism for setting the α_t. The value of the quantile is judged based on of the previous stage $t-1$, but with three major improvements:

- The distribution of $\{\rho_{\theta i}^{t-1}\}_{i=1,...,N}$ measure of samples set is not discrete. To keep the procedure flexible and support the interval $[0, 1]$ the best choice for continuous distributions seems to be the family of *Beta* distributions.
- The set $\{\rho_{\theta i}^{t-1}\}_{i=1,...,N}$ is extended by adding the values of $\rho(d, d_{obs})$ calculated in step t regardless whether the samples have been accepted or not.
- We propose to define the value of α_t to be dependent on the ratio between a number of currently accepted samples (AS) and the number of all hitherto generated samples (AGS) i.e.: $\alpha_t = 1 - AS/AGS$.

We chose transition kernel $K_{\lambda,t}(\lambda_i^t|\tilde{\lambda}_i)$ to be a three dimensional Gaussian kernel (e.g. [17]). The set of samples from $t-1$ is used to construct a kernel $K_{\lambda,t}(\lambda_i^t|\tilde{\lambda}_i)$, which will describe the probability mass. Additionally only samples keeping the parameter γ in a physically acceptable interval i.e., $0 < \gamma < 2$ are accepted.

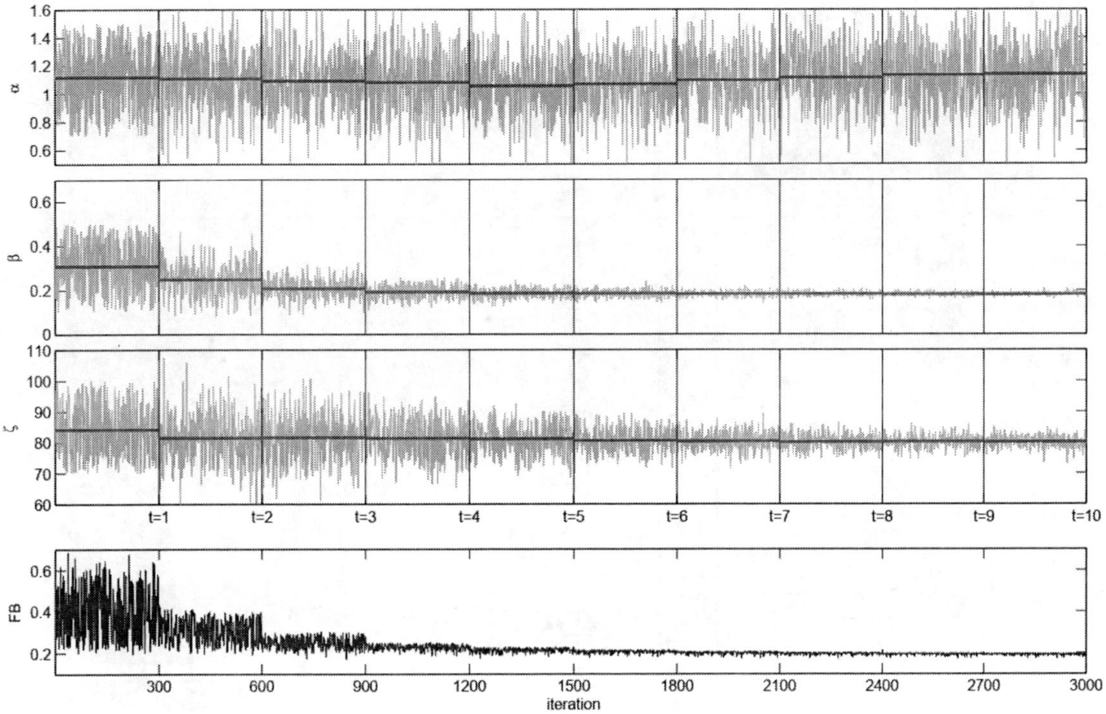

Figure 1. The trace plots for all searched parameters $\lambda \equiv (\alpha, \beta, \zeta)$ in subsequent iterations. Red line marks the mean value within 300 iterations; gray color represents simple trace plot of all samples. The bottom panel presents the FB value for all the accepted samples shown in top three panels.

4. The simulations results and discussion

We have employed the ABC SMC with Adaptive Weights method described in section 3 to find the diffusion coefficient parameters $\lambda \equiv (\alpha, \beta, \zeta)$ matching the Fd observed in March 2002 (Fig. 3). The code for the numerical realization of the described algorithm is realized in Julia environment [19]. The model of GCR stochastic transport in the heliosphere described in section 2 is easy to parallelize versus the number of simulated pseudoparticles using Distributed Arrays Package. This approach fallouts from the assumption that any random process is independent of the other realization, accordingly each pseudoparticles trajectory is independent on another. We performed the simulations applying the ABC SMC AW method setup described in section 3. Each calculation of the FB (Eq. 5) value requires one run of the model of stochastic transport of GCR in the heliosphere represented by the SDEs (Eq. 2). In a single run, the trajectories of $N = 3000$ pseudoparticles were traced backward in time in the spherical heliocentric coordinate system. The pseudoparticles with the rigidity R=10 GV were initialized in the point representing the Earth's orbit and traced backward in time until crossing the heliosphere boundary assumed at 100 AU. Details of the numerical solution of Eqs.2 are given in [13].

Fig. 1 illustrates the trace plots for all searched parameters in subsequent stages of the algorithm. The red line marks the mean value within 300 iterations. One can see, that with iterations the samples are more focused on the average value. This is the best visible in the case of β and ζ parameters. These parameters influence the most the amplitude of Fd. From the other side, we can see that the estimation of the parameter λ is not so unequivocal. The variance of this parameter is the largest. The bottom panel in Fig.1 presents the values of FB for accepted samples throughout the scanning procedure. One can see, that the FB value changes significantly in subsequent stages. In the initial iterations of given time, we observe the meaningful change of FB value. This is connected with adapting the threshold to

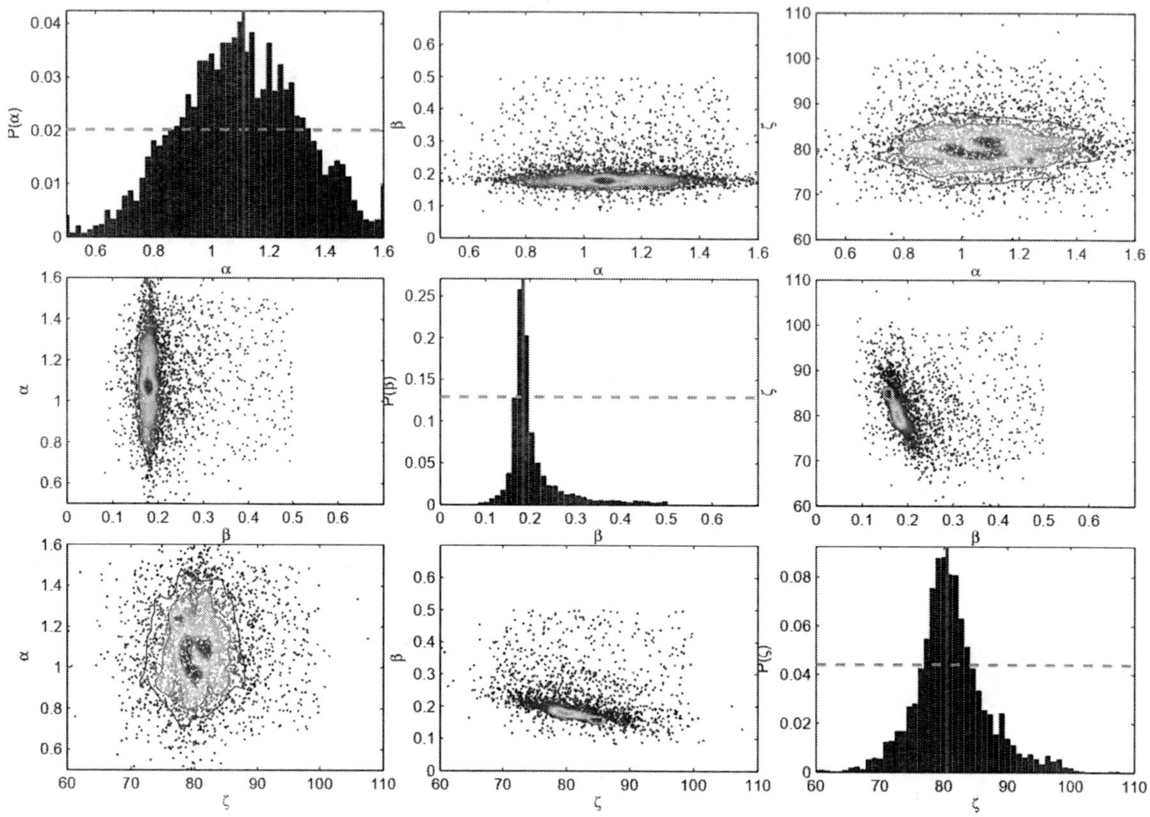

Figure 2. The bivariate and marginal posterior distributions for all searched parameters $\lambda \equiv (\alpha, \beta, \zeta)$. Probability density colors the plot, the more red regions, the higher is its probability. The red vertical lines in diagonal plots mark the value of the parameter with the higher probability. The green horizontal line cuts the value 50% of maximal probability.

the knowledge about the Fd amplitude, i.e., the value ϵ_t is modified according to the Acceptance Ratio (AR) . Starting from 1500 iteration, no significant improvement in FB value is observed. In result, ABC SMC AW algorithm accepts only samples with FB value lower than one estimated in previous stages. This reveals the strength of the proposed procedure, which keeps the Acceptance Ratio at the appropriate level in all stages.

The summarized results of reconstruction procedure are presented in Fig.2 as a trellis plot. Recall that a color pattern reflected in the Fig.2 corresponds to the empirical 2D probability distribution of all parameter combinations. The colored contour lines are enveloping higher probability of the joint posterior distributions. The diagonal plots are marginal empirical posterior distributions of the parameters (α, β, ζ).

The estimated as most probable values of parameters are highlighted with a vertical red line in diagonal subplots. The most probable values are: $P(\alpha = 1.11 \pm 0.01) = 0.0404$, $P(\beta = 0.1845 \pm 0.00635) = 0.2577$, $P(\zeta = 80.4545 \pm 0.45455) = 0.0880$. The green horizontal line cuts the half of the maximal probability distribution for each parameter, which we consider as a high probability region. Thus, as the values with high probability for each considered parameter we consider the interval $\alpha \in < 0.87, 1.35 >$, $\beta \in < 0.1718, 0.1973 >$ and $\zeta \in < 76.8182, 84.0909 >$.

We have incorporated the estimated parameters of diffusion coefficient into the stochastic model of the Fd. The expected amplitudes of the GCR intensity for the rigidity of 10 GV in comparison with the profile of the daily GCR intensity recorded by Apatity and Moscow neutron monitor during 18 March - 2 April presents Fig. 3. The shaded region denotes the amplitudes of the Fd obtained for the border

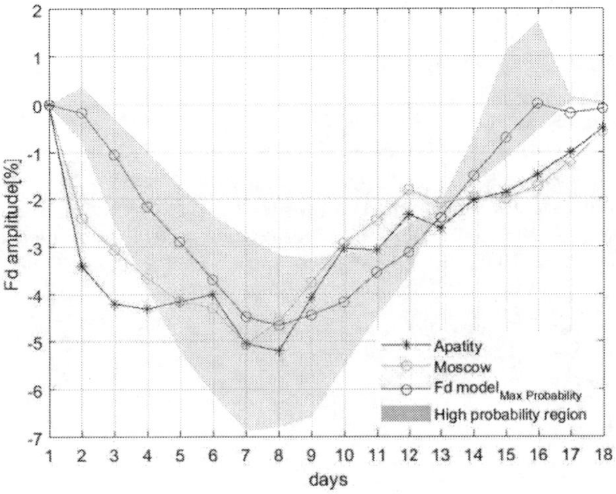

Figure 3. Temporal changes of the observed daily amplitudes of Fd in 18-March - 2 April 2002 for Apatity and Moscow neutron monitors and expected amplitude modeled with use of the parameters vector λ denoted by ABC SMC AW algorithm as the most probable. The shady region marks the expected amplitude for the minimal/maximal parameters indicated by ABC SMC AW with at least 50% of the maximal probability.

values of the high probability region of parameters. One can see that proposed model reconstructs the temporal changes of the GCR intensity observed in March 2002. However, in some periods, the model results do not coincide one to one with the experiment. The reasons are the assumed simplifications. The discrepancy results from the assumption that the Fd is modeled only by the change in diffusion coefficient, while the solar wind velocity was assumed as constant. However, one can see that proposed ABC SMC AW method in conjunction with the stochastic model of GCR transport was able to confirm that as the main cause of the Fd observed in March 2002 we can consider the decrease of the exponent γ causing the change of the diffusion coefficient. These results coincide with the experimental data for Fd in March 2002 presented in [2].

The obtained results confirm that the ABC methodology can be successfully applied to judge about the spatial and temporal changes of the parameters responsible for GCR modulation in the heliosphere. The future work will include testing other FB measures and summary statistics. The energy dependence of the diffusion coefficient and the spatial dependence of the drift velocity will also be fitted.

5. Acknowledgments

This work is supported by The Polish National Science Centre grant awarded by decision number DEC-2012/07/D/ST6/02488. Calculations were performed at the Interdisciplinary Centre for Mathematical and Computational Modelling (ICM) at Warsaw University within the computational grant no. G66-19. We thank the principal investigators of Apatity and Moscow neutron monitor for the ability to use their data.

6. References

[1] Forbush S E 1954 *J. Geophys. Res.* **59** 525
[2] Wawrzynczak A and Alania M V 2008 *Adv. Space Res.* **41** 325-334
[3] Wawrzynczak A and Alania M V 2010 *Adv. Space Res.* **45** 622-631
[4] Alania M V and Wawrzynczak A 2012 *Adv. Space Res.* **50** 6 725-730
[5] Wawrzynczak A and Alania M V 2015 *J. of Phys.: Conference Series* **632** 1742-6596
[6] Moraal H 2013 *Space Science Reviews* **176** 299-319
[7] Parker E N 1965 *Planetary and Space Science* **13** 9-49

IOP Conf. Series: Journal of Physics: Conf. Series **936** (2017) 012088 doi:10.1088/1742-6596/936/1/012088

[8] Jokipii J R 1971*Rev. Geophys. Space Phys.* **9** 21
[9] Toptygin I N 1985 *Cosmic rays in interplanetary magnetic fields*, Reidel Publishing Company, Dordrecht
[10] Potgieter M S 2013 *Space Science Reviews* **176** 165-176
[11] Alania M V 2002 *Acta Physica Pol. B* **33** 4 1149-1166
[12] Gardiner C W 2009 *Handbook of stochastic methods. For physics, chemistry and the natural sciences* Springer Series in Synergetics
[13] Wawrzynczak A, Modzelewska R and Gil A 2015 *J.of Phys.: Conference Series* **574** 012078
[14] Wawrzynczak A, Modzelewska R and Kluczek M 2015 *J. of Phys.: Conference Series* **633** 012058
[15] Wawrzynczak A and Modzelewska R *J. of Phys.: Conference Series* **738** 1
[16] Bonassi F V, West M et al. 2015 *Bayesian Analysis* **10** 1 171187.
[17] Filippi S, et.al. 2013 *Statistical applications in genetics and molecular biology* **12** 1 87107.
[18] Beaumont M A, Zhang W and Balding D J 2002 *Genetics* **162** 4 20252035.
[19] https://julialang.org/

IOP Conf. Series: Journal of Physics: Conf. Series **936** (2017) 012089 doi:10.1088/1742-6596/936/1/012089

An analysis of the influence of production conditions on the development of the microporous structure of the activated carbon fibres using the LBET method

Mirosław Kwiatkowski

AGH University of Science and Technology, Faculty of Energy and Fuels

al. A. Mickiewicza 30, 30-059 Krakow, Poland

kwiatkow@agh.edu.pl

Abstract. The paper presents the results of the research on the application of the new analytical models of multilayer adsorption on heterogeneous surfaces with the unique fast multivariant identification procedure, together called LBET method, as a tool for analysing the microporous structure of the activated carbon fibres obtained from polyacrylonitrile by chemical activation using potassium and sodium hydroxides. The novel LBET method was employed particularly to evaluate the impact of the used activator and the hydroxide to polyacrylonitrile ratio on the obtained microporous structure of the activated carbon fibres.

1. Introduction

At the present time, the activated carbon fibres occupy a major position among others carbonaceous adsorbents due to their specific properties [1-5]. The use of the mentioned materials requires their reliable characterization, which comprises determination of chemical and mechanical properties, description of porosity, surface texture and energetic heterogeneity [6-10]. Moreover, the right choice of the production process conditions for the given raw material which would ensure the production of the activated carbon fibres suitable for the particular adsorptive process requires experiments and comprehensive analysis [11,12]. In the experimental work to yield the anticipated results, mathematical tools have to be used with which the microporous structure of activated carbon fibres obtained under different conditions can be reliably verified. The tool in question should not only enable accurate evaluation of the structural parameters, but also permit an analysis of the microporous structure independently of the adsorbate used and an evaluation of the reliability of the parameters determined. Looking to meet these expectations, a new method with mathematical adsorption models were developed [13-21].

2. The method and materials

The LBET method, which has been described in detail in the previous works [16-22] is based on the original analytical LBET class models of adsorption on heterogeneous surfaces and the fast multivariant identification numerical procedure of adsorption systems. The LBET models have five adjusted parameters: V_{hA} [cm^3/g], Q_{Amax}/RT, α, β and B_C which can be adjusted by fitting LBET equation to the empirical adsorption isotherm, with a chosen variant of the surface energy distribution function [13-21].

The aim of the research was to applying the LBET method with the fast multivariant procedure of adsorption system identification to perform a comparative evaluation of the effect of the activating agent and the ratio of the weight of the activating agent to the weight of the raw material i.e. impregnation ratio R on the parameters of the microporous structure of carbonaceous adsorbents obtained by chemical activation. The calculations were carried out with the use of data taken from literature, namely the isotherms of nitrogen adsorption on high performance carbon fibre Hexcel obtained from polyacrylonitrile [22]. The results of the performed analyses are presented in Tables 1 and 2, as well as Figures 1 and 2.

Table 1. A compilation of the results of adsorption isotherm analyses carried out using LBET method for the activated carbon fibre Hexcel obtained from polyacrylonitrile by activation with NaOH at different impregnation ratios R.

R (wt:wt)	Q_{Amax}/RT	B_C	h	V_{hA} [cm^3/g]	α	β
4:1	-5.65	1.00	3	0.156	0.30	1.00
5:1	-5.57	1.00	3	0.279	0.28	1.00
8:1	-5.70	1.01	3	0.293	0.14	1.15

Table 2. Same as Table 1, but for the activated carbon fibre Hexcel obtained from polyacrylonitrile by activation with KOH.

R (wt:wt)	Q_{Amax}/RT	B_C	h	V_{hA} [cm^3/g]	α	β
4:1	-5.70	1.01	3	0.293	0.14	1.15
5:1	-5.71	1.01	0	0.359	0.15	1.00
8:1	-5.88	7.64	3	0.407	0.38	1.02

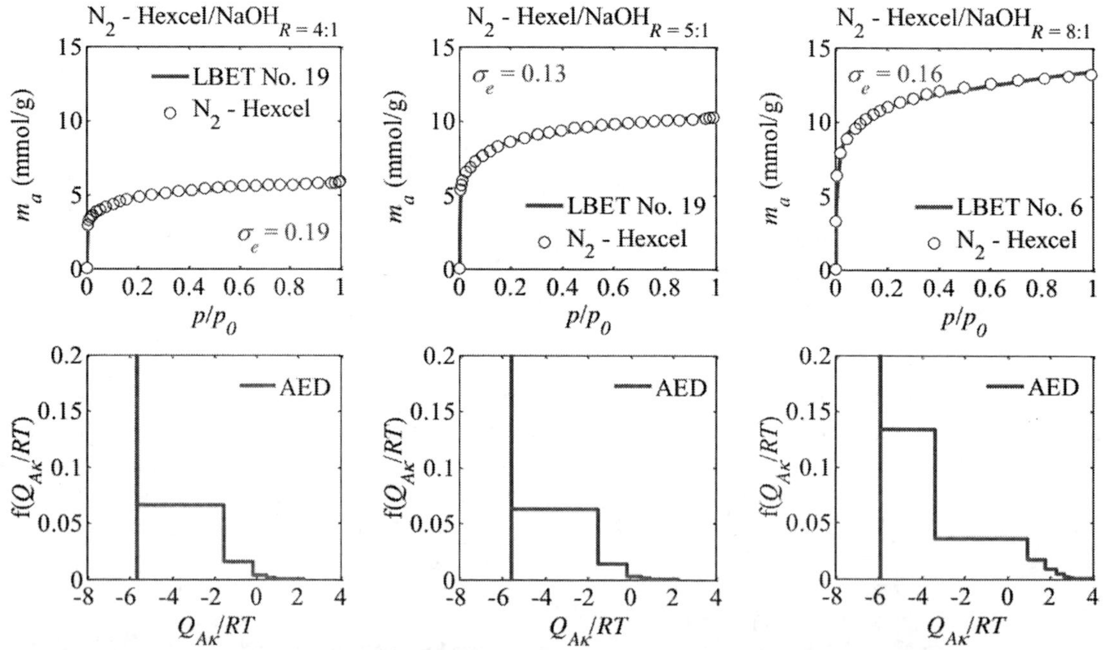

Figure 1. Results of multivariant identification of the nitrogen adsorption isotherms on activated carbon fibres Hexcel obtained from polyacrylonitrile with different impregnation ratios R; AED means adsorption energy distribution on the adsorbent surface.

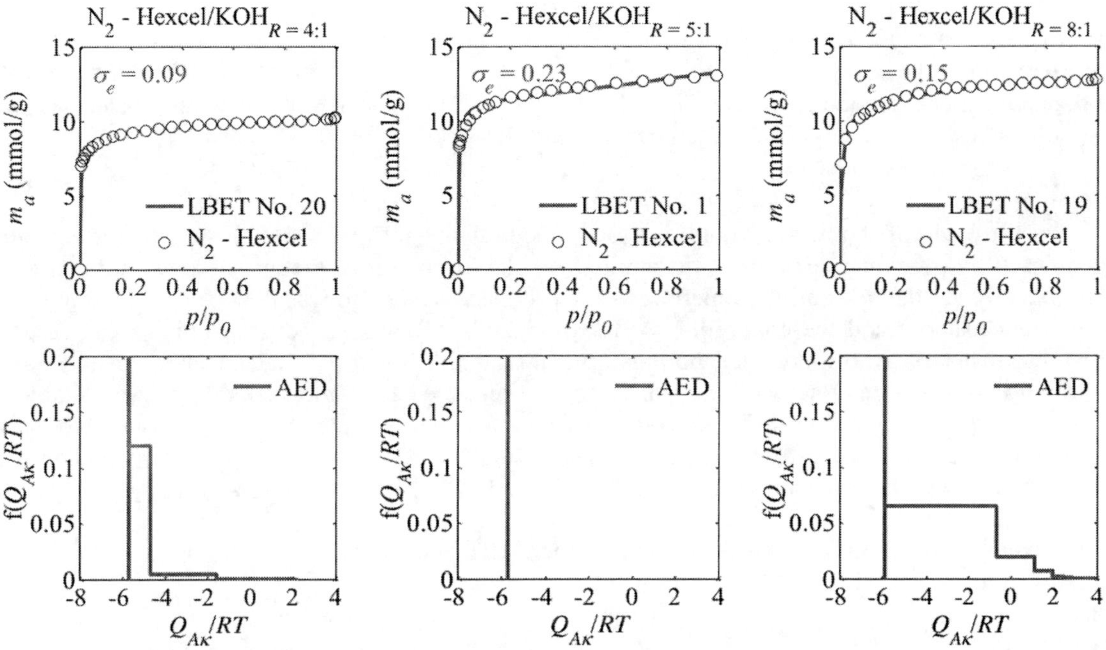

Figure 2. Same as Figure 1, but for the activated carbon fibres Hexcel obtained from polyacrylonitrile by activation with KOH.

3. Discussion of the obtained results

The analysis of the isotherms of adsorption of nitrogen on high performance activated carbon fibres Hexcel obtained from polyacrylonitrile [22] that were produced in the process of chemical activation for different impregnation ratios R with NaOH and KOH as the activating agents yielded interesting results. In the case of activation with potassium hydroxide, microporous carbon fibres with an identical degree of heterogeneity, namely $h = 3$, were obtained for all impregnation ratios. As regards the carbon fibre Hexcel in question, a rise in the ratio of the input raw material to the activating agent, or the impregnation ratio, resulted in an increase of the parameter V_{hA} denoting the volume of the first adsorption layer. Still, more attention should be paid to the changes of the geometrical parameters α and β as compared with the said volume of the first adsorption layer. The NaOH impregnation ratio of 4:1 yielded the geometrical parameters of $\alpha = 0.30$ and $\beta = 1.00$ and $V_{hA} = 0.156$ cm^3/g, which points to the formation of short, stack-like clusters of adsorbate particles in the micropores of the analysed material. It can be concluded from these data that narrow medium-depth micropores were formed in the analysed material during activation. An analysis of the results obtained for the NaOH impregnation ratio of 5:1 shows that the volume of the first adsorption layer is nearly twice as large for comparable geometrical parameters and almost identical distributions of adsorption energy on the first layer. Further increase in the NaOH impregnation ratio to 8:1 resulted in a minimum rise in the parameter V_{hA} with parallel changes to the geometrical parameters α and β, which points to the occurrence of shallow and flaring pores. Interestingly, the sample in question is characterised by a significantly narrower spectrum of energy of primary adsorption sites.

The analysis of the isotherms of adsorption of nitrogen on high performance activated carbon fibres Hexcel obtained from polyacrylonitrile that were produced by way of chemical activation for different impregnation ratios R with KOH as the activating agents also gave interesting results. For the KOH impregnation ratio of 4:1, a microporous material was obtained having the volume of the first adsorption layer $V_{hA} = 0.293$ cm^3/g and the structural parameters $\alpha = 0.14$ and $\beta = 1.15$.

For the KOH impregnation ratio of 5:1 a homogeneous material was produced whose geometrical parameters pointed to the formation of short, stack-like clusters of adsorbate molecules in the microporous structure of the said material, hence the conclusion that narrow pores with relatively small depth as compared with the volume of the adsorbate molecule occur in the structure of the analysed material. Interestingly, further rise in the impregnation ratio to 8:1 resulted in further increasing the volume of the first adsorption layer to $V_{hA} = 0.407$ cm^3/g and doubling the value of the geometrical parameter α denoting the height of the clusters of adsorbate molecules with a simultaneous increase of surface heterogeneity ($h = 3$).

4. Conclusions

On the basis of the research and analyses, a significant influence of the type of the activating agent used and the impregnation ratio applied on the formation of the microporous structure and, consequently, on the adsorptive properties of the produced adsorbate was observed. The right choice of the raw material and the conditions for the production of carbonaceous adsorbents can result in obtaining materials with a complex porous structure suitable for a particular technological process, provided that the optimal activating agent is applied and the preparation conditions are suitable for the raw material in question. Still, finding the right raw material, activating agent and production conditions requires detailed research, which cannot be carried out without access to precision tools with which to evaluate the microporous structure formed.

Acknowledgment: The research is led within the AGH University of Science and Technology in Krakow grant No. 11.11.210.373.

References:

[1] Alcañiz-Monge J, De La Casa-Lillo MA, Cazorla-Amorós D, Linares-Solano A 1997 *Carbon* **35** 291
[2] Fuertes AB, Marbán G, Nevskaia DM 2003 *Carbon* **41** 87
[3] Kunowsky M, Weinberger B, Lamari Darkrim F, Suárez García F, Cazorla-Amorós D, Linares-Solano A 2008 *International Journal of Hydrogen Energy* **33** 3091
[4] Vasiliev LL, Kanonchik LE 2010 *Chemical Engineering Science* **65** 2586
[5] Alcañiz-Monge J, Pérez-Cadenas M, Marco-Lozar JP 2012 *Adsorption Science and Technology* **30** 473
[6] Alcañiz-Monge J, Román-Martínez MC 2012 *Separation and Purification Technology* **85** 83
[7] Balanay JAG, Lungu CT 2012 *Adsorption Science and Technology* **30** 355
[8] Gliścińska E, Babeł K 2013 *Fibres and Textiles in Eastern Europe* **99** 42
[9] Vilaplana-Ortego E, Alcañiz-Monge J, Cazorla-Amorós D, Linares-Solano A 2002 *Fuel Processing Technology* **77-78** 445
[10] Maciá-Agulló JA, Moore BC, Cazorla-Amorós D, Linares-Solano A 2004 *Carbon* **42** 1367
[11] Carrott PJM, Nabais JMV, Ribeiro Carrott MML, Pajares JA 2001 *Carbon* **39** 1543
[12] Nahil MA, Williams PT 2011 *Journal of Analytical and Applied Pyrolysis* **91** 67
[13] Kwiatkowski M 2007 Langmuir 23 2569
[14] Kwiatkowski M 2007 Colloids and Surfaces A: Physicochemical and Engineering Aspects **294** 92
[15] Kwiatkowski M 2008 *Journal of Molecular Modeling* **14** 183
[16] Kwiatkowski M 2009 *Journal of Colloid and Interface Science* **330** 266
[17] Kwiatkowski M 2011 *Applied Surface Science* **257** 8912
[18] Kwiatkowski M, Policicchio A, Seredych M, Bandosz TJ 2016 *Carbon* **98** 250
[19] Kwiatkowski M, Kalderis D, Diamadopoulos E 2017 *Journal of Physics and Chemistry of Solids* **105** 81
[20] Kwiatkowski M, Fierro V, Celzard A 2017 *Journal of Colloid Interface Science* **486** 277
[21] Kwiatkowski M, Sreńscek-Nazzal J, Michalkiewicz B 2017 *Adsorption* **23** 551
[22] Maciá-Agulló JA, Moore BC, Cazorla-Amorós D, Linares-Solano A (2007) *Microporous and Mesoporous Materials* **101** 397

An analysis of burn-off impact on the structure microporous of activated carbons formation

Mirosław Kwiatkowski

AGH University of Science and Technology, Faculty of Energy and Fuels

al. A. Mickiewicza 30, 30-059 Krakow, Poland

kwiatkow@agh.edu.pl

Türkan Kopac

Department of Chemistry, Bülent Ecevit University , 67100 Zonguldak, Turkey

Abstract. The paper presents the results on the application of the LBET numerical method as a tool for analysis of the microporous structure of activated carbons obtained from a bituminous coal. The LBET method was employed particularly to evaluate the impact of the burn-off on the obtained microporous structure parameters of activated carbons.

1. Introduction

Microporous carbonaceous materials are widely used as adsorbents, filters, catalysts supports in industries technology [1-4]. In recent years, an increasing attention has been paid to the research into using activated carbon for the storage of gaseous fuels, including natural gas [5-7]. As has been concluded from the research, under specific conditions and with the same storage pressure, the volume of natural gas in a container filled with adsorbent can be even several times higher than the volume of that gas in a container without adsorbent. Consequently, the use of adsorbents for the storage of natural gas enables a reduction in the weight and size of gas containers and the compression of gaseous fuels at a reduced pressure. Another innovative method of energy storage in movable systems consists in hydrogen electrosorption by porous carbon cathode in the process of controlled water electrolysis at ambient temperature and atmospheric pressure [5-7]. Research results obtained so far permit a conclusion that the efficiency of hydrogen stored with the use of this method can exceed the one offered by high-pressure or cryogenic systems.

Since a few years ago, porous carbonaceous materials have been considered for use as the electrode of the electrochemical condenser in which electric charge is stored on the surface of polarised carbon electrode [8-10]. Work has been under way to use such systems to enhance the performance of fuel cells in electrically powered vehicles.

Another promising trend in the large-scale application of carbonaceous adsorbents is the use of carbon molecular sieves for Pressure Swing Adsorption (PSA), which takes advantage of changes in adsorptive properties at varying pressure and consists in repeated fast transient adsorption/desorption from adsorbent bed [11]. Through the suitable control of pressure and directions of the flow of gas through several adsorbent beds combined together, an effect of considerable gaseous mixture separation can be achieved, maintaining high efficiency of the production process. The PSA method is used in the processes of separating various gaseous mixtures with the use of carbon molecular sieves, including the separation of hydrogen from coal gas or refinery gases, the separation of air into particular components, the separation of carbon dioxide from biogas, the drying of gases, and the enrichment of methane from mine gases.

The middle of the 20th century saw a growing interest in hard coal, or anthracite, and bituminous coal as a raw material in the production of activated carbons. Coal is the most popular precursor owing to its cost-effectiveness and to better mechanical properties of the adsorbents obtained from coal compared to those produced with the use of lignocellulosic materials [12,13]. Interesting research results have been presented in a paper by Carrasco-Marin et al. [14], where bituminous coal from Puertollano (Spain) demineralized with HCl and HF was used to prepare microporous activated carbons. The demineralized coal was pyrolysed in an N_2 flow at 1,273 K for 30 min. and activated in a CO_2 flow at 1,123 K for different periods of time to achieve different degrees of activation [14]. The activated carbons are referred to in the text as BP followed by a number indicating the percentage burn-off. The particle size of the activated carbons was between 0.15 and 0.25 mm. The development of the microporosity was followed among others by adsorption of N_2 at 77 K.

The microporosity of the activated carbons was evaluated by the Dubinin-Astakhov (DA) [15], equation applied to the N_2 adsorption isotherms. The Dubinin-Astakhov equation (DA) [15], was more suitable in that case than the Dubinin- Radushkevitch equation [16] because of considerable surface heterogeneity of the activated carbons in question; however, this equation may be insufficient in a number of contexts to accurately determine the impact of burn-off on the formation of the porous structure of activated carbons. Considering the above, an idea has been conceived to analyze the structure of the said activated carbons on the basis of nitrogen adsorption isotherms using a more advanced method of structure analysis, namely the LBET method [17-25].

2. The LBET method

The LBET method, which has been described in detail in the previous works [17-25], was used for the calculations performed as part of this research. The LBET method is based on the original numerical LBET models of adsorption on heterogeneous surfaces, originating from an unique multilayer adsorption theory and the fast multivariant identification numerical procedure of adsorption systems [17-25]. The LBET models have five adjusted parameters: V_{hA} [cm^3/g], $Q_{Amax}/RT, \alpha, \beta$ and B_C which can be adjusted by fitting LBET equation to the empirical adsorption isotherm, with a chosen variant of the surface energy distribution function [17-25].

3 . Discussion of the obtained results

The results of the numerical analyses carried out for the isotherms of adsorption of nitrogen on the active carbons obtained for the various burn-off were presented in Figures 1–5and Table 1. In the top right-hand figure, the grey dots '•' mark the spots of the fitted empirical adsorption isotherm, while the black solid line '-' marks the theoretical isotherm plotted according to the best-fitted variant of the LBET class model. The dotted line represents the theoretical coverage of the first adsorption layer.

The heading of the top figure gives the name of the adsorption system, i.e. the name of the adsorbate (N_2) and the symbol of the adsorbent (BP). Next the name and number of the best-fitted LBET class model is given, preceded by a slash. The area of the top chart contains the parameters Q_{Amax}/RT, B_C, Z_A, h and the fitting error dispersion σ_e. The second figure presents adsorption energy distribution on the adsorbent surface; it contains the value of the volume of monolayer V_{hA} [cm^3/g] and the values of the geometrical parameters α, β.

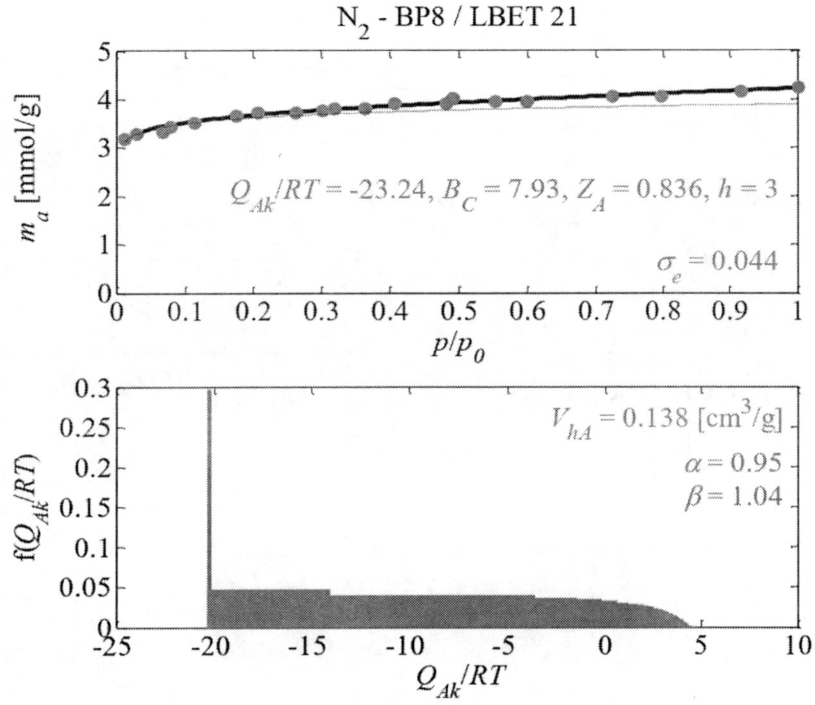

Fig. 1. The results of the nitrogen adsorption isotherm analyses on activated carbon BP, obtained with burn off 8% and adsorption energy distribution on first layers.

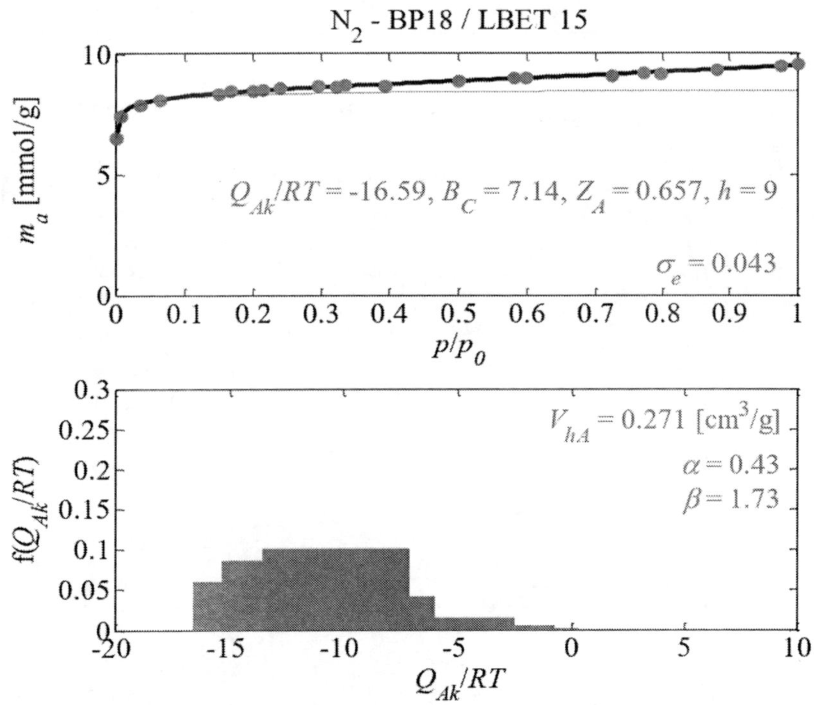

Fig. 2. The results of the nitrogen adsorption isotherm analyses on activated carbon BP, obtained with burn off 18% and adsorption energy distribution on first layers.

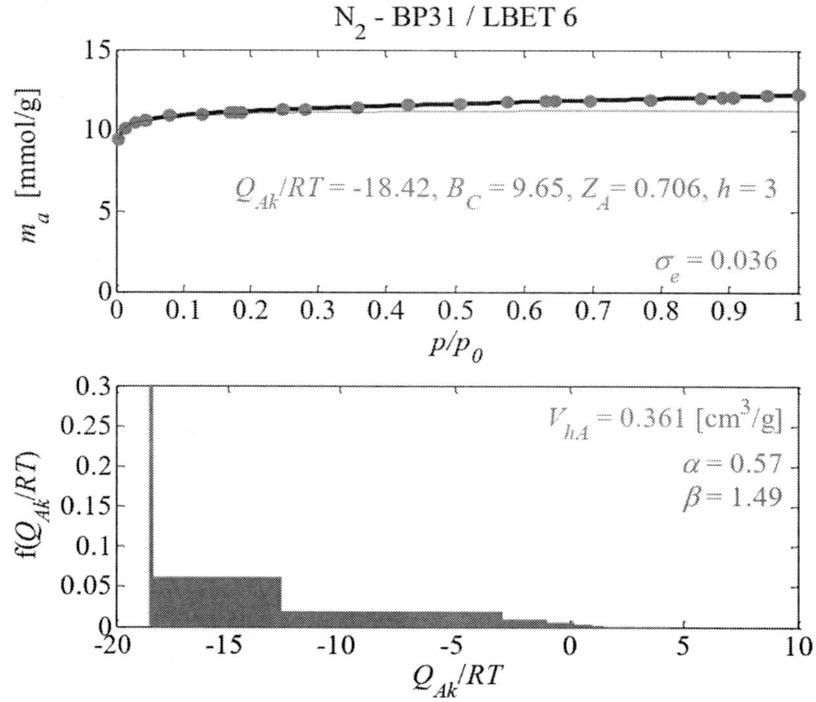

Fig. 3. The results of the nitrogen adsorption isotherm analyses on activated carbon BP, obtained with burn off 31% and adsorption energy distribution on first layers.

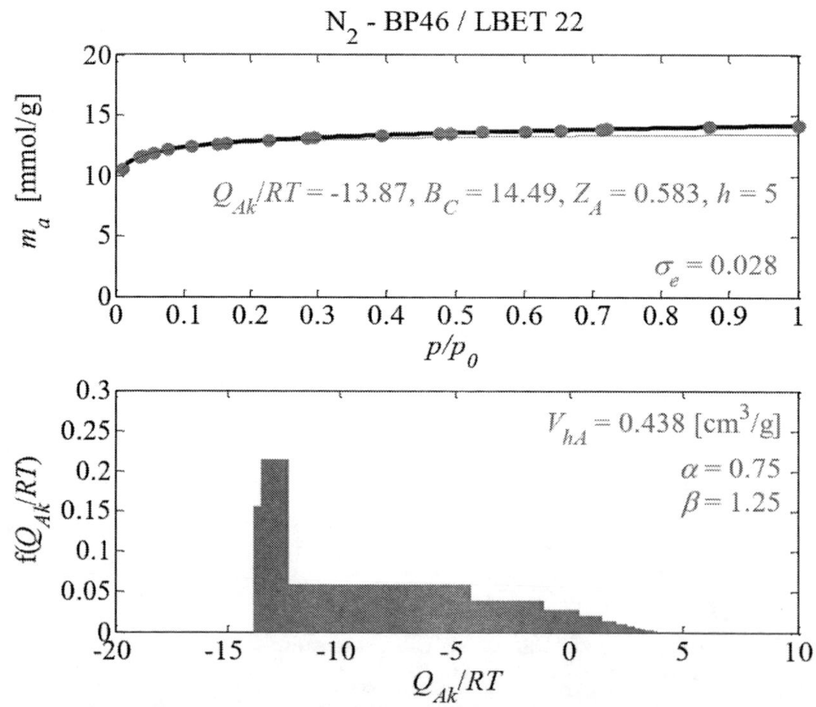

Fig. 4. The results of the nitrogen adsorption isotherm analyses on activated carbon BP, obtained with burn off 46% and adsorption energy distribution on first layers.

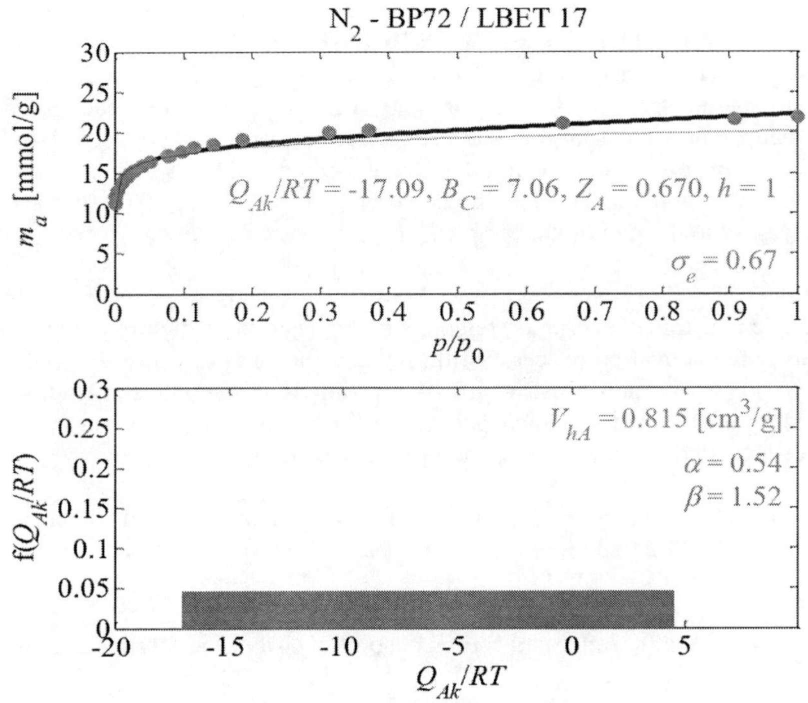

Fig. 5. The results of the nitrogen adsorption isotherm analyses on activated carbon BP, obtained with burn off 72% and adsorption energy distribution on first layers.

Table 1. A selection of results of the analyses carried out using the LBET method and by means of the BET and the DA equations [14].

	BP8	BP18	BP31	BP46	BP72
h	3	9	3	5	1
V_{hA} [cm^3/g]	0.138	0.271	0.361	0.438	0.815
α	0.95	0.43	0.57	0.75	0.54
β	1.04	1.73	1.49	1.25	1.52
S_{BET} [m^2/g]	306	697	924	1078	1632
W_0 [kJ/mol]	0.140	0.309	0.414	0.487	0.742

4. Discussion of the obtained results

Based on the research results compiled in Fig. 1–5 and in Table 1, it has been observed that the volume of the first adsorbed layer V_{hA} increases in proportion to a growth in the degree of burn-off; however, the values of the energetic and geometric parameters do not correlate with burn-off, implying that complex mechanisms take place during the activation process. More specifically, it has been evidenced with regard to the activated carbon BP8 that the surface of this material is moderately heterogeneous, with very high and unbranched clusters of adsorbate particles developing in its pores.

The results produced for the sample of BP18 reveal that the surface of this material is strongly heterogeneous, and the clusters formed in its pores are medium-sized and branched. This finding shows that an increase in the degree of burn-off from 8 to 18% resulted in the burning of the walls between adjoining micropores. Moreover, an increase in the degree of burn-off to 31% brought about further growth of V_{hA} concurrently with a significant decrease in surface heterogeneity and an increase in the height of the adsorbate particle clusters formed in the micropores. An increase in burn-off to 46% triggered further increase in the V_{hA} parameter in parallel to an increase in the h parameter. The adsorbate particle clusters formed in the micropores of the activated carbon BP46 were higher and thinner compared with those developed in the micropores of the activated carbon BP31.

The results obtained for the activated carbon BP72 showed that the value of the V_{hA} parameter of this material was the largest one compared with the remaining analyzed samples, while the value of the heterogeneity parameter h evidenced that the surface of this activated carbon was the most homogeneous one among the samples subject to the analysis. As concluded from a comparison between the V_{hA} parameters obtained using the LBET method and the S_{BET} and W_0 parameters obtained by means of the BET and the DA equations, respectively, the values of the said parameters are correlated with regard to the individual samples.

5. Conclusions

The research demonstrated a complex relationship between the structure parameters and the degree of burn-off during the activation process. Still, the tests carried out using the LBET method yielded precise information on the porous structure of the individual samples, including on their surface heterogeneity and pore geometry, which permit accurate determination of preparation conditions in order to produce activated carbons having anticipated properties.

Acknowledgment: The research is led within the AGH University of Science and Technology In Krakow grant No. 11.11.210.373.

References:

[1] Bansal RC, Goyal M 2005 Activated carbon adsorption. Taylor and Francis. London 1-3
[2] Sreńscek-Nazzal J, Michalkiewicz B 2011 *Polish Journal of Chemical Technology* **13** 63
[3] Młodzik J, Sreńscek-Nazzal J, Narkiewicz U, Morawski AW, Wróbel RJ, Michalkiewicz B 2016 *Acta. Phys. Pol. A* **129** 402
[4] Ahmed MB, Zhou JL, Ngo HH, Guo W, Chen M 2016 *Bioresource technology* **214** 836
[5] Menon VC, Komarneni S 1998 *Journal of Porous Materials* **5**, 43
[6] Policicchio A, Maccallini E, Agostino RG, Ciuchi F, Aloise A, Giordano G 2013 *Fuel* **104** 813
[7] Antoniou MK, Diamanti EK, Enotiadis A, Policicchio A, Dimos K, Ciuchi F, Maccallini E, Gournis D, Agostino RG 2014 *Microporous and Mesoporous Materials* **188** 16
[8] Kita H, Ye S, Aramata A, Fruya N 1990 *Journal of Electroanalytical Chemistry* **295** 317
[9] Jurewicz K, Frackowiak E, Béguin F 2002 *Fuel Processing Technology* **77-78** 213
[10] Kierzek K, Frąckowiak E, Lota G, Gryglewicz G, Machnikowski 2004 *Electrochimica Acta* **49** 515
[11] Ko D, Siriwardane R, Biegler LT 2005 *Industrial & Engineering Chemistry Rersearch* **44** 8084
[12] Kopac T, Toprak A 2007 *International Journal of Hydrogen Energy* **32** 5005
[13] Kopac T, Toprak A 2009 Korean *Journal of Chemical Engineering* **26** 1700
[14] Carrasco-Marin F, Alvarez-Merino MA, Moreno-Castilla C, 1996 *Fuel* **75** 966
[15] Dubinin MM, Astakhov VA 1971 *Advances in Chemistry* **102** 69
[16] Dubinin MM 1985 *Carbon* **23** 373
[17] Kwiatkowski M 2007 *Langmuir* **23** 2569
[18] Kwiatkowski M 2008 *Journal of Molecular Modeling* **14** 183
[19] Kwiatkowski M 2009 *Journal of Colloid and Interface Science* **330** 266
[20] Kwiatkowski M 2011 *Applied Surface Science* **257** 8912
[21] Kwiatkowski M, Policicchio A, Seredych M, Bandosz TJ 2016 *Carbon* **98** 250
[22] Kwiatkowski M, Fierro V, Celzard A 2017 *Journal of Colloid Interface Science* **486** 277
[23] Kwiatkowski M, Kalderis D, Diamadopoulos E 2017 *Journal of Physics and Chemistry of Solids* **105** 81
[24] Kwiatkowski M, Sreńscek-Nazzal J, Michalkiewicz B 2017 *Adsorption* **23** 551
[25] Kwiatkowski M, Broniek E 2017 *Colloids and Surfaces* A **529** 443

Modeling Interdisciplinary Notion "Substance" with Information Model

Tatyana Gnitetskaya[1], **Elena Ivanova**[2], **Natalya Kovalchuk**[3], **Yuliya Shutko**[4], **Larisa Dubovaya**[5], **Elena Karnaukhova**[6]
[1-6]The School of Natural Sciences, Far Eastern Federal University,
8 Sukhanova St., Vladivostok, 690950, Russia

E-mail: gnitetskaya.tn@dvfu.ru

Abstract. Wireless mobile technologies have caused a tendency to use mobile devices in the learning process. However, delivery mode should be changed. This paper describes a method of uploadable semantic structures for students' personal mobile devices that would ensure interdisciplinarity.

1. Introduction

Disciplinary education, evolving behind the sciences, contains subjects under names coinciding with those of the sciences and preserves specific features of descriptive framework pertinent to each of the sciences. Although nature is a common research subject for natural sciences, while explaining the same natural phenomena each of natural sciences uses different terminology and different theoretical methods of description. Different conceptual frameworks translated into contents of school subjects, such as physics and chemistry, results in student hesitation, or sometimes even inability to identify the same natural phenomenon discussed within courses in different disciplines.

Back in the XIX century, physics and chemistry competed for the title of a fundamental science with a theoretical basis offering groundwork for interpreting experiments conducted within the scope of other sciences. Both had short odds for being called a fundamental science, when quantum mechanics came around to explain an atomic model. Since then, physics has been enjoying its status of an absolutely fundamental science. However, nomenclatures of chemistry and physics courses vary enormously.

The issue of content interdisciplinarity goes back decades and has been a challenge for many famous researchers. In the 1960s, the Soviet Union awarded a public tender for investigation of interdisciplinary links. Under the contract, pedagogical community led by Institute of Pedagogics with the USSR Academy of Sciences performed a tremendous research to define properties, features and functions of interdisciplinary links [1], and to develop interdisciplinary maps and recommendations for teachers [2]. Yet, contents of textbooks on physics and chemistry remained the same, thus bringing the work done almost to naught. Today, authors of new textbooks on chemistry aim to refer to rendering certain issues in physics. However, the efforts prove to be insufficient for providing students with an insight into an integral notion of the surrounding world.

«Nature» devoted a special issue to interdisciplinarity. Authors of the journal articles ended in discussions on the importance of interdisciplinarity in a modern world. Mind Meld (Nature 2015, 289–

[1] To whom any correspondence should be addressed.

290) introduced the reader to the issue of interdisciplinarity. The article asked what interdisciplinary science was and proposed that, regardless of anything else, it would be a synthesis [3]. Integration of social science into research is crucial (Nature 525, 291) noted the disconnect between research and the general public, and called for the integration of social sciences into research projects to best represent that public [4]. Why interdisciplinary research matters (Nature 525, 305) demanded that hard and soft sciences work together to solve big social issues, and noted the pros, cons and difficulties addressed in the other articles in the issue [5]. Interdisciplinary research by the numbers (Nature 525, 306–307) acknowledged that interdisciplinarity was difficult to define, and presented citation research to demonstrate growth in interdisciplinary work [6]. How to solve the world's biggest problems (Nature 525, 308–311) claimed that large complex problems required a response from more than one discipline and that different skill sets must be assembled [7]. Grant giving: Global funders to focus on interdisciplinarity (Nature 525, 313–315) noted disciplinary resistance to collaboration, but pointed out three advantages to such sharing, mostly relating to complexity and liminality, and noted increases in global funding for such work [8]. Interdisciplinarity: How to catalyse collaboration (Nature 525, 315–317) noted disciplinary conflict and proposed the exploitation of five principles for interdisciplinarity (sharing, engaging difference, dialogue, support, and bridging), which were successfully tested in an institute [9].

2. Information Technologies and Information Model of Interdisciplinary Links

The main reason why Russian teachers neglected the results of the interdisciplinarity study was lack of tasks with interdisciplinary content in the textbooks, for students to analyse in class, or ponder over independently, or refer to sometime later and use for other educational objectives.

Introduction of wireless technologies into every sphere of human activity enhances possibilities for education and allows for a new solution of the old interdisciplinarity issue. Modern school students draw large amounts of information upon networks using their gadgets. They are embarrassed with neither fine print on a device display, nor intricate search for the information required. Moreover, teenagers have good understanding of search semantic hierarchy, like “follow a link”, and they are not surprised to find plenty of information in a word, which can be conveyed by other words. The above circumstances produced an idea to represent content of a paragraph in a mode that would be readily accommodated by a student’s phone display and stay at hand. Naturally, this would be interdisciplinary information provided with a path to where it could be used in the course of study.

To implement our idea, we exploited in information model of interdisciplinary links developed by T.N. Gnitetskaya [10].

A link is a philosophical category, which allows setting a comparison criteria for system elements and performing assessment of the category entirety. One of the varieties of links outlined in philosophy is a "substrate providing transfer of information.... This interpretation of a link is the basis of the interdisciplinary link information model, which is based on an interdisciplinary link graph model developed by the author. The interdisciplinary link graph model is a directed graph tree built between an element of the structure where this notion (or the law, or other element of knowledge) is introduced and an element of the structure where this concept (or other element of knowledge) is used. In fact, the intradisciplinary link graph model, where graph nodes are represented as semantic structures, updated by the calculation of informational characteristics, is considered by the author as an intradisciplinary link information model.

3. Substance as an Object of Interdisciplinary Link between Physics and Chemistry

Before you begin to Physics is a science about nature which introduces us to a wide range of types of substances. Meanwhile, chemistry is a science about substances. Not only the notion of a substance is characterized by high interdisciplinarity (used in contents of both physics and chemistry), but it also has a deep philosophical meaning. Being one of the matter species, a substance underlies the world structure. Therefore, it is critical, from methodological perspective, to provide students with a global view of a substance. Knowledge about a substance can be deepened and strengthened through wireless technology in the following way. We suggest shaping the data bank on a substance contained in textbooks on physics and chemistry into semantic structures [11]. The semantic structures can be handed out to students in physics and chemistry classes addressing a substance, and relied upon for covering the topic. The structures emphasize the nodes where a link can be executed to another discipline through a semantic state [12] of the notion of a substance. Structures for relevant semantic states of the notion of substance are distributed by the teacher among the students, e.g. within a dedicated Whatsapp Group. A task for the students is to insert a structure of the required semantic state of the notion of a substance studied in physics class into a structure related to a substance addressed in chemistry classes to independently identify the interdisciplinary kink and comprehend the integration of meanings of a substance within the contents of physics and chemistry.

The article illustrates the semantic structures by the notion of a substance introduced in the beginning of the chemistry course by O.S. Gabrielyan (8th grade, paragraph 2) The semantic states that the notion of a substance takes on in physics and chemistry were isolated from the contents of textbooks on physics and chemistry by N.S. Purysheva and above mentioned O.S. Gabrielyan . These books have been analyzed in our previous work [12, 13], and their contents were shown to have an adequate level of correlation.

Table1. Semantic states of the notion of a substance in physics and chemistry courses

Physics		
$Substance^1_{ph}$	As one of the matter species	§ 1, 7th grade
$Substance^2_{ph}$	As what physical bodies consist of	§ 1, 77th grade
$Substance^3_{ph}$	As what inert physical bodies with certain density consist of **(certain substance)**	§ 18, 7th grade
$Substance^4_{ph}$	As what consists of particles with spaces between them	§ 2, 8th grade
$Substance^5_{ph}$	As what consists of the same atoms and molecules **(certain substance)**	§ 3, 8th grade
Chemistry		
$Substance^1_{ch}$	As what physical bodies consist of (reference to physics)	§ 2, 8th grade
$Substance^2_{ch}$	As what consists of molecules, molecules consisting of atoms (reference to physics)	§ 2, 8th grade
$Substance^3_{ch}$	As what consists of atoms and molecules and possesses certain physical properties	§ 2, 8th grade
$Substance^4_{ch}$	As a molecule having a mass equal to the sum of masses of the chemical elements comprising the substance	§ 6, 8th grade

A fragment of the list of semantic states of the notion of a substance is given in Table 1, which shows that the semantics introduced in physics course is used in the chemistry course. For instance, one of the initial interpretations of the notion of a substance in the physics course correlates with its second semantic state in physics: "$substance^2_{ph}$ as what physical bodies consist of". This is the meaning used to introduce the notion of a substance in the chemistry course, while covering molecules and atoms helps broaden the meaning to the second semantic state in chemistry: "$substance^2_{ch}$ as what consists of molecules, molecules consisting of atoms".

4.Results and Discussion

Fig. 1 shows a semantic structure graph for the notion of a substance in its second semantic state in chemistry as mentioned above. An interdisciplinary link is established in nodes D_1 and D_2, wherein structures from physics must be ins

erted. It should be noted that in class, instead of the graphs students work with structures with the same number of nodes and links, but with notions specified in the nodes. For instance, Fig. 2 shows a semantic structure of the notion of a substance second semantic state in physics. Semantic structure of a paragraph is built up as follows: all physical concepts are picked up from the paragraph; the concepts are then placed onto the hierarchy levels so that the most simple ones, or those carried forward from previous paragraph, reside on the first level followed on upper levels by those with meaning formed by notions from the lower level(s), and so on, up to the top of the structure where a concept from the paragraph title resides.

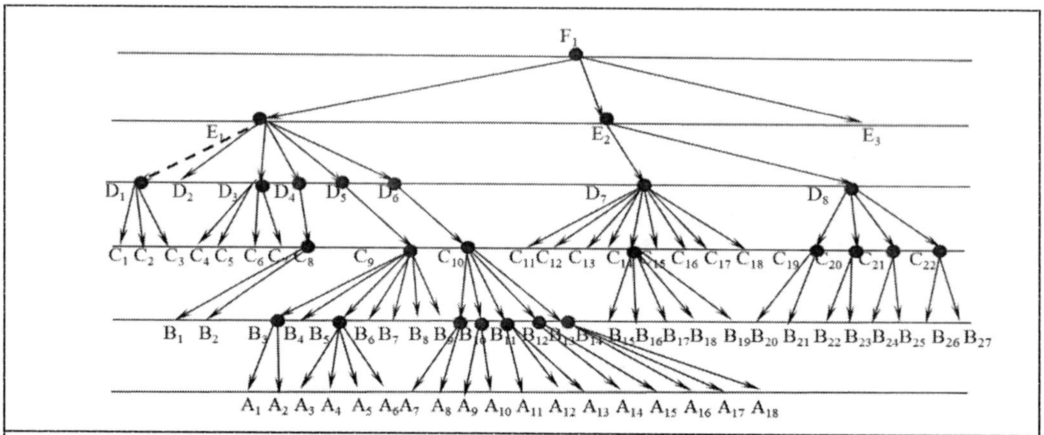

Figure 1. Semantic structure graph for the notion of a substance in its second semantic state in chemistry course for 8th grade

A_1 – type of atom, A_2 – atom, A_3 – neon, A_4 – argon, A_5 – krypton, A_6 – helium, A_7 – two atoms of hydrogen, A_8 – one atom of oxygen, A_9 – $\text{atom}^1_{(ph)}$, A_{10} – type of atom, A_{11} - oxygen, A_{12} - hydrogen, A_{13} – carbon, A_{14} – oxygen, A_{15} – carbon, A_{16} – oxygen, A_{17} – hydrogen, A_{18} – carbon, B_1 – type of atom, B_2 - $\text{atom}^1_{(ph)}$, B_3 – $\text{chemical element}^{(1)}_{ch}$, B_4 – inert gases, B_5 – hydrogen light gas (H_2), B_6 – iron, B_7 – oxygen gas (O_2), B_8 – aluminum Al, B_9 - copper Cu, B_{10} – water, B_{11} – chemical element, B_{12} – starch, B_{13} – carbon dioxide CO_2, B_{14} – sugar, B_{15} – light gas, B_{16} – density $9*10^{-5}$ g/cm^3, B_{17} – t_{boil} = -253^{0}C, B_{18} - t_{melt} = - 259^{0}C, B_{19} – colorless, tasteless, odorless, B_{20} – light weight, B_{21} – strength, B_{22} – light weight, B_{23} – high electrical conductivity, B_{24} – non-poisonous, B_{25} – plasticity, B_{26} – non-poisonous, B_{27} – high heat conductivity, C_1 – body composition, C_2 – glass, C_3 – water, C_4 – test tubes, C_5 – flasks, C_6 – glasses, C_7 – bottles, C_8 – chemical element $_{ch}^1$, C_9 – free hydrogen atoms on the Sun, C_{10} – (O) oxygen atom, C_{11} – atoms of the same chemical element, C_{12} – atoms of different chemical elements, C_{13} – example H_2, C_{14} – color, odor, gloss, C_{15} – density, C_{16} – hardness, C_{17} – plasticity, C_{18} - solubility, C_{19} – boiling temperature, C_{20} – melting temperature, C_{21}- heat-, electrical conductivity, C_{22} – aircraft and missile engineering, C_{23} – power lines, C_{24} – foil as packaging, C_{25} – aluminum dishes, D_1 – $\textbf{substance}^2_{(ph)}$, D_2 – $\textbf{substance}^3_{(ph)}$, D_3 – physical bodies made of glass, D_4 – free atoms, D_5 – elements, D_6 – compounds, D_7 – physical properties of substances, D_8 – example of aluminum applications, E_1 – $\textbf{substances}_{ch}^1$, E_2 – properties of substances, E_3 – transformations of substances, F_1 – subject of chemistry. $\textbf{Substances}_{ch}^2$.

Having comprehended contents of the chemistry course paragraph, students can easily find a point where an interdisciplinary link is established, and independently insert the structure distributed by the teacher, like doing the puzzles. The following training targets can be achieved this way:

1) interdisciplinarity of physics and chemistry courses is ensured;
2) mobile learning devices help cover the interdisciplinary link issues within the frames of self-directed learning and thus save class hours;
3) teenagers' undeniable desire to use gadgets for solving any issues whatsoever allows for the conclusion that the objective – building an interdisciplinary link in a structure using a gadget – is encouraged by the incentive.

Delivery in the form of a structure helps detect shortfalls of a paragraph content. For instance, the paragraph on chemistry illustrated by a graph in Fig. 1 introduces the notion of a substance in its second semantic state, namely, "Substance^2_{ch} as what consists of molecules, molecules consisting of atoms". The paragraph lacks a reference to the forth state of a substance, previously studied in the physics course, as "what consists of particles with spaces between them". Obviously, mentioning the state would have facilitated explanation of the notion of a substance by the notions of an atom and a molecule as ultimate particles comprising a substance, through establishment of an interdisciplinary link to physics.

We, therefore, recommend to O.S. Gabrielyan, author of the textbook on chemistry, 8th grade, to eliminate the detected shortfall. A graph for the notion of a substance in this particular semantic state is given in Fig. 3. The graph contains description of five experiments, which allow for two conclusions (see E_4 – experiment 1, E_5 – experiment 2, E_6 – experiment 3, E_7 – experiment in Fig. 3).

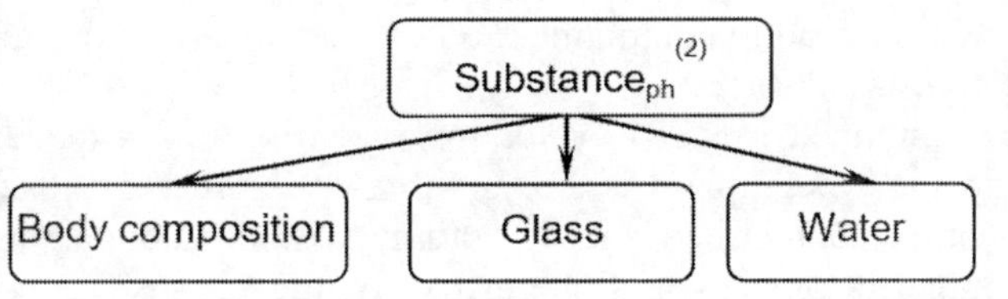

Figure 2. Semantic structure of the notion of a substance in the second semantic state in physics course

The structure contains notions instead of graph nodes and is therefore a good teaching aid both for chemistry students and teachers. It reveals the need for establishing the interdisciplinary links and offers a choice of self-guided leaning tasks. In fact, the students can be asked to analyse contents of the paragraph and to find the shortfalls, namely, the lack of interdisciplinary links.

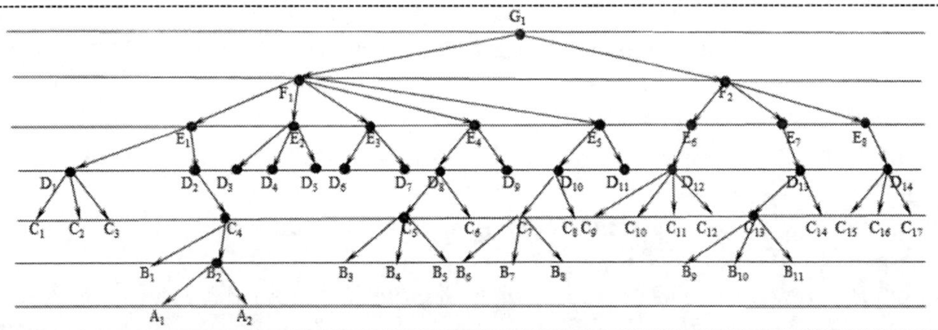

Figure 3. A graph for semantic structure of the notion of a substance in the fourth semantic state in physics course for grade 8th

A_1 –independence from creation, A_2 – independence from human will, B_1 – something existing, B_2 features, B_3 – chalk, B_4 – hammer, B_5 – strike, B_6 – water, B_7 – dye, B_8 – mixing in a glass, B_9 – metal ball, B_{10} – ring, B_{11} – passing a ball through a ring, C_1 – conduct electric current, C_2 – respond to light, C_3 – conduct heat, C_4 – matter, C_5 – small pieces of chalk, C_6 – strike, C_7 – light-colored water, C_8 – mixing in a glass of pure water, C_9 – test tubes, C_{10} – alcohol, 40 mL, C_{11} – water, 40 mL, C_{12} – mixing in a test tube, C_{13} – ball passes through a hole, C_{14} – heating a ball, C_{15} – water, C_{16} – flask with a glass tube in it, C_{17} – heating, D_1 – properties of substances, D_2 – matter species, D_3 – solids, D_4 – liquids, D_5 – gas, D_6 – stretching, D_7 – shrinking, D_8 – fine pieces of chalk, D_9 – strike, D_{10} – light-colored water, D_{11} – mixing in a glass of pure water, D_{12} – mixture volume below 40 mL, D_{13} – ball does not pass through a ring, D_{14} – raising water level in a flask, E_1 – substance$^2_{ph}$, E_2 – bodies, E_3 – changing shape, E_4 – experience 1, E_5 – experience 2, E_6 - experience 3, E_7 – experience 4, E_8 - experience 5, F_1 - conclusion 1, F_2 – conclusion 2, G_1 – substance$^4_{ph.}$

A colloquium, where every student is to present the interdisciplinary link he/she built using the archive of semantic structures of the notion semantic states, is another option. The number of tasks is unlimited, provided that physics and chemistry teachers cooperate in distributing e-structures of the notion semantic states to their students. Indeed, tracing changes in semantics of physical and chemical terms in the process of learning the subjects is reasonable for fundamental concepts only.

We developed the method of isolating fundamental concepts from contents of a course using the intradisciplinary link graph model [10]. Using the model, an interdisciplinary content cluster is formed for a group of notions. By calculation of quantitative characteristics of a model for each notion, a hierarchy of the notions can be built up. This method was applied to the subject chemistry course. The notion of a substance established itself in the top position, which emphasizes its high level of significance in the course.

Thus, the two problems – interdisciplinarity and integrity of physics and chemistry courses – can be solved through engagement of mobile communication devices in the learning process, as exemplified in the article by the notion of a substance.

The research can be continued towards evaluation of entropy as a quantitative characteristic of information contained in the paragraph structure graph. Availability of the entropy value will help determine whether the course content needs to be optimized, and enable a comparison of information content of the same notions in physics and chemistry for the purpose to choose the one with less entropy.

References

[1] I. Zverev, V. Maksimova 1981 *Intersubject links in the modern school.* (Moscow: Pedagogika) 160 p.

[2] V. Vershinin, S. Usova 1999 *Intersubject links In higher education: the mathematical support of the course of analytical chemistry.* (Bulletin of Omsk University, issue 2) p.32-34.

[3] Editorials. Mind Meld Nature 2015. V. 525: p. 289–290. doi:10.1038/525289b

[4] Editorials. Integration of social science into research is crucial Nature 2015. V.525: p. 291.. doi:10.1038/525291a

[5] Editorials. Why Interdisciplinary Research Matters, Nature 2015.V 525: p. 305, doi:10.1038/525305a

[6] Interdisciplinary research by the numbers *Nature* 2015. V.525: p.306–307. doi:10.1038/525306a

[7] Ledford, Heidi 2015 *How to Solve the World's Biggest Problems.* Nature V. 525: p.308–311. doi:10.1038/525308a

[8] Rick, Rylance. 2015 *Grant giving: Global funders to focus on interdisciplinarity.* (Nature. V 525) p. 313–315. doi:10.1038/525313a

[9] Rebekah R. Brown, Ana Deletic, 2015 *Interdisciplinarity: How to catalyse collaboration*. Nature. V. 525: p. 315–317. doi:10.1038/525315a

[10] T. Gnitetskaya 2014 *Graph Model of Intradisciplinary Connections in Example of General Physics Course.* (Journal of Physics: Conference Series. Volume 633, Issue 1, 21). DOI: 10.1088/1742-6596/633/1/012091.

[11] T. Gnitetskaya, E. Ivanova 2006 *Method of Semantic Structures on Chemistry Classes. J. Chemistry at School* №7. p. 45-51

[12] T. Gnitetskaya, E. Ivanova 2012 *Semantic Status of Chemical Concepts.* (Periodical of Advanced Materials Research Vols. 550-553) Pp. 3429 – 3432.

[13] Gnitetskaya T.N., Purisheva N.S., Ivanova E.B. 2014 *Connectedness of Physics and Chemistry Courses in the Group of Term.* (Advanced Materials Research Vols. 1033-1034) pp.1391-1394 doi: 10.4028/www.scientific.net/AMR. 1033-1034.1391

IOP Conf. Series: Journal of Physics: Conf. Series **936** (2017) 012092 doi:10.1088/1742-6596/936/1/012092

Modeling the Interdisciplinary in the Process of Training Physics

Tatyana Gnitetskaya[1], Elena Ivanova[2], Larisa Dubovaya[3], Yuliya Shutko[4], Natalya Kovalchuk[5], Elena Karnaukhova[6]
[1-6]The School of Natural Sciences, Far Eastern Federal University,
8 Sukhanova St., Vladivostok, 690950, Russia

E-mail: gnitetskaya.tn@dvfu.ru

Abstract. The information model of interdisciplinary links (Gnitetskaya T., 2016) allows to reach the interdisciplinary in the process of training physics. Presenting the fundamental physical term with semantic structure we can see that structure's node(s) where interdisciplinary link(s) from physics to chemistry and vice versa could be established. Method of optimization the interdisciplinary content was described in this paper. Method based on quantity estimation of the information – an entropy, as characteristic of semantic structure graph. Results of entropy calculation for interdisciplinary content of term "atom structure", corresponded to school physics' level, were discussed.

1. Introduction

Although the problem of increasing the interdisciplinary of educational material of such adjacent disciplines as Physics and Chemistry has been discussed for a long time, it still attracts universal interest. Multiple models have been developed and applied in the area of interdisciplinary research; however, all of them are qualitative. Qualitative models are easily available and can be applied to the interdisciplinary content thus proving their indisputable advantage. At the same time, introduction of interdisciplinary information into the educational material substantially increases its volume. Unfortunately, a qualitative model fails to evaluate the increase of educational information. Striving for the interdisciplinary content without restricting its volume poses a serious threat of overloading the students with information.

As a complex informational system, a human has at least three restrictions on transforming information. They involve restrictions in sensitivity, resolving power and carrying capacity. Carrying capacity is a maximal speed of transmitting information in a channel; it is measured in bps (bits per second) [2]. The greater the carrying capacity of a communication channel is, the bigger volume of information it can transmit per a time unit.

Carrying capacity of a human communication channel depends on its type. Thus, an optical channel can perceive and process 10^7 bps, thermal – $5*10^3$ bps, vestibular – 13 bps. The communication channel that transmits semantic information to consciousness has a carrying capacity of 16 bps. [7,8] (according to other sources - 100 bps) [9].

Theoretical restrictions on the carrying capacity in the theory of information are determined by the Shannon theory [5] that proves that if a channel has a capacity of P, then the amount of removable

[1] To whom any correspondence should be addressed.

IOP Conf. Series: Journal of Physics: Conf. Series **936** (2017) 012092 doi:10.1088/1742-6596/936/1/012092

uncertainty cannot exceed this value. In addition, according to the Ashby's Law of Requisite Variety, "the necessary choice is absolutely dependent on transmission of, at least, an equal amount of information". The possibility of applying the principles of information theory to describing the processes of information perception by consciousness has been studied by a research team from the RAS Institute of Psychology; the research, headed by B.F. Lomov, has been described in the paper [4]. These facts show an existing restriction of information perceived by consciousness; neglecting these restrictions results to the real danger of information overloads.

In this context, the interdisciplinary content can be modeled and optimized via an informational model of interdisciplinary links [1,3]; the qualitative features of this model allow evaluating the rate of content entropy growth in case of adding interdisciplinary information.

2. Informational model of intradisciplinary links.

The parameters that characterize the informational content of a portion of material are calculated via the method of semantic structures described in [1,3,10]. The method of sematic structures relies on the structuring of educational material (module, theme as part of the concept theme) with subsequent representation as a semantic structure graph, as well as on the calculation of information contained in this structure and the minimal time for acquiring it.

The semantic structure model of any portion of content has a hierarchic structure (refer to picture 1). Groups of sematic units that belong to the lower levels of the graph are connected with the sematic unit that lies at the following, higher level. The top of the graph contains the knowledge element with a semantic content reflected by the structure.

The graph reflects the collateral subordination of semantic elements in the structure. It registers their arrangement with levels and defines the degree of the structure's organization. The degree of abstraction calculated within this model characterizes the degree of semantic structure's organization. Sematic units located in the nods of the graph are the corresponding elements of knowledge (laws, theories, concepts, etc.), abilities and skilled included into the modeled portion of the content with the links represented by interdisciplinary connections. Informational characteristics of the model includes the volume of sematic information contained in the content portion, building the structure, as well as the coefficient of relative entropy that together represent the characteristics of both the quantity of information and the configuration of semantic units and interdisciplinary links in the structure, thus characterizing the complexity of the educational material content subject to research.

Construction of the semantic structure of a content portion starts with defining all concepts (physical, mathematical, etc.) contained in this structural element. The lower level of generalization holds the concepts that have been studied before (in the previous paragraph, section, different academic discipline, or introduced in this paragraph); the following level of generalization contains the concepts including the ones located below, and so on up to the top of the graph that holds the concept which is under study in this portion. One portion can carry several graphs; their amount is determined by the quantity of concepts that are studied in it.

In this article, we are going to show how changes in the paragraph's content can influence on the qualitative characteristics of the informational models of interdisciplinary links, in particular on the volume of information contained in the content of a Physics textbook paragraph and its entropy. We are also going to suggest a method of optimizing the interdisciplinary information based on the example of a very important interdisciplinary theme "Atom structure" studied in the 8th grade course of Physics of the high school.

3. Entropy as a criterion of optimizing the content of interdisciplinary information

The content of this theme definitely covers the concept of electron introduced within the course of Chemistry and Physics. The sequence of courses is such that the concept of electron in its various semantic states is introduced in the first paragraphs of the 8^{th}grade Chemistry textbook. We have mentioned the semantic states of concepts in the work [6, 10] suggesting eight semantic states of electron introduced in the Chemistry course – ranging from the first one, which defines an electron as

IOP Conf. Series: Journal of Physics: Conf. Series **936** (2017) 012092 doi:10.1088/1742-6596/936/1/012092

an atom particle that carries electricity to the eighth one where the concept means an electron of the external atomic level.

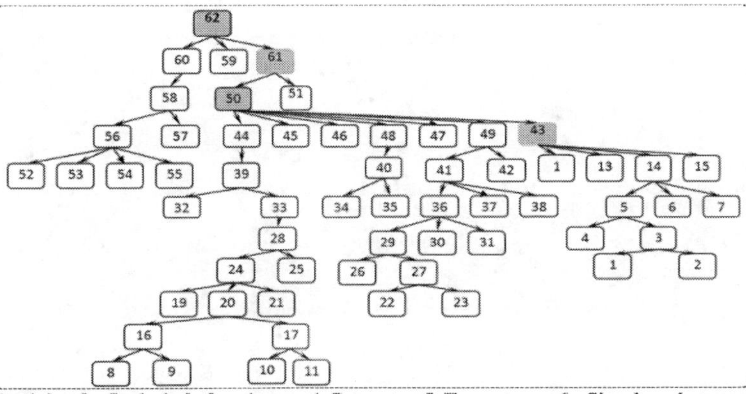

Notations: 1- Particle; 2- Secluded; 3 - Atom; 4 Separate, 5 Free atoms 6- Simple substances; 7- Complex substances; 8 - Negative charge; 9- Metal plate; 10 - Positive charge; 11 Metal plate; 12 Particle; 13 Electricity; 14 - Chemical element; 15 - Free atom; 16 - Cathode; 17 - Anode; 18 - Particle of an atom that carries electricity; 19 - Glass tube; 20 - Electrode; 21 - Vacuum; 22 - Particle; 23 - Secluded; 24 - Thomson tube; 25 - Voltage 26 - Separate; 27 - Atom; 28 - Cathode rays 29 - Free atoms; 30 - Simple substances; 31 - Complex substances; 32 - Fluorescent screen; 33 - Flow of free electrons; 34 - Metal. wire; 35 - Incineration; 36 - Chemical element; 37 - 1A p / g 38 Cesium; 39 - Glow of the screen; 40 - Flow of free electrons; 41 - Metal; 42 - Falling light; 43 - Electron (e_x^1); 44 - Thomson's experience; 45 – Mass; 46 - The speed of an electron; 47 - Negative charge; 48 - Method one; 49 - Method two; 50 - Free electron (e_x^2); 51 Electron, as the smallest negative charge; 52 - α - particles; 53 Source of α - particles; 54 - Gold foil; 55 - Screen; 56 - Dispersion; 57 - Screen; 58 - Reflection of a part of α-particles from the foil; 59 - β - particles; 60 - Rutherford's experiment; 61 - Electron as the smallest negative charge (e_x^3); 62 - Electron as a structural unit of atom (e_x^4).

Figure 1. Semantic structure of the fourth semantic state of electron as a structural atom unit, as introduced in the 8th grade Chemistry course.

The course of Chemistry starts with describing the structure and composition of an atom. The course of Physics introduces the concept of atom much later. It is logical to reduce the content of Physics by relying on the concepts already introduced in Chemistry. That is what the authors of the reviewed textbook do. However, there are no references to the paragraphs with necessary material in the Chemistry textbook. Lack of indication of the existence of interdisciplinary connection is a serious oversight that makes perception of the interdisciplinary topic more challenging.

The same topic of Physics introduces the concepts of electron in two semantic states - e^1_ϕ as negatively charged unit and e^1_ϕ as a structural unit of an atom. The first statement relies on the Millikan's experiment; the second is described via a planetary model for atoms of hydrogen, helium and lithium without describing the Rutherford's experiment. Perception of these abstractions is extremely complicated. Much more information than contained in the theme is required in order to form a consistent understanding of the atom structure. It is reasonable to establish interdisciplinary connections with Chemistry where the fourth semantic state of electron as atom's structural unit has already been studied. Picture 1 describes the corresponding semantic structure of electron in this semantic state included into the Chemistry course e^4_x. It is obvious that the structure has 11 levels of abstraction; its final elements are located at different levels. Configuration of the structure is very complex. Author of the Chemistry textbook has considered it necessary to describe the experiments of Thomson and Rutherford that require the knowledge of cathode rays, alpha- and beta- particles while introducing an electron as a structural unit of atom. This information into the 8th grade Physics course can be added in two places on the graph of the semantic structure in the topic of atom composition in paragraph 30 in the 8th grade Physics textbook. The first link is possible in nod K_1, where the second

IOP Conf. Series: Journal of Physics: Conf. Series **936** (2017) 012092 doi:10.1088/1742-6596/936/1/012092

semantic state of electron in Physics is introduced. The graph of semantic structure of this kind of presentation of an interdisciplinary topic is shown on Picture 2.

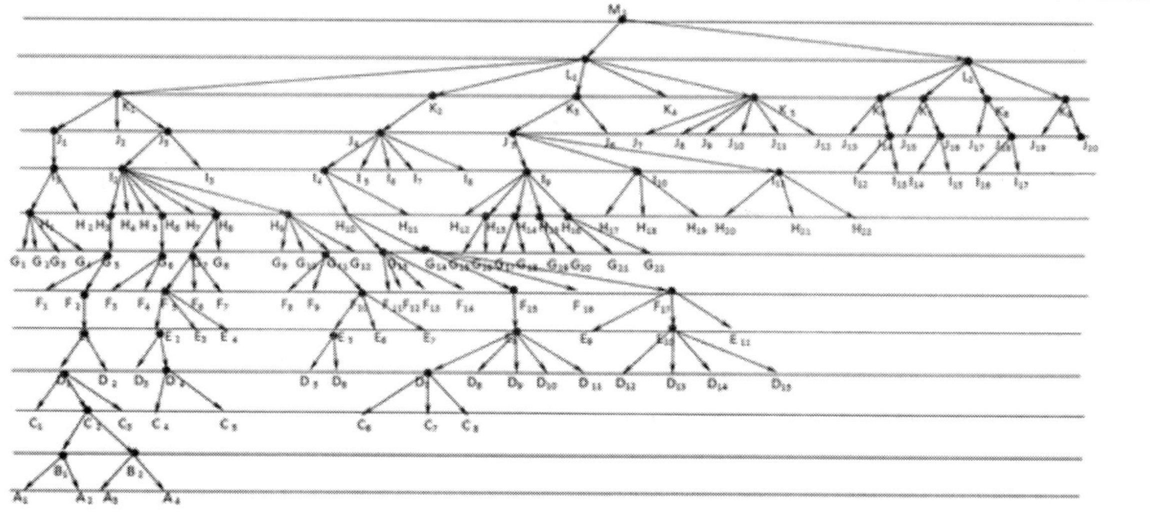

Notations: A_1 – negative charge; A_2 – Metal plate; A_3 – Positive charge; A_4 – Metal plate; B_1 – Cathode; B_2 - Anode; C_1 – Glass tube; C_2 – Electrode; C_3 - Vacuum; C_4 Particle; C_5 Isolated; C_6 – Neutral atoms ; C_7 – Neutral molecules; C_8 – Translocation; D_1 – Thomson's tube, D_2 - Voltage; D_3 – Isolated; D_4 – Atom D_5 – Particle; D_6 – Isolated; D_7 – No electrization; D_8 – Minimal electric charge; D_9 – Particle; D_{10} – Accumulation of minimal electric charge on the body; D_{11} – Removal of minima electric charge from the body; D_{12} – Charged electroscope q_1; D_{13} – Uncharged electroscope; D_{14} – Metallic wire with dielectric handle; D_{15} – Connection of electroscopes; E_1 – Cathode rays; E_2 – Free atoms; E_3 – Simple substances; E_4 – Complex substances; E_5 – Isolated; E_6 – Atom; E_7 – Electrization; E_8 – Connection of electroscopes; E_9 – Charged electroscope q/2; E_{10} – Metallic wire with dielectric handle; F_1 – Flow of free electrons; F_2 – Fluorescent screen; F_3 – Metallic wire; F_4 - Incandescent; F_5 – Chemical element; F_6 - 1A subgroups; F_7 – Cesium; F_8 – Free atoms; F_9 – Simple substances; F_{10} – Complex substances; F_{11} - Particle of zinc 1; F_{12} - Charge 1; F_{13} - Particle of zinc 2; F_{14} – Charge 2; F_{15} – Charge division; F_{16} – Particle; F_{17} – Charged electroscope q/4; G_1 - α - particles; G_2 – Source of α - particles; G_3 – Gold foil; G_4 – Screen; G_5 – Screen glow; G_6 – Flow of free electrons; G_7 – Negative charge; G_8 – Incident light; G_9 - Particle; G_{10} - Electricity; G_{11} – Chemical element; G_{12} – Free atom; G_{13} – Millikan's experiment; G_{14} – Minimal charge; G_{15} – Two electrons; G_{16} – Nuclear; G_{17} – eight electrons; G_{18} – Nuclear; G_{19} – Twenty six electrons; G_{20} – Nuclear; G_{21} – Ninety-two electrons; G_{22} – Nuclear; H_1 – Dissipation; H_2 – Screen; H_3 – Thomson's experiment; H_4 – Mass; H_5 – Velocity of electron; H_6 – Way 1 – Obtaining free electrons via incandescing the metallic wire; H_7 – Negative charge; H_9 – Obtaining free electrons in the incident light on alkali metal; H_{10} – Minimal non-divisible charge; H_{11} – Particle; H_{12} – Atom composition (Chemistry); H_{13} – Helium atom; H_{14} – oxygen atom; H_{15} – Iron atom; H_{16} – uranium atom; H_{17} – Chemical element 1; H_{18} – Reduction of the number of electrons; H_{19} – Reduction of the negative charge; H_{20} – Chemical element 1; H_{21} – Increase in the number of electrons; H_{22} – Increase of the negative charge; I_1 – Reflection of some α – particles off the foil; I_2 –Free electron; (e_x^2); I_3 - The electron charge equal to -1; I_4 – Electron e_ϕ^1; I_5 – Electron mass; I_6 – Unit of electron charge; I_7 – Electron charge; I_8 – q= $1{,}6*10^{-19}$ Кл; I_9 – Output1; I_{10} – Positive ion; I_{11} – Negative ion; I_{12} – One proton; I_{13} – Zero neutrons; I_{14}– Two protons; I_{15} – Two neutrons; I_{16} – Three electrons; I_{17} – Four neutrons; J_1 – Rutherford's experiment; J_2 - β – particles; J_3 – Electron as the smallest negative charge (e_x^3);J_4 – Electron e_ϕ^2; J_5 – Chemical element 1; J_6 – Change in the nucleus charge; J_7 – Proton; J_8 – Proton mass equals 1840 masses of electron; J_9 – Positive charge; J_{10} – Neutron; J_{11} – Proton mass equals to the neutron mass; J_{12} – Neutron charge equals to zero; J_{13} – One electron; J_{14} – Nucleus; J_{15} – Two electrons; J_{16} – Nucleus; J_{17}– Three electrons; J_{18} – Nucleus; J_{19} – Number of protons n_p; J_{20} – Number of neutrons n_n; K_1 – Electron as a structural unit of atom (e_x^4);K_2 – Electron; K_3 – Chemical element 2; K_4 – Nucleus charge defines the chemical element; K_5 – Nucleus; K_6 – Model of hydrogen atom; K_7 – Model of helium atom; K_8 –Model of lithium atom; K_9 – Atom is neutral (number of protons equals to the number of electrons); L_1 – Atom $_{\phi}^{1}$; L_2 – Atom model; M_1 – Atom composition.

Figure 2. Semantic structure graph of the topic “Structure of atoms” in the 8th grade course of Physics with the first method of adding interdisciplinary information from the course of Physics.

As we have mentioned before, this section of 8th grade Physics content of this author has a significant gap in the grounds that are available in the course of Chemistry. Therefore, the link is quite reasonable. Another place where the link can be established is congruent to the nod K_1 (Fig.2) , which introduces the first semantic state of atom in Physics in the planetary model concept but without describing Rutherford. The second variant of the topic’s graph structure with interdisciplinary information related to nod K_1 (Fig.3) is shown on Picture 3. This combination of interdisciplinary information of Chemistry and Physics will allow connecting the Rutherford’s experiment described in the Chemistry course with the planetary model of atom in the course of Physics.

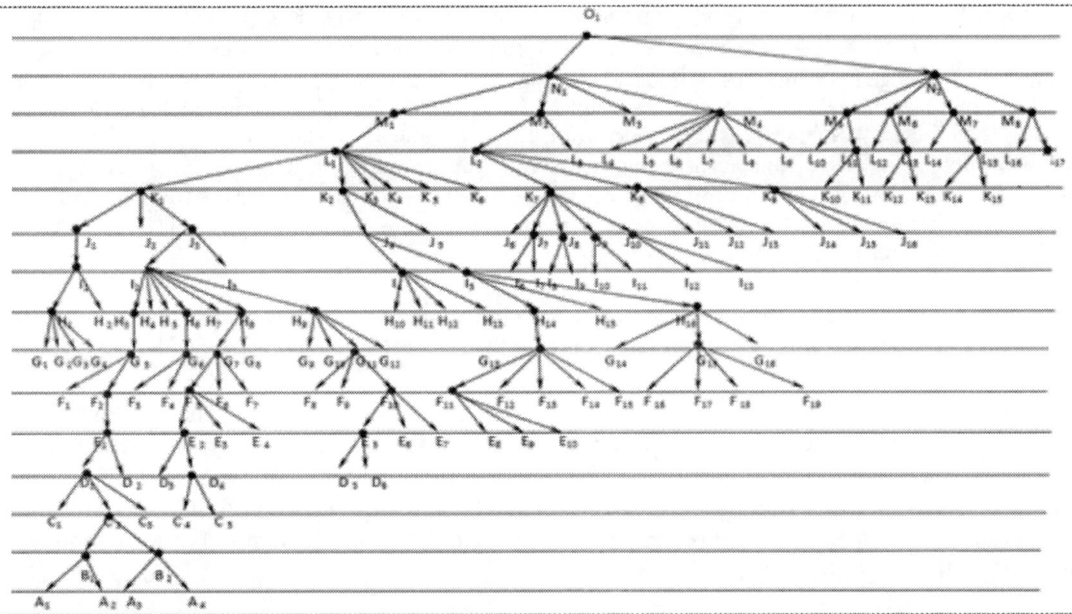

Notations: A_1 – negative charge; A_2 – Metal plate; A_3 – Positive charge; A_4 – Metal plate; B_1 – Cathode; B_2 - Anode; C_1 – Glass tube; C_2 – Electrode; C_3 - Vacuum; C_4 Particle; C_5 Isolated; D_1 – Thomson’s tube, D_2 - Voltage; D_3 – Isolated; D_4 – Atom; D_5 – Particle; D_6 – Isolated; E_1 – Cathode rays; E_2 – Free atoms; E_3 – Simple substances; E_4 – Complex substances; E_5 – Isolated; E_6 – Atom; E_7 – Electrization; E_8 – Neutral atoms ; E_9 – Neutral molecules; E_{10} – Translocation; F_1 – Flow of free electrons; F_2 – Fluorescent screen; F_3 – Metallic wire; F_4 - Incandescent; F_5 – Chemical element; F_6 - 1A subgroups; F_7 – Cesium; F_8 – Free atoms; F_9 – Simple substances; F_{10} – Complex substances; F_{11} – No electrization; F_{12} – Minimal electric charge; F_{13} – Particle; F_{14} – Accumulation of minimal electric charge on the body; F_{15} – Removal of minima electric charge from the body; F_{16} – Charged electroscope q_1; F_{17} – Uncharged electroscope; F_{20} – Metallic wire with dielectric handle; F_{19} – Connection of electroscopes; G_1 - α - particles; G_2 – Source of α - particles; G_3 – Gold foil; G_4 – Screen; G_5 – Screen glow; G_6 – Flow of free electrons; G_7 – Negative charge; G_8 – Incident light; G_9 - Particle; G_{10} - Electricity; G_{11} – Chemical element; G_{12} – Free atom; G_{13} – Electrization; G_{14} - Connection of electroscopes - G_{15} – Charged electroscope q/2; G_{16} - Metallic wire with dielectric handle; H_1 – Dissipation; H_2 – Screen; H_3 – Thomson’s experiment; H_4 – Mass; H_5 – Velocity of electron; H_6 – Way 1 – Obtaining free electrons via incandescing the metallic wire; H_7 – Negative charge; H_8 – Obtaining free electrons in the incident light on alkali metal; H_9– Millikan’s experiment; H_{10} – Minimal charge; H_{11} – Two electrons; H_{12} – Nuclear; H_{13} – eight electrons; H_{14} – Nuclear; H_{15} – Twenty six electrons; H_{16} – Nuclear; H_{17} – Ninety-two electrons; H_{18} –

Nuclear; I_1 – Reflection of some α – particles off the foil; I_2 –Free electron; (e_x^2); I_3 - The electron charge equal to -1; I_4 – Millikan's experiment; I_5 – Minimal charge; I_6 – Two electrons; I_7 – Nuclear; I_8 – eight electrons; I_9 – Nuclear; I_{10} – Twenty six electrons; I_{11} – Nuclear; I_{12} – Ninety-two electrons; I_{13} – Nuclear; J_1 – Rutherford's experiment; J_2 - β – particles; J_3 – Electron as the smallest negative charge (e_x^3); J_4 – Minimal non-divisible charge; J_5 – Particle; J_6 – Atom composition (Chemistry); J_7 – Helium atom; J_8 – oxygen atom; J_9 – Iron atom; J_{10} – uranium atom; J_{11} – Chemical element 1; J_{12} – Reduction of the number of electrons; J_{13} – Reduction of the negative charge; J_{14} – Chemical element 1; J_{15} – Increase in the number of electrons; L_{16} – Increase of the negative charge; I_4 – Electron e_Φ^1; I_5 – Electron mass; I_6 – Unit of electron charge; I_7 – Electron charge; I_8 – q= 1,6*10^{-19} Кл; I_9 – Output1; I_{10} – Positive ion; I_{11} – Negative ion; I_{12} – One proton; I_{13} – Zero neutrons; I_{14}– Two protons; I_{15} – Two neutrons; I_{16} – Three electrons; I_{17} – Four neutrons; J_4 – Electron e_Φ^2; J_5 – Chemical element 1; J_6 – Change in the nucleus charge; J_7 – Proton; J_8 – Proton mass equals 1840 masses of electron; J_9 – Positive charge; J_{10} – Neutron; J_{11} – Proton mass equals to the neutron mass; J_{12} – Neutron charge equals to zero; J_{13} – One electron; J_{14} – Nucleus; J_{15} – Two electrons; J_{16} – Nucleus; J_{17}– Three electrons; J_{18} – Nucleus; J_{19} – Number of protons n_p; J_{20} – Number of neutrons n_n; K_1 – Electron as a structural unit of atom (e_x^4); K_2 – Electron e_Φ^1; K_3 – Electron mass; K_4 – Unit of electron charge; K_5 – Electron charge; K_6 – q= 1,6*10^{-19} Кл; K_7 – Output1; K_8 – Positive ion; K_9 – Negative ion; K_{10} – One proton; K_{11} – Zero neutrons; K_{12}– Two protons; K_{13} – Two neutrons; K_{14} – Three electrons; K_{15} – Four neutrons; K_{16} – Proton; K_{17} – Neutron; L_1 – Electron e_Φ^2; L_2 – Chemical element 1; L_3 – Change in the nucleus charge; L_4 – Proton; L_5 – Proton mass equals 1840 masses of electron; L_6 – Positive charge; L_7 – Neutron; L_8 – Proton mass equals to the neutron mass; L_9 – Neutron charge equals to zero; L_{10} – One electron; L_{11} – Nucleus; L_{12} – Two electrons; L_{13} – Nucleus; L_{14}– Three electrons; L_{15} – Nucleus; L_{16} – Number of protons n_p; L_{17} – Number of neutrons n_n; M_1 – Electron; M_2 – Chemical element 2; M_3 – Nucleus charge defines the chemical element; M_4 – Nucleus; M_5 – Model of hydrogen atom; M_6 – Model of helium atom; M_7 –Model of lithium atom; M_8 – Atom is neutral (number of protons equals to the number of electrons); N_1 – $Atom_\Phi^1$; N_2 – Atom model; O_1 – Atom composition.

Figure 3. Semantic structure graph of the topic "Structure of atoms" in the 8th grade course of Physics with the second method of adding interdisciplinary information from the course of Physics.

It is obvious that one interdisciplinary material should not be discussed twice within one topic. Therefore, it has to be decided in which of the two places an interdisciplinary connection should be established. The answer to this question will be obtained if the entropy equivalence degree of the two variants of interdisciplinary connection is established. Our method of defining the informational characteristics of semantic structure graphs allows calculating the square of information entropy contained in the structure.

4. Conclusions

Table 1 shows the values of informational characteristics of semantic structure graphs, including the square of entropy, calculated for the two cases of adding interdisciplinary information from the 8th grade Chemistry course into the Physics course of the same grade.

Table 1. Informational characteristics of semantic structure graphs for the two methods of presenting interdisciplinary information.

Method	Square of entropy S^2 (Kb^2)
1	1792
2	1637

The table shows that the first method of presenting the interdisciplinary topic "Structure of atoms" has a bigger value of entropy square than the second method: 1792 Kb^2 for the first method, and 1637 Kb^2 for the second one respectively. Thus, without changing the volume of interdisciplinary information, we can influence on and optimize the content entropy by choosing the optimal place of including the

IOP Conf. Series: Journal of Physics: Conf. Series **936** (2017) 012092 doi:10.1088/1742-6596/936/1/012092

interdisciplinary information. This approach can help in compiling interdisciplinary courses while avoiding overloads.

References

[1] Gnitetskaya T.N. 2016 *The entropy estimation of the physics' course content on the basis of intradisciplinary connections' information model* (Journal of Physics: Conference Series) V. 738(1)
[2] Yaglom A.M. 1973 *Probability and information,* Moscow, 511
[3] Gnitetskaya T.N. 2015 *Information Model of Intradisciplinary Connections in the Context of a General Physics Course* (Proceedings 2015 International Conference on Simulation Modeling and Mathematical Statistics) Pp. 271-275
[4] Lomov B.F. 1966 *Humans and Technology. Articles on engineering psychology,* Moscow, 464
Shannon C. 1963 *Works of theory of information and cybernetics,* 830
[6] Gnitetskaya T.N, Ivanova E.B. 2012 *Semantic status of chemical concepts (*Advanced Materials Research) Volume 550-553, 2012, Pages 3429-3432
[7] Frank H. 1962 *Kybernetishe Grundlag der Pädagolik, AUGTS* – Verlag Baden – Baden, Gauthier – Villars Editeur, Paris
[8] Klaus G. 1966 *Kybernetic und Erkenntnistheorie, VEB – Deutschr Verlag der Wissenschaften,* Berlin 28, 175
[9] Steinbuch K. 1963 *Menschen oder Automaten im Weltraum? - "Natur-wissenschaft. Rundschau,"* 16, N 9, C. 341–349
[10] Gnitetskaya T.N., Ivanova E.B. 2006 *Method of semantic structures in the academic process.* (Chemistry in school) v.No7, Pp. 45-49

IOP Conf. Series: Journal of Physics: Conf. Series **936** (2017) 012093 doi:10.1088/1742-6596/936/1/012093

Impact of telecommunication technologies on the middle class formation

A Khusnullova[2], S Absalyamova[1,3,5], R Sakhapov[4] and Ch Mukhametgalieva[3]
[2]Kazan Cooperation Institute (branch) of the Russian University of Cooperation, Kazan, 420081, 58 Ershova street, Russia
[3]Kazan Federal University, Kazan, 420008, 18 Kremlyovskaya str., Russia
[4]Kazan State University of Architecture and Engineering, 420043, Kazan, Zelenaya str., 1, Russia

[5]E-mail: s.absalyamova@yandex.ru

Abstract. The article is devoted to the study of the impact of the information economy on the formation of the middle class. The paper identifies factors contributing to the increase in the share of the middle class in the transition to the information economy. The positive synergetic influence of telecommunication technologies on the formation of the middle class is considered through a possibility of using virtual spaces for labor and educational activities, a possibility of obtaining high returns in the form of dividends on intellectual capital, a qualitative change in the structure of needs, an access to new types of information services, etc. Authors develop a complex model of research of the middle class in the information economy, differing from those available using an expanded list of criteria. In addition to such widely used criteria as income level, level of education and self-identification, the criterion "degree of involvement in the information society" was introduced. The study substantiates that the transition to the information economy made an access to information and communication technologies one of the most significant criteria for social differentiation of society. On the basis of the model, an econometric estimate of the middle class has been carried out, which makes it possible to reveal the share of the middle class in modern society, dynamics of its development, as well as multicollinearity between spending on education, the Gini coefficient, access to information and telecommunication technologies and the size of the middle class.
Keywords: middle class, information economy, information society, social stratification

1. Introduction

Nowadays the formation of the middle class in Russia has become a priority task, which successful development largely determines the vector of the development of Russian society. The middle class, dominant in the social structure, is simultaneously proclaimed as the goal of economic reforms and as means of putting the state on a par with the developed European countries.

In the conditions of rapidly developing global information society, the middle class becomes the main initiator of innovative development of the economy. The availability and widespread use of information and communication technologies causes profound qualitative changes in it, contributes to

[1] To whom any correspondence should be addressed.

the emergence of new forms of employment and income sources, facilitates access to education, which leads to active growth of the middle class. Changes in the essential characteristics of the middle class in the information society were first noticed by such Western scientists as M. Castells and R. Florida, who introduced the concepts of "creative class" [1] and "information producers" into scientific circulation [2]. The American researcher T. Malone introduced the term "e-lancer" (electronic freelancer) to refer to a new category of workers. [3] In the conditions of the development of the information society, the quality of life increasingly depends on the level of consumption of information products and services, the middle class, and especially its creative layer, becomes both a mass user and a producer of high-tech and innovative products [4].

In this regard, not only the study of the features and regularities of the formation of the middle class in different countries, but also the development of new criteria for the social stratification of modern society acquires particular urgency.

2. Materials and methods.

The study was based on the analysis of modern trends in the formation of the middle class, conducted using the statistical base and methodological tools of UNDP [5], the Federal Service for State Statistics of the Russian Federation [6], the Territorial Body of the Federal State Statistics Service for the Republic of Tatarstan [7], including the data of a one-time sample survey "Survey of the population of the Republic of Tatarstan for the purposes of monitoring the level and quality of life" (with the technical assistance of the World Bank) [8], studies on the middle class by NISP under the leadership of T.M. Maleva [9] and the IS RAS under the leadership of N.E. Tikhonova [10].

Also, there were analyzed the results from numerous remote employment surveys, including the Sologig.com Freelance Survey, USA (Harris Interactive Marketing Research Company, commissioned by the Sologig.com remote workstation, surveyed 2400 companies using freelancers, And 5,600 self-employed) [11]; Global Freelancer Survey of 2016 USA, 2016 FreelanceSwitch, the study covered 3,700 freelancers from different professions from six continents [12]; Data of the "All-Russian Census of Freelancers", conducted in 2014 with the support of the Higher School of Economics [13].

In the process of research, such general scientific methods as system analysis, spectral analysis, correlation-regression analysis, synthesis, and graphical method were used. Each of these methods was used adequately to its functional capabilities and resolving abilities for solving the relevant stage-by-stage research tasks.

3. Results

The science has no universally accepted and empirically valid structural model of the middle class, despite the extensive scientific literature that offers its definitions of the middle class and various criteria for estimating its size [14].

Numerous works determine the structure and size of the middle class with the use of a combination of three characteristics: material and property status, social and professional status and self-identification.

The explored stratification criteria have high consistency: people with higher education are engaged in highly skilled work, they are the most competitive, have a high income, which in the end generates a relatively high self-esteem.

With the transition to the information economy, new criteria for stratification appear: the amount and quality of knowledge and information, the ability to accumulate them and use them to generate income and education are considered as the resources that differentiate the society. In our opinion, the main indicators of the emeregence of the middle class in the information economy are access to the widespread use of modern means of telecommunications for obtaining new sources of income and raising the level of education, the level of consumption of information products and services, and specific positions in the system of labor relations.

The use of indicators characterizing the involvement in the information society, in conjunction with generally accepted criteria for identifying the middle class, allows us to clarify its share in modern society and identify priority areas of public policy to increase its scale.

Considering the mentioned above, it can be concluded that in the conditions of the formation of the information economy, the scale of the middle class is derived from four main factors: income, education, self-identification and the degree of involvement in the information society.

Thus, the complex model of the middle class in the information economy can be represented in the following form:

$MC = f(P, E, S, ITT)$ (1),

where MC – middle class,

P – income level,

E – education level,

S – self-identification,

ITT – involvement in the information society.

The above criteria are not the same - in determining the middle class, some of them have a higher weight than others. The weight of these criteria depends on the level of the country's economic development. The transition to an information economy made an access to information and communication technologies one of the most significant factors of social differentiation of society.

According to the methodology of the World Bank, one of the criteria for referring to the middle class is not income, but household expenditure [8].

Taking into account the methodology of the World Bank, the above model can be presented in the following form:

$MC = f(D, E, S, ITT)$ (2),

where MC – middle class,

D – household expedintures,

E – education level,

S – self-identification,

ITT – involvement in the information society..

Theoretically, the scales of the middle class calculated using these models should be the same, but in practice they are significantly different, since there is a significant gap between the level of expenditure and household incomes. The reason for the gap can be significant extent of the shadow (non-observed economy), as well as the growing scale of consumer lending. Hidden income is determined by the balance sheet as the difference between total expenditures for households, including the growth of their economic assets, and formally registered income.

The introduction of the indicator of expenditures of households to the model makes it possible to calculate the real scales of the middle class taking into account the hidden incomes. Unfortunately, many types of distance employment and other sources of income obtained using telecommunication technologies are still not reflected in official statistics. In this case, they are more often manifested in the indicators of household expenditure.

The most accessible for the study of the involvement of the Russian population in the information society are official statistics published by UNDP in the Human Development Report, where in assessing the availability of information and telecommunication technologies in individual countries, their ranking was carried out according to the following criteria: the population covered by mobile networks ; Subscribers of broadband Internet access; Accessibility and cost of information and telecommunication technologies. The report presents data on 169 countries, so we have identified three groups of countries to simplify the analysis: with a very high level of human development, the USA (4th place), Germany (10th place) and Poland (41st place); With a high level of human development - Russia (65th place) and Ukraine (69th place), as well as the middle one - India (119th place).

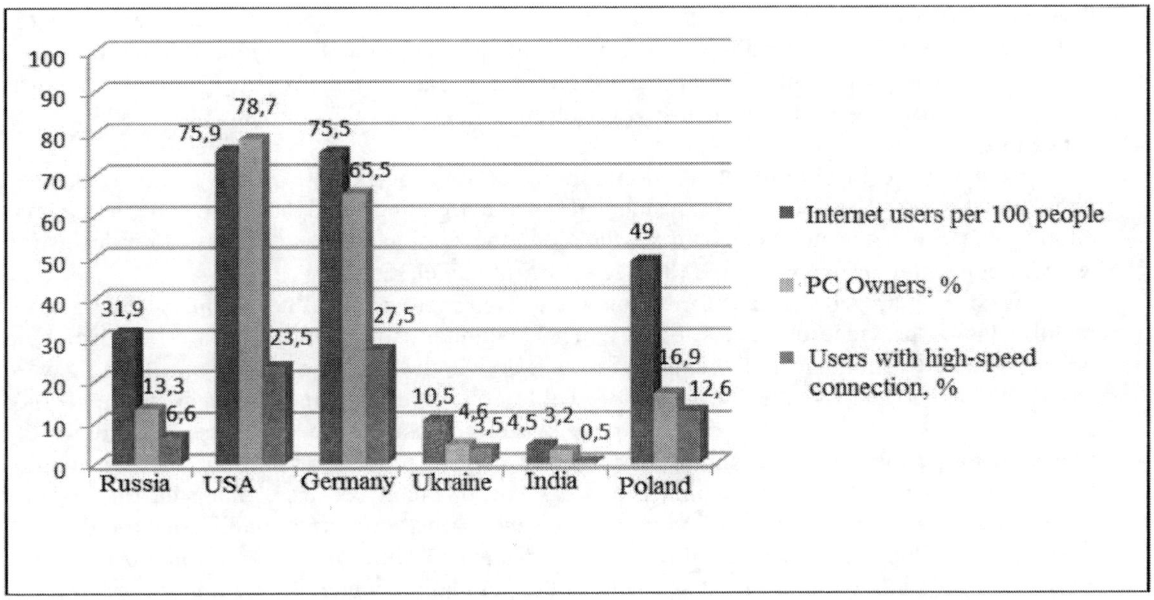

Figure 1. Access to the ITC among the population, %. (source: UNDP)

As can be seen from the given data (Fig. 1), the population of Russia, which has access to all the services provided by information and telecommunication technologies, and primarily to the broadband Internet, is only 6.6%, while 13.3% of the population are owners of personal computers and have access to the Internet, and 31.9% of the population use Internet. In our opinion, 6.6% of the population able to make maximum use of information technology, both for professional activities and for meeting the need for information services, can be considered the share of the intellectual core in the middle class. This, in general, is close to the results obtained by T. Maleva, according to which the intellectual core of the middle class today is 5.3% of the population of Russia.

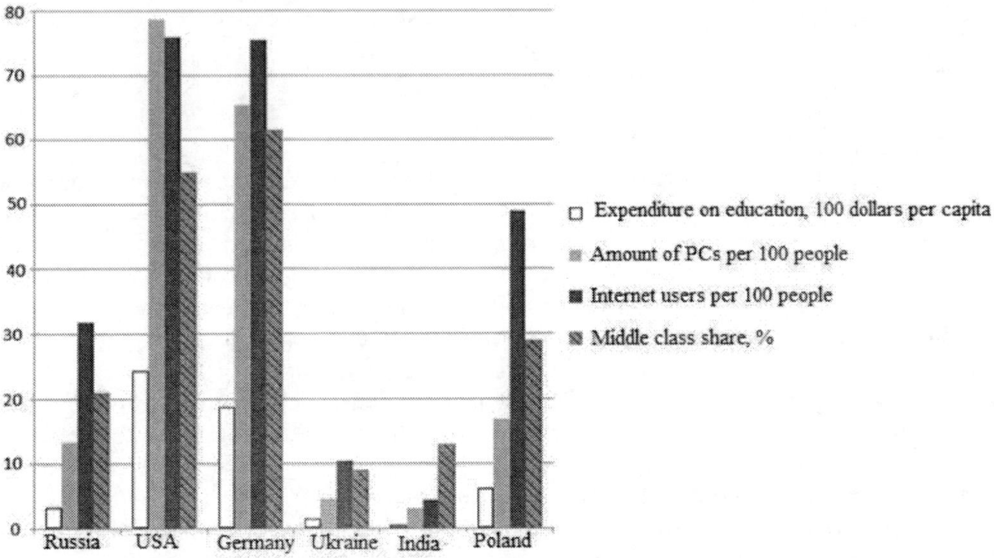

Figure 2. Correlation between expenditures of GDP on education, access to information and telecommunication technologies and the share of the middle class by countries (source: UNDP)

IOP Conf. Series: Journal of Physics: Conf. Series **936** (2017) 012093 doi:10.1088/1742-6596/936/1/012093

Analysis of personal computer availability, the availability of high-speed Internet per 100 people, as well as per capita GDP expenditure at various levels of education (Fig. 2) made it possible to reveal their positive correlation with the number of the middle class in the country. The higher the property stratification, the lower the availability of information and telecommunication technologies and education.

Studying the income differentiation using the Gini coefficient, on the contrary, the negative correlation is clearly seen. The results of the correlation analysis showed that there is a multicollinearity between the costs of education, the Gini coefficient, access to information and telecommunication technologies and the size of the middle class.

With the transition to an information society, an increasing influence on the formation of the middle class is provided by access to modern telecommunication technologies. The development of information and communication technologies and, above all, the Internet, has opened up new virtual spaces for labor activity, the formation of the Internet economy (e-economy), the corresponding electronic markets (e-markets, virtual markets) and e-business (e -business, e-commerce). The transition to the information economy has led to an active process of forming an elite middle class layer engaged in soft-tech and high-tech areas, in financial services, management, media, etc., characterized by a high standard of living based on the realization of their intellectual building.

The development of self-employment, the system of "information-based homework" makes it possible to get a job to a fairly large segment of the population, which for a number of reasons can not participate in traditional labor activity [15]. These are women with young children, disabled people, pensioners, who have limited mobility, but have a fairly large intellectual potential.

Information technologies have opened new opportunities for the development of new educational programs. In Europe, the spread of online learning is stimulated by the Bologna process and the Erasmus project, whereby students from different countries can participate in online seminars, learn any course of a foreign university via the Internet, and simultaneously study at several universities. The idea of developing a corporate system of distance learning is gaining more and more development. Up to 40% of large companies use online training technologies for retraining and upgrading their employees.

Factors contributing to the development of new forms of employment and changes in the system of labor relations of the middle class include: technological readiness at micro and macro levels, the development of financial markets and instruments, as well as the need for new services characteristic of the information economy.

4. Conclusion

The conducted researches have shown that formation of institutes of an information society and formation of an average class are closely interconnected. The growth of the share of the middle class in the structure of society leads to an increase in demand for institutions that ensure the transition to an information economy. This allows us to consider the middle class as a kind of resource for implementing the strategy of innovation and information development. The qualitative structure of the middle class, the nature and specificity of the groups existing in it, determine the possibilities of its influence on changes in the basic institutions of modern society.

The foregoing has led to the conclusion that the causal relationship within the triad "institutional environment" (a set of economic institutions) - the formation of the information economy - the formation of the middle class can have a direct character, when the change in the institutional environment follows the transition to an information society and the development of the middle class , And the reverse, when the growth of the scale of the middle class leads to a change in the institutional environment and acceleration of the transition to an information economy.

Thus, the conducted research allows to assert that involvement in the information society in the information economy becomes the most important criterion of social stratification of society.

References

[1] Castells, M The rise of the network society: The information age: Economy, society, and culture (Vol. 1). John Wiley & Sons. 2011.
[2] Florida R The Flight of the Creative Class. The New Global Competition for Talent. — HarperBusiness, HarperCollins, 2005.
[3] Malone T., Laubacher J. The Dawn of The E-Lance Economy // Harvard Business Review. Sep-Oct. 1998.
[4] Toffler, A. Revolutionary wealth. New Perspectives Quarterly, 30(4), 122-130. 2013
[5] United Nations e-Government Survey 2012: From e-Government to Connected Governance //URL: hllp:// unpanl. un. org/intradoc/groups/public/documcnls/un/unpan 028607. pdf.
[6] Rosstat data http://moscow.gks.ru/wps/wcm/connect/rosstat_ts/moscow/ru/statistics/ (accessed on May 25th 2017)
[7] Tatarstanstat data http://tatstat.gks.ru/wps/wcm/connect/rosstat_ts/tatstat/ru/statistics/ (accessed on May 25th 2017)
[8] Zaidi, S, Kandilov, V P, Krasnova, O M, & Semenova, O Y Issledovanie urovnja i kachestva zhizni naselenija Respubliki Tatarstan na osnove metodologicheskogo instrumentarija Vsemirnogo banka. Voprosy statistiki, (9), 42-47 2009..
[9] Maleva T M, Ovcharova L N Rossijskie srednie klassy na razlichnyh jetapah jekonomicheskogo razvitija. — INSOR, 2009
[10] Tikhonova, N E, & Mareeva, S V Srednij klass: teorija i real'nost' (pp. 001-320). Obshhestvo s ogranichennoj otvetstvennost'ju" Izdatel'skij Dom" Al'fa-M" 2009.
[11] http://www.creativepro.com/article/freelance-and-contract-opportunities-open-up-as-employers-face-qualified-labor-shortagessologigcom-survey-finds (accessed on May 25th 2017)
[12] http://freelanceswitch. com/general/3700-freelancers-surveyed-the-results-are-here (accessed on May 25th 2017)
[13] Strebkov, D O, & Shevchuk, A V Frilansery v informacionnoj jekonomike: kak rossijane osvaivajut novye formy organizacii truda i zanjatosti (po rezul'tatam Pervoj vserossijskoj perepisi frilanserov). 2009.
[14] Absalyamova, S, Absalyamov, T, Khusnullova, A, & Mukhametgalieva, C The impact of corruption on the sustainable development of human capital. In *Journal of Physics: Conference Series* (Vol. 738, No. 1, p. 012009). IOP Publishing. 2016, August
[15] Absalyamova, S G, & Absalyamov, T B Remote employment as a form of labor mobility of today's youth. *Mediterranean Journal of Social Sciences*, *6*(1 S3), 227, 2015

IOP Conf. Series: Journal of Physics: Conf. Series **936** (2017) 012094 doi:10.1088/1742-6596/936/1/012094

Geometric plasticity at leaves from *Ctenanthe oppenheimiana* probed by measure of distances between stomata

Humberto Antunes de Almeida Filho

Jeaneth Machicao

Odemir M. Bruno

Scientific Computing Group, Sao Carlos Institute of Physics, University of Sao Paulo, PO Box 369, 13560-970, Sao Carlos, SP, Brazil

E-mail: beto@ursa.ifsc.usp.br

Abstract. We measure the stomatal plasticity at *Ctenanthe oppenheimiana* leaves based on distance between stomata pairs in microscopic images. The distance measurements revealed that the plasticity induced by environmental adaptations are measurable using the distance between neighbor stomata. Plants exposed to extreme light irradiation times from 24 to 4 hours per day, presented discernible relations between stomatal distances. The average distances between stomata pairs revealed a powerful tool to predict changes of plasticity of stomata according with different light acclimation conditions at Ctenanthe openheimiana plants.

1. Introduction

Stomata is a cell that act as valve localized at plant leaves and stems performing the exchange of gas with the atmosphere [1]. They control the relation between the carbon gain and water loss to adjust photosynthetic needs with environmental and atmospheric conditions [2]. The plant epidermis is highly plastic during environmental adaptation, thus the number of stomata per leaf area can be highly variable and adapted according to the environmental conditions [3]. In this case the quantity and quality of light are also critical for stomatal plasticity on the leaf epidermis.

Therefore, the distribution of stomata in the epidermis can be a very valuable quantitative parameter in the analysis of plasticity as it provides many details of distance and angle between stomata that can change at plant adaptation to certain environmental conditions.

In this study, we attempt to develop computational metrics based on distance between stomata pairs located in the microscopic images from the live *Ctenanthe oppenheimiana* plants, exposed to different conditions of light acclimation.

2. Methods

2.1. Optic Microscopy and plant acclimation

We performed optic microscopy with *Ctenanthe oppenheimiana* leaves from plants artificially adapted at controlled growth chambers. The microscope used was Axio-Lab A1, Zeiss, adapted with cameras axiocam ERc5s. The magnification used was 100×. The plants used at microscopy were adapted with continuous 55% relative humidity, photon flux from $100\mu mol/m^2 \times s$, temperature 25^oc and light photo periods from 24 and 4 hours per day during a period of 60 days.

2.2. Measure of distances between stomata

The microscopic images were processed using Matlab image processing toolbox. The stomates were identified manually in the images, their centroids (s_i) were identified through an algorithm which found the central coordinates of each one of the manually marked dots, the cartesian centroid is represented by ($s_{ij} = [s_i, s_j]$). Thus, the Euclidean distance between a pair of stomates s_i and s_j is calculated as follows:

$$d(s_i, s_j) = \sqrt{(s_i - s_j)^2 + (s_i - s_j)^2}. \tag{1}$$

Images with magnifications of 100× were used. A distance matrix between each pair of centroid stomata at image is calculated as illustrated in Fig. 1.

A)

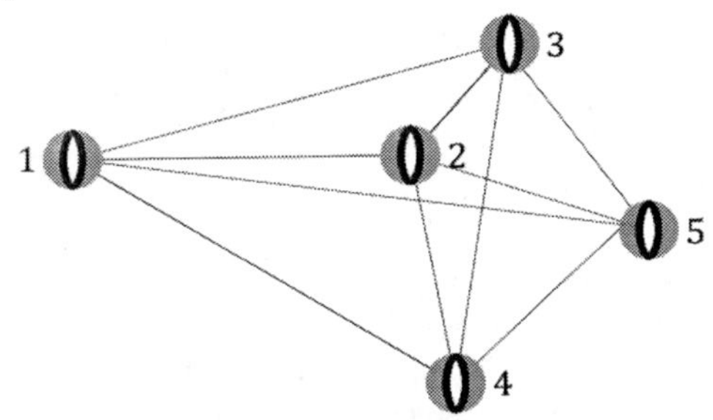

B)

	1	2	3	4	5
1	0	d_{12}	d_{13}	d_{14}	d_{15}
2	d_{21}	0	d_{23}	d_{24}	d_{25}
3	d_{31}	d_{32}	0	d_{34}	d_{35}
4	d_{41}	d_{42}	d_{43}	0	d_{45}
5	d_{51}	d_{52}	d_{53}	d_{54}	0

Figure 1. A) Illustration of five stomates (green) connected with each other by lines. B) Distance matrix of each stomata pair. The matrix indexes represent each stomata, and $d(i, j)$ is the distance between each pair of stomata found in A).

3. Results

The first evidence of the stomatal plasticity, observed in the leaves of the light acclimatized plants, was the great alteration of the stomatal density and the distance between the stomata of abaxial leaf epidermis according with the photo period . It was observed that the distance between stomata is inversely proportional to the stomatal density in the leaves. The stomatal density varied between the groups of plants studied at different light photo-periods of 4 and 24 hours per day under constant photon flux. Figure 2 depicts how the mean distance between the 20 nearest neighbor stomata varies in function of the stomatal density. Stomata from 3 images per leaf from 18 different plants submitted to photo periods of 4 and 24 hours per day were considered to distance measurements.

The graph above (Fig. 2C) shows that there is a range at the dispersion that has no points. It occurs in densities varying between 115 and 155 stomata/mm^2, which indicates that the

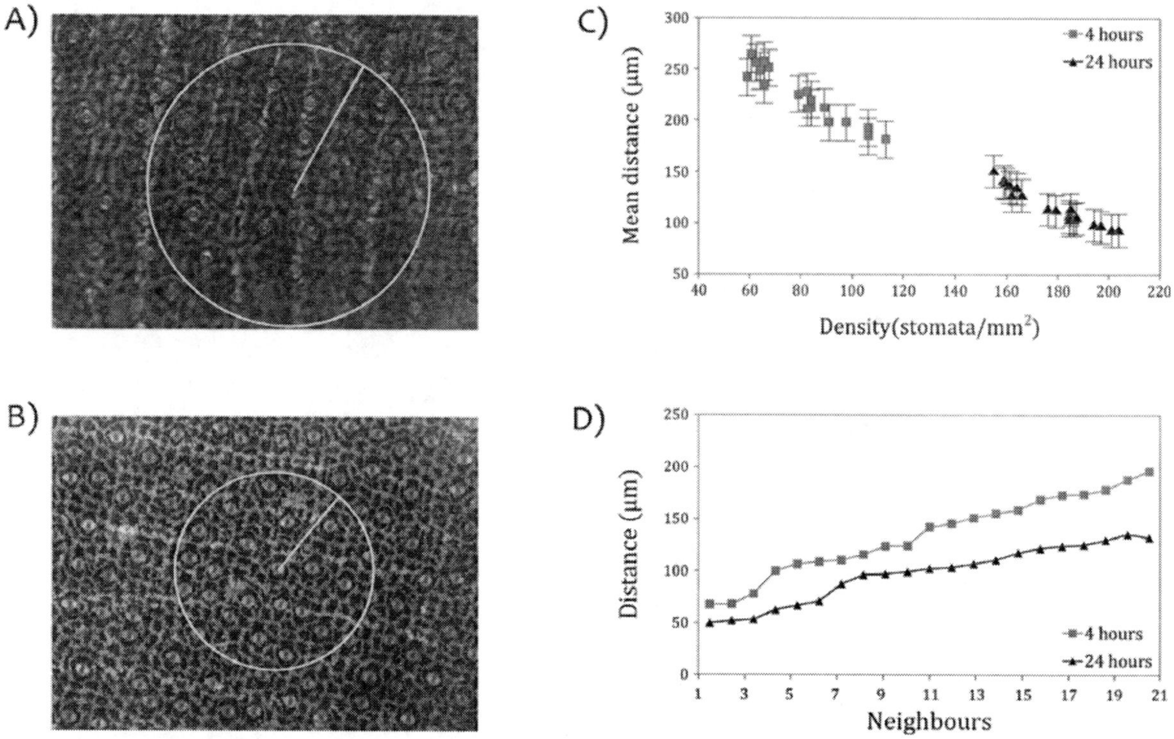

Figure 2. A) image of a plant exposed to photo period 4 hours per day. The generic stomata chosen is at center of the yellow a circumference with radius of the $200\mu m$ sweeping an area where 20 nearest stomata are located close to the central stomata B) Plant exposed to photo period of the 24 hours per day. A generic stomata is chosen over the surface and delimited by a circumference of 264 μm radius that cover the 20 nearest neighbors of this central stomata. The stomatic density observed at image A) is 82 stomata/mm^2, and at B, equal to 150 stomata/mm^2. C) Correlation between stomatal density and the mean distance between the 20 nearest neighbor stomata of each one of the stomata of the images from epidermis of the 18 plants acclimated to photo periods of the 24 (squares) and 4 hours per day (triangles), respectively. The error bars indicate the variation obtained in the average distance measurement from 3 images per plant. D) Graph showing the distance, in ascending order, of the central stomata of the circumference to each of the 20 nearest neighboring stomata for the plant image samples in A) and B).

intermediate acclimation conditions such as photo periods of 8, 12 and 16 hours per day can fill this interval.

4. Conclusions

We showed that the stomata distribution over the leaf surfaces (*Ctenanthe oppenheimiana*) varies according with the light acclimatization conditions. The changes at the stomatic plasticity over the leaves can be performed by measures of distances between stomata. Therefore, the precise measures of the stomatic distances, specifically at the (*Ctenanthe oppenheimiana*), can be useful as a sensor to environmental changes that plants can be exposed in general.

IOP Conf. Series: Journal of Physics: Conf. Series **936** (2017) 012094 doi:10.1088/1742-6596/936/1/012094

Acknowledgements
H. A. A. F. acknowledges support from CNPQ (Conselho Nacional de Desenvolvimento Cientfico e Tecnolgico) (Grants No.135137/2013-4). J. M. acknowledges a scholarship from the Brazilian agency CAPES (PROEX). O. M. B. acknowledges support from CNPq (Grant #307797/2014-7 and Grant #484312/2013-8) and FAPESP (grant #14/08026-1).

References
[1] COWAN, I. R.; TROUGHTON, J. H. The relative role of stomata in transpiration and assimilation. Planta, v. 97, n. 4, p. 325-336, 1971.
[2] KRAMER, Paul J.; BOYER, John S. Water relations of plants and soils. Academic press, 1995.
[3] HETHERINGTON, Alistair M.; WOODWARD, F. Ian. The role of stomata in sensing and driving environmental change. Nature, v. 424, n. 6951, p. 901-908, 2003.

IOP Conf. Series: Journal of Physics: Conf. Series **936** (2017) 012095 doi:10.1088/1742-6596/936/1/012095

On the decomposition of modular multiplicative inverse operators via a new functional algorithm approach to Bachet's-Bezout's Lemma

Luis A. Cortés–Vega
Mathematics Department, Antofagasta University, Antofagasta, Chile
E-mail: luis.cortes@uantof.cl

Abstract. In this paper, we consider modular multiplicative inverse operators (MMIO)'s of the form:

$$\mathscr{I}_{(m+n)} : (\mathbb{Z}/(m+n)\mathbb{Z})^* \longrightarrow \mathbb{Z}/(m+n)\mathbb{Z},\ \mathscr{I}_{(m+n)}(a) = a^{-1}.$$

A general method to decompose $\mathscr{I}_{(m+n)}(\cdot)$ over group of units $(\mathbb{Z}/(m+n)\mathbb{Z})^*$ is derived. As result, an interesting decomposition law for these operators over $(\mathbb{Z}/(m+n)\mathbb{Z})^*$ is established. Numerical examples illustring the new results are given. This, complement some recent results obtained by the author for (MMIO)'s defined over group of units of the form $(\mathbb{Z}/\varrho\mathbb{Z})^*$, with $\varrho = m \times n > 2$.

1. Introduction

In a general context, if $\mathbb{Z}/\varrho\mathbb{Z}$ denoted the residue system modulo ϱ, the modular multiplicative inverse (MMI) of $a \in \mathbb{Z}/\varrho\mathbb{Z}$, if it exists, is $a^{-1} \in \mathbb{Z}/\varrho\mathbb{Z}$, such that $a \times a^{-1} \equiv 1 \mod \varrho$, where $p \equiv q \mod \varrho$ is the usual modular representation of $q \in \mathbb{Z}/\varrho\mathbb{Z}$. This very special concept is a central element in fields of Public Key Cryptography, Cellular Automata, Computation Arithmetic, Elliptic Curves Cryptosystems and Particle Physics, as well as, in various branches of Electronic and Computer Engineering, see, for instance, [1, 2, 6, 7, 11, 13, 14, 17, 18].

Previously, the author in [4] introduce the notion of modular multiplicative inverse operator (MMIO): $\mathscr{I}_\varrho : (\mathbb{Z}/\varrho\mathbb{Z})^* \longrightarrow \mathbb{Z}/\varrho\mathbb{Z}$, such that $\mathscr{I}_\varrho(a) = a^{-1}$. and provide a general method for to decompose (MMIO)'s $\mathscr{I}_\varrho$ over $(\mathbb{Z}/\varrho\mathbb{Z})^*$, where $\varrho = m \times n \in \mathbb{N}$, $\varrho > 2$, which is closely related to the work in this paper. In this article, using a similar approach we extend and completed these results to a class of modular multiplicative inverse operator defined in group of units of the form $(\mathbb{Z}/(m+n)\mathbb{Z})^*$. In particular, under reasonable assumptions, we showed that:

$$\mathscr{I}_{m+n}(a) := \mathscr{I}_{m+\alpha}(a) + \mathscr{I}_{\beta+n}(a) - \mathscr{I}_{\alpha+\beta}(a) \tag{1.1}$$

is still valid, where $\phi_a(m) = \alpha$ and $\phi_a(n) = \beta$. Here, the operator $\phi_d(\cdot)$ have the following structure: $\phi_d : \mathbb{N}^* \to \mathbb{Z}/d\mathbb{Z}$, such that

$$\phi_d(a) = \begin{cases} a, \text{ if } 0 \leq a \leq d-1, \\ \mathbf{r}, \text{ if } a \geq d, \end{cases}$$

where $a \equiv \mathbf{r} \mod d$. In the same context, we shall also give several results relating which may become of an independent interest in the beautiful arena of the Number theory, Computer

Arithmetic and applicable in other fields as well. In order to establish Eq.(1.1), we need to show the following generalized algorithmic functional version of Bachet's-Bezout's Lemma:

$$a \times \mathscr{I}_b(a) = b \times \mathcal{L}_a(b) + 1, \tag{1.2}$$

where, the operators $\mathcal{L}_\beta(\cdot)$ have the following structure:

$$\gamma \in (\mathbb{Z}/\beta\mathbb{Z})^* \to \mathcal{L}_\beta(\gamma) \in \mathbb{Z}/\beta\mathbb{Z}, \text{ with } \mathcal{L}_\beta(\gamma) = \phi_\beta\left[(\beta - 1) \times \mathscr{I}_\beta(\gamma)\right].$$

An argument based on an algorithmic functional approach proposed by the author in [3, 4] confirms Eq. (1.2). From a mathematical point of view Eq. (1.1) is extremely interesting and naturally, this associated to (MMI). Furthermore, we should emphasize that from a computational point of view, the Eq. (1.1) offer opportunities of construct new architectures and schemes (problems that not is addressed in this work) which can complement to the different innovative algorithms mentioned in the references. It should, however, be noted that despite these recent and valuable contributions, to the author's knowledge, there is nothing in prior literature resembling Eq. (1.1).

Our work initially was motivated in part by some problems associated with certain inverse states defined about finite Galois Fields $(\mathbb{Z}/p_\alpha\mathbb{Z})^*$, where p_α is a large prime number, the which is key in the computation of Higgs mass - the particle related to the mechanism thought to be responsible for giving masses to all other particles - via a p-adic metric. A general survey on the state of the art of this last fascinating subject may be found in [6], [7] and relevant references.

This paper is organized as follows. Section §2 deals with preliminaries and notation. In §3, we introduce an algorithmic functional setting and we present in detail the main result of this work. Section §4 we offer some numerical experiments, the which displays and captures the richness algorithmic functional of all our theoretical results.

2. Preliminaries

To state our main results we shall give a rapid survey of those parts of the Number Theory that we shall need in what follows. Details can be found for instance in [15]. First, however, we need some notation. Here, and in the rest of the paper, $\mathbb{N}$ denote the set of the natural numbers, $a \in \mathbb{N}^*$ if $a \in \mathbb{N} \cup \{0\}$. To simplify the presentation we assume that $a \in \mathbb{N}^{\triangleright}$ if $a \in \mathbb{N}\backslash\{1\}$. We let $(\mathbb{Z}, +, \times)$ denote the ring of integers. The operation $\times$ in $\mathbb{Z}$ is usually called the product. Let b be a fixed positive integer. Two integers a and d are said to be congruent modulo b, written $a \equiv d \mod b$ if b divides $a - d$. Let $\mathbb{Z}/b\mathbb{Z}$ be the ring of residue classes modulo b, $a \in \mathbb{Z}/b\mathbb{Z}$ if $a \in \{0, 1, 2, \ldots, b-1\}$. Throughout this paper we use the convention that $(\mathbb{Z}/b\mathbb{Z})^* = \{a \in \mathbb{Z}/b\mathbb{Z} : \gcd(a, b) = 1\}$ denotes the group of unit of $\mathbb{Z}/b\mathbb{Z}$, the which under multiplication forms an abelian group. Let us emphasize that, as already mentioned above, the modular multiplicative inverse (MMI) of $a \in \mathbb{Z}/b\mathbb{Z}$, if it exists, is $a^{-1} \in \mathbb{Z}/b\mathbb{Z}$, such that $a \times a^{-1} \equiv 1 \mod b$. The symbol $\gcd(b, d)$ denotes the greatest common divisor between b and d (not both zero). In this notation, if $\gcd(a, b) = 1$, we say that a and b are relatively prime. The Bachet's-Bézout's Lemma, which states that: if a and b are positive integers, then there exist integers $\mathbf{s}$ and $\mathbf{t}$ such that $\gcd(a, b) = \mathbf{s} \times a + \mathbf{t} \times b$, is useful when a and b are relatively primes, in this case we have $\gcd(a, b) = 1$. Now following the arguments in [3], we can derive for the operators $\phi_b : \mathbb{N}^* \to \mathbb{Z}/b\mathbb{Z}$ and $\mathscr{C}_b : \mathbb{N}^* \to \mathbb{N}^*$, defined by

$$\phi_b(a) = \begin{cases} a, \text{ if } 0 \le a \le b-1, \\ \mathbf{r}, \text{ if } a \ge b \end{cases} \quad \text{and} \quad \mathscr{C}_b(a) = \frac{a - \phi_b(a)}{b},$$

where $a \equiv \mathbf{r} \mod b$ for any $b \in \mathbb{N}^{\triangleright}$, the following

Theorem 2.1 *Let $b \in \mathbb{N}^{\triangleright}$. Then, the following statements are true:*

(ϕ1) $\phi_b(0) = 0$,
(ϕ2) $\phi_b(d \times b) = 0$ *for every* $d \in \mathbb{N}^*$,
(ϕ3) $\phi_b(a) = \phi_b(\phi_b(a))$ *for every* $a \in \mathbb{N}^*$,
(ϕ4) $\phi_b(a + d) = \phi_b(\phi_b(a) + \phi_b(d)) = \phi_b(a + \phi_b(d)) = \phi_b(\phi_b(a) + d)$ *for every* $a,\ d \in \mathbb{N}^*$,
(ϕ5) $\phi_b(a \times d) = \phi_b(\phi_b(a) \times \phi_b(d)) = \phi_b(a \times \phi_b(d)) = \phi_b(\phi_b(a) \times d)$ *for every* $a,\ d \in \mathbb{N}^*$,
(ϕ6) $\phi_b(a + b) = \phi_b(a)$ *for every* $a \in \mathbb{N}^*$ *("periodicity" of* ϕ_b*)*,
(c1) $\mathscr{C}_b(0) = 0$,
(c2) $\mathscr{C}_b(b \times a) = a$ *for every* $a \in \mathbb{N}^*$. *In particular* $\mathscr{C}_b(b) = 1$,
(c3) $\mathscr{C}_b(\phi_b(a)) = 0$ *for every* $a \in \mathbb{N}^*$ *($\mathscr{C}_b$ is a "annihilator" of* ϕ_b*)*,
(c4) $\mathscr{C}_b(a + d) = \mathscr{C}_b(a) + \mathscr{C}_b(d) + \mathscr{C}_b(\phi_b(a) + \phi_b(d))$ *for every* $a,\ d \in \mathbb{N}^*$,
(c5) $\mathscr{C}_b(a \times d) = \mathscr{C}_b(a) \times d + \phi_b(a) \times \mathscr{C}_b(d) + \mathscr{C}_b(\phi_b(a) \times \phi_b(d))$ *for every* $a,\ d \in \mathbb{N}^*$,
(c6) $\mathscr{C}_b(a + b) = \mathscr{C}_b(a) + 1$ *for every* $a \in \mathbb{N}^*$ *($\mathscr{C}_b$ is quasi-periodic)*,
(c7) $\mathscr{C}_b(a + b \times \mu) = \mathscr{C}_b(a) + \mu$ *for every* $a,\ \mu \in \mathbb{N}^*$,
(e1) $a = \phi_b(a) + b \times \mathscr{C}_b(a)$ *for every* $a \in \mathbb{N}^*$,
(e2) $a < b$ *if and only if* $\mathscr{C}_b(a) = 0$ *for every* $a \in \mathbb{N}^*$,
(e3) $(\mathscr{C}_b \circ \mathscr{C}_d)(a) = \mathscr{C}_{b\times d}(a)$ *for every* $a \in \mathbb{N}^*$ *and every* $d \in \mathbb{N}^{\triangleright}$.

Remark 2.2 *Let us remark that in the Theorem 2.1 the compositions of the operators* $\mathscr{C}_d$ *with* $\mathscr{C}_b$*;* $\mathscr{C}_d$ *with* ϕ_b*,* ϕ_d *with* ϕ_b *and* ϕ_d *with* $\mathscr{C}_b$*, are defined by one usual way. We shall not prove this theorem here but assume its validity. A proof appear in [3].*

Another result coming out of Theorem 2.1, using the property $(e3)$ is the following decomposition law (Theorem 4 given in [3]):

Theorem 2.3 *For any* $b, d \in \mathbb{N}^{\triangleright}$ *and any* $a \in \mathbb{N}^*$*, we have*

$$\phi_{d\times b}(a) = \phi_d(a) + d \times \phi_b(\mathscr{C}_d(a)). \tag{2.1}$$

3. Decomposition-type theorems for modular multiplicative inverses operators in group of units of the form $(\mathbb{Z}/(m+n)\mathbb{Z})^*$

As already pointed out in the introduction, in this section we shall be concerned with the construction of a specific law of decomposition for the (MMI) over $(\mathbb{Z}/\varrho\mathbb{Z})^*$, with $\varrho = m + n$, and $m,\ n \in \mathbb{N}^{\triangleright}$. In order to investigate this, in the spirit of [4] we introduce first, the notion of modular multiplicative inverse operators (MMIO)'s. To be precise,

Definition 3.1 *If* $b \in \mathbb{N}^{\triangleright}$*, then the modular multiplicative inverse operator (MMIO) denoted by* $\mathscr{I}_b(\cdot)$ *is the mapping*

$$\mathscr{I}_b : (\mathbb{Z}/b\mathbb{Z})^* \to \mathbb{Z}/b\mathbb{Z},\ \textit{defined by } \mathscr{I}_b(a) = a^{-1},\ \textit{such that}$$

$$\phi_b(a \times \mathscr{I}_b(a)) = 1 \textit{ for every } a \in (\mathbb{Z}/b\mathbb{Z})^*. \tag{3.1}$$

Note that by the definition given, for any $a \in (\mathbb{Z}/b\mathbb{Z})^*$ *the (MMIO) always exist, and has the following additional property when acting on the natural numbers:*

$$\mathscr{I}_b(a) = \mathscr{I}_b(\phi_b(a)),\ a \in \mathbb{N} \textit{ with } \gcd(a, b) = 1,\ \textit{and } \mathscr{I}_b(1) = 1. \tag{3.2}$$

As examples, we have that $\mathscr{I}_7(5) = 3$ and $\mathscr{I}_7(13) = 6$, since

$$\phi_7(5 \times \mathscr{I}_7(5)) = \phi_7(5 \times 3) = \phi_7(15) = 1,$$

and similarly

$$\phi_7(13 \times \mathscr{I}_7(13)) = \phi_7(13 \times 6) = \phi_7(\phi_7(13) \times 6) = \phi_7(6 \times 6) = \phi_7(36) = 1.$$

In this last expression, we used property ($\phi 5$) of Theorem 2.1 given above. Further noting that Eq. (3.2) yields to $\mathscr{I}_7(13) = \mathscr{I}_7(\phi_7(13)) = \mathscr{I}_7(6) = 6$.

Another specific type of operator plays an important and particular role in what follows is given in the following

Definition 3.2 *If $b \in \mathbb{N}^{\triangleright}$, the operator $\mathcal{L}_b$ given by:*

$$a \in (\mathbb{Z}/b\mathbb{Z})^* \to \mathcal{L}_b(a) \in \mathbb{Z}/b\mathbb{Z}, \text{ with } \mathcal{L}_b(a) = \phi_b\left[(b-1) \times \mathscr{I}_b(a)\right] \tag{3.3}$$

is well defined. $\mathcal{L}_b(a)$ will be called the "predecessor operator modulo b", since

$$\phi_b\left[a \times \mathcal{L}_b(a)\right] = b - 1 \text{ for any } a \in (\mathbb{Z}/b\mathbb{Z})^*. \tag{3.4}$$

From this definition it is clear also that the predecessor operator $\mathcal{L}_b(a)$ satisfies the following additionals properties:

$$\mathcal{L}_b(1) = b - 1 \text{ and } \mathcal{L}_b(a) = \mathcal{L}_b(\phi_b(a)) \text{ if } a \in \mathbb{N}, \text{ with } \gcd(a, b) = 1. \tag{3.5}$$

Equipped with these operators, we can collect a number of important algebraic and functional properties, some of which, we shall use frequently in the current paper.

Theorem 3.3 (A algorithmic functional version of Bachet's-Bezout's Lemma) *Let be $m, n \in \mathbb{N}^{\triangleright}$ such that $gcd(m, n) = 1$. Then*

$$m \times \mathscr{I}_n(m) = n \times \mathcal{L}_m(n) + 1. \tag{3.6}$$

Proof. Indeed, as $\gcd(m, n) = 1$. The Bachet's-Bézout's Lemma, states that there exist integers $\mathbf{x}$ and $\mathbf{y}$ such that

$$1 = m \times \mathbf{x} + n \times \mathbf{y}. \tag{3.7}$$

To prove Eq. (3.6) we assume without loss of generality that $\mathbf{x} \geq 0$ and $\mathbf{y} < 0$. In this case, we obtain from Eq. (3.7) that:

$$\phi_n(\mathbf{x}) = \mathscr{I}_n(m) \tag{3.8}$$

and

$$\phi_m(-\mathbf{y}) = \phi_m[(m-1) \times \mathscr{I}_m(n)] = \mathcal{L}_m(n). \tag{3.9}$$

On the other hand, from the property ($e1$) of the Theorem 2.1, Eqs. (3.8) and (3.9) it follows that:

$$\mathbf{x} = \mathscr{I}_n(m) + n \times \mathscr{C}_n(\mathbf{x}) \tag{3.10}$$

and

$$-\mathbf{y} = \mathcal{L}_m(n) + m \times \mathscr{C}_m(-\mathbf{y}). \tag{3.11}$$

Now from Eqs. (3.7), (3.10), (3.11) together with the properties ($e2$) and ($e3$) of the Theorem 2.1 we obtain Eq. (3.6) as desired. ■

The following Theorem that I called of "modulus change", summarize one interesting property of $\mathscr{I}_b(\cdot)$, the which plays a role in this Section.

Theorem 3.4 (Modulus change) *Let b be in $\mathbb{N}^{\triangleright}$. Then, for any $a \in \mathbb{N}^{\triangleright}$, with* $\gcd(a, b) = 1$ *we get*

$$\mathscr{I}_b(a) = \mathscr{C}_a\{1 + \phi_a(b) \times \mathcal{L}_a(b)\} + \mathscr{C}_a(b) \times \mathcal{L}_a(b). \tag{3.12}$$

Proof. With the aid of Theorem 3.3, we obtain that

$$a \times \mathscr{I}_b(a) = 1 + b \times \mathcal{L}_a(b).$$

Note that using the properties (ϕ_4), (ϕ_5) and $(e1)$ of Theorem 2.1 we get:

$$\begin{aligned} a \times \mathscr{I}_b(a) &= 1 + b \times \mathcal{L}_a(b) - \underbrace{\phi_a\{1 + b \times (a-1) \times \mathscr{I}_a(b)\}}_{0} \\ &= 1 + b \times \mathcal{L}_a(b) - \phi_a\{1 + \phi_a(b) \times \phi_a((a-1) \times \mathscr{I}_a(b))\} \\ &= 1 + b \times \mathcal{L}_a(b) - \phi_a\{1 + \phi_a(b) \times \mathcal{L}_a(b)\} \\ &= 1 + b \times \mathcal{L}_a(b) - \{1 + \phi_a(b) \times \mathcal{L}_a(b) - a \times \mathscr{C}_a(1 + \phi_a(b) \times \mathcal{L}_a(b))\} \\ &= (b - \phi_a(b)) \times \mathcal{L}_a(b) + a \times \mathscr{C}_a(1 + \phi_a(b) \times \mathcal{L}_a(b)) \\ &= a \times \mathscr{C}_a(b) \times \mathcal{L}_a(b) + a \times \mathscr{C}_a(1 + \phi_a(b) \times \mathcal{L}_a(b)). \end{aligned}$$

This conclude the proof. ■

Theorem 3.4 have a important consequence which we now state.

Theorem 3.5 (A law of decomposition for (MMIO)'s over $(\mathbb{Z}/(m+n)\mathbb{Z})^*$**)**
Let $m, n \in \mathbb{N}^{\triangleright}$. Then for any $a \in (\mathbb{Z}/(m+n)\mathbb{Z})^* \cap \mathbb{N}^{\triangleright}$ we have

$$\mathscr{I}_{m+n}(a) = \mathscr{I}_{\alpha+n}(a) + \mathscr{I}_{m+\beta}(a) - \mathscr{I}_{\alpha+\beta}(a), \tag{3.13}$$

where $\alpha = \phi_a(m)$ and $\beta = \phi_a(n)$.

Proof. To prove this theorem we first of all note that for $a \in (\mathbb{Z}/(m+n)\mathbb{Z})^* \cap \mathbb{N}^{\triangleright}$, the identity

$$\mathscr{C}_a(n) \times \mathcal{L}_a(m+n) = \mathscr{I}_{\alpha+n}(a) - \mathscr{I}_{\alpha+\beta}(a)$$

follows from Theorem 3.4 and the properties $(\phi 4)$, $(c4)$ of the Theorem 2.1, respectively. Now as before, using Theorem 3.4 and the properties $(\phi 4)$, $(c4)$ of the Theorem 2.1 it is easy to check, repeating the previous argument, that if $a \in (\mathbb{Z}/(m+n)\mathbb{Z})^* \cap \mathbb{N}^{\triangleright}$, then

$$\mathscr{C}_a(n) \times \mathcal{L}_a(m+n) = \mathscr{I}_{m+n}(a) - \mathscr{I}_{m+\beta}(a).$$

These two last identities show the theorem, the proof is complete. ■

Remark 3.6 *We assume also that $\mathscr{I}_1(a) = 1$, for any $a \in (\mathbb{Z}/(m+n)\mathbb{Z})^* \cap \mathbb{N}^{\triangleright}$.*

4. Numerical experiments

The Chinese Remainder Theorem (CRT), is actually one of the main theorems of number theory [8]. Over the year, has been playing a prominent role due to its applicability in other fields of science, engineering and engineering genetic; see, for example [16] for Photo Radar, [15, 12] for Cryptography and Theory of Code, [10] for Matrix Theory, [3, 5, 15] for Acoustic Diffusers and [9] for DNA Sequencing.

We shall use some numerical results to illustrate most of the ideas and results of our approach. The discussion is limited only to Chinese remainder theorem in the spirit of [3], for the sake of completeness we outline it, emphasizing some specific facts.

Theorem 4.1 (Algorithmic Functional-CRT) *Let b_1 and b_2 relatively prime in $\mathbb{N}$, with $b_1, b_2 \geq 2$ and $\varrho = b_1 \times b_2$. Let γ and β be two arbitaries numbers such that $\gamma \in \mathbb{Z}/b_1\mathbb{Z}$ and $\beta \in \mathbb{Z}/b_2\mathbb{Z}$, respectively. Then we can find in the set $\mathbb{Z}/\varrho\mathbb{Z}$ one unique element a that satisfied the system:*

$$\begin{cases} \phi_{b_1}(a) = \gamma, \\ \\ \phi_{b_2}(a) = \beta. \end{cases} \tag{4.1}$$

A explicit version of the solution of (4.1), in the case $b_2 > b_1$ have the form (for more detaild, see [3]):

$$a = \beta + b_2 \times \phi_{b_1} \left\{ \mathscr{I}_{b_1} \left(\phi_{b_1}(b_2) \right) \times \left[\gamma + \beta \times (b_1 - 1) \right] \right\}. \tag{4.2}$$

To ilustrate some of our results, we analyze the numerical form of the solution of following system:

$$\begin{cases} \phi_{45}(a) = 31, \\ \\ \phi_{52}(a) = 47. \end{cases} \tag{4.3}$$

For tested it, we first recall that all the conditions of Theorem 4.1, can be verified. Now, $\min\{45, 52\} = 45$. As we have already mentioned, the expression (4.2) yeld to:

$$a = 47 + 52 \times \phi_{45} \left\{ \mathscr{I}_{45} \left(\phi_{45}(52) \right) \times \left[31 + 47 \times (45 - 1) \right] \right\}.$$

As, $\phi_{45}(52) = 7$, we get $a = 47 + 52 \times \phi_{45} \left\{ \mathscr{I}_{45}(7) \times [31 + 47 \times 44] \right\}$. In order to recover the solution a of (4.3) and to facilitate the computation of the (MMIO) $\mathscr{I}_{45}(7)$, one may works as follows: Using Eq. (3.13), we find that

$$\mathscr{I}_{45}(7) = \mathscr{I}_{24+21}(7) = \mathscr{I}_{3+21}(7) + \mathscr{I}_{24}(7) - \mathscr{I}_{3}(7).$$

Here, $m = 24$ and $n = 21$. So

$$\mathscr{I}_{45}(7) = 2 \times \mathscr{I}_{24}(7) - \mathscr{I}_{3}(\phi_3(7)).$$

It follows immediately from equation above that

$$\mathscr{I}_{45}(7) = 2 \times \mathscr{I}_{24}(7) - \mathscr{I}_{3}(1) = 2 \times \mathscr{I}_{24}(7) - 1. \tag{4.4}$$

Similarly, for $\mathscr{I}_{24}(7)$ we have that

$$\mathscr{I}_{24}(7) = \mathscr{I}_{12+12}(7) = \mathscr{I}_{5+12}(7) + \mathscr{I}_{12+5}(7) - \mathscr{I}_{10}(7).$$

So,

$$\mathscr{I}_{24}(7) = 2 \times \mathscr{I}_{17}(7) - \mathscr{I}_{10}(7).$$

Following this process for $\mathscr{I}_{17}(7)$ we find

$$\mathscr{I}_{17}(7) = \mathscr{I}_{10+7}(7) = \mathscr{I}_{3+7}(7) + \mathscr{I}_{10}(7) - \mathscr{I}_{3}(7) = 2 \times \mathscr{I}_{10}(7) - 1.$$

Now as $\mathscr{I}_{10}(7) = 3$ a simple computation yields

$$\mathscr{I}_{24}(7) = 3 \times \mathscr{I}_{10}(7) - 2 = 3 \times 3 - 2 = 7.$$

Finally, it follows immediately from Eq. (4.4) that

$$\mathscr{I}_{45}(7) = 2 \times \mathscr{I}_{24}(7) - 1 = 2 \times 7 - 1 = 13.$$

Consequently, the solution of (4.3) reduces to:

$$a = 47 + 52 \times \phi_{45}\left\{13 \times [31 + 47 \times 44]\right\}. \tag{4.5}$$

Now by the properties ($\phi5$), ($\phi4$) and ($\phi2$) of Theorem 2.1, we get

$$\begin{aligned} \phi_{45}\left\{13 \times [31 + 47 \times 44]\right\} &= \phi_{45}\left\{13 \times [31 + 2 \times 44]\right\} \\ &= \phi_{45}\left\{13 \times 74\right\} = \phi_{45}\left\{13 \times 29\right\}. \end{aligned} \tag{4.6}$$

Also, Theorem 2.3, the properties ($\phi5$) and ($c4$) of Theorem 2.1 yields to

$$\begin{aligned} \phi_{45}\left\{13 \times 29\right\} &= \phi_5(13 \times 29) + 5 \times \phi_9(\mathscr{C}_5(13 \times 29)) \\ &= \phi_5(\phi_5(13) \times \phi_5(29)) + 5 \times \phi_9(\mathscr{C}_5(13 \times 29)) = \phi_5(3 \times 4) \\ &+ 5 \times \phi_9\left[\mathscr{C}_5(13) \times 29 + \phi_5(13) \times \mathscr{C}_5(29) + \mathscr{C}_5(\phi_5(13) \times \phi_5(29))\right] \\ &= \phi_5(12) + 5 \times \phi_9\left[2 \times 29 + 3 \times 5 + \mathscr{C}_5(3 \times 4)\right] \\ &= 2 + 5 \times \phi_9\left[2 \times 2 + 3 \times 5 + 2\right] = 17. \end{aligned}$$

It then follows from (4.6) and (4.5) that $a = 931$ and $a < 2340$, like we should expect.

Acknowledgments
I am especially indebted to profesor Dr. Alvaro Restuccia, academic of Departament of Physics of the Antofagasta University by their support, via fondecyt Grant #1161192, Chile.

References

[1] H. M. AL-Matari, S. J. Aboud, N. F. Shilbayeh, Fast Fraction-Integer Method for Computing Multiplicative Inverse, *J. of Computing*, **1** (2009) 131–135.
[2] O. Arazi, H. Qi, On calculiting multiplicative inverses modulo 2^m, *IEEE Trans. Comput* **57** 1435–1438 (2008).
[3] L. A. Cortés–Vega, A functional technique based on the Euclidean algorithm with applications to 2-D acoustic diffractal diffusers, *J. Phys.: Conf. Ser* **633** 1–6 (2015).
[4] L. A. Cortés–Vega, About the decomposition of modular multiplicative inverse operators over group of units, *Submitted for publication* (2017).
[5] T. J. Cox, P. D'Antonio, *Acoustic Absorbers and Diffusers: Theory, Design and Application* Spon Press, 2004.
[6] Y. Dai, A. B. Borisov, K. Boyer, C. K. Rhodes, Computation with inverse states in a Finite Field F_{P_α}: The muon neutrino mass, the Unified-Strong-Electroweak coupling constant, and the Higgs mass, *Sandia National Laboratory, Report SAND2000-2043* 1–11 (2000).
[7] Y. Dai, A.B. Borisov, K. Boyer, C.K. Rhodes, A p-Adic metric for particle mass scale organization with genetic divisors, *Sandia National Laboratory, Report SAND2001-2903* 1–12 (2001).
[8] C. Ding, D. Pey, A. Salomaa, *Chinese remainder Theorem: Applications in Computing, Coding, Cryptography, Singapure; World Scientific*, (1999).
[9] Y. Elrich, K. Chang, A. Gordon, R. Ronen, O. Navon, M. Rooks, G.J. Hanon, DNA Sudoku-harnessing high-throughput sequencing for multiplexed specimen analysis, *Genome Res* **19** 1243–1253 (2009).
[10] M.A. Fiol, Finite Abelian groups and the Chinese remainder theorem, *Discrete Math* **67** 101–105 (1987).
[11] L. Hars, Modular inverse algorithms without multiplications for cryptographic applications, *J Embedded Systems* **032192** 1–13 (2006).
[12] D. E. Knuth, *The art of computer programming*, 2, Semi-Numerical Algorithms, 3rd Edition, Addison-Wesley, Reading, MA, (1997).
[13] W. H. Ko, Modular inverse and reciprocity formula, *arXiv:1304.6778v1* 1–7 (2013).
[14] R. Lórencz, New algorithm for classical modular inverse, in Kaliski, B.S., Jr., Koç, C.K., and Paar, C. (Eds.):CHES 2002, LNCS Springer-Verlag Berlin 2003, 57–70.
[15] M. R. Schroeder, *Number theory and in Science and comunication*, 3rd ed. Springer, Berlin (1997).
[16] R. J. Sullivan, *Microwave Radar Imaging and Advanced Concepts*, 2nd ed. Scitech Pub Inc. (2004).
[17] S. B. Verkhovsky, *Enhanced Euclid algorithm for modular multiplicative inverse and its applicattion in Cryptographic protocols*, Int. J. Communications, Network and System Sciences., 3, p.p., 901–906 (2010).
[18] S. Wei, Computation of modular multiplicative inverse using residue signed-digit additions, *IEEE Conf. Pub :2016 International SoC Design Conference (ISOCC)* 85–86 (2016).

IOP Conf. Series: Journal of Physics: Conf. Series **936** (2017) 012096 doi:10.1088/1742-6596/936/1/012096

The linear response function of a screw dislocation in a ferroelectrics

V V Dezhin

Voronezh State Technical University, Moskovsky prospect 14, Voronezh 394026, Russian Federation

E-mail: viktor.dezhin@mail.ru

Abstract. Small amplitude bending vibrations of a screw dislocation with arbitrary slip plane in an uniaxial ferroelectric is considered, the dislocation line is along the ferroelectric axis. To solve this problem the system of equations was made up. Fourier transform of the Peach-Koehler forces projection on the dislocation slip plane is obtained and a linear response function of a screw dislocation in a ferroelectric to an external force is found. The estimates of the additive to the dislocation effective mass and effective rigidity, the additional contribution to the damping of the dislocation vibrations are obtained.

1. Introduction

Dislocation bending vibrations in non-ferroelectric crystals using the self-consistent dynamic theory of dislocations in paper [1] were considered. The vibrations of the crystal lattice localized on dislocations were also considered in [2-4]. In [5], using the lagrangian formalism, a generalized susceptibility of dislocation and the equation of a crystal with a dislocation small oscillations are found, the results to the oscillations of a dislocation segment are applied [6]. A solution of the edge dislocation generalized susceptibility problem in a crystal with a structural phase transition was given in [7, 8], here also the natural frequencies and damping of bending dislocation vibrations were found. In ferroelectrics, dislocation oscillations will be accompanied by oscillations of electric fields and polarization, which are caused by variable elastic fields through a striction coupling. As a consequence, the effective dislocation mass and its effective rigidity are renormalized, an additional contribution to the damping of dislocation oscillations is appears. The problem of dislocation in a piezoelectric in the static case was solved in [9]. The case of uniform motion of an infinite rectilinear dislocation in an anisotropic piezoelectric medium was considered in [10]. The general theory of small bending vibrations of dislocations in ferroelectrics was developed in [11, 12, 8] and bending vibrations of an edge dislocation were investigated. The electric fields around an vibrating edge dislocation in paper [13] is found. In the present communication, as in [14], bending vibrations of a screw dislocation are studied, but the arbitrariness of the screw dislocation slip plane is taken into account.

2. Set of equations, describing screw dislocation vibrations in ferroelectrics

Let us consider a concrete case of a screw dislocation in an uniaxial ferroelectric. Let the ferroactive axis coincide with the axis Oz: $\boldsymbol{P}=(0,\,0,\,P)$ ($\boldsymbol{P}$ is polarization vector), the dislocation is also located along the axis Oz with the unit tangent vector to the dislocation line $\boldsymbol{\tau}_0=(0,\,0,\,-1)$, the Burgers vector $\boldsymbol{b}=(0,\,0,\,b)$ and the unit vector of the normal to the slip plane $\boldsymbol{n}=(n_x,\,n_y,\,0)$. Only small oscillations of the dislocation near the equilibrium position will be described. Then, in the linear dislocation displacement $u = u(z,t)$ approximation, we obtain $\tau = \left(-n_y\, \partial u/\partial z, -n_x\, \partial u/\partial z, -1\right)$. The presence of a dislocation in a ferroelectric crystal leads to a deviation of the polarization $P_1 = P - P_0$ from the equilibrium value P_0 in a homogeneous crystal [11, 12, 8]. Taking this into account and assuming the isotropy of the elastic and electrostriction properties $g_{lnkm} P_k P_m = g_1 \delta_{ln} P^2 + g_2 P_l P_n$, where g_{lnkm} is the tensor of the electrostriction coefficients, the generalization of the dynamic Beltrami equation [15] for a ferroelectric crystal with a screw dislocation in the linear approximation in the deviation of the polarization P_1 is:

$$\frac{1}{c_t^2}\frac{\partial^2\sigma_{ij}}{\partial t^2} - \frac{1}{2(1+\nu)}\frac{1}{c_t^2}\delta_{ij}\frac{\partial^2\sigma_{ll}}{\partial t^2} - \Delta\sigma_{ij} - \frac{1}{1+\nu}\left(\frac{\partial^2\sigma_{ll}}{\partial x_i \partial x_j} - \delta_{ij}\Delta\sigma_{ll}\right) - \rho(g_1+g_2)\delta_{ij}P_0\frac{\partial^2 P_1}{\partial t^2}$$

$$+2\mu(g_1+g_2)P_0\left(\delta_{ij}\Delta P_1 - \frac{\partial^2 P_1}{\partial x_i \partial x_j}\right) + \rho g_2\left(P_{0i}\frac{\partial^2 P_{1j}}{\partial t^2} + P_{0j}\frac{\partial^2 P_{1i}}{\partial t^2}\right)$$

$$+\mu g_2\left(P_{0j}\frac{\partial^2 P_1}{\partial x_i \partial z} + P_0\frac{\partial^2 P_{1j}}{\partial x_i \partial z} + P_{0i}\frac{\partial^2 P_1}{\partial x_j \partial z} + P_0\frac{\partial^2 P_{1i}}{\partial x_j \partial z} - P_{0i}\Delta P_{1j} - P_{0j}\Delta P_{1i} - 2P_0\delta_{ij}\frac{\partial^2 P_1}{\partial z^2}\right)$$

$$=\rho\frac{\partial}{\partial t}\left(j_{ij} + j_{ji} - \delta_{ij} j_{ll}\right) - 2\mu\eta_{ij} \tag{1}$$

Here $c_t = \sqrt{\mu/\rho}$ is transverse sound velocity, μ is shear modulus, ρ is density of the crystal, σ_{ij} is stress tensor, t is time, ν is Poisson's ratio, δ_{ij} is Kronecker symbol, $\Delta \equiv \partial^2/\partial x_i^2$, $j_{ij} = -n_i b_j V \delta(\xi) = n_i b_j\, \partial u/\partial t\, \delta(\xi)$ is the dislocation flux density tensor [16], V is the velocity vector projection of the dislocation line at a given point on the direction of its normal to the slip plane, ξ is the two-dimensional radius vector counted from the dislocation axis in a plane normal to the unit tangent vector to the dislocation line at a given point, $\delta(\xi)$ is two-dimensional delta function, $\eta_{ij} = e_{jkn}\frac{\partial}{\partial x_k}\left[\left(\tau_i b_n - \frac{1}{2}\boldsymbol{\tau}\cdot\boldsymbol{b}\delta_{in}\right)\delta(\xi)\right]$ is deformation incompatibility tensor [17] due to the presence of dislocations within the crystal, e_{jkn} is Levi-Civita symbol (unit antisymmetric tensor). Equation (1) must be multiplied by $n_i b_j$ to go over to the projection of the Peach-Koehler force onto the dislocation slip plane $f_\perp(\boldsymbol{r},t) = n_i \sigma_{ij} b_j$. For our case of locating a screw dislocation, taking into account the equality $p = -\sigma_{ll}/3$ (p is hydrostatic pressure), we have

$$n_i\sigma_{ij}b_j = n_x b\sigma_{xz} + n_y b\sigma_{yz} = f_\perp, \tag{2}$$

$$n_i\delta_{ij}b_j = n_i b_i = 0, \tag{3}$$

$$n_i\Delta\sigma_{ij}b_j = \Delta f_\perp, \tag{4}$$

$$n_i b_j\frac{\partial^2\sigma_{ll}}{\partial x_i \partial x_j} = -3n_x\frac{\partial^2 p}{\partial x \partial z} - 3n_y\frac{\partial^2 p}{\partial y \partial z}, \tag{5}$$

$$n_i b_j \frac{\partial^2 P_1}{\partial x_i \partial x_j} = n_x b \frac{\partial^2 P_1}{\partial x \partial z} + n_y b \frac{\partial^2 P_1}{\partial y \partial z}, \tag{6}$$

$$n_i b_j \left(P_{0i} \frac{\partial^2 P_{1j}}{\partial t^2} + P_{0j} \frac{\partial^2 P_{1i}}{\partial t^2} \right) = 0, \tag{7}$$

$$n_i b_j \left(P_{0j} \frac{\partial^2 P_1}{\partial x_i \partial z} + P_0 \frac{\partial^2 P_{1j}}{\partial x_i \partial z} \right) = 2 n_x b P_0 \frac{\partial^2 P_1}{\partial x \partial z} + 2 n_y b P_0 \frac{\partial^2 P_1}{\partial y \partial z}, \tag{8}$$

$$n_i j_{ij} b_j = -n_i n_i b_j b_j \frac{\partial u}{\partial t} \left[\delta(x)\delta(y) - n_x \delta(x)\delta'(y) u - n_y \delta'(x)\delta(y) \right]$$

$$\approx -n_x^2 b^2 \frac{\partial u}{\partial t} \delta(x)\delta(y) - n_y^2 b^2 \frac{\partial u}{\partial t} \delta(x)\delta(y) = -b^2 \frac{\partial u}{\partial t} \delta(x)\delta(y), \tag{9}$$

$$n_i j_{ji} b_j = -n_i n_j b_i b_j \frac{\partial u}{\partial t} \delta(\xi) = 0, \tag{10}$$

$$n_i \eta_{ij} b_j = n_x b \eta_{xz} + n_y b \eta_{yz} = \frac{1}{2} b^2 \left(-n_x \delta(x)\delta'(y) + n_y \delta'(x)\delta(y) + n_x^2 \delta(x)\delta''(y) u - n_y^2 \delta''(x)\delta(y) u \right). \tag{11}$$

After multiplying by $n_i b_j$ and taking into account equalities (2)-(11), equation (1) takes the form

$$\frac{1}{c_t^2} \frac{\partial^2 f_\perp}{\partial t^2} - \Delta f_\perp + \frac{3b}{1+\nu} \left(n_x \frac{\partial^2 p}{\partial x \partial z} + n_y \frac{\partial^2 p}{\partial y \partial z} \right) - 2\mu b g_1 P_0 \left(n_x \frac{\partial^2 P_1}{\partial x \partial z} + n_y \frac{\partial^2 P_1}{\partial y \partial z} \right)$$

$$= -\rho b^2 \frac{\partial^2 u}{\partial t^2} \delta(x)\delta(y) + \mu b^2 \left(n_x \delta(x)\delta'(y) - n_y \delta'(x)\delta(y) - n_x^2 \delta(x)\delta''(y) u + n_y^2 \delta''(x)\delta(y) u \right). \tag{12}$$

The convolution equation (1) given in equations

$$j_{ii} = n_i b_i \frac{\partial u}{\partial t} \delta(\xi) = 0, \qquad \eta_{ll} = b \frac{\partial u}{\partial z} \left(n_x \delta'(x)\delta(y) - n_y \delta(x)\delta'(y) \right),$$

received for the considered case of the screw dislocation location, the equation for the hydrostatic pressure we find

$$\frac{1}{c_l^2} \frac{\partial^2 p}{\partial t^2} - \Delta p - \frac{1}{3} \frac{1+\nu}{1-\nu} \frac{\mu}{c_t^2} (3g_1 + g_2) P_0 \frac{\partial^2 P_1}{\partial t^2} + \frac{2}{3} \frac{1+\nu}{1-\nu} \mu (2g_1 + g_2) P_0 \Delta P_1$$

$$- \frac{2}{3} \frac{1+\nu}{1-\nu} \mu g_2 P_0 \frac{\partial^2 P_1}{\partial z^2} = -\frac{2}{3} \frac{1+\nu}{1-\nu} \mu b \frac{\partial u}{\partial z} \left[n_x \delta'(x)\delta(y) - n_y \delta(x)\delta'(y) \right], \tag{13}$$

where c_l is longitudinal sound velocity.

As noted in [14], under the assumption of the elastic and electrostrictive properties isotropy, taking into account the chosen arrangement of the ferroactive axis, the polarization relationship with the stress tensor component σ_{zz} follows. In the case of an edge dislocation, this component is nonzero for a stationary dislocation and for a vibrating dislocation. In the case of a rectilinear screw dislocation $\sigma_{zz} = 0$, but for the screw dislocation performing bending vibrations, the components of the unit tangent vector to the dislocation line $\tau_x = -n_y\, \partial u/\partial z$ and $\tau_y = -n_x\, \partial u/\partial z$ appear, that is, the dislocation has an edge component of a dynamic nature that creates stresses $\sigma_{zz} \neq 0$. We find the equation for σ_{zz} from (1), believing $i = j = 3$:

$$\frac{1}{c_t^2}\frac{\partial^2 \sigma_{zz}}{\partial t^2} - \Delta\sigma_{zz} + \frac{3}{1+\nu}\left[\frac{1}{2c_t^2}\frac{\partial^2 p}{\partial t^2} - \left(\frac{\partial^2 p}{\partial x^2} + \frac{\partial^2 p}{\partial y^2}\right)\right]$$

$$- \mu(g_1 - g_2)P_0\left[\frac{1}{c_t^2}\frac{\partial^2 P_1}{\partial t^2} - \frac{2g_1}{g_1 - g_2}\left(\frac{\partial^2 P_1}{\partial x^2} + \frac{\partial^2 P_1}{\partial y^2}\right)\right] = 0\,. \tag{14}$$

We carry out the Fourier transform of equations (12)-(14) with respect to $\boldsymbol{r}$ and t. We obtain:

$$\left(k^2 - \frac{\omega^2}{c_t^2}\right)\widetilde{f}_\perp - \frac{3b}{1+\nu}(n_x k_x + n_y k_y)k_z \widetilde{p} + 2\mu b g_1 P_0 (n_x k_x + n_y k_y) k_z \widetilde{P}_1$$

$$= \mu b^2\left(n_x^2 k_y^2 - n_y^2 k_x^2 + \frac{\omega^2}{c_t^2}\right)\widetilde{u} + i(2\pi)^2 \mu b^2 (n_x k_y - n_y k_x)\delta(k_z)\delta(\omega)\,, \tag{15}$$

$$\left(k^2 - \frac{\omega^2}{c_t^2}\right)\widetilde{p} - \frac{1}{3}\mu\frac{1+\nu}{1-\nu}P_0\left[4g_1 k^2 + 2g_2 k_\perp^2 - (3g_1 + g_2)\frac{\omega^2}{c_t^2}\right]\widetilde{P}_1 = \frac{2}{3}\frac{1+\nu}{1-\nu}\mu b(n_x k_x - n_y k_y)k_z \widetilde{u}\,, \tag{16}$$

$$\left(k^2 - \frac{\omega^2}{c_t^2}\right)\widetilde{\sigma}_{zz} + \frac{3}{1+\nu}\left[k_\perp^2 - \frac{\omega^2}{2c_t^2}\right]\widetilde{p} - \mu(g_1 - g_2)P_0\left[\frac{2g_1}{g_1 - g_2}k_\perp^2 - \frac{\omega^2}{c_t^2}\right]\widetilde{P}_1 = 0\,. \tag{17}$$

In the equations (15)-(17) $\boldsymbol{k}$ is wave vector, $k_\perp^2 = k_x^2 + k_y^2$, ω is frequency; $\widetilde{f}_\perp = \widetilde{f}_\perp(\boldsymbol{k}, \omega)$, $\widetilde{p} = \widetilde{p}(\boldsymbol{k}, \omega)$, $\widetilde{P}_1 = \widetilde{P}_1(\boldsymbol{k}, \omega)$, $\widetilde{\sigma}_{zz} = \widetilde{\sigma}_{zz}(\boldsymbol{k}, \omega)$, $\widetilde{u} = \widetilde{u}(k_z, \omega)$ are Fourier-images. Supplement recorded system of equations (15)-(17) equation previously earlier in [11, 12, 8]

$$\widetilde{P}_1 = \chi(\boldsymbol{k}, \omega)P_0(-3g_1\widetilde{p} + g_2\widetilde{\sigma}_{zz})\,, \tag{18}$$

where $\chi(\boldsymbol{k}, \omega)$ is the susceptibility of a ferroelectric crystal.

3. Generalized susceptibility of screw dislocation

Solving the system of equations (15)-(18), the expressions for the Fourier transforms $\widetilde{f}_\perp = \widetilde{f}_\perp(\boldsymbol{k}, \omega)$, $\widetilde{p} = \widetilde{p}(\boldsymbol{k}, \omega)$, $\widetilde{P}_1 = \widetilde{P}_1(\boldsymbol{k}, \omega)$, $\widetilde{\sigma}_{zz} = \widetilde{\sigma}_{zz}(\boldsymbol{k}, \omega)$ we find. Among these expressions, the greatest interest is the expression that gives the connection between the force acting on the dislocation and the displacement of dislocation line:

$$\widetilde{f}_\perp(\boldsymbol{k}, \omega) = 2\frac{\mu b^2}{1-\nu}\frac{(n_x^2 k_x^2 - n_y^2 k_y^2)k_z^2}{k^2 - \omega^2/c_t^2}\left[k^2 - \frac{\omega^2}{c_t^2} - \mu g_2(g_1 - g_2)\chi P_0^2\left(\frac{2g_1}{g_1 - g_2}k_\perp^2 - \frac{\omega^2}{c_t^2}\right)\right.$$

$$\left. + 2(1+\nu)\mu\chi P_0^2 g_1^2\left(k^2 - \frac{\omega^2}{c_t^2}\right) + 2\mu\chi g_1 g_2 P_0^2\left(k_\perp^2 - \frac{\omega^2}{2c_t^2}\right)\right] \Big/ \left[\left(k^2 - \frac{\omega^2}{c_l^2}\right)\left(k^2 - \frac{\omega^2}{c_t^2}\right)\right.$$

$$+\mu\chi P_0^2\left[\frac{1+\nu}{1-\nu}g_1^2\left(k^2-\frac{\omega^2}{c_t^2}\right)\left(4k^2-3\frac{\omega^2}{c_t^2}\right)+\frac{g_1g_2}{1-\nu}\left(k_\perp^2-\frac{\omega^2}{2c_t^2}\right)\left(4k^2-3\frac{\omega^2}{c_t^2}\right)\right.$$

$$+\frac{1+\nu}{1-\nu}g_1g_2\left(k^2-\frac{\omega^2}{c_t^2}\right)\left(2k_\perp^2-\frac{\omega^2}{c_t^2}\right)+\frac{g_2^2}{1-\nu}\left(k_\perp^2-\frac{\omega^2}{2c_t^2}\right)\left(2k_\perp^2-\frac{\omega^2}{c_t^2}\right)$$

$$\left.\left.-g_2(g_1-g_2)\left(k^2-\frac{\omega^2}{c_t^2}\right)\left(\frac{2g_1}{g_1-g_2}k_\perp^2-\frac{\omega^2}{c_t^2}\right)\right]\right\}^{-1}\tilde{u}-\mu b^2\frac{n_y^2k_x^2-n_x^2k_y^2-\omega^2/c_t^2}{k^2-\omega^2/c_t^2}\tilde{u}$$

$$-i(2\pi)^2\mu b^2\frac{n_yk_x-n_xk_y}{k^2-\omega^2/c_t^2}\delta(k_z)\delta(\omega)\,. \tag{19}$$

Let us find the expression for the projection of the Peach-Koehler force on the dislocation line ($x=0$, $y=0$). For this we perform the inverse Fourier transform of (19) with respect to the variables k_x and k_y, and then set $x=y=0$:

$$\tilde{f}_\perp(x,y,k_z,\omega)=\int_{-\infty}^{\infty}\tilde{f}_\perp(\boldsymbol{k},\omega)e^{i(xk_x+yk_y)}\frac{dk_xdk_y}{(2\pi)^2}\,,$$

$$\tilde{f}_\perp(0,0,k_z,\omega)=\tilde{f}_\perp(k_z,\omega)=\int_{-\infty}^{\infty}\tilde{f}_\perp(\boldsymbol{k},\omega)\frac{dk_xdk_y}{(2\pi)^2}\,. \tag{20}$$

Passing to the polar coordinates in the integral (20) and performing integration over the polar angle, we obtain the equation of bending vibrations of a screw dislocation

$$\tilde{f}_\perp(k_z,\,\omega)=\alpha_D^{-1}(k_z,\,\omega)\tilde{u}(k_z,\,\omega)\,,$$

where $\alpha_D(k_z,\,\omega)$ is the linear response function (generalized susceptibility) of a screw dislocation in a ferroelectric, which is found by formula

$$\alpha_D^{-1}(k_z,\,\omega)=\frac{\mu b^2}{(2\pi)^2}\int k_\perp dk_\perp\left(\frac{2}{1-\nu}\frac{\pi k_\perp^2(n_x^2-n_y^2)k_z^2}{k_\perp^2+k_z^2-\omega^2/c_t^2}\left[k_\perp^2+k_z^2-\frac{\omega^2}{c_t^2}\right.\right.$$

$$-\mu g_2(g_1-g_2)\chi P_0^2\left(\frac{2g_1}{g_1-g_2}k_\perp^2-\frac{\omega^2}{c_t^2}\right)+2(1+\nu)\mu\chi P_0^2g_1^2\left(k_\perp^2+k_z^2-\frac{\omega^2}{c_t^2}\right)$$

$$\left.+2\mu\chi g_1g_2P_0^2\left(k_\perp^2-\frac{\omega^2}{2c_t^2}\right)\right]\left\{\left(k_\perp^2+k_z^2-\frac{\omega^2}{c_l^2}\right)\left(k_\perp^2+k_z^2-\frac{\omega^2}{c_t^2}\right)\right.$$

$$+\mu\chi P_0^2\left[\frac{1+\nu}{1-\nu}g_1^2\left(k_\perp^2+k_z^2-\frac{\omega^2}{c_t^2}\right)\left(4k_\perp^2+4k_z^2-3\frac{\omega^2}{c_t^2}\right)+\frac{g_1g_2}{1-\nu}\left(k_\perp^2-\frac{\omega^2}{2c_t^2}\right)\right.$$

$$\times\left(4k_\perp^2+4k_z^2-3\frac{\omega^2}{c_t^2}\right)+\frac{1+\nu}{1-\nu}g_1g_2\left(k_\perp^2+k_z^2-\frac{\omega^2}{c_t^2}\right)\left(2k_\perp^2-\frac{\omega^2}{c_t^2}\right)$$

$$+\frac{g_2^2}{1-\nu}\left(k_\perp^2-\frac{\omega^2}{2c_t^2}\right)\left(2k_\perp^2-\frac{\omega^2}{c_t^2}\right)$$

$$\left.\left.\left.-g_2(g_1-g_2)\left(k_\perp^2+k_z^2-\frac{\omega^2}{c_l^2}\right)\left(\frac{2g_1}{g_1-g_2}k_\perp^2-\frac{\omega^2}{c_t^2}\right)\right]\right\}^{-1}-\frac{\pi k_\perp^2(n_y^2-n_x^2)-\omega^2/c_t^2}{k_\perp^2+k_z^2-\omega^2/c_t^2}\right).$$

In the case of natural dislocation oscillations, it is necessary to solve the equation $\alpha_D^{-1}(q_z,\omega)=0$, from which the eigenfrequencies of the bending dislocation oscillations and their damping in the ferroelectric are found. Further investigation of the screw dislocation dynamic properties in a ferroelectrics is similar to the study carried out for an edge dislocation in [11, 12, 8]. Using these results, we give a qualitative description of the electrostriction coupling effect. Due to the interaction of the elastic field with the crystal polarization is an increase of effective mass, effective rigidity screw dislocation, velocity of bending waves propagating along it by a factor $1+\varepsilon$. Since $\sqrt{\tau_x^2+\tau_y^2}=|\partial u/\partial z|<<1$, we have $\varepsilon\sim 0{,}01\div 0{,}1$. An additional contribution to the attenuation of dislocation oscillations associated with the energy dissipation during polarization oscillations accompanying the dislocation oscillations will be of the same order.

4. Conclusion

The obtained result can be used to study the attenuation and scattering of ultrasound, electromagnetic waves and other external effects on a crystal with dislocations, to find the distribution of electric fields and polarization around an oscillating screw dislocation, to calculate the contribution of dislocations to the dielectric permittivity.

References

[1] Ninomiya T 1968 *J. Phys. Soc. Japan* **25** 830
[2] Lifshitz I M and Kosevich A M 1966 *Reports Progress Physics* **29** 217
[3] Lifshitz I M and Pushkarov H I 1970 *JETP Letters* **11** 310
[4] Tewary V K 1974 *J. Phys. C. Solid State Phys.* **7** 261
[5] Bataronov I L, Dezhin V V and Roshchupkin A M 1993 *Bulletin Russian Acad. of Sciences: Physics* **57** 1947
[6] Bataronov I L and Dezhin V V 2013 *Vestnik Tambovskogo Universiteta. Ser.: Yestestv. i Tekhnich. Nauki* **18** 1566 (In Russian)
[7] Dezhin V V, Nechaev V N and Roshchupkin A M 1990 *Fizika Tverdogo Tela* **32** 810 (In Russian)
[8] Dezhin V V, Nechaev V N and Roshchupkin A M 1990 *Z. für Kristallographie* **193** 175
[9] Kosevich A M, Pastur L A and Feldman E P 1967 *Kristallografiya* **12** 916 (In Russian)
[10] Minagawa S and Shintani K 1985 *Phyl. Mag. A* **51** 277
[11] Dezhin V V, Nechaev V N and Roshchupkin A M 1989 *Ferroelectrics Letters* **10** 155
[12] Dezhin V V, Nechaev V N and Roshchupkin A M 1990 *Fizika Tverdogo Tela* **32** 1148 (In Russian)
[13] Dezhin V V and Nechaev V N 2016 *J. Phys.: Conf. Ser.* **738** 012111.
[14] Nechaev V N and Dezhin V V 2017 *Fundamental'nye Problemy Sovremennogo Materialovedenia* **14** 34 (In Russian)
[15] Bross H 1964 *Physica status solidi* **5** 329
[16] Landau L D and Lifshitz E M 1986 *Theory of Elasticity* (*Course of Theoretical Physics* vol 7) (Butterworth-Heinemann, Oxford)
[17] de Wit R 1973 *J. Res. Nat. Bur. Stand.* **77A** 359

Behavior of the solution of the Cauchy problem for an inhomogeneous hyperbolic equation with periodic coefficients

Hovik A. Matevossian[1], **Anatoly V. Vestyak**[2]

[1] Federal Research Center "Informatics and Control", Dorodnitsyn Computing Center of the Russian Academy of Sciences, Vavilov str., 40, Moscow 119333 RUSSIA; and Department of Mathematical Modeling, Moscow Aviation Institute (National Research University), Volokolomskoe shosse, 4, Moscow 125993 RUSSIA

[2] Department of Mathematical Modeling, Moscow Aviation Institute (National Research University), Volokolomskoe shosse, 4, Moscow 125993 RUSSIA

E-mail: `hmatevossian@graduate.org`; `v.a.vestyak@mail.ru`

Abstract. We study the asymptotic behavior of the solution $u(x,t)$ of the Cauchy problem for a one-dimensional second-order hyperbolic equation with periodic coefficients (as $t \to \infty$) in which the initial data are zero, and the right-hand side of the equation has the form $f(x)\,exp(-i\omega t)$, where ω is real.

1. Introduction

Let $u(x,t)$ be the solution of the Cauchy problem

$$u_{tt}(x,t) - (p(x)\,u_x(x,t))_x + q(x)\,u(x,t) = f(x)e^{-i\omega t}, \quad (x,t) \in \mathbf{R}^1 \times \{t > 0\}, \tag{1}$$

with the initial conditions

$$u(x,t)|_{t=0} = 0, \quad u_t(x,t)|_{t=0} = 0, \quad x \in \mathbf{R}^1 \tag{2}$$

where the functions $p(x)$ and $q(x)$ are 1 - periodic,

$$p(x+1) = p(x) \geq const > 0, \quad q(x+1) = q(x) \geq 0.$$

Moreover, the functions $p(x)$ and $q(x)$ are continuous or have finitely many discontinuities of the first kind on the period. Eq. (1) describes the oscillation caused by a periodic force in t.

This paper is a continuation of research conducted in [10] and presented at the International Conference on "Mathematical Modeling in Physical Sciences (5th IC–MSQUARE)" [6], where for the solution $u(x,t)$ of the following Cauchy problem

$$u_{tt}(x,t - (p(x)\,u_x(x,t))_x + q(x)\,u(x,t) = 0, \quad (x,t) \in \mathbf{R}^1 \times \{t > 0\}, \tag{3}$$

$$u(x,t)|_{t=0} = 0, \quad u_t(x,t)|_{t=0} = \psi(x), \quad x \in \mathbf{R}^1 \tag{4}$$

the asymptotic expansion (as $t \to \infty$) was obtained, and, as above, the functions $p(x)$ and $q(x)$ are 1 - periodic, $p(x+1) = p(x) \geq const > 0$, $q(x+1) = q(x) \geq 0$, the functions $p(x)$ and $q(x)$ are continuous or have finitely many discontinuities of the first kind on the period, $\psi \in C_0^\infty(\mathbf{R}^1)$, $supp\ \psi \subset [0,1]$.

The behavior (as $t \to \infty$) of solutions of problems similar to problem (1), (2) (and also (3), (4)) and of the corresponding multi-dimensional problems under the condition that the potential differs from a constant by a finite function or tends to a constant sufficiently fast at infinity has been studied in many papers (see, e.g., [5],[7], [9]–[12]).

An asymptotic expansion as $t \to \infty$ of the solution $u(x,t)$ of the Cauchy problem

$$u_{tt}(x,t) - u_{xx}(x,t) + (\alpha_0 + q(x))u = 0, \quad (x,t) \in \mathbf{R}^1 \times \{t > 0\},$$

$$u(x,t)|_{t=0} = \varphi(x), \quad u_t(x,t)|_{t=0} = \psi(x), \quad x \in \mathbf{R}^1,$$

with finite initial functions, $\varphi(x) \in C^2(\mathbf{R}^1)$ and $\psi(x) \in C^1(\mathbf{R}^1)$ and under weaker restrictions on the potential $\alpha_0 + q(x)$, where $\alpha_0 = const$ and $q(x)$ satisfies, for some $k \geq 1$, the condition

$$\int_{-\infty}^{+\infty} |x|^k |q(x)| < \infty,$$

was obtained in [5]. For the equation of string oscillations, the Cauchy problem

$$u_{tt}(x,t) = (a(x)\, u_x(x,t))_x, \quad 0 < a_0 \leq a(x) \leq A_0 < +\infty, \quad (x,t) \in \mathbf{R}^1 \times \{t > 0\},$$

$$u(x,t)|_{t=0} = \varphi(x), \quad u_t(x,t)|_{t=0} = 0, \quad x \in \mathbf{R}^1,$$

was considered in [7]. Sufficient conditions for the stabilization (as $t \to +\infty$) of the solution $u(x,t)$ uniformly in x on any compact set and necessary and sufficient conditions for the stabilization of the solution $u(x,t)$ in the mean where obtained under some assumptions about the rate of tension $a(x)$.

In the case of periodic $p(x)$ and $q(x)$, the asymptotic behavior (as $t \to \infty$) of the solution of the Cauchy problem for a hyperbolic equation was first studied by the authors [10], [11]. This note contains the results announced in [11].

The Bloch principle for elliptic equations with periodic coefficients was obtained in [12].

We introduce the notation: $C_0^\infty(\Omega)$ is the space of the infinitely differentiable functions on the domain Ω with compact support in Ω; $L_2(\Omega)$ is the space of measurable functions in Ω suck that

$$||u(x); L_2(\Omega)|| = \left(\int_\Omega |u(x)|^2 dx\right)^{1/2} < \infty.$$

The Sobolev space $H^1(\Omega)$ in Ω is defined as follows:

$$H^1(\Omega) = \{u : u \in L_2(\Omega), \frac{\partial u}{\partial x_i} \in L_2(\Omega),\ i = 1, \ldots, n\},$$

equipped with the norm

$$||u(x); H^1(\Omega)|| = \left(\int_\Omega \left(|u|^2 + \sum_{i=1}^n \left|\frac{\partial u}{\partial x_i}\right|^2\right) dx\right)^{1/2}.$$

2. Definitions and auxiliary statements

Definition 1 *A function $u \in C^2(\boldsymbol{R}^1 \times \{t \geq 0\})$ is called a periodic (antiperiodic) solution of the Cauchy problem (1), (2), if it is satisfies the relation*

$$u(x+1,t) = (-1)^j u(x,t)$$

for any $x \in \boldsymbol{R}^1$, $t \geq 0$, and $j = 0$ ($j = 1$) in the case of periodic (antiperiodic) solutions, respectively.

By H_0 we denote the Hill operator

$$H_0 = -\frac{d}{dx}\left(p(x)\frac{d}{dx}\right) + q(x)$$

in the space $L_2(\mathbf{R}^1)$. The Hill operator has only a continuous spectrum which is located on the real axis and is left semibounded [8].

Let $y_n(x, \lambda_n)$ be an eigenfunction, normed by the condition $||y_n; L_2([0,1])|| = 1$, of the periodic Sturm-Liouville problem:

$$-(p(x)y')' + q(x)y = \lambda_n y, \quad x \in [0,1], \quad y(0) = y(1), \quad y'(0) = y'(1),$$

and let $y_n(x, \mu_n)$ be an eigenfunction, normed in $L_2([0,1])$, of the antiperiodic Sturm-Liouville problem:

$$-(p(x)y')' + q(x)y = \mu_n y, \quad x \in [0,1], \quad y(0) = -y(1), \quad y'(0) = -y'(1),$$

where λ_n $(\mu_n), n = 0, 1, 2, ..,$ are eigenvalues of the corresponding problems; these eigenvalues are numbered in ascending order with the multiplicity taken into account.

By $v_n(x, \lambda_n)$ $(v_n(x, \mu_n))$ we denote a periodic (antiperiodic) function in $\mathbf{R}^1$ whose restriction on the segment $[0,1]$ coincides with $y_n(x, \lambda_n)$ $(y_n(x, \mu_n))$.

Let

$$h(x,\omega) = \begin{cases} (H_0 - \omega^2)^{-1}(f), & if \quad \omega^2 \notin \sigma(H_0), \\ \lim\limits_{Imk>0, k\to\omega} (H_0 - k^2)^{-1}(f), & if \quad \omega^2 \in \sigma(H_0), \ and \\ \omega^2 \quad does\ not\ lie\ on\ the\ boundary\ of\ the\ spectrum \quad \sigma(H_0), \end{cases}$$

where the limit is understood in the uniform metric in each interval in $\mathbf{R}^1$. The function $h(x,\omega)$ is a solution of the equation

$$(p(x)y')' + (\omega^2 - q(x))y = f(x),$$

and belongs to $L_2(\mathbf{R}^1)$ if the point ω^2 does not belong to the spectrum $\sigma(H_0)$ of the operator H_0. Let $C > 0$ be an arbitrary fixed constant.

3. Main results

Theorem 1 *If the Hill operator H_0 is positive, $p(x) \geq const > 0$, $q(x) \geq 0$, and the point ω^2 does not lie on the boundary of the spectrum $\sigma(H_0)$ of the operator H_0, such that, for $|x| < C$ and $t > 0$, the solution of the problem (1), (2) has the form*

$$u(x,t) = -ie^{-i\omega t}h(x,\omega) + \frac{1}{\sqrt{t}}\{u_1(x,t) + u_2(x,t)\} + v(x,t),$$

where $u_1(x,t)$ *is the periodic solution of the Cauchy problem*

$$u_1(x,t) = \sum_{n=0}^{\infty} a_{\lambda_n}\psi_{\lambda_n} v_n(x,\lambda_n)\sin(\sqrt{\lambda_n}t + (-1)^n\frac{\pi}{4} + b_{\lambda_n}),$$

$u_2(x,t)$ *is the antiperiodic solution of the Cauchy problem*

$$u_2(x,t) = \sum_{n=0}^{\infty} a_{\mu_n}\psi_{\mu_n} v_n(x,\mu_n)\sin(\sqrt{\mu_n}t + (-1)^{n+1}\frac{\pi}{4} + b_{\mu_n}),$$

and the function $v(x,t)$ *satisfies the estimate*

$$|v(x,t)| \le \frac{C_0}{t}||f; L_2(\boldsymbol{R}^1)||, \quad C_0 = const(C),$$

for $|x| < C$ *and* $t > 0$*, where* $\psi_{\lambda_n}\,(\psi_{\mu_n})$ *are coefficients of the expansion of the function* $\psi(x)$ *in the Fourier series with respect to the system* $\{y_n(x,\lambda_n)\}$ $(\{y_n(x,\mu_n)\})$ *and* $a_{\lambda_n}\,(a_{\mu_n})$, $b_{\lambda_n}\,(b_{\mu_n})$ *are certain constants of order* $o(1/n^2)$ *as* $n \to \infty$.

Here the sums are taken only over the n for which $\lambda_n\,(\mu_n)$ are simple eigenvalues of the periodic (antiperiodic) Sturm-Liouville problem.

Let $\varepsilon > 0$ be a sufficiently small number and the point ω^2 lies on the boundary of the spectrum $\sigma(H_0)$ of the operator H_0. Then in the domain

$$\Omega_k \equiv \{k : Imk > 0, |k-\omega| < \varepsilon\}$$

the equality holds

$$(H_0 - k^2)^{-1}(f) = h_1(x,\omega)(k-\omega)^{-1/2} + h_2(x,\omega) + h_3(x,\omega)(k-\omega)^{1/2} + O(k-\omega).$$

Theorem 2 *If the Hill operator* H_0 *is positive,* $p(x) \ge const > 0$, $q(x) \ge 0$*, and the point* ω^2 *lies on the boundary of the spectrum* $\sigma(H_0)$ *of the operator* H_0*, then, for* $|x| < C$ *and* $t > 0$*, the solution of the problem (1), (2) has the form*

$$u(x,t) = C_1\sqrt{t}e^{-i\omega t}h_1(x,\omega) - C_2 ie^{-i\omega t}h_2(x,\omega) + \frac{1}{\sqrt{t}}\{u_1(x,t) + u_2(x,t) + C_3e^{-i\omega t}h_3(x,\omega)\} + v(x,t),$$

where the functions $u_1(x,t)$ *and* $u_2(x,t)$ *have the same form as in Theorem 1, only with the difference that in the expansions for* $u_1(x,t)$ *and* $u_2(x,t)$ *do not include a term corresponding to the* ω^2*, and for the function* $v(x,t)$*, for* $|x| < C$ *and* $t > 0$*, as in Theorem 1, the following estimate is holds*

$$|v(x,t)| \le \frac{C_0}{t}||f; L_2(\boldsymbol{R}^1)||, \quad C_i = const,\ i = 0,1,2,3.$$

It is well known [8] that the left end of the spectrum $\sigma(H_0)$ of the operator H_0 coincides with λ_0. Assume that $\lambda_0 \le 0$.

Remark 1 *If the left end of the spectrum* $\sigma(H_0)$ *of the Hill operator* H_0 *coincides with* $\lambda_0 = 0$ *and* $\omega > 0$*, then in Theorems 1 and 2 in the right-hand side in the expansions for the function* $u(x,t)$ *we add a term of the form* $h_0(x) = a_0\psi_0 v_0(x,\lambda_0)$*, where* a_0 *is some constant.*

Remark 2 *In the case of the finite-zone potential, the functions* $u_1(x,t)$ *and* $u_2(x,t)$ *can be represented as a finite sum of oscillating in* t*, because the spectrum of the operator* H_0 *has a zone structure and the ends of the zones coincide with the simple eigenvalues* λ_n *and* μ_n *[8].*

IOP Conf. Series: Journal of Physics: Conf. Series **936** (2017) 012097 doi:10.1088/1742-6596/936/1/012097

The validity of the principle of limiting amplitude, that is, the output of the solution (as $t \to \infty$) to the periodic regime with respect to t of the driving force, depends on whether ω^2 coincides with the spectrum boundary of the Hill operator.

The basic idea of the proof of theorems are the same as in the case of the finite function $q(x)$ [9]. Problem (1), (2) can be solved by using the Fourier transform, and the contour of integration in the inverse Fourier transform is shifted to the "nonphysical leave" of the Riemann surface of the spectral parameter. After this, the integral determining the solution is analyzed, i.e. the leading term of the asymptotics is distinguished, and then the remainder is estimated.

In the case, the integral in question and the methods for studying it differ strongly from those in the case of a finite function $q(x)$. The study of this integral is significantly based on the properties of the resolvent of the Hill operator H_0 ([1]–[4]).

References

[1] Firsova N E 1984, "Resonances of a Hill operator, perturbed by an exponentially decreasing additive potential", *Math. Notes*, **36**:5, 854–861

[2] Firsova N E 1987, "The direct and inverse scattering problems for the one-dimensional perturbed Hill operator", *Math. USSR–Sbornik*, **58**:2, 351–388.

[3] Firsova N E 1989, "On solution of the Cauchy problem for the Korteweg-de Vries equation with initial data the sum of a periodic and a rapidly decreasing function", *Math. USSR–Sbornik*, **63**:1, 257–265.

[4] Hochstadt H 1965, "On the determination of a Hill's equation from its spectrum", *Arch. Rational Mech. Analysis*, **19**:5, 353–362.

[5] Laptev S A 1975, "On the behavior for large values of the time of the solution of the Cauchy problem for the equation $\frac{\partial^2 u}{\partial t^2} - \frac{\partial^2 u}{\partial x^2} + \alpha(x)u = 0$", *Math. USSR–Sbornik*, **26**:3, 403–426

[6] Matevossian Hovik A and Vestyak Anatoly V 2016, "On the behavior of solutions of the Chauchy problem for the hyperbolic equation with periodic coefficients for large time value", 5th IC–MSQUARE: International Conference on Mathematical Modeling in Physical Sciences. Athens, Greece, May 23–26, 1–5.

[7] Peržan A V 1978, "Stabilization of the solution of the Cauchy problem for a hyperbolic equation", (Russian) *Differ. Uravnenija [Diff. Equations]*, **14**:6, 1065–1075.

[8] Titchmarsh E C 1958, *Eigenfunction expansions: associated with second-order differential equations*, Vol. 2, Oxford: Clarendon Press, 404 pp.

[9] Vainberg B R 1982, *Asymptotic methods in equations of mathematical physics*, Moscow: Moscow University Press, 296 pp. (Russian); English transl.: New York: Gordon and Breach, 1989, 498 pp.

[10] Vestyak A V and Matevossian O A 2016, "On the behavior of the solution of the Chauchy problem for the hyperbolic equation with periodic coefficients", *Math. Notes*, **100**:5, 126–129.

[11] Vestyak A V and Matevossian O A 2017, " On the behavior of the solution of the Chauchy problem for an inhomogeneous hyperbolic equation with periodic coefficients", *Math. Notes*, **102**:3.

[12] Zhikov V V and Pastukhova S E 2016, "Bloch principle for elliptic differential equations with periodic coefficients", *Russian J. Math. Phys.*, **23**:2, 257–277.

Conservation laws of the Hirota-Maxwell-Bloch system and its reductions

G.Shaikhova',K.Yesmakhanova', G.Bekova', S.Ybyraiymova"
'Eurasian International Center for Theoretical Physics and Department of General Theoretical Physics, Eurasian National University, Astana, 010000, Kazakhstan
"Department of Higher Mathematics, Al-Farabi Kazakh National university, Almaty, 050040, Kazakhstan

E-mail: g.shaikhova@gmail.com

Abstract. It is known that conservation law plays an important role in the study of nonlinear evolution equations and namely to integrability and constants of motion. In this paper, we construct infinitely many conservation laws for the Hirota-Maxwell-Bloch system and its reductions with symbolic computation from the Riccati form of the Lax pair.

1. Introduction

Conservation laws appear in various areas of the applied sciences, such as quantum physics, electromagnetism, plasma physics, physical chemistry, nonlinear optics [1-4]. For a nonlinear equation existence of the infinitely many conservation laws has been claimed to be a definition of the complete integrability [5, 6]. There are various methods [7] to compute conservation laws of nonlinear partial differential equations. A prevalent approach depends on the link between symmetries and conservation laws as stated in Noethers theorem [8-10].

In this work, we construct infinitely many conservation laws for the (1+1)-dimensional Hirota-Maxwell-Bloch system (HMBS) and its reductions with symbolic computation from the Riccati form of the Lax pair. The method that used in this paper was applied to several nonlinear evolution equations in mathematical physics [11-15]. Hirota-Maxwell-Bloch system and its reductions were studied in one and two dimensions [16-23]. Authors have found different kind of solutions but research regarding on infinitely many conservation laws of HMBS has not been presented in detail. Motivated by this reason we find the conservation laws for the (1+1)-dimensional Hirota-Maxwell-Bloch system, the (1 + 1)-dimensional Schrodinger-Maxwell-Bloch system (SMBS) and the (1 + 1)-dimensional complex modified Korteweg de Vries-Maxwell-Bloch equations (cmKdVMB).

The paper is organized as follows. In Section 2, we present the (1+1)-dimensional HMBS and its reductions. In Section 3, we construct infinitely many conservation laws for the (1+1)-dimensional HMBS, SMBS, cmKdVMB. Conclusion is given in Section 4.

2. The Hirota-Maxwell-Bloch system and its reductions

The (1+1)-dimensional Hirota-Maxwell-Bloch system reads as [17]

$$iq_t + \alpha(q_{xx} + 2\delta|q|^2 q) + i\beta(q_{xxx} + 6\delta|q|^2 q_x) - 2ip = 0, \quad (1)$$

$$p_x - 2i\omega p - 2\eta q = 0, \quad (2)$$

$$\eta_x + \delta(q^* p + p^* q) = 0, \quad (3)$$

where q, p are complex functions, η is real function, $\alpha, \beta, \delta, \omega$ are real constants and $\delta = \pm 1$. The symbol $*$ denotes the complex conjugate. The system of equations (1)-(3) admits the following integrable reductions:

2.1. *Case 1:* $\alpha = 1, \beta = 0$

The $(1 + 1)$-dimensional Schrodinger-Maxwell-Bloch equations result when $\alpha = 1, \beta = 0$ [17]:

$$iq_t + q_{xx} + 2\delta|q|^2 q - 2ip = 0, \quad (4)$$

$$p_x - 2i\omega p - 2\eta q = 0, \quad (5)$$

$$\eta_x + \delta(q^* p + p^* q) = 0. \quad (6)$$

2.2. *Case 2:* $\alpha = 0, \beta = 1$

The $(1 + 1)$-dimensional complex modified Korteweg de Vriese-Maxwell-Bloch equations are obtained for the choice $\alpha = 0, \beta = 1$ [17]:

$$iq_t + i(q_{xxx} + 6\delta|q|^2 q_x) - 2ip = 0, \quad (7)$$

$$p_x - 2i\omega p - 2\eta q = 0, \quad (8)$$

$$\eta_x + \delta(q^* p + p^* q) = 0. \quad (9)$$

2.3. *Case 3:* $\alpha = 1, \beta = 1, p = 0, \eta = 0$

The $(1 + 1)$-dimensional Hirota equations are obtained for the choice $\alpha = 1, \beta = 1, p = 0, \eta = 0$ [17]:

$$iq_t + (q_{xx} + 2\delta|q|^2 q) + i(q_{xxx} + 6\delta|q|^2 q_x) = 0. \quad (10)$$

3. Lax pairs and conservation laws

With the Ablowitz-Kaup-Newell-Segur scheme [1], the Lax pair associated with (1)-(3) can be derived as

$$\Psi_x = A\Psi, \quad (11)$$

$$\Psi_t = ((2\alpha\lambda + 4\beta\lambda^2)A + B)\Psi, \quad (12)$$

where

$$\Psi = \begin{pmatrix} \Psi_1 \\ \Psi_2 \end{pmatrix},$$

$$A = -i\lambda\sigma_3 + A_0,$$

$$B = \lambda B_1 + B_0 + \frac{i}{\lambda + \omega} B_{-1},$$

with

$$\begin{aligned}
B_1 &= 2i\beta\delta|q|^2\sigma_3 + 2i\beta\sigma_3 A_{0x}, \\
A_0 &= \begin{pmatrix} 0 & q \\ -r & 0 \end{pmatrix}, \\
B_0 &= (i\alpha\delta|q|^2 + \beta\delta(q^* q - q^* q_x))\sigma_3 + B_{01}, \\
B_{01} &= \begin{pmatrix} 0 & i\alpha q_x - \beta q_{xx} - 2\beta\delta|q|^2 q \\ i\alpha r_x + \beta r_{xx} + 2\beta\delta|q|^2 r & 0 \end{pmatrix}, \\
B_{-1} &= \begin{pmatrix} \eta & -p \\ -k & -\eta \end{pmatrix}, \\
\sigma_3 &= \begin{pmatrix} 1 & 0 \\ 0 & -1 \end{pmatrix},
\end{aligned}$$

and $r = \delta q^*, \quad k = \delta p^*$, where $\delta = \pm 1$. The compatible condition of system (11)-(12) is

$$A_t - B_x + [A; B] + (2\alpha\lambda + 4\beta\lambda^2)A_x = 0,$$

by direct calculation of above equation, we can yield the (1+1)-dimensional HMBS (1)-(3).

Making use of Lax pair (11)-(12), the infinetely many conservation laws for (1)-(3) could be derived. Introducing the function $\Gamma = \psi_2/\psi_1$ [24], we get the following Ricatti equation via Lax pair (11)-(12):

$$\Gamma_x = -r + 2i\lambda\Gamma - q\Gamma^2. \tag{13}$$

Expanding

$$\Gamma = \sum_{j=1}^{\infty} g_j \lambda^{-j} q^{-1} \tag{14}$$

in (13) and equating the same powers of λ to zero, we have

$$\begin{aligned}
g_1 &= -\frac{i}{2}\delta|q|^2, && (15) \\
g_2 &= -\frac{i}{2}(g_{1,x} - \frac{q_x}{q} g_1), && (16) \\
g_3 &= -\frac{i}{2}(g_{2,x} - \frac{q_x}{q} g_2 - g_1^2), && (17) \\
&\cdots && (18) \\
g_{j+1} &= -\frac{i}{2}(g_{j,x} - \frac{q_x}{q} g_j - \sum_{k=1}^{j-1} g_k g_{j-k}), (j = 3, 4, 5...), && (19)
\end{aligned}$$

where g_1, g_2 and g_n are the functions of x and t.

From Lax pair (11)-(12), we have

$$\begin{aligned}
\frac{\psi_{1,x}}{\psi_1} &= -i\lambda + q\Gamma, && (20) \\
\frac{\psi_{1,t}}{\psi_1} &= (2\alpha\lambda + 4\beta\lambda^2)a_{11} + b_{11} + ((2\alpha\lambda + 4\beta\lambda^2)a_{12} + b_{12})\Gamma. && (21)
\end{aligned}$$

Taking (20)-(21) into the compability condition $(ln\psi_1)_{xt} = (ln\psi_1)_{tx}$, we get

$$[-i\lambda + q\Gamma]_t = [(2\alpha\lambda + 4\beta\lambda^2)a_{11} + b_{11} + ((2\alpha\lambda + 4\beta\lambda^2)a_{12} + b_{12})\Gamma]_x, \tag{22}$$

where

$$\begin{aligned}
a_{11} &= -i\lambda, \\
a_{12} &= q, \\
b_{11} &= \lambda i\beta\delta|q|^2 + i\alpha\delta|q|^2 + \beta\delta(q_x^*q - q^*q_x) + \frac{i}{\lambda+\omega}\eta, \\
b_{12} &= \lambda 2i\beta q_x + i\alpha q_x - \beta(q_{xx} + 2\delta|q|^2q) - \frac{i}{\lambda+\omega}p.
\end{aligned}$$

Substituting (14) with (15)-(19) into (22), and collecting the coefficients of the same power of λ after multiplying both sides of (22) by $\lambda + \omega$, we obtain the infinitely many conservation laws for the equations (1)-(3)

$$\frac{\partial \rho_k}{\partial t} = \frac{\partial J_k}{\partial x} \tag{23}$$

with

$$\rho_1 = -\frac{i}{2}\delta|q|^2, \tag{24}$$
$$\rho_2 = -\frac{\delta}{2}(\frac{1}{2}q_x^*q + i\omega|q|^2), \tag{25}$$
$$\rho_3 = g_3 + \omega g_2, \tag{26}$$
$$\dots \tag{27}$$
$$\rho_n = g_n + \omega g_{n-1}, \quad (n = 4, 5, 6...) \tag{28}$$

and

$$J_1 = (b_{11}^0\omega + i\eta) + (2\alpha\omega + 2i\beta\omega\frac{q_x}{q} + b_{12}^0)g_1 + (2\alpha + 4\beta\omega + 2i\beta\frac{q_x}{q})g_2, \tag{29}$$
$$J_2 = (b_{12}^0\omega\frac{1}{q} - \frac{ip}{q})g_1 + (2\alpha\omega + 2i\beta\omega\frac{q_x}{q} + b_{12}^0)g_2 + (2\alpha + 4\beta\omega + 2i\beta\frac{q_x}{q})g_3, \tag{30}$$
$$J_3 = (b_{12}^0\omega\frac{1}{q} - \frac{ip}{q})g_2 + (2\alpha\omega + 2i\beta\omega\frac{q_x}{q} + b_{12}^0)g_3 + (2\alpha + 4\beta\omega + 2i\beta\frac{q_x}{q})g_4, \tag{31}$$
$$\dots \tag{32}$$
$$J_n = (b_{12}^0\omega\frac{1}{q} - \frac{ip}{q})g_{n-1} + (2\alpha\omega + 2i\beta\omega\frac{q_x}{q} + b_{12}^0)g_n + (2\alpha + 4\beta\omega + 2i\beta\frac{q_x}{q})g_{n+1}, \tag{33}$$

where

$$\begin{aligned}
b_{11}^0 &= i\alpha\delta|q|^2 + \beta\delta(q_x^*q - q^*q_x), \\
b_{12}^0 &= i\alpha q_x - \beta(q_{xx} + 2\delta|q|^2q).
\end{aligned}$$

Here ρ_k and $J_k(k = 1, 2, 3, ...)$ are the conserved densities and conserved fluxes, respectively.

Finally let us present the conservation laws for some particular cases:
i) If $\alpha = 1, \beta = 0$ in the equations (24)-(33) than we can receive conservation laws for SMBS

$$\begin{aligned}\rho_1 &= -\frac{i}{2}\delta|q|^2,\\ \rho_2 &= -\frac{\delta}{2}(\frac{1}{2}q_x^*q + i\omega|q|^2),\\ \rho_3 &= -\frac{i\delta}{2}(-q_{xx}^*q + (|q|^2)^2 - \omega q_x^*q),\end{aligned}$$

and

$$\begin{aligned}J_1 &= (i\delta|q|^2\omega + i\eta) + (2\omega + iq_x)g_1 + 2g_2,\\ J_2 &= (iq_x\omega\frac{1}{q} - \frac{ip}{q})g_1 + (2\omega + iq_x)g_2 + 2g_3,\\ J_3 &= (iq_x\omega\frac{1}{q} - \frac{ip}{q})g_2 + (2\omega + iq_x)g_3 + 2g_4.\end{aligned}$$

ii) If $\alpha = 0, \beta = 1$ in the equations (24)-(33) than we can get the conservation laws for cmKdVMB equations

$$\begin{aligned}\rho_1 &= -\frac{i}{2}\delta|q|^2,\\ \rho_2 &= -\frac{\delta}{2}(\frac{1}{2}q_x^*q + i\omega|q|^2),\\ \rho_3 &= -\frac{i\delta}{2}(-q_{xx}^*q + (|q|^2)^2 - \omega q_x^*q),\end{aligned}$$

and

$$\begin{aligned}J_1 &= (\delta(q_x^*q - q^*q_x)\omega + i\eta) + (2i\omega\frac{q_x}{q} - q_{xx} - 2\delta|q|^2q)g_1 + (4\omega + 2i\frac{q_x}{q})g_2,\\ J_2 &= (-(q_{xx} + 2\delta|q|^2q)\omega\frac{1}{q} - \frac{ip}{q})g_1 + (2i\omega\frac{q_x}{q} - q_{xx} - 2\delta|q|^2q)g_2 + (4\omega + 2i\frac{q_x}{q})g_3,\\ J_3 &= (-(q_{xx} + 2\delta|q|^2q)\omega\frac{1}{q} - \frac{ip}{q})g_2 + (2i\omega\frac{q_x}{q} - q_{xx} - 2\delta|q|^2q)g_3 + (4\omega + 2i\frac{q_x}{q})g_4.\end{aligned}$$

4. Conclusion

In the present paper, we construct infinitely many conservation laws for certain nonlinear evolution equations such as Hirota-Maxwell-Bloch system, Schrodinger-Maxwell-Bloch system, complex modified Korteweg de Vries-Maxwell-Bloch equations. The conservation laws were received from the Riccati form of the Lax pair. The existence of infinitely many conservation laws helpfully indicates the completely integrable property of these equations.

Acknowledgments

This work is supported by the Ministry of Education and Science of Republic of Kazakhstan (Grant No.0888/GF4)

References

[1] Ablowitz K J, Kaup D J, Newell A C, Segur H 1973 it Phys. Rev. Lett. **31** 125
[2] Kaup D J, Newell A C 1978 *J. Math. Phys.* **19** 798
[3] Wadati M, Konno K, Ichikawa Y H. 1979 *J. Phys. Soc. Jpn.* **46**

[4] Degasperis A. 1982 *Lett. Nuovo Cimento* **33** 425
[5] Fokas A. S. 1987 *Stud. Appl. Math* **77** 253
[6] Hereman W 2006 *Int. J. Quantum Chem* **106** 278
[7] Hereman W, Colagrosso M, Sayers R, Ringler A, Deconinck B, Nivala M, Hickman M. 2005 *In Differential Equations with Symbolic Computation* Basel
[8] Anderson I M. 2004 *The Variational Bicomplex D*, 318 pages
[9] Krasilshchik I S, Vinogradov A M 1998 *AMS: Providence*, Rhode Island
[10] Olver P J 1993 (Springer Verlag, New York)
[11] Lu X, Peng M 2013 *Commun Nonlinear Sci Numer Simulat* **18** 23042312
[12] Gao Zh, Gao Y, Su Ch, Wang Q and Mao B 2016 *Naturforsch* **71(1)a** 9-20
[13] Wang P, Tian B, Liu W, Qu Q, Li M and Sun K 2011 *Eur. Phys. J. D* **61**, 701-708
[14] Zhang H, Tian B, Meng X, Liu X, and Liu W 2009 *Eur. Phys. J. B* **72**, 233-239
[15] Guo R, Tian B, Lu X, Zhang H, and Liu W 2012 *Computational Mathematics and Mathematical Physics* **52** No. 4 565-577
[16] Yesmakanova K R, Shaikhova G N , Bekova G T, and Myrzakulova Zh R 2016 *Advances in Intelligent Systems and Computing* **441** 183-198
[17] Myrzakulov R, Mamyrbekova G K, Nugmanova G N, Lakshmanan M 2015 *Symmetry* **7(3)** 1352-1375
[18] Yesmakanova K R, Shaikhova G N , Bekova G T and Myrzakulov R 2016 *Journal of Physics: Conference Series* **738** 012018
[19] Yesmakanova K R, Shaikhova G N, Bekova G T 2016 *AIP Conference Proceedings* **1759** 020147
[20] Porsezian K and Nakkeeran K 1995 *Phys. Rev. Lett.* **74**, 2941
[21] Li Ch, He J *arXiv:1210.2501*
[22] Yang J, Li Ch, Li T, Cheng Zh 2013 *Chinese Physics Letters* **30**, 104201
[23] Li Ch, He J, Porsezian K 2013 *Physical Review E* **87** 012913
[24] Hisakado M and Wadati M 1995 *J. Phys. Soc. Jpn.* **64** 408

IOP Conf. Series: Journal of Physics: Conf. Series **936** (2017) 012099 doi:10.1088/1742-6596/936/1/012099

Aspects and Some Results on Passivity and Positivity of Dynamic Systems

M. De la Sen
Institute of Research and Development of Processes,
University of the Basque Country,
Campus of Leioa, Bizkaia , Barrio de Sarriena 48940,Spain

E-mail: manuel.delasen@ehu.eus

Abstract. This paper is devoted to discuss certain aspects of passivity results in dynamic systems and the characterization of the regenerative systems counterparts. In particular, the various concepts of passivity as standard passivity, strict input passivity, strict output passivity and very strict passivity (i.e. joint strict input and output passivity) are given and related to the existence of a storage function and a dissipation function. Later on, the obtained results are related to external positivity of systems and positivity or strict positivity of the transfer matrices and transfer functions in the time-invariant case. On the other hand, it is discussed how to achieve or how eventually to increase the passivity effects via linear feedback by the synthesis of the appropriate feed-forward or feedback controllers or, simply, by adding a positive parallel direct input-output matrix interconnection gain.

1. Introduction

This paper is devoted to discuss certain aspects of passivity results in dynamic systems and the characterization of the regenerative versus passive systems counterparts. In particular, the various concepts of passivity as standard passivity, strict input passivity, strict output passivity and very strict passivity (i.e. joint strict input and output passivity) are given and related to the existence of a storage function and a dissipation function. Basic previous background concepts on passivity are given in [1-4] , [10-12] and some related references therein. Later on, the obtained results are related to external positivity of systems and positivity or strict positivity of the transfer matrices and transfer functions in the time- invariant case. On the other hand, it is discussed and formalized how to achieve in case of passivity failing or how eventually to increase the passivity effects via linear feedback by the synthesis of the appropriate feed-forward or feedback controllers or, simply, by adding a positive parallel direct input-output matrix interconnection gain having a minimum positive lower-bounding threshold gain which is also an useful idea for asymptotic hyperstability of parallel disposals of systems, [10] . For the performed analysis, the concept of relative passivity index which is applicable for both passive and non-passive systems is addressed and modified to a lower number by the use of appropriate feedback or feed-forward compensators. Finally, the concept of passivity is discussed for switched systems which can have both passive and non-passive configurations which become active governed by switching functions. The passivity property is guaranteed by the switching law under a minimum residence time at passive active configurations provided that t he first active configuration of the switched disposal is active and that there are no two consecutive active non-passive configurations in operation.

Notation
- $\boldsymbol{R}_{0+}=\boldsymbol{R}_{+}\cup\{0\}$, where $\boldsymbol{R}_{+}=\{r\in\boldsymbol{R}:r>0\}$, $\bar{p}=\{1,2,....,p\}$,
- $D\succ 0$ denotes that the real matrix D is positive definite while $D\succeq 0$ denotes that it is positive semidefinite,
- $\lambda_{min}(.)$and $\lambda_{max}(.)$ denote, respectively, the minimum and maximum eigenvalues of the real symmetric $(.)$-matrix,
- $\hat{G}\in\{PR\}$denotes that the transfer matrix $\hat{G}(s)$ of a linear time-invariant system is positive real, i.e. $Re\,\hat{G}(s)\geq 0$ for all $Re\ s>0$ and $\hat{G}\in\{SPR\}$ denotes that it is strictly positive real, i.e. $Re\,\hat{G}(s)>0$ for all $Re\ s\geq 0$,
- A dynamic system is positive (respectively, externally positive) if all the state components (respectively, if all the output components) are non-negative for all time $t\geq 0$for any given non-negative initial conditions and non-negative input ,
- $\boldsymbol{i}=\sqrt{-1}$ is the complex unity,
- I_m is the m-th identity matrix,
- the superscript T stands for matrix transposition,
- $\boldsymbol{H}_\infty$ is the Hardy space of all complex-valued functions $F(s)$ of a complex variable s which are analytic and bounded in the open right half-plane $Re\,s>0$ of norm $\|F\|_\infty=sup\{|F(s)|:Re\,s>0\}=sup\{|F(i\omega)|:\omega\in\boldsymbol{R}\}$(by the maximum modulus theorem) and $\boldsymbol{RH}_\infty$ is the sub set of real-rational functions of $\boldsymbol{H}_\infty$.

2. Passivity and positivity

Consider a dynamic system $G:\mathrm{H}_e\to\mathrm{H}_e$ with state $x\in\boldsymbol{R}^n$, input $u\in\boldsymbol{R}^m$ and output $y\in\boldsymbol{R}^m$, where H_e is the extended space of the Hilbert space H endowed with the inner product $\langle .,.\rangle$ from $\mathrm{H}_e\times\mathrm{H}_e$ to $\boldsymbol{R}$ consisting of the truncated functions $u_t(\tau)=u(\tau)$ for $\tau\in[0,t]$ and $u_t(\tau)=0$; $\forall t,\tau(>t)\in\boldsymbol{R}_{0+}$ and $u:\boldsymbol{R}_{0+}\to\boldsymbol{R}^m$. If $u\in\mathrm{H}_e$ then

$$\|u_t\|_2^2=\langle u,u\rangle_t=\langle u_t,u_t\rangle=\int_0^t u^T(\tau)u(\tau)d\tau=\int_0^\infty u_t^T(\tau)u_t(\tau)d\tau\ ;\ \forall t\geq 0\,.$$

Definitions [2]. The above dynamic system is:. L_2*-stable* if $u\in L_2^m$ implies $Gu\in L_2^m$.

2. *Nonexpansive* if $\exists\lambda$ and $\exists\gamma>0$ s. t. for all $u\in\mathrm{H}_e$

$$\int_0^t(Gu)^T(\tau)u(\tau)d\tau\leq\lambda+\gamma^2\int_0^t u^T(\tau)u(\tau)d\tau\ ;\ \forall t\geq 0\,.$$

3. *Passive* if $\exists\varepsilon\geq 0$ such that $\int_0^t y^T(\tau)u(\tau)d\tau\geq-\varepsilon\ ;\ \forall t\geq 0$.

4. *Strictly- input passive* if $\exists\varepsilon\geq 0$ and $\exists\varepsilon_u>0$ s. t.

$$\int_0^t y^T(\tau)u(\tau)d\tau\geq-\varepsilon+\varepsilon_u\int_0^t y^T(\tau)u(\tau)d\tau\ ;\ \forall t\geq 0\,.$$

5. *Strictly- output passive* if $\exists\beta\geq 0$ and $\exists\varepsilon_y>0$ s. t.

$$\int_0^t y^T(\tau)u(\tau)d\tau\geq-\varepsilon+\varepsilon_y\int_0^t y^T(\tau)y(\tau)d\tau\ ;\ \forall t\geq 0\,.$$

6. *Strictly input/output passive* (or *very strictly passive*) if $\exists\beta\geq 0$, $\exists\varepsilon_u>0$ and $\exists\varepsilon_y>0$ s. t.

$$\int_0^t y^T(\tau)u(\tau)d\tau\geq-\varepsilon+\varepsilon_u\int_0^t y^T(\tau)u(\tau)d\tau+\varepsilon_y\int_0^t y^T(\tau)y(\tau)d\tau\ ;\ \forall t\geq 0\,.$$

The constants ε, ε_u and ε_y are respectively referred to as the passivity, input passivity and output passivity constants.

The following two results are relevant to mutually relate the time and frequency domains descriptions concerned with the passivity and positivity properties:

Theorem 1: Consider a linear time-invariant SISO (i.e. m=1) system whose transfer function $\hat{G} \in \{PR\}$. Then, the following properties hold:

(i) $\int_0^t y(\tau)u(\tau)d\tau \geq 0$ and $y(t)u(t) \geq 0$; $\forall t \geq 0$ and, furthermore, if $u \in L_2$ then $y \in L_2$. Then, the system is passive.

(ii) Assume, in addition, that $\hat{G} \in \{SPR\}$. Then $\gamma \geq \int_0^t y(\tau)u(\tau)d\tau \geq \varepsilon_u \int_0^t |u(\tau)|^2 d\tau - \varepsilon$ for any $t \in (0, \infty]$ and some $\gamma, \varepsilon \in \boldsymbol{R}_{0+}$.

(iii) If, furthermore, the system is externally positive then $\gamma \geq \int_0^t y(\tau)u(\tau)d\tau > 0$; $\forall t \geq 0$ for any given non-negative initial conditions and non-negative input.

(iv) Define $R_{\hat{G}} = \left\| \frac{1-\hat{G}(i\omega)}{1+\hat{G}(i\omega)} \right\|_{\infty} = \sup_{\omega \in \boldsymbol{R}_{0+}} \left| \frac{1-\hat{G}(i\omega)}{1+\hat{G}(i\omega)} \right|$ as the relative passivity index of the transfer function $\hat{G}(s) = \frac{\hat{N}(s)}{\hat{D}(s)} \in \boldsymbol{RH}_{\infty}$ ($\hat{N}(s)$ and $\hat{D}(s)$ being the numerator and denominator polynomials of $\hat{G}(s)$). Then, the constraint $a_G \leq R_{\hat{G}} = \left\| \frac{\hat{D}_G(i\omega) - \hat{N}_G(i\omega)}{\hat{D}_G(i\omega) + \hat{N}_G(i\omega)} \right\|_{\infty} \leq b_G$ is guaranteed for some $a_G, b_G (\geq a_G) \in \boldsymbol{R}_{0+}$ if

$$\frac{1-a_G^2}{2(1+a_G^2)}\left(Re^2\ \hat{D}_G(i\omega) + Re^2\ \hat{N}_G(i\omega)\right) \geq \frac{1-b_G^2}{2(1+b_G^2)}\left(Re^2\ \hat{D}_G(i\omega) + Re^2\ \hat{N}_G(i\omega)\right);\ \forall \omega \in \boldsymbol{R}_{0+}$$

If $b_G \leq 1$ (respectively, $b_G < 1$) then $\hat{G} \in \{PR\}$ (respectively, $\hat{G} \in \{SPR\}$).

Note that positivity is a very important property in some dynamic systems related to biological or epidemic-type models . See, for instance, [5-9] and references therein The generalization of Theorem 1 to the multi input multi-output (MIMO) case (i.e. $m > 1$) is direct by replacing the instantaneous power $y(t)u(t)$ by the scalar product $y^T(t)u(t)$ in the corresponding expressions. In particular, the subsequent two results discuss how the basic passivity property can become a stronger property as, for instance, strict-input passivity or very strict passivity, by incorporating to the input-output operator a suitable parallel static input-output interconnection structure.

Theorem 2. Consider a class $[G]_{\rho D}$ of dynamic systems $G(\rho, D): \mathrm{H}_e \rightarrow \mathrm{H}_e$, defined as $G(\rho, D) = G_0 + \rho D$ for given $\rho \in \boldsymbol{R}$, $D (\in \boldsymbol{R}^{m \times m}) \succeq 0$ and $G_0 : \mathrm{H}_e \rightarrow \mathrm{H}_e$, such that $G(\alpha, D) \in [G]_{\rho D}$ for any $\alpha \in [0, \rho]$. The following properties hold:

(i) Assume that $G(\rho, D)$ is very strictly passive, $D \succ 0$ and $\varepsilon_{up} > \frac{\rho \lambda_{max}\left(D + D^T\right)}{2}$, where ε_{up} is the input passivity constant for $G(\rho, D)$.Then, $G(\alpha, D)$ is very strictly passive for all $\alpha \in [0, \rho]$. If

$\varepsilon_{up} = \frac{\rho\lambda_{max}\left(D+D^T\right)}{2}$ then $G(\alpha, D)$ is very strictly passive for all $\alpha \in (0, \rho]$ while G_0 is strictly-output passive.

(ii) If G_0 is passive (respectively, strictly output passive) then $G(\alpha, D)$ is strictly-input passive (respectively, very strictly passive) if $D \succ 0$ for any $\alpha \in \boldsymbol{R}_+$.

Assume that G_0 is passive and non-expansive. Then:
(iii) $G_\rho = G_0 + \rho D$ is L_2 -stable and strictly-input passive if $\rho \in \boldsymbol{R}_+$,
(iv) G_0 is L_2 -stable if $\rho > -\frac{2\left(\lambda_0 + \gamma_0^2\right)}{\lambda_{max}\left(D+D^T\right) - \lambda_{min}\left(D+D^T\right)}$ if $D(\neq I_m, 0) \succeq 0$,
(v) G_0 is L_2 -stable if $D = I_m$ for any given $\rho \in \boldsymbol{R}$.

It turns out through simple mathematical derivations that Theorem 2 still holds with the replacement $D \to G_1$, where $G_1 : \mathrm{H}_e \to \mathrm{H}_e$ is passive with associated constant $\varepsilon_1 \le 0$ for the properties to be extended from the case that $D \succeq 0$ and strictly-input passive for those extended from the case when $D \succ 0$.

3. Control compensators

It is now discussed how the passivity properties can be improved via feedback with respect to an external reference input signal. Consider the following linear time-invariant SISO cases:

- The controlled plant transfer function $\hat{G}(s)$, whose relative passivity index [Theorem 1 (iv)] is $R_{\hat{G}} = \sup_{\omega \in \boldsymbol{R}_{0+}} \left| \frac{1-\hat{G}(i\omega)}{1+\hat{G}(i\omega)} \right|$, is controlled by a feedback controller of transfer function $\hat{K}_1(s)$ so that $\hat{M}_1(s) = \frac{1-\hat{T}_1(s)}{1+\hat{T}_1(s)} = \frac{1+\hat{G}(s)\left(\hat{K}_1(s)-1\right)}{1+\hat{G}(s)\left(\hat{K}_1(s)+1\right)}$, where $\hat{T}_1(s) = \frac{\hat{G}(s)}{1+\hat{G}(s)\hat{K}_1(s)}$ is the resulting closed-loop transfer function. The closed-loop relative passivity index is $R_{\hat{T}_1} = \sup_{\omega \in \boldsymbol{R}_{0+}} \left| \frac{1-\hat{T}_1(i\omega)}{1+\hat{T}_1(i\omega)} \right| = \sup_{\omega \in \boldsymbol{R}_{0+}} \left| \hat{M}_1(i\omega) \right|$. For any given $\hat{T}_1(s)$ and associated $\hat{M}_1(s)$, the controller transfer function is:

$$\hat{K}_1(s) = \frac{\left(2\hat{G}(s)-1\right)\hat{T}_1(s)+1}{2\hat{G}(s)\hat{T}_1(s)} = \frac{\hat{G}(s)\left(1+\hat{M}_1(s)\right)+\hat{M}_1(s)}{\hat{G}(s)\left(1-\hat{M}_1(s)\right)}.$$

- The controlled plant transfer function $\hat{G}(s)$ is controlled by a feed-forward controller of transfer function $\hat{K}_2(s)$ so that $\hat{M}_2(s) = \frac{1-\hat{T}_2(s)}{1+\hat{T}_2(s)} = \frac{1}{1+2\hat{G}(s)\hat{K}_2(s)}$, where $\hat{T}_2(s) = \frac{\hat{G}(s)\hat{K}_2(s)}{1+\hat{G}(s)\hat{K}_2(s)}$ is the resulting closed-loop transfer function. The closed-loop relative passivity index is $R_{\hat{T}_2} = \sup_{\omega \in \boldsymbol{R}_{0+}} \left| \frac{1-\hat{T}_2(i\omega)}{1+\hat{T}_2(i\omega)} \right| = \sup_{\omega \in \boldsymbol{R}_{0+}} \left| \hat{M}_2(i\omega) \right|$. For any given $\hat{T}_1(s)$ and associated $\hat{M}_1(s)$, the controller transfer function is:

$$\hat{K}_2(s)=\frac{\hat{T}_2(s)}{\hat{G}(s)\left(1-\hat{T}_2(s)\right)}=\frac{1-\hat{M}_2(s)}{2\hat{G}(s)\hat{M}_2(s)} .$$

The subsequent result is concerned with the fact that a positive real transfer function can be designed by using feedback or feed-forward control laws for the case when the plant transfer function is inversely stable even if it is not either positive real or stable.

Theorem 3. Assume that $\hat{G}(s)$ is inversely stable with relative degree 0 or 1 while non-necessarily in $\{PR\}$ (or even non-necessarily in $\boldsymbol{RH}_\infty$). Then, the following properties hold:

(i) A non-unique (state-space) realizable closed-loop transfer function $\hat{T}_1 \in \{PR\}$, or respectively $\hat{T}_1 \in \{SPR\}$, may be designed via a stable feedback controller of transfer function $\hat{K}_1(s)$ Eq. (14) which is realizable if $\hat{G}(s)$ and $\hat{T}_1(s)$ have respective zero relative degrees. In the above cases, $\hat{T}_1^{-1} \in \{PR\}$, or respectively, $\hat{T}_1^{-1} \in \{SPR\}$.

(ii) A non-unique realizable closed-loop transfer function $\hat{T}_2 \in \{PR\}$, or $\hat{T}_2 \in \{SPR\}$, may be designed via a feed-forward controller of transfer function $\hat{K}_2(s)$ via (16) which is realizable if the relative degree of the closed-loop transfer function $\hat{T}_2(s)$ is non less than that of the plant $\hat{G}(s)$. In the above cases, $\hat{T}_2^{-1} \in \{PR\}$, or respectively, $\hat{T}_2^{-1} \in \{SPR\}$.

4. Non-passive and passive systems

Note that passive systems are intrinsically stable and either consume or dissipate energy for all time. However unstable systems are always non-passive although some stable systems are also non-passive. Looking at Definition 3, we can give the next one:

Definition 7. A dynamic system is *Non-passive* (or *Active* or, so-called, *Regenerative*) if $\int_0^{t_i} y^T(\tau)u(\tau)d\tau+\varepsilon_{t_i}<0$ for some unbounded sequences $E=\{\varepsilon_{t_i}\}\subseteq \boldsymbol{R}_{0+}$, $T=\{t_i\}\subseteq \boldsymbol{R}_{0+}$ which satisfy the conditions:

a) $0<\delta_{i-1}\le t_{i+1}-t_i\le \delta_i<\infty$; $\forall i\in \boldsymbol{Z}_{0+}$ for some positive bounded sequence $\Delta=\{\delta_i\}$,

b) $0<\theta_{i-1}\le \widetilde{\varepsilon}_{t_i}=\varepsilon_{t_{i+1}}-\varepsilon_{t_i}\le \theta_i<\infty$; $\forall i\in \boldsymbol{Z}_{0+}$ for some positive bounded sequence $\Theta=\{\delta_i\}$,

c) $\varepsilon_i , t_i \to +\infty$ as $i \to +\infty$.

The following result follows for a non-passive system:

Theorem 4. If a dynamic system is non-passive then $\lim\limits_{t\to\infty} \int_0^t y^T(\tau)u(\tau)d\tau=-\infty$

Proof: Define $\widetilde{\varepsilon}_{t_i}\in \boldsymbol{R}_+$ such that $\int_0^{t_i} y^T(\tau)u(\tau)d\tau=-\varepsilon_{t_i}-\widetilde{\varepsilon}_{t_i}<-\varepsilon_{t_i}$. Thus,

$$\int_{t_i}^{t_{i+1}} y^T(\tau)u(\tau)d\tau=-\varepsilon_{t_{i+1}}-\widetilde{\varepsilon}_{t_{i+1}}-\int_0^{t_i} y^T(\tau)u(\tau)d\tau=\varepsilon_{t_i}+\widetilde{\varepsilon}_{t_i}-\left(\varepsilon_{t_{i+1}}+\widetilde{\varepsilon}_{t_{i+1}}\right)$$

$$\int_0^t y^T(\tau)u(\tau)d\tau=\int_0^{t_i} y^T(\tau)u(\tau)d\tau+\int_{t_i}^t y^T(\tau)u(\tau)d\tau=-\varepsilon_{t_i}-\widetilde{\varepsilon}_{t_i}+\int_{t_i}^t y^T(\tau)u(\tau)d\tau$$

$$\int_0^t y^T(\tau)u(\tau)d\tau=\int_0^{t_{i+1}} y^T(\tau)u(\tau)d\tau-\int_t^{t_{i+1}} y^T(\tau)u(\tau)d\tau=-\varepsilon_{t_{i+1}}-\widetilde{\varepsilon}_{t_{i+1}}-\int_t^{t_{i+1}} y^T(\tau)u(\tau)d\tau$$

Subtracting the two above ones:

$$\varepsilon_{t_{i+1}}-\varepsilon_{t_i}+\left|\widetilde{\varepsilon}_{t_{i+1}}-\widetilde{\varepsilon}_{t_i}\right|\ge\left|\int_{t_i}^{t_{i+1}} y^T(\tau)u(\tau)d\tau\right|$$

$$\ge 2\,max\left(\left|\int_t^{t_{i+1}} y^T(\tau)u(\tau)d\tau\right|,\left|\int_{t_i}^t y^T(\tau)u(\tau)d\tau\right|\right)$$

IOP Conf. Series: Journal of Physics: Conf. Series **936** (2017) 012099 doi:10.1088/1742-6596/936/1/012099

and

$$\left|\varepsilon_{t_i}+\widetilde{\varepsilon}_{t_i}-\varepsilon_{t_{i+1}}+\varepsilon_{t_i}-\left|\widetilde{\varepsilon}_{t_{i+1}}-\widetilde{\varepsilon}_{t_i}\right|\right| \le \left|\varepsilon_{t_i}+\widetilde{\varepsilon}_{t_i}-\left|\int_{t_i}^{t} y^T(\tau)u(\tau)d\tau\right|\right|$$
$$\le \left|\int_0^t y^T(\tau)u(\tau)d\tau\right| \le -\varepsilon_{t_i}-\widetilde{\varepsilon}_{t_i}+\left|\int_{t_i}^{t} y^T(\tau)u(\tau)d\tau\right| \le -\left(\varepsilon_{t_i}+\widetilde{\varepsilon}_{t_i}\right)+1/2\left(\varepsilon_{t_{i+1}}-\varepsilon_{t_i}+\left|\widetilde{\varepsilon}_{t_{i+1}}-\widetilde{\varepsilon}_{t_i}\right|\right)$$

since $\{\varepsilon_{t_i}\}$ is unbounded but its associated incremental sequence $\{\widetilde{\varepsilon}_{t_i}\}$ is bounded , $\{2\varepsilon_{t_i}-\varepsilon_{t_{i+1}}\}\to\infty$ as $t_i\to\infty$ then $\left|\int_0^{\infty} y^T(\tau)u(\tau)d\tau\right|<+\infty$ contradicts the above relations. □

Note that a non-passive system can reach an absolute infinity energy measure in finite time under certain atypical inputs as, for instance, a second-order impulsive Dirac input of appropriate component signs at some time instant $t_1<\infty$ with $u(t)=0$ for $t>t_1$. Then, $\int_0^{t_1} y^T(\tau)u(\tau)d\tau = \lim_{t\to\infty}\int_0^t y^T(\tau)u(\tau)d\tau=-\infty$.

Theorem 5. The following properties hold:
(i) A passive system cannot be non-passive in any time sub-interval. A non-passive system in some time interval cannot be a passive system.
(ii) A passive system is always stable and also dissipative (i.e. the dissipative energy function takes non-negative values for all time) including the conservative particular case implying identically zero dissipation through time.
(ii) A non-passive system can be stable or unstable (so, stable systems are non- necessarily passive).

Acknowledgements
This research is supported by the Spanish Government and by the European Fund of Regional Development FEDER through Grant DPI2015-64766-R and by UPV/EHU by Grant PGC 17/33.

References
[1] Zames G 1966 *IEEE Transactions on Automatic Control* **11** pp. 228-238
[2] N. Kottenstette N and Antsaklis P.J. 2010 *2010 American Control Conference*, Baltimore **1** pp 409-416
[3] Francis BA 1992 *A Course in H_∞ Control Theory*, Lecture Notes in Control and Information Sciences No. 88, Springer-Verlag, Heidelberg
[4] Dorato P , Fortuna L and Muscato G 1992 *Robust Control for Unstructured Perturbations- An Introduction*, Lecture Notes in Control and Information Sciences No. 168, Springer-Verlag, Heidelberg
[5] De la Sen M and Ibeas A 2005 *Discrete Dynamics in Nature and Society* **2005** pp. 31-67
[6] De la Sen M 2002 *Positivity*, **6** pp. 31-45
[7] De la Sen M 2008 *Applied Mathematical Modelling* **32** pp. 2312-2328
[8] De la Sen M and Alonso-Quesada S 2008 *Discrete Dynamics in Nature and Society* **2008** Article ID 793512, 21 pages
[9] De la Sen M and Alonso-Quesada S 2009 *Applied Mathematics and Computation* **215** pp. 2616-2633
[10] De la Sen M 1986 *International Journal of Control* **44** pp. 1769-1775
[11] Shen Z and Andersson SB, 2010 *49th IEEE Conference on Decision and Control* **1** pp. 4819-4824
[12] Fradkov A 2003 European Journal of Control **2003** pp. 577-586

IOP Conf. Series: Journal of Physics: Conf. Series **936** (2017) 012100 doi:10.1088/1742-6596/936/1/012100

Hipergeometric solutions to some nonhomogeneous equations of fractional order

Jorge Olivares[1], **Pablo Martin**[2] **and Fernando Maass**[2]
[1]Department of Mathematics, Universidad de Antofagasta, Antofagasta, Chile
[2]Department of Physics, Universidad de Antofagasta, Antofagasta, Chile

E-mail: jorge.olivares@uantof.cl, pablo.martin@uantof.cl, fernando.maass@uantof.cl

Abstract. In this paper a study is performed to the solution of the linear non homogeneous fractional order alpha differential equation equal to $I_0(x)$, where $I_0(x)$ is the modified Bessel function of order zero, the initial condition is f(0)=0 and 0< *alpha* < 1. Caputo definition for the fractional derivatives is considered. Fractional derivatives have become important in physical and chemical phenomena as visco-elasticity and visco-plasticity, anomalous diffusion and electric circuits. In particular in this work the values of alpha=1/2, 1/4 and 3/4. are explicitly considered . In these cases Laplace transform is applied, and later the inverse Laplace transform leads to the solutions of the differential equation, which become hypergeometric functions.

1. Introduction

Differential equations of fractional order has become important in Physics, Chemistry and several areas of Engineering as anomalous diffusion [1,2]. Gamma and Beta functions, and in general special functions playing an important role in Fractional Calculus [3,4]. Bessel functions are a kind of special functions, which play an important role in problems with cylindrical symmetry, and in this paper an investigation as carried out in the connection of these functions with Fractional Calculus. Here simple non-homogeneous fractional differential equation are considered and it is shown how its solution leads to hypergeometric functions. The procedure here used is through the Lapace transform, which is often used in this kind of calculations.
In this work an explanation of the differential equation and the theoretical treatment is done as well as its solution for different values of the fractional differential equations. The last section is devoted to the Conclusion.

2. Planning of the problem and its solution

Here the differential equation

$$D^{\alpha}G(x) = \frac{d^{\alpha}G(x)}{dx^{\alpha}} = I_0(x) \tag{1}$$

is considered with the condition $G_0 = 0$. The letter α is a fractional number between zero and one. Four values of α will be considered 1/4, 1/2 and 3/4, however other values de α could be considered in other works. The Bessel function $I_0(x)$ is taken as the non-homogeneous part of the differential equation, however most general treatment could be performed in future papers.

IOP Conf. Series: Journal of Physics: Conf. Series **936** (2017) 012100 doi:10.1088/1742-6596/936/1/012100

The procedure to be followed is to apply Laplace transform to both sides of the equation. The second step will be to separate the function F(x), followed by the application of the inverse Laplace transform.
Caputo definition of fractional derivatives will be considered in this paper. Let us considerer first the case of $\alpha = 1/4$, thus the differential equation will be

$$D^{1/4}G(x) = I_0(x) \tag{2}$$

$$\mathcal{L}D^{1/4}G_{1/4}(x) = s^{1/4}\tilde{G}_{1/4}(s) - s^{3/4}G_{1/4}(0) = \mathcal{L}I_0(x) = \frac{1}{\sqrt{s^2-1}} \tag{3}$$

where $\tilde{G}(s)$ is the Laplace transform of G(x), that is,

$$\tilde{F}(s) = \mathcal{L}F(x) = \int_0^\infty e^{-sx}F(x)\,dx \tag{4}$$

With the condition that G(0)=0, the Laplace transform $\tilde{G}(s)$ becomes

$$\tilde{G}(s) = \frac{1}{s^{1/4}\sqrt{s^2-1}} \tag{5}$$

The inverse transform of $\tilde{G}(s)$ is now

$$G(x) = \mathcal{L}^{-1}[\tilde{G}_{1/4}(s)] = \mathcal{L}^{-1}\left[\frac{1}{s^{1/4}\sqrt{s^2-1}}\right] = \frac{2\Gamma(3/4)x^{1/4}}{\pi}{}_2F_1\left(\frac{1}{2},\frac{5}{8};\frac{9}{8};\frac{x^2}{4}\right) \tag{6}$$

where ${}_2F_1\left(\frac{1}{2};\frac{5}{8},\frac{9}{8};\frac{x^2}{4}\right)$ is an hipergeometric function, F(a,a,b;c;z)
In similar way, the case of α=1/2 and α=3/4 can be also treated, giving

$$G_{1/2}(x) = \mathcal{L}^{-1}\left[\frac{1}{s^{1/2}\sqrt{s^2-1}}\right] = \frac{2}{\pi}x^{1/2}{}_2F_1\left(\frac{1}{2},\frac{3}{4};\frac{5}{4};\frac{x^2}{4}\right) \tag{7}$$

and

$$G_{3/4}(x) = \mathcal{L}^{-1}\left[\frac{1}{s^{3/4}\sqrt{s^2-1}}\right] = \frac{4}{3}x^{3/4}{}_2F_1\left(\frac{1}{2},\frac{7}{8};\frac{11}{8};\frac{x^2}{4}\right) \tag{8}$$

Futhermore it is known that

$$\int_0^x I_0(x)\,dx = x\,{}_2F_1\left(\frac{1}{2},1;\frac{3}{2};\frac{x^2}{4}\right) \tag{9}$$

In order to find some connections between the parameters of the tree hypergeometric functions, it is convenient to write the parameters as follows

$$\frac{3}{4} = \frac{6}{8} \quad ; \quad \frac{5}{4} = \frac{6}{8} \quad ; \quad 1 = \frac{8}{8} \quad ; \quad \frac{3}{2} = \frac{12}{8} \tag{10}$$

Now the characteristic parameters becomes 5/8; 6/8; 7/8 and 8/8, for α=1/4; 1/2=2/4; 3/4 and 1=4/4, respectively. In this way, all the a's parameters have the same denominator, and the numerators are obtained adding one to each previous numerator.

In the case of parameter b, this parameter can be also written as 9/8; 5/4=10/8; 11/8 and 3/2=12/8 for α=1/4; 1/2=2/4; 3/4 and 1=4/4, respectively. Here the rule is similar to case of the parameter "a".
The parameter c is equal for the four cases of α here considered, similarly the independent

variable for all, the case will be $x^2/4$. The factor with the hypergeometric is always x^α.
The rule for α=1/3, 2/3 and 3/3=1 is a little similar to the cases of 1/4; 2/4; 3/4 and 4/4, since now it is obtained

$$G_{1/3}(x) = \mathcal{L}^{-1}\left[\frac{1}{s^{1/3}\sqrt{s^2-1}}\right] = \frac{3}{2}\Gamma(2/3)x^{1/3}{}_2F_1\left(\frac{1}{2},\frac{2}{3};\frac{7}{6};\frac{x^2}{4}\right) \tag{11}$$

and

$$G_{2/3}(x) = \mathcal{L}^{-1}\left[\frac{1}{s^{2/3}\sqrt{s^2-1}}\right] = \frac{3}{2\Gamma(2/3)}x^{2/3}{}_2F_1\left(\frac{1}{2},\frac{5}{6};\frac{4}{3};\frac{x^2}{4}\right) \tag{12}$$

Now the parameter a and b can be written as a:2/3=4/6; 5/6 and 1=6/6. The parameter b will be 7/6; 4/3= 8/6 and 3/2=9/6, which correspond to α=1/3; 2/3 and 1 =3/3. The factor to go with the hypergeometric function, the independent variable and the third parameter c are obtained in similar to the cases of α=1/4, 1/2, 3/4 and 1.
Considering the two sets of values here analized it seems that the denominator of the parameters a and b, which are 6 and 8, can obtained as twice the denominator of α.

3. Conclusion

Here it is shown how to determine the solution of some simple linear fractional differential equation with some special functions as non homogeneous terms. The case of $I_0(x)$ is treated in detail, however it seems that the procedure could be applied to other cases.
Here the fractional derivative of order α is considered to be between zero and one. The solutions here obtained are always a combination of a fractional power of the independent variable and on hypergeometric function. The fractional power was always the fractional derivative in the cases here treated. Several values have been shown to determine the parameters of the hypergeometric function.

Acknowledgments

We thanks to the "Vicerrector de Investigación" and the "Decano de Ciencias Básicas", of the "Universidad de Antofagasta" for support to this research.

References

[1] H. M. Srivastava, R. K. Saxena, "Operators of fractional integration and their applications", Appl. Math. Comput. 118, 1-52 (2001).
[2] Baojin Wang, Zhiyuan Liu, Shengo Eben Li and Scott-Jason Moure, "State-of-Charge Estimation for Lithium-Ion Batteries Based on a Non linear Fractional Model", IEEE Trans. Control. Syst. Technol., 25, 3-11(2017).
[3] M. Ali Ozaeslan, Emine Ozergin, "Some generating relations for extended hypergeometric functions via generalized fractional derivative operator", Math. Comput. Modelliny 52, 1825-1833(2010).
[4] Igor Podlubny, "An introduction to Fractional Derivatives, Fractional Differential Equations, to Methods of their Solution and some of their Applications", (Academic Press, San Diego, Boston, New York, London, Sydney, Tokyo, Toronto, 1999).

Institute of Physics
Dirac House, Temple Back
Bristol BS1 6BE UK

ISSN: 1742-6588
ISBN 978-1-5108-5845-9

9 781510 858459

CURRAN ASSOCIATES INC.
proceedings.com